Fifteenth Asian Congress of Fluid Mechanics (15ACFM)

Journal of Physics: Conference Series Volume 822

Kuching, Malaysia
21 – 23 November 2016

ISBN: 978-1-5108-3907-6
ISSN: 1742-6588

Printed from e-media with permission by:

Curran Associates, Inc.
57 Morehouse Lane
Red Hook, NY 12571

Some format issues inherent in the e-media version may also appear in this print version.

Printed by Curran Associates, Inc. (2017)

Institute of Physics
Dirac House, Temple Back
Bristol BS1 6BE UK

Phone: 44 1 17 929 7481
Fax: 44 1 17 920 0979

techtracking@iop.org

Additional copies of this publication are available from:

Curran Associates, Inc.
57 Morehouse Lane
Red Hook, NY 12571 USA
Phone: 845-758-0400
Fax: 845-758-2633
Email: curran@proceedings.com
Web: www.proceedings.com

Table of contents

Volume 822

Fifteenth Asian Congress of Fluid Mechanics (15ACFM)

21–23 November 2016, Kuching, Sarawak, Malaysia

Accepted papers received: 13 March 2017
Published online: 11 April 2017

Preface

011001
OPEN ACCESS
Fifteenth Asian Congress of Fluid Mechanics (15ACFM)

C P Tso, T L Lim, L C Low, G C Lee and T C Yap

011002
OPEN ACCESS
Organising Committee of 15th Asian Congress of Fluid Mechanics

011003
OPEN ACCESS
Preface to the Proceedings

011004
OPEN ACCESS
Peer review statement

Papers

Aeroacoustics and aerodynamics

012001
OPEN ACCESS
Criterion for vortex acoustic lock-on in combustors with backward facing step

B Singaravelu and S Mariappan.....1

012002
OPEN ACCESS
Design of a novel and efficient lantern wind turbine

M D Ibrahim, L K Wong, M Anyi, Y S Yunos, M R A Rahman and M Z Mohta.....7

012003
OPEN ACCESS
Experimental investigation of boundary layer transition in flow past a bluff body

Rahul Deshpande, Aditya Desai, Vivek Kanti and Sanjay Mittal.....14

012004
OPEN ACCESS
Experimental investigation on the near flow field of dual stream nozzles

S Sudhakar, N Karthikeyan and S Ashwin Kumar.....20

012005
OPEN ACCESS
High-Alfa Aerodynamics with Separated Flow Modeled as a Single Nascent Vortex

Samuel B Antony and Rinku Mukherjee.....30

012006
OPEN ACCESS
Influence of non perfect impedance boundary on the bistable region in thermoacoustic interactions

B Mohan and S Mariappan.....37

012007
OPEN ACCESS
Numerical study of rigid and flexible wing shapes in hover

Aamer Shahzad, Fang-Bao Tian, John Young and Joseph C S Lai.....43

012008
OPEN ACCESS
Rear-roof spoiler effect on the aerodynamic drag performance of a simplified hatchback model

S Y Cheng and S Mansor.....49

012009
OPEN ACCESS
Some aspects of flow control over a NACA0015 airfoil using synthetic jets

T Parthasarathy and S P Das.....55

012010
OPEN ACCESS
Vortex Ring Formation Characteristics in Synthetic Jet due to Changes of Excitation Frequency in the ½-Ball Cavity Actuator

Engkos A. Kosasih, Harinaldi and Ramon Trisno.....62

Boundary layer flows

012011
OPEN ACCESS
Analysis of the coherent and turbulent stresses of a numerically simulated rough wall pipe

L Chan, M MacDonald, D Chung, N Hutchins and A Ooi.....70

012012
OPEN ACCESS
Application of shooting method on MHD thermally stratified mixed convection flow of non-Newtonian fluid over an inclined stretching cylinder

Khalil-Ur-Rehman and M.Y. Malik.....76

012013
OPEN ACCESS
CFD analysis for siting of wind turbines on high-rise buildings

K Veena, V Asha, C Arshad Shameem and T N Venkatesh.....88

012014
OPEN ACCESS
Development of streak instability modes excited by turbulent fluctuations

Huy Quang Ho, Masahito Asai and Shohei Takagi.....94

012015
OPEN ACCESS
Experiments in bypass boundary layer transition under a stream with and without shear

G Balamurugan and Alakesh C Mandal.....100

012016
OPEN ACCESS
Experiments on effective tripping device in a zero pressure gradient turbulent boundary layer

K Rengasamy and Alakesh Ch. Mandal.....106

012017
OPEN ACCESS
Generalized wall function and its application to compressible turbulent boundary layer over a flat plate

J Liu and S P Wu.....112

012018
OPEN ACCESS
Global stability analysis of axisymmetric boundary layer over a circular cone

Ramesh Bhoraniya and Narayanan Vinod.....119

012019
OPEN ACCESS
Numerical simulations of flow past a circular cylinder

Gaurav Chopra and Sanjay Mittal.....125

Computational fluid dynamics

012020
OPEN ACCESS
A comparative CFD study on the hemodynamics of flow through an idealized symmetric and asymmetric stenosed arteries

B Prashantha and S Anish.....131

012021
OPEN ACCESS
A hybrid method for improving the robustness of WENO scheme

Z H Ding, S P Wu and J Liu.....137

012022
OPEN ACCESS
Air flow optimization for energy efficient blower of biosafety cabinet class II A2

M D Ibrahim, M Z Mohtar, A A Alias, L K Wong, Y S Yunos, M R A Rahman, A Zulkharnain, C.S. Tan and R Thayan.....144

012023
OPEN ACCESS
Aqueous humour dynamics in anterior chamber under influence of cornea indentation

I Zuhaila, L Y Jiann, S Sharidan and A Fitt.....150

012024
OPEN ACCESS
Attributes of oscillatory physiological blood flow through 3-D geometry of single stenosed artery

Khairuzzaman Mamun, Mohammad Ali and Most. Nasrin Akhter.....159

012025
OPEN ACCESS
Blood-mimicking delivery in polygonal structure of inner quadrupletip microneedle with valveless micro-pump

M D Ibrahim, Y S Yunos, A R H Rigit, N A A Mohtadzar, N Watanabe, Y Sunami, M R A Rahman, L K Wong and M Z Mohtar.....166

012026
OPEN ACCESS
Complex fluid flow and heat transfer analysis inside a calandria based reactor using CFD technique

P. S. Kulkarni.....173

012027
OPEN ACCESS
Computational aerodynamic study of a supersonic reference projectile

S Zahir and Waseem Gul.....179

012028
OPEN ACCESS
Computational fluid dynamics simulation of laboratory scale reactor of fast pyrolysis fluidised bed

Seyed Amirmostafa Jourabchi, Hoon Kiat Ng, Suyin Gan and Zhong Jian Tan.....186

012029
OPEN ACCESS
Effect of mold designs on molten metal behaviour in high-pressure die casting

M D Ibrahim, M R A Rahman, A A Khan, M R Mohamad, M S Z M Suffian, Y S Yunos, L K Wong and M Z Mohtar.....198

012030
OPEN ACCESS
Implicit gradient reconstruction (IGR) method for compressible flow simulation

Manish K Singh, N Munikrishna, V Ramesh and N Balakrishnan.....204

012031
OPEN ACCESS
Influence of furnace cross section on the flow and velocity distribution in tangential firing furnace

A F Zainudin, H Hasini and S S A Fadhil.....211

012032
OPEN ACCESS
Investigation of flow inside pulverized coal (PC) pipes against coal particle size and air flow rate for a utility boiler

Savithry K. Thrangaraju, Kannan M. Munisamy and Rucgnes Apparao.....217

012033
OPEN ACCESS
Preliminary investigation on the effects of primary airflow to coal particle distribution in coal-fired boilers

N.A.W. Mohd Noor, H. Hassan, M.F. Hashim, H. Hasini and K.M. Munisamy.....230

012034
OPEN ACCESS
Nanofluid heat transfer between concentric independently rotating tubes with axial flow

Mohd Rosdzimin Abdul Rahman, Kin Yuen Leong, Azam Che Idris, Mohd Rashdan Saad and Mahmood Anwar.....236

012035
OPEN ACCESS
Novel fully Implicit Collocated grid incompressible flow solver on unstructured meshes

Jobin Joseph Unnupatt, N. Munikrishna and N. Balakrishnan.....241

012036
OPEN ACCESS
Numerical simulation of incompressible turbulent flows in presence of laminar to turbulent transition

G Satish, G A Vashista and Sekhar Majumdar.....249

012037
OPEN ACCESS
Numerical simulation of supersonic separating-reattaching flow through RANS

Rahul Kumar Soni, Nitish Arya and Ashoke De.....256

012038
OPEN ACCESS
Research in Varying Burner Tilt Angle to Reduce Rear Pass Temperature in Coal Fired Boiler

Savithry K. Thrangaraju, Kannan M. Munisamy and Saravanan Baskaran.....262

012039
OPEN ACCESS
The effect of nozzle diameter, injection pressure and ambient temperature on spray characteristics in diesel engine

Adiba Rhaodah Andsaler, Amir Khalid, Nor Sharifhatul Adila Abdullah, Azwan Sapit and Norrizam Jaat.....270

012040
OPEN ACCESS
Vein mechanism simulation study for deep vein thrombosis early diagnosis using cfd

Nabilah Ibrahim, Nur Shazilah Abd Aziz and Abreeza Noorlina Abd Manap.....277

012041
OPEN ACCESS
Vortex stretching in a homogeneous isotropic turbulence

M Hirota, Y Nishio, S Izawa and Y Fukunishi.....283

Environmental and natural fluid mechanics

012042
OPEN ACCESS
A numerical experiment on tidal river simplification in simulation of tide dominated estuaries

X Yin, L Jia and L Zhu.....289

012043
OPEN ACCESS
Buoyancy effect on atmospheric surface layer: measurements from the East Coast of Malaysia

Z. Harun, E. Reda and Rozli Zulkifli.....295

012044
OPEN ACCESS
CFD evaluation of erosion rate around a bridge near a sand dune

Wei He, Ning Huang, Hongchao Dun and Wenbo Wang.....302

012045
OPEN ACCESS
Drifting snow and its sublimation in turbulent boundary layer

Guang Li, Ning Huang and Zhengshi Wang.....308

012046
OPEN ACCESS
Effects of fractal grid on spray characteristics and flame development in burner combustion

S H Amirnordin, A Khalid, M Suardi, B Manshoor and M F Hushim.....314

012047
OPEN ACCESS
Investigation on terminal velocity and drag coefficient of particles with different shapes

Bin Xu, Ning Huang, Wei He and Youxing Chen.....319

012048
OPEN ACCESS
Propagation of landslide induced impulse wave in channel type reservoirs

An Yi, Liu Qingquan, Shi Chuanqi and Yang Jiaxiu.....325

012049
OPEN ACCESS
Study of a possible wake vortex encounter of an aircraft over the Arabian Sea

T N Venkatesh, C Arshad Shameem and J Saravanakumar.....331

012050
OPEN ACCESS
The effect of mountain wind on the falling snow deposition

Zhengshi Wang and Ning Huang.....338

Flow control and measurement

012051
OPEN ACCESS
Basic Coandă MAV Fluid Dynamics and Flight Mechanics

H Djojodihardjo and RI Ahmed.....344

012052
OPEN ACCESS
Blowing momentum and duty cycle effect on aerodynamic performance of flap by pulsed blowing

Ping Zhou, Yankui Wang, Jinjun Wang and Yongxiang Sha.....360

012053
OPEN ACCESS
Effect of nose perturbation on asymmetric vortices over a blunt-nose body at high angle of attack

Y.X. Sha, Y.K. Wang and Z.Y. Qi.....368

012054
OPEN ACCESS
Effect of splitter plate on fluid flow characteristics past a triangular cylinder

Sunakraneni Soumya and K Arul Prakash.....375

012055
OPEN ACCESS
Error propagation in the procedure of pressure reconstruction based on PIV data

Zhongyi Wang, Qi Gao, Runjie Wei and Jinjun Wang.....381

012056
OPEN ACCESS
Flow morphing by coaxial type plasma actuator

S Toyoizumi, H Aono and H Ishikawa.....387

012057
OPEN ACCESS
Solving Wave Flow Energy Propagated by Twin-Hull Ships (Catamarans)

Md Salim Kamil, Mazlan Muslim and Ishak Ali.....393

Flow visualization

012058
OPEN ACCESS
Comparision on dynamic behavior of diesel spray and rapeseed oil spray in diesel engine

Azwan Sapit, Mohd Azahari Razali, Mohd Faisal Hushim, Norrizam Jaat, Akmal Nizam Mohammad and Amir Khalid.....401

012060
OPEN ACCESS
Flow visualization around an apple with and without bagging

H Matsumoto, Y Kubota, M Ohishi and O Mochizuki.....408

Fluid machinery and industrial fluid mechanics

012061
OPEN ACCESS
Characteristics of a wind-actuated aerodynamic braking device for high-speed trains

H Takami and H Maekawa.....412

012062
OPEN ACCESS
Fluids vertical transfer utilizing VFD based centrifugal pumps

Vinesh Thiruchelvam and Lim Min Hong.....418

Heat and mass transfer

012063
OPEN ACCESS
Kinetic modeling of liquefied petroleum gas (LPG) reduction of titania in MATLAB

Tan Wei Yin, Sivakumar Ramakrishnan, Sheikh Abdul Rezan, Ahmad Fauzi Mohd Noor, Noor Izah Shoparwe, Reza Alizadeh and Parham Roohi.....426

012064
OPEN ACCESS
Numerical study on passive convective mass transfer enhancement

G P Aravind, K M Muhammed Rafi and M Deepu.....438

012065
OPEN ACCESS
Thermal cut-off response modelling of universal motors

Kashveen Thangaveloo and Yung Shin Chin.....444

Hydrodynamics and hydraulics

012066
OPEN ACCESS
Experimental study to the influences of rotational speed and blade shape on water vortex turbine performance

T C Kueh, S L Beh, Y S Ooi and D G Rilling.....454

012067
OPEN ACCESS
Freak Wave Formation from Phase Coherence

Arnida L. Latifah and E. van Groesen.....461

Unsteady flows and rotating flows

012068
OPEN ACCESS
The effects of the pulsatile period on the size of recirculation bubble in the vicinity of stent struts

B Jiang, V Thondapu, P Barlis, E K W Poon and A S H Ooi.....467

Others

012069
OPEN ACCESS
Analytical solution of subsonic gas flow past a circle

Kamyar Mansour.....473

012070
OPEN ACCESS
Average contra-rotation and co-rotation of line segments for flow field analysis

Jakub Šístek and Václav Kolář.....480

012071
OPEN ACCESS
Development of Flow over Blunt-Nosed Slender Bodies at Transonic Mach Numbers

Gireesh Yanamashetti, G. K. Suryanarayana and Rinku Mukherjee.....486

012072
OPEN ACCESS
Dynamical behaviour of non newtonian spiral blood flow through arterial stenosis

Mohammad Ali, Md. Mahmudul Hasan and Mahbub Alam Maruf.....492

012073
OPEN ACCESS
Numerical simulation of parametrically forced gravity waves in a circular cylindrical container

D Krishnaraja and S P Das.....498

012074
OPEN ACCESS
Performance of nickel-based oxygen carrier produced using renewable fuel aloe vera

NF Afandi, D Devaraj, A Manap and N Ibrahim.....505

012075
OPEN ACCESS
Rainfall forecast in the Upper Mahaweli basin in Sri Lanka using RegCM model

K M Muhammadh, M M M Mafas and S B Weerakoon.....512

012076
OPEN ACCESS
Streaming flows produced by oscillating interface of magnetic fluid adsorbed on a permanent magnet in alternating magnetic field

S Sudo, M Ito, Y Ishimoto and S Nix.....521

012077
OPEN ACCESS
Switching and transport mechanism in the branching tube network due to attached eddies

Yurika Nozaki, Kenta Ohhashi and Osamu Sano.....530

012078
OPEN ACCESS
Numerical simulation of vortex induced vibration of cylinders with flexible splitter plates

M Furquan, T R Sahu and S Mittal.....536

IOP Conf. Series: Journal of Physics: Conf. Series **822** (2017) 011001 doi:10.1088/1742-6596/822/1/011001

Asian Fluid Mechanics Committee

The Asian Fluid Mechanics Committee (AFMC) was founded in 1980 to advance research in Fluid Mechanics in Asia through interaction and exchange of information between workers in the field of Fluid Mechanics via the international conference series Asian Congress of Fluid Mechanics (ACFM). It is a member of the International Union of Theoretical and Applied Mechanics (IUTAM), and is the advisory board to ACFM.
Membership during 15ACFM is as below:

Chairman: Prof. G. S. Bhat (India)

Executive Committee:
Prof. Song Fu (China)
Prof. Duong Ngoc Hai (Vietnam)
Prof. Osamu Mochizuki (Japan)
Dr. C. S. Ow (Malaysia)

Members:
Prof. Y. T. Chew (Singapore)
Prof. Fazle Hussain (USA)
Prof. Hitoshi Ishikawa (Japan)
Prof. A. K. M. Sadrul Islam (Bangladesh)
Prof. Dr.-Ing. J. Ahmad Khan (Pakistan)
Prof. Victor V. Kozlov (Russia)
Prof. Sang-Joon Lee (Korea)
Prof. Qing-Quan Liu (China)
Dr. Sekhar Majumdar (India)
Prof. Sanjay Mittal (India)
Prof. Sri Redjeki Pudjaprasetya (Indonesia)
Prof. Ebrahim Shirani (Iran)
Prof. Chang Shu (Singapore)
Prof. Wei Shyy (HK, China)
Prof. C. P. Tso (Malaysia)
Prof. S. B. Weerakoon (Sri Lanka)
Prof. Jin-Jun Wang (China)

Organising Committee of 15th Asian Congress of Fluid Mechanics

Advisor: G. S. Bhat

Organizing Chairman: Kannan Munisamy

Committee Members (in alphabetical order):
Abreeza Noorliana Abd. Manap
Cheong Thiam Fook
Gopal Narian Kutty
Hasril Hasini
Kamal Ariffin Ahmad
Lim Chin Seong
Mohd. Zamri Yusoff
Nik Nazri Nik Ghazali
Noor Hisham Yahaya
Rizal Zahari
Savithry K. Thangaraju
Shahrir Abdullah
Siti Sarah Ain Fadhil
Srithar Rajoo
Tso Chih Ping
Yap Tze Chuen
Zambri Harun
Zulfikli Abdullah

Technical Committee of 15th Asian Congress of Fluid Mechanics

Technical Chairman: Tso Chih Ping

Secretary: Yap Tze Chuen

System Administrator: Lim Thong Leng

Web Master: Desmond Kho

Committee Members (in alphabetical order):
Chen Gooi Mee
Ervina Efzan
Hasril Hasini
Lee Gin Chong
Lim Boon Kian
Low Lay Chen
Mohd Zulkifly Abdullah
Shahrir Abdullah
Sim Hock Kheng

Than Cheok Fah
Toh Hoong Thiam
Zambri Harun

IOP Conf. Series: Journal of Physics: Conf. Series **822** (2017) 011003 doi:10.1088/1742-6596/822/1/011003

PREFACE

Through the mandate of the Local Organising Committee, the Local Technical Committee was formed to attend to the processing of the technical papers. Firstly I am thankful to advices from Chairman Dr. Kannan M. M. and Advisor Dr. C. S. Ow. I wish to thank each and every member of the Local Technical Committee for making possible this end result: the proceedings of a very prestigious Asian conference series on fluid mechanics. They are (in alphabetical order)

G. M. Chen	*Ervina Efzan*	*Hasril Hasini*
Desmond Kho	*G. C. Lee*	*B. K. Lim*
T. L. Lim	*L. C. Low*	*Shahrir Abdullah*
H. K. Sim	*C. F. Than*	*H. T. Toh*
T. C. Yap	*Zambri Harun*	*Zulkifly Abdullah*

Then the full credit for the standard of the Congress must go to the expert reviewers, both from within and outside Malaysia. They ensure that peer review was carried out through processes administered by the proceedings editors, in accordance with the high demand of IOP publication. The reviewers include (in alphabetical order)

AA Bidokhti	*Abdul Aziz Hairuddin*	*A. K. M. Sadrul Islam*	*Badrul Hisham*
G. S. Bhat	*S. Chang*	*G. M. Chen*	*Y. T. Chew*
K. H. Chuah	*J. J. Foo*	*Harijono Djojodihardjo*	*Hasril Hasini*
Hitoshi Ishikawa	*Y. M. Hung*	*Irfan Anjum Magami*	*Kahar bin Osman*
Kannan M. M.	*Kamarul Arifin Ahmad*	*Kenny Tan*	*C. S. Lim*
Q. Q. Liu	*Mehran Tadjfar*	*Mohd. Nazri*	*Mohd Radzi Abu Mansor*
Mohd Zamri Yusoff	*T. L. Mook*	*N. Balakrishnan*	*K. C. Ng*
Nik Nazri	*Noor Fadiya*	*Osamu Mochizuki*	*C. S. Ow*
Rahim Abu Talib	*Sanjay Mittal*	*Saqaff Ahmed Alkaff*	*Sekhar Majumdar*
Shahrir Abdullah	*Shuhaimi bin Mansor*	*Siti Aslina*	*Sivanandam Sivasankaran*
Sri Redjeki Pudjaprasetya	*Srithar Rajoo*	*Sudi Mungkasi*	*C. F. Than*
H. T. Toh	*C. P. Tso*	*Vinod Kumar Venkiteswaran*	*Wan Ahmad Najmi*
J. J. Wang	*H. K. Yeoh*	*Zambri Harun*	*Zulkifly Abdullah*

Last but not least, we wish to thank all the participants who contributed and attended the conference and made the event a success.

C. P. Tso

Editor -in -Chief

Peer review statement

All papers published in this volume of *Journal of Physics: Conference Series* have been peer reviewed through processes administered by the proceedings Editors. Reviews were conducted by expert referees to the professional and scientific standards expected of a proceedings journal published by IOP Publishing.

Criterion for vortex acoustic lock-on in combustors with backward facing step

B Singaravelu[1] and S Mariappan[2]

Department of Aerospace Engineering, Indian Institute of Technology Kanpur, Kanpur-208016, India
[1] Postdoctoral fellow
[2] Assistant Professor and Corresponding author

E-mail: [1] sbala@iitk.ac.in, [2] sathesh@iitk.ac.in

Abstract. In thermoacoustic systems vortex separation and breakdown are an important source of heat release rate fluctuations. The coupling between the acoustic field and the energy released by vortex breakdown can cause instability in these systems under certain operating conditions. Non-linear simulations show that combustion instability occurs after vortex acoustic lock-on. The objective of this work is to determine the criterion for vortex acoustic lock-on and thus the criterion for combustion instability as well. The 0-to-peak amplitude of the velocity fluctuations in the system are used as a criterion to determine the transition of Helmholtz number from vortex shedding mode to acoustic mode. Helmholtz number and acoustic pressure fluctuations computed using the model are compared with experimental investigations. The phenomenological model shows a good agreement with the experiments and the criterion determined for vortex lock-on provides a better understanding of the phenomenon.

1. Introduction

In thermoacoustic systems heat release rate fluctuations excite acoustic oscillations and the acoustic oscillations in turn affect the source of heat release. According to Rayleigh's criterion, thermoacoustic instability is encouraged if the heat release rate fluctuations are in phase with the pressure oscillations. Contemporary combustors use swirl flows, along with a backward facing step to stabilize the flame [1]. The problem of vortex shedding and breakdown are intrinsic to these combustors because of their geometrical characteristics. A number of experiments [2, 3, 4] have shown that during stable operation of a combustor there is low amplitude broadband noise and the dominant frequency is the vortex shedding frequency. After the onset of instability, there is a high amplitude discrete tone and the dominant frequency is found to be the acoustic frequency. The non-linear driving causes vortex shedding to occur at acoustic frequency. This phenomenon is termed as vortex acoustic lock-on. The objective of this work is to obtain the criterion for vortex acoustic lock-on and investigate the possibility of using this criterion as a precursor for instability.

A reduced-order model to describe the interaction among vortex separation, chamber acoustics and combustion process was given by Matveev & Culick[5]. Nair & Sujith[6] showed that the reduced order model can capture intermittent burst oscillations and subsequent flow acoustic lock-on observed in combustors. They postulated that the intermittency can be used as an early warning signal to an impending combustion instability.

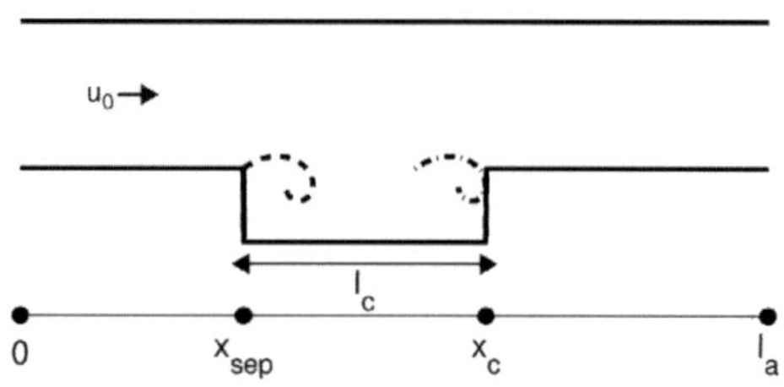

Figure 1. Geometry of a vortex shedding combustor with a backward facing step. The figure is not drawn to scale. Vortex formation and breakdown occur at x_{sep} and x_c respectively.

In the previous works [5, 6], the phenomenological model was used to study only a small range of operating conditions. Even though vortex acoustic lock-on was observed in both the model and the experiments for a given system configuration, the criterion for lock-on was not postulated. Such a criterion will allow us to define the safe operating regime for these combustors beforehand. The detailed objectives of this work are as follows: to explain the phenomenon of vortex acoustic lock-on, to find the criterion for lock-on as a function of all the system parameters and to establish the validity of this criterion by comparing the model with the experimental investigations available in the literature.

2. Theory

In this section we present the theory, governing vortex shedding combustors with a backward facing step. We deviate from the previous works [5, 6] by non-dimensionalising all the equations to reduce the number of free parameters in the system. The vortex shedding process is modelled using the method described by [5]. The present thermoacoustic system is a combustor with a backward facing step as shown in figure 1. The total length of the acoustic chamber is denoted by l_a. The backward facing step is located at x_{sep}. This is the point where vortex separation occurs. The separated vortex impinges and breaks down at the point denoted by x_c and the length of the cavity denoted by l_c. The ends of the acoustic chamber are assumed to be open.

The following scales are chosen for non-dimensionalising the equations governing the vortex shedding process and the acoustic flow field: $M = u_0/c_0, u' = \tilde{u}'/u_0, p' = \tilde{p}'/\gamma M p_0, t = c_0\tilde{t}/l_a$ and $x = \tilde{x}/l_a$. The steady state quantities are velocity (u_0), pressure (p_0) and the speed of sound (c_0). The other dimensional quantities are denoted with a *tilde* above them: velocity fluctuation ($\tilde{u}'$), pressure fluctuation ($\tilde{p}'$) and time ($\tilde{t}$). In case of non-dimensional quantities the tildes are dropped. The steady state vortex shedding frequency $f_{s0} = M\,St/d$, where St is the Strouhal number. Assuming uniform steady state density ρ_0, for low Mach number approximation, the linearised non-dimensional momentum and energy equations, which govern the acoustic field of this system are written as follows:

$$\frac{\partial u'}{\partial t} + \frac{\partial p'}{\partial x} = 0, \tag{1}$$

$$\frac{\partial p'}{\partial t} + \frac{\partial u'}{\partial x} + \zeta p' = \frac{(\gamma-1)}{\gamma}\frac{\beta}{M}\sum_m \Gamma_m \delta(t-t_m)\delta(x-x_c), \tag{2}$$

$$\frac{\mathrm{d}\Gamma_m}{\mathrm{d}t} = \frac{M}{2}\left[1+u'(x_{sep},t)\right]^2, \tag{3}$$

where β is the non-dimensional heat release coefficient, Γ_m is the circulation of the m-th vortex [5] and ζ is an artificially added acoustic damping term. The criterion for separation of the m-th vortex is $\Gamma_m \geq \Gamma_{sep}$, where $\Gamma_{sep} = [1+u'(x_{sep},t)]\,d/2St$. A separated vortex convects with velocity $dx_m/d\tilde{t} = M[\alpha + u'(x,t)]$, where α is the coefficient usually chosen in the range 0.4 to 0.6 [7].

The non-dimensional partial differential Eq. (1) and Eq. (2) are reduced to a set of ordinary differential equations (ODEs) by expanding the acoustic variables in terms of basis functions using the Galerkin technique. The mode shapes chosen for decomposing u' and p' are

$$u'(x,t) = \sum_{n=1}^{N} \cos(k_n x) U_n(t), \; p'(x,t) = \sum_{n=1}^{N} \sin(k_n x) P_n(t), \tag{4}$$

which ensures that the open-open boundary conditions of the acoustic system are satisfied. In the above expressions, U_n, P_n are the amplitude of the n-th Galerkin modes. The associated non-dimensional wave number of the nth mode $k_n = n\pi$, where $n = \{1, 2, \cdots N\}$. Substituting equations (4) in the partial differential equations (1) and (2) gives the following set of ODEs for the momentum and energy equation:

$$\dot{U}_n + k_n P_n = 0 \tag{5}$$

$$\dot{P}_n + 2\zeta_n k_n P_n - k_n U_n = 2\frac{(\gamma-1)}{\gamma}\frac{\beta}{M} \sin(k_n x_c) \sum_m \Gamma_m \delta(t - t_m). \tag{6}$$

where $\zeta_n = \frac{1}{2\pi}\left(c_1 \frac{\omega_n}{\omega_1} + c_2 \sqrt{\frac{\omega_1}{\omega_n}}\right)$ is the frequency dependent damping term [5]; c_1 and c_2 are the damping coefficients which are responsible for the end loses and loses due to boundary layers respectively. We rewrite the system of coupled first order differential equations (5) and (6) as a second order differential equation in U_n.

$$\ddot{U}_n + 2\zeta_n k_n \dot{U}_n + k_n^2 U_n = -2k_n \frac{(\gamma-1)}{\gamma}\frac{\beta}{M} \sin(k_n x_c) \sum_m \Gamma_m \delta(t - t_m), \tag{7}$$

is the second order differential equation which represents the dynamics of the thermoacoustic system presented. The dynamics is similar to that of a kicked oscillator, which behaves like a damped oscillator when the kicks are absent. The following jump conditions are obtained by integrating equation (7) in the time interval t_m^- to t_m^+, corresponding to time instants just before and just after the m-th kick.

$$\left.\begin{aligned} U_n^+ - U_n^- &= 0, \\ \dot{U}_n^+ - \dot{U}_n^- &= -2k_n \frac{(\gamma-1)}{\gamma}\frac{\beta}{M} \sin(k_n x_c)\Gamma_m \end{aligned}\right\} \tag{8}$$

gives the jump in pressure when a vortex impinges on the combustor wall. The strength of the kick for a fixed geometric configuration is given by $\beta\Gamma_m/M$.

2.1. Procedure to compute the fluctuations u' and p'

First, we set the geometric parameters $\{x_{sep}, x_c, d\}$ and the flow parameters $\{M, c_1, c_2, \alpha, \beta, St\}$. Next, we set the size of the time step dt, the number of time steps and the number of Galerkin modes N. Then, we solve for U_n and $\dot{U}_n$ from equations (5) and (7) using fourth order Runge-Kutta method. The above computations are solved in the absence of kicking ($t \neq t_m$) so the right hand side of equation (7) is set to zero. After that, we compute the circulation at x_{sep}, Γ_m and the critical circulation Γ_{sep}. If $\Gamma_m \geq \Gamma_{sep}$ the m-th vortex is shed and the circulation of the $(m+1)$-th vortex is computed in the next time step, else the circulation is incremented. All separated vortices in the cavity are convected for time step dt. If a vortex impinges at x_c, the jump conditions given by equation (8) are applied. Thus, the non-periodic Dirac delta function on the right hand side of equation (7) is solved. The same is repeated for desired number of time steps. Finally, the fluctuations at any location can be computed using equations (4).

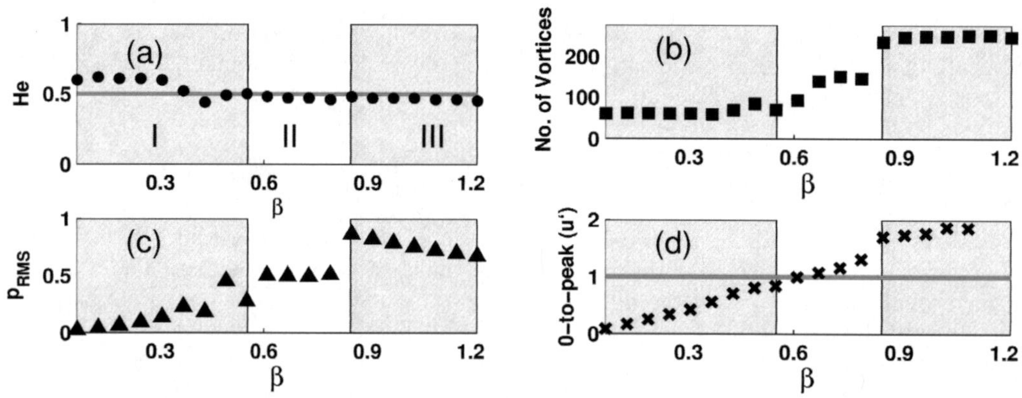

Figure 2. For $M = 0.03$: (a) Helmholtz number as a function of β, $He = 0.5$ corresponds to the acoustic mode. (b) Total number of vortices shed in 200 cycles as a function of β. (c) p_{RMS} as a function of β (computed after discarding the first 100 cycles), the vortex acoustic lock-on correlates with the sharp increase in the number of vortices shed and the p_{RMS}. (d) The 0-to-peak amplitude of u' as a function of β. As β is increased three regimes are noted, which are explained in detail in section 3.1.

3. Results

The geometrical parameters are set for all the case studies as follows $\{x_{sep} = 0.3,\ x_c = 0.45,\ d = 0.01\}$. The damping coefficients $c_1 = 0.315$, $c_2 = 0.015$[5], the vortex convection coefficient $\alpha = 0.4$ [7] and Strouhal number $St = 0.1$ are also fixed for all case studies. The system is studied for various values of M and β. For the non-linear simulations the size of the time step is set to be $\mathrm{d}t = 10^{-4}$ and the number of Galerkin modes is chosen as $N_{modes} = 32$. The number of modes chosen here is higher than the number chosen in earlier investigations [5, 6]. The quantities of interest, p' and u' are computed at the location x_{sep} for all the case studies. The dominant frequency in the Fourier transform of p' gives the Helmholtz number He. In this study values of He close to $\{0.5, 1, 1.5, \cdots\}$ correspond to the acoustic mode. If $He = mf_{s0}$, $m = \{1, 2, 3, \cdots\}$, the dominant mode is the vortex shedding mode.

3.1. Case study #1: Phenomenon of vortex lock-on

In this case study we explain the phenomenon of vortex lock-on using by varying β for a fixed $M = 0.03$. For the chosen system configuration the steady state frequency $f_{s0} = 0.3$. Figure 2 shows the computed values of Helmholtz number (He), number of vortices shed, root mean square value of the pressure fluctuations (p_{RMS}) and the 0-to-peak amplitude of u' as a function of β. In general the following sequence of events are observed for increasing values of β. In regime I, the $He = 0.6 = 2f_{s0}$, therefore the dominant mode is the vortex shedding mode. Next, in regime II, starting at $\beta = 0.55$, the number of vortices shed increases by more than 100% and $He = 0.5$ corresponds to the acoustic mode. Finally, in regime III, starting at $\beta = 0.85$, the number of the vortices shed increases 5-fold compared to the steady state value and there is a sharp increase in p_{RMS} indicating combustion instability. We also find that the vortex acoustic lock-on occurs (at $\beta = 0.55$) well ahead of the combustion instability (at $\beta = 0.85$). The same is true for most of other Mach numbers in the range $0.01 \leq M \leq 0.1$ for which the study was performed.

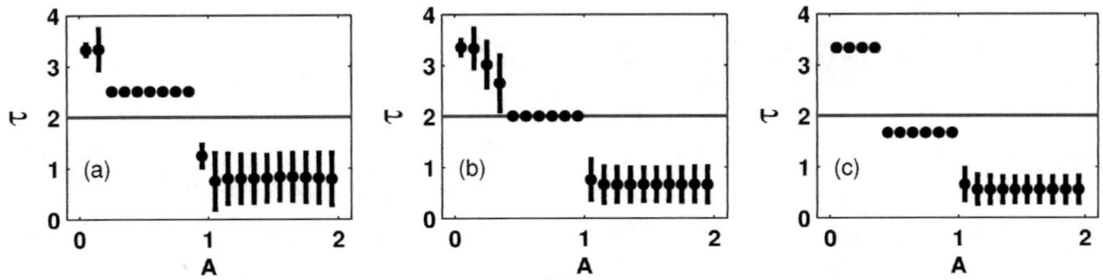

Figure 3. For $M = 0.03$ the time period between vortex breakdown (τ) is shown as a function of A for three different values of ω. (a) $\omega = 2\pi(0.4)$, (b) $\omega = 2\pi(0.5)$ and (c) $\omega = 2\pi(0.6)$. For small A, τ depends only on the steady state vortex shedding frequency. For intermediate values of A, τ locks-on to the forcing frequency and the error bars are negligible. Around $A = 1$, τ becomes aperiodic and is lower than both the steady state frequency and the forcing frequency. For A≥1, the error bars are significant in all the case studies. Here $A = 1$ is analogous to velocity fluctuations $u'(x_{sep}) = 1$.

3.2. Case study #2: Criterion for vortex lock-on

In this case study we find the criterion for lock-on as a function of the system parameters. In figure 2d, we find that the phenomenon of vortex lock-on correlates with the 0-to-peak amplitude of $|u'| > 1$ in the first 200 time cycles. The validity of this criterion can be tested by forcing the system using the function $u'(x_{sep}) = A\sin\omega t$. For $A \geq 1$, we expect the system change its dynamics. Figure 3 shows the mean and standard deviation of the time period between the kicks τ as a function of A for three different values of ω. For $M = 0.03$, $f_{s0} = 0.3$. Figure 3a shows the results for forcing frequency $\omega = 2\pi(0.4)$. For small A, vortex break down occurs at $\tau_{s0} = 1/f_{s0}$. Then it locks on to the forcing frequency of 0.4, which corresponds to $\tau = 2.5$ cycles. The 1σ standard deviation in τ is noticeably small. Around $A = 1$, the vortex breakdown becomes aperiodic and 1σ standard deviation in τ becomes large. Similar results are observed for $\omega = 2\pi(0.5)$ and $\omega = 2\pi(0.6)$. This change is dynamics correlates with vortex acoustic lock-on. Therefore, for a given system configuration the value of β for which 0-to-peak amplitude of $|u'(x_{sep})| > 1$, is designated as $\beta_{lock-on}$. For all the non-linear simulations performed, we observed that the vortex lock-on occurs at $\beta_{lock-on} \pm 0.2\beta_{lock-on}$.

3.3. Case study # 3: Comparison with experiments

In this case study we validate the criterion obtained in section 3.2 by comparing the phenomenological model with experimental investigations available in the literature. Chakravarthy et. al.,[3] showed the transition of He from vortex shedding mode to acoustic mode for a number of system configurations and found that this transition is marked by non-linearity in the p_{RMS} variations. In current study we reproduce the experimentally observed transitions using the model described in section 2. The experimental setup of [3] is as follows. The non-dimensionalised geometrical parameters of the configuration are $\{x_{sep} = 0.3, x_c = 0.45, d = 0.0224\}$. The air inlet velocity ranges from 3 ms^{-1} to 30 ms^{-1}, which corresponds to $0.004 \leq M \leq 0.04$. The speed of sound is assumed to be 750 ms^{-1}. In the experimental investigations of [3], the fuel flow rate is fixed as $\dot{m}_f = 229mg/s$. This information is not sufficient to determine β. Therefore, we perform all simulations for $\beta_{lock-on}$, which is a lower limit for combustion instability, according to our criterion.

Figures 4a & 4b show He and p_{RMS} as a function of M for both the experimental

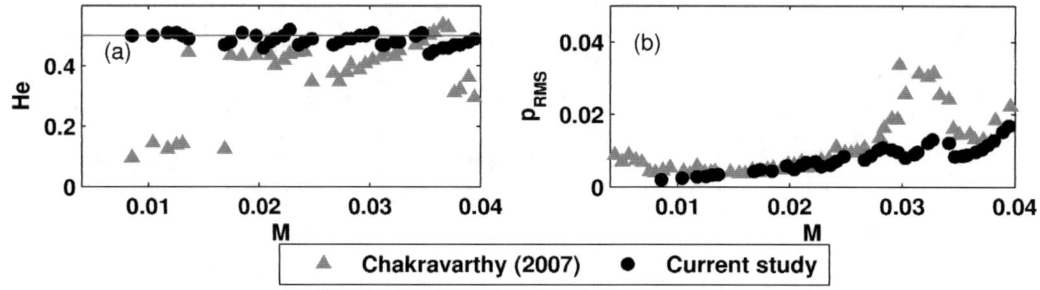

Figure 4. Comparison between experimental results of [3] and non-linear simulations in this study. (b) He as a function of M (c) p_{RMS} as a function of M.

investigations of [3] and the current study. In case of He there is a good agreement between the theory and the experiments for $0.017 < M < 0.037$. In both the experiments and non-linear simulations, we observe that He is close to the acoustic mode for $0.017 < M < 0.025$ and He is a multiple of the vortex shedding mode for $0.025 < M < 0.037$. The p_{RMS} given by [3] in units of Nm^{-2} is non-dimensionalised using the factor $1/\gamma M p_0$, where $\gamma = 1.4$ and $p_0 = 101325\mathrm{Pa}$. There is a good match in the trend between the theoretical and experimental values of p_{RMS} for all the M. The mismatch in the range $0.03 < M < 0.033$ can be attributed to the insufficient knowledge of all the free parameters. The role of other hydrodynamic phenomenon in the discrepancy can also be not ruled out.

4. Conclusion

Using a phenomenological model, we empirically find that the criterion for vortex acoustic lock-on highly correlates with the 0-to-peak amplitude of velocity fluctuations. The non-linear simulations show that the vortex acoustic lock-on occurs when the 0-to-peak velocity fluctuations are close to $u' > 1$. The same criterion is validated by externally forcing the system. We find that the system changes its dynamics when the amplitude of forcing is greater than one. After lock-on, vortex shedding occurs at the acoustic frequency and the low amplitude broadband noise becomes a high amplitude discrete tone, which is a characteristic of combustion instability. We use this criteria for vortex acoustic lock-on to predict the Helmholtz number (He) and the RMS fluctuations in pressure (p_{RMS}) during combustion instability. The predictions are compared with the available experimental results. A good agreement is observed in the trend of He and p_{RMS}.

The authors are grateful to the Indian Institute of Technology Kanpur for the financial support during the period of this research work, through *post-doctoral fellowship* and *Initiation grant (IITK/AE/20130179)*.

References

[1] Huang Y and Yang V 2009 *Progress in Energy and Combustion Science* **35** 293–364
[2] Yu K, Trouve A and Daily J W 1991 *J. Fluid Mech.* **232** 47–72
[3] Chakravarthy S R, Shreenivasan O J, Boehm B, Dreizler A and Janicka J 2007a *The Journal of the Acoustical Society of America* **122**
[4] Chakravarthy S R, Sivakumar R and Shreenivasan O J 2007b *Sadhana* **32** 145–154
[5] Matveev K and Culick F E C 2003 *Combust. Sci. Tech.* **175** 1059–1083
[6] Nair V and Sujith R I 2015 *Proceedings of the Combustion Institute* **3** 3193–3200
[7] Dotson A P, Koshigoe S and Pace K K 1997 *Journal of Propulsion and Power* **13** 197–206

Design of a novel and efficient lantern wind turbine

M D Ibrahim[1], L K Wong[1], and M Anyi[2], Y S Yunos[1], M R A Rahman[1] and M Z Mohta[1]

[1]Department of Mechanical and Manufacturing, Faculty of Engineering, Universiti Malaysia Sarawak, Malaysia
[2]Department of Electrical and Electronic, Faculty of Engineering, Universiti Malaysia Sarawak, Malaysia

E-mail: imdanial@unimas.my

Abstract. Wind turbine generates renewable energy when the forces acted on the turbine blades cause the rotation of the generator to produce clean electricity. This paper proposed a novel lantern wind turbine design compared to a conventional design model. Comparison is done based on simulation on coarse and fine meshing with all the results converged. Results showed that the pressure difference on the surface of novel design lantern wind turbine is much higher compared to the conventional wind turbine. Prototype is already manufactured and experimental result would be discussed in a separate future publication

1. Introduction

The lantern wind turbine is novel designed wind turbine that has multiple blades instead of one blade. Wind turbine is a type of renewable energy. It generates electricity when wind blow that acted as force on the blades of the wind turbine which causes the rotor to rotate produce electricity. United kingdom has the target to achieved 15% of energy consumption which came from renewable energy by 2020 [1]. The renewable energy reduces the use of natural gas (hydrocarbon fossil fuel) to generate electricity. The fossil fuel produces carbon dioxide and carbon monoxide during the incomplete combustion, which lead to environmental pollution [2]. Wind turbine reduces the usage of natural gas (hydrocarbon fossil fuel). In comparison, wind turbine system is capable of producing 5-8GWh of energy annually, which is equal to 1 tonne of burning coal to produce electricity.

As a result, the big and large wind turbine, known as wind farms with high renewable energy source had currently carried out by many countries [3]. However, the conventional wind turbine is huge consume a lot of space. It requires distance to be maintained when these wind turbines are placed. The diameter of a wind turbine can be larger than 124 m with the height of 200 m [4]. The huge and big conventional wind turbine has taken up lots of spaces and a distance is required to place another wind turbine next to it [5]. The development of wind power plant which taken up a lot of space have a direct impact towards citizen and road accessibility. Wind power plant project is usually time consuming therefore it requires the cleaning of lands for development in the particular area, which has the possibility of significant degradation and effects in quality of the ecosystem [6].

The purpose of this study is to reduce the large wind turbine to a smaller scale and improve its efficiency. The concept of this lantern wind turbine is obtained from the airplane nacelles[7]. This wind turbine has been produce with circular blades instead of just normal wind turbine blades. This increases the volume of wind turbines, which allow aerodynamic lifting. This study embarks on determining the pressure and velocity distribution over wind turbine on a smaller scale wind turbine. Table 1 shows the nomenclature of this paper.

Table 1: Nomenclature

Symbol	Name
HAWT	Horizontal axis wind turbine
VAWT	Vertical axis wind turbine
P	Power (W)
F	Force vector (N)
λ	Tip speed ratio
v	Wind speed (m/s)
ω	Blade revolution per minutes (rad/s)
R	Radius of the wind turbine (m)
$\bar{x}$	Mean (average values)

2. Aerodynamic consideration

The basics theory of lantern wind turbine is related to aerodynamics. Aerodynamics is the interaction between the motion of air and the airfoil design with a curved surface on the top and flat surface at the bottom. When the laminar flow of air passes through the airfoil, the speed above the airfoil is higher which produces lower pressure. The airfoil has a lower speed below it; these differences produce high pressure, which leads to lifting where the airfoil is lifted perpendicular to the direction of the wind flow. Lantern wind turbine is design with circular airfoil. High pressure outside the circular airfoil and lower pressure inside the airfoil create a boosting lifting effect for lantern wind turbine. The tip speed ratio is the ratio of the tip speed of the blade and the velocity of the wind flow [8].

$$\lambda = \frac{tip\ speed\ of\ the\ blade}{wind\ speed}$$

$$\lambda = \frac{\omega R}{v} \tag{1}$$

Where;

λ is tip speed ratio, ω is the blade revolution per minute, R is the radius of the wind turbine and v is the wind speed.

3. Betz's law

Albert Betz (1919) concluded that a wind turbine, which generates electricity through a rotor, the rotor of the wind turbine cannot convert more than 16 over 27 of kinetic energy to mechanical energy [9]. Betz limit results shows that utility-scale wind turbines can be achieve at peak 60% to 70%.

4. CFD simulation for Flow Simulation, fluid flow from one direction (wind turbine is rotating)

Table 2 shows the setting parameter of flow simulation. table 3 shows the setting for boundary condition of airflow over wind turbine. Velocity flow is set as 5, 6, 7, 8 and 9 (m/s) as the wind speed flowing through wind turbine. The rotation of the wind turbine is set as constant 2 rad per second, which is 19.1 (rpm). Solving mode, which is a serial and parallel mode. Serial mode is used in this case and the meshing process used for this study are auto meshing with the level of meshing set to the maximum of 8 (fine mesh) The minimum wall thickness and minimum gap size for meshing is set as

0.033m. All the result converged before 400 iteration. Table 4 shows the study condition which is the calculations end upon convergence and it is set to stop when iteration reached 1000. Coarse result is not discuss here, most of the journal has proven that fine meshing has more accurate reading compared to experimental result.

Table 2: Parameter for analysis setting

Detail	Setting
Analysis types	External flow
Fluid flow	Air
Flow type	Laminar and turbulence

Table 3: Boundary condition

Detail	Setting
Environment condition	101kPa
Velocity flow x-axis	0 m/s
Velocity flow y-axis	5 to 9 m/s
Velocity flow z-axis	0 m/s

Table 4: Study (run) finishing condition

Parameter	Value
Finish condition	All satisfied
Maximum iteration	1000
Goal convergence (Goal's criteria)	min velocity x-axis, max velocity x-axis, average velocity x-axis, min velocity y-axis, max velocity y-axis, average velocity y-axis, min velocity z-axis ,max velocity z-axis, average velocity z-axis, normal force x, normal force y, normal force z, force x, force y, force z, frictional force x, frictional force y, frictional force z, torque x, toque y and torque z

5. Design with AutoCAD®

The anemometers, rotor Darriues-H and lantern wind turbine are drawn with Auto-CAD® for numerical simulation comparison purposes. To determine the flow condition through simulation, the wind turbine has 60 cm length of radius. All model are design with the same length of radius.

5.1. Dimension

The figure 4 and figure 5 show the dimension of lantern wind turbine. The lantern wind turbine is designed with circular aerodynamic shape with a gap distance of 2 cm between each blade.

Figure 1. Market wind turbine design of Rotor Darrieus-H

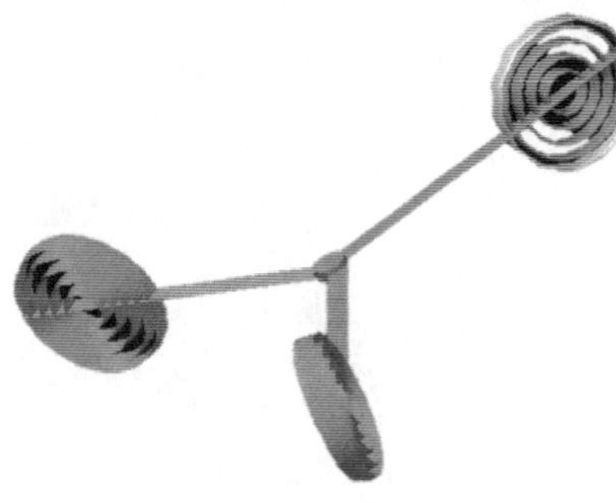

Figure 2. Novel Lantern wind turbine design

6. CFD simulation The Flow Simulation Solver fluid flow under rotational condition

Table 5: flow velocity and wind turbine rotation

Detail	Rotor Darrieus-H	Lantern wind turbine
Velocity of 5m/s with 2 rad/s rotation	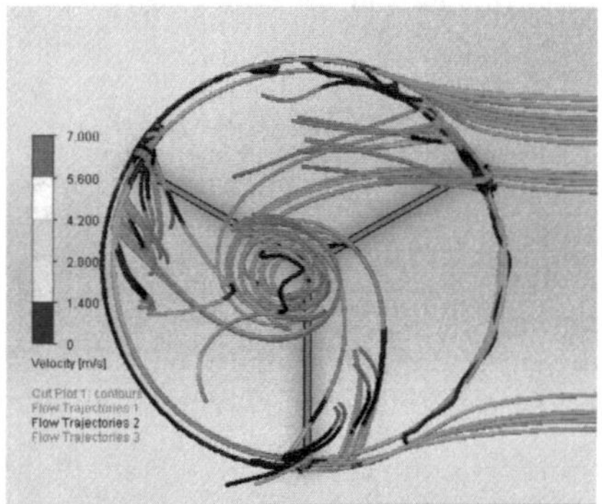	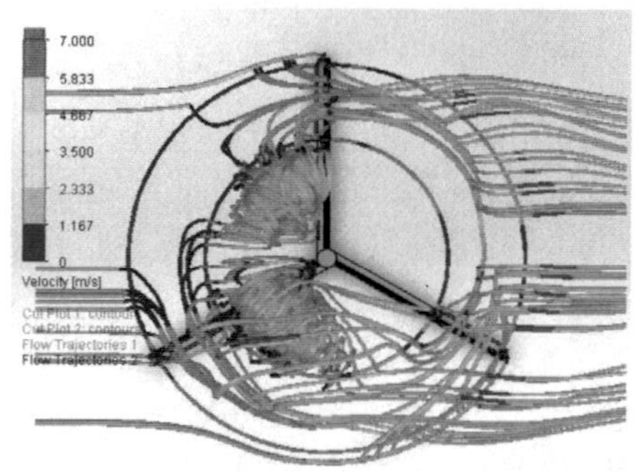
Velocity of 6m/s with 2 rad/s rotation	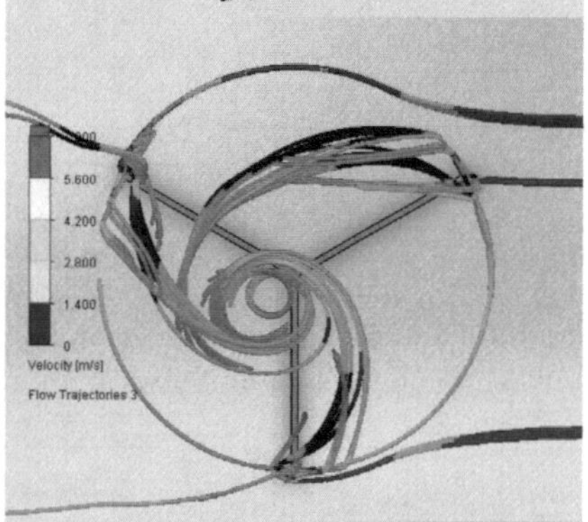	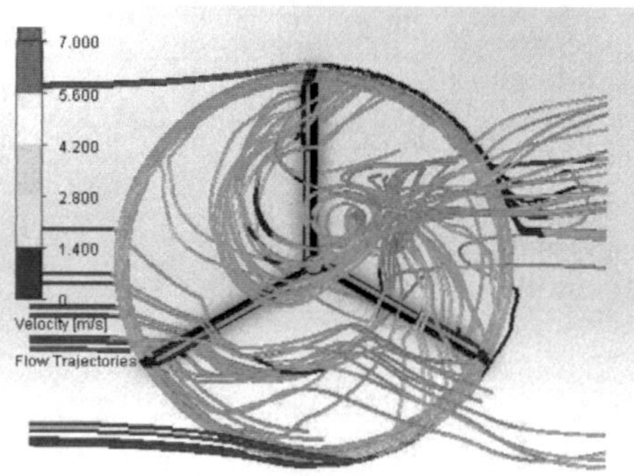
Velocity of 7m/s with 2 rad/s rotation	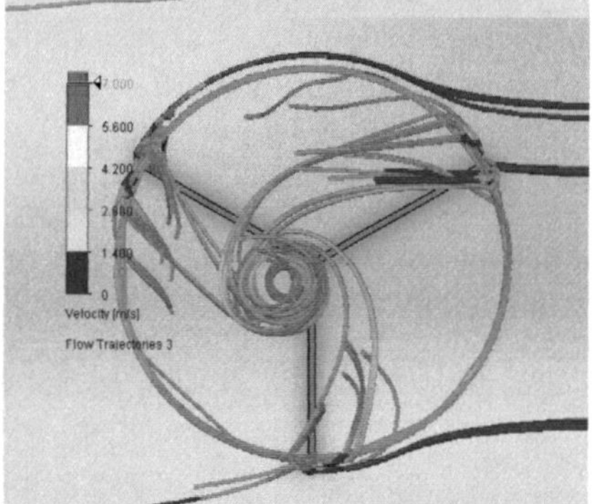	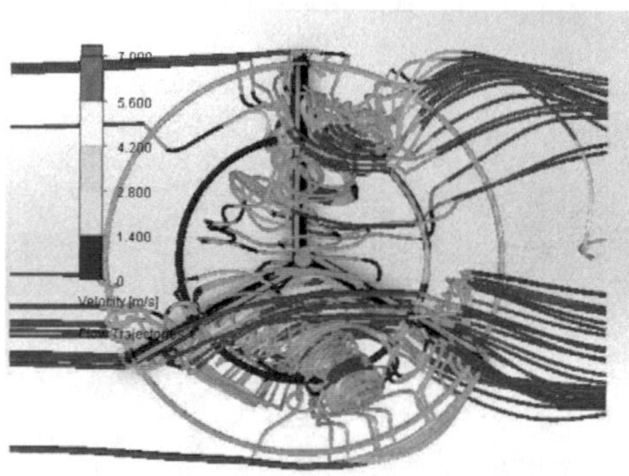
Velocity of 8m/s with 2 rad/s rotation	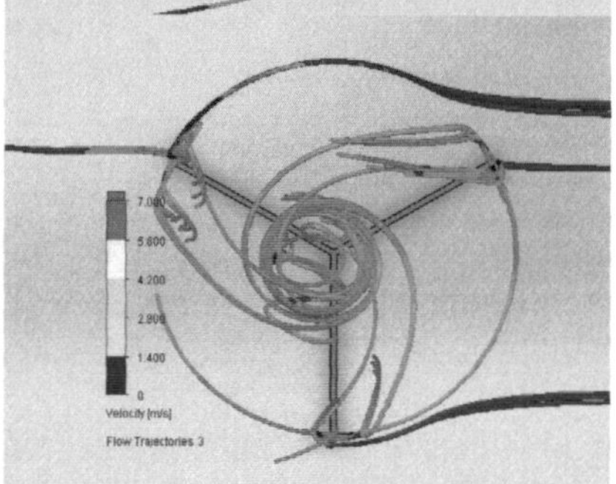	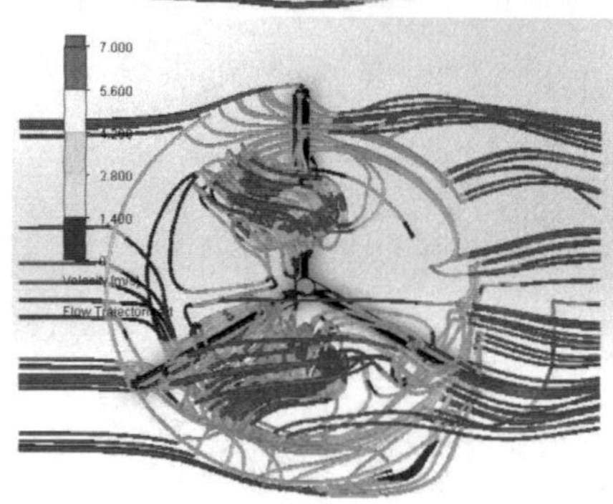

Velocity of 9m/s with 2 rad/s rotation	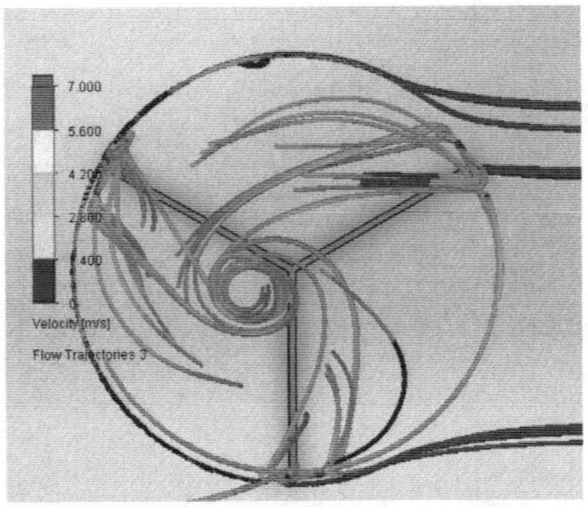	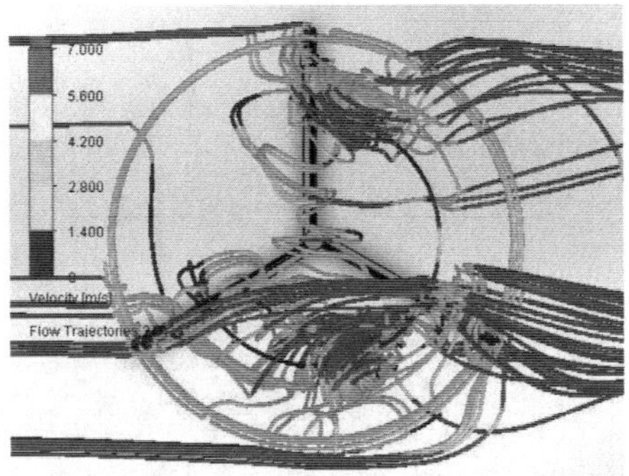

Based on the result obtained Velocity of 5m/s with 2 rad/s rotation from simulation the lantern wind turbine actually has very high shows the formation of unsteady vortices shed which form at the centre of the lantern wind turbine. This unsteady shed vortices leads to vibration and reduces the performance of the wind turbine.

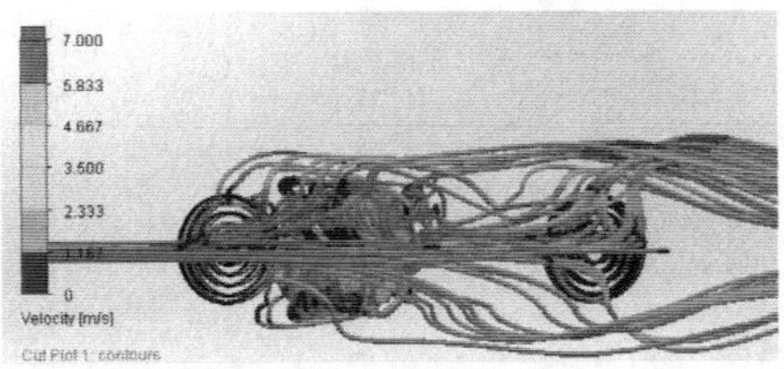

Figure 3. Lantern wind turbine design

In horizontal view, the flow condition with blue colour shows the airfoil vortex wake this increases the drag on the blade performance. The wind turbine design of Rotor Darrieus-H with velocity of 5m/s with 2 rad/s rotation shows a smooth pressure distributions flow on the blades and slight vortices wake at the centre of rotation. The velocity of 6m/s with 2 rad/s rotation shows that Lantern wind turbine the average wind speed rotating velocity of Lantern wind turbine is equal to the surrounding velocity this lead to the difference between the results compared to other results. The wind turbine design of Rotor Darrieus-H still shows a smooth pressure distributions flow on the blades and slight vortices wake at the centre of rotation.The wind turbine design of Rotor Darrieus-H shows an increase of vortices wake on the blade as the velocity increases and reduces in vortices wake at the centre of rotation of the wind turbine. Overall distributions still shows that it has a highly stable velocity flow distribution with low changes.

7. CFD simulation contour view

Table 6: flow velocity and wind turbine rotation contour view

Velocity of 5m/s with 2 rad/s rotation	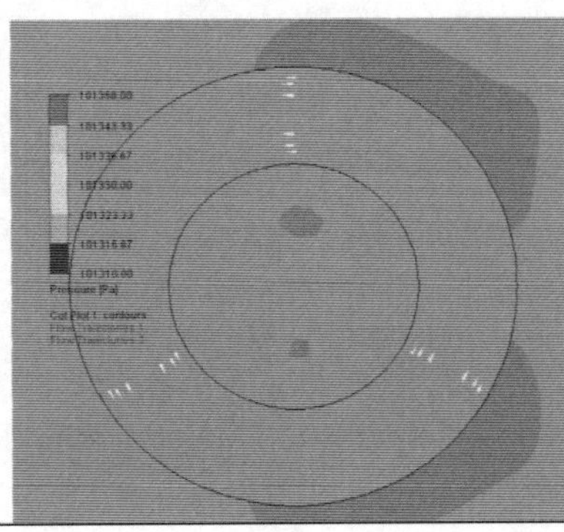	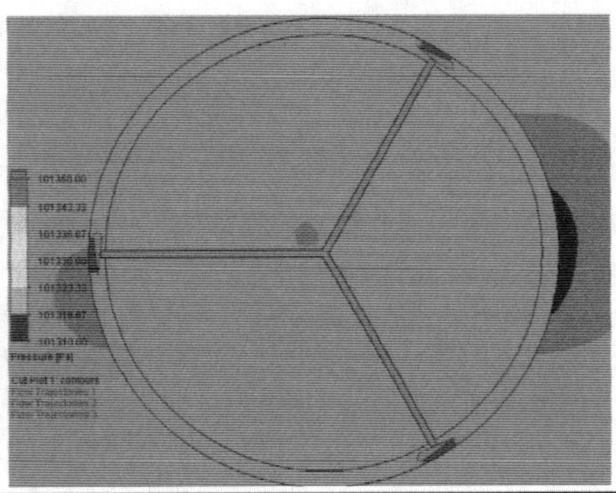

Velocity of 6m/s with 2 rad/s rotation	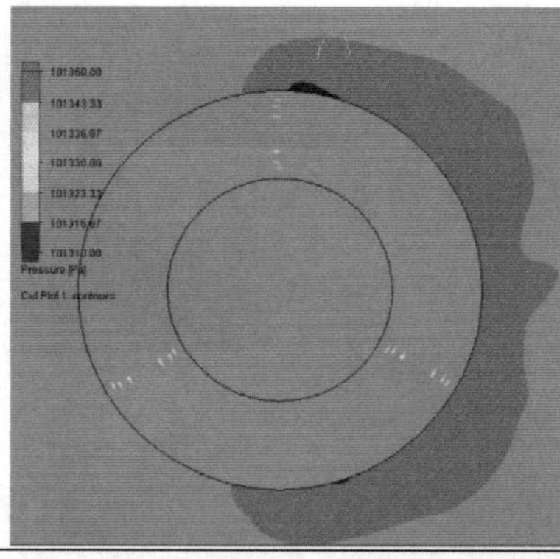	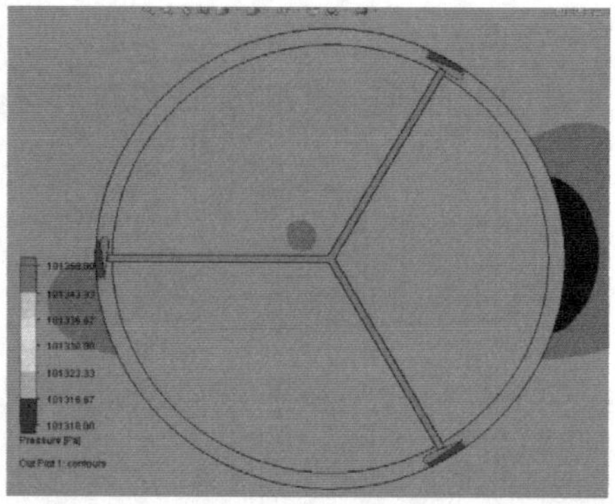
Velocity of 7m/s with 2 rad/s rotation	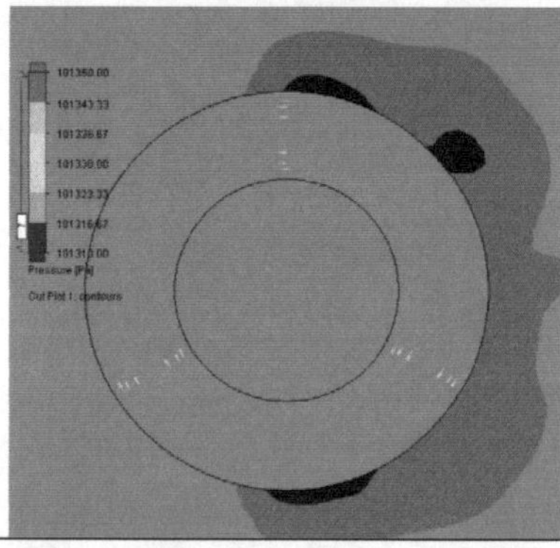	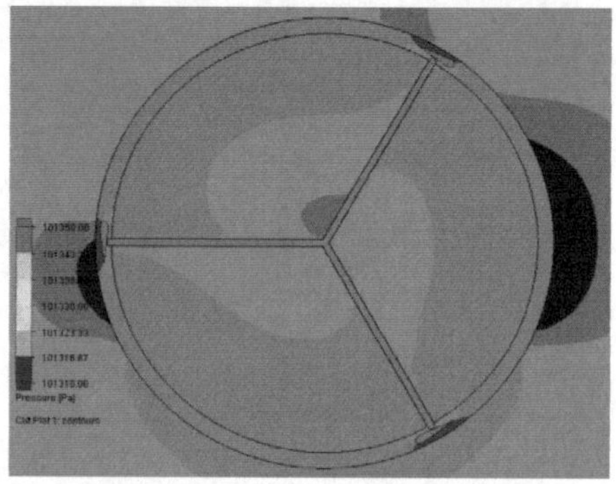
Velocity of 8m/s with 2 rad/s rotation	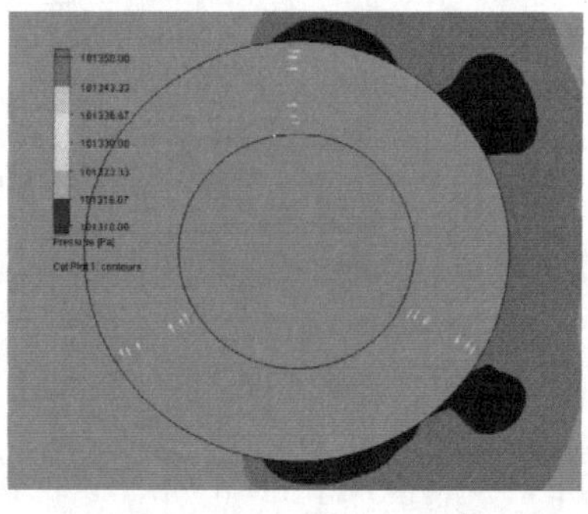	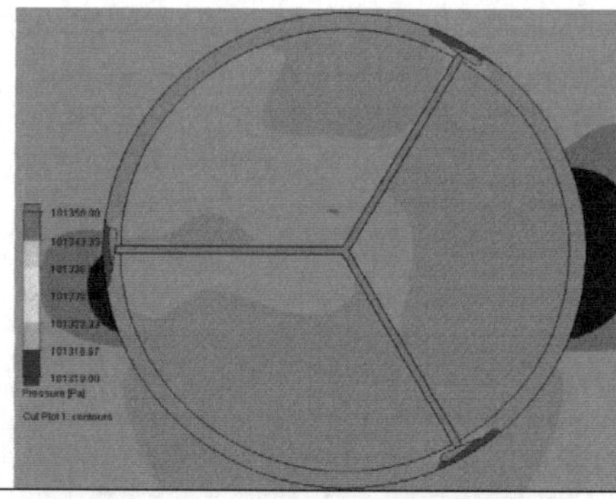
Velocity of 9m/s with 2 rad/s rotation	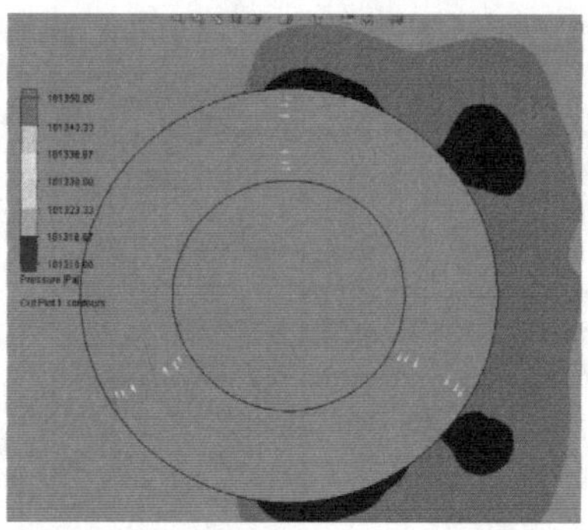	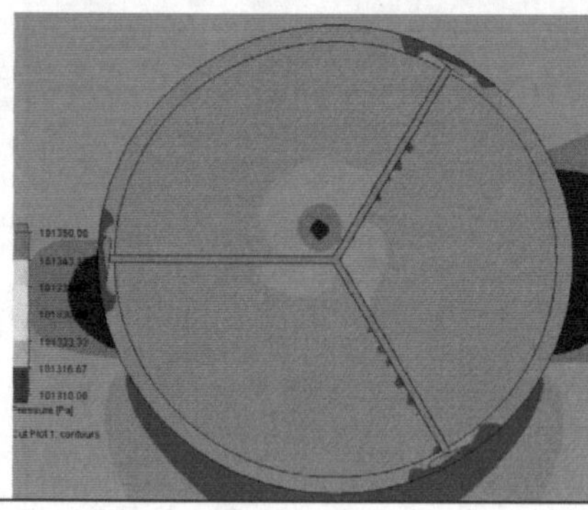

The pressure difference on the blade of lantern wind turbine is much lower in all the result shown above compared to darreis-H wind turbine. There is an obvious pressure difference on the darries H wind turbine which is red and green color. More likely for lifting to take place. Lantern wind turbine is more likely to have drag force acting on the blade compared to lifting force.

8. Conclusions and recommendations

The result generated based on rotational flow both shows that darries H wind turbine has higher tendency to create rotation compared to the lantern wind turbine since it has higher pressure difference. The study is done based on small scale wind turbine. Prototype is already manufactured, it should be compared simulation and experimental results.

Acknowledgement

The authors would like to thanks the Ministry of Education Malaysia which has financially supported this work through Fundamental Research Grant Scheme (FRGS) and Universiti Malaysia Sarawak for the facilities and support they provided to complete this project.

References

[1] Musgrove P 2010 *Wind power.* Cambridge: Cambridge University Press
[2] Demirbas A Combustion characteristics of different biomass fuels *Progress in energy and combustion science* **30** (2) 219-230
[3] Moroni S, Antoniucci V, and Bisello A 2016 Energy sprawl, land taking and distributed generation: towards a multi-layered density *Energy Policy* **98** 266-73.
[4] Dutton AG, Bonnet PA, Hogg P, and Lleong YL 2010 Novel materials and modelling for large wind turbine blades *Proc. of the Institution of Mechanical Engineers Part A: Journal of Power and Energy* **224** (2) 203-10
[5] Hand M, Jackson M, Ong S, Golden CO 2013 *National Renewable Energy Laboratory*
[6] Arnett EB, and Baerwald EF 2013 Impacts of wind energy development on bats: implications for conservation *InBat evolution, ecology, and conservation* Springer New York 435-456
[7] Ibrahim S Z, Miyanaga N, Ibrahim N, Ashari MF, Khan AA and Ibrahim MD 2015 *Proc. of the World Engineering Conf. and Convention 2015* WECC 2015 Engineering: Innovation and Society Kyoto Japan 1-6
[8] Çetin NS, Yurdusev MA, Ata R, Özdemir A 2005 Assessment of optimum tip speed ratio of wind turbines *Mathematical and Computational Applications* **10** (1) 147-54
[9] Manyonge AW, Ochieng RM, Onyango FN, and Shichikha JM 2012 Mathematical modelling of wind turbine in a wind energy conversion system: power coefficient analysis *Applied mathematical sciences* **6** (91) 4527-36

Experimental investigation of boundary layer transition in flow past a bluff body

Rahul Deshpande

Graduate Student, Department of Aerospace Engineering, Indian Institute of Technology Kanpur, Kanpur, Uttar Pradesh 208016, India

E-mail: raadeshpande@gmail.com

Aditya Desai

Postgraduate Student, Department of Aerospace Engineering, Indian Institute of Technology Kanpur, Kanpur, Uttar Pradesh 208016, India

E-mail: adityad@iitk.ac.in

Vivek Kanti

Project Engineer, National Wind Tunnel Facility, Indian Institute of Technology Kanpur, Kanpur, Uttar Pradesh 208016, India

E-mail: vkanti@iitk.ac.in

Sanjay Mittal

Professor, Department of Aerospace Engineering, Indian Institute of Technology Kanpur, Kanpur, Uttar Pradesh 208016, India

E-mail: smittal@iitk.ac.in

Abstract. We explore the phenomenon of *drag crisis* observed for the flow over bluff bodies at high Reynolds numbers. The drag coefficient reduces significantly beyond a certain Re due to the transition of the boundary layer from laminar to turbulent state. Flow past a smooth sphere and a circular cylinder is experimentally investigated for $1.0 \times 10^5 \leq Re \leq 5.0 \times 10^5$ via unsteady force, surface–pressure and 2–D Particle Image Velocimetry(PIV) measurements. In case of a smooth sphere, the *drag crisis* is observed for $Re > 3.3 \times 10^5$ .The unsteady force measurements reveal that the fluctuations in the force coefficients initially increase with Re in the high subcritical regime and then experience a steep fall in the critical regime. It is found from the PIV measurements that the normal Reynolds stresses in the separated shear layer from the sphere are one order lower in magnitude for the supercritical regime in comparison to the subcritical regime. In the case of flow past a smooth circular cylinder, a two–stage *drag crisis* is captured using surface–pressure measurements where the boundary layer over one side of the cylinder undergoes transition around $Re = 3.9 \times 10^5$ and that over the second side transitions around $Re = 4.8 \times 10^5$. The transition is accompanied with increased fluctuations in the surface–pressure coefficients near the shoulders of the cylinder.

1. Introduction

The phenomenon of boundary layer transition for the flow past a bluff body, popularly called the *drag crisis*, has been of continued interest for the fluid mechanics community. Several numerical and experimental studies have been carried out in the past to characterize the phenomenon and understand the underlying mechanism. During *drag crisis*, the transition of the boundary layer from a laminar to turbulent state causes a very significant reduction in the drag coefficient of the bluff body. The flow separation is delayed due to the increased near-wall momentum. This reduces the size of the wake and increases the base pressure. Here, the drag coefficient is the drag force non–dimensionalized with the dynamic pressure and frontal area of the bluff body. In this work, two bluff bodies are considered: sphere and cylinder. Achenbach[1] proposed dividing the entire $\overline{C}_D - Re$ variation into four flow regimes for flow past a sphere: (i) *subcritical regime* – $\overline{C}_D$ is almost independent of the Reynolds number and maintains a constant value of approximately 0.5, (ii) *critical regime* – a rapid drop in the drag coefficient,i.e., the drag crisis is observed with the $\overline{C}_D$ achieving its minimum value at the critical Reynolds number, (iii) *supercritical regime* – $\overline{C}_D$ increases gradually with Re while the point of transition of the laminar boundary layer to a turbulent state remains fixed at a particular azimuthal angle and (iv) *transcritical regime* – $\overline{C}_D$ increases due to the upstream shift of the point of boundary layer transition with an increase in Re.

Norman and McKeon[2] analyzed the statistics of the force coefficients for the flow over a sphere for $5 \times 10^4 \leq Re \leq 5 \times 10^5$ and observed that the standard deviation of the lateral forces increased rapidly with increasing Re in the subcritical regime. With further increase in Re into the critical regime, the fluctuations jump to a lower value and remain almost constant in the supercritical regime. Their study does not offer any explanation for the abrupt change in the force fluctuations.

One of the earliest and most comprehensive studies on drag crisis of a circular cylinder was carried out by Achenbach[3]. Similar to the flow past a sphere, the flow was classified in different flow regimes on the basis of the nature of the final flow separation from the cylinder surface. He also studied the variation of skin friction drag and pressure drag with Reynolds number and showed that the percentage of skin friction drag in total drag is less than 3% for $6 \times 10^4 \leq Re \leq 5 \times 10^6$. Several experimental and numerical studies have subsequently been carried out to understand the mechanism of drag crisis. Although experiments of different researchers report the onset of the drag crisis in roughly the same range of Reynolds numbers, the discrepancies in the comparison of the mean drag curves indicate that the phenomenon is extremely sensitive to flow conditions and surface roughness.

In the present study, we explore the mechanism for the phenomenon of *drag crisis* via unsteady force, surface–pressure and 2–D PIV measurements for the flow over a smooth sphere and a smooth circular cylinder. We define a sphere/cylinder to be *smooth* if $k/D < 10^{-5}$, where k is the height of the roughness elements on the sphere/cylinder surface and D is the diameter of the sphere/cylinder.

2. Results and Discussions

2.1. Flow past a smooth sphere

Figure 1(a) shows the variation of the time–averaged force coefficients with the Reynolds number. As has been reported in earlier studies, $\overline{C}_D$ is approximately constant in the subcritical regime followed by a rapid decrease in the critical regime. In the supercritical regime, $\overline{C}_D$ is again virtually constant. Although the comparison with results for earlier studies is not shown here, the present results are in excellent agreement with past results ([1],[3]). The lateral force coefficients, $\overline{C}_L$ and $\overline{C}_S$ are negligible in the early subcritical regime and gradually increase as Re approaches the critical Re. Non–zero lateral forces in the critical flow regime were also reported by Norman

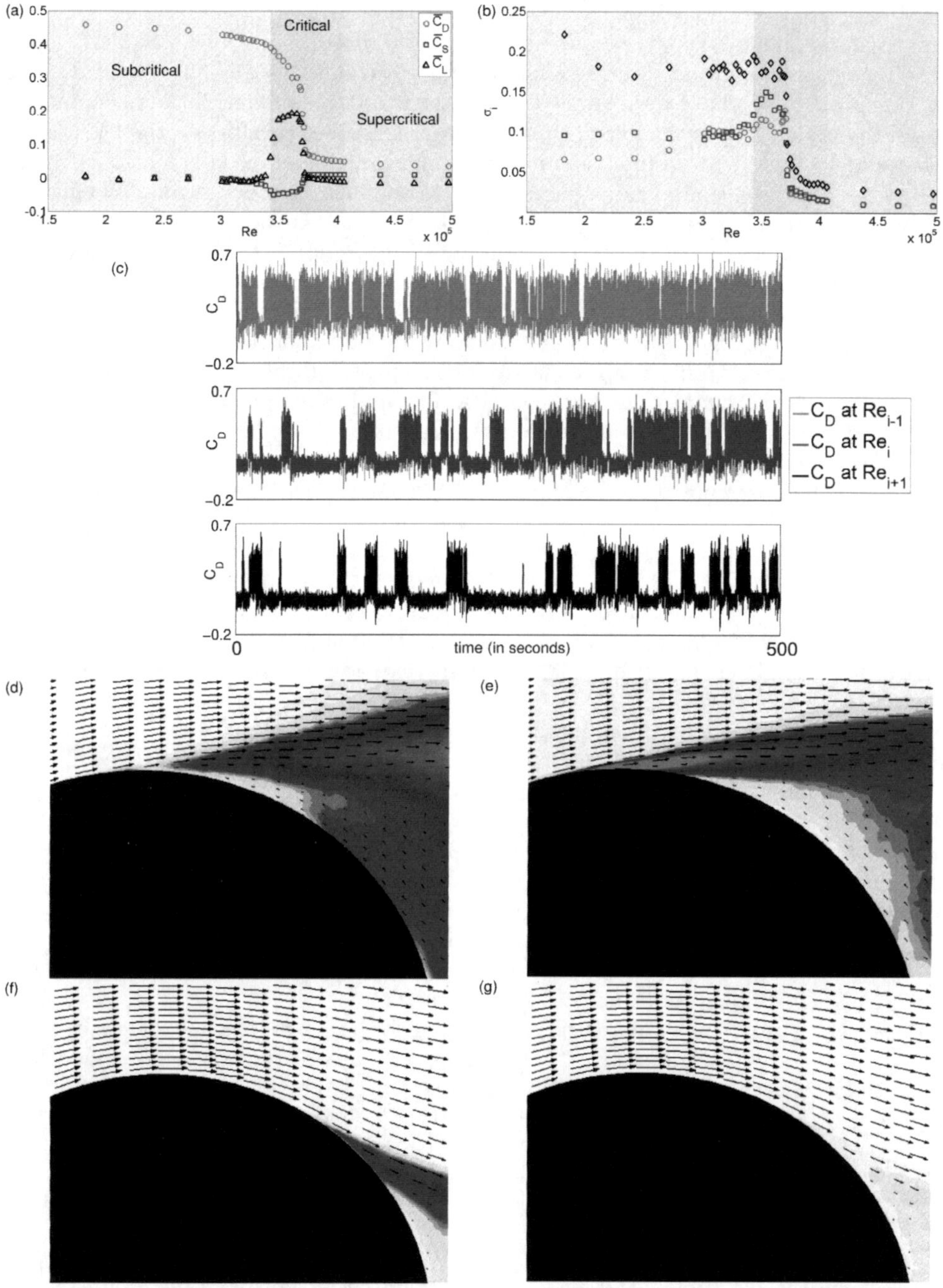

Figure 1. Flow past a sphere: (a)variation of the mean force coefficients with Re. Here, the mean force coefficients represent the time–averaged drag/lateral forces non–dimensionalized with the dynamic pressure and frontal area of the sphere. (b) variation of standard deviation of the force coefficients with Re. (c) C_D time series at three different Re. (d) $\overline{u_r' u_r'}/U^2_\infty$ and (e) $\overline{u_\theta' u_\theta'}/U^2_\infty$ for subcritical $Re = 3.08 \times 10^5$. (f) $\overline{u_r' u_r'}/U^2_\infty$ and (g) $\overline{u_\theta' u_\theta'}/U^2_\infty$ for supercritical $Re = 4.28 \times 10^5$.

and McKeon[3] who attributed it to the minute manufacturing imperfections on the sphere surface. These imperfections result in an earlier boundary layer transition over a particular region of the sphere resulting in a non-axisymmetric flow separation from the sphere surface. As Re approaches the end of the critical regime, the flow transition nears completion throughout the sphere surface due to which the lateral force coefficients decrease in magnitude. In order to check for a bias in the lateral plane, measurements were also carried out after rotating the sphere about the freestream axis at arbitrary angles. Even though the magnitude of $\overline{C}_L$ and $\overline{C}_S$ changed on rotation, the resultant lateral force coefficient remained same at a particular Re, irrespective of the rotations.

The standard deviation of the three force coefficients for various Re is depicted in Figure 1(b). In the subcritical regime, fluctuations in the lift force are almost three times larger than the drag fluctuations. Both the drag and lateral force fluctuations are maximum in the critical regime, with the maximum drag fluctuation occurring at a slightly higher Re as compared to the Re for maximum fluctuation in the lateral forces. On further increasing Re towards the end of the critical regime, these fluctuations decrease in magnitude. In addition, the decrease in the C_L and C_S fluctuations is large and gradual as compared to the jump observed for the fluctuations in C_D. This observation suggests a change in the flow physics near the shoulder of the sphere as the nature of the flow separation from the surface changes from a laminar to turbulent state. The force fluctuations remain nearly constant with Re in the supercritical regime.

To discuss the reason behind the jump in the C_D fluctuations in the critical regime, we show the time series of C_D at three different Re in Figure 1(c). Re_i denotes the Re where the standard deviation in C_D is found to be maximum. Re_{i-1} and Re_{i+1} represent the Reynolds numbers just before and after Re_i where the force measurement was carried out. The C_D time series depicts an intermittent nature at these Re due to the random shift in the C_D between the bistable states in the critical flow regime. For $Re_{i-1} \leq Re \leq Re_{i+1}$, the mean of the higher C_D state is 0.29, while that of the lower state is 0.09, approximately. It can be deduced from the time series data that the higher C_D state exhibits significantly high variation about its mean as compared to the lower C_D state. As we increase the Re from Re_{i-1} towards Re_{i+1}, the probability for the lower C_D state to exist increases as compared to the higher C_D state. The probability of existence of the two states is nearly equal at Re_i leading to the occurrence of maximum fluctuations in C_D. For $Re > Re_{i+1}$, the higher C_D state gradually diminishes. Beyond a certain Re, the lower C_D state is the only observed state. Hence, the standard deviation of C_D falls rapidly on increasing Re beyond Re_i.

To investigate the reason for decrease in lateral force fluctuations, we conducted 2–D Particle Image Velocimetry (PIV) measurements in a planar cut near the sphere surface for both subcritical and supercritical Re. Figures 1(d) and (e) depict the non-dimensionalized radial ($\overline{u_r'u_r'}/U^2{}_\infty$) and tangential Reynolds stresses ($\overline{u_\theta'u_\theta'}/U^2{}_\infty$) at a subcritical $Re = 3.08 \times 10^5$. The contour plots depict the unsteadiness after the flow separates as a laminar boundary layer. An increase in magnitude of the radial Reynolds stresses along the separated shear layer suggests increased unsteadiness near the surface of the sphere. Also, the maximum tangential Reynolds stress is observed to be near the shoulder, where the flow separates from the sphere. These stresses eventually decrease in magnitude along the shear layer downstream. The contour plots for the corresponding Reynolds stresses in supercritical regime are shown in Figure 1(f) and (g) at $Re = 4.28 \times 10^5$. In comparison to the laminar flow separation in the subcritical regime, the turbulent separation occurs at a greater azimuthal angle, the wake size is reduced and the unsteadiness close to the sphere surface decreases for the supercritical regime. Figure 1(f) shows that the values of radial Reynolds stresses is moderate along the separated shear layer, while it is low further downstream. Similarly, the tangential Reynolds stress distribution for the supercritical flow regime in Figure 1(g) depicts a decrease in the unsteadiness in the separated shear layer, when compared with the subcritical regime. The comparison between the radial and

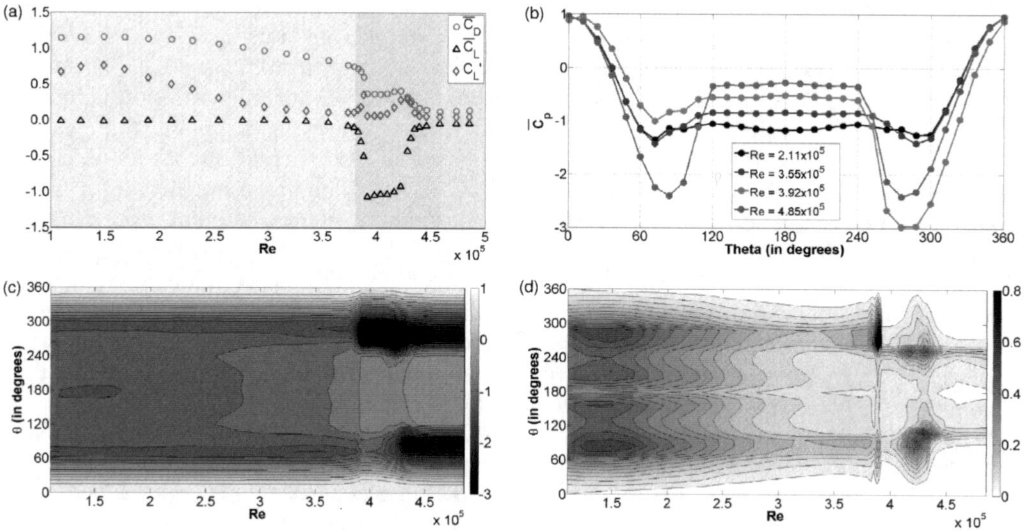

Figure 2. Flow past a circular cylinder: (a) variation of the mean force coefficients ($\overline{C}_D$ and $\overline{C}_L$) and rms of the lift coefficient ($C_L{}'$) with *Re*. Here, the force coefficients represent the drag/lateral forces non–dimensionalized with the dynamic pressure and frontal area of the cylinder. (b) $\overline{C}_P$ distribution at selected *Re*. (c) Angular distribution of $\overline{C}_P$ with *Re*. (d) Angular distribution of $C_P{}'$ with *Re*.

tangential Reynolds stresses shows that the unsteadiness near the sphere surface decreases when the laminar separation changes to a turbulent separation after the drag crisis. This decrease in the unsteadiness near the sphere shoulder can be correlated with the abrupt decrease in the lateral force fluctuations, shown in Figure 1(b), as the *Re* increases from subcritical to supercritical regime.

2.2. Flow past a smooth circular cylinder

Figure 2(a) shows the variation of aerodynamic force coefficients experienced by the cylinder with *Re*. The force is obtained by integrating the surface pressure distribution on the cylinder surface. The $\overline{C}_P$ distribution on the cylinder at selected *Re* are shown in Figure 2(b). Similar to what is observed for a sphere, $\overline{C}_D$ tends to be almost constant upto $Re = 2.3 \times 10^5$,i.e., in the lower subcritical regime. As is seen from the $\overline{C}_P$ distribution, at $Re = 2.11 \times 10^5$, the flow is subcritical over both halves of the cylinder and the flow separates at 84°, indicating a laminar flow separation from both the sides. As Reynolds number is increased to 3.55×10^5, the separation point moves downstream, to around 108° along with an increase in the base pressure. A characteristic of the higher subcritical regime for the flow past a circular cylinder is the decrease in the strength of the vortices being shed from the cylinder. This is confirmed by the significant decrease in $C_P{}'$ and an increase in the base pressure coefficient. Hence, the $\overline{C}_D$ in the higher subcritical regime decreases gradually with increase in *Re*.

For $Re > 3.8 \times 10^5$, the flow enters the critical regime as the $\overline{C}_D$ starts decreasing rapidly with increase in *Re*. In the case of a smooth circular cylinder, a two–stage drag crisis is observed along with a large asymmetric state (mean lift) in the critical regime. For convenience, the part of the cylinder between $0^o - 180^o$ is described as the top side while that between $180^o - 360^o$ is referred to as the bottom side. At $Re = 3.9 \times 10^5$, the boundary layer over the bottom side of the cylinder undergoes transition, thus delaying the flow separation on this side till 120°, while that over the top side remains laminar. This asymmetric flow separation leads to a large mean lift. For the Re where the asymmetric state persists, the circulation associated with this lift leads to a shift in the front stagnation point, as seen in figure 2(b). At $Re = 4.8 \times 10^5$, the boundary layer over the top side undergoes transition, leading to nearly identical $\overline{C}_P$ distribution on both the sides, with the flow resorting back to the symmetric state (near-zero lift). The two separation points in this regime are located at 120°, indicating a turbulent flow separation from both the sides.

The contour plots in Figures 2(c) and 2(d) depict the variation of $\overline{C}_P$ and C_P', the root-mean-square fluctuations in C_P with Re and θ. At the onset of the critical regime (around $Re \cong 3.8 \times 10^5$), C_P' near the bottom shoulder increases significantly, indicating the unsteady dynamics associated with the boundary layer transition. Following this, the -$\overline{C}_P$ (suction) increases on the bottom side. As Re is increased further, transition on the top side results in a significant rise in C_P' near top shoulder followed by an increase in $-\overline{C}_P$.

3. Conclusions

The present study attempts to investigate the *drag crisis* phenomena for flow past a 3-D bluff body(*sphere*) and a 2–D bluff body(*circular cylinder*). The first part of the study explores the variation in the flow physics in the vicinity of a sphere for $1.5 \times 10^5 \leq Re \leq 5.5 \times 10^5$ via unsteady force measurements and 2–D PIV. It was found that the intermittent behavior of C_D in the critical regime first results in the rise and eventually, in the fall of the fluctuations in the drag coefficient, as the Re reaches the end of critical regime. On increasing the Re from subcritical to supercritical regime, our measurements showed that the Reynolds stresses near the sphere surface decrease along with the reduction in the lateral force fluctuations in the critical regime. In the second part of the study, flow over a circular cylinder in the critical regime is investigated using surface pressure measurements. A two–stage drag crisis with each stage corresponding to the boundary layer transition on one side of the cylinder is observed. The boundary layer transition is observed to induce vigorous fluctuations in the surface pressure in the vicinity of the shoulder of the cylinder.

References

[1] Achenbach E.: Experiments on the flow past spheres at very high Reynolds numbers. J. Fluid Mech 54(03):565575, 1972.
[2] Norman A. K., McKeon B. J.: Unsteady force measurements in sphere flow from subcritical to supercritical reynolds numbers. Exp Fluids 51.5:1439-1453, 2011.
[3] Achenbach E.: Distribution of local pressure skin friction around a circular cylinder in cross flow up to $Re = 5 \times 10^5$. J. Fluid Mech 34(04):625-639, 1968.
[4] Norman A. K., Kerrigan E. C., McKeon B. J.: The effect of small amplitude time-dependent changes to the surface morphology of a sphere. J. Fluid Mech 675:268296, 2011.

Experimental investigation on the near flow field of dual stream nozzles

S Sudhakar, N Karthikeyan and S Ashwin Kumar
Experimental Aerodynamics Division, CSIR-National Aerospace Laboratories, Bangalore, India

E-mail: ssudha@nal.res.in

Abstract. An experimental investigation was carried out to investigate the effect of beveling of primary nozzle exit in the near field of a dual stream nozzle flow. Two exit geometry configurations of primary stream nozzle viz., (a) circular (b) bevel along with one exit geometry of the secondary stream-circular, were studied. Experiments were carried out at both subsonic and supersonic primary nozzle operating conditions M_p=0.96 and 1.2. The secondary nozzle exit Mach number was maintained at 0.65 and 0.85 respectively to maintain the velocity ratio of 0.7 between the primary and secondary jet. The by-pass ratio for this investigation is maintained at 2.0. Flow visualization using retro-reflective shadowgraph technique was used for the qualitative visualization of the near flow field at the Mach number of 1.2. The mean and turbulent quantities in near flow field were obtained using particle image Velocimetry (2D-PIV). The flow visualization and PIV investigations show significant change in mean and turbulent quantities brought about in the near field due to the beveling of the primary nozzle. PIV results show increase in the potential core length and reduction in turbulence levels in the potential core by the secondary flow regardless of the jet exit geometry. A differential trend is seen in the shear layer growth and the turbulence characteristics between the shorter and longer lips sides of the beveled nozzle. In the dual stream configurations, bevel nozzle shows lower Reynolds stress values than the circular one except in the shorter lip side at the larger downstream locations.

1. Introduction

Jet noise continues to be an significant component of aircraft noise. Several research and development efforts are underway around the globe to completely understand the physics behind generation of jet noise and develop a suitable technique to reduce it. Efforts towards passive methods of jet noise suppression – especially for civil aircraft using turbo-fan engines- have progressed in tandem with the efforts towards understanding noise generation [1]. In the past, numerous concepts such as the incorporation of vortex generators, tabs, chevrons and other mixing devices, offset and non-concentric nozzles, thermal shielding, etc., have been investigated. However, none of these devices has produced the level of noise reduction necessary to provide a substantial and noticeable relief for the communities living around airports. In addition, most of such techniques have resulted in thrust, weight and drag penalties. Recently, studies carried out by Viswanathan [2-5] have shown that beveling of primary jet nozzle exit has resulted in reduced noise levels both in single and dual stream configurations. This approach has proven to be the most effective technique for jet noise suppression, particularly with simulated flight conditions. The beveled nozzle concept has lesser thrust penalties without no additional weight or drag. Further, the extension of the concept to a working model in existing engines is much easier.

IOP Conf. Series: Journal of Physics: Conf. Series **822** (2017) 012004 doi:10.1088/1742-6596/822/1/012004

Computational simulation studies, using LES/DES, carried out by Viswanathan et al [6-7] on dual stream with beveled primary nozzle to relate the flow field changes to the acoustic characteristics. However, there has been no attempt to validate the result of these studies against any standard experimental database. An understanding of the physical mechanisms that may be responsible for the observed noise reduction can only be arrived, through detailed experimental flow-field investigations. The present study attempts to simulate exit flow conditions of commercial high by-pass ratio turbofan jet engines (velocity ratio and Mach number of core flow) using cold jets to investigate the effect of primary nozzle exit beveling on the flow fields of a dual stream nozzle. 2D-particle image Velocimetry and retro-reflective shadowgraph techniques are used to relate the subtle changes in the flow field (caused due to primary jet exit beveling) for better insight and understanding of noise generation in dual stream nozzles flow.

2. Experimental setup

2.1. Test Facility

The measurements are carried out in the jet aeroacoustics research facility at EAD, CSIR-NAL (Figure 1). The jet is supplied with dry compressed air from a 10 bar reservoir through a computerized globe valve (Make: Fisher, Type: ET) with a Whisper Cage Trim for cutting down valve pings. The settling chamber is provided with a perforated cone and three sets of screens for flow conditioning. An inline electric heater, provided in-between the valve and the settling chamber allows for studies on hot

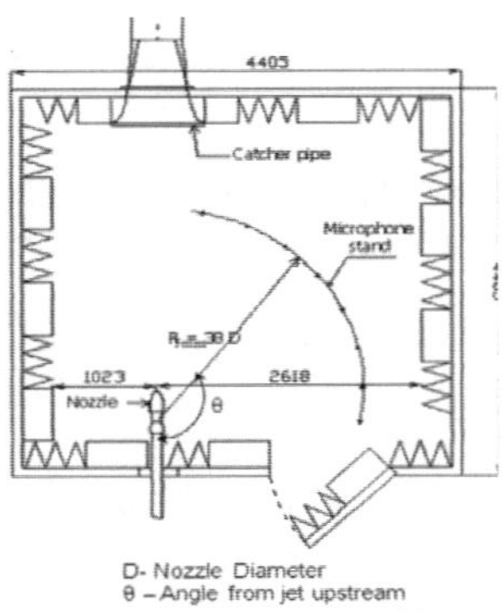

Figure 1.Schematic of jet acoustic research facility

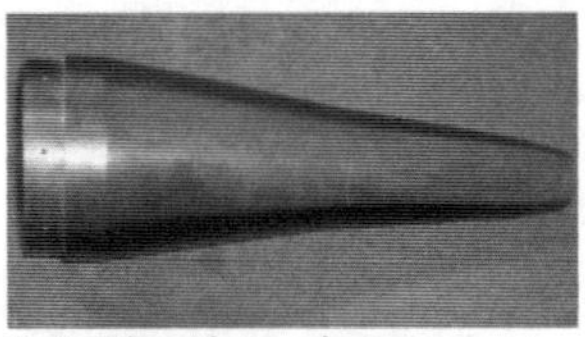

2a) Circular primary nozzle

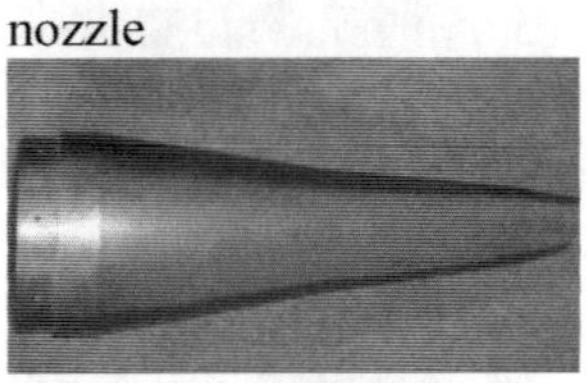

2b) Bevel primary nozzle

Figure 2. Primary nozzles

2c) Secondary Circular nozzle

Figure 3. Dual stream nozzle assembly

supersonic jets. The jet can be run continuously for up to 40 minutes. The driving pressure for the jet was maintained to within ±0.35 psi of set pressure during the run. The jet exhausts in to an anechoic chamber of inner dimensions 3.6m (L) x 3.6m (W) x 3m (H) for carrying out acoustic measurements. The anechoic chamber walls are mounted with fiberglass covered acoustic wedges, which are designed to achieve a low-frequency cutoff of 300 Hz. The chamber is provided with suitable acoustically treated intakes to allow for entrainment by the jet. During the run, the ambient pressure in the chamber dropped by less than 0.1psi demonstrating that adequate area has been provided for entrainment. A catcher of 0.6m x 0.6m outlet provided with proper acoustic termination collects the jet exhaust. Preliminary acoustic tests showed that the room is anechoic above 400Hz with a partial floor grating in place.

2.2. Nozzle configuration

Figure 2. Shows the nozzle configuration used for the present study. Two primary nozzles (one with circular and one with 24° bevel) of different exit geometry and one secondary nozzle with circular exit has been fabricated to study the effect of primary nozzle exit beveling on flow field of the dual stream nozzle flow. The primary nozzle extends beyond the secondary nozzle exit to simulate actual engine nozzle configuration. The exit diameter of the primary nozzle is 38.1mm. The secondary nozzle has exit diameter 90.58mm. The exit bevel angle for the modified primary nozzle is 24°. The increase in exit area of the beveled nozzles is 3%. Photograph of dual stream nozzle flow arrangement through combination of secondary and primary nozzle is shown in Figure 3. The exit area ratio between the secondary and the primary nozzle is 3.0.

Experiments planned on the selected nozzle configurations with different flow conditions are given in the following Table.1. The exit Mach number and velocity ratio for the present investigation is close to the jet Mach number of the aircraft during takeoff (M_P=0.96), climb and cruise (M_P=1.2) [1].

Table 1. Nozzle exit Mach numbers.

	Primary(M_P)	Secondary(M_S)
Circular primary nozzle	0.96	-
Circular primary nozzle	1.2	-
Beveled primary nozzle	0.96	-
Beveled primary nozzle	1.2	-
Dual stream with circular primary nozzle	0.96	0.65
Dual stream with circular primary nozzle	1.2	0.85
Dual stream with beveled primary nozzle	0.96	0.65
Dual stream with beveled primary nozzle	1.2	0.85

2.3. Shadowgraph Set up

A retro-reflective type shadowgraph was setup (Figure 4) inside the anechoic chamber to capture the flow field of dual stream nozzle configurations. The set-up was that of a coincident double pass shadowgraph as described in [8]. Light from high power LED illuminator was converged on to the mirrored surface of a rod mirror fitted on to a glass filter just ahead of an IDT Y5® Camera. The reflected light from the rod mirror illuminated a region large enough to capture the jet from the nozzle exit. The illuminator provided a short light pulse of high energy of sufficient enough to uniformly illuminate the designated area for 5μs and was synchronized with the exposure of the camera fitted with a Nikkor 70-210 mm lens.

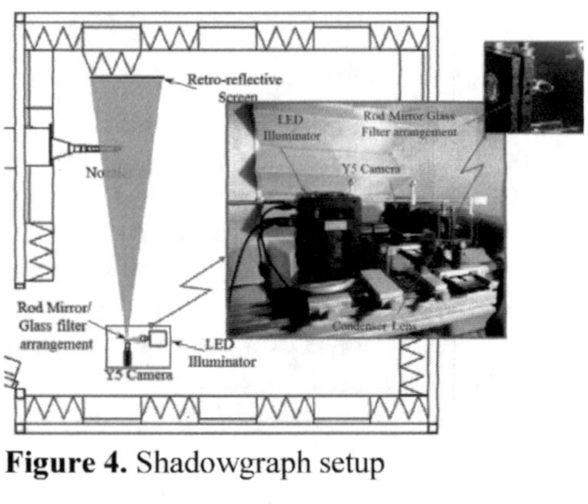

Figure 4. Shadowgraph setup

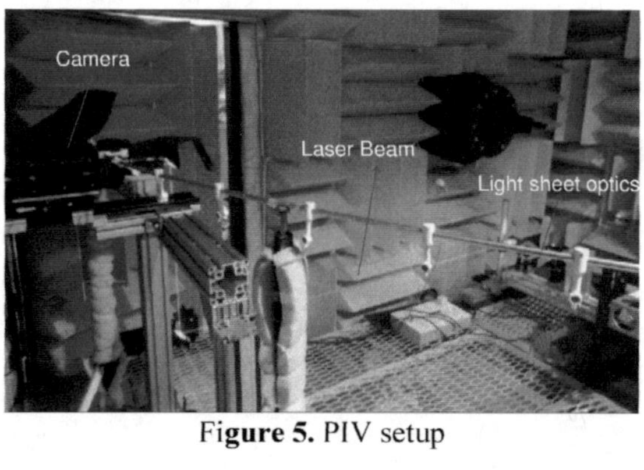

Figure 5. PIV setup

2.4. Particle Image Velocimetry

Figure 5 shows the setup for the 2-D PIV measurements inside the anechoic chamber of the jet acoustic research facility. The flow-field with tracer particles was illuminated by a double pulsed, frequency doubled dual Nd:YAG, PIV 400 laser, providing with a nominal 400 mJ of energy per pulse at a 532-nm wavelength. The optimum performance of the laser is at 15Hz (15 pairs per second).The light sheet optics are mounted to create sheet (380mmx300mm) in the streamwise plane at the center of the nozzle

IOP Conf. Series: Journal of Physics: Conf. Series **822** (2017) 012004 doi:10.1088/1742-6596/822/1/012004

cross section. The camera was mounted such that the light sheet and the camera image plane are parallel to acquire the image. The images are acquired using MotionPro® Y5 camera of 2314(H)X1728(V) pixels, with Nikkor® 75-210mm lens. The synchronization of laser and the camera image acquisition was achieved through IDT Motion-Pro® timing hub. The post processing of the acquired images were carried out with IDT ProVISION® XS software. Mean velocities were estimated from ensemble average of 2000 image pairs. The laser pulse separation time was kept at 3μs to allow the particle to move at least two pixels for the maximum velocity in the measurement. The velocity vectors were obtained from processing image with 48x48 pixel interrogation window with 50% overlap. A spatial resolution of 7.8mm in both directions(X and Y) was achieved. The flow field was seeded by injecting the fog of particle size 1μm in the settling chamber. The orientation of the nozzle for the PIV studies was such that shorter lip faced the floor of the anechoic chamber. As the light sheet for illumination was arranged from floor side of the chamber, this orientation prevents the obstruction of the illumination by the longer lip.

3. Results and Discussion

3.1. Shadowgraph

Figure 6a-d shows the flow field of the jet from circular and bevel nozzle with and without the presence of secondary stream captured using shadowgraph technique. Flow field for the primary jet from the circular nozzle at Mach number 1.2 is shown in figure 6a. The under expanded flow at the exit of the nozzle expands to adjust to the ambient. This is observed as a bulging jet at the primary nozzle exit. The ensuing flow features of shock cells are clearly captured in the shadowgraph. Three shock cells are

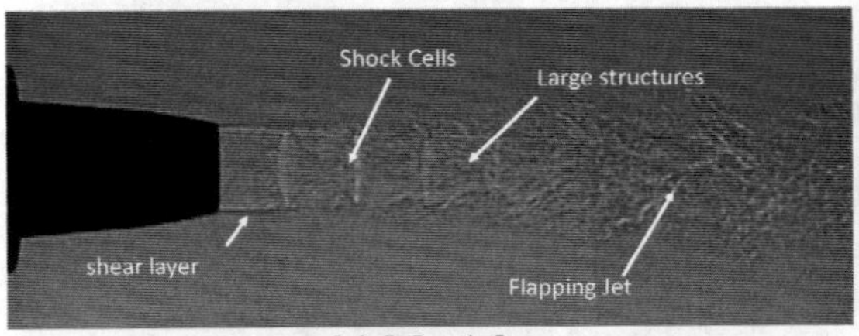

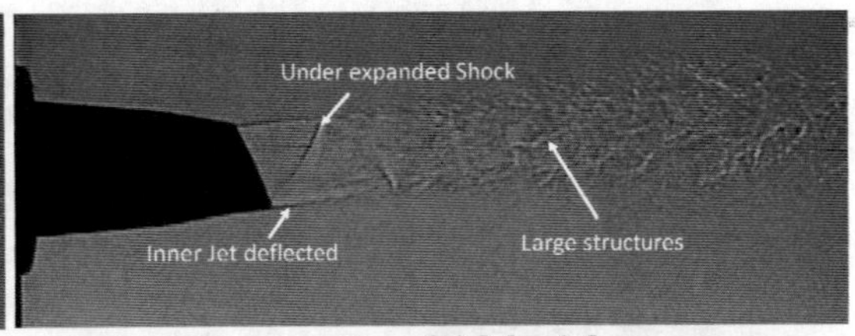

6a) M_P=1.2 6c) M_P=1.2

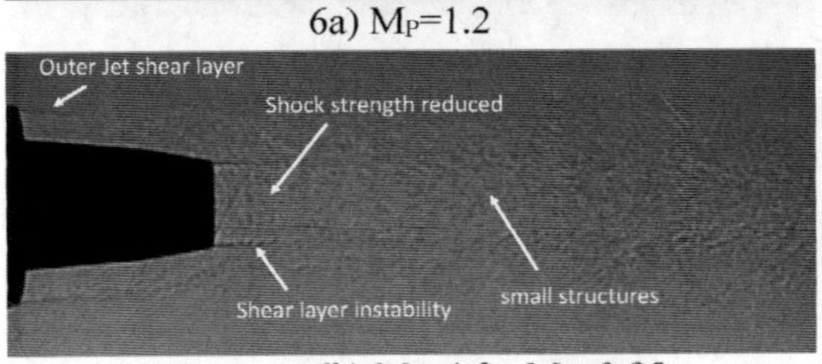

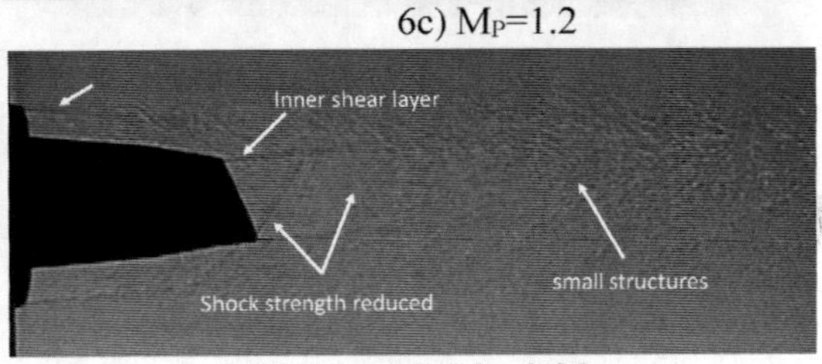

6b) M_P=1.2 ; M_S=0.85 6d) M_P=1.2; M_S=0.85

Figure 6. Shadowgraph images of the single and dual stream nozzle flow

discernable along with the shear layer emanating from the nozzle lip. Downstream of the third shock cell the jet dissipates into a flapping jet, the movement of which is captured in other shadowgraph realizations of the flow. In contrast the flapping of the jet is subdued when the secondary stream (M_S=0.85), is turned on as shown in figure 6b. Also introduction of secondary stream shows that the strength of the shock structures seen in the flow field of primary jet alone is heavily diminished. This is expected as the inner flow now exhausts into a different ambient dictated by the flow from the secondary stream. Since the under expanded jet from the inner nozzle cannot expand adequately at the exit due to the presence of secondary stream, it is deflected into itself resulting in oblique shocks at the exit of the primary jet. However these shocks are not as strong as the primary jet alone case and immediately dissipate, without forming any appreciable shock cells. Shear layers from the primary and secondary jets are captured clearly in the shadowgraph.

The shadowgraph of the flow field of the primary jet (M_P=1.2) from the beveled nozzle is shown in Figure 6c. Due to the extension of the lip on one side of the beveled nozzle, there is an azimuthal variation of the shear layer development, causing the flow to get inclined towards the shorter lip. The initial flow from the nozzle exit shows that a shock emanating from the longer lip and interacting with the shear developing from the shorter lip. The shock reflects from the opposite shear layer tilting away from the edge. The flapping of the jet was not observed in beveled nozzle case. The number of shock cells seen is also lesser than that seen in the jet from the circular nozzle. When the secondary stream is turned on (M_S=0.85) the strength of the shock at the beveled nozzle exit is reduced (Figure 6d). The flow inclination angle is found to be decreased by the presence of the secondary stream. Hence the overall flow inclination angle is limited thus reducing even the minor thrust loss due to beveling of the primary inner nozzle. It is observed that the size of the turbulent flow structures seen in the primary jet alone case was reduced in the presence of secondary stream for both the circular and bevel nozzle cases.

3.2. Particle image velocimetry

Comparison of mean streamwise velocity field of the jet from circular and beveled nozzle with and without the presence of secondary stream at subsonic and supersonic exit Mach numbers of the primary jet, is shown in figure 7a-h. The width of the jet potential core for the primary jet alone at exit condition of M_P=0.96 (figure 7a) for the circular nozzle is higher than that seen for jet from the beveled nozzle

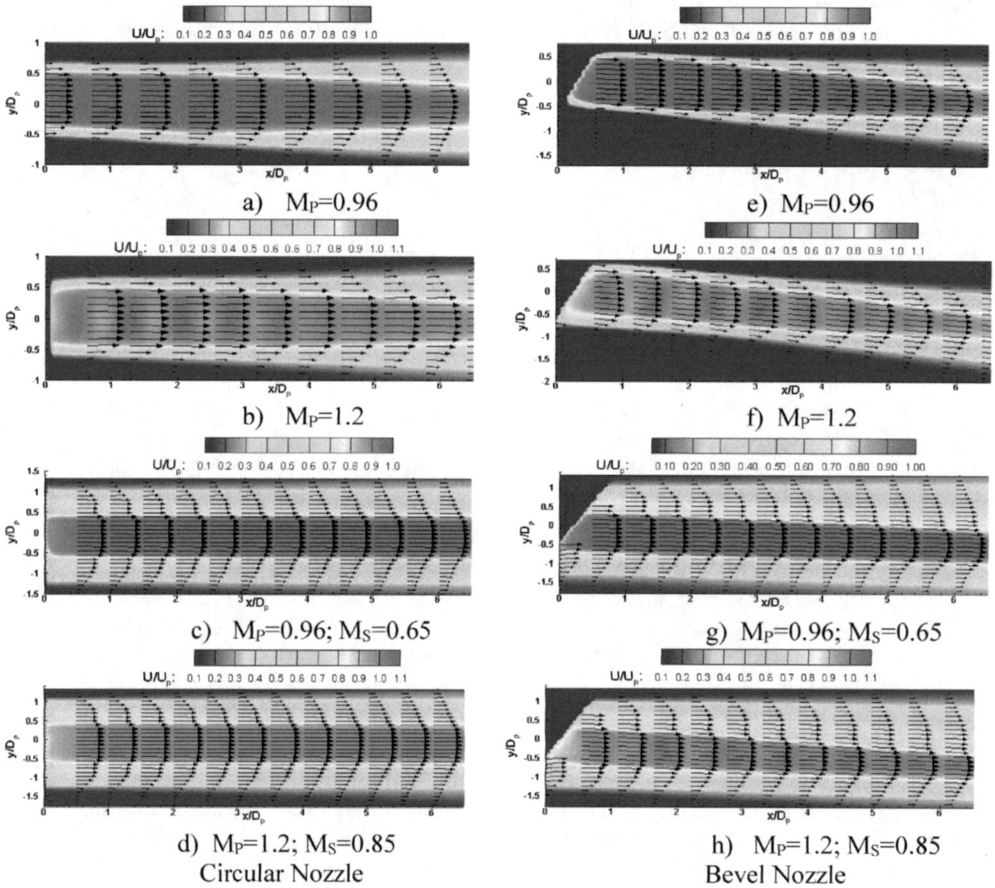

a) M_P=0.96 e) M_P=0.96
b) M_P=1.2 f) M_P=1.2
c) M_P=0.96; M_S=0.65 g) M_P=0.96; M_S=0.65
d) M_P=1.2; M_S=0.85 h) M_P=1.2; M_S=0.85
Circular Nozzle Bevel Nozzle

Figure 7. Mean streamwise velocity contour of single and dual stream nozzle flow

case (figure 7e). The jet flow field from the beveled nozzle is inclined towards the shorter lip of the nozzle at both subsonic and supersonic Mach number as seen in the shadowgraph. For the beveled nozzle case, the width of the shear layer in the shorter lip (bottom) is slightly higher than in the longer lip. The shock cells in the jet flow field is appeared as alternate high and low velocity contour levels in the potential core of the jet from both the circular and bevel nozzles (figure 7b and 7f). The number of shock cells present in the jet flow field looks similar for the both circular and bevel nozzle, though for the beveled nozzle case the latter shock cells are found to be weak. As expected, the width of the jet increases in the downstream locations with the introduction of secondary stream for both subsonic and supersonic cases. For the beveled primary nozzle, the secondary stream decays faster in the shorter lip side (bottom) than that in the longer lip side as seen from the velocity contours of figures 7g and 7h. *This could be due to early interaction of the secondary stream with the primary stream at the shorter lip than the longer lip.*

Contours of streamwise velocity fluctuations (u') normalized by jet exit velocity for both the cases of circular and bevel nozzle, for the subsonic and supersonic exit conditions are shown in figure 8a-h. At subsonic exit Mach number-M_P=0.96, both circular and bevel nozzles show the velocity fluctuations

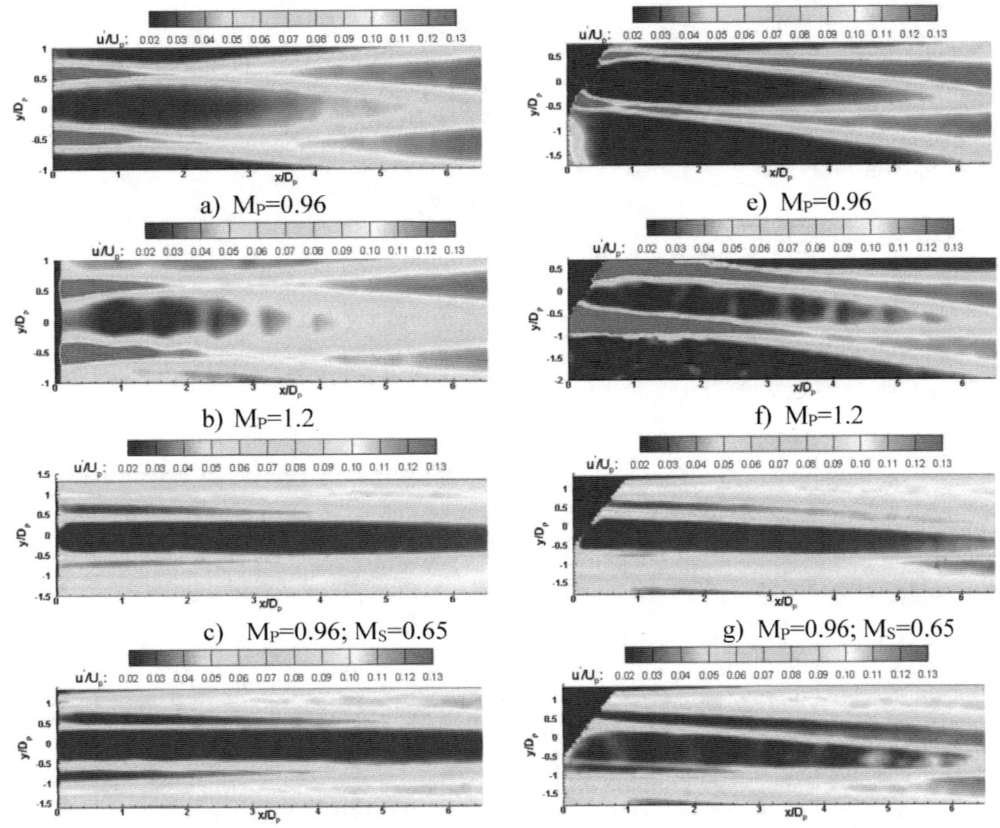

a) M_P=0.96 e) M_P=0.96

b) M_P=1.2 f) M_P=1.2

c) M_P=0.96; M_S=0.65 g) M_P=0.96; M_S=0.65

d) M_P=1.2; M_S=0.85 h) M_P=1.2; M_S=0.85

Circular Nozzle Bevel Nozzle

Figure 8. Mean streamwise velocity fluctuation contour of single and dual stream nozzle flow

higher in the shear layer and lower in the potential core. The bevel nozzle case shows (figure 8e) lower u_{rms}(u') levels in the potential core as compared to the circular nozzle (figure 8a). The extent of the shear layer is observed to be higher in the shorter lip side of the bevel nozzle than the longer lip. The u_{rms}(u') levels are higher in the shear layer at the longer lip side than that seen in the shorter lip. At M_P=1.2, the trend in u' levels for both the circular and beveled nozzles are similar to that seen in the subsonic case. However, u' values in the shear layer are higher than that seen in the subsonic condition (figure 8b and 8f). The increase in the u' value could be due to the presence of shock cells in the flow field. The presence of shock cells in the flow field is mildly visible in the streamwise velocity fluctuation contour as seen in figure 8b and 8f . Streamwise velocity fluctuation contour levels for the dual stream nozzle flow for the circular and bevel nozzle for M_P=0.96, M_S=0.65 is shown in figure 8c and 8g. The circular dual stream nozzle clearly shows lower u' in the potential cores of both the primary and secondary streams (figure 8c) while higher u' is seen in the outer shear layers. The inner shear layer between the primary and the secondary streams appear to have marginally higher values of u' as compared to the values in the primary stream core.

In the dual stream flow of beveled nozzle configuration(M_P=0.96, M_S=0.65), the secondary stream

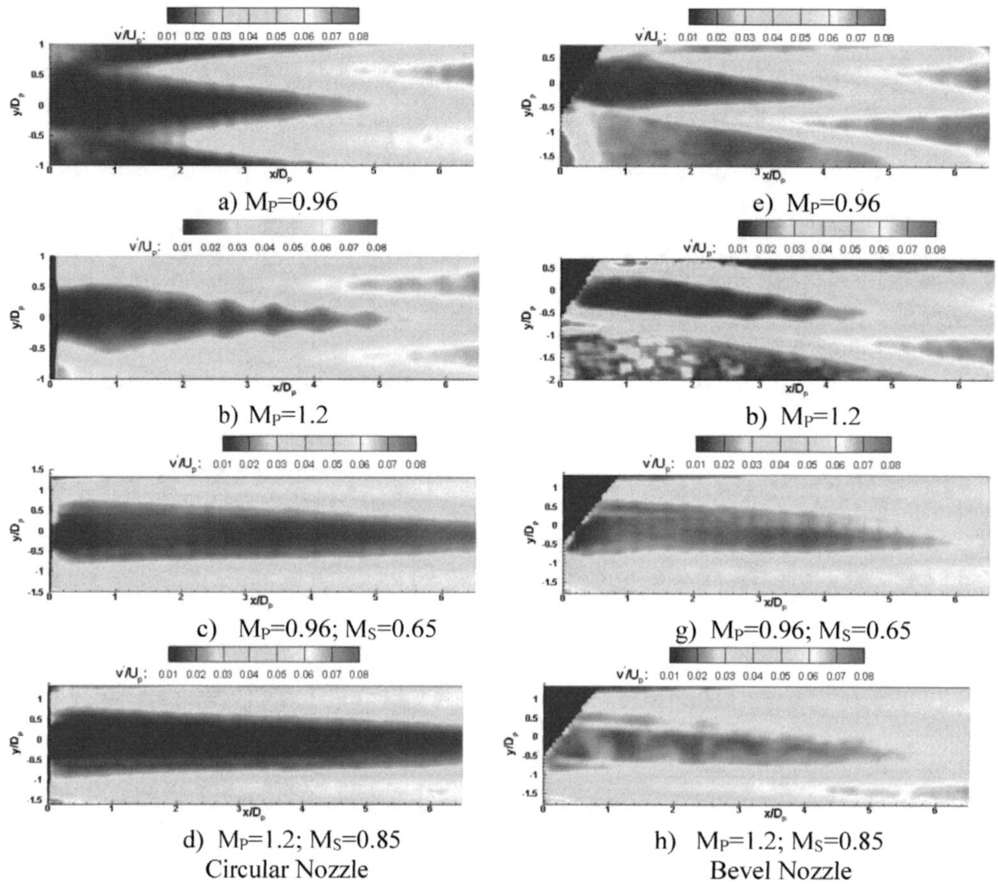

Figure 9. Mean transverse velocity fluctuation contour of single and dual stream nozzle flow

potential core is not clearly discernable at the shorter lip side as seen in the circular case wherein the streamwise velocity fluctuations (u') have lower values. However, at the longer lip side the secondary stream potential core is clearly seen since the velocity fluctuations are lower. Further, the axial extent of the low velocity fluctuation region, at the longer lip side, is larger than that seen in the circular case. The streamwise velocity fluctuation values in outer shear layer, as compared to the circular case, are lower in the longer lip side but higher in shorter lip in a small region beyond axial distance of x/D_P=5. At supersonic operating condition (M_P=1.2; M_S=0.85), the trend in the u' for the circular nozzle are similar to that seen in the subsonic case (figure 8c). However, the magnitude of streamwise velocity fluctuations (figure 8d) in the outer shear layer is higher than that seen in the subsonic case. In the bevel case the secondary stream potential core on the shorter lip side is seen up to an axial distance of x/D_P=3, while on the longer lip side it extends further (figure 8h). This is in contrast to the subsonic case where the secondary potential core on the shorter lip side is not clearly discernable (figure 8g). Weak shock patterns are visible in the primary stream potential core region similar to that seen in the mean streamwise velocity contour (figure 7).

Influence of primary nozzle exit geometry on the transverse velocity fluctuation (v′) of the dual stream nozzle flow is shown in figure 9. The transverse velocity fluctuation is normalized by the primary stream jet exit velocity. Comparison of transverse velocity fluctuation contour of the primary stream from the circular and bevel nozzle at subsonic Mach number of M_P=0.96 is shown in figure 9a and 9e. The transverse velocity fluctuation in the shear layer is higher for the bevel nozzle compared to the circular

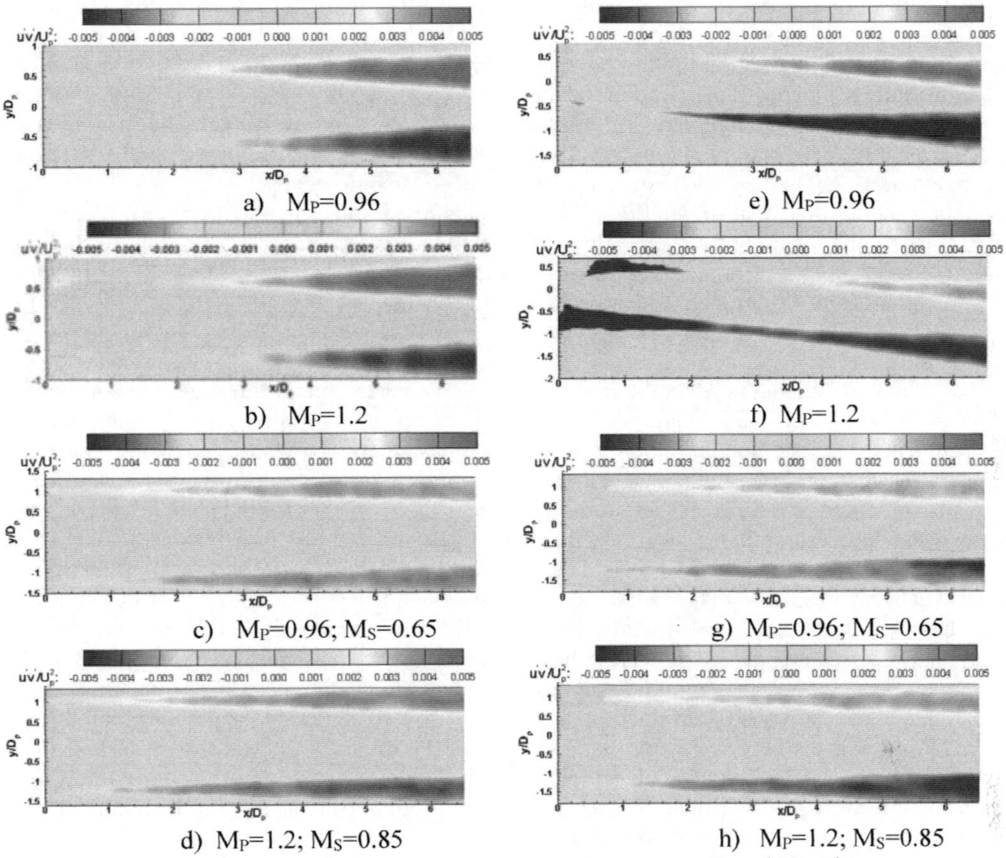

a) M_P=0.96 e) M_P=0.96

b) M_P=1.2 f) M_P=1.2

c) M_P=0.96; M_S=0.65 g) M_P=0.96; M_S=0.65

d) M_P=1.2; M_S=0.85 h) M_P=1.2; M_S=0.85

Circular Nozzle Bevel Nozzle

Figure 10. Mean Reynolds stress contour of single and dual stream nozzle flow

nozzle because of the transverse deflection of the flow. In the bevel nozzle case, the axial extent as well as the magnitude v′ in the shear layer is higher on the shorter lip as compared to the longer lip side. At supersonic operating condition M_P=1.2, the trend in the transverse velocity fluctuations at the shear layer is similar to that seen in the subsonic case. A wavy pattern is seen in the v′ levels at the boundary of the core of the circular and bevel nozzle jet (figure 9b) indicating the presence of shock in the jet flow field. However, the amplitude of this wavy pattern for the bevel nozzle is (figure 9f) lower than that seen in circular. Influence of secondary stream on the transverse velocity fluctuations at subsonic Mach number M_P=0.96; M_S=0.65 for the circular and bevel nozzle is shown in figure 9c and 9g. In the circular nozzle case, the axial extent of the low transverse velocity fluctuation region in the jet core is longer for dual stream case as compared to the primary stream alone case. In the bevel case, similar to the circular one the axial extent of the low fluctuation region in the potential core is increased (figure 9c) by the secondary stream. However the value is slightly higher than in the circular case. The magnitude of the v′ is lower in the outer shear layer at the longer lip side as compared to the circular dual stream. Further, the fluctuations are higher in the outer shear layer at the shorter lip side than that in the longer lip side. At supersonic condition M_P=1.2; M_S=0.85, the trend of velocity fluctuation in the core and outer shear layer is similar to that seen in the subsonic case (figure 9h). Additionally, the fluctuation values in the core and in the outer shear layer at the shorter lip are higher than dual stream flow with circular nozzle. Effect of primary nozzle exit geometry modification on the average Reynolds stress (u'v') of the dual stream nozzle flow at subsonic and supersonic exit condition is shown in figure 10a-h. The Reynolds stress values are normalized by square of the jet exit velocity of the primary stream. The Reynolds stress contour for the circular nozzle shows (figure 10a and 10b) that the values in the shear layer is antisymmetric about jet axis for the primary stream alone case at Mach numbers of 0.96 and 1.2. Whereas in the bevel nozzle, the Reynolds tress has higher magnitude in the shorter lip shear layer and lower in the longer lip shear layer(figure 10e and 10f). The Reynolds stress sign change is seen in the longer lip shear layer at Mach number 1.2. Further the Reynolds stress value is higher near the exit (up to $x/D_P=2$) and lower beyond $x/D_P=2$ in both shear layers of bevel nozzle in comparison with the circular at Mach number 1.2. The Reynolds stress values are reduced significantly by the secondary stream for both circular and beveled dual stream flow. The Reynolds stress values in the outer shear layer are lower at longer lip side and higher in the shorter lip as compared to the circular nozzle for both subsonic (figure 10c and 10g) and supersonic condition (figure 10d and 10h). The values of u'v' have direct influence on noise generation. Hence examination/comparison is important.

4.Conclusions

Influence of primary stream nozzle exit modification on the mean and turbulence fluctuation quantities in the near flow field of dual stream nozzle flow was experimentally investigated using 2D particle image Velocimetry. Additionally, the flow field was qualitatively visualized using shadowgraph technique. The results show that the beveling of the nozzle causes a reduction in the axial extent of shock cells compared to the circular jet, in the supersonic case. Further for the beveled nozzle case, the number of shock cells seen in the jet is less than the circular nozzle case and they are weakened by the presence of secondary stream. The flow field measurement from the 2D-PIV shows the differential shear layer growth between shorter and longer lip sides of the beveled nozzle for both supersonic and subsonic cases. The co-flowing secondary stream reduces the flow inclination angle of bevel nozzle. For the beveled nozzle case while the streamwise velocity fluctuations in the shear layer is higher at the longer lip and lower at the shorter lip, the transverse velocity fluctuation in the shear layer is lower at the longer lip and higher at the shorter lip. Introduction of secondary stream increased the potential core length due to the reduction in shear between the primary and secondary stream. The beveling of the nozzle results in Reynolds stress values being much lower for both single and dual stream configurations, when compared to the corresponding circular nozzle cases. This could be the reason for the reduction of noise level reported in the literature for the beveled nozzle.

Acknowledgments
Authors wish to acknowledge assistance of EAD Model Shop in fabricating the components of the co-flow facility.

References

[1] Dimitri Papamoschou 2004 New method for jet noise reduction in turbofan engines *AIAA Journal* **42** 11
[2] K.Viswanathan 2004 Parametric study of noise from dual-stream nozzles *Journal of Fluid Mechanics* **521** 35-68
[3] Viswanathan, K 2005 Nozzle shaping for reduction of jet noise from single jets *AIAA Journal* **43** 5 1008–1022
[4] Viswanathan, K. 2006 An elegant concept for reduction of jet noise from turbofan engines *Journal of Aircraft* **43** 3 616–626
[5] Viswanathan, K. 2006 Noise of dual-stream beveled nozzles at supercritical pressure ratios," *Journal of Aircraft* **43** 3 627–638
[6] Viswanathan, K., Shur, M., Spalart, P. R., and Strelets, M. 2008 Flow and noise predictions for single and dual-stream beveled nozzles *AIAA Journal* **46** 3 601–626
[7] K. Viswanathan, P. R. Spalart, and M. J. Czech 2011 Tailored nozzles for jet plume control and

noise reduction *AIAA Paper* 2011-2011
[8] Hargather, M. J., and Settles, G. S. 2009 Retroreflective shadowgraph technique for large-scale flow visualization *Applied optics* **48** 22 4449-4457

IOP Conf. Series: Journal of Physics: Conf. Series **822** (2017) 012005 doi:10.1088/1742-6596/822/1/012005

High-Alfa Aerodynamics with Separated Flow Modeled as a Single Nascent Vortex

Antony Samuel B
MS Scholar
Dept.of Applied Mechanics, Indian Institute of Technology Madras, India 600036.
E-mail: antonysamuelb@gmail.com

Rinku Mukherjee
Assistant Professor
Dept.of Applied Mechanics, Indian Institute of Technology Madras, India 600036.
E-mail: rinku@iitm.ac.in

Abstract. A numerical iterative vortex lattice method is developed to study flow past wing(s) at high angles of attack where the separated flow is modelled using NY nascent vortex filaments. The wing itself is modelled using $NX \times NY$ bound vortex rings, where NX and NY are the number of sections along the chord and span of the wing respectively. The strength and position of the nascent vortex along the chord corresponding to the local effective angle of attack are evaluated from the residuals in viscous and potential flow, i.e. $(C_l)_{visc} - (C_l)_{pot}$ and $(C_m)_{visc} - (C_m)_{pot}$. Hence, the 2D airfoil viscous $C_l - \alpha$ and $C_m - \alpha$ is required as input (from experiment, numerical analysis or CFD). Aerodynamic characteristics and section distribution along span are predicted for 3D wings at a high angle of attack. Effect of initial conditions and existence of multiple solutions in the post-stall region is studied. Results are validated with experiment.

1. Introduction

The interest in extending linear methods to include post-stall regimes of flow has existed since Prandtl's Lifting Line Theory. A vortex-lattice method algorithm based on a novel decambering approach [1] uses an estimate of the reduction in camber at post-stall angles of attack to account for the change in C_l and C_m from the inviscid case.

Sarpakaya [2] in his review paper, discusses shedding of discrete vortices from the points of separation. He mentions that vortex methods are the weakest when used for simulating separated flows. Various methods, using both variable and fixed positions of the nascent vortex have been discussed while the strength of the circulation of the discrete shed vortex at each time step is found out from the velocity at a certain point, e.g. at the position of the nascent vortex.

Katz [3] places the separated vortex at the half way point of the distance covered in a given time step, by the panel LE in which the separation point lies. The separation point in this case is previously known. The magnitude of circulation of Nascent Vortex shed is calculated using velocities across the shear layer. Antonini et al [4] uses C_l value of the airfoil to find the point of separation and the separated nascent vortex is placed at a certain predetermined distance from the point of separation. Sarpakaya [5] estimates the velocity which is used for calculating the strength of the shed vortex by taking the average velocity of 4 previously shed vortices. The position of the nascent vortex is selected in such a way that the kutta condition is satisfied. Chorin [6] has set a different path, wherein nascent vortices are introduced throughout the surface. No-slip condition is applied at the boundary in order to find the strength and the core radius of the Nascent Vortices. The method has been used to predict the temporal loading on a cylinder.

Separated flow, especially post stall is an unsteady phenomenon. The motivation of the current work is to incorporate viscous effect into a 2D discrete vortex method and a 3D vortex lattice method, while also being able to predict the position of and strength of Nascent Vortex. This work can then be extended to unsteady cases.

2. Mathematical Model

In the present work, separated flow is assumed to be steady and is modelled by a single nascent vortex in the separated region as shown in Fig. 1, whose strength and the position along the chord are unknown.

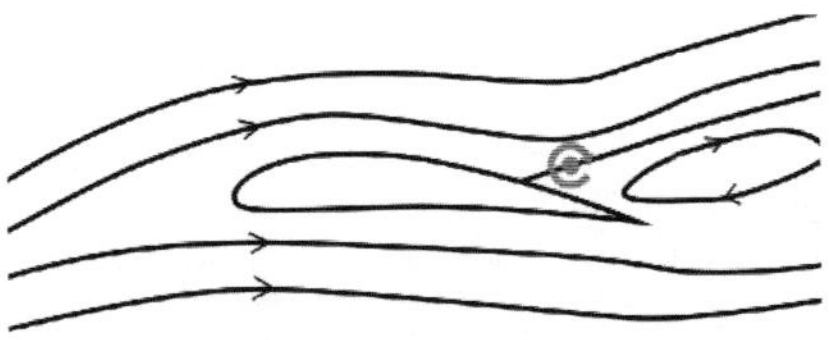

(a) Separated Flow in an airfoil

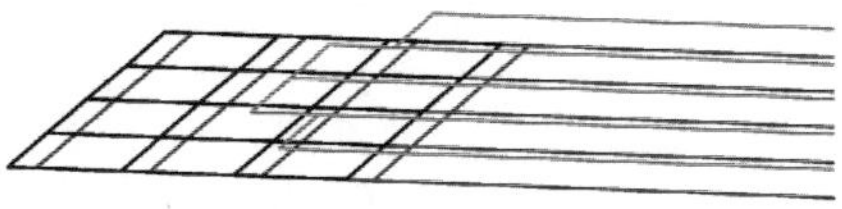

(b) Separated Flow model for a 3D wing

Figure 1: Separated Flow

The airfoil characteristics in the inviscid regime are predicted by a discrete vortex element method (DVEM) as in Katz et al [8]. The inviscid model is extended to account for post-stall flow using a single nascent vortex with a fixed finite core as shown in Fig 1a in addition to DVEM.

The idea is that when viscous flow deviates from the inviscid flow $C_l - \alpha$ curve due to flow separation from the lifting surface, it essentially causes a deficit in the lift-creating vorticity around the airfoil. Therefore, if we can compensate for this loss in vorticity, we should be able to get back the missing lift. The present method proposes to account for this missing vorticity by a nascent vortex of strength Γ_{nv} and position, x_{nv} along chord and z_{nv} perpendicular to the airfoil chord such that its effect on the airfoil is able to recover the loss in the coefficients of lift and pitching moment.

However, we have three unknowns (Γ_{nv}, x_{nv}, z_{nv}) and only two pieces of information ($\Delta C_l, \Delta C_m$) to evaluate these unknowns. Hence, one of the values must be assumed or found empirically so that a unique solution is obtained. In the present method, z_{nv} is assumed. Therefore, the effect of the nascent vortex (Γ_{nv}, x_{nv}) is expected to be such that it accounts for the residuals: $(C_l)_{visc} - (C_l)_{pot}$ and $(C_m)_{visc} - (C_m)_{pot}$. The unknowns are then solved iteratively so that they satisfy Eqn. 1.

$$J \cdot \delta x = -F \tag{1}$$

where

$$F = \begin{pmatrix} (C_l)_{pot} - (C_l)_{visc} \\ (C_m)_{pot} - (C_m)_{visc} \end{pmatrix} = \begin{pmatrix} \Delta C_l \\ \Delta C_m \end{pmatrix}$$

$$J = \begin{pmatrix} \frac{\delta \Delta C_l}{\delta \Gamma_{nv}} & \frac{\delta \Delta C_l}{\delta x_{nv}} \\ \\ \frac{\delta \Delta C_m}{\delta \Gamma_{nv}} & \frac{\delta \Delta C_m}{\delta x_{nv}} \end{pmatrix}; \qquad \delta x = \begin{pmatrix} \Delta \Gamma_{nv} \\ \Delta x_{nv} \end{pmatrix}$$

The $C_l - \alpha$ and $C_m - \alpha$ from the present methodology are plotted along with the input experimental data from Abbott [10] in Fig. 2.

The core radius (cr) of the Nascent Vortex is an important parameter. The effect of variation of the size of the core on the strength and position of the NV for NACA0009 is shown in Fig. 3. It is found that for $cr = 0.025$ and $cr = 0.05$, convergence does not occur for some α, while for some α the nascent vortex is located beyond the trailing edge, which is physically not possible.

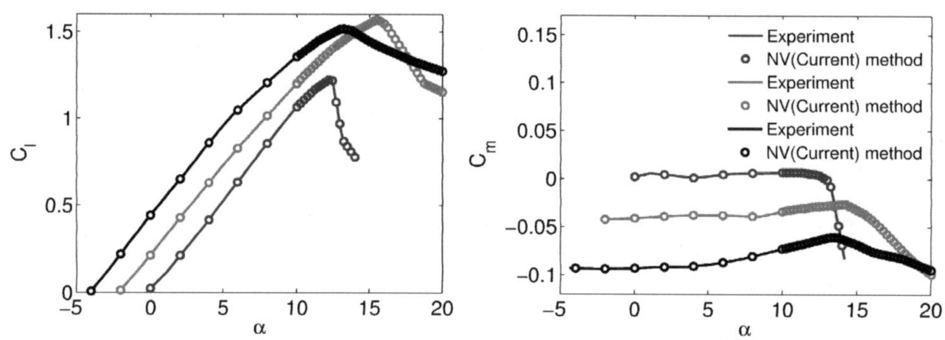

Figure 2: Aerodynamic Characteristics. Blue: NACA0009, Red: NACA2412, Black: NACA4412.

With increase in cr, the NV moves closer to the leading edge and the predicted circulation also increases. This trend is similar to the one observed when the z_{nv} is increased. Hence, a core radius of at least 0.1 has to be used. For all of the current calculations, a $cr = 0.2$ is used.

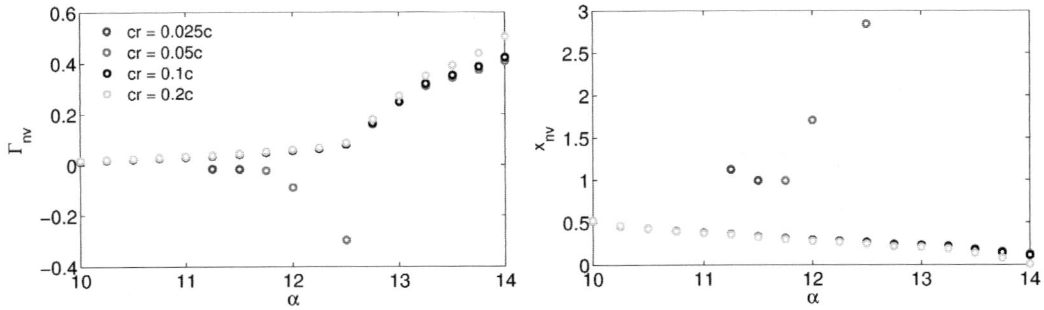

Figure 3: Variation of Γ_{nv} & x_{nv} with core radius (cr)

2.1. *Implementation of Methodology for a 3D Wing*

The inviscid solution of a 3D wing is obtained by using a Steady Vortex Lattice Method(VLM) [8]. The nascent vortex concept illustrated for a 2D airfoil is incorporated into the solution for a 3D wing, where the separated flow is modelled by horse-shoe vortices as shown in Fig. 1b. For the 3D solution, Eqn. 1 transforms to Eqn. 2.

$$J_{(2NY\times 2NY)} \cdot \delta x_{(2NY\times 1)} = -F_{(2NY\times 1)} \tag{2}$$

The local effective angle of attack $(\alpha_{eff})_i$ is computed using Eqn. 3.

$$(\alpha_{eff})_i = -sin^{-1}\left(\frac{K+G_i}{NX\times\left|\overrightarrow{U}_\infty\right|}\right); \qquad i = 1, NY \tag{3}$$

Where,

$$K_i = \sum_{k=1}^{NX}\sum_{l=1}^{NX}\left(\left(\overrightarrow{V_b}\right)_{lk}\cdot\hat{n_k}\right)_i; \qquad G_i = \sum_{k=1}^{NX}\left(\left(\overrightarrow{V}_{nv}\right)_{ik}\cdot\hat{n_k}\right)_i;$$

$\left(\overrightarrow{V_b}\right)_{lk}$ is the velocity induced by the bound vortex of panel l on the collocation point of panel k; $\left(\overrightarrow{V}_{nv}\right)_{ik}$ is the velocity induced by the Nascent Vortex on section i on collocation point of panel k; $\hat{n_k}$ is unit normal vector to panel k and $\overrightarrow{U}_\infty$ is the free stream velocity.

Solve Eqn. 2 for the entire wing and find $\Delta\Gamma_{nv_i}$ & Δx_{nv_i}

$$\begin{pmatrix} \frac{\delta\Delta C_{l_j}}{\delta\Gamma_{nv_i}} & \frac{\delta\Delta C_{l_j}}{\delta x_{nv_i}} \\ \frac{\delta\Delta C_{m_j}}{\delta\Gamma_{nv_i}} & \frac{\delta\Delta C_{m_j}}{\delta x_{nv_i}} \end{pmatrix}\begin{pmatrix}\Delta\Gamma_{nv_i} \\ \Delta x_{nv_i}\end{pmatrix} = -\begin{pmatrix}\Delta C_{li} \\ \Delta C_{mi}\end{pmatrix}$$

Where $i = 1, NY$; $j = 1, NX$

3. Results

The present work is compared with the experimental work of Naik and Ostowari [11] for a rectangular wing of aspect ratio 12 at $Re = 0.5\times 10^6$ & 0.75×10^6 as shown in

IOP Conf. Series: Journal of Physics: Conf. Series **822** (2017) 012005 doi:10.1088/1742-6596/822/1/012005

Fig. 4. Airfoil data is required as input, which is also available from the above reference. NACA4415 airfoil is used. No thickness correction is applied. $NY = 16$ is used.

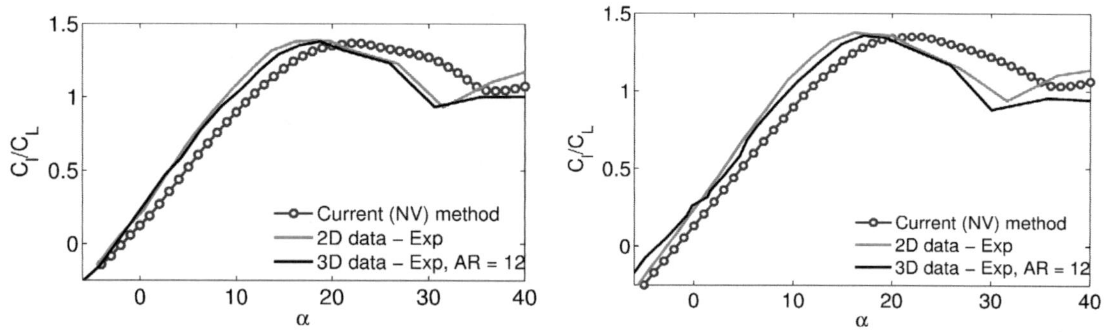

(a) Re $= 0.5 \times 10^6$ (b) Re $= 0.75 \times 10^6$

Figure 4: $C_L - \alpha$ for a finite wing (section: NACA 4415, $AR = 12$).

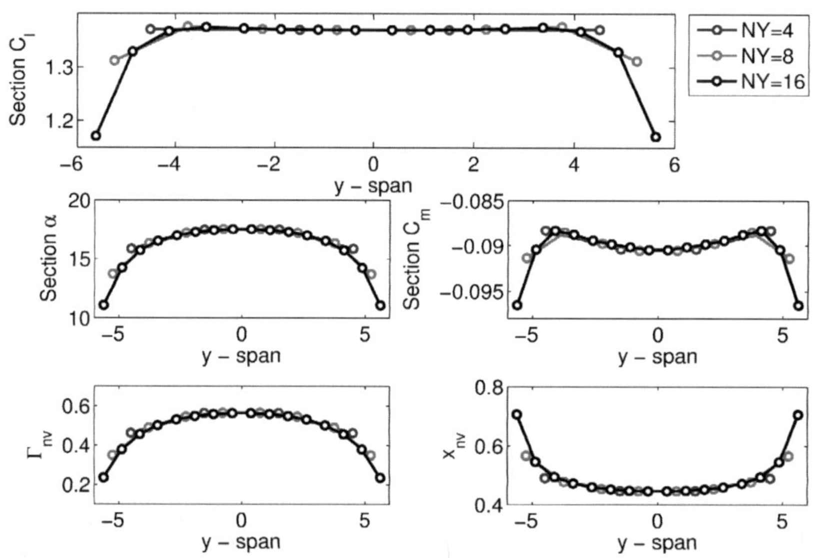

Figure 5: Variation along span of rectangular wing (section: NACA 4415, $AR = 12$, α=20°).

It is seen from the 2D and 3D experimental plots in Fig. 4 that the airfoil and wing stall at $\alpha \approx 18^o$, which corresponds to a maximum $C_l \approx 1.4$ while the present results show that the wing is near-stall at $\alpha \approx 20^o$ for a similar maximum C_l. It is seen in Fig. 5 that in the root section of the wing, section $\alpha \approx 18^o$, which corresponds to a maximum $C_l \approx 1.4$. This is confirmed by the section C_l distribution. The high section C_l causes very high induced velocities near the root-section and hence there is an increase in the negative C_m. Conversely, the sectional distribution of the strength and location along

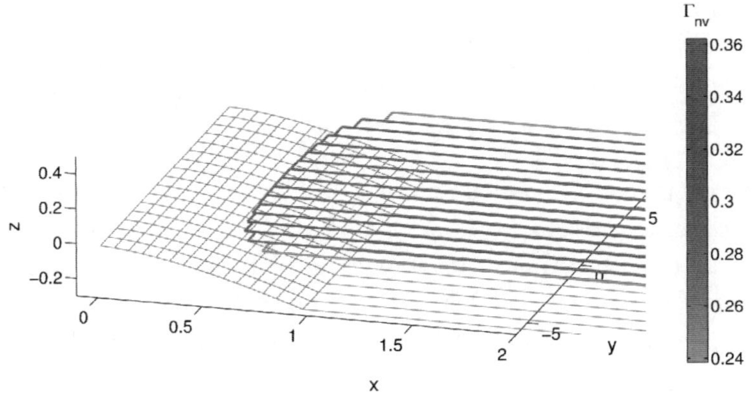

Figure 6: Distribution of circulation over Rectangular Wing(section: NACA 4415, $AR = 12$, α=15°)

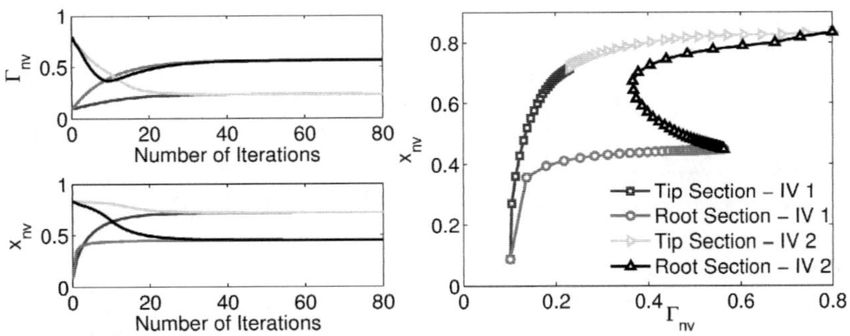

Figure 7: Effect of number of iterations and initial conditions on rectangular wing (section: NACA 4415).

chord of the nascent vortex show that Γ_{nv} is stronger and x_{nv} is nearly 0 near the root, suggesting that the vortex moves closer to the leading edge. It is interesting to note that even at the wing tips, Γ_{nv} is not close to zero. A particular case of the distribution of Γ_{nv} along the wing span at $\alpha = 15^o$ is shown in Fig. 6, where it clearly shows that the strength of the nascent vortex is stronger near the root and weakest near the tips for a rectangular wing.

The section C_l distribution is clearly devoid of any sawtooth and this result is obtained without the use of any *smoothing algorithm.* This is definitely a clear advantage over the post-stall predictive tool developed by Mukherjee et. al. [1].

The variation in Γ_{nv} & x_{nv} at the root and tip sections as the iterations progress is shown in Fig. 7 for a wing with NACA 4415 section, $\alpha = 20^o$, $AR = 12$ & $NY = 16$. Effect of initial conditions on Γ_{nv} & x_{nv} is studied at the tip and root sections for two different initial values for a final converged result. The initial values used are as follows: IV 1: Γ_{nv}=0.1 & x_{nv}=0.1 and IV 2: Γ_{nv}=0.8 & x_{nv}=0.8. For initial values of Γ_{nv}=0.9

& $x_{nv} = 0.9$ the solution did not converge. Hence, different initial values did not result in multiple solutions for a post-stall angle of attack.

4. Conclusion

A multi-dimensional post-stall predictive tool is developed for wing(s) using VLM, which uses a nascent vortex to account for flow separation. The tool is robust for wings with cambered airfoil sections but some convergence issues exist for wings with symmetric airfoil sections. For wings with cambered airfoil sections, section C_l distributions are completely devoid of sawtooth. Different initial conditions do not result in multiple solutions.

References

[1] Mukherjee, R., and Gopalarathnam, A., *Post-Stall Prediction of Multiple-Lifting-Surface Configurations Using a Decambering Approach*, Journal of Aircraft, Vol. 43, No. 3, pp. 660-668, May-June 2006.

[2] Sarpkaya T., *Computational Methods With Vortices-The 1988 Freeman Scholar Lecture.* ASME. J. Fluids Eng., 111(1):5-52, 1989. doi:10.1115/1.3243601.

[3] Katz J., *Large-scale vortex-lattice model for the locally separated flow overwings.* AIAA Journal 20:12, 1640-1646, 1982.

[4] Enrico G. A. Antonini, Gabriele Bedon, Stefano De Betta, Luca Michelini, Marco Raciti Castelli, and Ernesto Benini, *Innovative Discrete-Vortex Model for Dynamic Stall Simulations.* AIAA Journal 2015 53:2, 479-485.

[5] Sarpkaya T., *An inviscid model of two-dimensional vortex shedding for transient and asymptotically steady separated flow over an inclined plate.* Journal of Fluid Mechanics, 68, pp 109-128, 1975.

[6] Alexandre Joel Chorin. *Numerical study of slightly viscous flow.* Journal of Fluid Mechanics, 57, pp 785-796, 1973.

[7] A Leonard, *Vortex methods for flow simulation.* Journal of Computational Physics, Volume 37, Issue 3, Pages 289-335, ISSN 0021-9991, October, 1980.

[8] Katz, Joseph and Allen Plotkin. *Low-Speed Aerodynamics.* 2nd ed. Cambridge: Cambridge University Press, 2001.

[9] Leishman, J. G. *Principles of helicopter aerodynamics.* Cambridge: Cambridge University Press, 2006.

[10] Ira H. Abbott, Albert E. von Doenhoff, Louis S. Stivers, *Summary of airfoil data.* NACA Report, No. 824, 1945.

[11] Ostowari, C. and Naik, D., *Post Stall Studies of Untwisted Varying Aspect Ratio Blades with an NACA 4415 Airfoil Section Part 1*, Wind Engineering, Vol. 8, No. 3, 1984,pp. 176194.

IOP Conf. Series: Journal of Physics: Conf. Series **822** (2017) 012006 doi:10.1088/1742-6596/822/1/012006

Influence of non perfect impedance boundary on the bistable region in thermoacoustic interactions

B Mohan[1] **and S Mariappan**[2]

Department of Aerospace Engineering, Indian Institute of Technology Kanpur, Kanpur-208016, India

[1] Graduate Student and Corresponding author

[2] Assistant Professor

E-mail: [1] mbala@iitk.ac.in, [2] sathesh@iitk.ac.in

Abstract. We investigate the influence of non perfect impedance boundary on the bistable zone in thermoacoustic interactions of a horizontal Rijke tube. A wave based approach is used to obtain the nonlinear dispersion relation with frequency dependent impedance boundary condition. The location and the time delay in the response of the heater are considered as bifurcation parameters to obtain the stability boundaries. In the presence of non perfect impedance boundary condition, we find that the extent of globally unstable regime reduces and the bistable zone significantly increases. The quantitative changes in the stability boundaries and the bistable zone are investigated for different time lags. However, the nature of bifurcation remains sub critical and unaltered for the range of time delays considered in the present study.

1. Introduction

Thermoacoustic instability is one of the major issue which has been faced by scientist, engineers and industrialist over the last five decades. Starting from rocket engines to gas turbines, boilers, afterburners and incinerators, thermoacoustic instability has influenced their operational limits. Controlling this instability during design phase is still a challenging task. Detailed review on the dynamics and control of this phenomenon can be found in [1, 2]. Rijke found that when heated gauze is placed at quarter length of a vertical tube in the bottom half, produced sound and this was later known as thermoacoustic instability [3].

Most of the study in thermoacoustic interactions concerned about different types of heat sources and their influence on the stability. The amplitude of the limit cycle oscillations is determined by the balance between the acoustic driving caused due to the unsteady heat release rate and damping in the system. Energy losses due to acoustic radiation and dissipation at the walls are the major sources of damping. Only few included radiation losses directly into analysis [4, 5, 6] and a few others added the effects of damping implicitly in energy equation [7, 8]. However, no detailed investigation on the effects of radiation losses on the stability regimes is performed yet. Thermoacoustic system possess three different stability regimes namely globally stable, globally unstable and bistable regions [5, 10, 11, 8]. In the globally stable zone, all the disturbances decay exponentially irrespective of the amplitude of oscillation and eventually, the system reaches its steady state. In the globally unstable zone, disturbances grow even if it is infinitesimally small; there is no linearly stable state and finite amplitude oscillation (limit cycle) prevails. In the bistable zone, the system dynamics relies completely on the initial perturbation

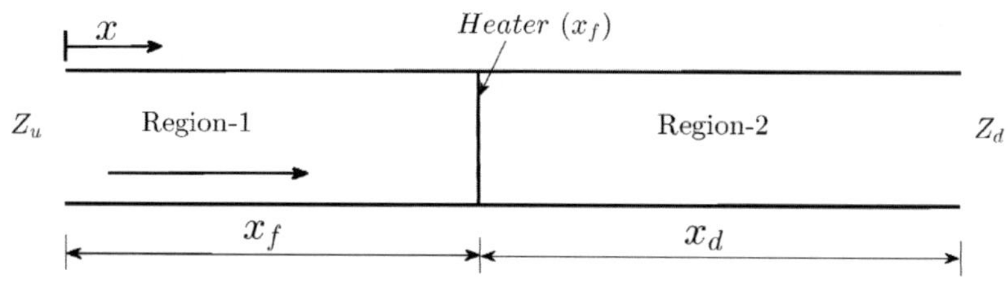

Figure 1. Schematic of Rijke tube.

given to the system [12]. This is manifested as an hysteresis in the experiments. Furthermore, classical linear stability analysis is not valid in this zone, making the prediction on the dynamics difficult. Since the bistable zone possesses crucial dynamics such as triggering and lies in between the linearly stable and linearly unstable zones, it is interesting to explore it in detail.

The presence of noise on the dynamics of the bistable zone was investigated exhaustively in the recent past. Experimental investigation in a simulated gas turbine combustor reported that inherent noise in the combustor can make nominally stable combustor into unstable [13]. Waugh et al. [15] found that high amplitude low frequency (pink) noise has strong influence on triggering. Experimental investigation of noise induced transition in a diffusion flame was performed by inducing random disturbances in fuel flow. A reduction in the amplitude of limit cycle oscillation was observed [16]. In reference [14], it was found that the system exhibits completely different type of bifurcation above certain threshold noise level. All the above investigations indicate that the nature of bifurcation or in turn the extent of the various stability regimes were altered significantly in the presence of noise.

We believe that the extent of bistable zone is not only affected by the noise, but also by the presence of acoustic losses in the system. Therefore our main objective in this study is to investigate the effects of radiation losses on the bistable zone, which are not given due attention in the past. Radiation losses in flanged and unflanged tubes, their dependence on frequency and amplitude were investigated in references [17, 18]. In this paper, describing function (DF) technique is used to predict the nonlinear stability of thermoacoustic oscillations. This technique is preferred, as it is easily possible to incorporate non perfect impedance boundary conditions. In the past, this framework was used to predict limit cycle, triggering and frequency of oscillations [10, 11, 4].

This paper is arranged in the following manner. Section 2 deals with the theoretical modeling of Rijke tube and obtaining the dispersion relation, followed by section 3 interpreting the results. In the end, section 4 provides a summary of the present work.

2. Modeling system dynamics using wave approach

One of the most simple and elegant model to capture the salient dynamics associated with thermoacoustic interactions is a Rijke tube and it is best suited for our study. Linearized acoustic momentum and energy equations govern the flow properties in a Rijke tube. Figure 1 shows the schematic of a Rijke tube with a heat source located at x_f. The heat source is assumed to be compact compared to the length of the tube l. The steady flow is from left to right and the associated Mach number is small. The mathematical model for heat source is obtained from Heckl [6]. As shown in figure 1, region 1 and 2 has constant flow properties. Linearized momentum and energy equations are combined to obtain the wave equation as described in reference[5],

$$\frac{\partial^2 p'}{\partial t^2} + \zeta \frac{\partial p'}{\partial t} - c^2 \frac{\partial^2 p'}{\partial x^2} = \frac{(\gamma - 1)}{S_c} \frac{\partial \dot{Q}'}{\partial t} \delta(x - x_f) \tag{1}$$

IOP Conf. Series: Journal of Physics: Conf. Series **822** (2017) 012006 doi:10.1088/1742-6596/822/1/012006

where p', $\dot{Q}'$, S_c, γ, ζ and δ denotes acoustic pressure in the duct, fluctuating heat release rate from the heat source, cross sectional area, ratio of specific heat capacities, acoustic damping and Dirac delta function respectively. The above equation (1) is solved using wave based approach [19]. The pressure and velocity fluctuations are represented as $p' = \Re\left[\hat{p}(x)exp(i\omega t)\right]$ and $u' = \Re\left[\hat{u}(x)exp(i\omega t)\right]$, $i = \sqrt{-1}$, where ω is the complex frequency with $\omega = \omega_r + i\omega_i$. ω_r and ω_i represent the oscillating circular frequency and the growth/decay rate of the oscillations respectively. These expressions are then inserted in the above equation to obtain an Helmholtz equation. Solving Helmholtz equation results in a solution for $\hat{p}$ and $\hat{u}$ as follows,

$$\hat{p}_n(x) = A_n \exp\left[is_n\left(x - x_f\right)\right] + B_n\, exp\left[-is_n\left(x - x_f\right)\right] \tag{2}$$

$$\hat{u}_n(x) = -\frac{s_n}{\omega\rho_n}\left\{A_n\, exp\left[is_n\left(x - x_f\right)\right] - B_n\, exp\left(-is_n\left(x - x_f\right)\right)\right\} \tag{3}$$

For locations in the up $(0 \leq x < x_f)$ and downstream $(x_f < x \leq l)$ of the heat source, the subscripts n is replaced with u and d respectively. c_n is the local speed of sound and $s_n^2 = (\omega^2 - i\zeta\omega)/c_n^2$. Note that the above solutions are valid for all $x \neq x_f$. At x_f, the acoustic jump conditions, $p'_u(x_f^-) = p'_d(x_f^+)$ and $u'_d(x_f^+, t) - u'_u(x_f^-, t) = (\gamma - 1)\dot{Q}'(t)/(\gamma S_c P)$ are applied. The superscripts $'$, $\hat{}$, $^-$, $^+$ denote fluctuating quantities, complex amplitude, quantities just up and downstream of the heat source respectively. In the velocity jump condition, $\dot{Q}'(t)$ is replaced with describing function. This describing function assumes weak nonlinearity, meaning that only the response in the excitation frequency is used for the analysis and the higher harmonics are neglected. Furthermore, A_n, B_n are all constants and are eliminated to obtain the following nonlinear dispersion relation.

$$\xi\left[G_d\cos(s_d x_d) - iH_d\sin(s_d x_d)\right]\left[H_u\cos(s_u x_f) + iG_u sin(s_u x_f)\right] - \left[G_u\cos(s_u x_f) + iH_u\sin(s_u x_f)\right]\left[H_d\cos(s_d x_d) - iG_d\sin(s_d x_d)\right]\left[1 + \frac{(\gamma-1)\dot{Q}}{\gamma S_c P u_u}\mathcal{F}\right] = 0 \tag{4}$$

Here $x_d = l - x_f$, $G_n = I_n - 1$, $H_n = I_n + 1$ and $I_n = (s_n/k_n - 1/Z_n)\,/\,(s_n/k_n + 1/Z_n)$. Further $\xi = \rho_u c_u/\rho_d c_d$ and $\mathcal{F}\left(u_u^-, \omega\right) = (\dot{Q}'/\dot{Q})/[u'_u(x_f^-)/u_u] = \mathcal{G}\ exp\left(i\phi\right)$. Z, P, $\mathcal{F}$, $\mathcal{G}$, ϕ represent acoustic impedance at the boundary associated with n (for instance when $n = u$, $Z_n = Z_u$(at $x = 0$)), steady state pressure, describing function, gain and phase. The expression for damping is taken from references [7, 18], $\zeta = \omega_r/\pi\left(c_1\omega_r/\omega_{r1} + c_2\sqrt{\omega_{r1}/\omega_r}\right)$. c_1 and c_2 are damping co-efficients which accounts for the losses. ω_{r1} denotes fundamental oscillation frequency of an organ pipe.

In equation (4), $\mathcal{F}$ represents the response of the heat source to various input frequencies and amplitude of velocity fluctuations. In this paper, the heat source is considered to be an electrical heater. The instantaneous heat transfer (release) rate $(\tilde{\dot{Q}})$ for a given unsteady velocity fluctuation is obtained from modified King's law [6].

$$\tilde{\dot{Q}} = L_w\left(T_w - T_u\right)\left\{X + 2\sqrt{\pi X c_v \rho u_u \frac{d}{2}}\left[\left(1 - \frac{1}{3\sqrt{3}}\right) + \frac{1}{\sqrt{3}}\sqrt{\left|\frac{1}{3} + \frac{u'_u\left(x_f^-, t - \tau_w\right)}{u_u}\right|}\right]\right\} \tag{5}$$

where L_w, T_w, T_u, d, X, c_v and ρ represent the wire length, wire surface temperature, upstream temperature, wire diameter, thermal conductivity, specific heat at constant volume and steady state density of the flow respectively. From equation (5), steady state heat transfer rate $(\dot{Q})$ is found by putting $u'_u = 0$ and the fluctuating heat transfer rate$(\dot{Q}')$ is obtained from $\dot{Q}' = \tilde{\dot{Q}} - \dot{Q}$.

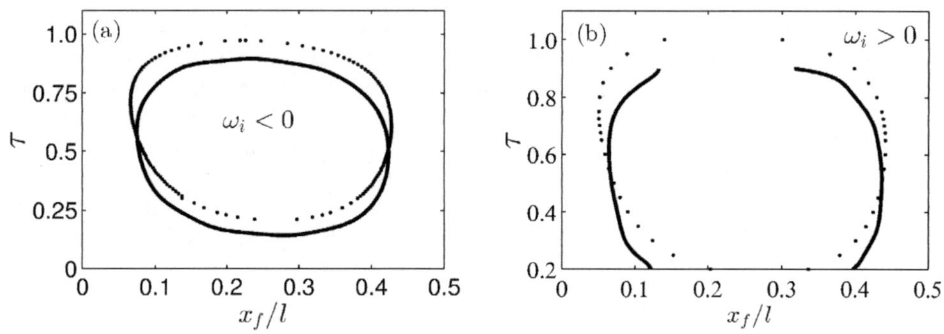

Figure 2. (a) Linear stability boundary and (b) nonlinear stability boundary in the $x_f/l - \tau$ plane. ω_i shown, is obtained in the limit $|u'_u(x_f^-)| = 0$. Further —— and $\cdots\cdots$ lines denotes the study of Subramanian et al. [8] and present analysis respectively. The non dimensional heat release rate $Q_{ND} = 0.8$.

Non dimensionalized fluctuating heat transfer rate is written as $Q_{ND} = (\gamma - 1)\dot{Q}/(\gamma S_c P u_u)$. τ_w is the time lag between the input u'_u and the response $\dot{Q}'$. It is non dimensionalized as $\tau = \tau_w c_u/l$.

Table 1. Parameters used in this investigation.

T_w=700 K	T_u=295 K	u_u=0.24 m/s	γ =1.4	R=288.35 J/kg K	l= 1 m
d=0.45 mm	L_w=13 m	r_t=52.5 mm	P=1atm	$c_1 = 0.1$	$c_2 = 0.06$

3. Results and discussion

In this section, results from the above solution procedure are first compared with the previous analysis [8], where the solution was obtained in the time domain using continuation methods. Acoustically perfect boundary conditions (open-open in this case) were used for the comparison. In perfect boundary conditions, the time averaged acoustic intensity is zero and hence there are no radiation losses. Later, the effects of inclusion of non perfect impedance boundary condition, which is the main objective of the present paper is discussed.

For a given x_f, acoustically open-open boundary condition ($Z_n = 0$) and other parameters shown in table 1, equation 4 is solved to obtain ω. From the complex frequency, both stability boundary and frequency of oscillations are obtained. To facilitate a one to one comparison with reference [8], ξ is kept 1 (although not true in reality) and other parameters are chosen as the same value given in the above reference. The comparison is shown in figure 2 for the linear and nonlinear stability boundaries, where the non dimensional time lag τ and heater location x_f/l are the parameters. The system is linearly stable, when $\omega_i > 0$ in the limit $|u'_u(x_f^-)| = 0$ and linearly unstable when $\omega_i < 0$. The demarcation line is the linear stability boundary. On the other hand, in the bistable zone, ω_i equals zero for two values of $|u'_u(x_f^-)| > 0$. At these two locations, when $d\omega_i/d(u'_u(x_f^-)) < 0$, there is an unstable limit cycle. Otherwise, a stable limit cycle occurs. As x_f is varied, these two solutions merge to form a *fold point* (in the parlance of dynamical systems' theory). The locus of this fold point in the $x_f/l - \tau$ plane forms the nonlinear stability boundary. The region between the linear and nonlinear stability boundaries forms the bistable zone.

We compare the boundaries obtained from this paper (dotted lines), with those from reference [8] (full lines) and are shown in figure 2. It is expected to have a perfect match between the two

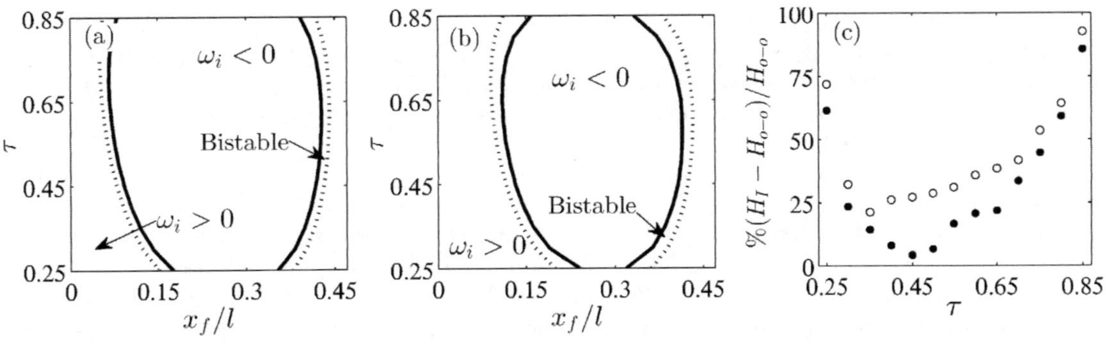

Figure 3. Stability boundaries for (a) perfect (acoustically open) (b) non perfect boundary conditions. —— and $\cdots\cdots$ lines indicate linear and nonlinear stability boundaries respectively. (c) Relative change in the extent of the bistable zone between figures 3 (a) and (b). ○ and ● denote left and right envelopes of stability boundaries respectively. $Q_{ND} = 0.95$.

results, as the equations, boundary conditions and the parameters are the same. We only observe that the stability boundaries are analogous qualitatively, they are different quantitatively. This difference is attributed to the different analysis methods used in both the investigations. In reference [8], Galerkin approach was used to convert the partial differential equation (1) to a set of ordinary differential equations and later continuation algorithm was used to obtain the stability boundaries. In the present investigation, we use wave based approach to predict the same. Both the methods use different approximations. In the former, since Galerkin spatial decomposition is used, the acoustic velocity jump across the heat source is excluded, while in the latter, higher harmonics in the describing function for the heat source are neglected.

Furthermore, for a given location of the heat source (say $x_f/l = 0.25$), the system is linearly and nonlinearly stable for low values of τ. As τ is increased, it crosses the stability boundaries, enters initially the bistable region and then linearly unstable zone. Further, as τ is increased to large values, the system again becomes globally stable. τ is related (approximately) inversely to the mass flow rate in the system. This phenomenon of a initially globally stable system becoming unstable and then becoming again stable, as the mass flow rate is increased is also observed in the experiments [9]. Hence, the present solution procedure is able to capture the observed trends in the experiments and compares qualitatively with the previous numerical investigations.

We can now use this model and the analysis procedure to investigate the effects of non perfect impedance boundary conditions. Frequency dependent impedance boundary conditions for an unflanged pipe is used from reference [17] for our investigation.

$$Z_n = \left.\frac{\hat{p}}{\hat{u}}\right|_n = \left(\frac{\omega_r r_t}{2c}\right)^2 - 0.6i\frac{\omega_r r_t}{c} \tag{6}$$

where, r_t is the radius of tube. This expression is included as Z_n in the dispersion relation (4) and solved for the complex frequency, thereby obtaining the stability boundaries. The obtained stability boundaries are then compared with those obtained by using perfect (open) boundary conditions (figure 3(a,b)). First, one could observe a reduction in the globally unstable zone (boundaries in full lines shown in both the figures). Non perfect impedance boundary conditions lead to additional losses due to acoustic radiation and hence the size of the unstable zone is decreased. For the same reason, the size of the globally stable zone, indicated by the nonlinear stability boundary (shown as dotted lines) is also increased. However, non trivially, region between the two boundaries (bistable zone) is increased.

One can observe two branches of the bistable zone in figure 3(a,b). These are indicated as left and right envelopes. The percentage relative change in the bistable zone due to the inclusion of the non perfect boundary conditions for both the envelopes are shown in figure 3(c). H_I and H_{o-o} denote the size of the bistable zone corresponding to non perfect impedance and acoustically open boundary conditions respectively. The increase in the size of bistable zone is significant and it varies with τ. The non monotonic trend with τ is because of the phase relation between heat release rate and pressure oscillation which in turn is related to the stability boundaries. The changes are more noticeable for the left envelope, where the values of x_f/l is close to the end, where a non perfect boundary condition is applied. However, the nature of bifurcation remained a sub critical Hopf in both the cases. The results indicate that the inclusion of non perfect impedance boundary condition leads to significant (on an average 25%) increase in the bistable zone. Hence, incorporation of correct impedance boundary conditions is essential to obtain accurate stability maps and thus the operating regimes of practical gas turbine combustors.

4. Conclusion

Stability analysis of thermoacoustic interactions in an electrically heated horizontal Rijke tube is performed. The dynamics of the heat source is modeled using modified King's law. Describing function analysis is then used to obtain the stability boundaries. A good qualitative comparison in the results with the previous investigations corresponding to the acoustically open-open boundary condition is observed. The analysis is extended to include non perfect impedance boundary conditions and the results are compared with those from acoustically open boundary conditions. From the obtained linear and nonlinear stability boundaries, we found that the inclusion of non perfect impedance boundary condition leads to an increase in the extent of globally stable regime and a decrease in the globally unstable regime. The major finding of this investigation is the *non trivial* increase in the extent of the bistable zone while incorporating non perfect impedance boundary conditions. However, there is no change in the observed nature of bifurcation. Relative change in the extent of the bistable zone is significant and exhibits a non monotonic variation with time lag. In conclusion, it is essential to include the correct non perfect impedance boundary condition to accurately predict the stability maps of thermoacoustic oscillations.

References

[1] Dowling A P and Morgans A S 2005 *Annu. Rev. Fluid Mech.* **37** 151-82
[2] Candel S 2002 *Proc. Combust. Inst.* **29** 1-28
[3] Rijke P L 1859 *Philo. Mag.* **17** 419-22
[4] Li J and Morgans A S 2015 *J. Sound Vib.* **346** 345-60
[5] Matveev K I 2003 *J. Fluid. Struct.* **18** 783-94
[6] Heckl M A 1990 *Acustica* **72** 63-71
[7] Matveev K I and Culick F E C 2003 *Combust. Sci. Tech.* **175** 1059-83
[8] Subramanian P, Mariappan S, Wahi P and Sujith R I 2010 *Int. J. Spray Combust.* **2** 325-56
[9] Mariappan S, Sujith R I and Schmid P J 2015 *Int. J. Spray Combust.* **7** 315-352
[10] Noiray N, Durox D, Schuller T and Candel S 2008 *J. Fluid Mech.* **615** 139-67
[11] Boudy F, Durox D, Schuller T and Candel S 2011 *Proc. Combust. Inst.* **33** 1121-28
[12] Strogatz S H 2000 *Nonlinear Dynamics and Chaos: with applications to Physics, Biology, Chemistry and Engineering* (Westview Press)
[13] Lieuwen T C 2002 *J. Prop. Power* **18** 61-7
[14] Gopalakrishnan E A and Sujith R I 2015 *J. Fluid Mech.* **776** 334-53
[15] Waugh I, Geuss M and Juniper M 2011 *Proc. Combust. Inst.* **33** 2945-52
[16] Jegadeesan V and Sujith R I 2013 *Proc. Combust. Inst.* **34** 3175-83
[17] Levine H and Schwinger J 1948 *Phys. Rev.* **73** 383-406
[18] Peters M C A M, Hirschberg A, Reijnen A J and Wijnands A P J 1993 *J. Fluid Mech.* **256** 499-534
[19] Dowling A P and Stow S R 2003 *J. Prop. Power* **19** 751-64

Numerical study of rigid and flexible wing shapes in hover

Aamer Shahzad, Fang-Bao Tian, John Young and Joseph C S Lai

School of Engineering and Information Technology, the University of New South Wales, Canberra, ACT, 2600, Australia

a.shahzad@student.adfa.edu.au

Abstract. This study is focused on evaluating the aerodynamic performance of rigid and isotropic flexible wing shapes defined by the radius of the first moment of wing area ($\overline{r_1}$) at Reynolds number of 6000. An immersed boundary method was used to solve the 3D, viscous, incompressible Navier-Stokes equations, and coupled with an in-house non-linear finite element solver for fluid structure interaction simulations. Numerical simulations of flexible $\overline{r_1}$ = 0.43, 0.53 and 0.63 wing shapes performed with a single degree of freedom flapping shows that thrust and peak lift coefficients increase with $\overline{r_1}$. Higher thrust in the $\overline{r_1}$ = 0.63 wing is attributed to the large induced pitch angle, and higher peak lift (compared to the rigid counterpart) results from an increase in the stroke amplitude and spanwise deformation of the wing that anchors the leading edge vortex.

1. Introduction

There has been a significant increase in the research interest recently to understand the inspiring flight characteristics of insects for developing insect-like flapping wing micro aerial vehicles (MAVs). Typical applications include search and rescue operations, situational awareness in the combat field, aerial reconnaissance, counter drug operations, sensing biological, nuclear, chemical agents, sports broadcasting and many more. The aerodynamic performance of flapping wings is governed by wing geometric parameters such as shape, area, aspect ratio (AR), leading edge sweep and location of pivot point. Of these, the wing shape and AR have been studied extensively in the past. Here AR is defined as the ratio of the wing span (R) to the mean chord length (c) of a wing. Phillips et al. [1] and Ozen and Rockwell [2] observed similar flow structures on generic wing shapes. While Luo and Sun [3] found a less than 10% difference in the lift of ten wing shapes based on a fruitfly's wing, clear trends have been reported by others [4, 5] . It was found that the wings with more area outboard (distributed towards the wing tip) produce higher lift at the cost of efficiency.

For MAV applications, Ansari et al. [4] recommended high AR wings with straight or almost straight leading edge (LE), and more area outboard to produce high lift. However, their quasi-steady model had limitations, as it did not take into account the spanwise flow, tip vortex and aspects of viscosity, for example, vortex diffusion and stretching. In addition, the generic wing shapes in their study have different LE sweep. For instance, rectangle, ellipse and reverse ellipse have straight, swept back and swept forward LE respectively. The LE orientation is expected to influence the formation and positioning of the leading edge vortex (LEV), and thus, the lift. The limitations of the study of Ansari et al. [4] warrant further investigation of wing shapes.

In our previous work [6], the Reynolds number (Re) effects on the performance of $\overline{r_1}$= 0.43, 0.53, 0.63 and rectangular wings with AR = 2.96 were assessed at Re = 12, 400 and 13500. Here $\overline{r_1}$, the radius of the first moment of inertia, defines the distribution of the wing area along the span of the wing. The $\overline{r_1}$= 0.43 wing is found to be the most efficient in terms of power economy (PE) and the finding is Re independent. PE represents the ratio of the mean lift coefficient to the aerodynamic power coefficient. Since previous studies [1-5] have been restricted to rigid wing shapes, this study extends the previous work [6] to investigate the effect of introducing flexibility on the performance of different wing shapes. The study is carried out at Re = 6000, representative of a hawkmoth, based on the mean chord length (c) and tip velocity (U_{tip}).

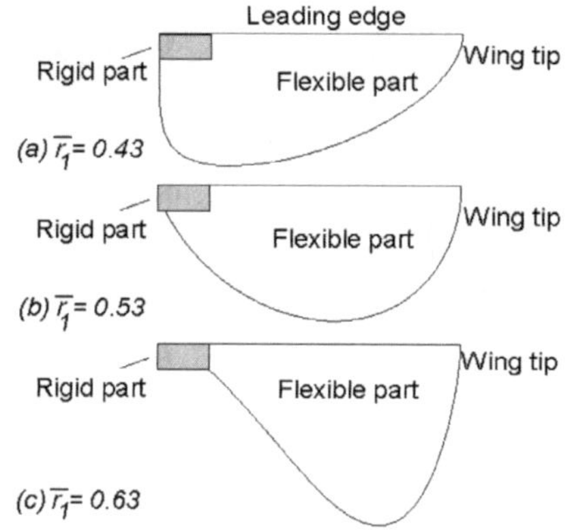

Figure 1. Flexible wing shapes with AR = 2.96.

2. Wing geometry and kinematics

The insect wings have a variety of shapes that can be reasonably approximated by the radius of the first moment of wing area ($\overline{r_1}$) using the beta distribution [7]. This distribution generates shapes with straight leading edge, constant span and area. The beta distribution shapes with AR = 2.96 for fluid structure interaction (FSI) simulations are given in figure 1.

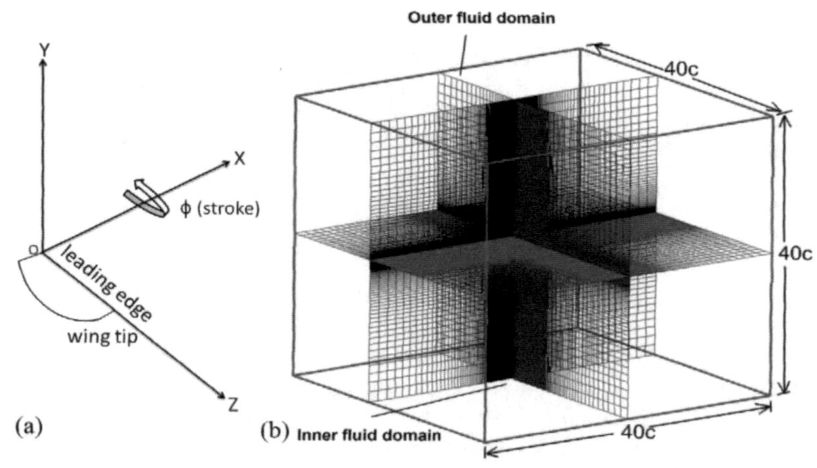

Figure 2. (a) Schematics of the wing kinematics and (b) Computational domain.

Unlike real insect wings, the $\overline{r_1}$ = 0.63 wing produced by the beta distribution has pointed wing root. The wing root is slightly modified, which brings an insignificant change (less than 3%) in AR and the area of the wing. In addition, the modification of the wing root is not expected to influence the aerodynamic performance significantly because it is closer to the pivot point. The flexible wing shapes in figure 1 show rigid rectangular area at the intersection of wing root and LE, which constitutes approximately 4.3% of the total wing area. This is a reasonable approximation because, in insect wings, this part being closer to the pivot point is about 10^2-10^3 times more rigid than the distal area [8]. In the absence of this rigid part, $\overline{r_1}$= 0.63 wing may collapse as it will not be able to withstand high stresses at the narrow section of wing root. For rigid wing simulations, the flexible part of the wing in figure 1 is also considered rigid.

For the study, one degree of freedom (DOF) kinematics is used, where the wing flaps in a vertical stroke plane (figure 2 (a)). The flapping angle (ϕ) is defined by the sinusoidal function. The pitch and deviation angles are kept zero. The up stroke and down stroke motions are between t/T = 0-0.5 and 0.5-1.0, respectively. The wing flaps with a frequency (f) of 20 Hz and stroke amplitude of 56.35° [1].

For flexible wing shapes, we have assumed isotropic flexibility. These wings have same mass ratio (m^*, in equation (1)), Poisson's ratio (ν, the ratio of transverse to axial strain) and effective stiffness (π_1, in equation (2)).

$$m^* = \rho_s h_s/\rho_f c \tag{1}$$

$$\pi 1 = E{h_s}^3/[12c^3(1-\nu^2)\rho_f {U_{tip}}^2] \tag{2}$$

Where ρ_s and ρ_f are the structural and fluid densities respectively, h_s the wing thickness, E the Young's modulus and U_{tip} the mean tip velocity. While m^* = 4.0 is representative of Hawkmoth wing [9], π_1 = 100 has been chosen to keep the frequency ratio, the ratio of the flapping frequency to the first natural frequency, of all the wings less than 0.3 so that the wing deformation is not dictated by the structural resonance in any case. It is reasonable to use ν = 0.3 for flapping wings [10].

3. Computational method

Numerical simulations are performed using a second order sharp interface immersed boundary method [11] to solve viscous, 3D, incompressible Navier–Stokes equations and the same solver is coupled with an in-house non-linear finite element solid mechanics solver for FSI simulations [11] of flexible wing cases. The computational domain is a cube with sides of 40c and the flow computed on a non-uniform Cartesian grid shown in figure 2 (b). This grid is kept fine (30 grid points per chord) in the inner fluid domain around a wing where it flaps in one complete cycle. The wing surface is meshed with triangular elements. The flow is assumed to be laminar. Neumann boundary condition for pressure and velocity (zero gradients) is applied on domain boundaries and no slip or penetration boundary condition on the wing surface. The performance of wing shapes is assessed using the time histories of lift coefficient ($C_L = 2F_L/\rho_f U^2 S$), thrust coefficient ($C_T = 2F_T/\rho_f U^2 S$), and aerodynamic power coefficient ($C_P = 2P/\rho_f U^3 S$) and their average values over 4th, 5th and 6th flapping cycles. Here F_L, F_T and P are the vertical lift force, horizontal thrust force and aerodynamic power respectively and U the mean tip velocity. The wing area (S) of approximately 0.0023m^2 has been used for all the wings. The aerodynamic power ($P = -\int \boldsymbol{F}.\boldsymbol{V}\, dS$) is determined by integrating the dot product of fluid force and local velocity over the wing surface.

4. Validation

After performing grid and time step independence studies (details not mentioned here), the validation of the fluid solver is performed against the experiment of Nagai et al. [12] using a single hind wing and the computational study of the same case by Naidu et al. [13]. Time histories of C_L given in figure 3 (a) show reasonable agreement, noting that the slightly higher peak values in Nagai et al. [12] are potentially attributable to the flexibility of the aluminium wing in the experiment. Rigid wing is used in this validation and the case of Naidu et al. [13]. There is also a reasonable agreement with the time history of C_D that was obtained from Nagai through private communication.

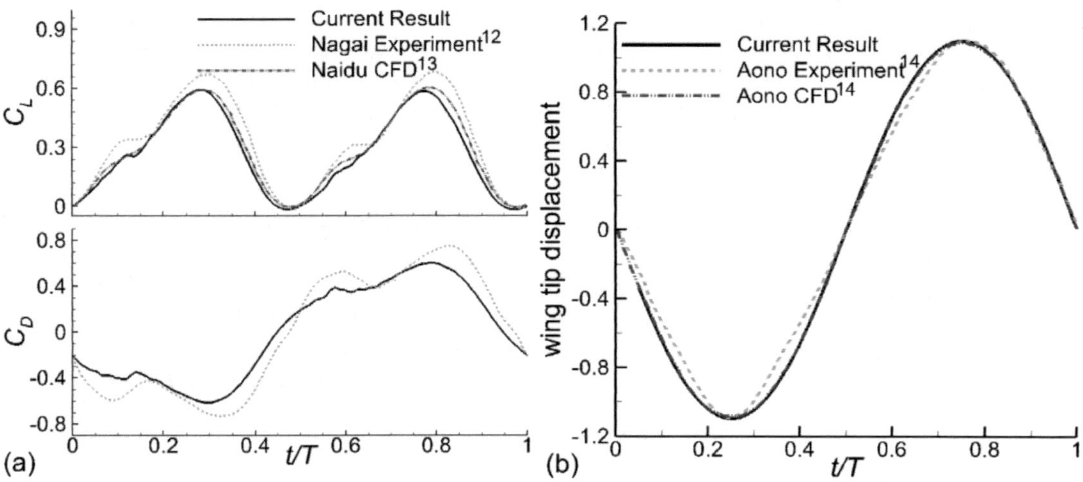

Figure 3. (a) Fluid solver validation of the time histories of C_L and C_D and (b) FSI solver validation of normalized vertical displacement of the wing tip.

FSI validation is performed against the experiment and computational fluid dynamics (CFD) result of Aono et al. [14] using a Zimmerman wing prescribed with a single DOF flap rotation at the wing root. Figure 3 (b) shows that the amplitude and phase of the vertical displacement of wing tip normalized by the root chord are in reasonable agreement with the baseline experimental and CFD results [14]. The tip displacement shows good symmetric response between the up stroke and down stroke.

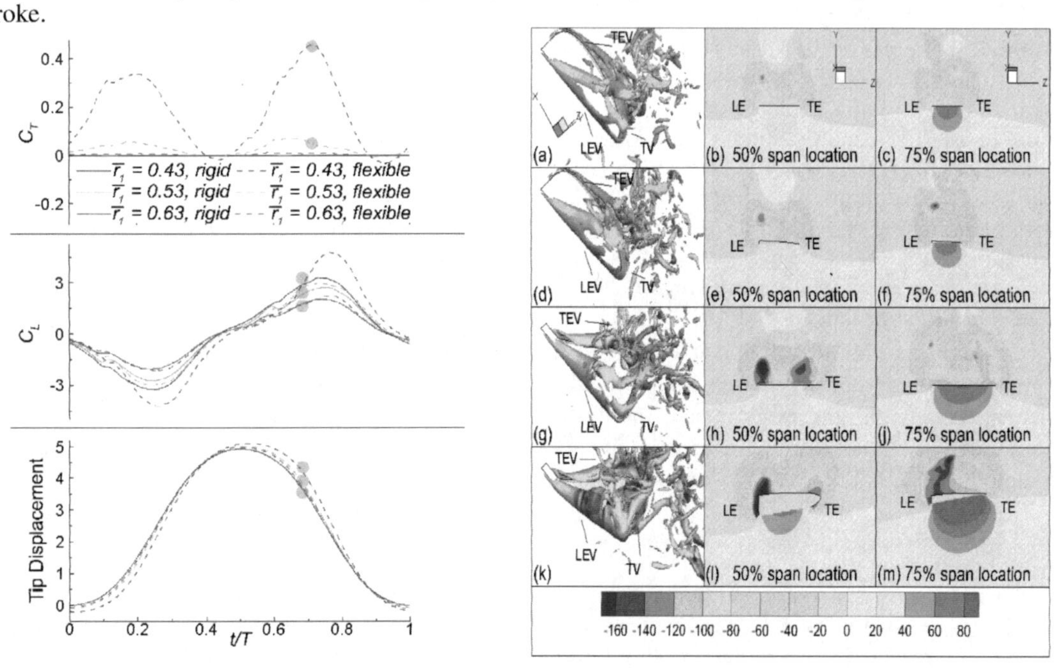

Figure 4. (Left) Time histories of C_T, C_L, and wing tip displacements of rigid and flexible wings . (Right) Iso-Q surfaces and pressure coefficients at t/T = 0.7: (a-c) $\bar{r}_1$= 0.43, rigid, (d-f) $\bar{r}_1$= 0.43, flexible (g-j) $\bar{r}_1$= 0.63, rigid and (k-m) $\bar{r}_1$= 0.63, flexible wing.

5. Results and discussions

This section includes the details of time history of force coefficients and wing tip displacement, and the flow features of rigid and flexible $\overline{r_1}$ = 0.43 and 0.63 wings at t/T = 0.7 (figure 4). For $\overline{r_1}$ = 0.43, the rigid (figure 4 (b) and (c)) and flexible wings (figure 4 (e) and (f)) show similar pressure coefficient distributions at 50% and 75% span locations. As a result, the C_T of flexible $\overline{r_1}$ = 0.43 wing, although non-zero, is very small. For $\overline{r_1}$ = 0.63, at the same spanwise locations, the pressure distributions of the rigid (figure 4 (h) and (j)) and flexible wings (figure 4 (l) and (m)) show marked differences. There is an induced pitch angle in the flexible wing that orients the pressures and resultant of the aerodynamic force for favourable thrust production. This indicates that, with more area distributed outboard towards the tip, the high $\overline{r_1}$ wing shows high chordwise flexibility in the distal part of the wing that result in higher thrust. At this time instant (t/T = 0.7), the flexible $\overline{r_1}$ = 0.63 wing produces the peak C_T. Like the thrust, the time history of C_L in figure 4 also shows that flexibility improves the peak lift and it increases with increasing $\overline{r_1}$. For $\overline{r_1}$ = 0.63 wing, there is a 30.3% increase in the peak lift with the introduction of flexibility and this peak is registered during the down stroke at t/T = 0.75, when the flapping velocity is maximum. Careful examination of the wing tip displacements in figure 4 shows that the flapping amplitude of the flexible wing increases and this increase is more pronounced for high $\overline{r_1}$ (3.4% increase for $\overline{r_1}$= 0.63 wing). Since the time period remains unchanged, the velocity is bound to increase with flexibility, and this in turn results in higher suction pressures and more lift. Although the flexible $\overline{r_1}$= 0.43 wing produces high peak lift than its rigid counterpart, their lift coefficients are similar at t/T = 0.7 as evident from the iso-Q surfaces of the rigid (figure 4 (a)) and flexible (figure 4 (d)) wings. For $\overline{r_1}$ = 0.63, the iso-Q surfaces of rigid (figure 4 (g)) and flexible (figure 4 (k)) wings show that the LEV on the rigid wing starts to break down (at approximately 60%) after the mid span. On the contrary, the spanwise deformation in the flexible wing anchors the LEV on the wing as it joins the tip vortex. This results in an increase in the peak lift because of flexibility. The degree of spanwise deformation in $\overline{r_1}$ = 0.43 and 0.63 wings can be observed in figure 5. While the flexibility in $\overline{r_1}$= 0.43 wing is small, there is significant deformation in the distal parts of the flexible $\overline{r_1}$= 0.63 wing.

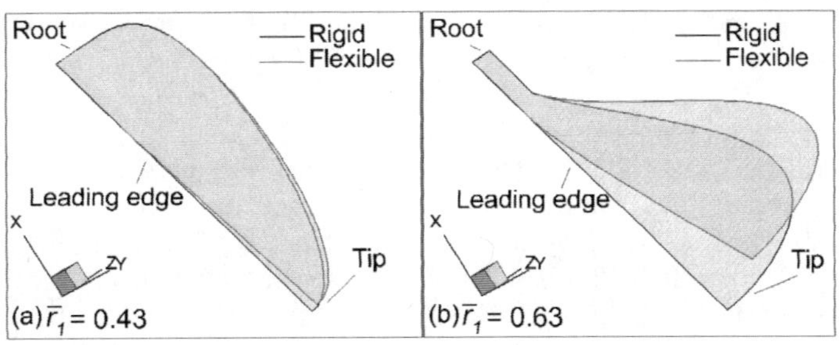

Figure 5. Spanwise deformation in flexible wings at t/T = 0.7.

6. Conclusion

Numerical simulations were performed at Re = 6000 using an IBM solver to compare the aerodynamic performance of rigid and flexible wing shapes. The results show that, while rigid wings do not produce any thrust, the flexible wings generate thrust whose magnitude and peak value increases with increasing $\overline{r_1}$ because high $\overline{r_1}$ wings induce pitch angles, which changes the orientation of the resultant aerodynamic force to produce more thrust. In addition, the peak lift forces of flexible wings are also higher than their rigid counterparts. This is because lift scales as the square of the wing tip velocity which increases with $\overline{r_1}$ due to an increase in the stroke amplitude (maximum of 3.4% increase in stroke amplitude in $\overline{r_1}$ = 0.63 wing). The spanwise deformation of flexible wings also anchors the LEV on the wing and contributes to more lift. Among flexible wings, $\overline{r_1}$ = 0.63 wing produces maximum peak lift. In conclusion, flexibility plays a key role in defining the aerodynamic performance of wing shapes.

Acknowledgments
This work was supported under the Australian Research Council's Discovery Projects funding scheme (project number DP130103850), and was conducted with the assistance of resources from the National Computational Infrastructure (NCI), which is supported by the Australian Government. Aamer Shahzad acknowledges the support of NUST, Islamabad, Pakistan via the Prime Minister's Gold Medal Scholarship and SEIT, UNSW, Canberra, which provided a top-up scholarship.

References
[1] Phillips, N., Knowles, K. & Lawson, N., 2010, "Effect of wing planform shape on the flow structures of an insect-like flapping wing in Hover", *27th International Congress of the Aeronautical Sciences. ICAS. Nice, France.*
[2] Ozen, C. & Rockwell, D., 2013, "Flow Structure on a Rotating Wing: Effect of Wing Aspect Ratio and Shape", *51st AIAA Aerospace Sciences Meeting including the New Horizons Forum and Aerospace Exposition. Grapevine, Texas.*
[3] Luo, G. & Sun, M., 2005, "The effects of corrugation and wing planform on the aerodynamic force production of sweeping model insect wings", *Acta Mechanica Sinica*, 21(6), pp 531-541.
[4] Ansari, S.A., Knowles, K. & Zbikowski, R., 2008, "Insectlike flapping wings in the hover part II: Effect of wing geometry", *Journal of Aircraft*, 45(6), pp 1976-1990.
[5] Wilkins, P.C., 2008, "Some unsteady aerodynamics relevant to insect-inspired flapping-wing micro air vehicles", Cranfield University, UK.
[6] Shahzad, A., Tian, F., Young, J. & Lai, J., 2014, "Reynolds Number Effects on the Aerodynamic Performance of Wing Planform Shapes in Hover", *19th AFMC. RMIT University, Melbourne, Australia.*
[7] Ellington, C., 1984, "The aerodynamics of insect flight. II. Morphological parameters", *Phil. Trans. R. Soc. Lond. B*, 305, pp 17-40.
[8] Combes, S.A. & Daniel, T.L., 2003, "Flexural stiffness in insect wings I. Scaling and the influence of wing venation", *Journal of Experimental Biology*, 206(17), pp 2979-2987.
[9] Shyy, W., Aono, H., Kang, C.-k. & Liu, H., 2013, "*An introduction to flapping wing aerodynamics*", Vol 37, Cambridge University Press.
[10] Combes, S. & Daniel, T., 2003, "Flexural stiffness in insect wings II. Spatial distribution and dynamic wing bending", *Journal of Experimental Biology*, 206(17), pp 2989-2997.
[11] Tian, F.-B., Dai, H., Luo, H., Doyle, J.F. & Rousseau, B., 2014, "Fluid–structure interaction involving large deformations: 3D simulations and applications to biological systems", *Journal of Computational Physics*, 258, pp 451-469.
[12] Nagai, H., Isogai, K. & Fujimoto, T., 2010, "Experimental study on flow interaction between fore-and hindwings of dragonfly in hovering and forward flight", *27th Congress of International Council of the Aeronautical Sciences. Nice, France.*
[13] Naidu, V., Young, J. & Lai, J., 2013, "Effect of wing flexibility on the performance of tandem wings during hover flight", *31st AIAA Applied Aerodynamics Conference. San Diego, CA.*
[14] Aono, H., et al., 2010, "A computational and experimental study of flexible flapping wing aerodynamics", *48th AIAA aerospace sciences meeting including the new horizons forum and aerospace exposition*. pp. 4-7.

Rear-roof spoiler effect on the aerodynamic drag performance of a simplified hatchback model

S Y Cheng[1,2] and S Mansor[3]

[1]Centre for Advanced Research on Energy, Universiti Teknikal Malaysia Melaka, Hang Tuah Jaya, 76100 Durian Tunggal, Melaka, Malaysia

[2]Faculty of Mechanical Engineering, Universiti Teknikal Malaysia Melaka, Hang Tuah Jaya, 76100 Durian Tunggal, Melaka, Malaysia

[3]Faculty of Mechanical Engineering, Universiti Teknologi Malaysia, 81310 UTM Skudai, Johor, Malaysia

E-mail: cheng@utem.edu.my

Abstract. The effect of rear-roof spoiler on the aerodynamic drag performance of simple hatchback model has been investigated for a range of spoiler pitch angles. RANS based CFD method was utilized. Validation was done by comparing the CFD results of Ahmed model without spoiler to the experimental data. The spoiler has a beneficial influence at positive inclination angle below 5°. Above that angle, the aerodynamic drag increases with pitch angle. The mechanism of how the spoiler influences the aerodynamic drag was identified through extensive flow visualization.

1. Introduction

A rear-roof spoiler is an external structure fitted to the trailing edge of the roof of a vehicle to modify the airflow going over it. The purpose is to improve the vehicle's aerodynamic performance. Spoiler designs fall into two categories: strips and free-standing wing. The latter could have a simple flat plate profile or an inverted airfoil. The effectiveness of wing-type spoiler has been reported in numerous studies. For instance, five different wing-type spoiler configurations had been tested on a simplified car model based on a HONDA S2000 [1]. All of them had achieved lift reduction. However, the drag coefficient (C_d) of all the cases had increased tremendously as compared to the case without a spoiler.

On a similar note, a numerical study has shown up to 75% reduction in the lift coefficient C_l of a passenger car model when properly manipulated the inclination angle of the flat-plate-type spoiler, i.e. at 50° relative to the ground [2]. However, the C_d had increased by 8%. Meanwhile, a simplified-passenger-car model mounted with an inverted-airfoil type spoiler has experienced 80% reduction in the C_l with an increase of 3% C_d [3]. The increase in drag is a drawback as it is directly linked to fuel consumption. This performance is important because the emission from automobiles is accounting for about 10% of global carbon dioxide emissions, a major contributor to the greenhouse effect [4]. On the other hand, there are also cases where a spoiler is found to reduce the aerodynamic drag of vehicles (e.g. [5]; [6]).

As for the strip-type rear-roof spoiler, it was reported that the highest drag reduction of around 5 to 6% has been achieved on a simple hatchback model, namely, an Ahmed model at 35° slant angle [7].

The spoiler was configured at the length of about 32% of the slant length, and at 0° spoiler angle. However, the study did not cover the range of positive inclination angles, which are more commonly found in real life applications. To fill the gap, hence, the present paper aims to investigate the effect of pitch angle of rear-roof spoiler, in particular, at positive inclination, on the aerodynamic performance of simplified hatchback model.

2. Simplified hatchback model and spoiler configurations

Rear-roof spoilers are commonly found in hatchback-type vehicles. Hence, the Ahmed body with the slant angle of 35°, which is typical for most hatchback cars, was adopted. Figure 1 shows the Ahmed model fitted with a 0° pitch angle spoiler. Also shown in the figure are the spoiler configurations tested, i.e. pitch angles of -15°, 0°, 5°, 10°, and 15°. The length of the spoiler was 66.6 mm, which corresponds to 30% of the length of the slant section. The trailing edge of the spoiler was filleted (5 mm radius) for avoiding highly skewed cells during meshing. For details of the Ahmed model dimensions, the reader is referred to [8].

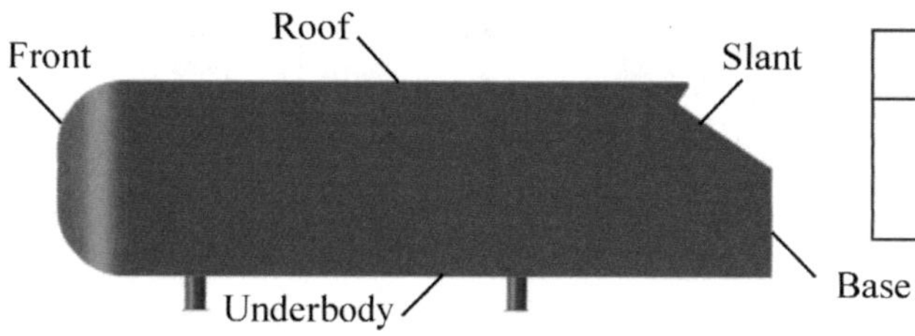

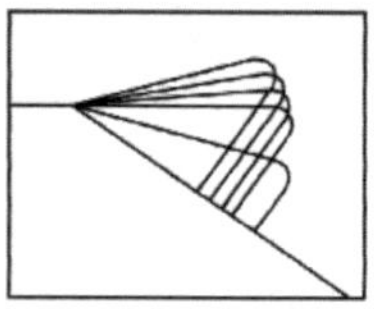

Figure 1. Side view of Ahmed model fitted with a rear-roof spoiler (left), and the spoiler configurations (right).

3. Numerical Method

The present paper employed numerical simulations to investigate the influence of pitch angle of rear-roof spoiler on the aerodynamics of a simplified hatchback model. All results were obtained using the commercial finite-volume solver ANSYS Fluent 16. The Reynolds-averaged Navier-Stokes (RANS) approach was used. The turbulence model was the widely used two-equation model, namely, k-epsilon realizable model with enhanced wall treatment. The steady, pressure-based solver was utilized to achieve steady-state simulations. The effects of compressibility was neglected due to the relatively low Mach numbers (about 0.12). All reported results were obtained using a second-order node-based upwinding discretization scheme.

The inlet boundary condition was set as uniform flow at 40 m/s with the turbulence intensities of 0.2%. The corresponding Reynolds number (Re) was 768,000 based on the model height. The pressure outlet at zero gauge pressure was imposed on the outlet boundary. As for the side and top walls of the domain, symmetry boundary condition was applied. Meanwhile, the ground and model surfaces were set as no slip wall.

The computational domain resembled a rectangular box. Since the model is symmetric, all simulation cases were run for half of the flow domain with a symmetry plane placed at the centerline location. The cross sectional area of the half flow domain was 1738 mm height x 1129.5 mm width. The corresponding blockage ratio was less than 1.5%, which is well within the typically accepted range of 5% in automotive aerodynamic testing [9]. The upstream and downstream extends of the domain were $1.4l$ and $11.4l$, respectively, where l is the vehicle length.

The computational domain comprises unstructured and prismatic cells (see figure 2). The latter was employed around the model and the ground for obtaining better boundary layer resolutions. The first prismatic cell layer thickness was at 0.5 mm. The corresponding y^+ around the model surface was within 1 to 57, which is in the appropriate range for k-epsilon turbulence model. The total nodes and cells for all simulation cases were around 315,000 and 890,000, respectively.

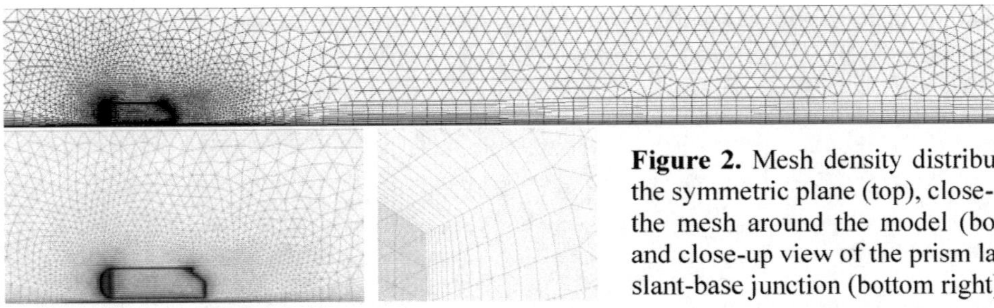

Figure 2. Mesh density distribution along the symmetric plane (top), close-up view of the mesh around the model (bottom left), and close-up view of the prism layers at the slant-base junction (bottom right).

4. Validation

To validate the numerical method, the present study carried up a case of flow simulation for the Ahmed model without the spoiler. The Re was the same as the experimental value by [10], i.e. 768,000 based on the model height. Table 1 summarizes the validation results. As depicted, the total C_d value is in excellent agreement with the experiment, i.e. percentage difference of 2.9%. The drag breakdown shows that the C_d of the base and slant sections are well predicted. Good prediction of C_d in these two sections is important because the influence of rear-roof spoiler is expected to be most prominent at the rear part of the model.

Table 1. Comparison of the experimental C_d.

	CFD	Experiment [10]	Percentage difference (%)
Slant	0.095	0.097	2.3
Base	0.094	0.096	4.0
Front	0.039	0.015	161.8
Viscous	0.049	0.055	11.4
Total	0.0265	0.0258	2.9

5. Result and Discussion

Figure 3 indicates that the rear-roof spoiler could have a positive or negative impact on C_d, depends on its pitch angle. At 0° and 5° pitch, its effect is favorable. Figure 4 compares the pressure coefficient (C_p) at the rear section of the model between the with- and without-spoiler cases. As seen, in the latter, the flow accelerates near the rear end of the roof. The higher airflow produces lower surface C_p at the slant section. On the other hand, the one with the 0° pitch spoiler is showing a relatively higher C_p at both the slant and base sections.

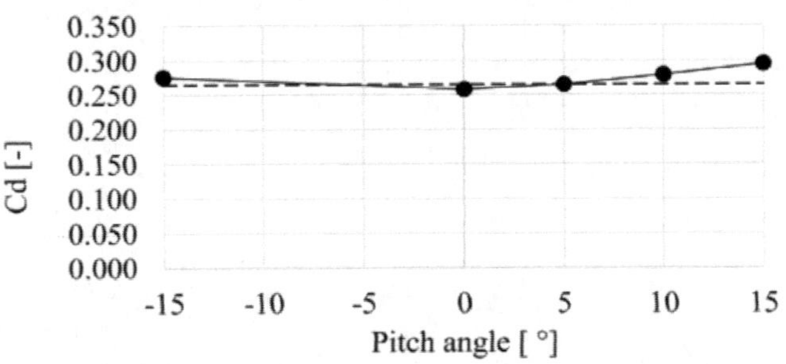

Figure 3. Graph of drag coefficient vs pitch angle of rear-roof spoiler; Dashed-horizontal line is without spoiler.

Another factor that contributes to the higher C_d value in the without-spoiler case is the shedding of longitudinal vortex structure at the side edge of the slant section (see figure 5). This vortex structure which initiated at the upper corner of the slant has induced lower surface C_p along the side edge of the slant. In the with-spoiler case, however, the onset of the vortex structure has been prevented by the spoiler geometry.

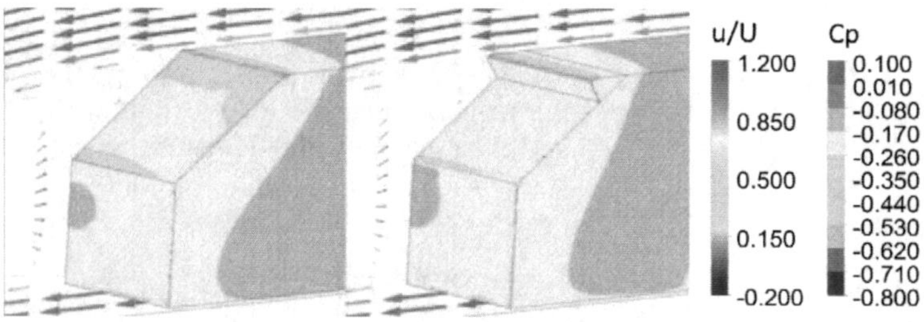

Figure 4. Rear section surface C_p distribution and centerline velocity vectors; Without spoiler (left); With 0° pitch angle spoiler (right).

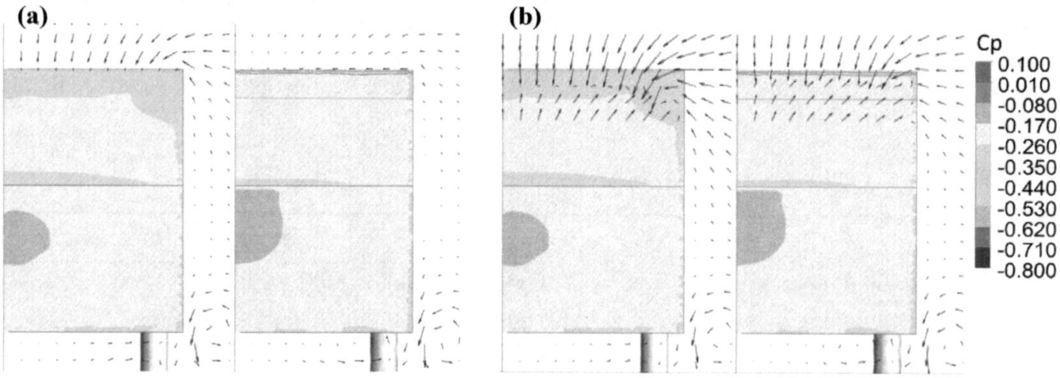

Locations of visualization planes

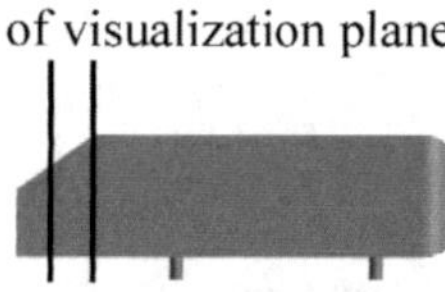

Figure 5. Rear section surface C_p distribution and cross flow velocity vectors; Without spoiler (left); With 0° spoiler (right); Visualization plane locations (a) x = -80 mm from the base, and (b) -180 mm.

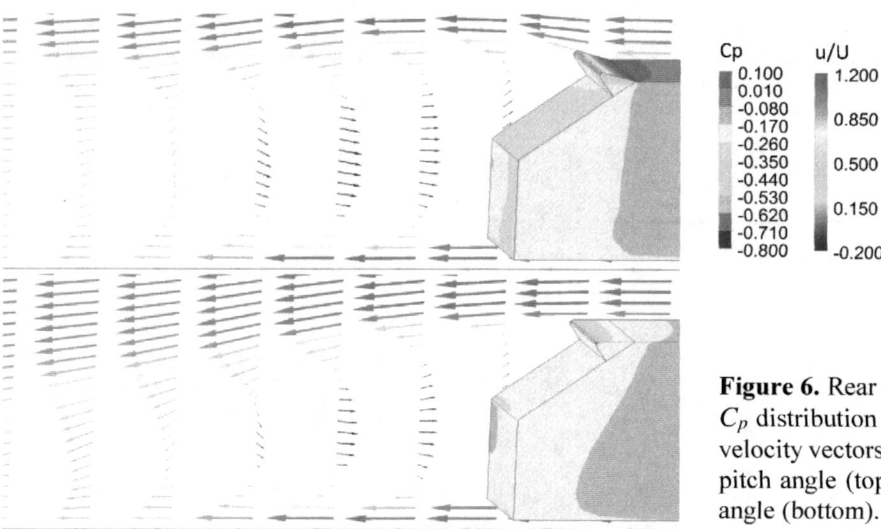

Figure 6. Rear section surface C_p distribution and centerline velocity vectors; Spoiler of 15° pitch angle (top), and 0° pitch angle (bottom).

In general, the C_d of the model is found to increase with larger spoiler angle for the angle range tested. However, the trend is relatively steeper from 5° onwards. The reason for the higher C_d value at larger pitch angle was partly due to the larger separation bubble behind the model (see figure 6). The larger separation bubble comprises stronger recirculating flow and upwash (see figure 7). Hence, lower surface C_p is induced on the slant, particularly around the lower part. Also notable in Figure 7 is the formation of the trailing vortices at the side edge of the spoiler at higher pitch angle. These vortices only occurred for spoiler angle larger than 5°, and are associated with strong upwash at the middle part of the slant.

The present study included a case of spoiler at negative inclination angle, i.e. -15°. This test case was carried out to verify the results found in the literature [7]. As shown in figure 3, similar tendency was observed in which the case of negative pitch angle does exhibit a higher C_d value than the 0° case.

The reason for the higher C_d value in the case of negative spoiler angle is to be twofold. First, the declining spoiler surface augments the flow near its leading edge (see figure 8). Consequently, the C_p at the entire upper surface of the spoiler drops. Secondly, similar to the slant section of the model, the declining spoiler surface generates side edge vortices. Strong side edge vortices are the characteristic of a high drag generating flow [8].

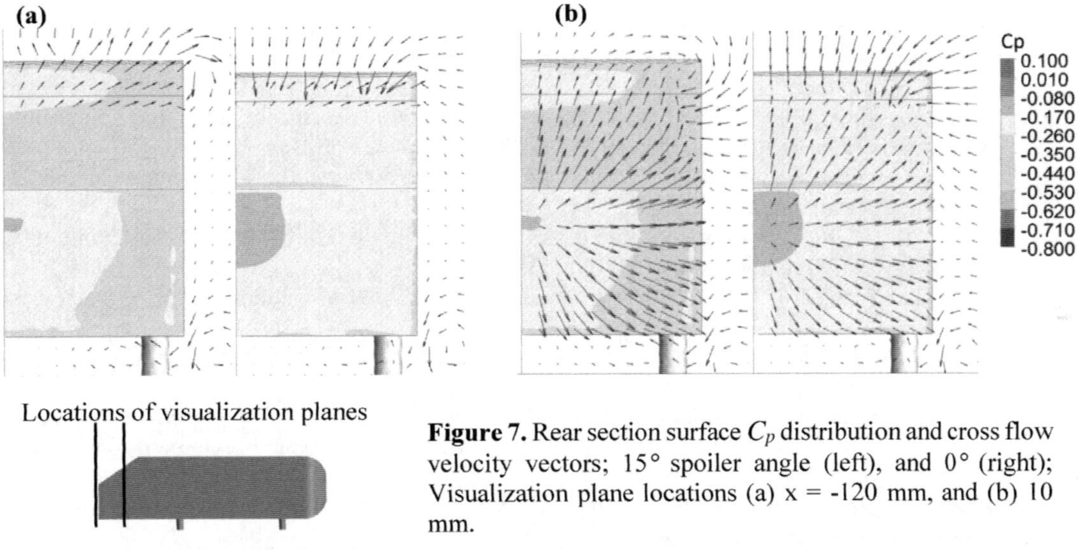

Figure 7. Rear section surface C_p distribution and cross flow velocity vectors; 15° spoiler angle (left), and 0° (right); Visualization plane locations (a) x = -120 mm, and (b) 10 mm.

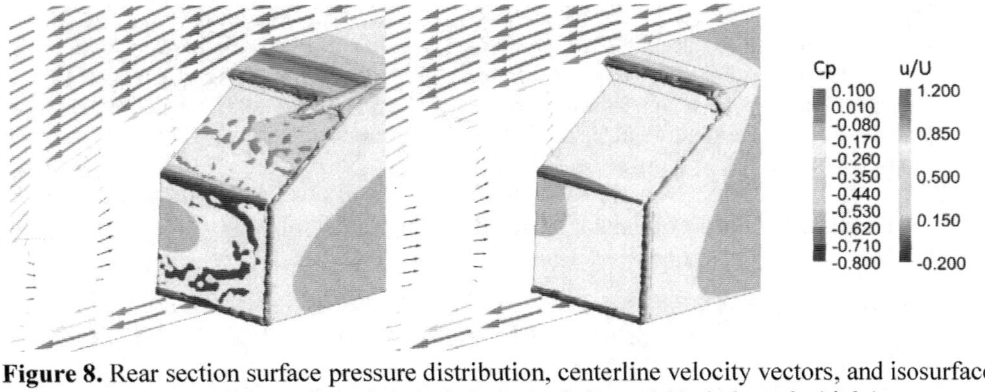

Figure 8. Rear section surface pressure distribution, centerline velocity vectors, and isosurface of Q criterion; Spoiler of -15° pitch angle (left), and 0° pitch angle (right).

6. Conclusions

The RANS-based CFD simulations have been carried up to investigate the effect of rear-roof spoiler on the aerodynamic drag performance of a simplified hatchback model. The results show that the simple strip-type spoiler could have a beneficial influence from 0° to 5° pitch angle. Beyond this range, the C_d increases with larger pitch angle. The spoiler relies on two main mechanisms to reduce aerodynamic drag. First, by preventing the airflow from accelerating at the leading edge of the slant section. Second, by preventing the formation of longitudinal vortices at the side edge of the slant section.

The present results were obtained from stationary simulations in which the motion of the vehicle body has not been considered. In practice, motion of vehicle body is common, and could change the inclination angle of the spoiler when the motion mode is of pitching. Hence, it helps to improve the realism of the simulation if the numerical method that incorporates the pitching mode of motion into the flow simulation [11] could be employed in the future studies.

Acknowledgement

Authors would like to thank Universiti Teknikal Malaysia Melaka (UTeM) and Ministry of Higher Education for supporting this research under FRGS FRGS/1/2015/TK03/FKM/02/F00273.

References

[1] Tsai C H, Fu L M, Tai C H, Huang Y L and Leong J C 2009 Computational aero-acoustic analysis of a passenger car with a rear spoiler *J. Appl. Math. Model.* **33** 3661-73
[2] Daryakenari B, Abdullah S, Zulkifli R, Sundararajan E and Mohd Sood A B 2013 Numerical study of flow over ahmed body and a road vehicle and the change in aerodynamic characteristics caused by rear spoiler *Int. J. Fluid Mech. Res.* **40** 354-72
[3] Kodali S P and Bezavada S 2012 Numerical simulation of air flow over a passenger car and the influence of rear spoiler using CFD *IJATP* **01** 6-13
[4] Metz N 2001 Contribution of passenger cars and trucks to CO2, CH4, N2O, CFC and HFC Emissions *SAE Technical Paper* 2001-01-3758
[5] Hu X X and Wong T T 2011 A numerical study on rear-spoiler of passenger vehicle *WASET* **57** 636-41
[6] Sharma R B and Bansal R 2013 Drag and lift reduction on passenger car with rear spoiler *IJAuERD* **3** 13-22
[7] Menon D P, Kamat G S, Mukkamala Y S and Kulkarni P S 2014 To improve the aerodynamic performance of a model hatchback car with the addition of a rear roof spoiler *16th Annual CFD Symposium* (Bangolore)
[8] Ahmed S R 1981 An experimental study of the wake structures of typical automobile shapes *J. Wind Eng. Ind. Aerodyn.* **9** 49-62
[9] Hucho W-H and Sovran G 1993 Aerodynamics of road vehicles *Annu. Rev. Fluid Mech* **25** 485-537
[10] Lienhart H, Stoots C and Becker S 2000 Flow and turbulence structures on the wake of a similified car model (Ahmed model) *DGLR Fach. Symp. der AG ATAB* (Stuttgart University)
[11] Cheng S Y, Tsubokura M, Nakashima T, Nouzawa T and Okada Y 2011 A numerical analysis of transient flow past road vehicles subjected to pitching oscillation *J Wind Eng Ind Aerod* **99** 511-22
Cheng S Y, Tsubokura M, Nakashima T, Okada Y and Nouzawa T 2012 Numerical quantification of aerodynamic damping on pitching of vehicle-inspired bluff body *J Fluid Struct* **30** 188-204

Some aspects of flow control over a NACA0015 airfoil using synthetic jets

T Parthasarathy* and S P Das

Department of Mechanical Engineering, Indian Institute of Technology Madras, Chennai 600036, India

E-mail: * tejaswin.sarathy@gmail.com

Abstract. Flow control for performance enhancement over airfoils has become an increasingly important topic. This work details the characteristics of flow control using synthetic jets over a NACA0015 airfoil at a Reynolds number of 896,000 based on the chord length and free stream velocity, and at 20° angle of attack wherein the flow is separated. Numerical simulations were performed to help understand the behaviour of the controlled flow for a range of synthetic jet parameters. Analysis of key flow parameters such as phase averaged pressure and streamline profiles indicate that the synthetic jet is efficient in increasing the lift coefficient; more so for larger jet amplitudes and at smaller angles of jet injection. Behaviour of the flow characteristics for controlled cases has been analysed from the flow structures obtained from the same. This work serves as a platform to qualitatively and quantitatively understand the effects of the jet parameters on the separated flow over the airfoil.

1. Introduction

Unsteady active flow control has been an actively pursued area of research in the past decades. A major advantage of such an unsteady control technique is that it can exploit the existing instabilities in the flow [1]. Periodic excitation accelerates and regulates the generation of large coherent structures with the flow and hence transfers high momentum fluid across the mixing layer. Of particular interest in periodic excitation control is the use of zero net mass flux devices, also called synthetic jets [2]. Synthetic jets are formed by oscillatory flow through an orifice. Hence, while similar to continuous jets (used for suction and blowing) in form and function, they vary in the fact that they offer no mass addition (or removal), entrain low momentum fluid [3] and promote mixing through the formation of vortical structures. The performance of the synthetic jets in controlling a flow greatly relies on the amplitude, frequency, inclination angle and location of the synthetic jets [4,5,6].

Several experimental and numerical works in characterizing the flow control over an airfoil have been done in the past. Experimental data for the same airfoil at different angles of attack for a Reynolds number ($Re = UC\nu^{-1}$, where U is the free stream velocity, C is the chord length and ν is the kinematic viscosity) of 896,000 [7] suggests that the maximum lift coefficient value can be enhanced by 80% and the stall angle can be pushed back from 12° to 18°. The effectiveness of using synthetic jets for stall control over airfoils was also demonstrated experimentally [8,9]. Numerical studies also demonstrate and characterize the effect of synthetic jets on a wide range of airfoils at different angles of attack [10,11,12]. Duvigneau and Visonneau [13] explore the effect of jet parameters on the lift and drag coefficients for a NACA0015 airfoil at the same Reynolds number, and suggest an increase in the lift by 34%.

In the present study, numerical investigations have been carried out to understand the effects of the synthetic jet and its parameters on the external flow over a NACA0015 airfoil at an angle of attack of 20° for a Reynolds number of 896,000, based on the free stream velocity and chord. Thrust is also given in understanding the reason behind the performance enhancement on using the synthetic jets in such separated flows. The jet location and width are fixed based on previous studies [12,14]. The important jet parameters chosen are summarised in table 1.

Table 1. Summary of parameters

Non dimensional Frequency ($F = fC\,U^{-1}$)	Amplitude Ratio at jet exit ($VR = V_j\,U^{-1}$)	Angle of jet injection (in °)
0.6446	1.50	30
0.9670	2.00	45
1.1281	2.25	60
1.2893	2.50	90
1.6117	3.00	-

2. Basic Flow

An external flow over an airfoil has been studied, with suitable modifications to include the model of a synthetic jet. The equations to be solved are the incompressible two-dimensional continuity and Navier Stokes equations, elaborated as follows (where $\vec{V}$, P, ρ and μ are velocity, pressure, density and dynamic viscosity respectively)

$$\nabla.\vec{V}=0 \tag{1}$$

$$\frac{d\vec{V}}{dt}+(\vec{V}.\nabla)\vec{V}=-\frac{1}{\rho}\nabla P+\frac{\mu}{\rho}\nabla^2\vec{V} \tag{2}$$

3. Case Setup

3.1. Geometrical setup and Boundary conditions

The geometrical setup and corresponding boundary conditions are as shown in figure 1. An inlet velocity of 34.9 ms^{-1} is specified, corresponding to a Reynolds number of 896,000 for air, with very low turbulent intensity. The chord length is taken as 0.375m, with a single slot of width 0.0053C at 0.12C from the leading edge. Other geometrical details are avoided for the sake of brevity. The steady flow Neumann boundary condition is imposed at the exit. The airfoil wall is given a no-slip boundary condition. A time varying sinusoidal boundary condition is given at the sinusoidal velocity inlet shown (figure 1) in such a way that at the jet exit, the flow is fully developed and has an exit velocity of V_j .

$$V_i=VR\times U\times\sin(2\pi ft) \tag{3}$$

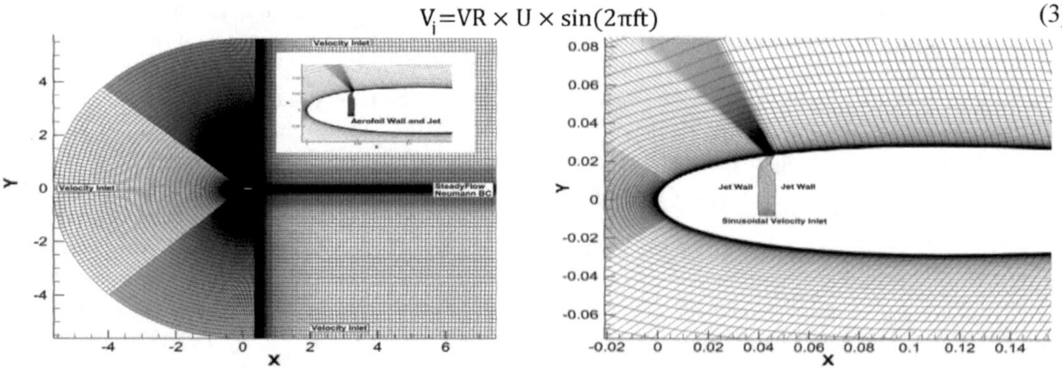

Figure 1. Representative mesh with geometry & boundary conditions, for an injection angle of 30°

3.2. Numerical Scheme

The simulations were set up in the commercial CFD solver Fluent, at 20° angle of attack. This particular configuration is studied as the baseline flow is completely separated, and hence the effect of the jet will be predominant. The velocity at the base of the jet is taken such that the average velocity at the jet exit is VR × U. A time implicit second order upwind discretization scheme with the SIMPLE coupling, was used to solve the equations that were derived based on the Spalart-Allmaras model. The simulations were run till a quasi (dynamic)- steady state is reached for the system.

3.3. Grid Setup

Different meshes were generated for different angles of injection, and the details of the mesh also change. A grid independence test was done for all the aforesaid grids, with cells numbering 35,200, 74,879 and 144,000. Based on the same, the mesh with 74,879 cells is selected. The spacing values ($\Delta c/C$) near the jet exit is 2.01×10^{-4}, at the point of maximum near the airfoil wall is 7.36×10^{-3} and normal to the wall is 2.6×10^{-5}(corresponding to a y^+ value of 1). The grid size for the semi-circular inlet, horizontal and vertical edges is $363\times120\times240$.

3.4. Validation

The validation of the current setup is done by comparing the obtained values with the results of [13] and XFoil data as shown in figure 2 (experimental validation is avoided due to reasons stated in [13]), which suggests a reasonable match. The same can be said for the streamlines in the uncontrolled flow.

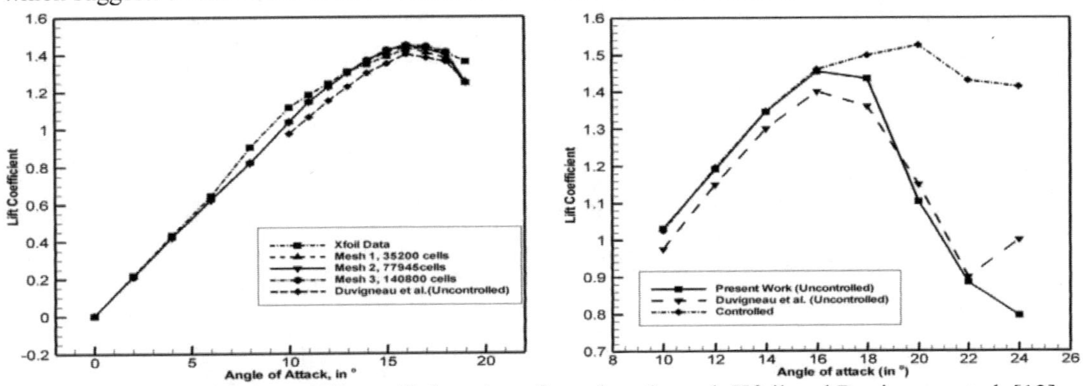

Figure 2. Validation of lift coefficient data, from data through Xfoil and Duvigneau et al. [13]

4. Results and Discussions

4.1. Effect of jet parameters

Extensive simulations were performed to identify the effect of the parameters on the external flow. The lift characteristics were observed for all the cases. The effects of the parameters on the flow become apparent on analysing the same, many of which are avoided here for the sake of brevity.

The slope of the lift coefficient graph (figure 3 (a)) with injection angle changes abruptly, except for the case of 150 Hz frequency, wherein it decreases linearly. The nature of the graph seems to suggest a local maxima and minima at around 45° - 60 ° angles of injection, which becomes less pronounced at higher frequencies. While at lower amplitudes, the lift versus angle curve is concave upwards, this slowly changes to a concave downwards curve at higher amplitudes. However, this effect is barely visible at the highest frequency tested. Lift also increases substantially at lower angles of injection, if other parameters are kept the same.

The lift values in the case of 90° angle of injection do not display any trend with different amplitudes (figure 3(b)), and is also among the least measured values. Lift characteristics plotted against amplitude ratios suggest an upward concave monotonic curve for all angles of attack except at 90°. The graph flattens out as the angle of injection increases. Lift increases substantially at higher amplitudes, with other parameters kept the same. As amplitude is the most important factor tested, the values of lift/drag are also represented in figure 3(d).

At higher amplitudes and smaller angles, the frequency of operation becomes increasingly irrelevant for lift enhancement, as visible in the figure 3(c). At lower angles and lower amplitudes, the frequency must be kept at a minimum for desirable characteristics. The sensitivity of the lift values to the frequency of operation seems to be minimal at lower angles of injection. However, it is interesting to note that the graph changes concavity from almost zero to concave upwards as the angle increases.

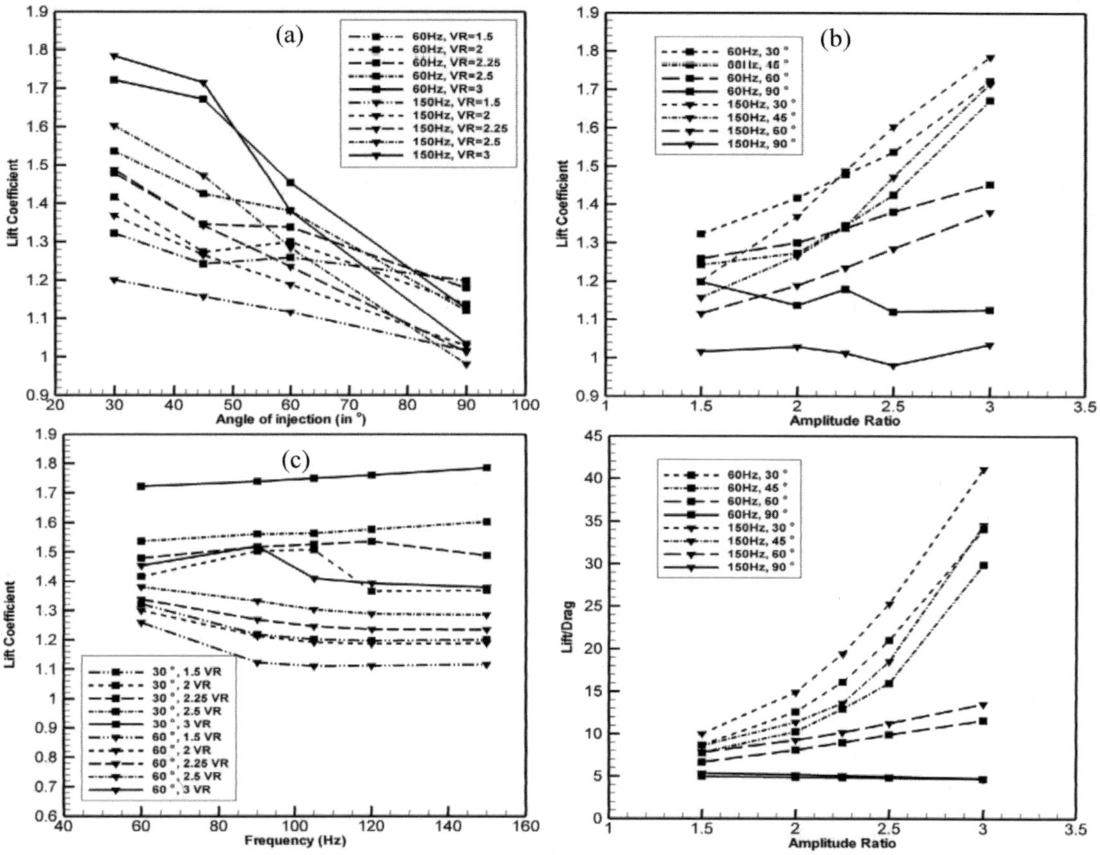

Figure 3. Performance characteristics of the parametric study

4.2. Flow physics

While the previous section details the characteristics and behaviour of the performance of a controlled airfoil with variation in the input parameters, the physics behind the flow is still left unexplained. The controlled flow exhibits different characteristics and flow structures for different frequencies, amplitudes and angles of injection. A comparison to the steady baseline flow at the same angle of attack is also given so as to delineate the cause of increased lift in the controlled cases (figure 4). While the characteristics associated with the change in input parameters are discussed, the detailed flow structure for a representative case is only given.

The controlling aspect of a synthetic jets stems from the fact that it adds momentum flux and vorticity flux to the external flow, without any mass flux addition. The momentum flux is instrumental in separation control, which is done by adding high momentum fluid to the boundary layer during the blowing stroke and by removing the proximal low momentum fluid during the suction stroke, leading to stifled separation. A synthetic jet in a quiescent flow is characterized by the vortex rings (which are

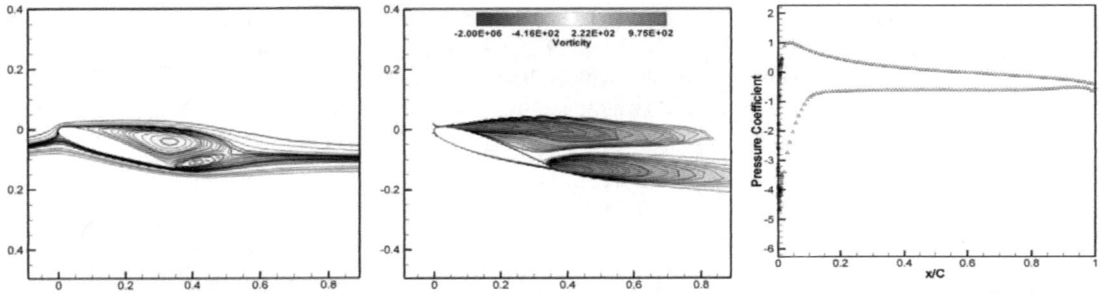

(a) Streamline Contour (b) Vorticity Contour (c) Pressure Distribution

Figure 4. Flow patterns for the baseline(uncontrolled) case

manifested as a pair of counter-rotating vortices in 2D) emanating into the surroundings, which carries the aforesaid momentum, and vorticity flux along with it.

Optimal flow control (using low angles of injection and feasible amplitudes) involves the convective transport of these vortices downstream. Vortices are generated during the blowing stroke, while the suction stroke is marked by coalescence of the external flow. While the strength of the anticlockwise rotating vortex (from the left edge of the synthetic jet wall here) is diminished, the other vortex is sustained and can be seen as a thick blob of vorticity (figure 5). During the suction stroke, the boundary layer seems to have lesser thickness compared to the baseline flow case. While the said vortex is generated due to the jet, its point of manifestation (physical significance) depends on the flow parameters selected. This vortex grows in size during the suction stroke, travelling downstream at the same time (This is seen in both the vorticity and streamline contours of figure 5). This vortex is finally shed (the location of separation depends on the flow parameters) and in case of moderate parameters, present a typical vortex shedding pattern – with an opposite rotating vortex generated

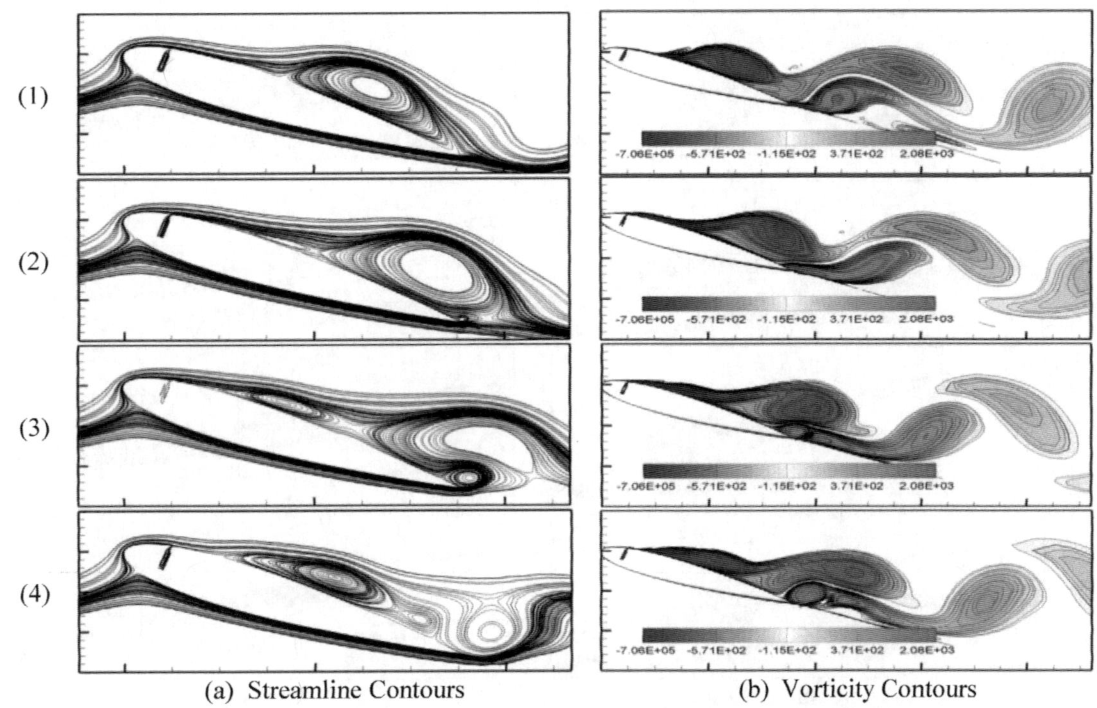

(a) Streamline Contours (b) Vorticity Contours

Figure 5. Phase averaged flow patterns for a controlled case (60Hz, 30° injection, 1.5 VR) at ϕ = (1) 0°, (2) 90°, (3) 180°, (4) 270°

from the pressure side of the airfoil. On noticing the phase average pressure coefficient plots, shown in figure 6, it is clearly seen that the vortex emanating from the synthetic jet acts like a low pressure "wave". This wave increases the area bound within the pressure coefficient curve (thus increasing the lift), while at the same time travelling downstream from its point of creation. While this forms the basis for the controlling aspect of the flow, some differences do creep in based on the parameters selected.

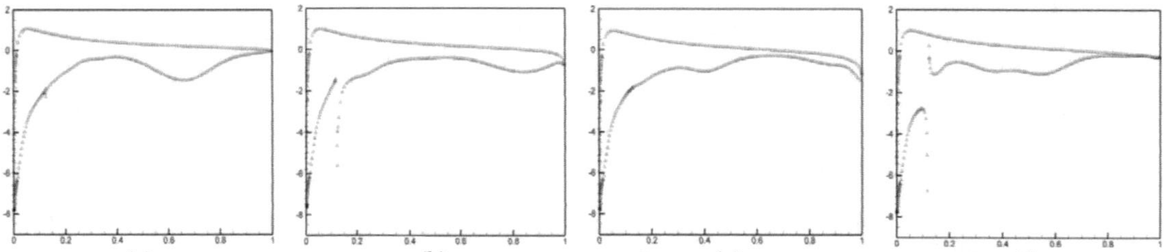

(a) (b) (c) (d)

Figure 6. Phase averaged pressure coefficient distribution for a controlled case (60Hz, 30° injection, 1.5 VR) at ϕ = (1) 0°, (2) 90°, (3) 180°, (4) 270°

Amplitude adversely affects the flow characteristics. With an increase in amplitude, the shed vortices are stretched and continuous (from both the suction side and pressures side of the airfoil). Furthermore, the leading edge separation observed in the baseline case shifts to a trailing edge type separation at higher amplitudes (figure 7). This is attributed to the fact that the vortex formed from the jet does not manifest to significant proportions and is shed almost immediately from the airfoil wall.

Major changes in the flow structures can be witnessed on changing the frequency. For the same amplitudes, vortices appear stretched and continuous at higher frequencies. Changes in the streamline patterns corresponding to the same indicate a constriction of streamlines forming a thinner separation

(1)
(2)
(3)
(4)

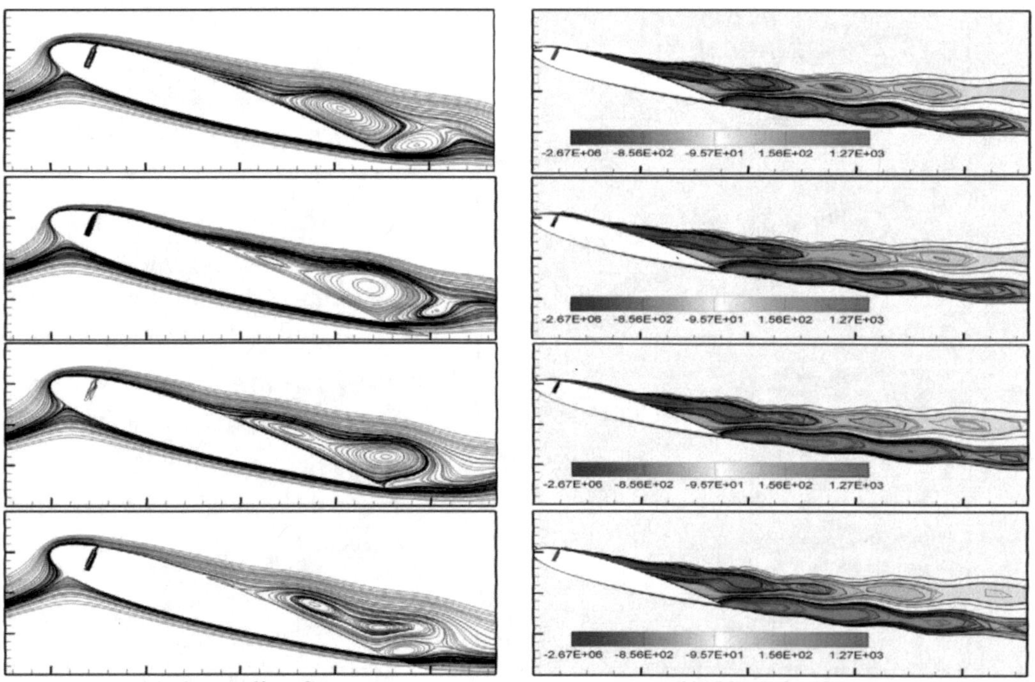

(a) Streamline Contours (b) Vorticity Contours

Figure 7. Phase averaged flow patterns for a controlled case (150Hz, 30° injection, 2.25 VR) at ϕ = (1) 0°, (2) 90°, (3) 180°, (4) 270°

bubble at higher frequencies (figure 7). As the vortices correspond to the frequency of operation, it is understandable that the pressure waves mentioned before correspond to the frequency of operation - more the frequency, more such waves per unit time. However the amplitudes of these waves also decrease, probably owing to the nature of the inherent instabilities in the flow. Cumulatively, a marginal increase in the area under the graph is seen for higher frequencies.

Less surprising is the fact that to maintain optimal control, the angle of injection is to be restricted to a minimal value. With an increase in the angle of injection, the normal component of the velocity increases for the same amplitude. The manifestation of the jet vortex occurs upstream as the angle is increased. As a result, the vortex from the synthetic jet is thus shed earlier from the airfoil, leading to a larger separation region. While the pressure "waves" have higher amplitudes at the jet vortex inception, it quickly diminishes to very low values. A comparison between figure 5 and figure 7 highlights the differences creeping in to the flow structures as the jet parameters are varied.

5. Conclusion

The effect of synthetic jet parameters on the flow control over a NACA0015 airfoil has been studied in depth. Qualitative characteristics for a representative case are given to highlight the effect of the jet on the performance enhancement of the airfoil. Thrust was placed on understanding the physics of the flow and the reason for effective control. Several remarks and explanations were given based on the observed characteristics, paving way for the better understanding of the turbulent physics associated with such jets. While this work serves as a platform to understand the effect of the synthetic jet, several modifications to the same can be made that can better enhance desired characteristics. A study on such parameters will be carried out in the near future.

References

[1] Collis S S, Joslin R D, Seifert A, and Theofilis V 2004 Issues in active flow control: theory, control, simulation, and experiment *Prog. Aerosp. Sci.* **40** 237-89

[2] Glezer A and Amitay M 2002 Synthetic jets *Annu. Rev. Fluid Mech.* **34** 503-29

[3] Smith B L and Swift G W 2003 A comparison between synthetic jets and continuous jets *Exp. Fluids.* **34** 467-72

[4] Mittal R and Rampunggoon P 2002 On the virtual aeroshaping effect of synthetic jets *Phys. Fluids.* **14** 1533-36

[5] Okada K, Fujii K and Miyaji K 2009 Computational study of frequency and amplitude effects on separation flow control with the synthetic jet *Int. Mechanical Engineering Congress and Exposition (Buena Vista)* **9** 289-298

[6] Akcayoz E and Tuncer I H 2009 Numerical Investigation of flow control over an airfoil using synthetic jets and its optimization *Int. Aerospace Conference (Ankara)*

[7] Gilarranz J L, Traub L W and Rediniotis O K 2005 A new class of synthetic jet actuators—part II: application to flow separation control *J. Fluids. Eng.* **127** 377-387

[8] Goodfellow S D, Yarusevych S and Sullivan P E 2009 Momentum coefficient as a parameter for aerodynamic flow control with synthetic jets *AIAA J.* **51** 623-31

[9] Mccormick D C 2000 Boundary layer separation control with directed synthetic jets 38th *Aerospace Sciences Meeting and Exhibit (Reno) AIAA* 519

[10] Raju R, Mittal R and Cattafesta L 2008 Dynamics of airfoil separation control using zero-net mass-flux forcing *AIAA J.* **46** 3103-15

[11] You D and Moin P 2008 Active control of flow separation over an airfoil using synthetic jets *J. Fluid. Struct.* **24** 1349-57

[12] Lasagna D, Donelli S, De Gregorio F, Orazi M and Iuso G 2013 Separation delay on thick airfoil using multiple synthetic jets *XXI Congresso Aimeta (Torino)*

[13] Duvigneau R and Visonneau M 2004 Simulation and optimization of aerodynamic stall control using a synthetic jet *2nd AIAA Flow Control Conf. (Portland)* 2315

[14] Dannenberg RE and Weiberg JA 1952 Section characteristics of a 10.5-percent thick airfoil with area suction as affected by chordwise distribution of permeability *NASA TN 2847*

Vortex Ring Formation Characteristics in Synthetic Jet due to Changes of Excitation Frequency in the ½-Ball Cavity Actuator

Ramon Trisno[1,a*], Harinaldi[2,b*] and Engkos A. Kosasih[3,c]
[1]Mechanical Engineering Department, Faculty of Engineering, University of Pancasila, Jagakarsa, Jakarta 12640, Indonesia
[2,3]Mechanical Engineering Department, University of Indonesia, Depok, West Java, Indonesia 16424
E-mail : [a]ramontrisno@gmail.com, [b]harinald@eng.ui.ac.id, [c]kosri@eng.ui.ac.id

Abstract. A jet flow that contains vortex ring has a large energy compared to a regular jet. As one of the causes of the aerodynamic drag to the vehicle, the flow separation that occurs behind the bluff body must be controlled, so that aerodynamic drag can be significantly reduced. This study is a basic work on the development of turbulent flow separation control for aerodynamic purpose, especially in the design of the vehicle body. The main objective of this study is to analyze the performance of the synthetic jet (SJA) as one of flow control tool to reduce separation area.
To get the maximum performance of the synthetic jet actuator, the research starts by characterizing the actuator. Characterization of ½ ball-shaped cavity is done with excitation frequency changes and orifice diameter of 3, 5 and 8 mm. The study was conducted using computational and experimental methods. The experimental data was obtained by testing synthetic jet actuator with providing sinusoidal signal to drive the membrane and at the orifice end a hotwire probe that is set and plugged into a CTA (Constant Temperature Anemometry) to obtain the speed velocity of the exhaust jet. Computational methods used a commercial CFD software (FLUENT 6.3) with a Reynolds Stress Model as a model of turbulence. Each of these calculations or measurements was conducted under the same conditions.
The research result is displayed in frequency testing curve to get the maximum velocity of the jet stream. The results are further indicative of the synthetic jet actuator capability to generate vortex rings. In the experimental results, the determination of ring vortex formation taken from the calculation of the flow velocity, while the CFD simulations, the formation of vortex rings can be seen from the visualization of the flow contour. Vortex ring formed from this ½ -ball cavity, occurred at 3 mm and 5 mm orifice diameter, while the 8 mm orifice diameter cavity cannot form a ring vortex.

Keyword: synthetic jet actuators, reversed Ahmed body, vortex ring, bluff body

1. Introduction

The latest research in vehicle aerodynamics is carried out with a design standpoint related to fuel economy. For this reason, ground vehicle aerodynamics have been studied experimentally and numerically by many researchers. Most previous studies have used simple models of vehicles that can generate relevant features of the flow around a real vehicle [1] [2] [3] [4] [5] [6]. The wake around the flow field is characterized by a pair of horseshoe vortices and trailing vortices which comes from the edge of the sloping side of the body

Experiments of passive flow control in the wind tunnel model or prototype vehicle has been widely performed in various research [7] [8]. However, passive flow control give a weak effect in vehicle aerodynamic because the properties of the flow that involved is turbulent and the contribution of friction on aerodynamic drag is still small, only about 10% [9].

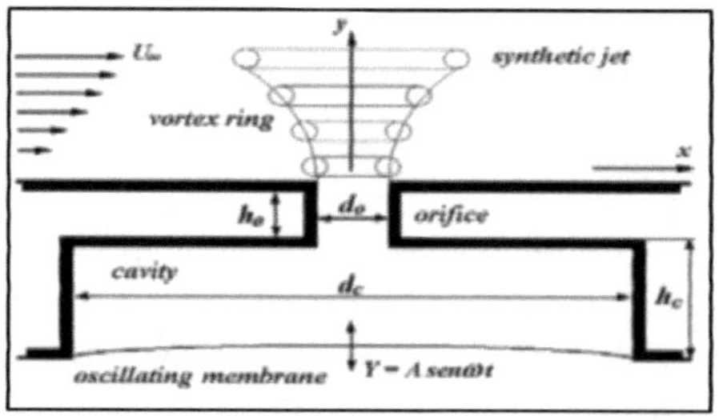

Figure 1 Sketch of a synthetic jet actuator in a cavity formed by the outlet orifice

Flow control on the bluff body in order to reduce drag and noise is one of the main issues in aerodynamics. The pressure difference between the front and the back side of the bluff body is a major contributor to the overall drag, the difference is primarily due to flow separation at the rear region of the body [10].

The need for an effective drag reduction encourage the automotive design to be more creative in developing innovative active control model. Active control method makes it possible to modify the topology of the flow without changing the shape of the vehicle. Within the scope of academic and industrial laboratories, active control methods have been developed with computational and experimental methods, and significant results have been obtained in an academic framework [7]. Suction active control that are placed on top of the rear window is able to eliminate the separation on a simplified geometry fastback cars where aerodynamic drag reduction obtained was about 17% [11]. Numerical study conducted by Kuorta and Gillieron using a synthetic jet active control placed at the top of the rear window Ahmed model showed a drag reduction of 13% [12].

Characteristics of the flow velocity on synthetic jet is taken from the center axis, U_{CL}, as the assumption of the highest velocity profile at the exit orifice. Since the spatial velocity profile can show significant deviation from the top-hat shape, a more general velocity scale is the time-and-spatially averaged exit velocity over the expulsion stroke, [13]

$$U = \frac{1}{T/2}\frac{1}{A_n}\int_{A_n}\int_0^{T/2} U(t, A_n)\,dt\,dA_n \qquad [1]$$

Where U is the average velocity of the axial phase, T is the period of excitation and A_n is the cross-sectional area of the exit nozzle.

The average velocity can be used to determine the Reynold number of the jet. The important thing at the length scale of synthetic jet flow is the stroke length L_o / D, that given by:

$$L_o = \int_0^{T/2} U_{ave}(t)\,dt \qquad [2]$$

U_{ave} is a spatial average flow velocity. Stroke length is the distance that the fluid taken during the pushing process as a part of the cycle. The excitation frequency f can be non-dimensionalized as a Stokes number, S, that given by :

$$S = \sqrt{\frac{2\pi f D^2}{\nu}} \qquad [3]$$

The non-dimensional parameter uniquely determines the operating conditions of the synthetic jet and greatly affects its ability to transfer linear momentum. In particular, Holman et al. [19] showed that in order to achieve vortex escape after the expulsion stroke (defined as synthetic jet formation)

$$\frac{\mathrm{Re}_U}{S^2} > C \qquad [4]$$

where C is a constant which equal to 0.16 for the axis symmetric and equal to 1 for the rectangular nozzles. According Utturkar et al [14] [15], the condition for a vortex ring to be formed is $Re_U / S^2 > 1$. Meanwhile, according to Smith and Swift opine the ring vortex can occur if: [16] [17] [18]

$$\frac{L_0}{D} = St > 6 \qquad [5]$$

The purpose of this study was to obtain the characteristics of the SJA with half ball-shaped cavity by providing the oscillation frequency of the membrane, in order to obtain the best performance from changes in the diameter of the orifice in the cavity.

2. Experiment Settings

Before starting the study of synthetic jet performance to the drag reduction on the vehicle, it need to know about the performance of SJA itself, in order to get the maximum results for drag reduction. The purpose of this research on the performance of SJA is to choose the type of actuator that is good in forming vortex ring during the operation of SJA. Selection was based on the SJA actuator cavity shape, orifice diameter and comparison of performance through experiments and CFD simulations. In this research, ½ ball cavity shape was studied with a diameter of 3 mm (B3), 5 mm (B5) and 8 mm (B8) (Figure 2).

Figure 3 is a schematic diagram of the synthetic jet performance test through the experiment. Synthetic jet component consisted of cavity and *piezoelectric* membrane component, mounted as a whole and driven by the function generator. This tool unleashed the desired wave signal. A signal which used in this experiment was a sinusoidal wave. The movement of the membrane made the air pushed out of the cavity and sucked into the cavity back. This tool had a mechanical system with a very small mass with the stem and the spring move. This mass was hanged just two microns (two millionths of a meter) from the electronic circuit.

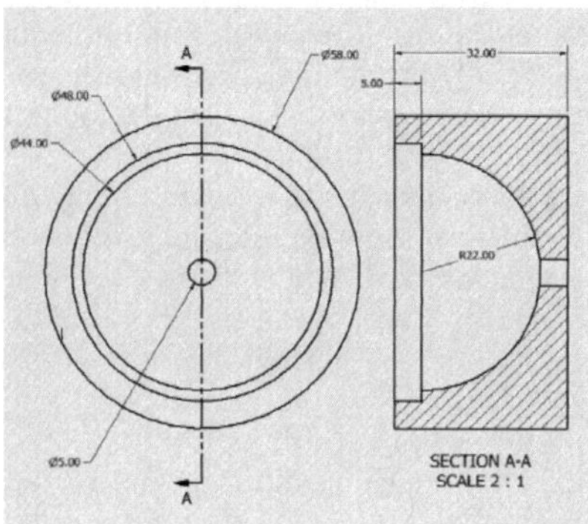

Figure 2. ½ -ball shaped SJA cavity

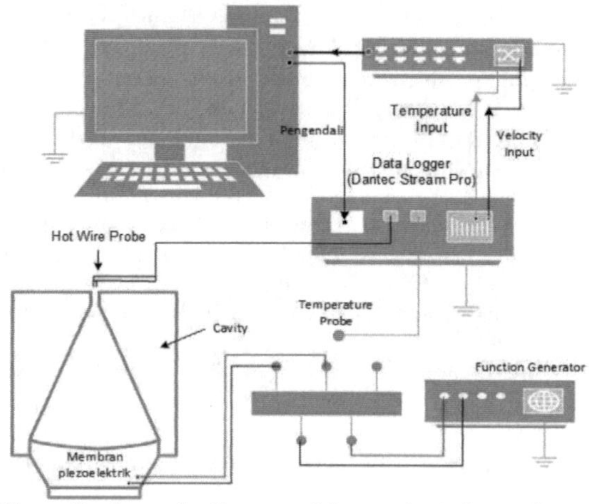

Figure 3 schematic diagram of the synthetic jet performance testing

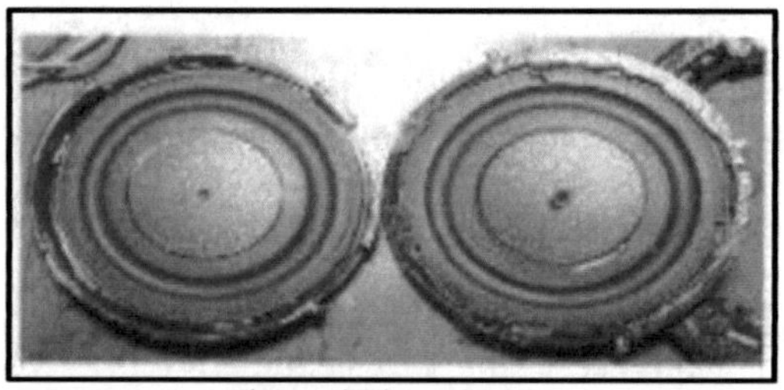

Figure 4 Piezoelectric

Furthermore, the speed of the air that come out of this cavity through the orifice was measured using a hot wire probe 55P14 Gold Plate. This single wire probe had a wire diameter $d_w = 5$ μm and the distance of the branch 1.2 mm, placed on a SJA as near as possible, in this case the possible distance was equal to ± 0.5 mm above the orifice of synthetic jet actuator, so that the movement of the probe along the axis x did not touch the cavity. These probes were connected to the Data Logger, brand DANTEX Stream pro, type 91C10, CTA module that served to record the incoming data from hotwire in the form of low voltage. Then the data was entered into the tool Data Acquisition Board of National Instrument BNC-2110. This tool was used to change the analog data into digital data. The tool was connected to the computer via a data cable. This computer was also connected to a data logger which serves to control the speed of recording desired data. Before recording the data, the probe had to be calibrated using a calibration unit Dantec Streamline Pro Automatic Calibrator System S / N 9091H0013445 and King's Law that suitable for the job to interpret the data speed below the calibration range. The range of the calibration performed at 0.02 ÷ 30 m / s before each experiment performed. The calculation result of uncertainty due to calibration for all data retrieval speed was about 2.5%. Radial velocity profile was measured with a hot wire probe which across the hole using the tool transfer (vise) and the displacement distance was measured using a dial indicator with 0.01 mm accuracy.

Velocity data collection procedure was done under two conditions. The first was the velocity measurement for determining the frequency that produces the maximum speed. The process of data collection was intended to get the maximum speed that can be generated SJA and the amount of data allowed (uncertainty) by varying the sinusoidal frequency issued by the function generator. The given frequency is 20 Hz ÷ 200 Hz. Data taken as much as 60,000 data at a data rate of 30,000 data/s. The second procedure was the jet velocity measurements at points determined by the Cartesian coordinates (*X/D*, *Y/D*). This jet velocity was measured only in the half part region, because the jet was considered symmetrically formed. The value of *X/D* is 0, 0.25, 0.5, 0.75, ... ,4, while the value of *Y/D* is 0, 1, 2, 3, ... , 10. Velocity data were recorded at 10,000 data/s for 6 seconds.

The purpose of this simulation was to compare the results obtained with the results of experiments conducted as well as to visualize the vortex ring that formed by SJA. The test condition used in the experiment was similar to that used in the CFD simulation. Determination of the frequency used in CFD simulations are based on experimental results that were achieved, such that the use of User Defined Function (UDF) in the simulation matched with experimental results.

Simulations performed using software FLUENT computational fluids dynamic software version 6.3.2.6 and Gambit version 2.4.6 to form geometry, defining the role of the components, and define meshing. Meshing is used for CFD simulations in the study had determined its density is coarse (5000 mesh), intermediate (20 000 mesh) and fine (60 000 mesh). Mesh independencies done to find amount of mesh are most suitable for use in the simulation of fluid flow as well as the convergence correcting numerical calculation by computer. This study uses two approaches to mesh independencies. First, look for the minimum total error with the same parameters on the frequency and the same wavelength as well as at 1 time step using the turbulence models of Reynolds Stress Model. Second, match the simulation results (using the turbulence models of Reynolds Stress Model) with experimental results. Mesh is closest to the result of experiments that would be used for computational stages.

Turbulence models is a computational procedure for calculating the main equations of flow with a certain flow conditions. To obtain the solution of turbulent flow, the flow should be modeled mathematically. Turbulence models used in this computing method is RSM (Reynolds Stress Model). The Reynolds stress model involves calculation of the individual Reynolds stresses, $\overline{u_i'u_j'}$, using differential transport equations. The individual Reynolds stresses are then used to obtain closure of the Reynolds-averaged momentum equation. The exact transport equations for the transport of the Reynolds stresses, $\rho\,\overline{u_i'u_j'}$, may be written as follows:

$$\begin{aligned}
&\frac{\partial}{\partial t}\rho\overline{u_i'u_j'}+\frac{\partial}{\partial x_k}\left(\rho u_k\overline{u_i'u_j'}\right)=-\frac{\partial}{\partial x_k}\left[\overline{\rho u_i'u_j'u_k'}+\overline{p}\left(\delta_{kj}u_i'+\delta_{ki}u_j'\right)\right]\\
&+\frac{\partial}{\partial x_k}\left[\mu\frac{\partial}{\partial x_k}\left(\overline{u_i'u_j'}\right)\right]-\rho\left(\overline{u_i'u_k'}\frac{\partial u_j}{\partial x_k}+\overline{u_j'u_k'}\frac{\partial u_i}{\partial x_k}\right)-\rho\beta\left(g_i\overline{u_j'\theta}+g_j\overline{u_i'\theta}\right)\\
&+\overline{p\left(\frac{\partial u_i'}{\partial x_j}+\frac{\partial u_j'}{\partial x_i}\right)}-2\mu\overline{\frac{\partial u_i'}{\partial x_k}\frac{\partial u_j'}{\partial x_k}}-2\rho\Omega_k\left(\overline{u_j'u_m'}\in_{ikm}+\overline{u_i'u_m'}\in_{jkm}\right)+S_{user}
\end{aligned}\quad [6]$$

3. Results and Discussion

3.1 Frequency Testing and Mesh Independencies

The early stages of the experiment was to determine the frequency that can produce a maximum flow rate so that the performance of this SJA as a reference in subsequent experiments. This testing is done by varying the frequency of the wave with function generator tool which use sinusoidal wave.

The test results on the SJA frequency shows that the type B3 and B8 experience the maximum velocity at the frequency of 120 Hz, while at the B5, maximum velocity occurs at a frequency of 130 Hz. The flow velocity produced by SJA types B3 is higher than other types of SJA, where the average jet velocity is equal to 7.786 m / s.

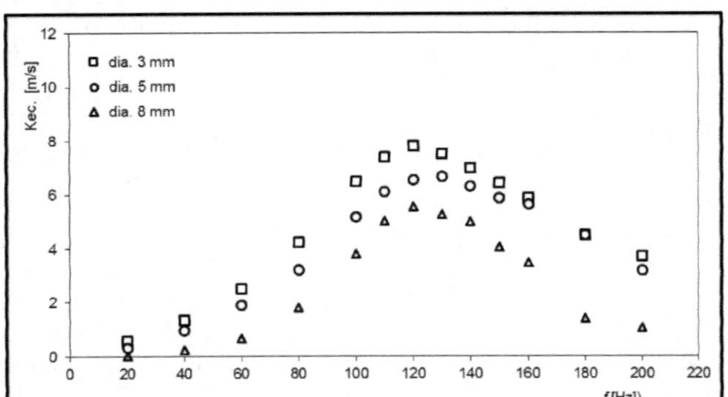

Figure 5. The graph of test results on the synthetic jet actuator half-ball cavity with sinusoidal amplitude types.

In the computational study, meshing testing needs to be done in advance, using software Gambit 2.4.6. This meshing testing aims to get the meshing residues that are not changed to deformation and meshing journals. Then the predefined meshing is used later as input material in CFD simulation processing. The type of the cavity and orifice diameter adapted to the experimental conditions. Each meshing is then processed using software *Fluent 6.3.26*. Conditions of iterations taken on the position of $X/D = 0$ and $Y/D = 0$ with 200 data at the speed of data recording 10,000/s.

To determine the certainty and reliability of the data obtained from the experiment, the uncertainty values used in the equation below, applies for a confidence level of 95%, given by:

$$U_x \propto \sigma_{\bar{x}} \Rightarrow U_x \approx k(convidence\ limit)\,\sigma_{\bar{x}}$$

$$\sigma_{\bar{x}} = \frac{2S_x}{\sqrt{N}} \qquad [7]$$

As for determining the percentage level of measurement uncertainty (U_x), the following equation was used:

$$U_x = \frac{\sigma_{\bar{x}}}{\bar{x}} \times 100\% \qquad [8]$$

The calculation of uncertainty in each cavity was taken based on the number of sampling data. The data are processed on the condition $Y/D = 0$ and $X/D = 0$, the point (0,0) which is the onset of the synthetic jet. This point has a greater uncertainty value than the other points.

For first 10,000 data, uncertainty calculation using equations [7] and [8], the uncertainty calculation result in B3 cavity type at 1.61%, B5 cavity types at 1.58 % and B8 cavity types at 0.98%. Based on the value of this uncertainty, the B3 cavity type have greater uncertainty value than the other cavity types. Hence, the value of uncertainty in B3 should be aware of its magnitude, so that the amount of data captured on the experimental conditions can qualify a data confidence level at 95%, then the value of the uncertainty should be below 5%. But the results of the calculations shows that the data obtained by the experimental method provides that the uncertainty value does not exceed 2%. The value is below the required value.

3.2 Average Jet Velocity

Membrane on synthetic jet actuator (SJA) oscillates with a sinusoidal frequency, so that the SJA generates air flow out through the orifice. The orifice tip, i.e. the (0,0) point experienced a higher air flow rate compared to other points. U_{CL} is the airflow velocity at the point (0,0) coming out form the of the tip orifice, while U_x is the local air flow rate at a certain point.

Figure 6 is a comparison of U_x/U_{CL} at the point (0,0) produced by synthetic jet actuator with a half-ball cavity shaped and orifice diameter of 3 mm (B3). The left graph is an experimental results graph, while the right graph is a CFD simulation results graph. Both of these graphs looks similar in the data distribution. On the $0 < X/D < 1$, it appears that the value of U_x/U_{CL} is above 0.2. This means at the $0 < X/D < 1$ condition, the velocity of the jet flow is high. While at the $1 < X/D < 2$, the value of the speed ratio U_x/U_{CL} decreased quite dramatically, so that the possible occurrence of velocity difference is quite large in this area. The author notes, that in regions $1 < X/D < 2$ there has been a flow separation between areas affected by jet to areas not affected by the jet. This makes it possible to form a kind of the boundary layer between the area experiencing jet to areas not affected by the jet. But for areas $2 < X/D < 4$ speed differences tend to be smaller and more constant. It is influenced by the jet center which is farther to the area. The presence of the flow velocity in the area is not caused by a jet of air that comes out of the SJA but rather caused by turbulence around the mouth of the orifice. Turbulence is caused by the collision of air flow into the cavity with the flow of air that had not sufficient time to get out. So that the air flow is partially not enter into the cavity but being around the orifice of the SJA.

Figure 7 depicts the distribution of air flow velocity at B5 obtained from experiments and CFD simulations. The graph of experimental results (7a) shows that for $0 < X/D < 2$ speed ratio U_x/U_{CL} showed a significant difference due to the change of Y/D with a value range $0.1149 < U_x/U_{CL} < 1$. The smallest value (0.1149) and the largest value (1) is located along the center line. The velocity change in the central line is experiencing considerable fluctuation. As for $2 < X/D < 4$, U_x/U_{CL} did not show a big difference. Figure 8, shows the experimental results that U_x/U_{CL} is very significant at $0 < X/D < 1$. A very sharp slope of the U_x/U_{CL} occurs in this area. It shows that the area where the jet is possibly formed is only at $0 < X/D < 1$. While at $X/D > 1$, it tends to form a horizontal chart, meaning that the speed of air flow in that region is unlikely to change significantly. Likewise with the graph from the CFD simulation results, shows similar results to the experiment. Only in $Y/D = 0$ gives a greater change of velocity compared with other Y/D.

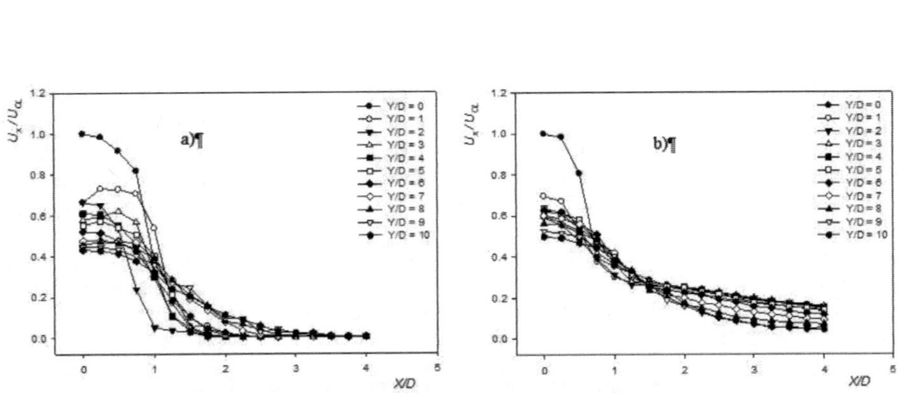

Figure 6 the comparison of U_x/U_{CL} graph on the B3, a) experiment, b) CFD simulation

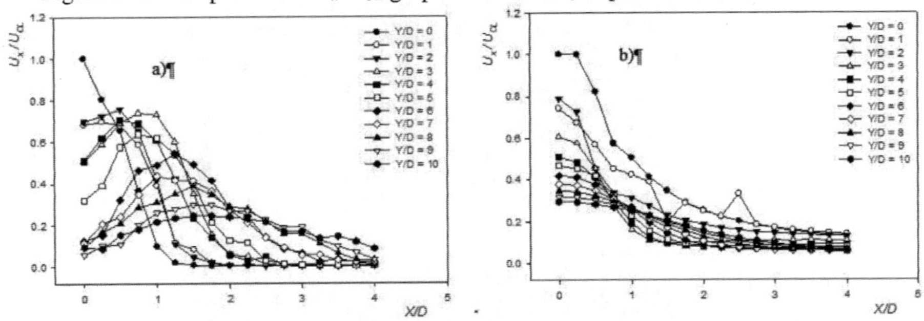

Figure 7 the comparison of U_x/U_{CL} graph on the B5, a) experiment, b) CFD simulation

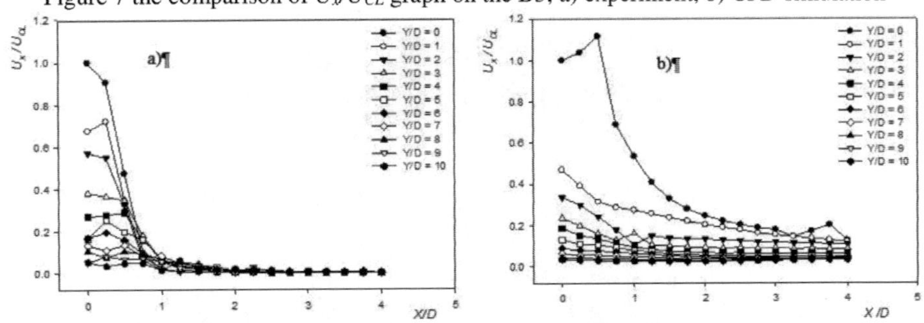

Figure 8 the comparison of U_x/U_{CL} graph on the B8, a) experiment, b) CFD simulation

3.3 Determining the occurrence of vortex Ring

When the bursts of flow come out through the orifice, the bursts is not necessarily form a vortex ring. Vortex ring formation is preferred so that the thrust of the bursts is more powerful and could protect the bursts to remain solid. The un-formation of a vortex ring can be caused by the flow rate, the frequency of the excitation and the diameter of the orifice. Jet is formed when the ratio between the Reynolds numbers with squared Stoke number is > 1 and the ratio between the stroke length and orifice diameter is > 6 filled. Here, Table 3 is the result of calculations based on equations [4] and [5] at the B3, B5 and B8 cavity model, at exactly above the orifice and with the maximum frequency.

Table 3 Results of calculation of the occurrence of vortex rings in the half ball cavity

Cavity	*Lo* [mm]	*Uo* [m/s]	Stroke ratio	Re_U/S^2	Explanation
B3	48.74	5.85	16.25	2.77	Occur
B5	26.25	3.41	5.25	0.84	Not-Occur
B8	22.25	2.67	2.78	0.44	Not-Occur

The next step is the determination of the occurrence of vortex rings on each type of cavity, and the visualization of computational simulation results can be seen in Figure 9a, 9b and 9c. Figure 9a shows the velocity contours due to the movement of the membrane in the cavity B3. When the membrane moves at $t/T = 1/8$, the membrane is at blowing step, and it shows the airflow out through the orifice. Then at $t/T = 1/4$, the contour of the flow is beginning to show the formation of vortex rings. This position is the culmination expulsion step by the movement of the membrane. Membrane starts to move downwards and SJA perform suction step. After reaching at $t/T = 3/8$, then the vortex ring will be seen more clearly. Then at $t/T = 1/2$, the vortex ring is split off and move up. Suction step continued to stage $t/T = 5/8$. In this position, some of the air under the vortex ring will be sucked back into the cavity. This can affect the shape of the vortex ring which has been formed by the blowing of SJA. But on B3 cavity, the vortex ring which is formed remains intact and further intensified at $t/T = 3/4$. Then at $t/T = 7/8$, the membrane moves up, and the SJA start to blow. In this condition, the vortex ring which has been formed is ready to break away by the formation of a new vortex ring. At $t/T = 1$, vortex ring have released itself from the tip of the orifice.

a)

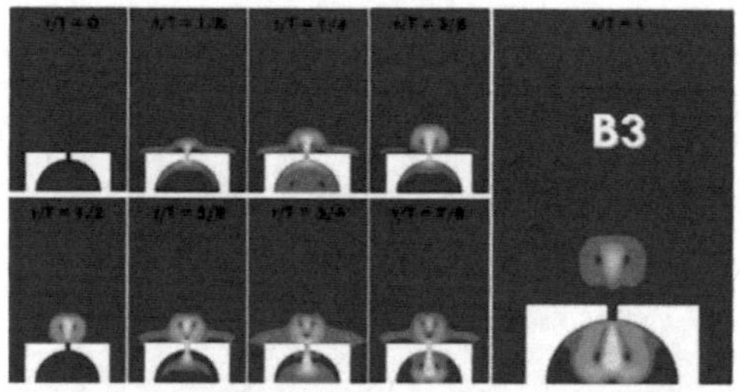

b)

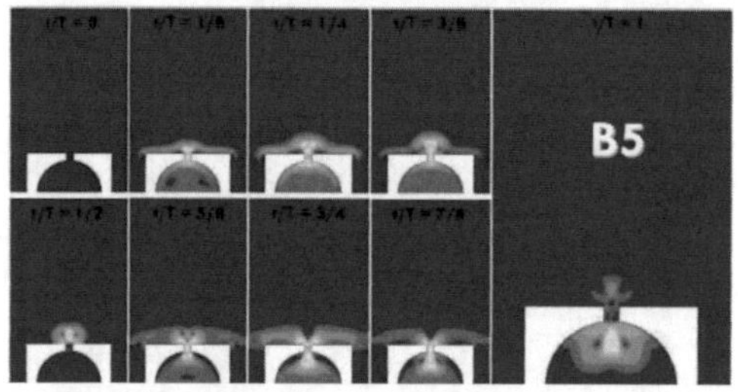

c)

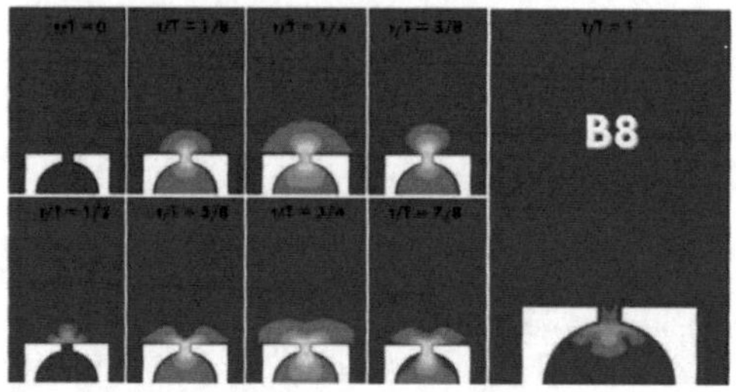

Figure 9. Formation visualization of the vortex rings at B3, B5 and B8

Figure 9b shows the velocity contours produced by B5 cavity SJA. The vortex ring began to take shape at $t/T = 1/4$. At $t/T = 3/5$, the vortex ring begin to separate from the tip of the orifice. It clearly shows that the vortex ring is leaving the tip orifice at $t/T = 1/2$. But when the membrane moves down at $t/T = 5/8$, most of the air around the mouth of the orifice began to be sucked into the cavity, and the vortex ring that has been formed over the orifice is immediately affected by the suction of SJA, as shown by the figure at $t/T = 3/4$. Consequently, the vortex ring is rupture (at $t/T = 7/8$) and failed to form, as shown at $t/T = 1$.

Figure 9c shows the velocity contours produced by B8 cavity SJA. In the initial condition of blow step which generated by SJA, velocity contours do not shown the formation of the vortex ring. As also at the suction conditions, at $t/T = 5/8$, most of the air at the end orifice, seen flowing into the cavity and filled the cavity. This situation shows that there is no vortex ring formed by the process of SJA on B8. Thus, it can be said that both method, experiment and CFD simulation give the same result.

4. Conclusion

From the results of the research on the characteristics of synthetic jet actuator, some conclusions can be taken. The frequency obtained to get the maximum velocity jet flow of B3 and B8 cavity is at 120 Hz and B5 is at 130 Hz. Based on the number of data taken, the value of uncertainty has been reached below 2%. Results of flow velocity measurements in absolute terms through experiments and CFD simulations shows near close results. Distribution of the jet stream average speed occurs higher in $X/D < 1$, then decreased after $X/D > 1$. The flow rate at absolute maximum jet streams occur on B3. In determining whether the occurrence of vortex ring due to movement of SJA, the experimental results were processed in the existing equation and the results obtained from CFD simulation, gives the same results. Vortex ring can occur only at B3, whereas at B5 and B8, SJA cannot form a vortex ring.

5. References

[1] Ahmed S.R., G. Ramm and G. Faltin., 1984, SAE paper, Detroit, Michigan, USA, 8400300-01
[2] Hintenberger C., Villalba M. G. &Rodi W., 2004, Large eddy simulation of flow around the Ahmed body, Institute for Hydromechanics, University of Karlsruhe, Germany
[3] Fares E., 2006, Unsteady flow simulation of the Ahmed reference body using a lattice Boltzmann approach, Computers and Fluids, 35, pp. 940-950
[4] Minguez M., Pasquetti R. &Serre E., 2008, High-order Large Eddy Simulation of Flow over the Ahmed Body' Car Model', Physics of Fluids, 20
[5] Uruba V., and Hladik O., 2009, On the Ahmed Body Wake, Colloquium Fluid Dynamics, Institute of Thermomechanics AS CR, v.v.i., Prague
[6] Conan B., Anthoine J., and Planquart P., 2011, Experimental Aerodynamic study of a car-type bluff body, Experimental of Fluids, 50, pp. 247 – 259
[7] Gak-El-Hak, M., 1996, Modern Development in Flow Control, Appl. Mech. Rev., 9, pp. 365-379
[8] Hucho W.H., 1998, Aerodynamics of Road Vehicle, Annu. Rev. Fluid Mech., 25, pp 285-537
[9] Kourta A. &Gillieron P., 2009, Impact of automotive Control on the Economic Issues, Journal of Applied Fluid Mechanics, vol. 2, no. 2, pp. 69 – 75
[10] Hucho W. H., 2002, Aerodynamik der stumpfer K''orper – PhysikalischeGrundlagen und Anwendung in der Praxis, Vieweg-Verlag
[11] Roumeas M., Gillieron P., and Kuorta A., 2009, Analysis and Control of near wake flow over a square back geometry, Computers & Fluids, 38, pp. 60 – 70
[12] Roumeas M., Gillieron P., and Kuorta A., 2009, Drag Reduction by Flow Separation Control on a Car after Body, International Journal for Numerical Method in Fluids, 60, pp. 1222 – 1240
[13] Feero, Mark A., Lavoie, Philippe, Sullivan, Pierre E, Influence of cavity shape on synthetic jet performance. Sensors and Actuators A: Physical 223 (2015) page 1 – 10, Elsavier, 2014.
[14] R. Holman, Y. Utturkar, R. Mittal, B.L. Smith, L. Cattafesta, Formation criterionfor synthetic jets, AIAA J. 43 (10) (2005) 2110–2116.
[15] Utturkar, Y., "Numerical Investigation of Synthetic Jet Flow Fields," M.S. Thesis, Department of Mechanical Engineering, University of Florida, 2002.
[16] Smith, B. and Glezer, A. "The formation and evolution of synthetic jets." Phys. Fluids, Vol. 10, No. 9, 1998, pp. 2281-2297
[17] Glezer, A., "The Formation of Vortex Rings," Phys. Fluids, Vol. 31, No. 12, 1998, pp. 3532-3542.
[18] Smith, B. and Swift, G. "Synthetic Jets at Large Reynolds Number and Comparison to Continuous Jets," AIAA Paper 2001-3030, 2001.

Analysis of the coherent and turbulent stresses of a numerically simulated rough wall pipe

L Chan[1], M MacDonald[2], D Chung[2], N Hutchins[2] and A Ooi[2]

[1] Department of Mechanical Engineering, Universiti Tenaga Nasional, Jalan IKRAM-UNITEN, 43000 Kajang, Selangor, Malaysia

[2] Department of Mechanical Engineering, The University of Melbourne, Victoria 3010, Australia

E-mail: leonchan@uniten.edu.my

Abstract. A turbulent rough wall flow in a pipe is simulated using direct numerical simulation (DNS) where the roughness elements consist of explicitly gridded three-dimensional sinusoids. Two groups of simulations were conducted where the roughness semi-amplitude h^+ and the roughness wavelength λ^+ are systematically varied. The triple decomposition is applied to the velocity to separate the coherent and turbulent components. The coherent or dispersive component arises due to the roughness and depends on the topological features of the surface. The turbulent stress on the other hand, scales with the friction Reynolds number. For the case with the largest roughness wavelength, large secondary flows are observed which are similar to that of duct flows. The occurrence of these large secondary flows is due to the spanwise heterogeneity of the roughness which has a spacing approximately equal to the boundary layer thickness δ.

1. Introduction

Typically, a turbulent quantity can be decomposed into its mean and fluctuating component

$$u_i = \overline{u}_i + u_i'' \tag{1}$$

where $\overline{u}_i$ is the time-averaged mean and u_i'' is the fluctuation about the time-averaged mean (also known as the turbulent fluctuation) for component i. This decomposition is known as the Reynolds decomposition and is commonly used to analyse the turbulent fluctuations in a flow. However, in a rough wall flow, variations due to the unevenness of the spatial geometry must be taken into account when analysing the turbulent flow data. Therefore, a triple decomposition is applied to the turbulent quantity where it is decomposed into three components [1, 2],

$$u_i = U_i + \tilde{u}_i + u_i''. \tag{2}$$

Here, $U_i = \langle\overline{u_i}\rangle$ is the spatial and temporal averaged mean, which is also known as the global mean and $\tilde{u}_i = \overline{u}_i - U_i$ is the spatial variation of the time-averaged flow around individual roughness elements (also known as the coherent or dispersive component). The sum of $\tilde{u}_i + u_i'' = u_i'$ is the total fluctuation and contains both the turbulent and coherent fluctuations. For a smooth wall, $\tilde{u}_i = 0$ and therefore $U_i = \overline{u_i}$. The triple decomposition is a result from double averaging, where the governing equations are averaged over time and then in space. This technique has been used to analyse plant canopy flows (urban-canopy model) [3, 4, 5] and also in open-channel flow over rough beds [6, 7, 8] where the rough surfaces are spatially inhomogeneous.

When conducting laboratory experiments, point measurement techniques such as hot wire anemometry and laser Doppler velocimetry gather data locally in space and only measures the turbulent fluctuations of the flow. This can result in misinterpretation of the data especially when conducting rough wall experiments where large spatial variation of velocity might exist in the flow. In this paper, we will investigate the importance of the coherent and turbulent fluctuations in a turbulent rough wall flow.

2. Computational setup

DNS is conducted in a pipe using CDP which is a second-order, energy conserving finite volume code. The rough wall pipe has a Cartesian 'O-grid' mesh and a body conforming grid is used to explicitly represent the roughness elements. The pipe has a length of $L_x = 4\pi R_0$ where R_0 is the reference radius. Further details of the numerical method used can be found in [9].

Throughout the paper, the roughness cases are identified by the following identifying code

$$\underbrace{\boxed{1}\boxed{0}}_{h^+}\text{-}\underbrace{\boxed{1}\boxed{4}\boxed{1}}_{\lambda^+} \qquad (3)$$

where the first two digits represent the roughness height and the last three digits represent the streamwise or spanwise wavelength of the roughness elements ($\lambda_x = \lambda_s$). Two groups of cases were simulated where in group $S1$, the roughness elements are geometrically scaled while maintaining a fixed h/λ ratio and in group $S2$, the wavelength of the roughness λ is varied while the roughness semi-amplitude h is held constant. The computational details and mean flow properties of the rough cases are tabulated in table 1.

Table 1. Details of the computational mesh and mean flow properties of the cases simulated. $N_{r,\theta}$ is the number of elements in an (r, θ) plane, N_x the number of elements in the streamwise direction and N_{λ_x} the number of elements per roughness wavelength. Δr^+ and Δx^+ are the radial and streamwise mean viscous grid spacings at the wall calculated using the local friction velocity U_τ. The largest cells are located at the centre of the pipe where $\Delta r^+ \approx \Delta r\theta^+ \approx \Delta x^+$. k_a^+ is the average roughness height and ES is the effective slope of the roughness (see [9]). Re_τ and Re_D are the friction and bulk Reynolds numbers and ΔU^+ is the Hama roughness function.

$S1$: Geometrically increasing roughness (fixed h/λ)											
Case	Symbol	$N_{r,\theta}$	N_x	N_{λ_x}	Δr^+	Δx^+	k_a^+	ES	Re_τ	Re_D	ΔU^+
Smooth	○	94752	1152	-	0.23	5.8	0.0	0.00	538	18729	0.0
20_141	●	104400	1152	24	0.14	4.4	8.1	0.36	541	12114	6.4
40_283	●	104400	1152	48	0.13	4.1	16.2	0.36	534	9027	9.1
60_424	●	108720	1152	72	0.15	4.0	24.3	0.36	543	7870	10.5
$S2$: Decreasing roughness wavelength λ^+ (fixed h^+)											
Case	Symbol	$N_{r,\theta}$	N_x	N_{λ_x}	Δr^+	Δx^+	k_a^+	ES	Re_τ	Re_D	ΔU^+
60_848	▲	109440	1152	144	0.15	4.4	24.3	0.18	543	9609	8.5
60_424	▲	108720	1152	72	0.15	4.0	24.3	0.36	543	7870	10.5
60_212	▲	109440	1152	36	0.08	3.5	24.3	0.72	544	7620	10.8

3. Mean velocity profile

The mean velocity profile of the rough cases are plotted in figure 1. The downward shift in the mean velocity profile of the rough wall is quantified by the roughness function ΔU^+ which is measured 200 viscous wall units above the crest of the roughness elements. The roughness function increases with increasing roughness height and also with decreasing roughness wavelength. However decreasing the roughness wavelength from $\lambda^+ = 424$ to 212 only results in a marginal increase in the ΔU^+ as the surface is in the dense regime ($ES = 2\Lambda = 0.72$). From [9], it is found that case 60_424 is in the fully rough regime. Therefore, it is deduced that case 60_212 which has a higher ES (or solidity) is also in the fully rough regime.

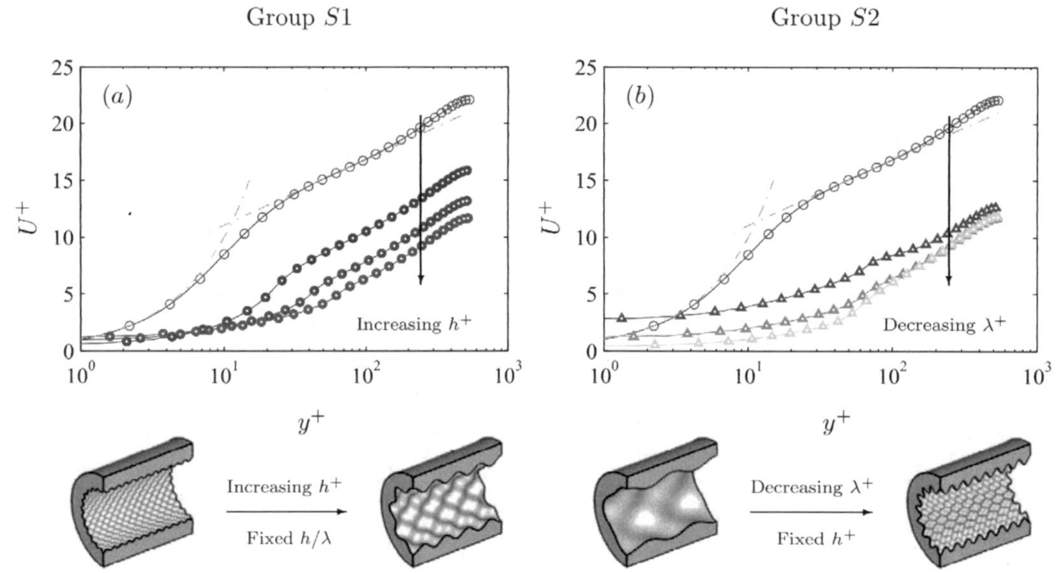

Figure 1. Streamwise velocity profile for (a) geometrically scaled roughness cases (group $S1$) and (b) roughness cases with fixed h^+ and varying wavelength λ^+ (group $S2$). Dash-dotted lines show $U^+ = y^+$ and $U^+ = (1/\kappa)\log(y^+) + C$, $\kappa = 0.40$ and $C = 5.3$. Symbols are as in table 1.

4. Stress profiles

The various fluctuating terms will be compared in this section using the triple decomposition method introduced in § 1. The stress profiles for the azimuthal turbulent $u''^+_{\theta,rms}$, total $u'^+_{\theta,rms}$ and coherent $\tilde{u}'^+_{\theta,rms}$ fluctuations are plotted in figure 2.

In figures 2 (a, d), the occurrence of a secondary peak is observed within the roughness canopy ($y \leq h$). This peak increases with increasing h^+ (group $S1$) and also with increasing roughness wavelength λ^+ (group $S2$). The secondary peak is contributed by the coherent (or dispersive) fluctuating component of the flow which is time independent (see figures 2 (c, f)).

The peak of the azimuthal coherent stress profile is located at a wall-normal location which ranges from $y^+ \approx 10-20$ and has a value which is comparable to the azimuthal turbulent stress. The peak in the azimuthal coherent stress profile is due to the channelling of the flow around the three-dimensional sinusoidal roughness elements. However, the coherent stress profiles quickly reduces to zero above the crest of the roughness elements (expect for case 60_848). This indicates that the effects of the azimuthal coherent stresses to the flow are small. However, for case of

60_848, the azimuthal coherent stress profile extends more than 200 viscous wall units above the crest of the roughness elements. This is because there are large stationary features occurring in the flow and this will be investigated further in § 5.

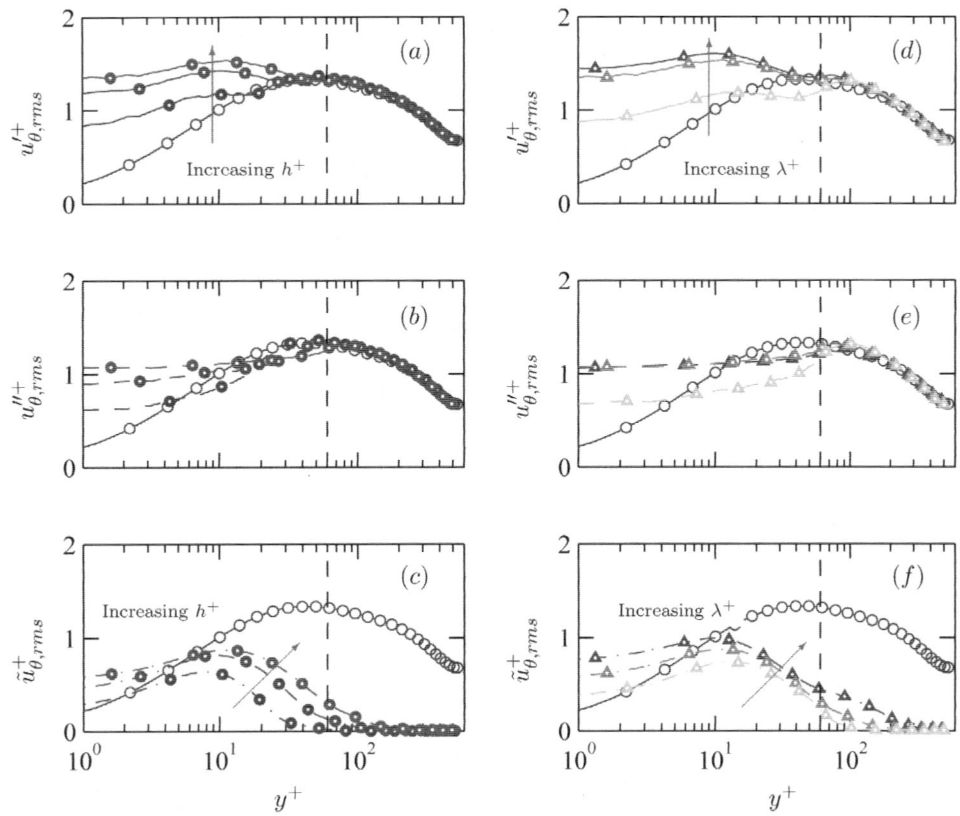

Figure 2. Azimuthal velocity (a, d) total (solid lines), (b, e) turbulent (dash lines) and (c, f) dispersive (dash-dotted lines) stress for group $S1$ (left) and $S2$ (right). Dashed vertical line shows the wall-normal location of the crest of the largest roughness in the group ($y^+ = h^+$).

The stress profiles for the total (figures 2 (a, d)) and turbulent (figures 2 (b, e)) fluctuation in the azimuthal direction for the rough cases collapses with the smooth wall in the outer region of the flow. Good collapse in the outer region of the flow is also obtained for the stress profiles of the total and turbulent fluctuations in the streamwise and radial direction (not shown here). Good collapse is observed as the total fluctuation and turbulent stresses scale with the friction Reynolds numbers. Therefore, our findings support Townsend's outer-layer similarity hypothesis when considering the second-order statistics. However, when analysing the contours of the streamwise premultiplied energy spectra, it is observed that the collapse occurs at a much larger wall-normal location as energy is redistributed to the wavelength corresponding to the roughness elements and its harmonics [10].

In general, the maximum coherent stresses in all three dimensions (streamwise and radial components, not shown here) increases with increasing geometrical scale ($S1$) and increasing wavelength ($S2$). In addition, with increasing h^+ and λ^+, the influence of the coherent component is felt further away from the roughness and for case 60_848, extends even into the edge of the outer layer. As for the turbulent stresses, the profiles for the rough cases above the roughness elements are not too dissimilar with the smooth case profiles as the turbulent stresses scales predominantly with the friction Reynolds number.

5. Streamwise vorticity

The azimuthal coherent fluctuation stress profile for case 60_848 extends far above the crest of the roughness elements due to large secondary flows. Isosurfaces of the time-averaged streamwise vorticity $\overline{\omega}_x^+$ of the unwrapped surface for case 60_848 are plotted in figures 3 (a, b). There is strong streamwise vorticity near the surface of the roughness which alternates between positive and negative values in the streamwise direction due to the topological changes of the three-dimensional sinusoidal roughness (figure 3 (a)). Plotting isosurfaces of the time-averaged streamwise vorticity where $|\overline{\omega}_x^+| = 0.01$ in figure 3 (b), we observe large secondary flows which extend up to 200 viscous units from the wall. Figures 3 (c, d) shows the contours of the time-averaged streamwise vorticity overlaid with the time-averaged azimuthal and radial velocity vectors along cross sectional slice I where the surface is locally rough and II where the surface is locally smooth. The zoom in view of these contours clearly shows the occurrence of these secondary flows which are similar to that observed in duct flows [11]. The sketch in figure 3 (e) illustrates the orientation of the secondary flow which alternates in the spanwise direction. Fluid is pushed upwards from the wall along the locally rough streamwise plane and is pushed downwards in the locally smooth streamwise plane. Secondary flows are also observed in the other rough cases (not shown here) but is confined to regions near the crest of the roughness. Large secondary flows occur in case 60_848 due to the spanwise heterogeneity of the roughness elements which are spaced approximately equal to the boundary layer thickness $\delta^+ = R_0^+ = 540 \approx L_s^+ = 424$ (for a pipe, the δ is fixed and is equal to the reference radius of the pipe R_0). This is consistent with the findings of [12] who found that δ-spaced roughness elements induces secondary flows on the scale of δ.

6. Conclusion

DNS in a rough wall pipe with explicitly gridded three-dimensional sinusoidal roughnesses is simulated at moderate Reynolds numbers where the height h^+ and the wavelength λ of the roughness elements are systematically altered. Increasing the roughness height and/or roughness wavelength results in an increase in the coherent fluctuation stresses. For case 60_848, the coherent fluctuation stress profile extends much further above the crest of the roughness elements compared to the other rough cases due to the occurrence of large time invariant secondary flows. These large secondary flows develop in the pipe due to the spanwise heterogeneity of the roughness which has a spanwise spacing approximately equal to the boundary layer thickness.

References

[1] Reynolds W C and Hussain A K M F 1972 *J. Fluid Mech.* **54** 263–288
[2] Coceal O and Belcher S E 2004 *Quart. J. Roy. Meteorol. Soc.* **130** 1349–1372
[3] Raupach M R and Shaw R H 1982 *Bound-Layer Meteor.* **22**(1) 79–90
[4] Finnigan J J 1985 *The Forest-Atmosphere Interaction* (Springer) pp 443–480
[5] Finnigan J J 2000 *Annu. Rev. of Fluid Mech.* **32** 519–571
[6] Nikora V, Goring D, McEwan I and Griffiths G 2001 *J. Hydraul. Eng.* **127**(2) 123–133
[7] Nikora V, McEwan I, McLean S, Coleman S, Pokrajac D and Walters R 2007 *J. Hydraul. Eng.*
[8] Mignot E, Barthelemy E and Hurther D 2009 *J. Fluid Mech.* **618** 279–303
[9] Chan L, MacDonald M, Chung D, Hutchins N and Ooi A 2015 *J. Fluid Mech.* **771** 743–777
[10] Chan L, MacDonald M, Hutchins N, Chung D and Ooi A 2015 *International Symposium on Turbulence and Shear Flow Phenomena (TSFP-9)* pp 2A–3
[11] Brundrett E and Baines W D 1964 *J. Fluid Mech.* **19**(3) 375–394
[12] Vanderwel C and Ganapathisubramani B 2015 *J. Fluid Mech.* **774** R2–1–R2–12

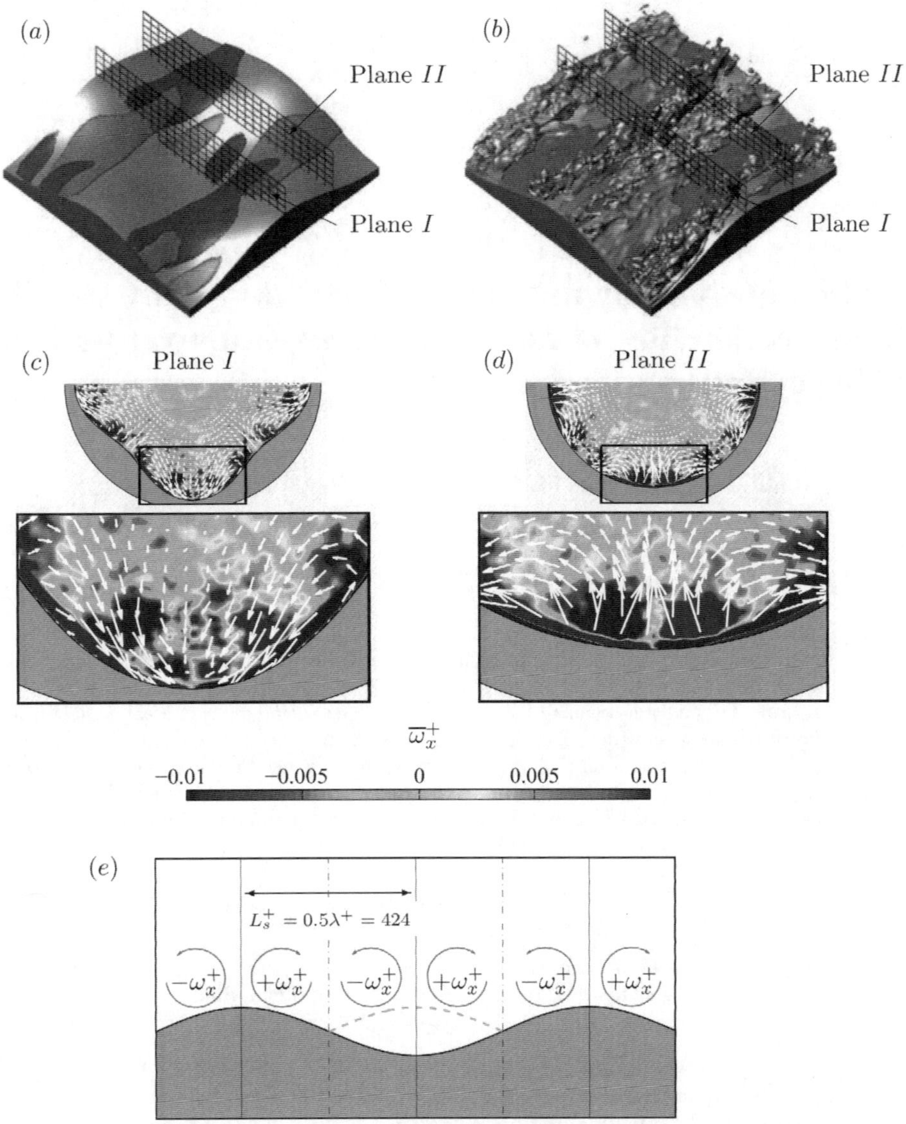

Figure 3. Isosurfaces of the time-averaged streamwise vorticity of the unwrapped rough wall pipe for case 60_848 are shown in (a) and (b) where $|\overline{\omega}_x^+| = 0.1$ and $|\overline{\omega}_x^+| = 0.01$ respectively. Contours of the time-averaged streamwise vorticity $\overline{\omega}_x^+$ for case 60_848 at cross-sectional (a) plane I (locally rough), (b) plane II (locally smooth). Overlaid on top of the contour is the time-averaged azimuthal and radial velocity vectors. Black solid box shows the zoom in view. Flow is into the page. The cross-sectional sketch of the unwrapped pipe in (e) illustrates the secondary flow occurring between the roughness elements where the solid vertical and dash-dotted lines are streamwise planes which are locally rough and smooth respectively. Grey dashed line represents the roughness element located downstream.

Application of shooting method on MHD thermally stratified mixed convection flow of non-Newtonian fluid over an inclined stretching cylinder

Khalil-Ur-Rehman and M.Y. Malik

Department of Mathematics, Quaid-i-Azam University Islamabad 44000, Pakistan

E-mail: krehman@math.qau.edu.pk

Abstract: An analysis is made to examine the magnetohydrodynamic mixed convection boundary layer flow of Eyring-Powell fluid brought by an inclined stretching cylinder. Flow field analysis is accounted by thermal stratification phenomena. The temperature is assumed to be higher across the surface of cylinder as compared to ambient fluid. The arising mathematical model regarding Eyring-Powell fluid is governed by interesting physical parameters which includes mixed convection parameter, thermal stratification parameter, heat generation/absorption parameter, curvature parameter, fluid parameters, magnetic field parameter and Prandtl number. The numerical solutions are computed through the application of shooting technique conjunction with fifth order Runge-Kutta algorithm. In addition, numeric values for two unlike geometries namely, plate and cylinder for skin friction coefficient and Nusselt number are presented with the aid graphs and some particular cases are discussed. The present study is validated by establishing comparison with previously published works, which sets a benchmark of quality of shooting method.

Keywords: Shooting Method; MHD; Thermal Stratification; Mixed Convection; Eyring-Powell Model; Heat Generation/Absorption, An Inclined Stretching Cylinder.

1. Introduction

The mutual interaction of fluid flow and magnetic field relates MHD. Such type of analogy comprises those fluids which are electrically conducting and nonmagnetic comprise liquid metals, strong electrolytes and hot ionized gases. Magnetohydrodynamic fluid flows have various applications in the area of polymer and metallurgical industry. The cooling of filaments or continuous strips confirms the metallurgical use, for example process involves drawing of and thinning of copper wires. The characteristics of final products highly depends on cooling rate. Drawing such strips under magnetic field, we can control stretching and rate of cooling. In a same way many other properties of such materials can be adjusted by iterating magnetic field. Pavlov [1] studied the magnetohydrodynamic fluid flow past over a stretching surface and provided an exact solution of momentum equation. Kumari and Nath [2] examined the magnetohydrodynamic effects on a non-Newtonian fluid flow over a continuously moving surface. The researchers are still showing interest to study magnetohydrodynamic flows over a plane geometry, as far as cylindrical geometry is concern, Ishak et al. [3] explored the magnetohydrodynamic heat mass transfer flow induced by stretching cylinder. Mukhopadhyay et al. [4] identified the characteristics of magnetohydrodynamic boundary layer flow by way of stretching cylinder. Malik et al. [5] reported the magnetohydrodynamic flow of tangent hyperbolic fluid past over a stretching cylinder.

Convective heat transfer involves both heat diffusion known as conduction and by bulk heat transfer of fluid known as advection. Convection heat transfer may be natural, forced or mixed. Convection phenomena play a key role in practical life. The chemical transportation in packs bed reactors, drying of porous solid and solar power collectors to mention just a few are the practical applications of mixed convection phenomena. So that, the mixed convection was identified by many scientists as Heckel et al. [6] studied mixed convection flow along slender vertical cylinder by considering variable surface temperature. The characteristics of heat transfer along mixed convection flow of non-Newtonian fluid past a vertical plate was taken by Wang [7]. Chen [8] examined the laminar mixed convection flow over a vertical continuously stretching sheet. Seddeek et al. [9] reported mixed convection flow past a continuously stretching vertical cylinder by considering variable viscosity and thermal diffusivity. The properties of mixed convection flow past a permeable vertical cylinder along surface heat flux was provided by Bachok and Ishak [10]. Hsiao [11] examined the magnetohydrodynamic mixed convection flow of viscoelastic fluid over a porous wedge. Mukhopadhyay and Ishak [12] explored the mixed convection flow past a stretching cylinder by means of thermal stratification. The mixed convection flow of viscoelastic fluid over a stretching cylinder was discussed by Hayat et al. [13] . The mixed convection nanofluid flow over an inclined stretching cylinder with thermal slip effects by way of Buongiorno's model was considered by Dhanai et al. [14].

Stratification has an abundant applications in real world. For example environmental heat rejection such as seas, lakes and rivers, thermal energy storage systems similar to solar ponds etc. So, due to huge implementations in fluid mechanics many researchers probed stratification phenomenon. Whereas geology, biology, oceanography, agriculture, astrophysics and many chemical processes also enclosed double diffusion occurrence. In addition, closed containers, ecological heated walls chambers are supported by convectional environments with thermal and solutal stratification. Such a stratification of the medium is due to temperature and concentration differences, which results density variation in the medium. In short, stratification contribute a vital role in many industrial and natural phenomena. The double stratification flow of nanofluid over a vertical plate was taken by Ibrahim and Makinde [15]. The radiative Jeffrey fluid flow near a stretching sheet with double stratification effects was examined by Hayat et al. [16]. The influence of magnetohydrodynamic flow of micropolar fluid was presented by Srinvasacharya and Upendra [17].

In 1944, Eyring and Powell introduced a fluid model known as Eyring-Powell fluid model [18]. Even though it is complex model but it has certain advantages over a non-Newtonian models in this sense, it is derived from kinetic theory of liquid rather than the experimental relation and turn into Newtonian mode of behaviour for both low and high shear rates. The Eyring-Powell model with heat transfer play a important role in different industrial, natural and geophysical processes. Such a processes contains dispersion and formulation of fog, groves of fruit trees, crops damaging period due to freezing, thermal insulation and underground energy transport etc. A countable attempts has been made to study Eyring-Powell fluid flow over a stretching cylinder. To mention just a few, Malik et al. [19] examined the Eyring-Powell fluid past a stretching cylinder along variable viscosity effect. Recently, Hayat et al. [20] studied the Eyring-Powell magnetohydrodynamic nanofluid flow due to stretching cylinder by considering thermal radiation effect. More recently, Rehman et al. [21] identified stagnation point Eyring-Powell fluid flow over a vertical cylinder.

The above mentioned literature assessment indicates that the boundary layer flow of non-Newtonian fluid (Eyring-Powell) over an inclined stretching cylinder is not investigated widely, as yet. Therefore, the main focus of present study is to analyse magnetohydrodynamic mixed convectional thermally stratified flow of Eyring-Powell fluid induced by an inclined stretching cylinder by way of heat generation/absorption effect. The flow directing differential equations are converted into a system of non-linear ordinary differential equations by using suitable transformations. The numerical solution is obtained by shooting method with the aid of fifth order Runge-Kutta algorithm. The effect logs regarding achieved results for velocity and temperature profiles are presented graphically. The numeric values for local skin friction coefficient and Nusselt number against two different geometries namely, plate and cylinder are reported with the aid of graphs, which is very essential for industrial application point of view. It is also trusted here, that the results found will not only offer convenient information for applications, but also serve as a complement to the preceding studies.

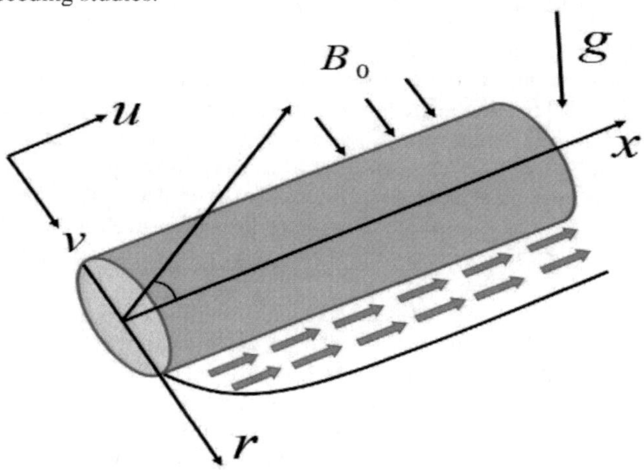

Figure 1. The physical model and coordinate's system.

2. Flow Field Analysis

In this study, we have consider steady two dimensional, an incompressible boundary layer flow of Eyring-Powell fluid induced by an inclined stretching cylinder. Flow field analysis is taken with thermal stratification phenomena. The temperature is supposed to be higher near the cylinder surface as compare to ambient fluid. For the Eyring-Powell fluid model, the rheological equation of state is given as:

$$\Gamma = [\, \mu + \frac{1}{\beta \gamma^1} \sinh^{-1}\left(\frac{1}{c}\gamma^1\right) \mathbf{A}_1, \tag{1}$$

$$\gamma^1 = \sqrt{\frac{1}{2} tr(\mathbf{A}_1)^2}, \tag{2}$$

The second order approximation for $\sinh^{-1}(\backslash.)$ function is estimated as:

$$\sinh^{-1}(\frac{1}{c}\gamma^1) \cong \frac{\gamma^1}{c} - \frac{\gamma^{1^3}}{6c^3}, \text{ where } \left|\tfrac{1}{c}\gamma^1\right| << 1. \tag{3}$$

The steady two dimensional equations under boundary layer approximation for Eyring-Powell fluid flow, in usual notation are given as follows:

$$\frac{\partial(ru)}{\partial x} + \frac{\partial(rv)}{\partial r} = 0, \tag{4}$$

$$u\frac{\partial u}{\partial x}+v\frac{\partial u}{\partial r}=\left(\nu+\frac{1}{\beta\rho c}\right)\frac{\partial^2 u}{\partial r^2}-\frac{1}{2\beta c^3\rho}\left(\frac{\partial u}{\partial r}\right)^2\frac{\partial^2 u}{\partial r^2}+\frac{1}{r}\left(\nu+\frac{1}{\beta\rho c}\right)\frac{\partial u}{\partial r}-\frac{1}{6\beta r\rho c^3}\left(\frac{\partial u}{\partial r}\right)^3$$
$$-\frac{\sigma B_0^{\,2}}{\rho}u+(g\beta_T\left(T-T_\infty\right))\cos\alpha, \quad (5)$$
$$u\ \frac{\partial T}{\partial x}+v\ \frac{\partial T}{\partial r}=\frac{\alpha'}{r}\frac{\partial}{\partial r}\left(r\frac{\partial T}{\partial r}\right)+\frac{Q_0}{c_p\rho}(T-T_\infty), \quad (6)$$

The axial line of cylinder is consider as x-axis and radial direction perpendicular to axial line is taken as r-axis. So that, u, v, ν, g, β_T, α, c_p, ρ, Q_0, and α', denotes velocity components in the x and r direction, kinematic viscosity, gravity, coefficient of thermal expansion, inclination, specific heat at constant pressure, fluid density, heat generation/absorption coefficient and thermal diffusivity respectively. Further, A_1, tr, c, β and μ denotes first Rivlin-Ericksen tensor, trace, fluid parameters and dynamic viscosity respectively.

The endpoint conditions are given as:

$$u\left(x,r\right)=U(x)=\frac{U_0}{L}x\ ,\ \ v(x,r)=0\ \text{ at } r=R,$$
$$u\left(x,r\right)\to 0\ \text{ at }\ \ r\to\infty,$$
$$T\left(x,r\right)=T_w(x)=T_0+\frac{b\,x}{L}\ ,\quad \text{at } r=R, \quad (7)$$
$$T\left(x,r\right)\to T_\infty(x)=T_0+\frac{c\,x}{L}\ ,\quad \text{as }\ r\to\infty.$$

To trace out the solution of Eqs. (5)-(6) under end point conditions Eq. (7), we have considered following transformations:

$$u=\frac{U_0 x}{L}F'(\eta),\ v=-\frac{R}{r}\sqrt{\frac{U_0\nu}{L}}F\left(\eta\right),\ \eta=\frac{r^2-R^2}{2R}\left(\frac{U_0}{\nu L}\right)^{\frac{1}{2}},$$
$$\psi=\left(\frac{U_0\nu x^2}{L}\right)^{\frac{1}{2}}R\,\mathrm{F}(\eta),\quad T\left(\eta\right)=\frac{T-T_\infty}{T_w-T_0}, \quad (8)$$

where $T_w(x)$, $T_\infty(x)$, T_0, U_0, L, $F(\eta), F'(\eta)$ denotes prescribed surface temperature, variable ambient temperature, reference temperature, free stream velocity, reference length, dimensionless variable, velocity of fluid over an inclined stretching cylinder because prime denotes differentiation with respect to η (similarity variable) and b, c, are positive constants, ψ is the stream function, which identically satisfies the continuity Eq. (4) and is defined as:

$$u=\frac{1}{r}\left(\frac{\partial\psi}{\partial r}\right),\quad v=\frac{-1}{r}\left(\frac{\partial\psi}{\partial x}\right), \quad (9)$$

by substituting Eqs. (8)-(9) in Eqs. (5)-(6), we get:

$$FF''-(F')^2+(1+2K\eta)(1+M)F'''+2K(1+M)F''-\frac{4}{3}\lambda MK(1+2K\eta)(F'')^3- \\ M\lambda(1+2K\eta)^2(F'')^2F'''-\gamma^2F'+\lambda_T T(\eta)\cos\alpha=0, \tag{10}$$

$$2KT'+(1+2K\eta)T''+\Pr(FT'-F'T-F\delta_1+\delta T)=0, \tag{11}$$

the transformed endpoint conditions are:

$$F'=1\ ,\ F=0\ ,\ T=1-\delta_1 \text{ at } \eta=0, \\ F'\to 0\ ,\ T\to 0 \text{ as } \eta\to\infty, \tag{12}$$

where γ, K, M, λ, λ_T, Pr, δ_1, δ, denotes magnetic field parameter, curvature parameter, fluid parameters, mixed convection parameter, Prandtl number, thermal stratification parameter and heat generation/absorption parameter respectively and defined as follows:

$$\gamma=\sqrt{\frac{\sigma B_0^{\,2}}{\rho a}},\ K=\frac{1}{R}\sqrt{\frac{\nu}{a}},\ M=\frac{1}{\mu\beta c},\ \lambda=\frac{a^3x^2}{2c^2\nu}, \\ \lambda_T=\frac{Gr}{\mathrm{Re}_x^{\,2}},\ \Pr=\frac{\nu}{\alpha},\ \delta_1=\frac{c}{b},\ \text{and } \delta=\frac{LQ_0}{U_0\rho c_p}. \tag{13}$$

Furthermore, Gr denotes thermal Grashof number which is defined as:

$$Gr=\frac{g\beta_T(T_w-T_0)x^3}{\nu^2}. \tag{14}$$

The skin friction coefficient at the surface of cylinder is considered as:

$$C_f=\frac{\tau_w}{\rho\frac{U^2}{2}},\quad \tau_w=\left[\mu\left(\frac{\partial u}{\partial r}\right)+\frac{1}{\beta c}\frac{\partial u}{\partial r}-\frac{1}{6\beta c^3}\left(\frac{\partial u}{\partial r}\right)^3\right]_{r=R}, \tag{15}$$

where τ_w and μ denotes shear stress and viscosity of fluid respectively. The dimensionless form of skin friction coefficient is prearranged as:

$$\frac{1}{2}C_f\,\mathrm{Re}_x^{1/2}=(1+M)F''(0)-\frac{M\lambda}{3}(F''(0))^3, \tag{16}$$

with $\mathrm{Re}_x=\dfrac{U_0x^2}{\nu L}$ as a local Reynolds number.

The local Nusselt number is defined as:

$$Nu_x=\frac{xq_w}{k(T_w-T_\infty)},\quad q_w=-k\left(\frac{\partial T}{\partial r}\right)_{r=R}, \tag{17}$$

in dimensionless form it can be written as:

$$Nu_x\,\mathrm{Re}_x^{-1/2}=-T'(0). \tag{18}$$

3. Special Cases

3.1. Case-A

A mathematical formulation for steady two dimensional flow of Eyring-Powell fluid brought by an inclined stretching cylinder is given by Eq. (10). In the absence of magnetic field parameter and curvature parameter we obtained steady, two dimensional an incompressible mixed convection flow of Eyring-Powell fluid past over a inclined stretching plate. So that, the comparative study of cylindrical and plane geometry is examined and plotted through figures 11-12 for skin friction coefficient and heat transfer rate respectively. Mathematically by using $\gamma = 0$ and $K = 0$, Eq. (10) takes the form:

$$FF''-(F')^2+(1+M)F'''+M\lambda(F'')^2F'''+\lambda_T T(\eta)\cos\alpha=0,$$
$$T''+\Pr(FT'-F'T-F\delta_1+\delta T)=0, \tag{19}$$

with boundary conditions:

$$F'=1\ ,\ \ F=0\ ,\ \ T=1-\delta_1 \quad \text{at} \quad \eta=0,$$
$$F'\to 0\ ,\ \ T\to 0 \quad \text{as} \quad \eta\to\infty. \tag{19a}$$

3.2. Case-B

In addition, by incorporating inclination $\alpha=0$ with mixed convection parameter $\lambda_T=0$, we achieved two dimensional steady, an incompressible flow of Eyring-Powell fluid past over a stretching palte. Mathematics for such a physical problem is given by:

$$FF''-(F')^2+(1+M)F'''+M\lambda(F'')^2F'''$$

with boundary conditions

$$F'=1\ ,\ \ F=0\ ,\ \ \text{at} \quad \eta=0,$$
$$F'\to 0\ ,\ \ \text{as} \quad \eta\to\infty. \tag{20}$$

Which was discussed by Javed at al. **[24]**

4. Shooting Method

The non-linear boundary value problem (BVP) given in Eq. (10) and (11) subjected to endpoint conditions Eq. (12) has been computed numerically by employing shooting method with the aid of fifth order Runge-Kutta algorithm. We reduce Eq. (10) and (11) into a system of five first order simultaneous equations by appropriate letting i-e

$x_2 = F'$,

$x_3 = x'_2 = F''$,

$x_5 = T'$,

the equivalent form of Eqs. (10) and (11) under newly signed variables is given by:

$$\begin{bmatrix} x'_1 = x_2 \\ x'_2 = x_3 \\ x'_3 = \dfrac{(x_2)^2 - x_1x_3 - (2K)(1+M)x_3 + \frac{4}{3}\lambda MK(1+2K\eta)x_3^{\ 3} + \gamma^2 x_2 - \lambda_T x_4 \cos\alpha}{(1+2K\eta)(1+M) - M\lambda(1+2K\eta)^2 x_3^{\ 2}} \\ x'_4 = x_5 \\ x'_5 = \dfrac{\Pr(x_2x_4 + \delta_1 x_2 - x_1x_5 - \delta x_4) - 2Kx_5}{1+2K\eta} \end{bmatrix}, \tag{21}$$

the corresponding endpoint conditions regarding new variables are given as follows:

$$\begin{aligned} &x_1(0) = 0, \\ &x_2(0) = 1, \\ &x_3(0) = unknown, \\ &x_4(0) = 1-\delta_1, \\ &x_5(0) = unknown. \end{aligned} \tag{22}$$

For the integration of first order system i-e Eq. (21) as a IVP with the aid of fifth order Runge-Kutta scheme, we required values for $x_3(0)$ i.e. $F''(0)$, and $x_5(0)$ i.e. $T'(0)$. Newton's method is than used to estimate the values of $F''(0)$ and $T'(0)$ till the solution approaches zero with the desired efficiency of 10^{-4} along step size $\Delta\eta = 0.025$ by way of additional conditions given as:

$$\begin{aligned} &x_2(\infty) = 0, \\ &x_4(\infty) = 0. \end{aligned} \tag{23}$$

5. Results and Discussion

5.1. Velocity profiles

Table 1 includes the assessment sample of our results for heat transfer rate corresponding to different values of Prandtl number *Pr* with results obtained in **[20], [22]-[23].** An excellent match has been found among existing work, which yields a conformity of present results. The influence of magnetic field parameter on velocity profile is presented in figure 2. It is seemed that the velocity profile decreases for higher values of magnetic field parameter γ. Physically, when we increase magnetic field parameter γ the Lorentz force increases, which is resistive force. As a result fluid velocity decreases. In figure 3**,** velocity profile is evaluated at different values of thermal stratification parameter δ_1. It is evident that the velocity profile decreases for higher values of thermal stratification parameter δ_1. Essentially, this effect is due to decline in convective potential between surface of cylinder and ambient temperature. Figure 4 includes the effect of mixed convection parameter λ_T on velocity profile. It is noticed that for higher iterations of mixed convection parameter λ_T, velocity of the fluid increases. Physically, it is due to inciting of thermal buoyancy force. The impact of fluid parameter *M* is described through figure 5**.** It is found that the velocity of fluid increases for increasing values of fluid parameter M, because fluid parameter M has inverse relation towards viscosity of fluid, so increase in fluid parameter M turns fluid to be less viscous and hence deformation rate increases which yields increase in velocity. Figure 6 presents the influence of curvature parameter K on velocity profile. It is observed that the higher values of curvature parameter K is the cause of increase in velocity profile. Curvature parameter K is inversely proportional to radius of curvature. So when we increase curvature parameter, the radius of cylinder decreases and hence contact surface area of cylinder with fluid reduces which offers less resistance to fluid flow. So increase in curvature parameter K causes increase in velocity profile.

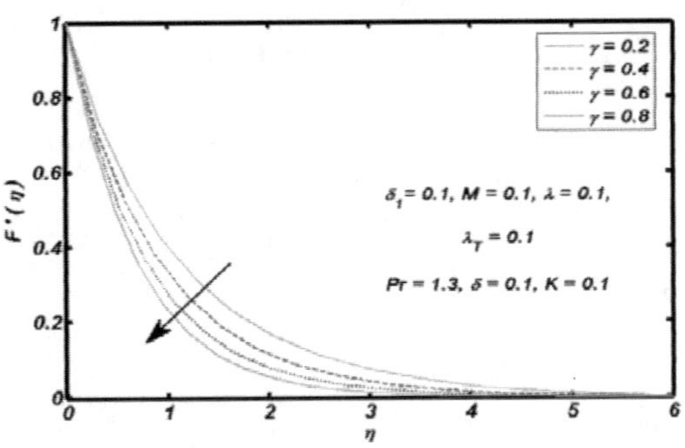

Figure 2. Influence of γ on velocity profile.

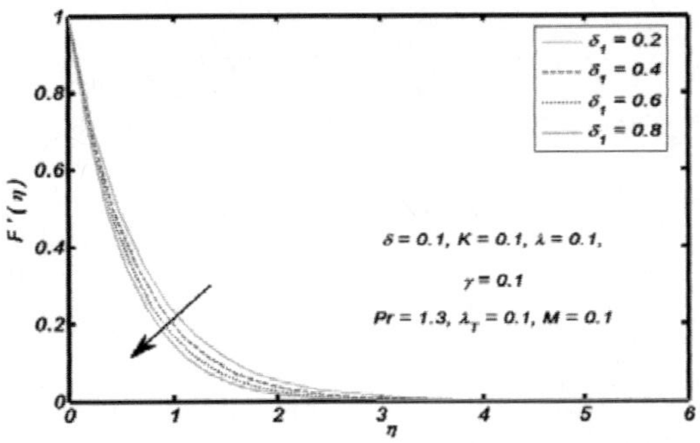

Figure 3. Influence of δ_1 on velocity profile.

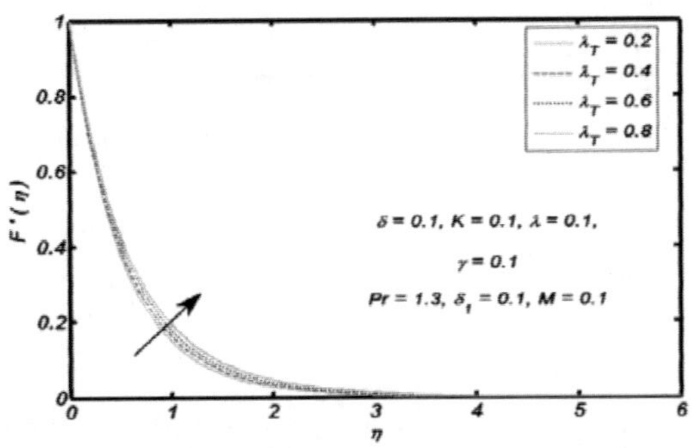

Figure 4. Influence of λ_T on velocity profile.

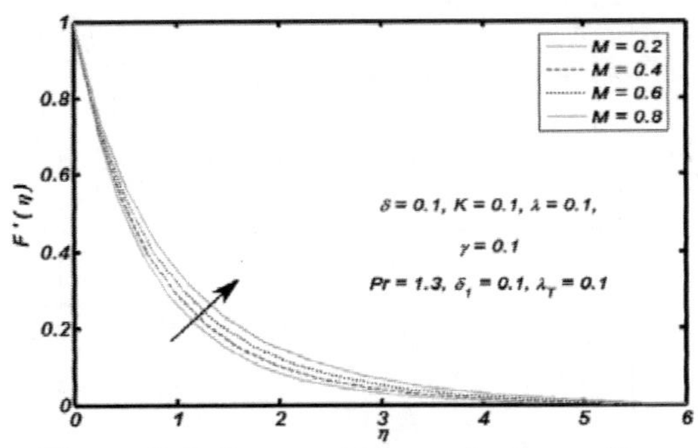

Figure 5. Influence of M on velocity profile.

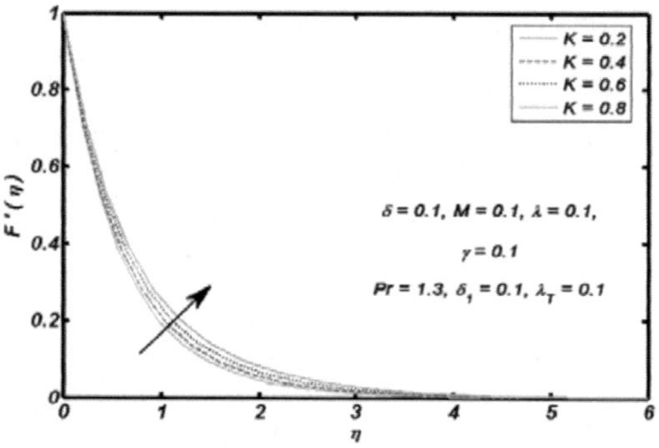

Figure 6. Influence of K on velocity profile.

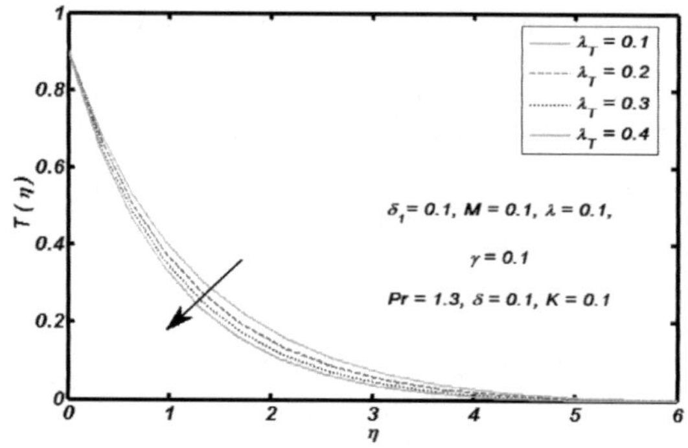

Figure 7. Influence of λ_T on temperature profile.

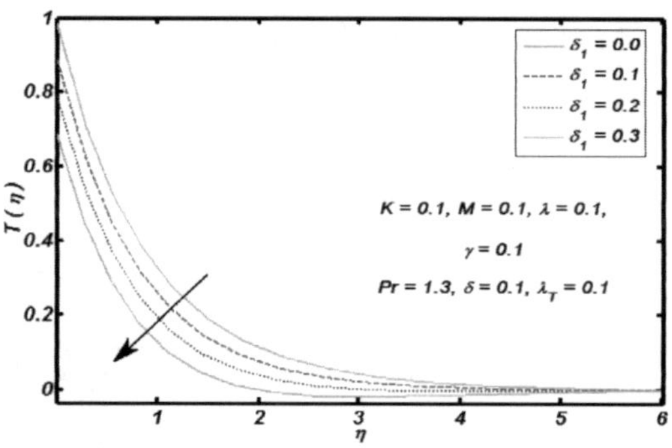

Figure 8. Influence of δ_1 on temperature profile.

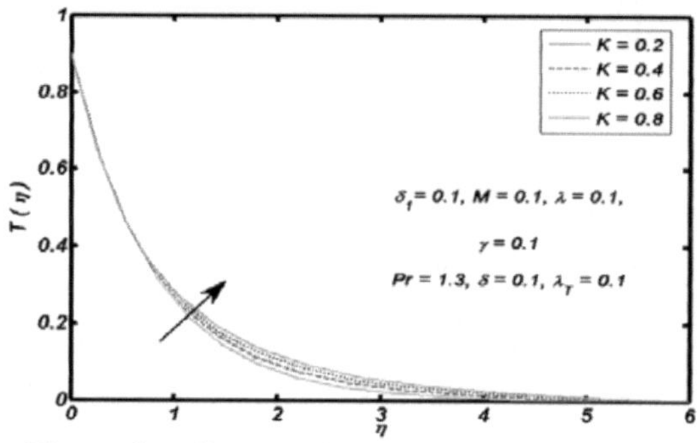

Figure 9. Influence of K on temperature profile.

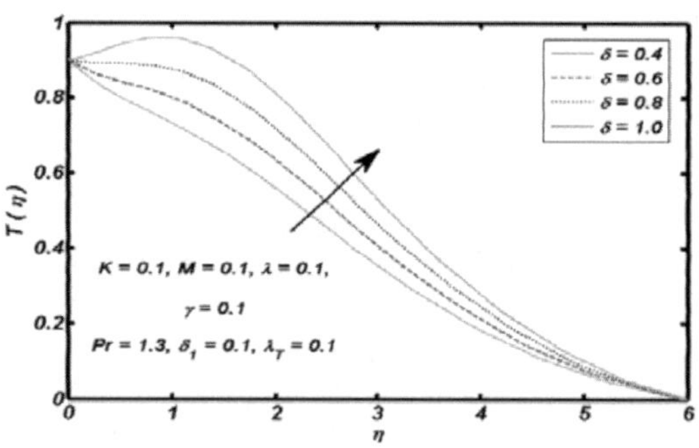

Figure 10. Influence of δ on temperature profile.

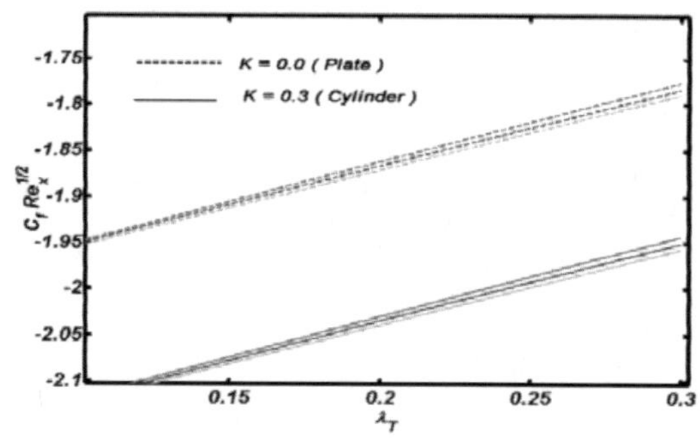

Figure 11. Influence of λ_T and δ on skin friction coefficient.

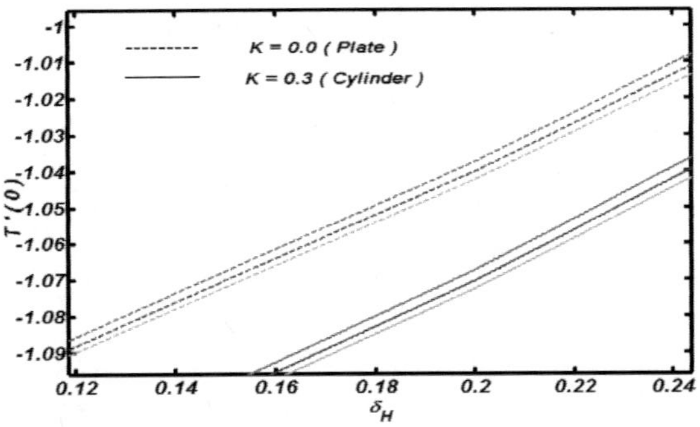

Figure 12. Effect of α and δ on local Nusselt number.

Table 1. Comparison value of $-T'(0)$ against *Pr*.

Pr	*Hayat et al. [20]*	*Ishak and Nazar [22]*	*Vajravelu et al. [23]*	*Current results*
1	*1.00000*	*1.0000*	*1.000002*	*1.0021*
1.1	-	-	-	*1.0924*
1.2	*1.120662*	-	-	*1.1589*
1.4	*1.2135*	-	-	*1.2235*

5.2. Temperature profiles

In figure 7, the effect of mixed convection parameter λ_T on temperature profile is presented. It is found that increasing values of mixed convection parameter λ_T corresponds decrease in temperature profile. When mixed convection parameter λ_T increases, thermal buoyancy forces enhanced which results inciting in heat transfer rate and hence temperature profile decreases. The influence of thermal stratification parameter δ_1 is depicted in figure 8, it is observed that the temperature profile decreases for increase in thermal stratification parameter δ_1. This is due to drop of temperature difference between surface of cylinder and ambient fluid therefore, temperature profile decreases. Figure 9 paints that the temperature profile increases for increase in curvature parameter K. As Kelvin temperature is defined as an average kinetic energy so when we increase curvature of cylinder, velocity of the fluid increases, resultantly kinetic energy increases and due to this temperature increases. Note that temperature of fluid decreases near the surface of cylinder and increases away from the surface. The effects of heat generation/absorption parameter δ on temperature profile can be seen through figure 10. It is clearly seen that for altered values heat generation/absorption parameter δ temperature of fluid increases. During heat generation process energy is produced which brings enhancement in temperature distribution. In addition, for higher values of heat generation/absorption parameter δ, over shoot in temperature profile is observed. Such type of overshoot can be controlled by introducing heat sink which helps to reduce temperature of fluid. Figure 11 includes the plots of skin friction coefficient verses mixed convection parameter and heat generation\absorption parameter ($\delta = 0.2, 0.3, 0.4$). It is worth pointing here the skin friction decreases (in absolute sense) for larger values of mixed convection parameter λ_T and shows opposite attitude towards heat generation\absorption parameter δ. It is also observed that the magnitude in case of cylinder is higher than the plate. Figure 12 witnessed that the heat transfer rate decreases (absolute sense) for higher values of inclination (i-e $\alpha = 30^0, 45^0, 60^0$) and heat generation/absorption parameter δ. The significant enhancement of local Nusselt number is observed in case of cylinder as compare to plate.

6. Concluding Remarks

The shooting method conjunction with fifth order Runge-Kutta algorithm has been successfully applied to perform a comparative study of mixed convectional thermally stratified boundary layer flow of Eyring-Powell fluid induced by an inclined stretching cylinder by means of heat generation/absorption effect. The behaviour of dimensionless velocity and temperature profiles are identified under different physical parameters. The key results of this study are summarized and itemized as follows:

- The velocity profile increases for higher values of mixed convection parameter λ_T, curvature parameter K and fluid parameter M, while it shows opposite behaviour towards magnetic field parameter γ and thermal stratification parameter δ_1.
- The temperature profile increases significantly for larger values of heat generation/absorption parameter δ and curvature parameter K, whereas it shows decline via mixed convection parameter λ_T and thermal stratification parameter δ_1.
- The skin friction coefficient shows decreasing behaviour towards mixed convection parameter λ_T and increasing via heat generation/absorption parameter δ.
- The local Nusselt number is decreasing function of mixed convection parameter λ_T and heat generation/absorption parameter δ.
- It is seemed that magnitude of skin friction as well as heat transfer rate is remarkably large for cylinder as compare to plane geometry (i-e plate).
- The computed results are validated for heat transfer rate by iterating Prandtl number *Pr,* which leads to surety of the present work.

Acknowledgment

The authors would like to thank the reviewers for their valuable suggestions and also the financial support provided by Higher Education Commission (HEC), Pakistan.

References

[1] Pavlov K B 1974 Magnetohydrodynamic flow of an incompressible viscous fluid caused by deformation of a plane surface *Magnitnaya Gidrodinamika* **4** 146-147
[2] Kumari M and Nath G 2001 MHD boundary-layer flow of a non-Newtonian fluid over a continuously moving surface with a parallel free stream *Acta Mech* **146** 139-150
[3] Ishak A, Nazar R, Pop I 2008 Magnetohydrodynamic (MHD) flow and heat transfer due to a stretching cylinder *Energy Convers Manage* **49** 3265–3269
[4] Mukhopadhyay S, Golam Arif M, Wazed Ali Pk M 2013 MHD boundary layer flow along a stretching cylinder *Acta Technica* **58** No. 3.

[5] Malik M Y, Salahuddin T, Hussain A, Bilal S 2015 MHD flow of tangent hyperbolic fluid over a stretching cylinder: Using Keller box method *Journal of Magnetism and Magnetic Materials* **395** 271–276
[6] Heckel J J, Chen T S, Armaly B F 1989 Mixed convection along slender vertical cylinders with variable surface temperature *International Journal of Heat and Mass Transfer* **32** 431-1442
[7] Wang T Y 1995 Mixed convection from a vertical plate to non-Newtonian fluids with uniform surface heat flux *International Communications in Heat and Mass Transfer* **22** 369-380
[8] Chen CH 1998 Laminar mixed convection adjacent to vertical, continuously stretching sheet, *International Journal of Heat and Mass Transfer* **33** 471-476
[9] Seddeek M A, Salem A M 2005 Laminar mixed convection adjacent to vertical continuously stretching sheet with variable viscosity and variable thermal diffusivity *Heat and Mass Transfer* **41** 1048-1055
[10] Bachok N and Ishak A 2009 Mixed convection boundary layer flow over a permeable vertical cylinder with prescribed surface heat flux *European Journal of Scientific Research* **34** 46-54
[11] Hsiao K L 2011 MHD mixed convection for viscoelastic fluid past a porous wedge *International Journal of Non-Linear Mechanics* **46** 1-8
[12] Mukhopadhyay S and Ishak A 2012 Mixed Convection Flow along a Stretching Cylinder in a Thermally Stratified Medium *Journal of Applied Mathematics* doi:10.1155/2012/491695
[13] Hayat T, Anwar M S, Farooq M and Alsaedi A 2015 Mixed Convection Flow of Viscoelastic Fluid by a Stretching Cylinder with Heat Transfer *Plos One* **10** doi:10.1371/journal.pone.0118815.
[14] Dhanai R, Rana P, Kumar L 2016 MHD mixed convection nanofluid flow and heat transfer over an inclined cylinder due to velocity and thermal slip effects: buongiorno's model *Powder Technol* **288** 140–150
[15] Ibrahim W, Makinde O D 2013 The effect of double stratification on boundary layer flow and heat transfer of nanofluid over a vertical plate *Computers and Fluids* **86** 433-441
[16] Hayat T, Hussain T, Shehzad S A, Alsaedi A 2014 Thermal and concentration stratifications effects in raditaive flow of Jeffrey fluid over a stretching sheet *Plos One* **9** 1-15 doi:10.1371/journal.pone.0107858
[17] Srinivasacharya D, Upendar M 2013 Effect of double stratification on MHD free covection in a micropolar fluid Journal of the *Egyptian Mathematical Society* **21** 370-378
[18] Powell R E and Eyring H 1944 Mechanisms for the relaxation theory of viscosity Nature **154** 427-428
[19] Malik M Y, Hussain A, Nadeem S 2013 Boundary layer flow of an Eyring–Powell model fluid due to a stretching cylinder with variable viscosity *Scientia Iranica* **20** 313–321
[20] Hayat T, Gull N, Farooq M, Ahmad B 2015 Thermal Radiation Effect in MHD Flow of Powell-Eyring Nanofluid Induced by a Stretching Cylinder *Journal of Aerospace Engineering* doi:10.1061/(ASCE)AS.1943-5525.0000501
[21] Rehman A, Achakzai S, Nadeem S, Iqbal S 2016 Stagnation point flow of Eyring Powell fluid in a vertical cylinder with heat transfer *Journal of Power Technologies* **96** 57–62
[22] Ishak A and Nazar R 2009 Laminar boundary layer flow along a stretching cylinder *European Journal of Science* **36** 22-29
[23] Vajravelu K, Prasad K V, Santhi S R 2012 Axisymmetric magneto-hydrodynamic (MHD) flow and heat transfer at a nonisothermal stretching cylinder *Applied Mathematics and Computation* **219** 3993-4005
[24] Javed T, Ali N, Abbas Z, Sajid M 2013 Flow of an Eyring-Powell nonNewtonian fluid over a stretching sheet *Chemical Engineering Communications* **200** 327–36

CFD analysis for siting of wind turbines on high-rise buildings

K Veena, V Asha, C Arshad Shameem, T N Venkatesh

Aerospace Weather Research Group, CTFD Division, CSIR National Aerospace Laboratories, Bengaluru, 560017

E-mail: aarya.vjs@gmail.com, tnv@nal.res.in

Abstract. Extracting energy from wind by placing wind turbines over tall buildings is an emerging idea for "greener" cities. This paper discusses the best location for placing wind turbines based on numerical simulations. Different building geometries are considered by varying the dimension and number of buildings. Here real atmospheric boundary layer profiles are created with the help of the Weather Research and Forecasting model and also compared with the theoretical profiles from power law equation. Outputs from the simulations show the location where wind velocity is maximum and the possibility of placing wind turbines on the top as well as in the gap between the buildings. The paper also describes the significance of using realistic profiles instead of theoretical profiles.

1. Introduction

Construction of tall buildings with more than ten-twenty floors has become a part of urbanization all over the world. Nowadays attempts for optimal use of energy from renewable energy resources is gaining ground and wind energy is one of the sources which can be tapped in urban environments. An emerging idea is to place wind turbines over the high-rise buildings to extract wind energy in the form of electricity. Prominent examples include, the Bahrain World Trade Center, which has large scale wind turbines (See Figure 1).

The location where it has to be placed for maximum utilization of wind in the most efficient way is still an area of research. For that computational simulations can help to find out how the wind flows over building/buildings, its direction of flow, turbulence nature and the interference effects if more than one building is present. Among the previous works on energy harvesting using updrafts of wind over buildings using Computational Fluid Dynamics(CFD) are [1–4]. Analysis by [3] shows the increase in wind velocity at the building with different wind direction and size and shape of the building. Dependence of wind direction and size and shape of the building in exploiting the wind energy are analyzed with a series of experiments by [3, 4].

Our approach is to use realistic atmospheric boundary layer (ABL) profiles of wind to ascertain the suitable location for wind-turbines over high-rise buildings. Realistic ABL profiles are derived from simulations of the WRF [5, 6] public domain weather prediction model. Building simulations are done using the OpenFOAM CFD package.

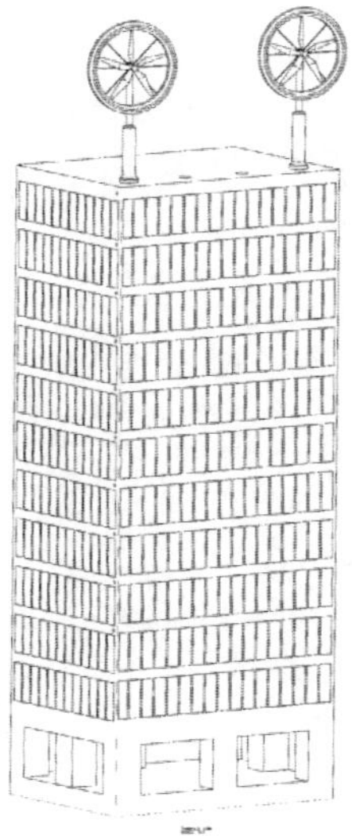

(a)

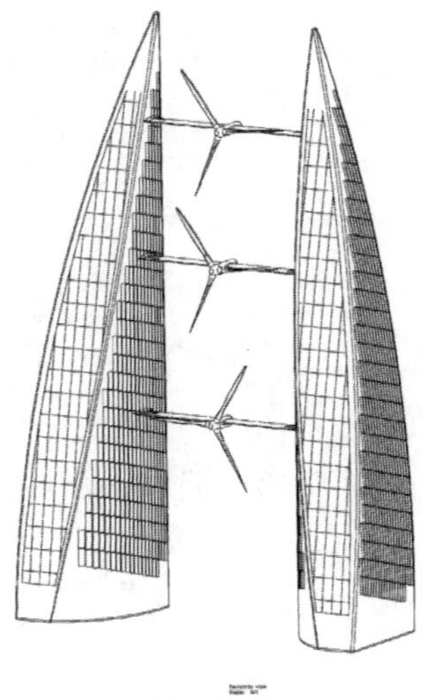

(b)

Figure 1. Schematic diagrams showing (a) building with roof-top wind turbines and (b) wind turbines in-between twin towers as in World Trade Center of Bahrain

2. Computational simulation model

2.1. Inlet profiles and boundary conditions

The wind which acts on a building can be recreated in software by introducing a velocity profile at the front of the building. It can be achieved in two ways. One is by using equations and the other with real atmospheric boundary conditions obtained from meteorological sources. Here both the types of profiles are used in the CFD simulation. Theoretical velocity profile derived from power law equation and the atmospheric boundary layer profile from WRF model. The power law equation used here is

$$\frac{U}{U_{max}} = \left(\frac{y}{y_{max}}\right)^{\alpha} \tag{1}$$

where U is the flow velocity at height y and y_{max} is the maximum flow velocity at the maximum height. Here all sets of profiles are normalized to a maximum height of 500m and maximum velocity of 4.5m/s with average velocity in the range of 3.6-3.9m/s. To have average velocities of the same order the power law profile exponent α is taken as 0.2 which is similar to the case done by [2] in his works.

The terrain locations selected for creating atmospheric boundary layer profiles were the three Indian cities with lat-lon 13N-77E (Bangalore), 13.07N-80.27E (Chennai) and 9.97N-76.28E (Cochin). These realistic velocity profiles are named as Profile A, B and C respectively for Bangalore, Chennai and Cochin.

In an earlier study [7], the effect of these profiles on Coefficient of Drag for two dimensional and three dimensional buildings was compared. We found that there was a significant difference in drag coefficient between theoretical and realistic profiles and also found that the variations within the realistic profiles are comparatively small. Among them Profile B (Profile of Chennai) is considered here for all simulation. As mentioned above theoretical and realistic inlet velocity profiles are normalized to same height and average velocity and plots are as shown in Figure 2. Average velocity is integrated for further calculations so that we make sure the comparison is meaningful for all cases.

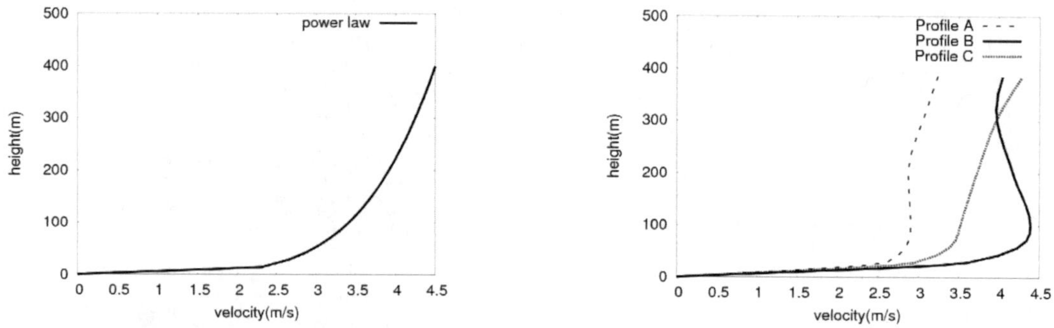

(a) Idealized wind profiles

(b) Atmospheric boundary layer profiles over some south Indian cities

Figure 2. Inlet velocity profiles

Reynolds number 10,000 was kept constant for all simulation cases and ν value is calculated in accordance with the Reynolds number. Since the fluid medium is air the density was set as 1.225 kg/m^3 itself. The other boundary conditions such as inlet and exit pressure values were taken as 2 bar and 0 bar respectively.

2.2. Geometry and grid

There are three different geometries included in this paper according to the three sets of analysis works. The ultimate aim of the work is to propose the best locations where wind turbines can be placed on a high-rise building. According to the number of buildings the geometries can be divided into two sets.

(i) Single building (Building A and Building B)
Two different dimensional sets of single building are tabulated in Table 1.

Table 1. Geometry

Geometry	Building dimension	Domain size	Number of points
Building A	100 × 100 × 100	1200 × 400 × 700	351421
Building B	20 × 100 × 20	1260 × 500 × 620	3510015

Here the building considered is in rectangular block shape with a constant height(h) of 100m. Building A is much larger in area and looks like cubic in shape. The Building B is more or less similar to the Public Utility Building of Bangalore. Geometries were done using Gambit software and the blockMesh utility of OpenFOAM software. The domain exit was made at least at a distance of 8h from the building.

(ii) Twin buildings
Two buildings are constructed with same dimension and height kept the same as for single building. Breadth and width of both the buildings are of 20m in length and the gap between the buildings is varied. The domain size is 1260 × 500 × 620 and Figure 3 represents the model.

IOP Conf. Series: Journal of Physics: Conf. Series **822** (2017) 012013 doi:10.1088/1742-6596/822/1/012013

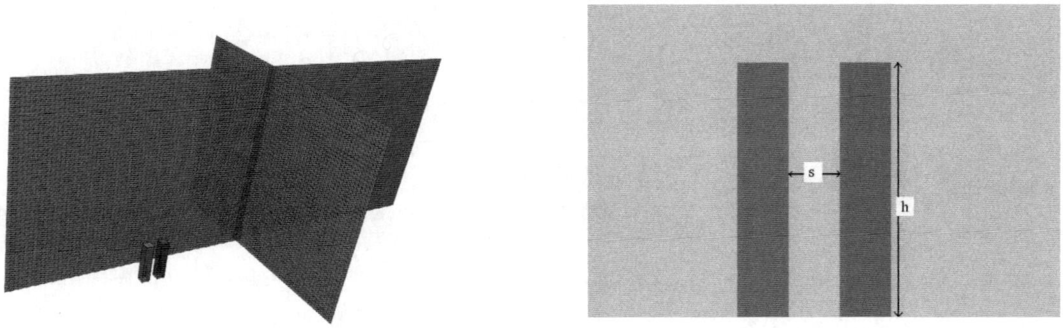

(a) Grid (b) Schematic representation of twin buildings

Figure 3. Twin buildings with dimension $20 \times 100 \times 20$ and domain size $1260 \times 500 \times 620$, and schematic representaion

2.3. Governing equations and simulation setup

The incompressible Navier-Stokes equation used in the simulation is

$$\frac{\partial \mathbf{u}}{\partial t} + \mathbf{u}.\nabla \mathbf{u} = -\frac{1}{\rho}\nabla p + \nu \nabla^2 \mathbf{u} \tag{2}$$

where u is the flow velocity, ρ density of the flow, p pressure and ν the kinematic viscosity of the fluid. Incompressible, unsteady solver called icoFoam of OpenFoam was used in the test cases. OpenFOAM uses pressure-implicit splitting operator (PISO) algorithm to solve the Navier Stokes equations iteratively and the method of parallel computing used by OpenFOAM is known as domain decomposition, in which the geometry and associated fields are broken into pieces and allocated to separate processors for solution. The parallel version uses the public domain openMPI implementation of the standard message passing interface (MPI). The current simulations were done using the HPC systems at the CSIR Fourth Paradigm Institute and up to 64 nodes were used.

3. Computational results

3.1. Evaluation of wind velocity around the Building A and Building B

The flow gets separated at the front corner of the building and this flow stagnation creates a reversed flow at the base of the building. The vortices appear near to the corner of the building and the accumulation of these vortices results in a re-circulation zone behind the building. This region of turbulence has less intensity of wind and is not suitable for placing wind turbines. From the Figure 4 it can be inferred that just above this wake region there is a region of maximum wind speed created. The magnitude of this maximum wind velocity is much higher to that of mean wind speed. The area covering maximum wind speed depends on the cross sectional area of the building.

3.2. Comparison between theoretical wind profile with realistic profiles

As there is a drag coefficient variation due to the use of realistic boundary layer profiles [7], it is likely that it has an influences on the maximum velocity at the building roof top. Interestingly , the trend of variation is similar to that of drag coefficient change. When compared to theoretical power law profile, by the use of realistic profile an increase of 20% in maximum velocity near the building can be achieved from simulations. Both Building A and Building B are tested and

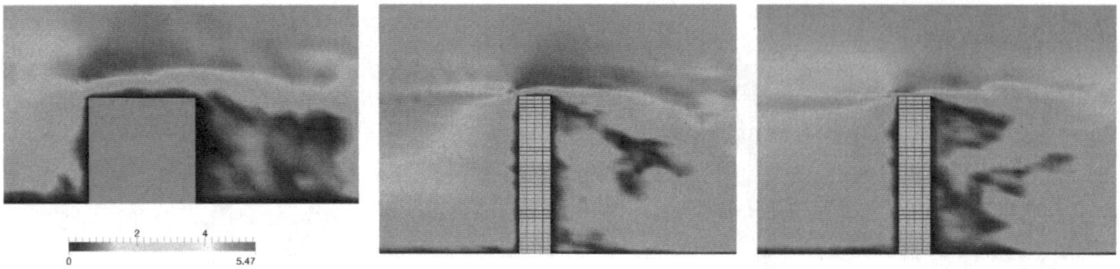

(a) Building A, Realistic profile (b) Building B, Realistic profile (c) Building B, Power law profile

Figure 4. Comparison of different inlet profiles

the results are as tabulated in Table 2. From the results, Building A and Building B showed an increase in velocity of 30% and 22% respectively.

The velocity contours obtained for this quantitative analysis is depicted in Figure 4.

Table 2. Comparison of theoretical and realistic wind profiles

Geometry	Profile	Maximum velocity (U_{max} in m/s)
Building A	Theoretical	4.36
	Realistic	5.68
Building B	Theoretical	4.53
	Realistic	5.56

3.3. Feasible location for placing wind turbines

From the plots of velocity contours (Figure 4) it can be seen that, the area covering maximum wind velocity is at a distance of 15m from the building top. Placing wind turbines at this height will lead to maximum power generation, Even at a lesser height (upto 10m) there is sufficient wind for power generation. The other regions to place the turbines are at the sideways of the building below to the rooftop, where velocity is still high in magnitude.

When more than one building is present, the interference effect also has to be considered. Figure 5 shows the simulation over twin buildings and a schematic representation in which s is the distance of separation between the buildings and U_{max_s} is the maximum wind speed in between the buildings. The Table 3 indicates the influence of separation distance on the wind speed.

Table 3. Influence of separation distance between the buildings on wind velocity

Twin building cases	s/h	U_{max_s}/U_∞
Case I	50/100	0.973
Case II	20/100	0.351

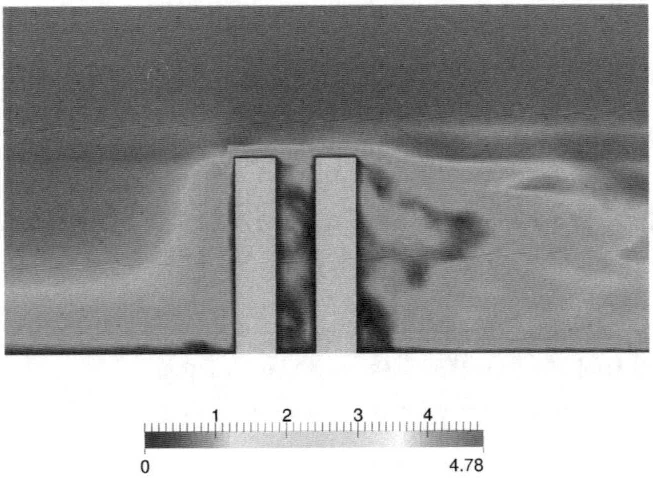

Figure 5. Simulation of wind flow over twin buildings-velocity contours

4. Conclusion

We have carried out a study on the optimal siting of wind turbines over high rise buildings using computational simulations using OpenFOAM. A feature of this study is the use of realistic wind profiles derived from a numerical weather prediction model. While it is true that results depend on the exact geometry of the building, the general conclusions we can draw are as follows. We find that for a single tall building, there is a significant fraction over which winds are sufficient to be used for wind power generation.

For twin buildings, wind turbines can be placed not only at the roof top but also in between the buildings. A velocity of 2m/s can be obtained for a separation distance of fraction 0.2 and 2-2.8m/s for 0.5.

5. Acknowledgments

This work was carried out as part of the ARIEES project, under the twelfth Five year plan of the Government of India. The support is gratefully acknowledged. The computational facilities at CSIR-4PI were used for the simulations.

References

[1] Park J, Jung H J, 2 S W L and Park J 2015 *energies* **8** 11846
[2] Mohamed A, White C and Watkins S 2012 *15th Australasian Wind Engineering Society Workshop*
[3] Cai Y, Zhao F and Liu S 2015 *International Conference on Information Sciences, Machinery, Materials and Energy* 591
[4] Toja-Silva F, Peralta C, Lopez-Garcia O, Navarro J and Cruz I 2015 *computation* **3** 299
[5] Michalakes J, Dudhia J, Gill D, Henderson T, Klemp J B, Skamarock W C and Wang W 2005 *Proceedings of the Eleventh ECMWF Workshop on the Use of High Performance Computing in Meteorology*
[6] Klemp J B, Skamarock W C and Dudhia J 2007 *Monthly Weather Review* **135** 2897
[7] Veena K, Asha V, Shameem C A and Venkatesh T N 2015 *17th CFD Annual Symposium CFD Division - Aeronautical Society of India 11-12 August, Bangalore* 164

Development of streak instability modes excited by turbulent fluctuations

Huy Quang Ho, Masahito Asai and Shohei Takagi

Department of Aerospace Engineering, Tokyo Metropolitan University, 6-6 Asahigaoka, Hino, Tokyo 191-0065, Japan

E-mail: ho-huyquang@ed.tmu.ac.jp

Abstract. Development of streak instability modes excited by disturbances whose frequency-spectrum was similar to that of wall turbulence was examined experimentally for a single screen-generated low-speed streak. The disturbances were introduced into the low-speed streak through two small holes made on both the sides of the low-speed streak. The result showed that even when the low-speed streak was forced by turbulent fluctuations with zero spanwise correlation, sinuous instability modes were excited with distinct frequency selectivity around the most unstable frequency of the linear streak instability and dominated streak breakdown in the downstream region. The magnitude of the sinuous modes excited was compared to the case when the anti-symmetric forcing with the same turbulence spectrum was applied.

1. Introduction

Breakdown of low-speed streak(s) into quasi-streamwise vortices, is considered to be a key event for generation and sustenance of wall turbulence structure. We have been studying the streak instability and breakdown experimentally for well-controlled low-speed streak(s) under various disturbance conditions/environments. The experiments examined the sinuous and varicose instabilities of a single low-speed streak [1] and the fundamental and subharmonic instabilities in spanwise-periodic low-speed streaks [2, 3]. In these experiments, the low-speed streak/streaks were generated artificially in a laminar boundary layer by using a small piece/pieces of screen set normal to the boundary-layer plate, and the streak instability was triggered by well-controlled sinusoidal disturbances with mono-frequency. These experiments showed that sinuous instability could continue to grow downstream and lead to breakdown of low-speed streak(s) into quasi-streamwise vortices not unlike those observed in near-wall turbulence [4, 5, 6]. Furthermore, streak instability and breakdown in a re-transition caused by suction-surviving residual turbulent fluctuations was examined [7], and streak breakdown due to growth of sinuous instability waves was observed in spite that disturbances exciting the streak instability were frequency-rich.

In experiments on bypass boundary-layer transition, laminar boundary layer was subjected to free-stream turbulence with relatively high intensity which has often been controlled by turbulent grid upstream of the boundary-layer plate [8, 9, 10]. On the other hand, in numerical computations simulating bypass boundary-layer transition under free-stream disturbance, initial disturbances were supplied using continuous-spectral modes of Orr-Sommerfeld and Squire equations [11, 12]. These experimental and computational studies showed that occurrence and growth of sinuous instability modes dominated the streak breakdown.

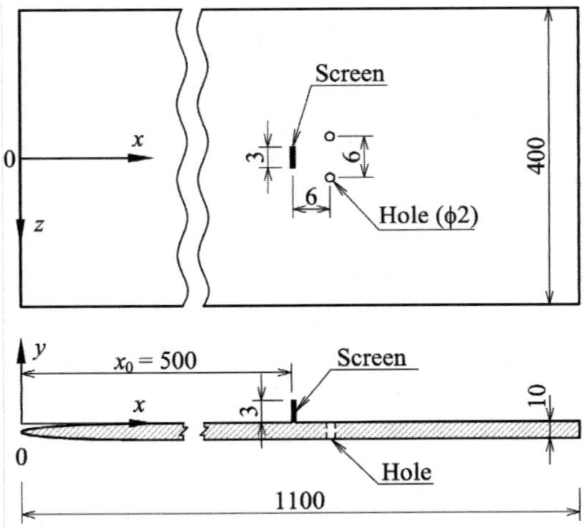

Figure 1. Schematic of boundary-layer plate (dimensions is mm).

In the present experimental study, aiming at giving further insights on instability and breakdown of low-speed streak(s) in natural or turbulent environments, a low-speed streak is forced by disturbances with a continuous spectrum similar to that of developed wall turbulence. A particular attention is paid to how the growth of sinuous instability modes depends on nature of disturbances triggering the instability.

2. Experimental setup and procedure

The experiment was conducted in a low-turbulence wind tunnel with an exit cross-section of 400×400 mm^2. As illustrated in Fig. 1, a boundary-layer plate (10 mm thick and 1100 mm long) was set parallel to the oncoming uniform flow in the test section. This facility is the same as that used in experiments [1, 2, 3]. The free-stream velocity U_∞ was fixed at 4 m/s throughout the experiment. A single low-speed streak was produced in the boundary layer by using a small piece of 40-mesh screen (wire-gauze) set normal to the boundary-layer plate at a station 500 mm downstream of the leading edge. The screen height and width were both 3 mm. Note that the displacement thickness δ^* of Blasius boundary layer was about 2.5 mm at the screen location in the absence of the screen. The coordinate system was such that x was the streamwise distance, measured from the leading edge, y the normal-to-wall distance and z the spanwise distance. The screen location (x = 500 mm) is denoted by x_0.

The laminar streak flow was forced by external disturbances which were introduced through two small holes which are drilled on the wall 6 mm downstream of the screen, with a spanwise interval of 6 mm and were connected to two loudspeakers as has been done in [1] where the streak instability was excited at a single frequency. In the present experiment, the instability was excited by a turbulent fluctuation signal which was deduced by a hot-wire measurement in a buffer region ($y^+ = 23$) of turbulent boundary layer generated by gluing cylinder roughness elements on the boundary layer plate, at an upstream location (x = 100 mm) under the same free-stream condition. The screen width (3 mm) was about 50 ν/u_τ where u_τ was the friction velocity of the turbulent boundary layer and ν was the kinematic viscosity. We applied two kinds of forcing to excite streak instability. One was an anti-symmetric forcing (Case 1) in which two loudspeakers were driven with turbulent signals that were 180° out of phase, and the other was a forcing due to two turbulent signals with zero correlation (Case 2). We of course expected that the anti-symmetric forcing was the most effective for excitation of sinuous instability modes. A constant temperature hot-wire anemometer was used to measure time-mean and fluctuating velocities in the streamwise direction, U and u.

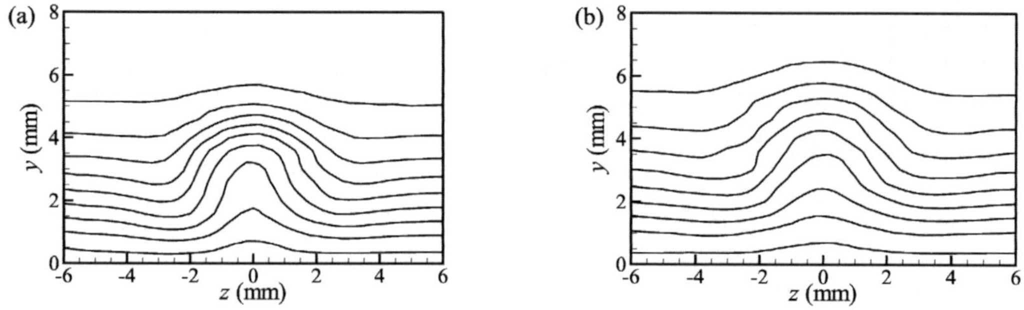

Figure 2. Contours of mean velocity (U/U_∞) in the (y, z) plane. (a) x - x_0 = 60 mm, (b) x - x_0 = 140 mm. Contour levels range from 0.1 to 0.9.

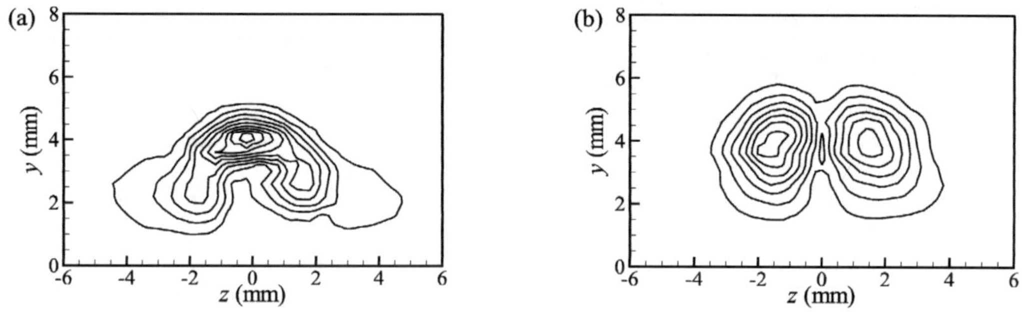

Figure 3. Amplitude distributions of u-fluctuations (u'_f/U_∞) in the (y, z) plane. (a) Varicose mode of f = 110 Hz at x - x_0 = 50 mm. Contour levels range from 0.002 to 0.018. (b) Sinuous mode of f = 90 Hz at x - x_0 = 100 mm. Contour levels range from 0.005 to 0.035.

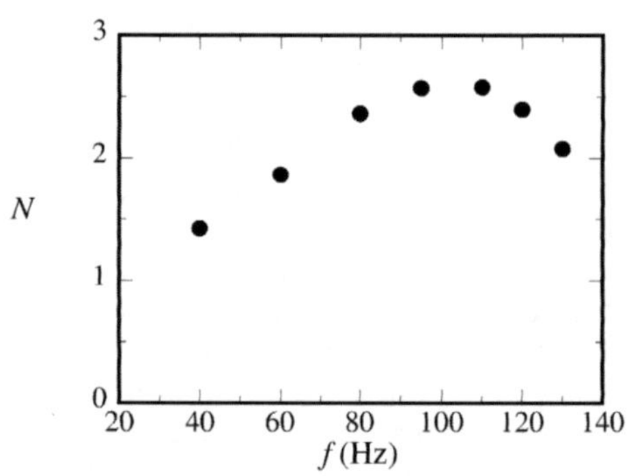

Figure 4. Amplification of sinuous mode from x - x_0 = 40 mm to 180 mm. N vs. f.

3. Result and discussion

Figs. 2(a) and (b) illustrate distributions of the mean velocity U/U_∞ in the (y, z) plane at x - x_0 = 60 mm and 140 mm, respectively. A laminar low-speed streak developed downstream of the screen. The velocity difference across the low-speed streak was about 35% and 20%, at x - x_0 = 60 mm and 140 mm, respectively. In the absence of any artificial disturbances, the low-speed streak maintained laminar completely in the observation region up to x - x_0 = 300 mm.

Before examining frequency-selective growth of disturbances initially consisting of turbulent spectral components, we examined the instability characteristics of the present low-speed streak by driving the loudspeakers at a single frequency. By introducing symmetric and anti-symmetric disturbances, we could easily excite varicose and sinuous instability modes as shown in Figs. 3(a) and (b), respectively, which illustrate the amplitude distributions of u-fluctuation (of the forcing frequency) in the (y, z) plane. The symmetric varicose modes could not continue to grow downstream in the present low-speed streak because the span width of the horizontal ($\partial U/\partial y$) shear layer away from the wall in the low-speed streak was too small compared to the shear-layer thickness: Varicose modes decay downstream beyond $x - x_0 = 60$ mm. Whereas the anti-symmetric sinuous modes could continue to grow downstream almost exponentially. Fig. 4 illustrates the frequency selective growth of sinuous disturbances in terms of N-value defined as $N = \ln[u'_m/u'_{m0}]$ where u'_{m0} is the maximum rms value of u-fluctuation component at $x - x_0 = 40$ mm. The maximum amplification of the sinuous mode occurred at and around 100 Hz in the present streak flow condition. Then we examine the response of the low-speed streak to disturbance with turbulence spectrum mentioned above.

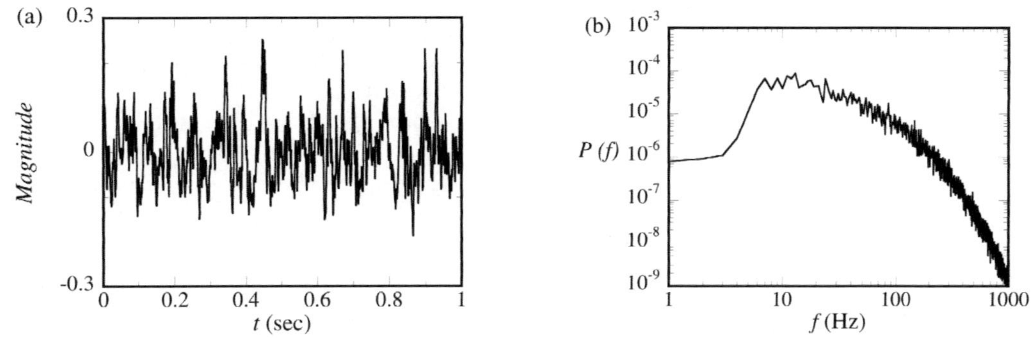

Figure 5. Waveform and power spectrum of the forcing signal. (a) Waveform, (b) power spectrum.

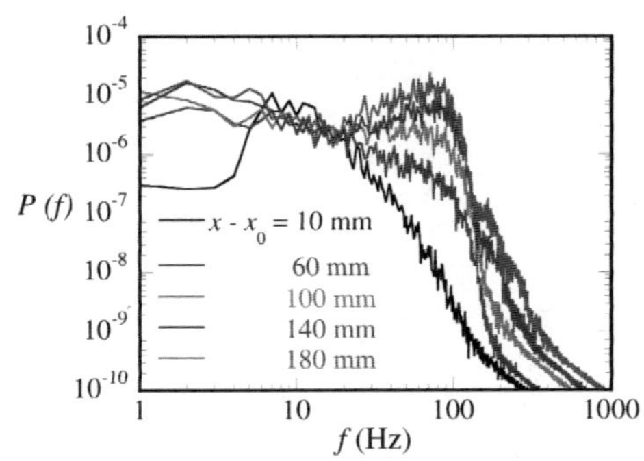

Figure 6. Power spectra of u at $x - x_0 = 10$ mm, 60 mm, 100 mm, 140 mm and 180 mm for anti-symmetric excitation (Case 1).

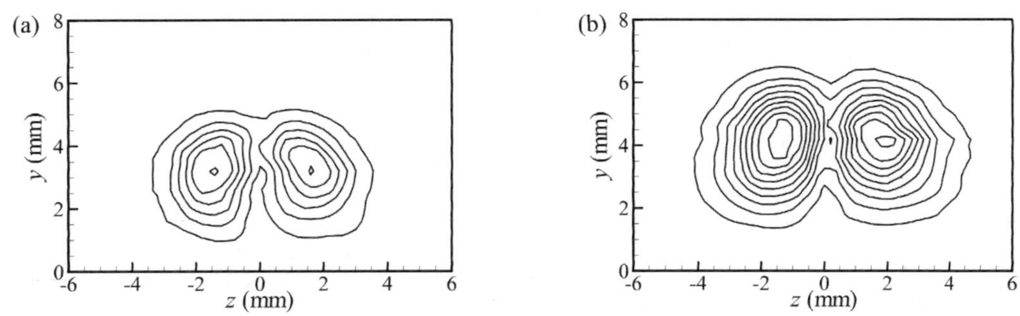

Figure 7. Amplitude distribution of u (u' over 30-120 Hz) in the (y, z) plane for anti-symmetric excitation (Case 1). (a) $x - x_0 = 60$ mm. Contour levels range from 0.0015 to 0.009. (b) $x - x_0 = 140$ mm. Contour levels range from 0.003 to 0.03.

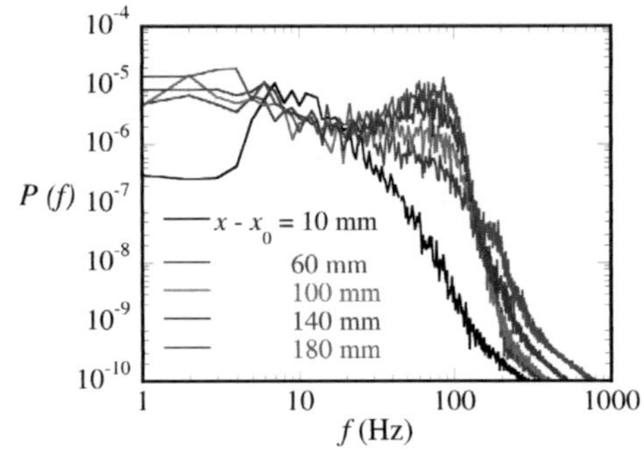

Figure 8. Power spectra of u at x - x_0 = 10 mm, 60 mm, 100 mm, 140 mm and 180 mm for Case 2.

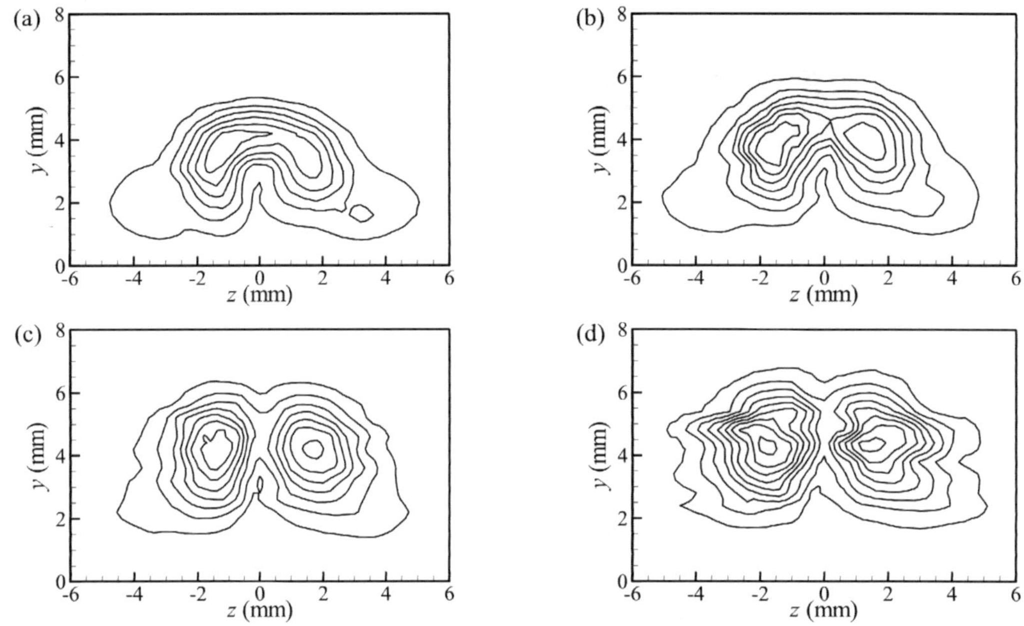

Figure 9. Amplitude distribution of u (u' over 30-120 Hz) in the (y, z) plane for anti-symmetric excitation (Case 2). (a) x - x_0 = 60 mm. Contour levels range from 0.0015 to 0.009. (b) x - x_0 = 100 mm. Contour levels range from 0.002 to 0.014. (c) x - x_0 = 140 mm. Contour levels range from 0.003 to 0.021. (d) x - x_0 = 180 mm. Contour levels range from 0.003 to 0.027.

Figs. 5(a) and (b) display the waveform and power spectrum of turbulent signals supplied to the two loudspeakers, where the spectral components below 5 Hz were filtered out for the input signals to the loudspeakers. Fig. 6 illustrates power spectra of u at x - x_0 = 10 mm, 60 mm, 100 mm, 140 mm and 180 mm for the anti-symmetric excitation (Case 1); they were measured at a vertical shear-layer ($\partial U/\partial z$) position where the disturbance took a maximum in amplitude at each x station. At the initial stage, at x - x_0 = 10 mm, the spectra were close to those of the input signals to the loudspeakers, and the rms value of u-fluctuations were calculated from the spectra to be about 1.3% of U_∞. In the downstream stations, we can see disturbances are amplified over the frequency range of 30-120 Hz. The growing disturbance over the frequency range of 30-120 Hz was of course a sinuous instability mode as shown in Fig. 7 which illustrates amplitude distributions of u in the (y, z) plane at x - x_0 = 60 mm and 140 mm.

Next we examined the disturbance development when the streak flow was excited by the zero-correlation fluctuation signals (Case 2). Fig. 8 illustrates power spectra of u at x - x_0 = 10 mm, 60 mm, 100 mm, 140 mm and 180 mm for Case 2; they were measured at a vertical shear-layer similarly to Case

1. At the downstream locations, we can see disturbances are amplified over the frequency range of 30-120 Hz in this case, too, suggesting that the sinuous instability mode was also amplified. Fig. 9 shows the disturbance development in terms of the amplitude distributions in the (y, z) plane at four x stations: $x - x_0 = 60$ mm, 100 mm, 140 mm and 180 mm. The structure of the disturbance excited was like that of the varicose mode at $x - x_0 = 60$ mm and gradually changed to be that of sinuous instability mode beyond $x - x_0 = 100$ mm.

4. Conclusions

Growth of streak instability mode excited by initial disturbances whose frequency-spectrum was similar to that of wall turbulence was examined experimentally for a single low-speed streak. Two kinds of driving signals were used for the excitation of streak instability. One was a forcing using a pair of turbulent signals with zero correlation and the other was an anti-symmetric forcing which was the most effective for anti-symmetric sinuous modes. Even when the low-speed streak was forced by turbulent signals with zero spanwise correlation, the sinuous instability mode occurred dominantly in the downstream region around the most amplified frequency of the linear instability. The amplitude of the excited sinuous mode was found to be only by 25% smaller than that in the case of the anti-symmetric forcing.

Acknowledgments
This work was, in part, supported by the Grant for Scientific Research from Tokyo Metropolitan Government.

References

[1] Asai M, Minagawa M and Nishioka M 2002 The instability and breakdown of a near-wall low-speed streak. *J. Fluid Mech.* **455** 289-314.
[2] Konishi Y and Asai M 2004 Experimental investigation of the instability of spanwise-periodic low-speed streaks. *Fluid Dyn. Res.* **34** 299-315.
[3] Konishi Y and Asai M 2010 Development of subharmonic disturbance in spanwise-periodic low-speed streaks. *Fluid Dyn. Res.* **42** 035504.
[4] Jeong J, Hussain F, Schoppa W and Kim J 1997 Coherent structures near the wall in a turbulent channel flow. *J. Fluid Mech.* **332** 185-214.
[5] Jiménez J and Pinelli A 1999 The autonomous cycle of near-wall turbulence. *J. Fluid Mech.* **389** 335-359.
[6] Kawahara G and Kida S 2001 Periodic motion embedded in plane Couette turbulence: regeneration cycle and burst. *J. Fluid Mech.* **449** 291-300.
[7] Asai M, Konishi Y, Oizumi Y and Nishioka M 2007 Growth and breakdown of low-speed streaks leading to wall turbulence. *J. Fluid Mech.* **586** 371-396.
[8] Matsubara M and Alfredsson P H 2001 Disturbance growth in boundary layers subjected to free stream turbulence. *J. Fluid Mech.* **430** 149-168.
[9] Mans J, Kadijk E C, de Lange H C and van Steenhoven A A 2005 Breakdown in a boundary layer exposed to free-stream turbulence. *Exp. Fluids* **39** 1071-1083.
[10] Mans J, de Lange H C and van Steenhoven A A 2007 Sinuous breakdown in a at plate boundary layer exposed to free-stream turbulence. *Phys. Fluids* **19** 088101.
[11] Brandt L, Schlatter P and Henningson D S 2004 Transition in boundary layers subject to free-stream turbulence. *J. Fluid Mech.* **517** 167-198.
[12] Hack M J P and Zaki T A 2014 Streak instabilities in boundary layers beneath free-stream turbulence. *J.Fluid Mech.* **741** 280-315.

IOP Conf. Series: Journal of Physics: Conf. Series **822** (2017) 012015 doi:10.1088/1742-6596/822/1/012015

Experiments in bypass boundary layer transition under a stream with and without shear

G Balamurugan, Alakesh C Mandal

Department of Aerospace Engineering, Indian Institute of Technology, Kanpur, India

E-mail: alakeshm@iitk.ac.in

Abstract. Effect of mean flow shear on bypass boundary layer transition is investigated experimentally in a low-speed wind tunnel. To initiate bypass transition on a flat plate, the level of freestream turbulence is intensionally enhanced using a uniform parallel rod grid and a non-uniform parallel rod grid. The non-uniform grid produces a mean velocity profile with shear, and therefore, turbulence production due to this shear will be non-zero compared to a uniform grid which produces a mean velocity profile without shear. The growth rate of the disturbance energy for the non-uniform grid is found to be higher compared to the uniform grid. The intermittency values are also found to be higher for the non-uniform grid under same flow conditions. Particle Image Velocimetry (PIV) measurements in the spanwise plane reveal instability characteristics similar to the one reported in the literature.

1. Introduction

Boundary layer transition is influenced by several factors such as freestream turbulence (FST), pressure gradient, wall roughness, leading edge geometry, etc. Rapid boundary layer transition that occurs at an elevated level of turbulence is often called as bypass transition. Various studies [1, 2, 3, 4, e.g.] were carried out to understand bypass transition in a boundary layer. These studies clearly reveal that, at an elevated level of freestream turbulence, low- and high-velocity streaks appear in the flow. The experimental work of Matsubara and Alfredsson [1] shows that the growth of these streaks are algebraic, and these streaks eventually break down into a turbulent spot due to instability. The numerical simulation of Jacobs and Durbin [2] also shows that localized instability on a lifted low-speed streak can lead the flow to the turbulent state. The PIV measurements of Mandal et al. [4] also indicate breakdown due to similar instability mechanism.

Generally two types of secondary instabilities are reported in the literatures, i.e. sinuous and varicose instabilities. These instabilities are often held responsible for generation of incipient turbulent spots in the flow. Inviscid secondary instability studies on optimal streaks were performed by Andersson et al. [5]. They have reported that instability develops beyond a critical streak amplitude of 26%. Asai et al. [6] studied the breakdown characteristics of an artificially generated low speed streak and observed that sinuous modes are more unstable, if the streak width is comparable with the boundary layer thickness. In their numerical simulation, Brandt et al.[7] observed similarity between their spot precursors and the instabilities reported earlier in various controlled studies [5, 6, e.g.]. Dye flow visualizations of Mans et al. [8] in an actual bypass transition clearly show the presence of these sinuous and varicose instabilities. A comprehensive and comparative study by Schlatter et al. [9] indicates that these instabilities are perhaps triggered by localized perturbations which are present in the freestream. To elucidate the role of these freestream perturbations, a comparative study has to be performed between two cases differing only in the amount of perturbation produced. In our present work, we attempt to investigate this aspect by using a uniform and a non-uniform turbulent generating grids in the freestream.

2. Experimental Setup

Experiments were performed in an open circuit wind tunnel available at the Department of Aerospace Engineering, Indian Institute of Technology, Kanpur, India. The tunnel has a 3000 mm long test section with a square cross section of 610 mm × 610 mm. The tunnel has a square three dimensional contraction section with 16:1 contraction ratio. Measurements were carried out on a flat plate placed horizontally in the mid-plane of the tunnel test section. The plate of length 1200 mm, was made of transparent acrylic material with thickness of 12 mm. The plate has an asymmetric modified super elliptical leading edge which was designed following the guideline of Hanson et al. [10]. This leading edge shape is chosen to reduce the receptivity of the boundary layer to external disturbances by eliminating the leading edge and flat plate juncture curvature discontinuity which can affect the stability properties of the flow [11]. The streamwise turbulent intensity of the tunnel was found to be 0.1% of the freestream velocity.

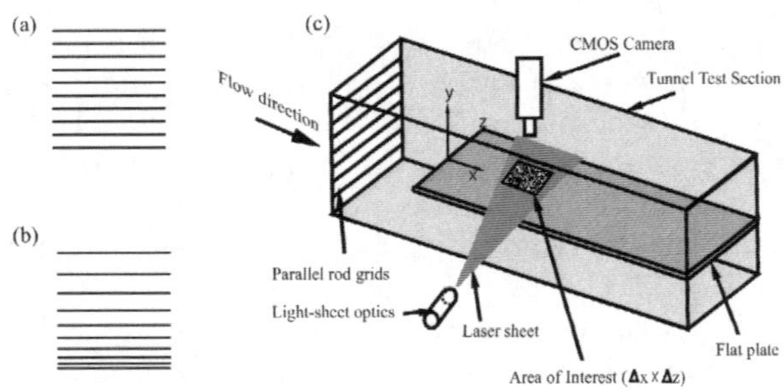

Figure 1. Simple sketches for the uniform grid (a), the non-uniform grid (b) and the PIV setup for the spanwise plane measurements (c).

However, the freestream turbulence level in the tunnel test section was intensionally enhanced using two types of passive grids, as shown schematically in figures 1(a) and (b). The grid with equal spacing of parallel rods is called as uniform grid (see figure 1(a)) and the grid with non-uniform spacing of the rods are called as non-uniform grid (see figure 1(b)). Both the uniform and non-uniform grids in figures 1 (a) and (b) are made of 10 mm diameter circular rods and the numbers of rods were kept same for both the grids. The non-uniform grid was fabricated by varying the spacing among the parallel rods linearly, whereas the uniform grid has a uniform spacing between two consecutive rods. These grids were designed such that the solidity (σ, defined as the ratio of the blockage area by the grid to the tunnel test section area), remains the same for both the grids. The solidity value was smaller than 0.45 as suggested by Kurian and Fransson [12] to avoid interaction of the jets that gets created between the rods. The distance between the grid and the leading edge was maintained more than 20 mesh widths to ensure the development of isotropic turbulence [3] for both the grids. For a same rotational rpm of the tunnel, the measured turbulent intensities for both the grids at the plate leading edge position are shown in table 1, along with the integral length scale and Taylor micro scale.

A hotwire anemometry system (procured from Dantec Dyanmics) along with a National Instruments data acquisition card and Labview software was used for data acquisition. A single

Table 1. Free stream parameters at the leading edge.

Grid	u_{rms}/U_0 (%)	$\Lambda(mm)$	$\lambda(mm)$
Uniform grid	2.3	14.8	7.02
Non-uniform grid	2.3	13.95	6.88

wire miniature hotwire probe with 5μm diameter tungsten wire as the sensing element was used to measure the instantaneous streamwise velocity, U_I, and the fluctuating streamwise velocity, u. The length to diameter ratio of the probe was 250. The probe was calibrated using a Pitot-static tube connected to a Furness FC012 digital manometer and King's law fit.

We also utilize the particle image velocimetry (PIV) technique to measure the instantaneous velocity fields in the spanwise plane, i.e. in the $x-z$ plane, where x and z denote the streamwise and the spanwise distances. The wall normal distance in this work is denoted by y. The experimental setup for the PIV measurements is schematically shown in figure 1(c). A double pulsed Nd:YAG laser (Innolas Spitlight Compact 400 PIV, 180 mJ/pulse, 10 Hz) is used for illuminating the flow field. The flow is seeded with the help of a fog generator placed at the rear end of the tunnel entrance. The fog is distributed uniformly across the tunnel using a small fan in front of the fog generator. A CMOS camera (model Motion pro Y5, IDTpiv, USA) of 2336×1728 pixel resolution was used to capture the particle images. The laser and the camera were synchronized with the Motion pro timing unit (procured from IDTpiv, USA). The field of view of the PIV measurements in the $x-z$ plane was 100 mm× 70 mm. For the PIV measurements, the laser sheet was made parallel to the flat plate at $y = 2.5$ mm. We acquired 700 image pairs for both the grids and these images were analyzed using Provision-XS software developed by IDTpiv. The software utilizes a mesh free cross-correlation algorithm (see Lourenco and Krothapalli [13], for details). While processing these images, we used a 32 pixel×32 pixel correlation window, and the spatial resolution based on this correlation window is found to be 1.4 mm. Further details of image processing using this software is given in our previous work [4].

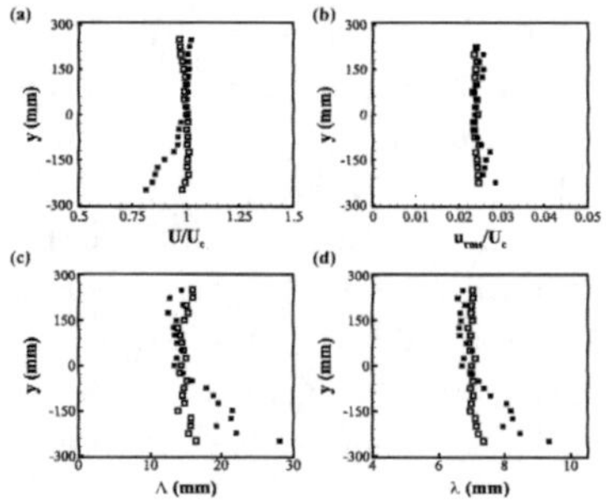

Figure 2. Transverse distribution of the normalized mean velocity (a) and rms velocity (b) along with the integral scale, Λ, (c) and Taylor micro scale, λ, (d), for the uniform grid (unfilled symbol) and the non-uniform grid (filled symbol) at the leading edge.The length scales are found by using the autocorrelation function, R_{uu} and the local mean velocity, U

3. Results and Discussion

The freestream turbulence characteristics along the transverse direction (i.e. along y direction) are shown in figure 2, for both the grids. These measurements are carried out at $x = 0$ in an empty test section; that is at the streamwise location where the plate leading edge starts if the plate is placed inside the tunnel. These measured quantities describe the conditions imposed by these grids on the developing boundary layer over the plate after its installment in the test section. The normalized mean velocity profile in figure 2(a) shows nearly uniform variation for the uniform grid, whereas the velocity profile for the non-uniform grid shows a increasing trend, although the difference is small after the plate leading edge. Here, the mean velocity, U, and the root-mean-squared velocity, u_{rms}, are normalized by the center-line velocity, U_c, measured at $y = 0$ location. The variation in the mean velocity for the non-uniform grid produces a mean shear in the freestream which in turn keep alive the turbulence production, and thereby the turbulence produced by the non-uniform grid is comparatively more active. The variation of u_{rms}/U_c, the integral length scale, Λ and the Taylor micro scale, λ, is shown in figures 2(b), (c) and (d), respectively, for both the grids. One may notice that the values of these parameter in the working side of the plate are nearly of same magnitude.

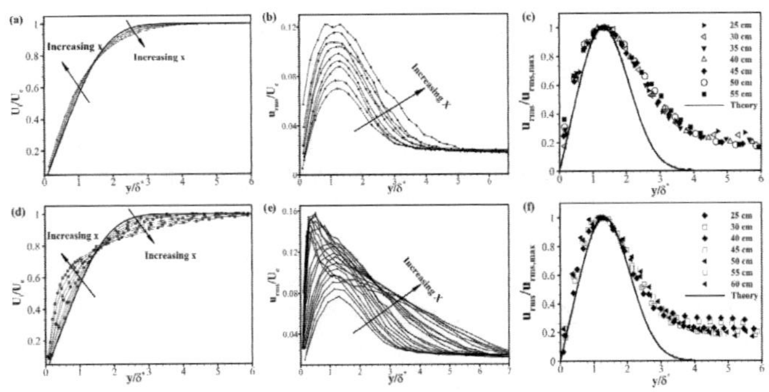

Figure 3. Mean (a) and u_{rms} (b) velocity profiles in the boundary layer, for the uniform grid at $x = 300, 400, 500, 600, 700$ mm. Mean (d) and u_{rms} (e) velocity profiles, for the non-uniform grid at $x = 250, 300, \ldots\ldots, 1250$ mm. Comparison with the non-modal theory of Luchini [14], for uniform (c) and non-uniform (f) grids. Solid line in (a) & (d) corresponds to the Blasius profile.

The boundary layer flow, subjected to the above flow condition (figure 2), was measured on the flat plate after placing it at y=0 location in the tunnel. Figures 3(a) and (d) show the mean velocity profiles in the boundary layer measured at different x locations, for the uniform and the non-uniform grids, respectively. The Blasius profile is also shown in these figures for comparison purpose. We may note that the mean and the u_{rms} velocities in the boundary layer are normalized by the boundary layer edge velocity, U_e, and the displacement thickness, δ^*, is estimated based on the edge velocity. The u_{rms} velocity profiles normalized by U_e and $u_{rms,max}$ are shown in figures 3(c-d) and (e-f), respectively, for the uniform and the non-uniform grids. Figures 3(c) and (f) show that the u_{rms} distribution compares well with the optimal perturbation theory of Luchini [14].

The variation of $u^2_{rms,max}/U^2_e$ with the streamwise distance is shown in figure 4, for both the grids. A linear growth region can clearly be seen in this figure, for both the grids. This is similar to the measurements by Matsubara & Alfredsson [1]. However, we find that the growth rate is higher for the non-uniform grid compared to the uniform grid.

The boundary layer transition is often quantified by flow intermittency, which indicates the fraction of time a flow resides in turbulent state at a spatial location. Here, we use the 'Dual

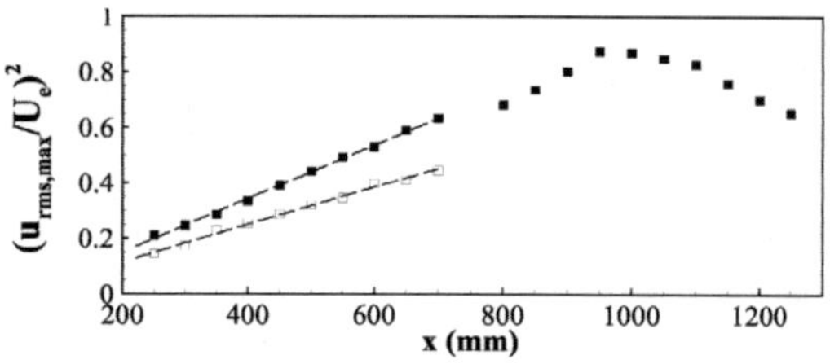

Figure 4. Growth of disturbance energy inside the boundary layer. □, Uniform grid; ■, Non-uniform grid

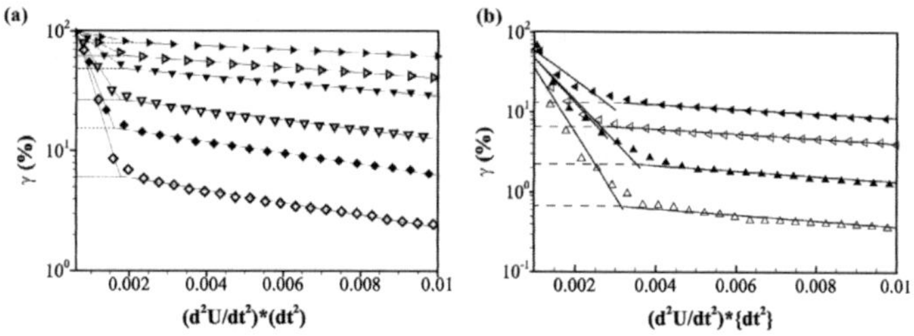

Figure 5. Intermittency for uniform grid (unfilled symbol), and non-uniform grid (filled symbols). (a) At x = 700 mm for U_e = 7, 8, and 9 m/s. (b) At x = 550 mm (triangle), and x = 700 mm (left triangle) location, for U_e = 7 m/s.

Slope Method' of Kuan and Wang [15] for estimating the flow intermittency. Briefly, a double derivative of the velocity signal is used as a criterion function. After estimating intermittency with various threshold value of the double derivative function, a plot of intermittency vs the double derivative is obtained, and the curve shows a sharp change in the slope when plotted in a semi-log graph. By fitting two straight lines for these two regions, an intersection point is found which corresponds to the actual estimated intermittency value. The intermitttency plots at x = 750 mm for three velocities, i.e. 7, 8 and 9 m/s are compared for both the grids. It is clearly seen in figure 5(a) that the value of intermittency function for the non-uniform case is always greater than the uniform grid for a corresponding velocity. The intermittency comparison at same velocity but at different streamwise locations also shows that the flow intermittency is higher for the non-uniform grid, as may be seen in figure 5(b). This again confirms that the presence of shear increases the turbulent activity in the freestream.

PIV measurements have been carried out mainly to identify the spanwise flow structures in the boundary layer. The secondary instabilities, as mentioned in the introduction, can clearly be seen in figure 6. These are distinguished as sinuous and varicose modes depending on the nature of streak oscillation. This confirms that the nature of transition is the same for both the grids, while the non-uniform grid only accelerates the transition process.

4. Summary and concluding remark

Bypass transition scenario under a stream with and without shear in the incoming flow is experimentally investigated using two parallel rod grids. For both the cases, measurements are carried out at the same rotational speed of the tunnel motor. Transverse distribution of the streamwise mean velocity at $x = 0$, without the flat plate inside the tunnel, reveals a moderate shear profile for the non-uniform grid and a nearly uniform velocity profile for the uniform grid. The u_{rms} values and the length scales in the test section are also found to be nearly equal for

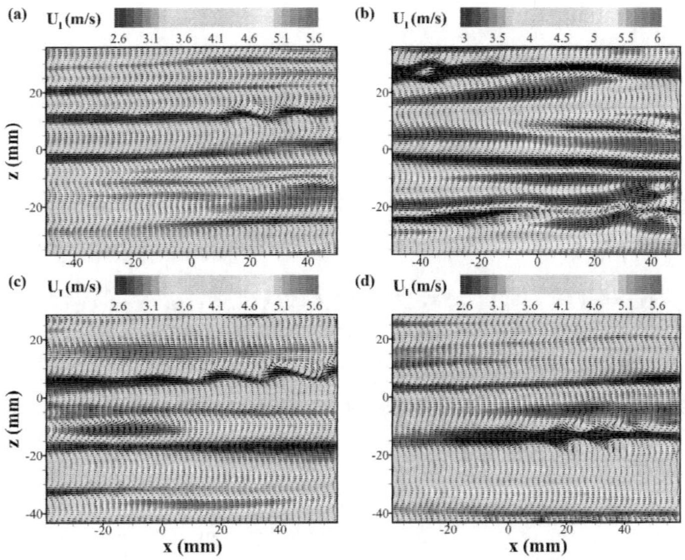

Figure 6. Sinuous (a) and varicose (b) instabilities, for the uniform grid. Sinuous (c) and varicose (c) instabilities, for the non-uniform grid. Origin is at (x, y, z)=(700 mm, 2.5 mm ,0 mm).

both grids in the working side of the flat plate. Boundary layer measurements over the flat plate revealed similar flow characteristics reported in the literature [1]. The wall normal distribution of the normalized u_{rms} profile follows the non-modal theory of Luchini [14], for both the cases. The linear growth rate of the disturbance energy and the flow intermittency levels are found to be higher, for the non-uniform grid case (see figure 5), although similar secondary instability features are seen in the spanwise PIV realizations for both the grids (see figure 6). This is perhaps because of the fact that shear in the freestream contributes to turbulence production, however small it may be, and enhances the turbulent activity in the freestream. This enhanced turbulent activity in turn increases the generation of instability wave packets in the boundary layer, which enhance the energy growth rate and the flow intermittency.

References

[1] Matsubara M and Alfredsson P H 2001 *J. Fluid Mech.* **430** 149–169
[2] Jacobs R G and Durbin P A 2001 *J. Fluid Mech.* **428** 185–212
[3] Fransson J H M, Matsubara M and Alfredsson P H 2005 *J. Fluid Mech.* **527** 1–25
[4] Mandal A C, Venkatakrishnan L and Dey J 2010 *J. Fluid Mech.* **660** 114–146
[5] Andersson P, Brandt L, Bottaro A and Henningson D 2001 *J. Fluid Mech.* **428** 29–60
[6] Asai M, Minagawa M and Nishioka M 2002 *J. Fluid Mech.* **455** 289–314
[7] Brandt L, Schlatter P and Henningson D S 2004 *J. Fluid Mech.* **517** 167–198
[8] Mans J, Kadijk E C, de Lange H C and van Steenhoven A A 2005 *Exp. Fluids* **39** 1071–1083
[9] Schlatter P, Brandt L, de Lange H C and Henningson D S 2008 *Phys. Fluids* **20** 101505–1–101505–15
[10] Hanson R E, Buckley H P and Lavoie P 2012 *Experiments in fluids* **53** 863–871
[11] Saric W S, Reed H L and Kerschen E J 2002 *Annual review of fluid mechanics* **34** 291–319
[12] Kurian T and Fransson J H M 2009 *Fluid dynamics research* **41** 021403
[13] Lourenco L M and Krothapalli A 2000 TRUE resolution PIV: a mesh-free second order accurate algorithm *Proceedings of the International Conference in applications of lasers to fluid mechanics, Lisbon, Portugal*
[14] Luchini P 2000 *J. Fluid Mech.* **404** 289–309
[15] Kuan C L and Wang T 1990 *Experimental Thermal and Fluid Science* **3** 157–173

Experiments on effective tripping device in a zero pressure gradient turbulent boundary layer

Rengasamy K, Alakesh Ch. Mandal

Deparment of Aerospace Engineering, Indian Institute of Technology Kanpur, Kanpur, Uttar Pradesh, India

E-mail: alakeshm@iitk.ac.in

Abstract. Laminar-turbulent transition for flow over a flat plate can artificially be excited by introducing a tripping device. This will promote early transition in the boundary layer; hence a turbulent boundary layer can be realized at low Reynolds number. In this work, we consider three different types of tripping device, i.e. a circular rod, a square rod and a zigzag trip. Height of these three trips has been maintained same, i.e. 2 mm. Hotwire measurements have been carried out at several streamwise locations behind these trips. Measured flow characteristics behind these tripping devices have been presented in terms of recently suggested diagnostic plots along with the streamwise normal stress scaled by inner variables, and an effort has been made to characterize the best tripping device.

1. Introduction

A laminar boundary layer over a flat plate is often tripped to promote early transition-to-turbulence in the boundary layer. This process of promoting early transition-to-turbulence by introducing a tripping device in the boundary layer is known as boundary layer tripping. It is also used to prevent flow separation and to reduce drag in bluff bodies. Boundary layer tripping can be done by various methods such as vibrating ribbons, blowing/suction through walls, passive roughness elements on the wall. Various tripping devices can be used to achieve fully turbulent flow. For example, Elsinga and Westerweel [1] have carried out measurements using zigzag trip. Height of the zigzag trip used was 1.6 mm, which corresponds to the critical Reynolds number of 200. They have chosen the spanwise wavelength for the zigzag trip to be 6 mm, as suggested by Andersson et al. [2]. They reported hairpin like structures due to spanwise vortices behind the trip, and streaks in the separated flow region due to waviness in the spanwise vortices. Undulating spanwise vortices from the shear layer separating from the trip were seen in their instantaneous flow measurements. They also reported that the spanwise vortices could be seen in cylindrical wire, as well, but at a higher Reynolds number than zigzag trip. Adrian et al. [3] studied hairpin vortices in the outer region of the turbulent boundary layer. In their experiments, they carried out measurements for three different Reynolds number for flow over a flat plate. They used a circular trip of diameter 4.7 mm in all three cases. For their measurements at $Re_\theta = 2370$ and 6845, they placed the trip at 110 mm from the leading edge. Based on their data for these two Reynolds numbers, the estimated boundary layer thicknesses (δ) based on the Blasius boundary layer are found to be 3.4 mm and 2 mm at the trip location, respectively. For $Re_\theta = 930$, trip wire placed at 1520 mm from the leading edge which corresponds to $\delta = 19.4$

mm. This shows that the ratio $\frac{k}{\delta}$ (trip height/boundary layer thickness) is 1.4 and 2.35 for Re_θ = 2370 and 6845 cases, respectively. Even though the trip height is higher than the boundary layer thickness, they mentioned that there is no discernible differences in the boundary layer structure from the PIV measurements that they carried out without the trip.

Comparing a wide range of low Reynolds number data from the literature along with their simulated data, Schlatter and Örlü [4] found a considerable differences in the boundary layer parameters, such as skin-friction coefficient, shape factor and also in the mean and fluctuation profiles. However, they speculated that these differences might be from the simulation set up like inflow conditions, turbulence generation and boundary conditions. Schlatter and Örlü [5] spent an effort to explain the reason for differences in the boundary layer parameters given in Schlatter and Örlü [4]. They found that the upstream flow conditions and the tripping effects are main reasons for the differences in the boundary layer integral properties. They suggested that the upstream memory due to the presence of tripping device can be removed if the transition is initiated at the Reynolds number range of $Re_\theta < 300$, and no differences in the flow properties due to the tripping effects could be seen in the boundary layer for $Re_\theta > 2000$. Hutchins [6] recommended some caution remainder for using tripping devices in transition to turbulent boundary layer. He suggested that for flow to become fully turbulent, trip should be introduced inside the boundary layer. He also recommended that further work is needed for over-stimulating case where upstream velocity to the trip is increased.

In this work, three different tripping devices are used to excite early transition in the boundary layer, and the measured data behind these trips are utilized to find the effective tripping device for a turbulent boundary layer.

2. Experimental setup

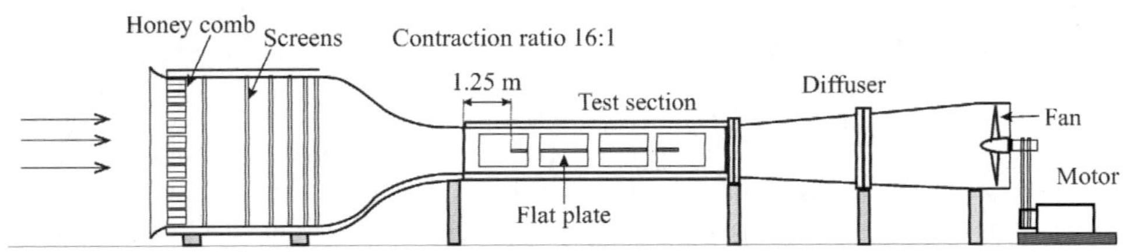

Figure 1. A simple sketch of the low turbulence and low-speed wind tunnel.

The experiments reported in this work were carried out in the low turbulence wind tunnel available at the low speed aerodynamics laboratory, Indian Institute of Technology, Kanpur, India. A simple sketch of this tunnel is shown in figure 1. It is an open suction type tunnel with a test section of length 3000 mm and 610 mm × 610 mm square cross section. The freestream turbulence level in the test section estimated based on the streamwise fluctuating velocity is about 0.1 % in the bandwidth of 2 Hz to 3 kHz. The contraction ratio of the tunnel is 16:1. The flat plate used for the measurements is 12 mm thick, 610 mm in width and 2000 mm in length. The leading edge of the plate is an asymmetrical modified super elliptic leading edge, which was designed following the work of Hanson et al. [7]. This leading edge is chosen because it reduces the receptivity of the boundary layer to disturbance by eliminating the leading edge and flat plate juncture curvature discontinuity and also maintains a larger region of zero pressure gradient over the working section of the plate. The flat plate is mounted horizontally at the mid-plane of the tunnel test section. The distance from the starting of the test section to the leading edge of the flat plate is 1250 mm. Tripping devices have been attached over the upper side of the plate at a distance of 150 mm from the leading edge.

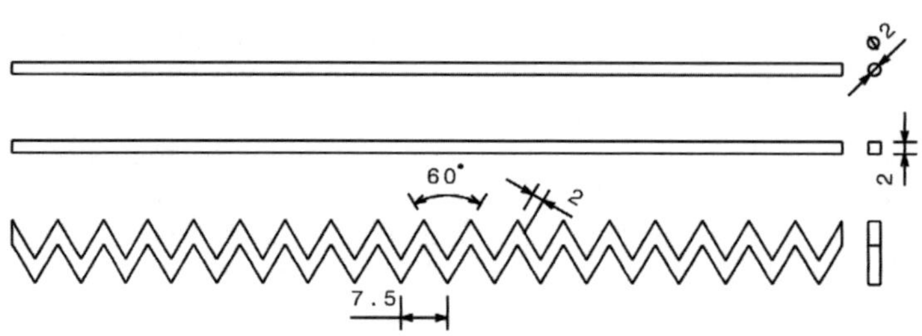

Figure 2. Schematic of three different types of trips: circular, square and zigzag; all dimensions are in mm.

Three different trips which are used in the present experimental work are circular, square and zigzag trip. A schematic of the three different trips, i.e. circular, square and zigzag trip, are shown in figure 2. All these trips are made of brass material. Spanwise wavelength of the zigzag trip is 7.5 mm. This has been chosen based on the suggestion of Andersson et al. [2]. Circular and square trip of 2 mm diameter and thickness, respectively, have been used directly for the measurements. The zigzag trip has been made from a rectangular plate of 2 mm thickness. Height of these trips has been chosen such that the Reynolds number based on the trip should be greater than the minimum critical Reynolds number, Re_c. This has been done to make sure that the flow becomes turbulent in the nearest downstream location behind the trip. Different critical Reynolds number values are suggested in the literature. For example, Gibbings [8] suggested that the critical Reynolds number based on the trip height and the freestream velocity should be more than 826, i.e. $\frac{U_\infty k}{\nu} \geq 826$, whereas Braslow and Knox [9] suggested that the critical Reynolds number based on the trip height and the local velocity at the height of the trip has to be more than 600, i.e. $\frac{U_k k}{\nu} \geq 600$; here U_∞ and U_k denote the freestream velocity and the velocity at the height of the trip location in the undisturbed laminar boundary layer, respectively, whereas the wire diameter/thickness and the kinematic viscosity are denoted by k and ν, respectively. For our all three trips, the value of the first critical Reynolds number, $\frac{U_\infty k}{\nu}$, is found to be 1032, and the second critical Reynolds number, $\frac{U_k k}{\nu}$, is found to be 973; here U_∞ is same for all three trips. However, both Re_c values are above the critical values suggested by these authors. Moreover, we find that the Reynolds number based on the momentum thickness and the freestream velocity velocity, Re_θ, is 175 at the trip location. This value is less than 300, which is suggested by Schlatter and Örlü [5].

Measurements in the boundary layer have been carried out using a hotwire anemometry system procured from the Dantec Dynamics. Data acquisition have been carried out using the Dantec multichannel CTA system along with a 16 bit NI-6034E data acquisition card and the Labview software. The maximum frequency response attainable with the system is 10 kHz. The sensor is made of 5 μm diameter tungsten wire soldered between the two prongs. The hotwire is calibrated with the aid of a Pitot-static probe connected to a digital manometer and using the King's law fit. Hotwire measurements were carried at a sampling rate of 6 kHz with a low pass filter of 3 kHz and without any gain.

3. Results and Discussions

Placing the tripping devices at x = 150 mm from the leading edge of the flat plate, boundary layer measurements have been carried out at eight different streamwise locations behind the

Table 1. Details of boundary layer parameters measured at a fixed U_∞ for all three trips; 'cir', 'sqr' and 'zz' refer to the circular, square and zigzag trip, respectively.

Trip	x_1(mm)	Re_θ	δ(mm)	$\delta^\star$(mm)	θ(mm)	H	$C_f\times10^5$	$u_\tau(ms^{-1})$
cir	250	716	12.97	2	1.37	1.46	508	0.408
sqr	250	639	13.26	1.8	1.22	1.48	530	0.419
zz	250	803	13.49	2.3	1.55	1.47	468	0.388
cir	400	793	14.97	2.31	1.56	1.48	483	0.387
sqr	400	742	13.16	2.18	1.47	1.48	486	0.386
zz	400	937	16.09	2.68	1.83	1.47	467	0.384
cir	600	1030	19.42	2.97	2.02	1.47	437	0.369
sqr	600	993	17.07	2.85	1.95	1.46	444	0.372
zz	600	1102	18.73	3.12	2.15	1.45	426	0.367
cir	1000	1364	22.57	3.81	2.67	1.43	407	0.357
sqr	1000	1344	23.33	3.85	2.67	1.44	404	0.35
zz	1000	1398	23.11	3.96	2.76	1.44	402	0.352

three trips, i.e. at x_1 = 40 mm, 100 mm, 170 mm, 250 mm, 400 mm, 600 mm, 800 mm and 1000 mm, where x and x_1 are the streamwise distances from the leading edge of the plate and from the location of the trip, respectively. However, measured data at x_1 = 250 mm, 400 mm, 600 mm and 1000 mm are only presented in this paper, for brevity. Boundary layer parameters at those locations are summarized in table 1. Here, the Reynolds numbers are based on the momentum thickness, θ; boundary layer and displacement thicknesses are denoted by δ and $\delta^\star$, respectively; shape factor, skin-friction and the friction velocity are denoted by H, C_f and u_τ, respectively. We may note that the skin-friction, C_f, has been estimated using the Clauser chart method [10]. Accurate wall identification has been done using the local peak probability distribution function method proposed by Orlu et al. [11]. For the present data, this method is also discussed in details in [12]. In the following, we also compare our data with the available direct numerical simulation (DNS) data of Schlatter and Örlü [4].

Our measured data are presented in the form of a diagnostic plot proposed by Alfredsson et al. [13] in figure 3, for all the three tripping devices. Here, we plot u_{rms}/U_∞ against the normalized local mean velocity U/U_∞. Figures 3 (a), (b), (c) and (d) show the comparison of our experimental data at x_1 = 250 mm, 400 mm, 600 mm and 1000 mm, respectively, with the DNS data from Schlatter and Örlü [4] corresponding to Re_θ = 1000 and 1410. At x_1 = 250 mm, data for the zigzag trip compare well with the DNS data compared to the data of other trips. As we go downstream, our measured data closely compare with the DNS data, for all the three trips. This indicates that early turbulence can perhaps be achieved using a zigzag trip. We may mention that diagnostic plot has also been used for checking the quality of our measured data in the near wall locations, as suggested by Orlu et al. [11].

Based on the diagnostic plot, Alfredsson et al. [14] suggest another plot where u_{rms} is normalized by the local mean velocity U and plotted against U/U_∞. The parameter u_{rms}/U is called as the local TI which is found to increase with decreasing value of U/U_∞. Local TI is shown in the figure 4. Similar to the figure 3, zigzag trip data collapsed with the reference data for x_1 = 250 mm downstream location. Our measured data for all three trips correspond to the reference data well, as we go downstream location.

The wall-normal distribution of the streamwise component of the normal Reynolds stresses is shown in figure 5, for all the three trips; here the normal stress is normalized by the friction velocity, u_τ. De Graaff and Eaton [15] suggested that the peak Reynolds stress in inner coordinates occurs at y^+ = 14, as is also seen in our present data. Although, figures 3 and 4 show

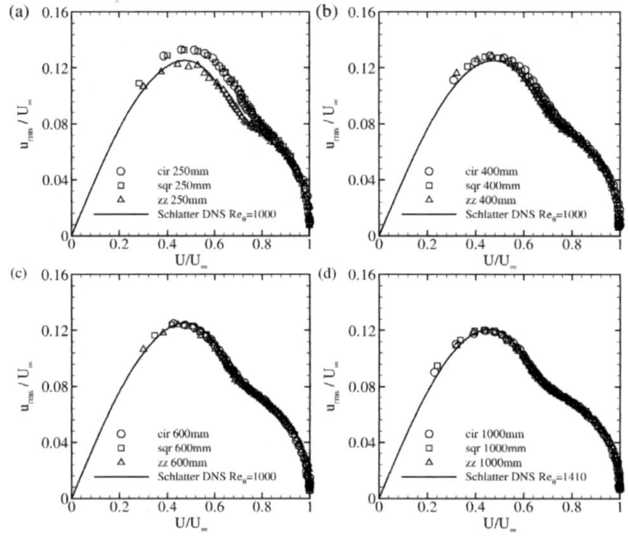

Figure 3. Diagnostic plot at different downstream locations. (a) $x_1 = 250$ mm. (b) $x_1 = 400$ mm. (c) $x_1 = 600$ mm. (d) $x_1 = 1000$ mm

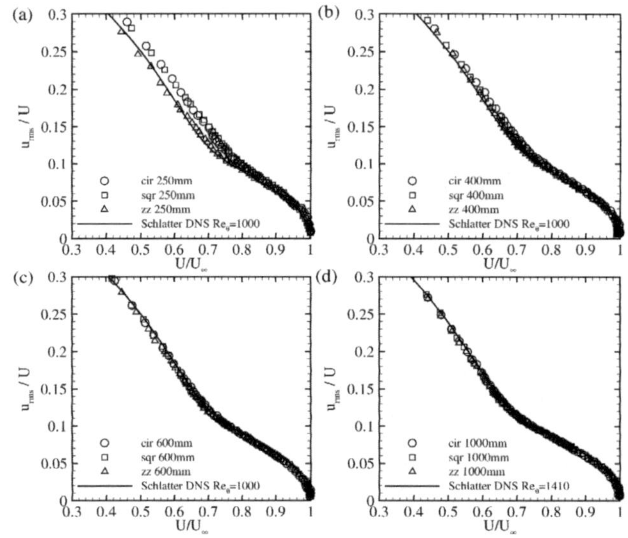

Figure 4. Local TI plot at different downstream locations. (a) $x_1 = 250$ mm. (b) $x_1 = 400$ mm. (c) $x_1 = 600$ mm. (d) $x_1 = 1000$ mm.

that our measured data collapse with the reference data from $x_1 = 400$ mm, the normal stress in inner coordinates, as shown in figure 5, indicates that the flow is yet to be fully turbulent even at $x_1 = 1000$ mm.

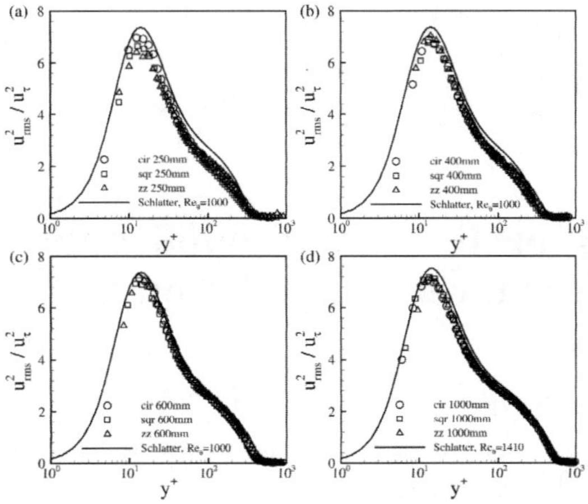

Figure 5. Streamwise component of the normal Reynolds stresses in inner coordinates. (a) x_1 = 250 mm. (b) x_1 = 400 mm. (c) x_1 = 600 mm. (d) x_1 = 1000 mm

4. Conclusion

To achieve turbulent boundary layer over a flat plate at a low Reynolds number, various tripping devices have been used. Measured data are presented at various downstream locations, i.e. from x_1 = 40 mm to 1000 mm behind a particular trip. While comparing the measured data with the available data in the literature, the present diagnostic and local TI plots (see figures 3 and 4) show that the flow may be considered to be fully turbulent at x_1 = 1000 mm. On the other hand, the wall-normal distribution of the streamwise component of the normal Reynolds stresses (figure 5) shows that the flow is yet to be fully turbulent, as compared to the reference data. This may be due to the memory effect of the tripping devices, as the Reynolds numbers, Re_θ, in the present measurements are less than the value of 2000. Because, in their DNS simulation, Schlatter and Örlü [5] found that the upstream memory effect due to a tripping device may not be there if the Reynolds number, Re_θ, is more than 2000. Although, closer to the trip location, comparatively better turbulence could be achieved using the zigzag trip with respect to the other trips, there may not be any significant effect of different trips if someone measures turbulent quantities far downstream, i.e. more than 1000 mm downstream from the trip location for the present freestream velocity.

References

[1] Elsinga G and Westerweel J 2012 *Exp. Fluids* **52** 865–876
[2] Andersson P, Berggren M and Henningson D S 1999 *Phys. Fluids.* **11** 134–150
[3] Adrian R, Meinhart C and Tomkins C 2000 *J. Fluid Mech.* **422** 1–54
[4] Schlatter P and Örlü R 2010 *J. Fluid Mech.* **659** 116–126
[5] Schlatter P and Örlü R 2012 *J. Fluid Mech.* **710** 5–34
[6] Hutchins N 2012 *J. Fluid Mech.* **710** 1–4
[7] Hanson R E, Buckley H P and Lavoie P 2012 *Experiments in fluids* **53** 863–871
[8] Gibbings J 1959 *On boundary-layer transition wires* (HM Stationery Office)
[9] Braslow A L and Knox E C 1958 *NACA TN 4363*
[10] Clauser F H 1956 *Advances in applied mechanics* **4** 1–51
[11] Örlü R, Fransson J H and Alfredsson P H 2010 *Progress in Aerospace Sciences* **46** 353–387
[12] Rengasamy K 2015 Master's thesis Indian Institute of Technology Kanpur, India
[13] Alfredsson P H and Örlü R 2010 *Eur. J. Mech. B/Fluids* **29** 403–406
[14] Alfredsson P H, Segalini A and Örlü R 2011 *Phys. Fluids.* **23** 041702
[15] De Graaff D B and Eaton J K 2000 *J. Fluid Mech.* **422** 319–346

Generalized wall function and its application to compressible turbulent boundary layer over a flat plate

J Liu[1] and S P Wu[2,3]

[1] Sino-French Engineer School, Beijing University of Aeronautics and Astronautics, 37 Xueyuan Road, Beijing 100191, China
[2] School of Aeronautic Science and Engineering, Beijing University of Aeronautics and Astronautics, 37 Xueyuan Road, Beijing 100191, China
[3] Ministry-of-Education Key Laboratory of Fluid Mechanics, Beijing University of Aeronautics and Astronautics, China

E-mail: jun.liu@buaa.edu.cn

Abstract. Wall function boundary conditions including the effects of compressibility and heat transfer are improved for compressible turbulent boundary flows. Generalized wall function formulation at zero-pressure gradient is proposed based on coupled velocity and temperature profiles in the entire near-wall region. The parameters in the generalized wall function are well revised. The proposed boundary conditions are integrated into Navier-Stokes computational fluid dynamics code that includes the shear stress transport turbulence model. Numerical results are presented for a compressible boundary layer over a flat plate at zero-pressure gradient. Compared with experimental data, the computational results show that the generalized wall function reduces the first grid spacing in the directed normal to the wall and proves the feasibility and effectivity of the generalized wall function method.

1. Introduction

The high-Reynolds-number turbulence models, which do not include wall correction terms in the differential equations, require modification for near-wall flow to account for fluid viscous effects and for damping of velocity fluctuations normal to the wall. Two main approaches for near-wall flows are near-wall simpler models and wall functions. The first approach is to patch in a simpler model near the wall, such as an algebraic model or a one-equation model. This method requires fine grid resolution near the wall, resulting in a relatively large total number of grid points. Although this method has been shown to yield good results for some flowfields[1], more numerical experiments are needed before it may be fully assessed. A disadvantage of the patched simpler models is that they also require fine grid spacing at the wall.

The second approach is to choose wall functions. Empirically derived algebraic models of the near-wall region of the turbulent boundary layer, high-Reynolds-number turbulence models provide boundary conditions to the mean flow Navier-Stokes equations at the first point off the wall. Researchers[2,3] demonstrated that the computational results of friction coefficient and heat transfer highly rely on the divided ways of grid in the directed normal to the wall, especially the grid spacing. These models of the near-wall region are called wall functions. Because the high-gradient near-wall region for velocity and temperature is modeled with these empirical relationships, the first point off the wall may be placed much farther away from the viscous sublayer. Wall functions that do not constraint the position of the first grid point between the wall and logarithmic layer are called generalized wall function. Generalized wall function method reduces the number of points required to discretize a flowfield and increases the maximum allowable time step. Both of these advantages are especially useful in modeling unsteady viscous flows.

Based on analysis of Millikan[4] for turbulent channel flows, most of the early wall function boundary conditions did not consider the pressure gradient or heat transfer. Numerous formulas have been proposed to describe the universal turbulent velocity profile, called by Coles[5] "the law-of-the-wall". Spalding[6] introduced a single formula over the whole range of the wall which represented adequately the experimental data for the universal turbulent velocity profile when the viscosity and density of the fluid are uniform. White and Christoph[7] proposed a new approach for analyzing the compressible turbulent boundary layer by using the Crocco-Busemann energy approximation. Nichols and Nelson[8] described a wall function boundary condition upon coupled velocity and temperature boundary-layer profiles based on White's work. The novel wall function considering heat transfer and compressibility was applicable to incompressible flows, compressible flows and flows with heat transfer.

Based on Nichols's work, the paper includes mainly three sections. In Section 2, wall function boundary conditions are improved in the aspect of compressibility and heat transfer. The proposed wall functions are formulated for zero-pressure gradient flows. The validation of the proposed generalized wall function for the compressible turbulent boundary layer over a flat plate is performed in section 3. By integrating the wall function boundary conditions with the shear stress transport turbulence model, detailed numerical results for flow over a flat plate at zero-pressure gradient are also presented in this section. The conclusions of this paper are given in section 4.

2. Generalized wall function including heat transfer and compressibility

For turbulent flow conditions with constant molecular viscosity, the velocity profile can be split into three distinguished region: the viscous sublayer, the logarithmic layer and the defect layer. In a quasi-equilibrium boundary layer, such as flow over a flat plate at zero-pressure gradient, the region between the wall and the outer edge of the logarithmic layer is universal.

Near the wall, derivatives in streamwise direction can be neglected and the turbulence variables depend only on the coordinate y, which is directed normal to the wall. Essentially, this is a turbulent Couette flow approximation. To derive solutions for the viscous sublayer and logarithmic layer, the equations are recast in dimensionless form:

$$u^+ = \frac{u}{u_\tau},\quad y^+ = \frac{\rho_w u_\tau y}{\mu_w},\quad u_\tau = \left(\frac{\tau_w}{\rho_w}\right)^{1/2} \tag{2.1}$$

u, velocity parallel to the wall; τ_w, wall shear stress; ρ_w, wall density; μ_w, wall molecular viscosity; y, the distance to the wall; u_τ, friction velocity, and subscript w is a label meaning the wall. Nichols proposed the velocity profiles near the wall concerning heat transfer and compressibility:

$$y^+ = u^+ + y^+_{\text{White}} + e^{-\kappa B}\left[-1-\kappa u^+ - \frac{(\kappa u^+)^2}{2!} - \frac{(\kappa u^+)^3}{3!}\right] \tag{2.2}$$

where

$$y^+_{\text{White}} = \exp\{(\kappa/\gamma)[\sin^{-1}(2\gamma u^+ - \beta/Q) - \phi] - \kappa B\} \tag{2.3}$$

$$\gamma = \frac{r u_\tau^2}{2C_p T_w},\quad \beta = \frac{q_w \mu_w}{\rho_w T_w k_w u_\tau},\quad Q = (\beta^2 + 4\gamma)^{1/2},\quad \varphi = \sin^{-1}\left(\frac{-\beta}{Q}\right) \tag{2.4}$$

with r, recovery factor; C_p, pressure coefficient; T_w, wall temperature; q_w, wall heat transfer; k_w, thermal conductivity; heat transfer $q_w = -k_w \partial T_w / \partial y$; the constant κ and B are generally taken as 0.41 and 5.5, respectively. Non-dimensional parameter γ and β model compressibility effects and heat transfer effects. The wall heat transfer q_w appears only in the definition of β. The two parameters γ

and β arise from writing the Crocco-Busemann assumption for the energy equation $T = a + bu + cu^2$ [9]. The Crocco assumption appears to be quite reasonable for arbitrary compressible turbulent flows. For example, Lee[10] expressed the density distribution in terms of velocity, hence uncoupling the energy and momentum equations. The constants (a,b,c) can be related to wall conditions as follows. First, at the wall, $u = 0$, which is the no-slip condition. Hence $a = T_w$. Second, the temperature gradient at the wall must reflect the wall heat flux. Hence $b = q_w\mu_w / k_w\tau_w$. Finally, $c = -r/2C_p$ for the adiabatic wall. It is desirable to use the wall variable $u^+ = u/u_\tau$ in the final form of the Crocco-Busemann relation. With the above considerations, for a perfect gas Crocco-Busemann relation can now be written as:

$$\rho_w / \rho = T / T_w = 1 + \beta u^+ - \gamma u^{+2} \Rightarrow$$
$$T = T_w (1 + \beta u^+ - \gamma (u^+)^2) \tag{2.5}$$

Equation (2.5) is used to formulate the equation (2.3). For adiabatic wall cases $q_w = 0$, that is $\beta = 0$, with the definition of γ in equation (2.4) and the wall variable $u^+ = u/u_\tau$ in equation (2.1), the Crocco-Busemann equation reduces to:

$$T = T_w - ru^2 / (2C_p) \tag{2.6}$$

2.1. *Improved wall function formulation*

Equation (2.5) is derived from the Crocco-Busemann energy approximation. For the compressible turbulent boundary layer at zero-pressure gradient, the expression of Crocco-Busemann equation is:

$$T = T_w + (T_{aw} - T_w)\frac{u}{U_e} - \frac{ru^2}{2C_p} \tag{2.7}$$

where

$$T_{aw} = T_e(1 + \frac{\Gamma - 1}{2} rMa_e^2) \tag{2.8}$$

U_e, the free stream velocity; T_e, the free stream temperature and Ma_e^2, the free stream Mach number; T_{aw}, adiabatic wall temperature; $\Gamma = 1.4$ for the perfect gas; subscripts of $(\,)_{aw}$ and $(\,)_e$ mean the adiabatic wall and the free stream. T_w has a special relation with q_w as follows[11]:

$$\frac{T_{aw} - T_w}{U_e} = \frac{q_w \mu_w}{k_w \tau_w} \tag{2.9}$$

Integrating equation (2.7) and equation (2.9), we obtain:

$$T = T_w + \frac{q_w \mu_w}{k_w \tau_w} u - \frac{ru^2}{2C_p} \tag{2.10}$$

Equation (2.10) becomes equation (2.5) by supposing the parameters β and γ. Moreover, the above derivation process shows that parameter β comes from equation (2.9). In other words, the formulation of β is expressed implicitly in the equation (2.9). Utilizing $\Pr = \mu C_p / k_w$, Stanton number St [12] is formulated by:

$$St = \frac{q_w}{\rho_e U_e C_p (T_{aw} - T_w)} = \frac{2C_f}{\Pr} \tag{2.11}$$

$$\Rightarrow \frac{St}{2C_f} = \frac{1}{\Pr} \approx 1.39 \tag{2.12}$$

where C_f, skin friction coefficient; ρ_e, free stream density; $\Pr = 0.72$, laminar Prandtl number. Therefore the equation (2.9) is rewritten to:

$$\frac{T_{aw}-T_w}{U_e}=\frac{q_w\mu_w}{k_w\tau_w}=\frac{q_w}{\rho_e U_e^2 C_p St}=\frac{\Pr}{2C_f}\cdot\frac{q_w}{\rho_e U_e^2 C_p} \tag{2.13}$$

According to a large number of supersonic experimental data with skin friction $C_f \approx 3\times10^{-3}$ and CFD numerical simulation results, the empirical Reynolds analogy factor $St/(2C_f)$ [13] should be 1.16. Hence the equation (2.10) must be revised by:

$$T=T_w+\frac{1.2q_w\mu_w}{k_w\tau_w}u-\frac{ru^2}{2C_p} \tag{2.14}$$

with $1.2\approx 1.39/1.16$.

We denominate the improved wall function "Method 1", with heat transfer parameter β:

$$\beta=\frac{1.2q_w\mu_w}{\rho_w T_w k_w u_\tau} \tag{2.15}$$

2.2. Generalized wall function based on Walz relation

In order to revise the Crocco-Busemann energy approximation on condition of $\Pr\approx 1$, Walz[14] proposed:

$$\frac{T}{T_e}=\frac{T_w}{T_e}+(\frac{T_{aw}-T_w}{T_e})f(\frac{u}{U_e})+\frac{T_e-T_{aw}}{T_e}(\frac{u}{U_e})^2 \tag{2.16}$$

where

$$f(\frac{u}{U_e})=(1-\alpha)(\frac{u}{U_e})^2+\alpha\frac{u}{U_e},\quad \alpha\approx 0.82$$

(2.17)

Equation (2.16) is called Walz function. According to equation (2.17), we have:

$$\frac{T}{T_e}=\frac{T_w}{T_e}+\alpha(\frac{T_{aw}-T_w}{T_e})\frac{u}{U_e}+[\frac{T_e-T_{aw}}{T_e}+(1-\alpha)\frac{T_{aw}-T_w}{T_e}](\frac{u}{U_e})^2 \tag{2.18}$$

Then

$$T=T_w+\alpha\frac{q_w\mu_w}{k_w\tau_w}u-[\frac{r}{2C_p}-(1-\alpha)\frac{q_w\mu_w}{k_w\tau_w U_e}]u^2 \tag{2.19}$$

The new dimensionless parameters are in the following form:

$$\beta=\frac{\alpha q_w\mu_w}{\rho_w T_w k_w u_\tau},\quad \gamma=\frac{ru_\tau^2}{2C_p T_w}-\frac{(1-\alpha)q_w\mu_w}{\rho_w T_w k_w U_e},\quad \alpha\approx 0.82 \tag{2.20}$$

We dominate the generalized wall function method based on Walz equation "Method 2".

In this section, the generalized wall function, as a semi-empirical approach, has been corrected and proposed by two different ways. In the section 3, we will compare the improved wall function "Method 1" and the generalized wall function method based on Walz relation "Method 2" by means of numerical simulation of compressible turbulent boundary layer over a flat plate.

3. Implication to compressible boundary layer over a flat plate

The generalized wall functions described above are implemented in a two-dimensional compressible RANS flow and SST turbulence model. Inflow Conditions[15] of flow over a flat plate are defined by $M_\infty=3$, $T_\infty=288K$, $\rho_\infty=0.17527kg/m^3$, wall temperature $T_w=294.44K$. In the numerical simulation, discretized format is Roe scheme[16]. SST model[17] is solved on different first grid spacing $\Delta y=5\times10^{-7}$, and $2.5\times10^{-5}, 5\times10^{-5}$, which corresponds the dimensionless distance $y^+=0.1$, 25, 50 by making use of the relation $y^+=\rho_w u_\tau \Delta y/\mu_w$. The numerical simulation is performed using a finite difference CFD code developed by us. Reliability of the code has been validated and verified by a variety of simulations[18,19]. The computational domain is a rectangle with a length of 1m and a width of 0.8m.

The main work for CFD code coupled with wall function boundary condition is to replace the corresponding terms of momentum and energy equations in CFD code respectively by wall shear stress τ_w and heat transfer q_w obtained from the wall function. Firstly, we use velocity law-of-the-wall equation (2.2) and the velocity u_1 at the first point off the wall obtained by CFD code to solve τ_w iteratively. Then, we replace the shear stress term $\tau_{1/2}$ of the momentum equation in CFD viscous subroutine by τ_w. Based on the constant stress assumption in the lower part of the turbulent boundary layer[20], it is reasonable to let $\tau_{1/2}$ be equal to τ_w. Secondly, we utilize u^+ obtained from the above velocity law-of-the-wall and the temperature T_1 at the first point off the wall computed by CFD code to calculate parameter β by temperature law-of-the-wall equation (2.5) and further solve the wall heat transfer q_w. Similarly, we replace the heat transfer term $q_{1/2}$ of the energy equation in CFD viscous subroutine by q_w. According to the assumption by Nichols[8] that heat transfer term keeps constant in the lower part of the turbulent boundary layer, in other words $q_{1/2}$ equals q_w. Finally, the results of τ_w and q_w can be returned back to main CFD code.

Nichols's wall function formulation is described from equation (2.2) to (2.4). The improved wall function "Method 1", equation (2.15) and the generalized wall function method based on Walz function "Method 2", equation (2.20) are integrated into de boundary conditions of the flow over a flat plate.

Figure 1 shows the numerical simulation solutions for u^+ and y^+ in logarithmic coordinates when $y^+ = 0.1, 25, 50$. Linear in the legend represents $u^+ = y^+$ for the viscous sublayer near the wall, and log-law of wall stands for $u^+ = \kappa^{-1} \cdot \mathrm{Ln}(y^+) + B$, the logarithmic layer[21].

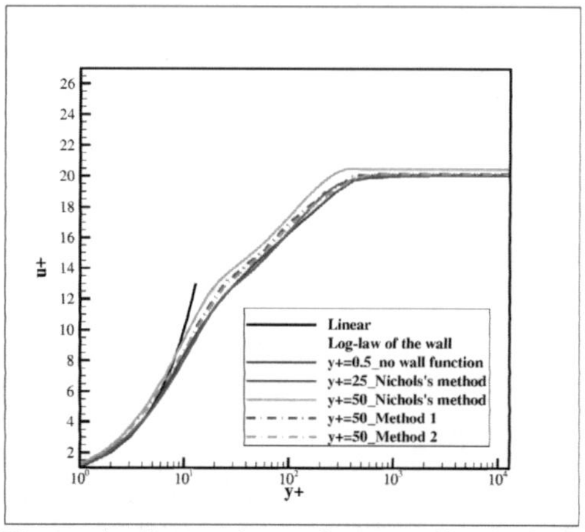

Figure 1. Comparison of near-wall velocity profiles among different grid spacing in directed normal to the wall

Three results could be obtained from the figure 1. When $y^+ = 0.1$, the first grid notes in the directed normal to the wall are located in viscous sublayer, the simulation do not need use the wall function; when the first grid spacing is broadening to $y^+ = 50$, the generalized wall function become crucial in order to get the better results compared with the theoretical solutions. Generalized wall functions are designed to be used with coarse near-wall grids; the results of Method 1 and Method 2 presented with dashed line are much better than the Nichols's wall function method with solid lines.

Figure 2 demonstrates comparisons of skin friction coefficient $C_f = 2\tau_w / (\rho_\infty U_\infty^2)$ and Van Driest theoretical data[13] under the above inflow conditions and different grid spacing. Figure 3 presents the computational results for heat transfer q_w and x. It can be seen from the two figures that computational results agree with the theoretical solutions well, which illustrates that the simulations are reliable.

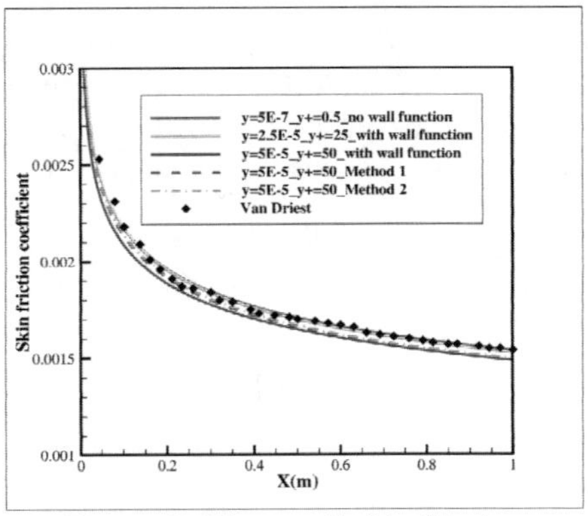

Figure 2. Near-wall skin friction profiles of different grid spacing

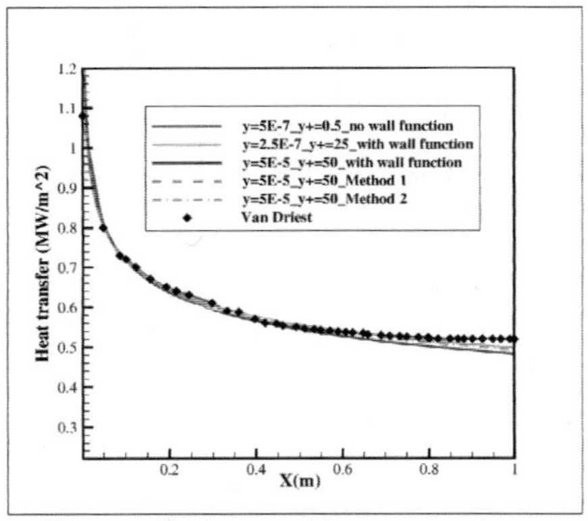

Figure 3. Near-wall heat transfer profiles of different grid spacing

We define the deviation[22] between the numerical result and experimental data ε by

$$\varepsilon = \frac{1}{N}\sum_{j=1}^{N} |\phi(r)_j - \phi(d)_j| \tag{3.1}$$

with N, number of discrete points; ϕ, turbulent quantity; r and d mean the numerical result and theoretical data. The deviation defined can indicate well the reliability of method. Finally, we can compare the deviation between Method 1 and Method 2 according to equation (3.1) with the same first grid spacing $y^+ = 50$.

Table 1. Deviation of velocity, skin friction and heat transfer

	$\varepsilon(u^+)$	$\varepsilon(C_f)$ $(\times 10^{-5})$	$\varepsilon(q_w)$ $(\times 10^{-3})$
Method 1	1.240	3.912	9.789
Method 2	1.149	3.267	9.142

From the table 1, the results of Method 2, C_f and q_w, are more reliable than Method 1 on account of the much smaller deviation of Method 2. Therefore Method 2 is chosen as the generalized wall function. We formulate finally the generalized wall function based on Walz function as follows:

$$y^+ = u^+ + y^+_{\text{White}} + e^{-\kappa B}[-1 - \kappa u^+ - \frac{(\kappa u^+)^2}{2!} - \frac{(\kappa u^+)^3}{3!}] \tag{3.2}$$

with $$y^+_{\text{White}} = \exp\{(\kappa/\gamma)[\sin^{-1}(2\gamma u^+ - \beta/Q) - \phi] - \kappa B\}$$

$$\beta = \frac{\alpha q_w \mu_w}{\rho_w T_w k_w u_\tau},\quad \gamma = \frac{r u_\tau^2}{2C_p T_w} - \frac{(1-\alpha) q_w \mu_w}{\rho_w T_w k_w U_e},\quad \alpha \approx 0.82,\ Q = (\beta^2 + 4\gamma)^{1/2},\quad \varphi = \sin^{-1}(\frac{-\beta}{Q}) \tag{3.3}$$

4. Conclusions

A general set of wall function boundary conditions has been developed and corrected for compressible flows with heat transfer near the wall. The generalized wall functions are based on coupled velocity profiles and Walz relation in the near-wall of the boundary layer. The proposed boundary conditions are integrated into Navier-Stokes computational fluid dynamics code that includes the shear stress transport turbulence model. The formula of the heat transfer term in the near-wall region was derived so that the coupling between the wall function boundary condition and CFD code is realized more accurately.

Through the analysis of numerical results of the near-wall region, we obtained the velocity, skin friction and heat transfer focusing on compressible turbulent boundary layer over a flat plate at zero-pressure gradient. Moreover, the results verify that the proposed generalized wall functions are designed to be used with coarse near-wall grids. For the flat plate case, the modified compressible wall function can produce reasonable solutions of skin friction and heat transfer. The successful numerical results powerfully show that generalized wall function and near-wall grid adaptation ensure both an appropriate resolution of near-wall flow physics and take into account the range of validity and stability of the wall-function formulation.

Acknowledgements
The research was sponsored by the National Natural Science Foundation of China (NSFC 73024802).

References

[1] Horstman C 1992 *AIAA J* 30 pp 1480–1
[2] Alter S J 2004 *AIAA Paper* p 612
[3] Thivet F Besbes O and Knight D D 2000 *38 th ASME*(Reno, NV)
[4] Millikan C B 1938 *Proc. 5th Int. Cong. Appl. Mech* (New York-London: Wiley/Chapman and Hall) pp 386–92
[5] Coles D 1956 *J. Fluid Mech.* **1** pp 191–226
[6] Spalding D B 1961 *J. Appl. Mech.* **28** pp 455–8
[7] White F M and Christoph G H 1972 *J. Fluids Eng.* **94** pp 636–42
[8] Nichols R H and Nelson C C 2004 *AIAA J* **42** pp 1107–14
[9] White F M and Christoph G H 1971 *Final Technical Report* pp 70–133
[10] Lee R E, Yanta W J and Leonas A C 1969 *Naval Ordnance Laboratory Report* (White Oak, Md)
[11] Wilcox D C 1998 *Turbulence modeling for CFD*(Canada, CA: DCW industries)
[12] Achaichia A and Cowell T A 1988 *Exp. Therm Fluid Sci.* **1** pp 147–57
[13] Chen M Z 2002 *Fundamentals of viscous Fluid Dynamics*(Beijing: Higher Education Press)
[14] Walz A 1962 Compressible turbulent boundary layers *Ed. du CNRS*
[15] Boris J P, Oran E S, Gardner J H et al 1985 *Ninth International Conf. on Numerical Methods in Fluid Dynamics* Springer Berlin Heidelberg pp 98–102
[16] Roe P L 1981 *J Comput Phys* **43** pp 357–72
[17] Menter F R 1994 *AIAA J* **32** pp 1598–605
[18] Lee P F and Wu S P 2010 *J. of Aerospace Power* **8** pp 1705–10
[19] Xie J R, Wu S P and Wang H 2005 *J. of Beijing University of Aeronautics and A* **31** pp 274–7
[20] Tennekes H and Lumley J L 1974 *A First Course in Turbulence*(Cambridge, MA: MIT Press)
[21] Eça L and Hoekstra M 2010 *Int. J. Numer. Methods Fluids* **63** pp 953–88
[22] Sandham N D and Yee H C 2001 *Computational Fluid Dynamics* Springer Berlin Heidelberg pp 361–6

Fifteenth Asian Congress of Fluid Mechanics (15ACFM) IOP Publishing
IOP Conf. Series: Journal of Physics: Conf. Series **822** (2017) 012018 doi:10.1088/1742-6596/822/1/012018

Global stability analysis of axisymmetric boundary layer over a circular cone

Ramesh Bhoraniya[1], **Narayanan Vinod**[2]

[1] Assistant professor,Department of Mechanical Engineering, Faculty of Engineeering, Marwadi Education Foundation,Rajkot,INDIA

[1] Research Scholar, Department of Mechanical Engineering,Indian Institute of Technology, Gandhinagar, INDIA

[2] Assistant Professor, Department of Mechanical Engineering,Indian Institute of Technology, Gandhinagar, INDIA

E-mail: rameshbhoraniya@iitgn.ac.in

Abstract. This paper presents linear biglobal stability analysis of axisymmetric boundary layer over a circular cone. An incompressible flow over a sharp circular cone is considered with zero angle of attack. The base flow velocity profile is fully non-parallel and non-similar. Linearized Navier-Stokes (LNS) equations are derived for disturbance flow quantities using the standard procedure. The LNS equations are discretized using Chebyshev spectral collocation method. The governing equations along with boundary conditions form a general eigenvalues problem. The numerical solution of general eigenvalues problem is obtained using ARPACK subroutine, which uses Arnoldis iterative algorithm. The global temporal modes are computed for the range of Reynolds number and semi-cone angles(α)for the axisymmetric mode(N=0). The flow is found temporally and spatially stable for 1° semi-cone angle and the range of Reynolds numbers considered. However, flow becomes temporally unstable and spatially stable with the increase in semi-cone angle(α). The wave-like behaviour of the disturbances is found at small semi-cone angles (α).

1. Introduction

The laminar-turbulent transition of the boundary layer is important to study in aerodynamics and marine hydrodynamics applications i.e. submarines, aircrafts etc. While it is important to study the physics of the problem as it has practical applications in aerodynamics and marine. The growth of small disturbances is the first step towards the transition process and it is studied through linear stability analysis. The linear stability analysis of parallel flow is well understood from the Orr-Sommerfeld equation and its solutions [1]. Parallel flow assumption is not valid when the base flow is varying in wall normal and streamwise direction. Thus, the mean flow is non-parallel. The stability analysis of such flows is performed using global stability analysis [2].

In the present work, we study the stability analysis of an incompressible boundary layer over a circular cone. The incoming flow is parallel to the axis of the cone. A pressure gradient is generated in the streamwise direction due to the cone angle. The flow is developing in the streamwise direction and is nonparallel. In this case the transverse curvature and pressure gradient play a crucial role in the stability of boundary layer formed. The literature on the stability analysis of incompressible flow over a circular cone is very sparse. The objectives of the present work are to understand the effect of transverse curvature and pressure gradient on the boundary layer stability .

2. Problem formulation

Linearized Navier-Stokes equations are derived for disturbance flow quantities in spherical coordinates. The equations are normalized by free stream velocity (U_∞) and cone radius (a) at the inlet of the domain respectively. The angle of attack is zero and hence the base flow is axisymmetric. Base flow in the boundary layer over a circular cone is axisymmetric and developing in streamwise direction. The recent development in the global stability analysis has demonstrated that wave-like behaviour of the disturbance amplitudes is not only the solution [3]. The Reynolds number is defined based on cone radius at the inlet of domain and free stream velocity.

$$R_e \equiv \frac{U_\infty a}{\nu} \tag{1}$$

We follow the standard procedure for stability analysis with three dimensional perturbation to mean flow. Disturbances are assumed to be in normal mode form with the amplitudes are varying in r and θ direction.

$$\overline{U_r} = U_r + u_r, \overline{U_\theta} = U_\theta + u_\theta, \overline{P} = P + p \tag{2}$$

$$q(r,\theta,t) = \hat{q}(r,\theta)e^{[i(N\phi-\omega t)]} \tag{3}$$

where,
r = radial coordinate
θ = polar coordinate
ϕ = azimuthal coordinate
ω = frequency of waves
N = azimuthal wavenumber
a = cone radius at the inlet
The Linearized Navier-Stokes equations for disturbance flow quantities are obtained as,

$$U_r\frac{\partial u_r}{\partial r} + \frac{\partial U_r}{\partial r}u_r + \frac{U_\theta}{r}\frac{\partial u_r}{\partial \theta} + \frac{\partial U_r}{\partial \theta}\frac{u_\theta}{r} - \frac{2U_\theta}{r}u_\theta + \frac{\partial p}{\partial r} - \frac{1}{Re}[\nabla^2 u_r - \frac{2}{r^2}u_r - \frac{2}{r^2}\frac{\partial u_\theta}{\partial \theta} - \frac{2cot\theta}{r^2}u_\theta] = i\omega u_r \tag{4}$$

$$U_r\frac{\partial u_\theta}{\partial r} + u_r\frac{\partial U_\theta}{\partial r} + \frac{U_\theta}{r}\frac{\partial u_\theta}{\partial \theta} + \frac{u_\theta}{r}\frac{\partial U_\theta}{\partial \theta} + u_\theta\frac{U_\theta}{r} + u_r\frac{U_\theta}{r} + \frac{1}{r}\frac{\partial p}{\partial \theta} - \frac{1}{Re}[\nabla^2 u_\theta + \frac{2}{r^2}\frac{\partial u_r}{\partial \theta} - \frac{u_\theta}{r^2 sin^2\theta}] = i\omega u_\theta \tag{5}$$

$$\frac{\partial u_r}{\partial r} + \frac{2u_r}{r} + \frac{1}{r}\frac{\partial u_\theta}{\partial \theta} + \frac{u_\theta cot\theta}{r} = 0 \tag{6}$$

where,

$$\nabla^2 \equiv \frac{\partial}{\partial r^2} + \frac{2}{r}\frac{\partial}{\partial r} + \frac{1}{r^2}\frac{\partial}{\partial \theta^2} + \frac{cot\theta}{r^2}\frac{\partial}{\partial \theta} \tag{7}$$

2.1. Boundary conditions

At the solid surface of cone no slip and no penetration conditions are applied to all velocity disturbance amplitudes. At the free stream, far away from a solid wall, it is expected to vanish all velocity and pressure disturbances. The boundary conditions for pressure do not exist physically at the wall. However, compatibility conditions derived from the LNS equations have collocated at the solid wall [2].

$$u_r(r,\theta_{min}) = 0,\ u_\theta(r,\theta_{min}) = 0, \tag{8}$$

$$u_r(r,\theta_{max}) = 0,\ u_\theta(r,\theta_{max}) = 0,\ p(r,\theta max) = 0 \tag{9}$$

$$\frac{\partial p}{\partial r} = \frac{1}{Re}[\nabla^2 u_r - \frac{2}{r^2}\frac{\partial u_\theta}{\partial \theta}] - U_r\frac{\partial u_r}{\partial r} - \frac{U_\theta}{r}\frac{\partial u_r}{\partial \theta} \tag{10}$$

$$\frac{1}{r}\frac{\partial p}{\partial \theta} = \frac{1}{Re}[\nabla^2 u_\theta + \frac{2}{r^2}\frac{\partial u_r}{\partial \theta}] - U_r\frac{\partial u_\theta}{\partial r} - \frac{U_\theta}{r}\frac{\partial u_\theta}{\partial \theta} \tag{11}$$

At inlet and outlet linearly extrapolated boundary conditions are applied. The disturbances are defined in the form of unknown values of the disturbances within the interior of the domain itself [3, 4].

$$u_{r_1}[r_3 - r_2] - u_{r_2}[r_3 - r_1] + u_{r_3}[r_2 - r_1] = 0 \tag{12}$$

$$u_{r_{n-2}}[r_n - r_{n-1}] - u_{r_{n-1}}[r_n - r_{n-2}] + u_{r_n}[r_{n-1} - r_{n-2}] = 0 \tag{13}$$

LNS equations are discretized using Chebyshev spectral collocation method (**n** points in the streamwise direction and **m** points in the wall normal direction). The clustering of collocation points is employed in wall normal direction near the wall to improve resolution and hence to capture real flow dynamics [5]. In the streamwise direction grid mapping is applied to remove the Gibbs oscillation [6]. The governing equations together with the boundary conditions forms a generalized eigenvalue problem, of the form,

$$\begin{bmatrix} A_{11} & A_{12} & A_{13} \\ A_{21} & A_{22} & A_{23} \\ A_{31} & A_{32} & A_{33} \end{bmatrix} \begin{bmatrix} u_r \\ u_\theta \\ p \end{bmatrix} = i\omega \begin{bmatrix} B_{11} & B_{12} & B_{13} \\ B_{21} & B_{22} & B_{23} \\ B_{31} & B_{32} & B_{33} \end{bmatrix} \begin{bmatrix} u_r \\ u_\theta \\ p \end{bmatrix} \tag{14}$$

$$[A][\phi] = i\omega[B][\phi] \tag{15}$$

where the matrices A and B are real and of size 3×n×m , $i\omega$ is an eigenvalues and ϕ is eigenvector.

2.2. Solution of general eigenvalues problem

The eigenvalues problem described above is very large to solve for all the eigenvalues. However flow becomes unstable due to very few nuber of eigenmodes with its largest imaginary part. Hence, we considered the few number of dominant eigenvalues and associated eigenvectors only. The QZ algorithm computes the full spectrum of eigenvalues and hence it requires large computational efforts. The iterative method based on Arnoldi's algorithm is a proper choice to compute few selected eigenvalues. The Krylov subspace provides the possibility of extracting major part of the spectrum using shift and invert strategy. The computations of Krylov subspace along with Arnoldis algorithm applied to eigenvalues problem become easy.

$$(A - \lambda B)^{-1}B\phi = \mu\phi \qquad where, \quad \mu = \frac{1}{i\omega - \lambda} \tag{16}$$

Where, λ being the shift parameter and μ is the eigenvalues of the converted problem. Sometimes it is also called spectral transformation, which converts generalized eigenvalues problem to a standard eigenvalues problem. The Krylov subspace may be computed by successive resolution of a linear system with matrix (A-λB), using LU decomposition. Full spectrum method is employed for this small subspace to get a good approximate solution to the original general eigenvalues problem [7]. The major part of the spectrum can be extracted with the large size of Krylov subspace. The computed spectrum is always nearby the shift parameter. The good approximation of shift parameter reduces the number of iterations to converge the solution to required accuracy level. However, the larger size of subspace makes the solution almost independent from the shift parameter λ. We tested the code for several values of shift parameter. The convergence of the solution depends on the value of shift parameter. A good approximation of the shift value needs less number of iterations. However, we have taken a maximum number of iterations equal to 300. Hence the convergence of the solution may not be affected by shift parameter. Given the large subspace size of $k = 250$, the part of the spectrum for our instability analysis could be recovered in the single computation that took about 4 hours on Intel Xenon(R)CPU E5 26500@2.00GHz × 18.

IOP Conf. Series: Journal of Physics: Conf. Series **822** (2017) 012018 doi:10.1088/1742-6596/822/1/012018

3. Results and discussions

In the present analysis, Reynolds number is varying from 174 to 1047 and semi-cone angle from 1° to 4° for axisymmetric mode(N=0). The stream wise length is normalized by cone radius at the inlet of the domain. The domain length in wall normal direction is taken as $L_\theta=12^o$. The number of collocation points taken in radial and polar directions are n=121 and m=121 respectively. The general eigenvalues problem is solved using Arnoldi's iterative algorithm. Eigenvalues with largest imaginary part is selected. It has been also checked for spurious mode. The sponging is applied with 90 percentage sponge strength over 20 percentage domain to prevent spurious reflection at the outlet.

Figure 1 shows the eigenspectrum for semi-cone angle $\alpha=1^o$, Re=174 and azimuthal wave

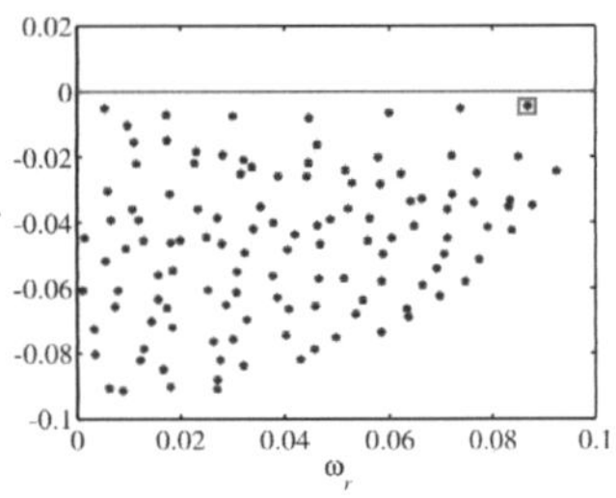

Figure 1. Eigenspectrum for N=0, $\alpha=1^o$ and Re=174 .

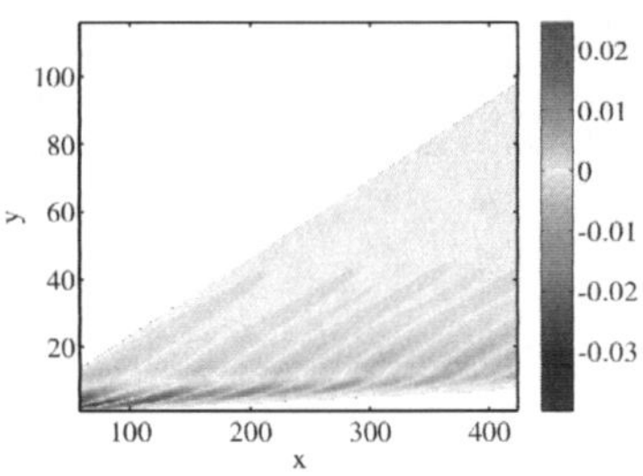

Figure 2. Contour plot of u_r for N=0, $\alpha=1^o$ and Re = 174.

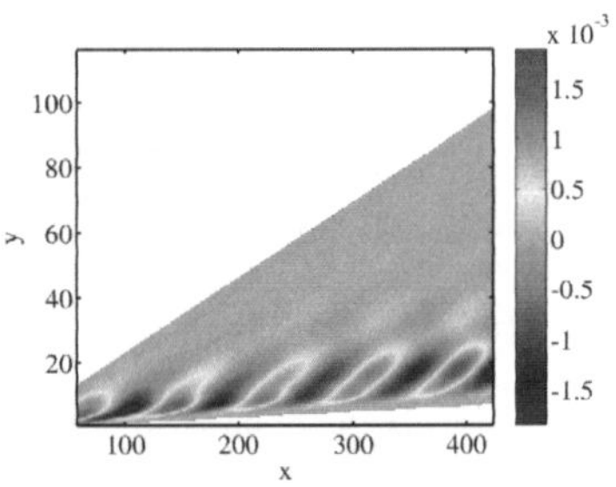

Figure 3. Contour plot of u_θ for N=0, $\alpha=1^o$ and Re = 174.

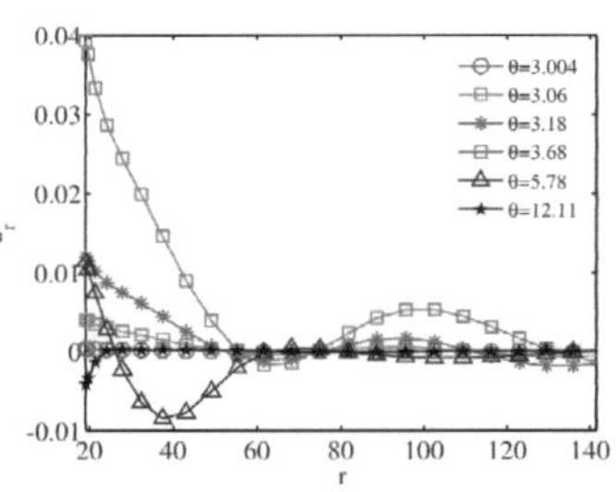

Figure 4. Streamwise variation of u_r for N=0, $\alpha=3^o$ and Re=174.

number N=0. The most unstable oscillatory mode has an eigenvalues, $\omega=0.0869-0.0045i$ and it is temporally stable because $\omega_i < 0$. The computed eigenmodes (all are not shown here) in the range of Reynolds number and $\alpha=1^o$ are temporally stable because all eigenmode have largest $\omega_i < 0$. Figure 2 and 3 shows the global mode structure for radial (u_r) and polar (u_θ) disturbances for eigenmode $\omega=0.0869-0.0045i$, as marked by square in Figure 1. The magnitude of the u_r disturbance amplitudes is one order higher than that of u_θ disturbances. The spatial structure of the disturbances is also elongated in spatial directions. The radial and polar disturbance amplitudes show positive and negative wavelets in the streamwise direction. This proves wave like nature of the disturbance amplitudes in the stream wise direction for Re = 174. Figure 4 shows variation of u_r in streamwise direction for 3° semi-cone angle. It shows that amplitudes of disturbances decay in streamwise direction towards the downstream. Increase in semi-cone angle makes the flow spatially stable.

Figure 5 shows the eigenspectrum for semi-cone angle α=2° and Re=872. The most unstable oscillatory mode has an eigenvalue $\omega = 0.1460 + 0.0012i$. The largest $\omega_i > 0$, proves that the flow is temporally unstable. Figure 6 and 7 shows the spatial structure for radial (u_r) and polar (u_θ) disturbances for most unstable temporal mode as marked by square in figure 5. Figure 6 and 7 shows that the region of contamination is comparatively less for u_r and u_θ for higher semi-cone angle (α=2°) and higher Reynolds number(Re=872).

Figure 9 shows the normalized spatial amplification rate (A_x) in the streamwise direction for various Reynolds numbers for 1° semi-cone angle . A_x shows overall effect of all the disturbances together in the streamwise direction towards the down stream [8]. The A_x reduces initially

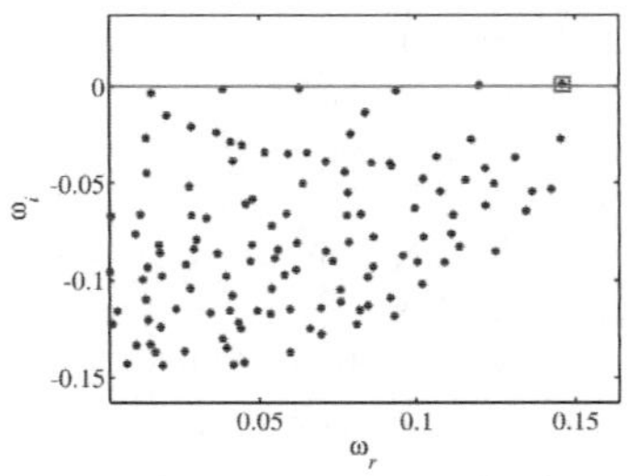

Figure 5. Eigenspectrum for N=0, α=2° and Re=872.

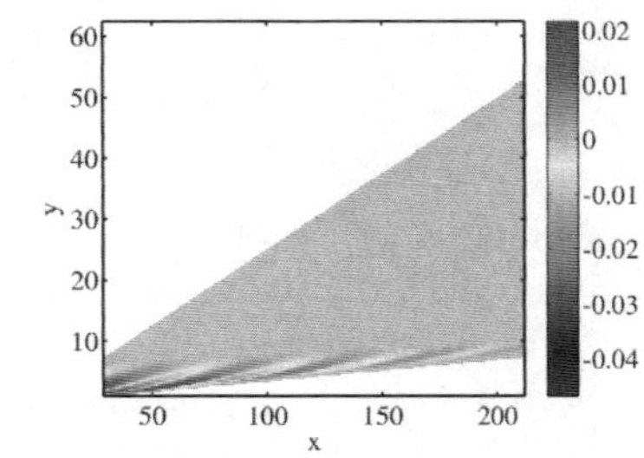

Figure 6. Contour plot of u_r for N=0, α=2° and Re = 872.

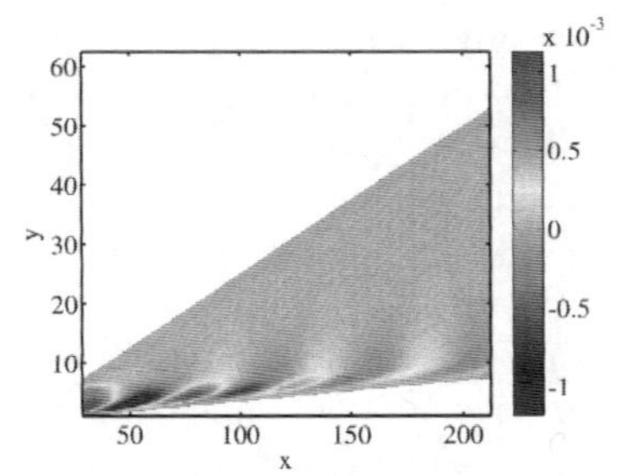

Figure 7. Contour plot of u_θ for N=0, α=2° and Re=872.

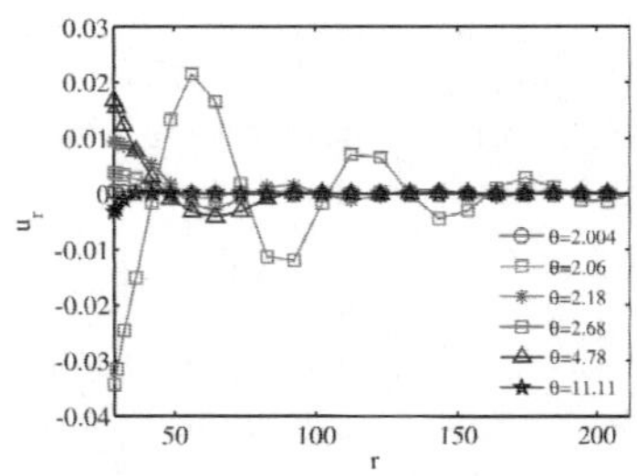

Figure 8. Streamwise variation of u_r for N=0, α=2° and Re=872.

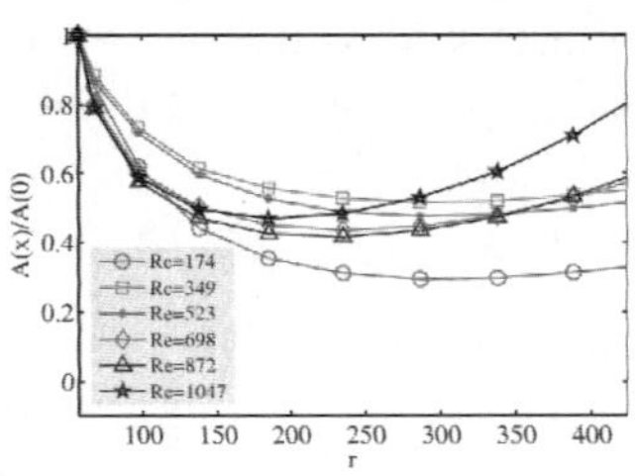

Figure 9. Normalized spatial amplification rate for α=1°.

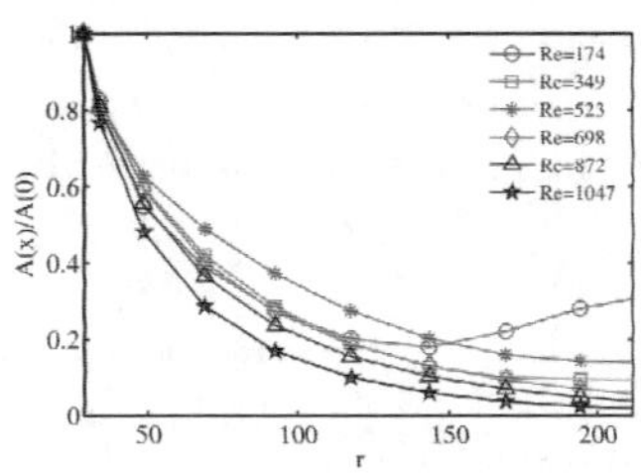

Figure 10. Normalized spatial amplification rate for α=2°.

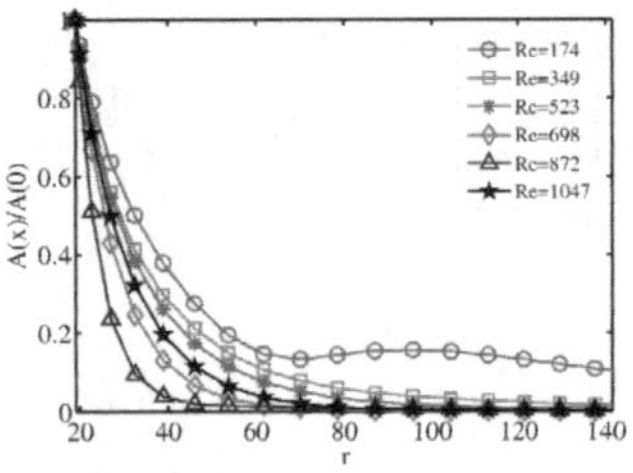

Figure 11. Normalized spatial amplification rate for α=3°.

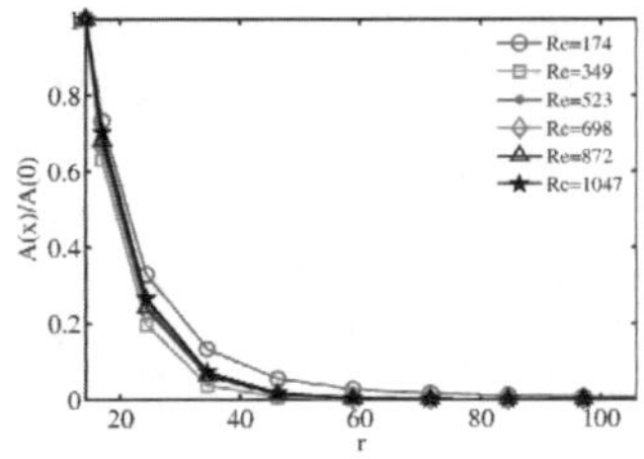

Figure 12. Normalized spatial amplification rate for α=4°.

and then gradually increases for all Reynolds number in the streamwise direction towards the downstream. A_x also increases with the increase in Reynolds number. However, flow is spatially stable for all the Reynolds number as $A_x < 1$. The decay rate reduces with the increase in Reynolds number. The computations for stability analysis are performed for semi-cone angles α=1°,2°, 3° and 4°. It has been observed that with the increase in semi-cone angle the flow becomes temporally unstable and wave-like behaviour disappears. Figure 10, 11 and 12 shows A_x for semi-cone angle α=2°, 3° and 4° for various Reynolds number. Spatial amplification rate reduces with increase in semi-cone angles at given Reynolds number. The flow is spatially stable for the range of Reynolds number for α=1°,2°,3° and 4° for all Reynolds number.
The transverse curvature and favourable pressure gradient have overall damping effect on the disturbances [9]. Effect of pressure gradient is comparatively less for small semi-cone (1°) angles. A_x initially reduces and then increases in streamwise direction for α=1° due to decrease in transverse curvature. However, A_x reduces in the streamwise direction for α=2^0 and higher values of α due to the development of larger favourable pressure gradient.

4. Conclusions
Global stability analysis of an incompressible axisymmetric boundary layer over a circular cone is performed. The global temporal modes are computed for different semi-cone angles for the axisymmetric mode. The flow is found temporally stable for small semi-cone angle (α=1°), however with the increase in α, ω_i also increases and makes the flow temporally unstable. Spatial amplification rate of the modes are computed. Spatial amplification rate reduces with the increase in semi-cone angle α from 1° to 4°, which makes the flow spatially stable. Overall, the disturbances are found spatially stable in the streamwise direction. This is due to the transverse curvature and favourable pressure gradient present in the boundary layer. The flow is spatially stable for the range of Reynolds number and semi-cone angles α considered. The wave-like nature of the disturbances observed in the streamwise direction for small semi-cone angles and it disappears as the semi-cone angle(α) increases.

References
[1] Drazin P and Raid W 2002 *Hydrodynamic stability* (Cambridge University Press)
[2] Theofilis V 2003 *Progress in Aerospace Science.* **39** 3229
[3] Swaminathan G and Govindrajan R 2011 *Theor. Comput. Fluid Dyn.* **64** 25
[4] Alizard F and Robinet J 2007 *Physics of fluid.* **19** 3229
[5] Malik M 2007 *Journal of Computational Physics.* **86** 376
[6] B Costa W D and Simas A 2007 *Proc. SCPDE: Recent Progress in Scientific Computing.* 179
[7] Theofilis V 2011 *Annu. Rev. Fluid Mech.* **43** 319
[8] Ehrenstein U and Gallaire F 2005 *J.Fluid. Mech.* **536** 209
[9] Narayanan V and Bhoraniya R 2015 *68th meeting of APS, DFD* **60** 21

IOP Conf. Series: Journal of Physics: Conf. Series **822** (2017) 012019 doi:10.1088/1742-6596/822/1/012019

Numerical simulations of flow past a circular cylinder

Gaurav Chopra

Postgraduate Student, Department of Aerospace Engineering, Indian Institute of Technology Kanpur, Kanpur, Uttar Pradesh 208016, India

E-mail: gchopra@iitk.ac.in

Sanjay Mittal

Professor, Department of Aerospace Engineering, Indian Institute of Technology Kanpur, Kanpur, Uttar Pradesh 208016, India

E-mail: smittal@iitk.ac.in

Abstract. Flow past a circular cylinder is numerically investigated by solving the incompressible Navier-Stokes e uations with a stabilized finite element method. The computations are carried out over the Reynolds number (Re) range 1×10^4 to 4×10^5. In this Re range, drag on cylinder drops rapidly with increase in Re. This phenomena, known as *drag crisis*, is associated with transition of boundary layer from laminar to turbulent state. The computations successfully capture the *drag crisis* phenomena and the accompanying flow field features like Laminar Separation Bubble (LSB). The computed flow field was used to study the statistical properties of the flow as well as its time-dependent features. It is found that appearance of LSB and the overall process of transition is intermittent. The fre uency of appearance of LSB as well as the duration of its stay, in the critical Re regime, increases with increase in Re. The LSB is tracked and its intermittency is evaluated using fluctuations in surface pressure and Reynolds stress.

1. Introduction

Flow past a circular cylinder can be divided into four different regimes: subcritical, critical, supercritical and transcritical [1]. In the subcritical regime, the boundary layers separate in the laminar state at about 80° from the front stagnation point. The early separation of boundary layer leads to high a value of drag coefficient ($\overline{C}_D$) of about 1.2. In the critical regime, the boundary layer is laminar when it first separates from the surface of cylinder. Due to the instabilities, the separated shear layer undergoes a transition to turbulent state and reattaches to the surface of cylinder [2, 3]. The boundary layer separates in turbulent state at a higher angle compared to laminar boundary layer. As a result, the wake is narrower and base pressure higher. This leads to a very rapid reduction in the drag coefficient ($\overline{C}_D$) with increase Reynolds number (Re). This phenomena is popularly known as *drag crisis*. The turbulent reattachment of shear layer leads to the formation of LSB. [4] and [5] experimentally established that formation of LSB is responsible for the *drag crisis*. In the supercritical regime, the boundary layer immediately transitions from laminar to turbulent state at a critical angle from front stagnation point. Also, the $\overline{C}_D$ increases with increase in Re. In the transcritical regime, the formation of LSB ceases, the boundary layer separates in turbulent state, and the $\overline{C}_D$ remains nearly constant with increse in *Re*.

In an experimental investigation [6] found that the variation of $\overline{C}_D$ with *Re* during the drag crisis, in the critical regime, takes place in two stages. In the first stage, at the lower *Re* of the critical regime, the LSB is present only one of the shoulders of the cylinder. At higher *Re*, the LSB appears on both sides of the cylinder. [3] carried out three-dimensional large eddy simulations (LES) for smooth cylinder and with a with trip located at $\theta = 55^o$ from the stagnation point on one of the sides. They observed that the smooth cylinder undergoes a single stage drag crisis while the one with a trip experiences a two-staged drag crisis similar to what is seen in experimental studies. This suggests that the asymmetry in the transition of the flow on the two sides of the cylinder is a consequence of a slight asymmetry in either the geometry or the flow conditions at the inlet. [ ] reported two kinds of boundary layer reattachment during the critical regime: symmetric and asymmetric. Weakening of vortex shedding activity was also observed during drag crisis. The probability of appearance of one and two bubble states was presented for various scenarios of transition. It was observed that for asymmetric reattachment, one bubble state has a higher probability at the onset of the critical regime, while the two bubble state shows increased probability at *Re* close to the end of critical regime.[8] performed three-dimensional large eddy simulations with high span wise grid resolution. They observed asymmetric reattachment in the critical regime.

Despite the various studies in the critical regime, the exact dynamics of the formation of the LSB on the circular cylinder is not well understood. Does the LSB appear suddenly at a certain critical *Re*? Or does it have an intermittent nature at the lower *Re* end of the critical regime and later exists at all times. We explore this in the present work. For the same we use the concept of *intermittency* which is extensively used in the description of transition from laminar to turbulent states. We denote the intermittency factor by I_f. It is the fraction of time during which the flow is in a turbulent state [9]. I_f lies between 0 and 1; $I_f = 0$ implies that the flow is laminar while $I_f = 1$ means that the flow is turbulent all the time. In the present work, we utilize I_f to quantify the fraction of time during which LSB exists in the flow.

2. Numerical method and computational details

2.1. The finite element formulation

The flow is modeled by the incompressible Navier-Stokes equations in the primitive variables. A stabilized finite element formulation [10] is utilized to discretize the equations. The second-order-accurate-in-time, Crank-Nicholson scheme is employed for time integration. In this work the streamline-upwind/Petrov- alerkin (SUP) and pressure-stabilizing/Petrov- alerkin (PSP) method is utilized [10] to stabilize the computations. The linear algebraic equations resulting from the finite element discretization are solved using a matrix-free implementation of the eneralized Minimal RESidual (MRES) technique [11] in conjunction with diagonal preconditioners. To handle large scale computations, the solution method is implemented on a distributed memory parallel computing machine. For more details on the finite element formulation and its parallel implementation the reader is referred to the article by [12].

Computations are carried out in both two- and three-dimensions. Detailed results for the 2D computations were presented in our earlier study [2]. In this work, we focus on 3D computations; the results for 2D computations are presented mostly for comparison, and to bring out the three-dimensional effects. At the *Re* considered in this study, the flow involves small scale flow structures. It is not possible to resolve the flow at all the scales via a Direct Numerical Simulation (DNS), with the present computational resources, as the number of grid points required is extremely large. We present two sets of computations. In the first set, the smaller unresolved flow structures are modeled using a sub-grid scale model. Large Eddy Simulation (LES) is employed; the effect of small scales is represented by a constant coefficient ($C_s = 0.1$)

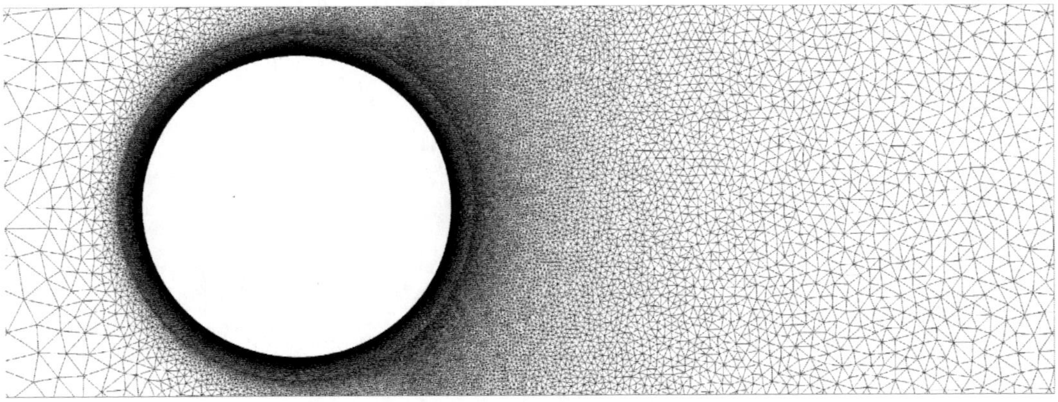

Figure 1. Flow past a cylinder: close-up view of the finite element mesh in the $x - y$ plane near the cylinder.

Smagorinsky model. The implementation of the model is similar to that used by [13].

2.2. Problem set-up and the finite element mesh

The two-dimensional mesh employed for computations is shown in figure 1. The mesh consists of a structured part close to the cylinder. The remaining domain consists of an unstructured mesh, which is created using Delaunay triangulation. The structured part of the mesh is adequately fine to resolve the boundary layer, its separation and the subsequent rolling up of the separated shear layer into vortices, in the range of Re being considered. The height of the first element from the surface of the cylinder is $5 \times 10^{-6}D$ in the radial direction, where D is the diameter of the cylinder. The 2D mesh consists of 115,84 nodes and 230,840 triangular elements.

The mesh for the three-dimensional computations is generated by stacking sections of the two-dimensional mesh along the span. The spanwise extent of the domain is $L_z = 1D$. Fifty one uniformly spaced sections of the two-dimensional mesh, described above, are utilized to construct the three-dimensional mesh. The resulting 3D mesh consists of 5,908,19 nodes and 11,542,000 6-noded wedge elements. A time step of $t = 5 \times 10^{-4}$ is utilized for all the computations. All the results presented in the paper are in terms of the non-dimensional time.

3. Results

Results are presented for the flow past a cylinder for $1 \times 10^4 \leq Re \leq 4 \times 10^5$. In all the cases the unsteady computations are carried out for sufficiently long time such that the flow achieves a fully developed unsteady state. For estimating the statistical quantities, the data for the transient during the unsteady development of the flow, following the initiation of the computation, is discarded.

Figure 2 shows the variation of $\overline{C_D}$ with Re from the present computations along with their comparison from earlier studies. We observe that both two- and three-dimensional simulations predict drag crisis. The range of Re for which the flow lies in the critical regime is very similar from the two sets of computations. This suggests that the phenomenon itself is largely two-dimensional. Of course, the three-dimensional effects are quite significant as indicated by the difference in $\overline{C}_D$ in the sub- and super-critical regime from the 2D and 3D computations. However, the mechanism of transition in a span-averaged sense appears to be two-dimensional. Compared to the measurements from experiments, the present results predict the drag-crisis at a slightly lower Re. Results are also shown in Figure 2 for the Large Large Eddy Simulation

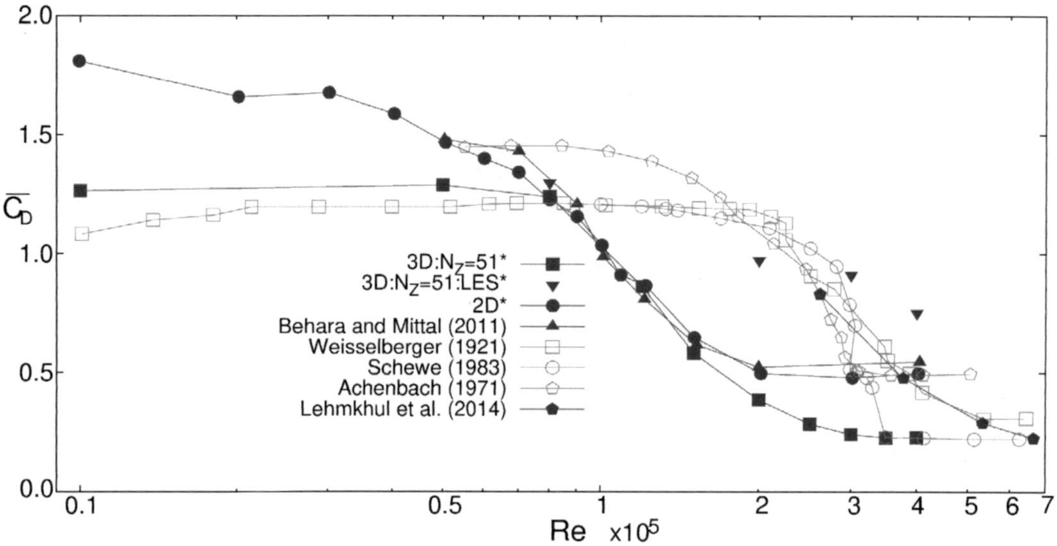

Figure 2. Flow past a cylinder: variation of the mean drag coefficient with Re. Also shown are results from earlier studies. The data from experiments by [14] has been taken from [15].

(LES) with a static Smagorinsky model to account for the subgrid scales. Compared to results from model-free computations, these results are in closer agreement with the experimental data. Since the model-free computations also capture the essential trend, therefore, all the analysis in this work is carried out for data from model-free computations. Both, the computational and experimental data in Figure 2 shows that the variation of $\overline{C}_D$ with Re during the transition is smooth, and not abrupt. We explore the changes in the flow during transition.

The instability of the separated shear layer and its subsequent rolling up into vortices plays a major role in the reattachment of the boundary layer and formation of LSB. To study this we need to filter out the activity due to the Karman shedding that has a much lower frequency compared to the activity of the shear layer. For the same, consider the time variation of the span-averaged surface coefficient of pressure $C_P(\theta, t)$ at different Re. We compute a moving average of $C_P(\theta, t)$: $\widetilde{C_P}(\theta, t)$ over a time window of size $T_k/10$, where T_k is the time period of the Karman shedding. This averaging filters out the shear layer activity while retaining the vortex shedding effects. By subtracting $\widetilde{C_P}(\theta, t)$ from the original signal, we compute the rms of the fluctuations that correspond to shear layer activity: $\widetilde{C'_P C'_P}(\theta, t)$. Figure 3 shows a space-time diagram of the variation of $\widetilde{C'_P C'_P}(\theta, t)$ for $Re = 1.5 \times 10^5$. The flow is in the transitional regime at this Re and undergoing drag-crisis. Figure 3 shows the streamlines for the moving time-averaged flow that are also span averaged at two time instants that are marked in figure 3. These figures reveal the co-relation between the presence/absence of LSB with the level of fluctuations due to shear layer activity as observed in $\widetilde{C'_P C'_P}(\theta, t)$. At $t = t_1$ the fluctuations in the figure 3 are high at $\theta \approx 105^o$ but relatively low at $\theta \approx 250^o$. Correspondingly a LSB can be seen in the time-averaged streamlines shown in figure 1(d) on the upper shoulder of the cylinder. At $t = t_2$, an LSB appears on both the upper and lower shoulder of the cylinder. Correspondingly, high fluctuations in $\widetilde{C'_P C'_P}(\theta, t)$ can be seen at $\theta \approx 105^o$ and $\theta \approx 250^o$. This demonstrates that high fluctuations in $\widetilde{C'_P C'_P}(\theta, t)$ correspond to increased shear layer activity and formation of LSB. As seen from figure 1(b), the high fluctuations in $\widetilde{C'_P C'_P}(\theta, t)$ appear in bursts. The space-time

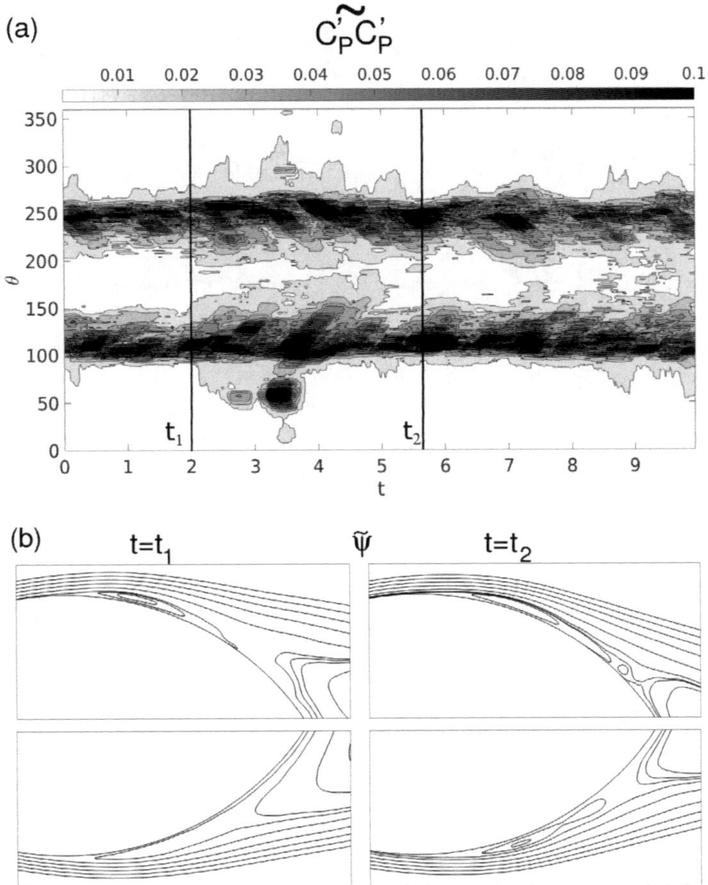

Figure 3. $Re = 1.5 \times 10^5$ flow past a cylinder: space-time diagram of the *rms* of the fluctuations subjected to high pass filter, (a) $\widetilde{C'_P C'_P}(\theta, t)$. Shown in (b) are the close-up views of the streamlines for the span- and moving time-averaged flow ($\tilde{\psi}$) at the time instants $t_1 = 2.0$ and $t_2 = 5.6$ marked in (a).

diagram is utilized to compute the Intermittency factor, I_f. An LSB exists on the shoulder if $\widetilde{C'_P C'_P}(\theta = 105^o$ or $250^o, t)$ exceeds a certain threshold value. Figure 4 shows the variation of $1 - I_f$ and C_D with Re. This figure clearly shows the correlation between $1 - I_f$ and C_D. In the subcritical regime $I_f \approx 0$, indicating LSB does not appear in the flow. In the early stages of critical regime, I_f starts increasing from a zero value and reaches unity towards the end of the critical regime. This shows that LSB appears intermittently during the regime of drag crisis and its frequency of appearance increases towards the end of drag crisis. Thus it can be said that in the transitional regime during the drag crisis, the flow fluctuates between laminar and turbulent states. The laminar state is associated with larger drag and is devoid of LSB. The turbulent state has a lower drag and is associated with an LSB. The average drag at any Re depends on the relative time spent in the two states.

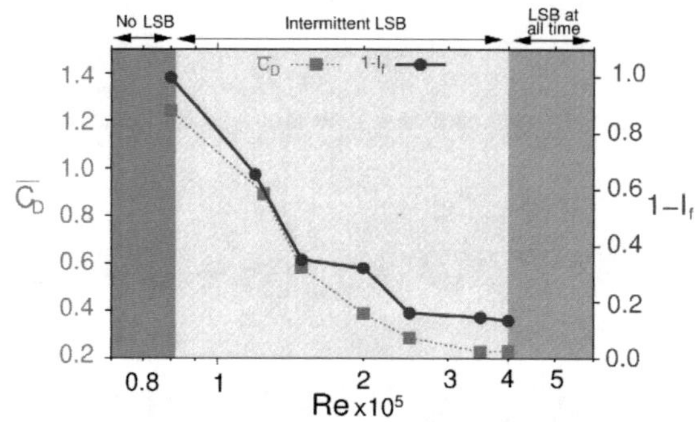

Figure 4. Flow past a cylinder: variation of the intermittency factor with Re. The scale for $(1 - I_f)$ is shown on the right and its variation is displayed in solid line. Also shown in broken line is the span- and time-averaged drag coefficient. Its scale is shown on the left. I_f is calculated from the *rms* of the high pass filtered pressure time series on the surface of the cylinder $(\widetilde{C'_P C'_P}(\theta, t))$.

4. Conclusions

The numerical simulations capture the drag-crisis. The Laminar separation bubble (LSB) is found to exhibit an intermittent behaviour. This nature of LSB is utilized to explain the gradual change in $\overline{C}_D$ in the transition regime during drag crisis as opposed to a sudden decrease in drag at a certain Re.

Acknowledgement

The authors are grateful to Mr. Nishchal Agrawal and Mr. Aditya Desai for their help in carrying out the two-dimensional simulations as well as their enthusiastic participation in the discussion related to the statistical analysis in this work. Much of the computations were carried out in the High Performance Computing facility at the Computer Center, Indian Institute of Technology Kanpur that has been set-up by the generous assistance from Department of Science and Technology (DST), India.

References

[1] Roshko A 1961 *Journal of Fluid Mechanics* **10** 345–356
[2] Singh S and Mittal S 2005 *International Journal for Numerical Methods in Fluids* **47** 75–98
[3] Behara S and Mittal S 2011 *Journal of Fluids and Structures* **27** 702–715
[4] Tani I 1964 *Progress in Aerospace Sciences* **5** 70–103
[5] Achenbach E 1968 *Journal of Fluid Mechanics* **34** 625–639
[6] Bearman P 1969 *Journal of Fluid Mechanics* **37** 577–585
[7] Cadot O, Desai A, Mittal S, Saxena S and Chandra B 2015 *Physics of Fluids (1994-present)* **27** 014101
[8] Lehmkuhl O, Rodríguez I, Borrell R, Chiva J and Oliva A 2014 *Physics of Fluids (1994-present)* **26** 125110
[9] Tennekes H and Lumley J L 1972 *A first course in turbulence* (MIT press)
[10] Tezduyar T E, Mittal S, Ray S and Shih R 1992 *Computer Methods in Applied Mechanics and Engineering* **95** 221–242
[11] Saad Y and Schultz M H 1986 *SIAM Journal on Scientific and Statistical Computing* **7** 856–869
[12] Behara S and Mittal S 2009 *Parallel Computing* **35** 195–212
[13] Johari H and Stein K 2002 *Physics of Fluids (1994-present)* **14** 3459–3474
[14] Wieselsberger C 1922
[15] Schlichting H, Gersten K and Gersten K 2000 *Boundary-layer theory* (Springer Science & Business Media)

A comparative CFD study on the hemodynamics of flow through an idealized symmetric and asymmetric stenosed arteries

B Prashantha[1], S Anish[1]
[1] Mechanical Engineering Department,
National Institute of Technology, Karnataka, PIN- 575025, India

E-mail: prashanthb.mech@gmail.com

Abstract. The aim of the present study is to numerically evaluate the hemodynamic factors which affect the formation of atherosclerosis and plaque rupture in the human artery. An increase of atherosclerosis in the artery causes geometry changes, which results in hemodynamic changes such as flow separation, reattachment and adhesion of new cells (chemotactic) in the artery. Hence, geometry plays an important role in the determining the nature of hemodynamic patterns. Influence of stenosis in the non-bifurcating artery, under pulsatile flow condition has been studied on an idealized geometry. Analysis of flow through symmetric and asymmetric stenosis in the artery revealed the significance of oscillating shear index (OSI), flow separation, low wall shear stress (WSS) zones and secondary flow patterns on plaque formation. The observed characteristic of flow in the post-stenotic region highlight the importance of plaque eccentricity on the formation of secondary stenosis on the arterial wall.

1. Introduction

Atherosclerosis is a common dangerous disease death in many countries because of pulsatile blood behaviour in the human artery [1]. Blood is an inhomogeneous fluid consisting of red cells, platelets and a number of white cells. Several autopsy study have demonstrated 70-80% of cardiovascular diseases (atherosclerosis plaque) occurs at complex geometries [2]. This continuous development causes narrowing of the vessel lumen (known as stenosis) creating significant flow disturbances [3]. These disturbances and turbulences in the flow influence the variations in the mechanical parameters such as stress acting on wall and wall elongation. Large plaque formation in the artery changes the wall shear stress distribution on the arterial walls [4, 9]. Flow behaviour in the stenotic arteries influences the stress on the plaque and thereby stimulates the plaque rupture also.

Quite often, the adhesion of new cells on the wall occurs at regions where the wall shear stress is lower [5]. In a normal healthy person, blood flow through the artery is laminar. However, the presence of new cells in the blood flow and its adhesion on the arterial wall make the flow transitional or turbulent [6] and change of pulse frequency causes laminar to turbulence flow in the artery. Certain combination of pulsatile flows and geometries will generate vortices and secondary flows in the flow and it promotes accumulation of new cells on to the arterial wall [7]. For the accurate prediction of plaque formation and development a thorough understanding of the flow behavior at the post stenotic region is essential.

Objective of the present study is to understand the effect of pulsatile blood flow behavior through an idealized symmetric and asymmetric stenosed channel. The analysis is carried out for severe stenosis case (75% of area reduction). During pulsatile flow, distributed streamlines, velocity vector fields and oscillatory shear stresses in the post stenotic region clearly defines the influence of stenosis location (eccentricity) in the arterial wall. Symmetric case results were validated with experimental results of [8].

2. Methodology

Real artery geometry is very difficult to reconstruct due to its complex internal structure and the difficulty in obtaining real patient data. Hence, present computational study is an ideal geometry model has been considered [8]. The computational fluid model is set up for 75% reduction in area (severe case study [8]) and was modelled using CATIA. The geometry of an idealized artery model has been scaled

up keeping the dynamic similarity. Inlet and outlet diameter has been considered as 2 inch and upstream length will be considered as 100 times of diameter, downstream length will be considered as 14 times of the diameter from the throat section as shown in the figure 1(a).
Asymmetric stenosis model will be constructed by varying eccentricity of the minimum constricted flow diameter in the passage (stenosis) as shown in the figure 1(b). Three-dimensional smooth geometries are developed by using CATIA and meshing has been done through ICEM CFD. Grid independence study has been carried out to arrive at an optimum number of grid elements for the computational study. This optimizes the computational time without compromising with accuracy of the solution. Computational simulations have been carried out using Finite volume method (FVM) by commercial software ANSYS FLUENT. Two equation turbulence modelling (K-omega Standard) Naiver strokes equations has been solved by using second order upwind discretization method and Pressure Implicit Splitting Operator algorithm (PISO).

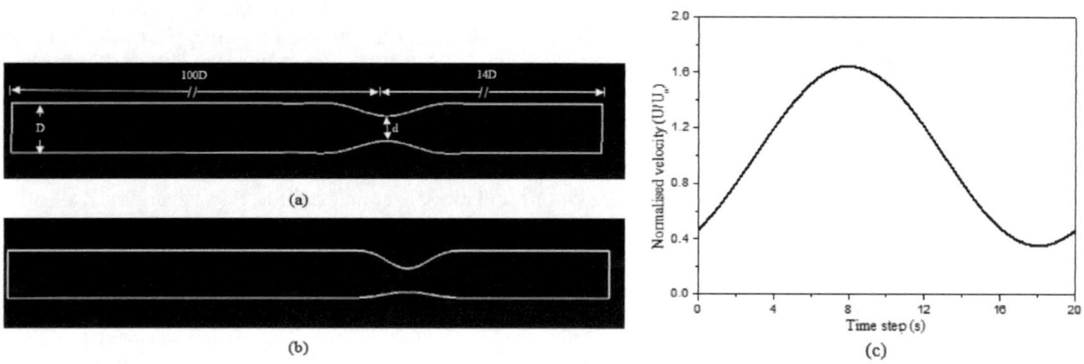

Figure 1. Geometric representation of (a) symmetric and (b) asymmetric stenosis (c) Inlet velocity profile Boundary condition

Blood is assumed as Newtonian fluid with density of 1060 kg/m^3 and viscosity of 0.0036104 Pa s [8]. Velocity wave form figure 1(c) taken out from the Ahmed and Giddens experimental result and created as a User Defined Function (UDF) using C language, was applied as an input velocity boundary condition and outlet can be defined as outflow boundary condition within the FLUENT. After the completion of three inlet velocity wave form, the simulation results were extracted from third cycle [6].

3. Results and Discussion

This study is an extension of symmetric stenosis case study carried out by J. Banks [6] simulation work to asymmetric case of deposited plaque on the arterial wall, to examine the detailed hemodynamic flow behaviour during the presence of symmetric and asymmetric stenosis in the rigid arterial wall. Previous Ahmed and Giddens [8] literature data describes hemodynamic behaviour in the different degrees of symmetric stenosis in an artery. Present study focuses on the hemodynamic behaviour in the presence of protruded asymmetric plaque over the lumen area. In the post stenotic region flow becomes turbulent when the stenosis exceeds 50% of area [8], hence, extended our study for the severe case (75%). In the present study, analysis has been done at three axial locations in the post stenotic region.

3.1. Validation

The present computational work has been validated with Ahmed and Giddens [8] experimental result as shown in figure 2. These plots show that computed values are in good agreement with the experimental results. Available experimental velocity profile at -1.5D axial location, figure 1(c), has been considered for the simulation inlet boundary condition for the analysis, equation (1) shows the similar velocity profile as taken out from the experimental velocity profile at -1.5D location. Where U/U_m the normalized inlet velocity along centreline axis and U_m is the mean velocity of 0.04254 ms^{-1}.

$$V(t) = [0.042517+0.027969*\cos(2*3.1415*t/20-2.53)] \quad (1)$$

where 't' is the time step.

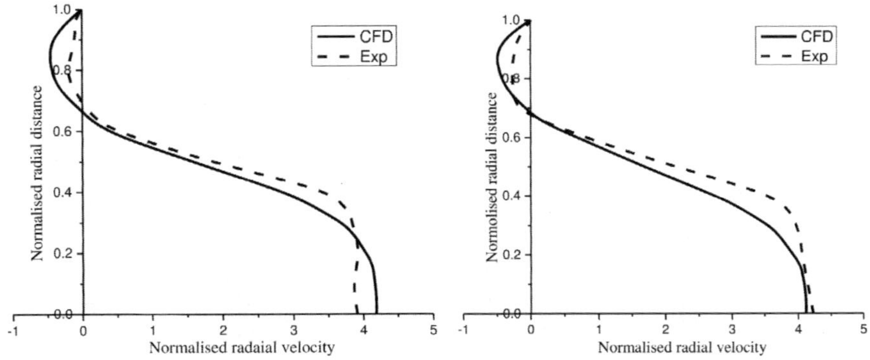

(a) (b)

Figure 2. Comparison of radial velocity profile at peak time step in the post stenosis region (a) Plane at 1D location (b) Plane at 1.5D location

3.2 Effect of Stenosis Nature on Hemodynamic Flow Field

Flow streamlines at two different instance of the pulsating flow is shown in figure 3. One is in the accelerating and other is in the deceleration phase of the systole and diastole. It is observed that the recirculation zone is quite large in the deceleration phase (t=14s). The reattachment occurs at 5D distance for the symmetric model whereas for the asymmetric model it occurs at 9D during the deceleration phase. During deceleration diastolic time phase, asymmetric stenosis distributes strong recirculated secondary vortices up to 9D locations, however in the case of symmetric stenosis only mild fluctuated turbulent flow pattern can be observed along the stream wise distance.

In the case of peak systolic time phase (t=8s), flow reattachment in both the cases have nearly same length (3D) but vortices are unevenly distributed. Whereas symmetric stenosed atrial wall generates a set of symmetric vortices immediate next to the stenosis, those generated vortices will not continue their growth. Hence, after small region of stream distance fully developed (reattachment) flow can be observed. But in the asymmetric case strong vortices will grow with the time step and more recirculation region length occurs along the stream wise distance (9D).

Recirculation region length of velocity stream lines clearly signifies developed strong recirculating vortex patterns and with time steps. In asymmetric cases primary vortex are detached as soon and reattach next opposite to the wall at a certain region of length as shown in figure 3(c) (d). Hence, clear picture of these results defines there will be more chances of deposition of new cells (plaque) irregularly in the post stenotic region as compare with the symmetric case. This signifies importance of the lumen geometry in the distribution of wall shear stress.

3.3. Effect of Wall Shear Stress

Flow streamlines at two different instance of the pulsating flow is shown in figure 3. One is in the accelerating and other is in the deceleration phase of the systole and diastole. It is observed that the recirculation zone is quite large in the deceleration phase (t=14s). The reattachment occurs at 5D distance for the symmetric model whereas for the asymmetric model it occurs at 9D during the deceleration phase. In the case of peak systolic time phase (t=8s), flow reattachment in both the cases have nearly same length (3D) but vortices are unevenly distributed. Whereas symmetric stenosed arterial wall generates a set of symmetric vortices immediate next to the stenosis. These generated vortices will not continue their growth but after small distance downstream reattachment can be observed. In the asymmetric case strong vortices will grow with the time step and more recirculation region length occurs along the stream wise distance (9D).

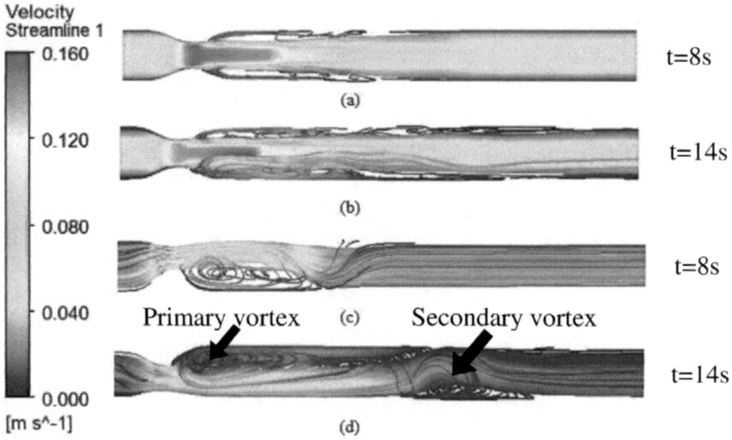

Figure 3. Effect of stenosis eccentricity allots recirculation regions in the post stenotic region towards outer wall (a) Symmetric case at peak time step (b) Symmetric case at deceleration time step (c) Asymmetric case at peak time step (d) Asymmetric case at deceleration time step.

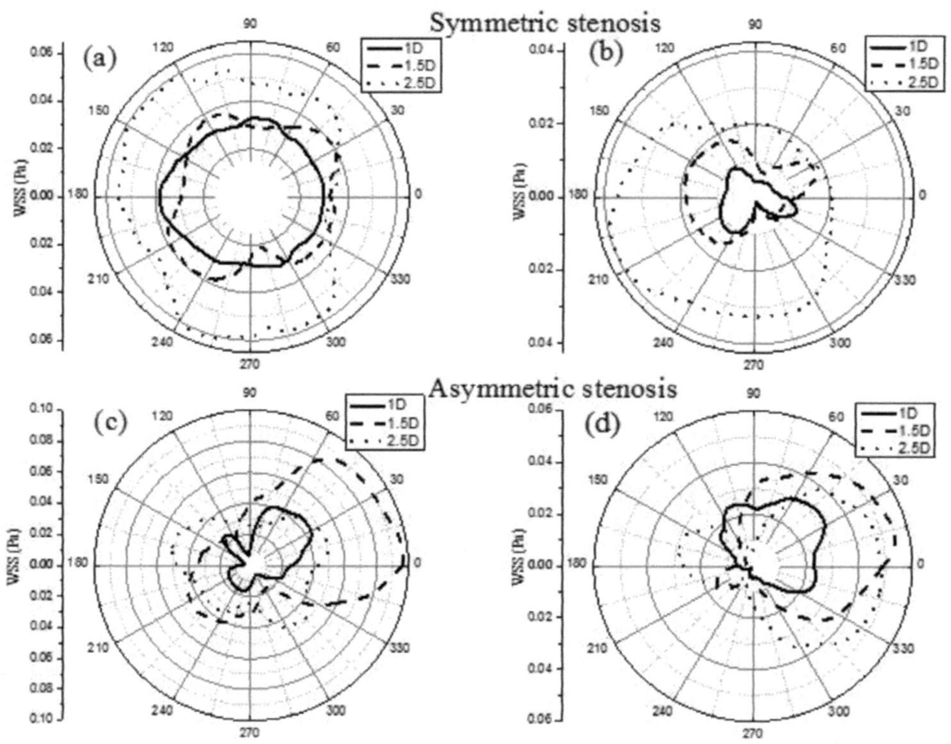

Figure 4. Angular variation of WSS at different axial locations (a) Symmetric case at peak time step (b) Symmetric case at deceleration time step (c) Asymmetric case at peak time step (d) Asymmetric case at deceleration time step.

The length of the recirculation region differs in both cases and so the wall shear stress distribution figure 4. Asymmetric channel develops strong recirculating vortex patterns which evolves with time step. In asymmetric cases the detached primary vortex soon reattaches opposite to the wall at a certain length which may depends on the degree of stenosis. Hence with asymmetric stenosis there will be more

IOP Conf. Series: Journal of Physics: Conf. Series **822** (2017) 012020 doi:10.1088/1742-6596/822/1/012020

chances of deposition of new cells (plaque) irregularly in the post stenotic region as compare with the symmetric case, signifying the importance of lumen geometry in the atherogenesis progression.

3.3. Effect of Oscillatory Shear Stress

Oscillatory shear index, is an important parameter in the pulsatile blood flow through the arteries, which is used to quantify the multiple sign changes of WSS (unsteady) in the cardiac cycle. OSI can be calculated as,

$$\text{OSI}= \frac{1}{2}\left(1 - \frac{|\int_0^n \tau|}{\int_0^n |\tau|}\right) \qquad (2)$$

where 'τ' is the wall shear stress derived from velocity fields. It ranges from 0 to 0.5, small values of OSI indicates flow separation and reattachment, whereas large values of index highlights fully reversed steady flow [7].

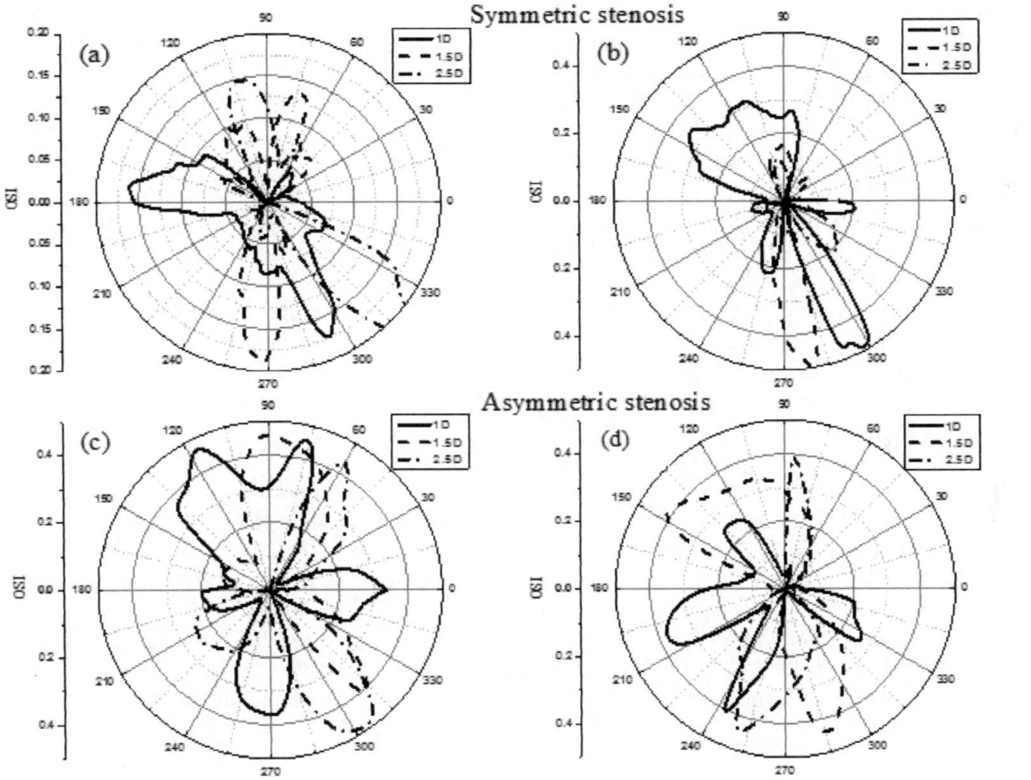

Figure 5. Variation of OSI in the post stenotic region (a) Symmetric case at peak time step (b) Symmetric case at deceleration time step (c) Asymmetric case at peak time step (d) Asymmetric case at deceleration time step.

The OSI values are localized to a comparatively small angular space for the symmetric stenosis case. Especially for the decelerating phase the OSI values are very small at 1.5D and almost nil at 2.5D. On the contrary, for the asymmetric case the oscillatory shear index exhibits a significantly strong presence over a wide range of angular location. For each axial location, the peak OSI value occurs at almost similar angular location for the symmetric stenosis. Whereas for the asymmetric stenosis the angular location of the peak OSI keeps on shifting along the axial distance as shown in the figure 5. This is possibly due to the formation of the secondary vortex along the downstream.

4. Conclusion

In this CFD analysis, a comparison of hemodynamics features through two idealized stenosed artery models are carried out. These two models differ in such a way that one has symmetric stenosis and the other has asymmetric stenosis.

The symmetric model has been validated with the established experimental results. From the analysis of streamlines, it is understood that pulsatile flow nature and complex geometries are responsible for the asymmetric plaque formations in the artery. Even though the degree of stenosis are same for both symmetric and asymmetric case the pulsatile flow and the eccentricity creates a complex secondary vortex generation in the asymmetric channel. As the adhesion of new cells and subsequent plaque formation happens at low wall shear stress region, an increase in the irregularity of the artery lumen may be conducive for secondary plaque formation and atherogenesis. The angular variation of WSS and OSI values show that asymmetric stenosed channel is more prone to atherogenesis than the symmetric channel.

References

[1] Beaglehole R, Irwin A, and Prentice T, "*Facts and Figures: The World Health Report 2003—Shaping the Future,*" World Health Organization, Tech. Rep.

[2] Constantinides P 1984 "Atherosclerosis–A General Survey and Synthesis". *Pathology and Immunopathology Research*, vol. *3*(**6**), pp 477-98.

[3] Marshall I, Zhao S, Papathanasopoulou P, Hoskins P, Xu X Y 2004 "MRI and CFD studies of pulsatile flow in healthy and stenosed carotid bifurcation models". *Journal of Biomechanics, vol.* **37(5)**, pp 679-87.

[4] Tang D, Yang C, Zheng J, Woodard P K, Saffitz J E, Sicard G A, and Yuan C 2005 "Quantifying effects of plaque structure and material properties on stress distributions in human atherosclerotic plaques using 3D FSI models", *Journal of biomechanical engineering*, vol. **127(7)**, pp 1185-94.

[5] Razavi A, Shirani E, Sadeghi M R 2011 "Numerical simulation of blood pulsatile flow in a stenosed carotid artery using different rheological models", *Journal of Biomechanics*, vol. **44(11)**, pp 2021-30.

[6] Banks J, Bressloff N W 2007 "Turbulence modelling in three-dimensional stenosed arterial bifurcations", *Journal of biomechanical engineering*, vol. **129(1)**, pp 40-50.

[7] Buchanan J R, Kleinstreuer C, and Comer J K 2000 "Rheological effects on pulsatile hemodynamics in a stenosed tube," *Computers and Fluids*, vol. **29(6)**, pp 695–724.

[8] Ahmed S A, Giddens D P 1984 "Pulsatile post stenotic flow studies with laser Doppler anemometry", *Journal of biomechanics,* vol. **17(9)**, pp 695-705.

[9] Basavaraja P, Surendran A, Gupta A, Saba L, Laird J R, Nicolaides A, ... and Suri J S (2016) "Wall shear stress and oscillatory shear index distribution in carotid artery with varying degree of stenosis: a hemodynamic study". *journal of mechanics in medicine and biology*, vol. **17(4)**, 1750037.

A hybrid method for improving the robustness of WENO scheme

Ding Z H[1], Wu S P[1, 2], Liu J[*3]

[1]School of Aeronautic Science and Engineering, Beijing University of Aeronautics and Astronautics, 37 Xueyuan Road, Beijing 100191, China
[2]Ministry-of-Education Key Laboratory of Fluid Mechanics, Beijing University of Aeronautics and Astronautics, China
[3]Sino-French Engineer School, Beijing University of Aeronautics and Astronautics, 37 Xueyuan Road, Beijing 100191, China

Email: lantingjixulj@126.com

Abstract. In this paper, a hybrid scheme based on fifth-order finite difference WENO scheme is proposed for solving spurious oscillation in strong shock-capturing. In this scheme, we consider smoothness indicator as a measurement of gradient and construct a function of indicator, using which the new hybrid can switch between WENO scheme and Lax-Friedrichs (LxF) scheme. Numerical experiments demonstrate that the hybrid scheme dominates the accuracy of the solution in smooth region and non-oscillation can be promised around discontinuity. The new hybrid scheme could essentially preserve positive density and positive pressure. The idea of constructing the hybrid can be easily extended to other high resolution schemes (TVD, ROE etc.) coupled with WENO scheme. These ideas will be illustrated in further study.

1. Introduction

Exact solution of scalar conservation law

$$\begin{aligned} u_t + \nabla \cdot F(u) = 0 \\ u(x,0) = u_0(x) \end{aligned} \qquad (1.1)$$

has an important property that it satisfies a strict maximum principle, namely: $u(x,t) \in \left[\min_x u_0(x), \max_x u_0(x)\right]$ [1], the same property is desired to be satisfied in numerical simulation.

Weighted essentially non-oscillatory (WENO) schemes, which are based on essentially non-oscillatory (ENO) schemes, perform well in high accuracy solution but still have some problems in shock-capturing. [2, 3, 4, 5] develop advanced smoothness indicators of WENO scheme since Jiang and Shu proposed classical WENO-JS [6], among which WENO-M and WENO-Z can measure smoothness indicator in the vicinity of critical points, and assign substantially larger weights to discontinuous stencils [3], which improves the robustness of smoothness indicator. Due to the Gibbs' phenomenon, high order Lagrangian interpolation always occur oscillation. Numerical experiment demonstrate that classical WENO-JS shows mild oscillation while Flux Vector Splitting (FVS) WENO-

Z is sensitive to discontinuity and performs severe oscillation which might lead to negative density and pressure during calculation in some complex problem.

The spurious oscillation of fifth-order WENO scheme comes from following reasons: (a) Odd rank truncation error term ($\partial^{(2n-1)}u/\partial x^{(2n-1)},(n=1,2,......)$) in WENO scheme, because the dispersion always comes from $(2n-1)^{th}$ -derivative and the dissipation from $(2n)^{th}$ -derivative $(n=1,2,......)$, detail analysis of the problem in [7]; (b) The solution of unbalanced weights in advanced WENO schemes ($\mathrm{WENO-M, WENO-Z, WENO-NS, WENO}-\eta$) may do not strictly satisfy maximum-principle in scalar conservation law, which lead to severe oscillation when using characteristic splitting.

In this paper, we further deal with the spurious oscillation in the vicinity of discontinuity. There are mainly three techniques for solving the problem: (a) Add higher order term to WENO scheme. Then counteract the fifth-derivative ($\partial^{(5)}u/\partial x^{(5)}$) of the truncation error, then the main term of truncation error is a sixth-derivative which mainly perform dissipation; (b) Direct limitation: solution $u(x,t)$ that does not satisfy maximum-principle will be forced into $\left[\min_x u_0(x), \max_x u_0(x)\right]$, this technique perform well in some mildly oscillatory solution; (c) As will be discussed in following part, a hybrid of WENO and low accuracy scheme that satisfies maximum principle.

It is well-known that Lax-Friedrichs (LxF) schemes solving equation (1.1) strictly satisfies the maximum principle [1]. Both explicit and implicit 1st-order LxF scheme keep positive density and internal energy [8]. And some Positive-Preserving WENO scheme is a combination of WENO and LxF and a good behaviour of such schemes are demonstrated [9]. Thus we use LxF scheme to construct a Hybrid based on discontinuity recognition.

In following sections, we review the WENO-Z scheme and LxF scheme, then we propose the hybrid scheme. Finally, numerical experiments can demonstrate the robustness of the scheme.

2. Review of WENO and LxF

In this section, we will describe the construction of the two schemes applied on one dimension Euler system:

$$\frac{\partial \boldsymbol{U}}{\partial t}+\frac{\partial \boldsymbol{F}}{\partial x}=0 \tag{2.1}$$

Where

$$\boldsymbol{U}=\begin{pmatrix}\rho\\ \rho u\\ E\end{pmatrix},\boldsymbol{F}=\begin{pmatrix}\rho u\\ \rho u^2+p\\ u(E+p)\end{pmatrix} \tag{2.2}$$

Spatial-discrete: on the cell $\boldsymbol{I}_j=[x_j-\Delta x/2, x_j+\Delta x/2]$, we have:

$$\frac{\partial \boldsymbol{U}}{\partial t}+\frac{\boldsymbol{H}_{j+1/2}-\boldsymbol{H}_{j-1/2}}{\Delta x}=0 \tag{2.3}$$

Numerical flux $\boldsymbol{F}_{j+1/2}$ is used for approximating flux $\boldsymbol{H}_{j+1/2}$.

2.1. FVS WENO-Z scheme

On stencils $S=\{\boldsymbol{I}_{j-2},\boldsymbol{I}_{j-1},\boldsymbol{I}_j,\boldsymbol{I}_{j+1},\boldsymbol{I}_{j+2}\}$, $\boldsymbol{F}_{j+1/2}$ of WENO-Z scheme can be obtained by:

$$\boldsymbol{F}_{j+1/2}=\boldsymbol{F}^+_{j+1/2}+\boldsymbol{F}^-_{j+1/2} \tag{2.4}$$

Hereinafter, only the positive part $\boldsymbol{F}^+_{j+1/2}$ is described for brevity, the formulas for the negative part of the split flux are symmetric.

$$\boldsymbol{F}^{+}_{j+1/2} = R_{j+1/2}\sum_{k=0}^{2}\omega_k \overline{\boldsymbol{F}}^{+}_{k} \tag{2.5}$$

In the right hand term of (2.5), $\overline{\boldsymbol{F}}^{+}_{k}$ can be obtained on sub stencil $S_k = \{I_{j+k-2}, I_{j+k-1}, I_{j+k}\}$, namely:

$$\overline{\boldsymbol{F}}^{+}_{k} = \Lambda^{+}_{j+1/2} L_{j+1/2}\sum_{l=0}^{2}\boldsymbol{c}_{k,l}\boldsymbol{U}_{j+k+l-2} \tag{2.6}$$

where $\Lambda^{+}_{j+1/2}, L_{j+1/2}, R_{j+1/2}$ is eigenvalues, left eigenvectors and right eigenvectors of Jacobi matric $A_{j+1/2} = \left(\partial F / \partial x\right)_{j+1/2}$ respectively, which can be obtained by Roe-average.

$$\omega_k = \frac{\alpha_k}{\alpha_0 + \alpha_1 + \alpha_2}, \quad \alpha_k = \frac{C_k}{IS_k}. \tag{2.7}$$

Where C_k is a constant named ideal weight and denominator IS_k :

$$IS_k = \frac{IS_k^{JS} + \epsilon}{IS_k^{JS} + \epsilon + \tau_5} \tag{2.8}$$

Where IS_k^{JS} is smoothness indicator of classical WENO-JS [6], the role of ϵ is avoid denominator being zero.

And τ_5 in fifth-order WENO-Z scheme:

$$\tau_5 = | IS_2^{JS} - IS_0^{JS} | \tag{2.9}$$

2.2. LxF scheme

$\boldsymbol{F}_{j+1/2}$ of LxF scheme can be directly obtained by:

$$\boldsymbol{F}^{LF}_{j+1/2} = \frac{1}{2}\left(\boldsymbol{F}_j + \boldsymbol{F}_{j+1} - (|u| + c)_{\max j}(\boldsymbol{U}_{j+1} - \boldsymbol{U}_j)\right) \tag{2.10}$$

3. Construction of hybrid scheme

The basic idea of the hybrid is distinguishing discontinuity from flow region and constructing a function θ to construct modified numerical flux:

$$\boldsymbol{F}^{m}_{j+1/2} = \theta\left(\boldsymbol{F}^{LF}_{j+1/2} - \boldsymbol{F}_{j+1/2}\right) + \boldsymbol{F}_{j+1/2} \tag{3.1}$$

Where θ can vary from 0-1, and θ can be a function of measurement of discontinuity.

3.1. A choice for recognizing discontinuity

In this section, we devise a measurement for recognizing discontinuity, which is the variable for function θ.

Generally, smoothness indicators (IS) play an important role in constructing weights of WENO scheme on sub stencils. IS measure the smoothness of the numerical solution on a stencil to form a weight. We define the standard deviation of IS:

$$SDIS = \sum_{k=0}^{2}\left(IS_k - \overline{IS}\right)^2 \tag{3.2}$$

Where

$$\overline{IS} = \frac{\sum_{k=0}^{2} IS_k}{3} \tag{3.3}$$

When the stencil $S = \{I_{j-2}, I_{j-1}, I_j, I_{j+1}, I_{j+2}\}$ is a flat region, namely

$$\boldsymbol{U}_{j-2} = \boldsymbol{U}_{j-1} = \boldsymbol{U}_j = \boldsymbol{U}_{j+1} = \boldsymbol{U}_{j+2} \tag{3.4}$$

We can easily find $SDIS = 0$, and when the standard deviation of IS become great, we consider these points as a discontinuity. Though these stencils with huge gradient may not discontinuity, we both know that WENO scheme will degenerate to low order in complex region, so a low accuracy LxF scheme used to replace in these regions is sustainable.

3.2. Modified term limiter θ

Based on discussion above all, we find that $SDIS$ can be an effective indicator of huge gradient. In order to limit the scheme to low order accuracy, we add a term $\theta\left(\boldsymbol{F}_{j+1/2}^{LF} - \boldsymbol{F}_{j+1/2}\right)$ to flux, in which θ is a function of huge gradient measurement, so $SDIS$ can be the variable for function θ, namely $\theta = \theta(SDIS)$. θ satisfies the following requirements:

- Varying from 0 to 1.
- θ must be positive correlation with $SDIS$.
- Once $SDIS$ ensure a smoothness region, $\boldsymbol{F}_{j+1/2}^{LF}$ cannot influence flux $\boldsymbol{F}_{j+1/2}^{m}$ in case of losing excessive accuracy.

There are two possible choices of $\theta(SDIS)$ as displayed in figure 1:

$$f_1 = 1 - \frac{1}{1+x} \tag{3.5}$$

$$f_2 = 1 - e^{-x} \tag{3.6}$$

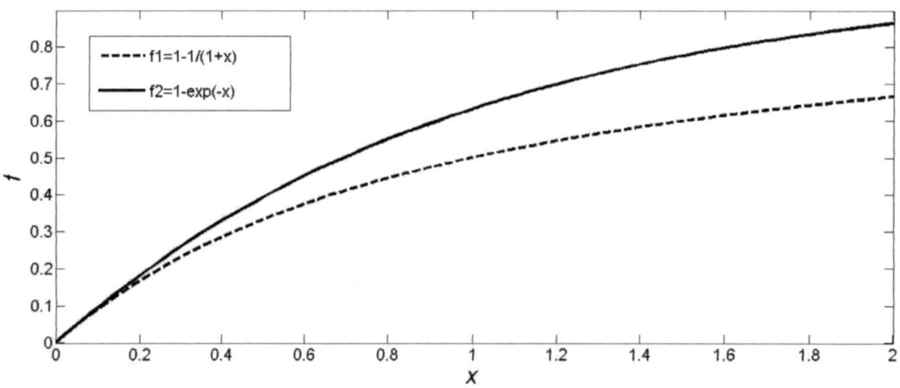

Figure 1. Functions: $f_1 - x$ and $f_2 - x$

f_2 shows some better properties comparing to f_1: when huge gradient is recognized, a bigger weight is assigned to modify term $\theta\left(\boldsymbol{F}_{j+1/2}^{LF} - \boldsymbol{F}_{j+1/2}\right)$, which could keep the scheme more robust and a steep gradient of f_2 can easily distinguish smooth region. Above all, we use

$$\theta = 1 - e^{-SDIS} \tag{3.7}$$

In practical application, we use a modified function for a better robust:

$$\theta^m = \begin{cases} 1 - e^{-SDIS \times Ac}, when: SDIS \geq threshold \\ \Delta x^5, when: SDIS < threshold \end{cases} \tag{3.8}$$

where Ac is an amplified coefficient for adjust SDIS's effects to θ^m, and when SDIS is smaller than threshold, we consider the stencils are relative smooth, and modified term show little effects on flux. Because $\boldsymbol{F}_{j+1/2}^{WENO} - \boldsymbol{H}_{j+1/2} = O(\Delta x^5)$, we assign θ the same order as Δx^5.

4. Numerical test

We solve Euler system (2.1) as test:

Test 1: Lax problem with initial condition:

$$(\rho, u, p) = \begin{cases} (0.445, 0.698, 3.548), -1 < x < 0 \\ (0.500, 0.000, 0.571), 0 \le x < 1 \end{cases}$$

The numerical solution of density of LxF, FVS WENO and Hybrid schemes are shown in figure 2, figure 3 and figure 4 respectively.

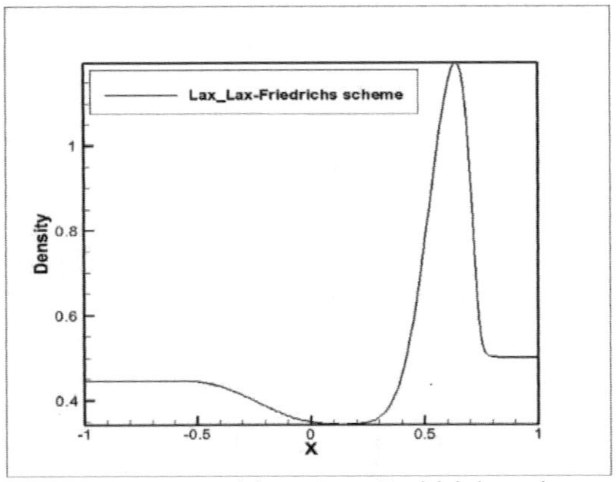

Figure 2. Lax problem: Lax-Freidrichs scheme, One-order Runge-Kutta; CFL=0.1; 200points; T=0.2;

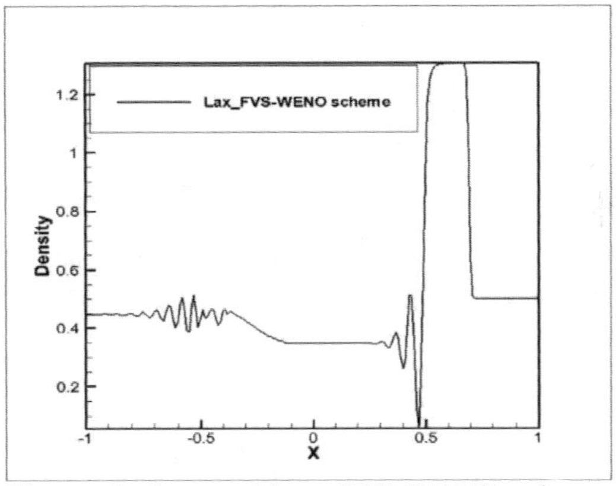

Figure 3. Lax problem: FVS-WENO scheme, One-order Runge-Kutta; CFL=0.1; 200points; T=0.2;

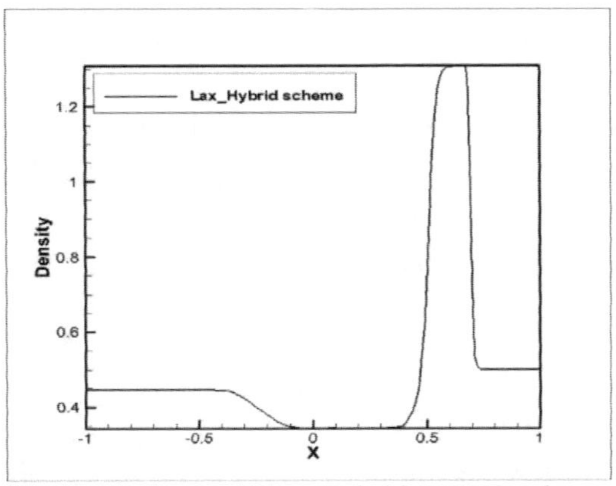

Figure 4. Lax problem: Hybrid-WENO scheme, One-order Runge-Kutta; CFL=0.1; 200points; T=0.2;

As shown in figure 2, figure 3 and figure 4, we can find that LxF scheme does not bring in oscillation but a large dissipation about discontinuity, WENO scheme show a high resolution of the discontinuity but a huge dispersion which would lead negative density that may cause blow up. Hybrid show a middle performance, an acceptable resolution, which is better than LxF scheme, and restrain the oscillation.

Test 2: Sod problem with initial condition:

$$(\rho, u, p) = \begin{cases} (1.000, 0.750, 1.000), -1 < x < 0 \\ (0.125, 0.000, 0.100), 0 \le x < 1 \end{cases}.$$

The numerical solution of density of LxF, FVS WENO and Hybrid schemes are shown in figure 5, figure 6 and figure 7 respectively. Hybrid show the same property of resolution and oscillation restrained as Lax problem but a mild oscillation which means the hybrid scheme still need some correction.

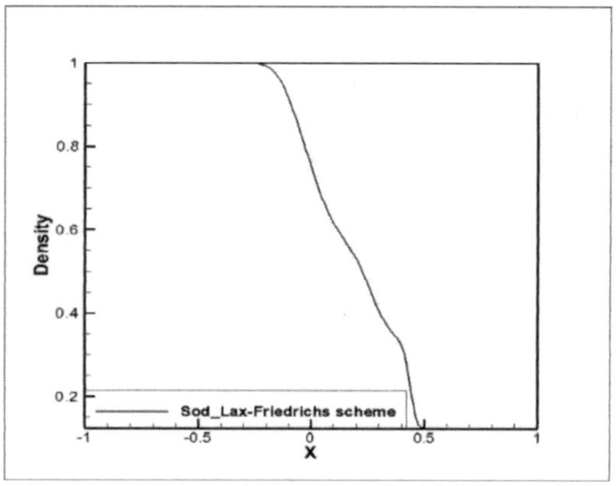

Figure 5. Sod problem: Lax-Freidrichs scheme, One-order Runge-Kutta; CFL=0.1; 200points; T=0.2;

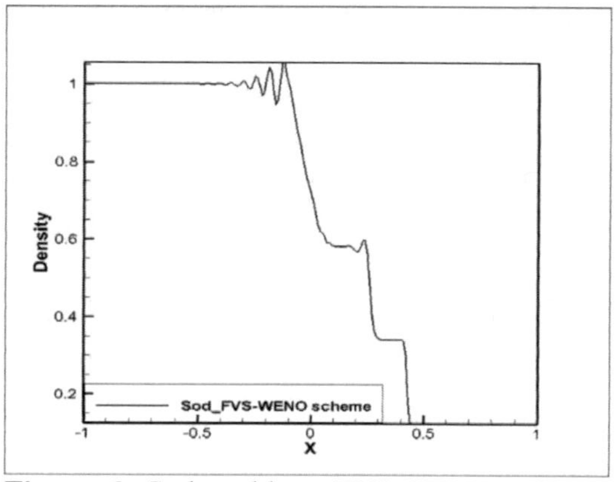

Figure 6. Sod problem: FVS-WENO scheme, One-order Runge-Kutta; CFL=0.1; 200points; T=0.2;

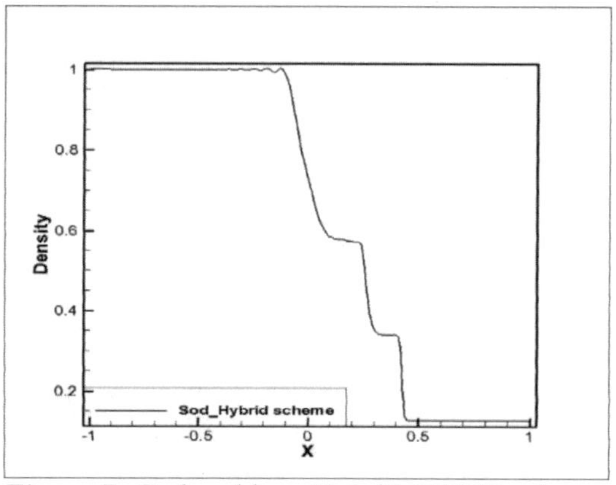

Figure 7. Sod problem: Hybrid-WENO scheme, One-order Runge-Kutta; CFL=0.1; 200points; T=0.2;

5. Conclusion

The property of FVS WENO scheme we have analysed in this paper show high dispersion because of its odd-order truncation error. We construct a hybrid scheme with the help of Lax-Friedrichs to overcome it. A limiter based on smoothness indicator is constructed for connecting WENO scheme with Lax-Friedrichs. Numerical simulations demonstrate that the hybrid show a high resolution and the ability to restrain oscillation. Further study will focus on a higher resolution of the hybrid of WENO and TVD and extend the hybrid scheme to multi-dimensions, more cases will be tested.

Acknowledgements
The research was sponsored by the National Natural Science Foundation of China (NSFC 91530325).

References

[1] Zhang X X and Shu C W 2011 *Proc. R. Soc. A.* **467** pp 2752–76.
[2] Henrick A K, Aslam T D and Powers J M 2005 *J. Comput. Phys.* **207** pp 542–67.
[3] Borges R, Carmona M, Costa B and Don W S 2008 *J. Comput. Phys.* **227** pp 3191–211.
[4] Ha Y, Kim C H, Lee Y J and Yoon J H 2013 *J. Comput. Phys.* **232** pp 68–86.
[5] Fan P, Shen Y Q and Tian B L 2014 *J. Comput. Phys.* **269** pp 329–354.
[6] Jiang G S and Shu C W 1996 *J. Comput. Phys.* **126** pp 202–28.
[7] Anderson J D(translators: Wu S P, Liu Z M) 2007 *Computational Fluid Dynamics* (Peking: China Machine Press) pp 163.
[8] Tang H Z and Xu K 2000 *J. Sci. Comput.* **15** pp 19-28.
[9] Zhang X X and Shu C W 2012 *J. Comput. Phys.* **231** pp 2245–58.
[10] Hu X Y, Adams N A, Shu C W 2013 *J. Comput. Phys.* **242** pp 169–80.

Air flow optimization for energy efficient blower of biosafety cabinet class II A2

M D Ibrahim[1], M Z Mohtar[1], A A Alias[1], L K Wong[1], Y S Yunos[1], M R A Rahman[1], A Zulkharnain[2], C.S. Tan[3] and R Thayan[4]

[1]Department of Mechanical and Manufacturing, Faculty of Engineering, Universiti Malaysia Sarawak, Malaysia
[2]Department of Molecular Biology, Faculty of Resource Science & Technology, Universiti Malaysia Sarawak, Malaysia
[3] Department of Para-Clinical Sciences, Faculty of Medicine & Health Sciences, Universiti Malaysia Sarawak, Malaysia
[4]Virology Unit Department, Institute for Medical Research, Malaysia

E-mail: AbgZaidi87@gmail.com

Abstract. An energy efficient Biosafety Cabinet (BSC) has become a big challenge for manufacturers to develop BSC with the highest level of protection. The objective of research is to increase air flow velocity discharge from centrifugal blower. An aerodynamic duct shape inspired by the shape of Peregrine Falcon's wing during diving flight is added to the end of the centrifugal blower. Investigation of air movement is determined by computational fluid dynamics (CFD) simulation. The results showed that air velocity can be increased by double compared to typical manufactured BSC and no air recirculation. As conclusion, a novel design of aerodynamic duct shape successfully developed and proved that air velocity can be increase naturally with same impeller speed. It can contribute in increasing energy efficiency of the centrifugal blower. It is vital to BSC manufacturer and can be apply to Heating, Air Ventilation and Air Conditioning (HVAC) industries.

1. Introduction

The World Health Organisation (WHO), the U.S. Centers For Disease Control and Prevention (CDC) and other organisation classified BSC into 3 classes. Each classes are distinguished in two ways; level of personal and environmental protection and level of product protection provided. This research focuses on BSC Class II Type A2. This is because it is commonly used due to their versatility and economic designs [1].

One of the important equipment inside BSC is the centrifugal blower. The purpose of this centrifugal blower is to maintain air circulation in certain air speed inflow to BSC and downflow to enclosure work area. Air flow movement generated by centrifugal blower needs to provide a non turbulent airflow distribution inside enclosure work area to ensure that the material inside the enclosure are undisturbed by air flow [2]. Generally, the best design of a BSC is the ones with a non-recirculated air flow. Recirculation of air flow at work area can cause concentrated of contaminate airborne. It will also increase pressure drop due to the increase of flow velocity. Recirculation normally happened caused by turbulent flow. Acceleration of air by centrifugal blower with very less turbulent flow is one of objective in this research. Normally, BSC uses forward curved centrifugal blower type that has high efficiency, low noise level and relatively small air flow with a high increase of static pressure [3].

IOP Conf. Series: Journal of Physics: Conf. Series **822** (2017) 012022 doi:10.1088/1742-6596/822/1/012022

A guideline from Air Movement and Control Association (AMCA) stated that each blower needs an extended length of ductwork specified as a 100 percent effective length for the velocity pressure to be fully converted to static pressure [4]. It should be long enough so that the air velocity becomes uniform across the face of the duct that is called static to regain its velocity pressure. Therefore, the design of the duct at the outlet has great effect on the system performance. Unfortunately, current BSC design in the market does not follow the guideline highlighted by AMCA. Centrifugal blowers are used without any duct connected at the end of the blower outlet and does not effectively convert velocity pressure to static pressure to gain uniform air velocity. Thus, a non-uniformed velocity pressure can cause chaos air flow and have potential to generate turbulent flow.

From another perspective, Ultra Low Particulate Air (ULPA) filter lifetime highly depends on the characteristic of the inlet velocity flow. A non-uniform velocity creates non-distributed air pressure towards filter inlet area. As a result, certain area in the BSC gains higher pressure while some areas gain lower pressure. Area which received higher pressure frequently has less filter effectiveness and shorter lifetime compared to area with lower pressure. The necessity to create a uniform distribution of air inlet velocity is important to extend longer lifespan for ULPA filters.

This research will focus on the development of a novel shape design of a duct outlet for forward-curved centrifugal blower of BSC in order to improve air flow performance with less turbulent, uniform velocity and less power usage for biosafety cabinet class II A2. Theoretically, surface modification at the outlet or any surfaces has potential to increase air flow velocity [5] and improve its aerodynamic characteristics.

Recent study areas mostly covered on chemical fume hood rather than biosafety cabinet. There is no research study on improvement of aerodynamic design of components of the biosafety cabinet itself. CFD software was used to analyze flow characteristic inside a chemical fumehood using k-ω turbulence model [6]. Then, risk management system was developed for high level biosafety laboratory in China due to outbreak of diseases such as foot and mouth disease, SARS, and highly pathogenic avian influenza [7]. Outcome from research showed that flow characteristic of air inside a biosafety cabinet is influenced by jet velocity, suction flow velocity and descending flow velocities [8]. A new way to conduct an airflow assessment was introduced in a biosafety building using CFD application [9]. Successful improvement of air flow for an enclosed work area with robotic equipment was conducted to provide an unidirectional air flow [2]. On the other hand, CFD application was used in order to improve performance test for chemical fume hood [10]. Aerodynamic design for chemical fume hood was studied by safety cabinet manufacturer, ESCO to show significant impact on air flow in minimizing reverse and turbulence flow [11]. This paper research will induce a novel research area by improving air movement and pattern inside biosafety cabinet from improvements of centrifugal blower performance.

2. Methodology

Basic 3D model of biosafety cabinet was designed similar to an actual size with nominal width size of 1219 mm. Configuration of the biosafety cabinet are described in the figure 1.

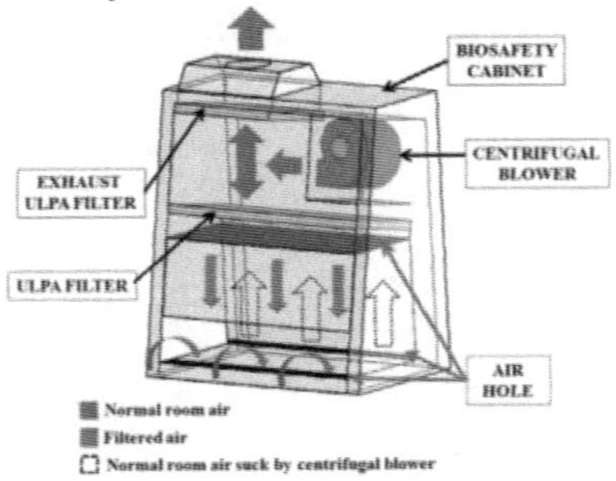

Figure 1. Basic 3D model of biosafety cabinet

Two types of biosafety cabinet configuration are simulated as can be seen in comparison from figure 2 and figure 3. The first configuration is configuration A with a ductless centrifugal blower outlet as can be seen from figure 2. This is a typical manufactured BSC configuration. Then, figure 3 shows duct after centrifugal blower.

Duct length was calculated complying to the guideline by Air Movement and Control Association (AMCA) [4]. Equation (1) was used to calculate hundred percent effective duct length, $L_{100\%}$ with flow rate less 12.7 m/s.

$$L_{100\%} = 2.5 \times D \tag{1}$$

Equivalent duct diameter, D is calculated using equation (2) with a and b are the width and the height of rectangular shape for outlet duct. The values were taken same as rectangular shape dimension for centrifugal blower outlet.

$$D = \sqrt{\frac{4ab}{\pi}} \tag{2}$$

By solving equation (1) and equation (2) with $a = 287$ mm and $b = 141$ mm, yields D = 227 mm and $L_{100\%} = 567$mm. A 3D model for configuration B was drawn based on this length.

Simplified aerodynamic shape of Peregrine Falcon was applied at middle of duct for configuration B. This aerodynamic shape is a Cupped-Shape and V-Shape design as shown in figure 4 [12]. The authors previously designed geometries based on nature's inspired showed tremendous improvements in obtaining drastic results [13].

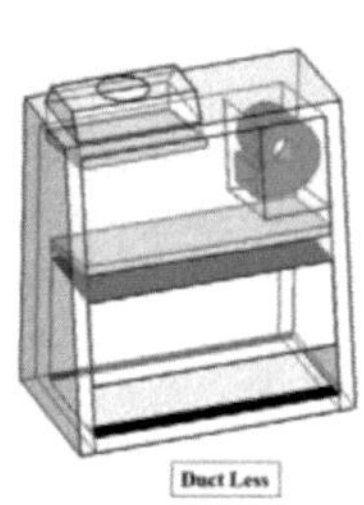

Figure 2. Centrifugal blower of configuration A

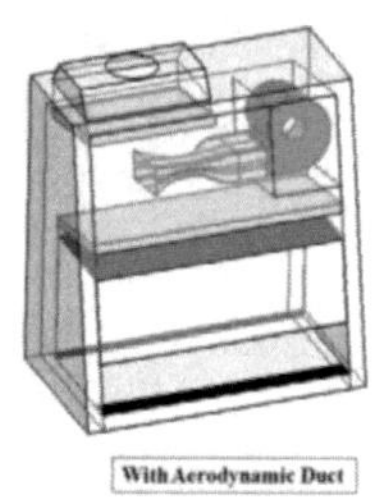

Figure 3. Centrifugal blower of configuration B

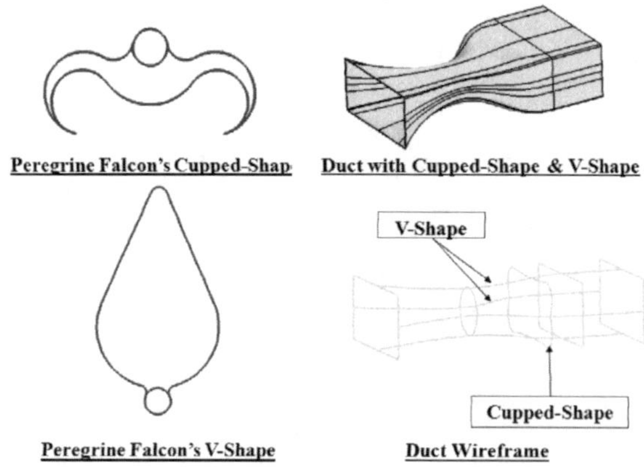

Figure 4. Peregrine Falcon's aerodynamic shape

Solid 3D of configuration A and B are imported to simulation software and mesh generated for each configuration contain approximately two million elements.

Numerical simulation using Fluid Flow (Fluent) analysis system with a standard of k-ε turbulence model has been used. Impeller rotational speed is being set to 66.9 rad/s for all configurations according to target of inflow air speed, V_i equal to 0.53 m/s as specified in BSC's catalog [14]. Concerning numerical simulation parameters, pressure outlet boundary conditions have been applied at outlet, respectively. A gauge pressure, P_o = 101325 Pa been applied to the environment. Simulation was conducted until achieved maximum residual value.

Average air velocity output, V_o are determined at the end of discharge area for configuration A and B. Then, the percentage of air velocity increment, η is determined using equation (3).

$$\eta = \left(\frac{V_o - V_i}{V_i}\right) \times 100 \quad (3)$$

Velocity uniformity between both configurations also to be analyzed as determined from this research objective.

Air flow pattern at top compartment and work area compartment was observed to check the degree of severity of recirculation of air. Besides, the pressure at two inlet surface of ULPA filter was determined in order to confirm the pressure distribution towards the ULPA filter.

3. Results and Discussions

Results of simulation shown in figure 5 and figure 6 shows that is the discharge area with duct has positive impact on air velocity pattern. The theory and numerical prediction converged as the result of Configuration A had a non-uniform velocity at discharge area. High air velocity at bottom region appeared compared to top region with a less velocity distribution. Configuration B with the duct successfully converts the non uniform velocity to a uniformed velocity pressure. Figure 5 and figure 6 below shown velocity comparison contour for both configurations.

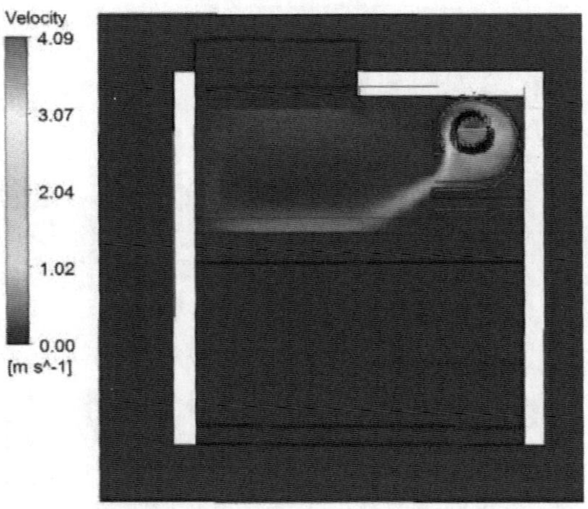

Figure 5. Velocity contour for configuration A

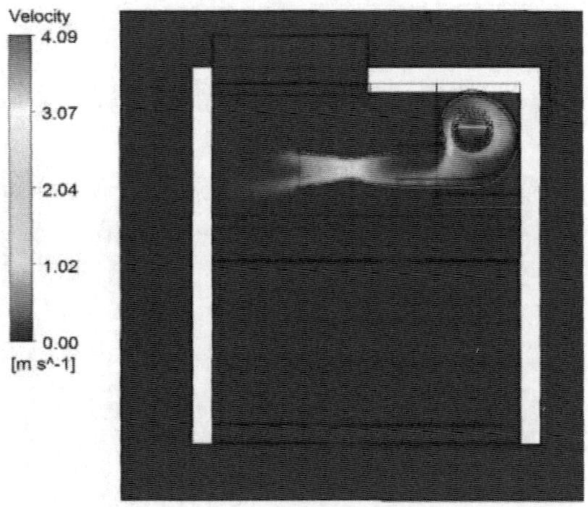

Figure 6. Velocity contour for configuration B

Furthermore, configuration B had greater average velocity value at duct discharge area compared to configuration A. The average velocity for configuration A is 0.32 m/s and configuration B is 0.77 m/s. Configuration A is set as a reference value as shown in Table 1 for comparison in percentage. It is because Peregrine Falcon shape at middle of duct for configuration B increased about 3 times of air average velocity from 0.32 m/s to 1.07 m/s. This supports the fact that air velocity accelerated naturally in configuration B was formed when air passes through the aerodynamic duct shape.

Figure 7 below shows bad air recirculation happened at the top compartment of configuration A caused by non-uniform velocity output from ductless centrifugal blower. Compared to configuration B the air flow had a smooth movement and less air recirculation can be seen from figure 8.

Local pressure distribution at inlet ULPA filter for configuration B was almost distributed evenly as can be seen from figure 10 below. While for configuration A as can be seen from figure 9, had higher concentrated pressure at one area, promoting backflow and recirculation inside the work area.

Table 1. Comparison of percentage of average air velocity (configuration A as reference)

Configuration	Average air velocity outlet, V_o (m/s)	Percentage comparison to configuration A,η (%)
A	0.32	-
B	0.77	+ 141 %

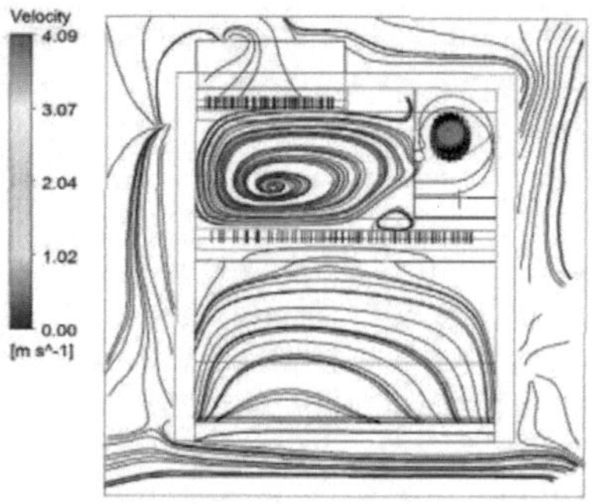

Figure 7. Streamline for configuration A

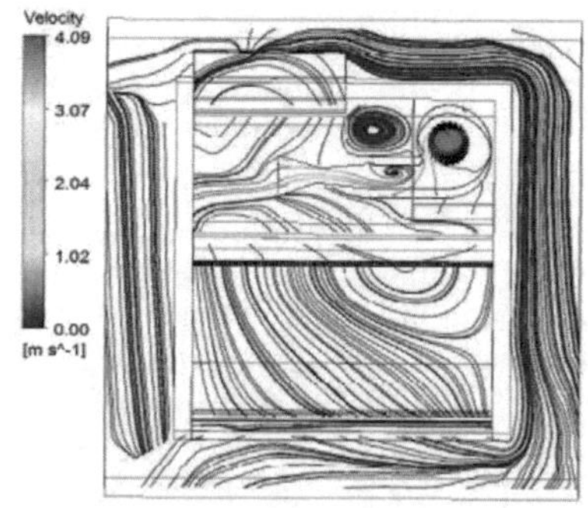

Figure 8. Streamline for configuration B

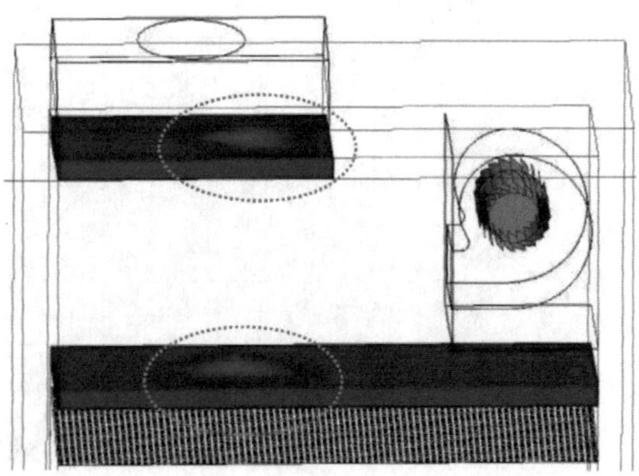

Figure 9. Pressure ditribution at ULPA filter for configuration A

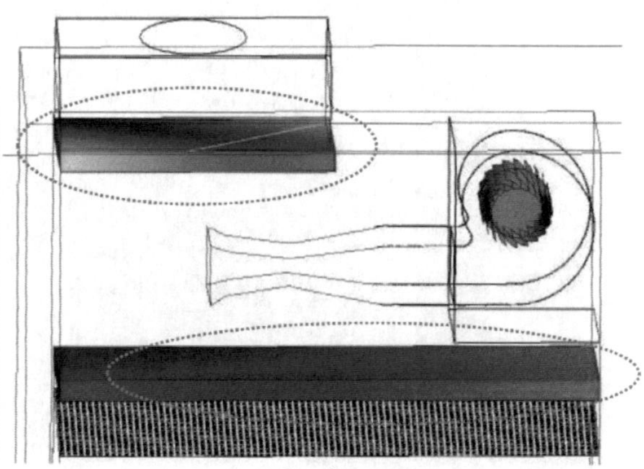

Figure 10. Pressure ditribution at ULPA filter for configuration B

Table 2 shows that the minimum and maximum pressure gauges inside BSC. Configuration B had greater pressure different on the maximum value compare to configuration A. It showed that air movement faster in configuration B instead of configuration A.

Table 2. Comparison of Pressure Gauge for configuration A and configuration B

Configuration	Min Pressure Gauge, ΔP_{min} (Pa)	Max Pressure Gauge, ΔP_{max} (Pa)
A	-1.32	0.89
B	-1.32	1.07

4. Conclusion

Biosafety cabinet with aerodynamic duct shape at discharge area had better performance compared to a ductless centrifugal blower. The study showed that uniform velocity generated manages to distribute pressure evenly to ULPA filter inlet area and less air recirculation happened. It also proved that an accelerated air can be introduced naturally without increase impeller rotational speed. As a result, rotational speed of impeller can be reduced to obtain similar performance and power consumption can be saved. It will help to reduce carbon emission by reducing power consumption as Kyoto Protocol requirement to reduce greenhouse gases emission. This research can be propose to all BSC manufacturer for create an energy efficient BSC.

Acknowledgments

The present work was partially funded by Zamalah Scholarship UNIMAS (ZSU), Universiti Malaysia Sarawak (UNIMAS) and Special Short Term Grant from UNIMAS, Grant No: F02/SpSTG/1381/12/23.

References

[1] Dondelinger R 2013 *The fundamental of biological safety cabinet, Biomedical instrumentation & technology* (Kansas: Allen Press Publishing Services Inc.) 333–338
[2] Kolesnikov A, Ryan R and Walters D B 2004 *Applied Biosafety* **9** 188–199
[3] EBM-PAPST 2007 *Centrifugal fans and blowers-version 7* (Mulfingen: EBM-PAPST Mulfingen GmbH & Co.KG) 2
[4] Air Movement And Control Association (AMCA) 2007 *AMCA Publication 201-02 (R2007) Fan & Systems* (Arlington Height: AMCA Inc.) 29–33
[5] Ibrahim M D et al 2015 *Proc. of the World Engineering Conference and Convention 2015* (Kyoto: WECC 2015 Engineering: Innovation and Society) 1–6
[6] Omar M A A, Wisnoe W and Bakri A 2013 *Applied Mechanics and Materials* **393** 753–758
[7] Ding F et al 2011 *Proc. Environmental Sciences* vol 10 (Beijing: Elsevier Ltd.) 1193-1198
[8] Rong F H and Chun I C 2009 *Ann. Occup. Hyg.* **53** 425–440
[9] Scott T, Banks D and Mishra A M 2006 *Applied Biosafety* **11** (4) 188–196
[10] Nicholson P G, Clark R P and Calcina-goff M L D 2000 *Ann. Occup. Hyg.* **44** 203–217
[11] Manik C and Peh J 2000 *High performance low flow fume hood design (Singapore:* Laboratory Fume Hood Division) 1–14
[12] Ponitz B, Triep M and Brucker C 2014 *Open Journal of Fluid Dynamics* **4** 363–372
[13] Ibrahim M D, Asmara A, Abdullah S N, Ashaari M F, Ibrahim N, Miyanaga N 2014 *J. Appl. Sci. & Agric.* **7(13)** 20-27
[14] Esco Micro Pte. Ltd. 2008 *Labculture Class II, Type A2 Biological Safety Cabinet User and Service Manual* (Hatboro: ESCO Technologies, Inc.) 95–119

IOP Conf. Series: Journal of Physics: Conf. Series **822** (2017) 012023 doi:10.1088/1742-6596/822/1/012023

Aqueous humour dynamics in anterior chamber under influence of cornea indentation

I Zuhaila[1], L Y Jiann[1], S Sharidan[1] and A Fitt[2]

[1] Department of Mathematical Sciences, Faculty of Science, Universiti Teknologi Malaysia, 81310 UTM Johor Bahru, Johor, Malaysia.
[2] Faculty of Technology, Design and Environment, Oxford Brookes University, Headington Campus, Gipsy Lane, Oxford, OX3 OBP, United Kingdom.

E-mail: zuhaila@utm.my

Abstract. The existing temperature different between the cornea and the pupil induces the aqueous humour (AH) to circulate in the anterior chamber (AC). The buoyancy forces produced by the temperature gradient has driven the AH to flow. Previous studies have shown that cornea indentation changes the structure of the AC. This imply that the cornea indentation may change the fluid flow behaviour in the AC. A mathematical model of AH flow has been developed in order to analyse the fluid mechanics concerning the indentation of the cornea. Naiver-Stokes equations is used to describe the flow of AH in the AC. The governing equations have been solved numerically using finite element method. The results show that the cornea indentation has slow down the circulation the AH in the AC.

1. Introduction

Buoyancy force due to temperature differences, secretory flow from the ciliary body through the pupil aperture and ultimately to the trabecular meshwork, and movement of the eye itself are the suggested mechanisms that drive the AH to flow in the AC [1]. A simple fluid dynamical models was used by the authors to study each of the mechanisms and the solutions was obtained by the aiding of asymptotic analysis methods. The authors concluded that the buoyancy induced by temperature gradient in the eye is the main mechanism for causing AC flow. Besides, [2] applied a simple standard fluid dynamical model to examine the flow in the AC which is driven by thermal processes. The formation of a hyphema or a Krukenberg spindle were predicted by them with the aiding of the model. They had justified that the buoyancy driven flows in the AC play a significant role and indicates that only very small temperature gradient are required to drive such flows. Further, a Boussinesq model of natural convection in the AC was developed by [3] in order to predict the magnitude of the natural convection flow of the AH. The model also used to estimate the movement of pigment particles in the AC. The lubrication theory was used by [2] in order to determine the analytical solutions of the governing equations, was not applied by [3]. However, [3] solved the model numerically with the help of FIDP computer package. The governing equations were formulated using the finite element method and the domain was divided into 27-node brick elements. The authors found a same results as shown in [2] that was the temperature gradients within the AC of the human eye are sufficient to create significant natural convection currents of AH. [4] used a commercial CFD package FLUENT to study the fluid dynamics of the AH in the AC of eye. In addition, the drainage mechanisms through the Trabecular Meshwork (TM) was also considered. The authors concluded that the buoyancy force is the dominant source to drive the AH flow.

[5], developed a 3D model of the human eye which include the TM, Schlemm's canal and collecting channels structures. The AH flow characteristics under physiological and pathological conditions such as primary open-angle glaucoma was simulated by ANSYS CFX (ANSYS Inc., Canonsburg, PA, USA). Recently, [6] developed a fluid mechanical model of buoyancy driven AH flow in the AC around Descemet Membrane Detachment (DMD) to investigate the spontaneous reattachment phenomena of the DMD. They concluded that, the clinical outcomes for DMDs were depended on the temperature difference across the eye.

On the other hand, a research was conducted by [7] to investigate the alteration of the angle between the iris and cornea and the change of the iris contour during indentation of cornea or sclera. The indentation of the cornea or sclera consequently rotate the iris root. Moreover, [8-11] had shown that the AC angle can be wider by indent the cornea. [12] extended their research to investigate the fluid structure interaction between the iris and AH, driven by iris root rotation due to the indentation of cornea or sclera. These indicate that the structure of the AC is changed when the cornea is indented, then it may alter the behaviour of the AH flow in the AC. However, there is limit discussion on the effect of the cornea indentation to the circulation of the AH driven by the temperature gradient in the AC. Therefore, the objective of this paper is to study the influence of the cornea indentation to the AH flow in AC numerically by using finite element method (FEM). Matlab software is utilized to compute and visualize the numerical solutions.

2. Mathematical formulation

AC is an approximately ellipsoidal shape region that bounded by the cornea, iris and lens, which has depth in range 2.5 to 3 [mm] and diameter in the plane of the iris root is in range 9 to 12 [mm] ([2, 4, 6, 13]). Hence, the AC is modelled as a "half-moon" shape as illustrated in figure 1 in this study. Besides, the diameter of the AC is assumed to be 11 [mm] and the depth to be 2.75 [mm] as follow [2, 6]. The iris is assumed as a rigid surface at the bottom and the curvature of the iris surface is neglected and assumed it as a flat plane. It is justified by [1, 2, 4, 6] that those assumptions incorporate insignificant error in the results. The circulation of AH in the AC is assumed mainly driven by the buoyancy forces. The temperature gradient between the iris and the cornea has generated the buoyant force to drive the AH. The temperature drop between the cornea and the iris is taken to be 2-4 [^{0}C] for the open eye ([2, 4, 6]). 98 % of the AH is built up by water and the other 2 % is made up by protein, several of amino acids, oxygen, lactate and glucose ([13, 14]). Thus, the density and viscosity of the AH is similar to water ([15, 16]).

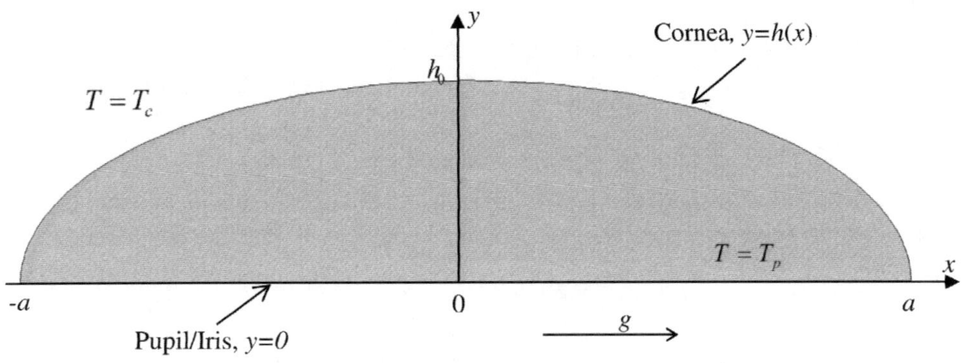

Figure 1. Schematic diagram of the AC.

In the plane z = 0, a two-dimensional AH flow driven by buoyancy effects in the AC has been considered, as shown in figure 1. The temperature at the back of the AC is assumed close to core body temperature (37 [^{0}C]). Due to the ambient temperature (say 25 [^{0}C]), the cornea is assumed to has temperature 35 [^{0}C]. Therefore in Cartesian coordinate system (x, y), the AH is supposed flows between the plane formed by pupil aperture and the iris, $y = 0$ and the anterior surface of the cornea, $y = h(x)$. At the iris, the temperature is fixed at T_p which is close to the human body temperature, 37 [^{0}C], and the temperature at the cornea is assumed to be T_c, around 35 [^{0}C]. The gravity, g is acted along the positive x-axis as seen in figure 1. To be realistic, a set of typical values for human eye is used such as $h_0 = 2.75$ [mm], $a = 5.5$ [mm], gravity $g = 9.8$ [ms^{-2}] and the AH has a typical velocity of $U = 10^{-4}$[ms^{-1}], density $\rho = 1000$ [kgm^{-3}], viscosity $\mu = 0.9 \times 10^{-3}$ [kgm^{-1}s^{-1}], specific heat $c_P = 4200$ [Jkg^{-1}K^{-1}], thermal conductivity $k = 0.57$ [Wm^{-1}K^{-1}] and coefficient of thermal expansion $\beta = 3 \times 10^{-4}$ [K-1]. The AH is assumed to be Newtonian, viscous and incompressible. The governing equations are

$$\begin{gathered}
\frac{\partial u}{\partial x} + \frac{\partial v}{\partial y} = 0, \\
\rho\left(u\frac{\partial u}{\partial x} + v\frac{\partial u}{\partial y}\right) + \frac{\partial P}{\partial x} - \mu\left(\frac{\partial^2 u}{\partial x^2} + \frac{\partial^2 u}{\partial y^2}\right) + g\rho\beta\left(T - T_\infty\right) = 0, \\
\rho\left(u\frac{\partial v}{\partial x} + v\frac{\partial v}{\partial y}\right) + \frac{\partial P}{\partial y} - \mu\left(\frac{\partial^2 v}{\partial x^2} + \frac{\partial^2 v}{\partial y^2}\right) = 0, \\
c_P\rho\left(u\frac{\partial T}{\partial x} + v\frac{\partial T}{\partial y}\right) - k\left(\frac{\partial^2 T}{\partial x^2} + \frac{\partial^2 T}{\partial y^2}\right) = 0.
\end{gathered} \tag{1}$$

Following [2], the boundary conditions are

$$\begin{gathered}
u(x,0) = v(x,0) = 0, \\
u(x,h(x)) = v(x,h(x)) = 0, \\
T(x,0) = T_p,\ T(x,h(x)) = T_c, \\
P(a,0) = P_a,
\end{gathered} \tag{2}$$

where P_a is the normal atmospheric pressure and equal to 101325 [Pa]. FEM is applied to solve equation (1) with subjected to the boundary conditions (2). To the best of authors' knowledge, the application of the finite element analysis to solve this problem is not yet adequately provided. Initially, the domain is discretized into nonuniform and unstructured triangular Taylor-Hood elements ([17-19]) and in each finite element e_i, $i = 1, 2, \ldots n$, the variables u^{e_i}, v^{e_i}, T^{e_i} and P^{e_i} are approximated as

$$\begin{gathered}
u = u^{e_i} = \phi_j^{e_i}(x,y)\,u_j^{e_i},\ j = 1,2,\ldots,6. \\
v = v^{e_i} = \phi_j^{e_i}(x,y)\,v_j^{e_i},\ j = 1,2,\ldots,6. \\
P = P^{e_i} = \psi_k^{e_i}(x,y)\,P_k^{e_i},\ k = 1,2,3. \\
T = T^{e_i} = \phi_j^{e_i}(x,y)\,v_j^{e_i},\ j = 1,2,\ldots,6.
\end{gathered} \tag{3}$$

where, $\phi_j^{e_i}$ and $\psi_k^{e_i}$ are the shape functions for the dependent variables velocity, temperature and the pressure respectively. Galerkin finite element methods is practiced in this study. After discretises the equations (1) by the method, the equations (1) can be represented in a matrix form as

$$\begin{bmatrix} S_{11} & S_{12} & S_{13} & S_{14} \\ S_{21} & S_{22} & S_{23} & S_{24} \\ S_{31} & S_{32} & S_{33} & S_{34} \\ S_{41} & S_{42} & S_{43} & S_{44} \end{bmatrix} \begin{Bmatrix} \{u_j^{e_i}\}^T \\ \{v_j^{e_i}\}^T \\ \{P_k^{e_i}\}^T \\ \{T_j^{e_i}\}^T \end{Bmatrix} = \begin{bmatrix} r_1 \\ r_2 \\ r_3 \\ r_4 \end{bmatrix}, \tag{4}$$

where

$$S_{11} = \left[\int_{A_{e_i}} \begin{matrix} \{\phi_l^{e_i}\}\rho\left(u^{e_i} \dfrac{\partial\{\phi_j^{e_i}\}}{\partial x} + v^{e_i} \dfrac{\partial\{\phi_j^{e_i}\}}{\partial y} \right) \\ +\mu\left(\dfrac{\partial\{\phi_l^{e_i}\}}{\partial x}\dfrac{\partial\{\phi_j^{e_i}\}}{\partial x} + \dfrac{\partial\{\phi_l^{e_i}\}}{\partial y}\dfrac{\partial\{\phi_j^{e_i}\}}{\partial y} \right) \end{matrix} dA_{e_i} \right]$$

$$S_{12} = S_{21} = S_{24} = S_{41} = S_{42} = [0]_{6\times 6},\ S_{13} = \left[\int_{A_{e_i}} \{\phi_l^{e_i}\} \frac{\partial\{\psi_k^{e_i}\}}{\partial x} dA_{e_i} \right],$$

$$S_{14} = \int_{A_{e_i}} \{\phi_l^{e_i}\} g\rho\beta \{\phi_j^{e_i}\} dA_{e_i},\ S_{22} = S_{11},$$

$$S_{23} = \left[\int_{A_{e_i}} \{\phi_l^{e_i}\} \frac{\partial\{\psi_k^{e_i}\}}{\partial y} dA_{e_i} \right],\ S_{31} = \left[\int_{A_{e_i}} \{\psi_m^{e_i}\} \frac{\partial\{\phi_j^{e_i}\}}{\partial x} dA_{e_i} \right],$$

$$S_{32} = \left[\int_{A_{e_i}} \{\psi_m^{e_i}\} \frac{\partial\{\phi_j^{e_i}\}}{\partial y} dA_{e_i} \right],\ S_{33} = [0]_{3\times 3},\ S_{34} = [0]_{3\times 6},\ S_{43} = [0]_{6\times 3}$$

$$S_{44} = \left[\int_{A_{e_i}} \begin{matrix} \{\phi_l^{e_i}\} c_P \rho\left(u^{e_i} \dfrac{\partial\{\phi_j^{e_i}\}}{\partial x} + v^{e_i} \dfrac{\partial\{\phi_j^{e_i}\}}{\partial y} \right) \\ +k\left(\dfrac{\partial\{\phi_l^{e_i}\}}{\partial x}\dfrac{\partial\{\phi_j^{e_i}\}}{\partial x} + \dfrac{\partial\{\phi_l^{e_i}\}}{\partial y}\dfrac{\partial\{\phi_j^{e_i}\}}{\partial y} \right) \end{matrix} dA_{e_i} \right]$$

$$r_1 = \oint_{\Gamma_{e_i}} \mu\left(\{\phi_l^{e_i}\} \frac{\partial\phi_j^{e_i} u_j^{e_i}}{\partial x} n_x + \{\phi_l^{e_i}\} \frac{\partial\phi_j^{e_i} u_j^{e_i}}{\partial y} n_y \right) d\Gamma_{e_i} + \int_{A_{e_i}} \phi_l^{e_i} g\rho\beta T_\infty dA_{e_i},$$

$$r_2 = \oint_{\Gamma_{e_i}} \mu\left(\{\phi_l^{e_i}\} \frac{\partial\phi_j^{e_i} v_j^{e_i}}{\partial x} n_x + \{\phi_l^{e_i}\} \frac{\partial\phi_j^{e_i} v_j^{e_i}}{\partial y} n_y \right) d\Gamma_{e_i},\ r_3 = [0]_{3\times 1},$$

$$r_4 = \oint_{\Gamma_{e_i}} k\left(\{\phi_l^{e_i}\} \frac{\partial\phi_j^{e_i} T_j^{e_i}}{\partial x} n_x + \{\phi_l^{e_i}\} \frac{\partial\phi_j^{e_i} T_j^{e_i}}{\partial y} n_y \right) d\Gamma_{e_i},$$

and

$$\{\phi_j^{e_i}\} = \{\phi_1^{e_i} \quad \phi_2^{e_i} \quad \phi_3^{e_i} \quad \phi_4^{e_i} \quad \phi_5^{e_i} \quad \phi_6^{e_i}\},$$
$$\{\psi_k^{e_i}\} = \{\psi_1^{e_i} \quad \psi_2^{e_i} \quad \psi_3^{e_i}\},$$
$$\{u_j^{e_i}\}^T = \{u_1^{e_i} \quad u_2^{e_i} \quad u_3^{e_i} \quad u_4^{e_i} \quad u_5^{e_i} \quad u_6^{e_i}\}^T,$$
$$\{v_j^{e_i}\}^T = \{v_1^{e_i} \quad v_2^{e_i} \quad v_3^{e_i} \quad v_4^{e_i} \quad v_5^{e_i} \quad v_6^{e_i}\}^T,$$
$$\{T_j^{e_i}\}^T = \{T_1^{e_i} \quad T_2^{e_i} \quad T_3^{e_i} \quad T_4^{e_i} \quad T_5^{e_i} \quad T_6^{e_i}\}^T,$$
$$\{P_j^{e_i}\} = \{P_1^{e_i} \quad P_2^{e_i} \quad P_3^{e_i}\}^T,$$
$$\{\phi_l^{e_i}\} = \{\phi_j^{e_i}\}^T, \ \{\psi_m^{e_i}\} = \{\psi_k^{e_i}\}^T.$$

Assembly the element matrix (4) for each finite element into a global matrix and then determine the solutions by solving the global matrix with subjecting to the boundary conditions (2). Nevertheless, the global matrix is a nonlinear system since equation (4) is nonlinear, thus it is needed to solve the system in an iteration manner. Newton-Raphson methods is applied to solve the global nonlinear system.

On the others hand, in [7] four kind of cornea indentation had considered. As an early stage of the study, only one of the instance is proposed in this paper which is the manual indentation of the central cornea. In addition, this nature of indentation obey the assumption of the iris surface as a flat plane. Generally, squint hook, cotton bud, glass rod, and gonioprism are the instruments used by the ophthalmologist to indent the central or peripheral surface of the cornea. However in this paper, a glass rod, which has 7 [mm] diameter is used to indent the central of cornea as shown in figure 2(a). This specific diameter is used so that the indentation formed with depth *d*, in range 0 to 0.5 [mm] always produces a flat indention on the surface of the cornea (see figure 2(b)). The consequence of the indentation to the fluid flow of the AH in the AC is investigated.

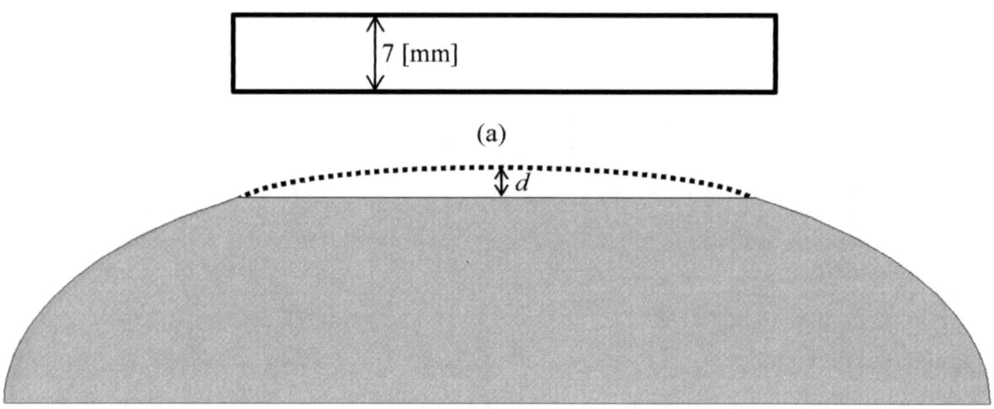

(b)

Figure 2. (a) A glass rod and (b) the indentation of the central cornea by the rod.

3. Results and discussion

A Matlab coding was developed to compute the numerical results and a personal computer with a processor speed of 2.30 GHz and a RAM of 8GB was used to run all the computation in the study. A mesh test was conducted first in order to show that the results is independent from the total number of elements used to mesh the domain. Figure 3 shows the *u*-velocity profiles along a vertical line passing through the geometric center of the cornea with total number of elements 6413, 7374, 8546, 9266 and 10224 respectively. The figure illustrates that the results are excellent match. Therefore, we have shown that the results do not depend on the number of elements used. For the following study, the domain is meshed with 8000-10000 elements, 19000-20000 nodes and contained 60000-70000 degreed of freedoms. In other word, the global stiffness matrix that produced by the sources code are in size between $60000 \times 60000 - 70000 \times 70000$.

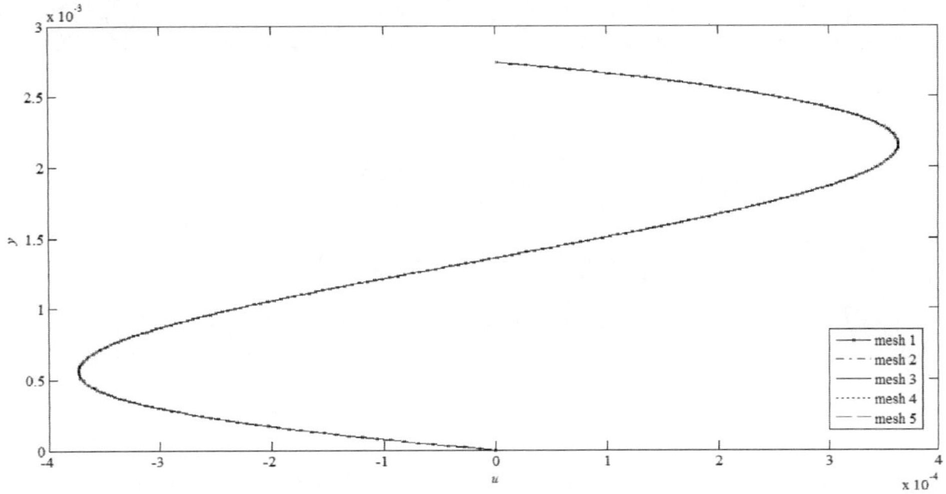

Figure 3. The *u*-velocity profiles along a vertical line passing through the geometric centre of the cornea with mesh 1 = 6413 elements, mesh 2 = 7374 elements, mesh 3 = 8546 elements, mesh 4 = 9266 elements and mesh 5 = 10224 elements.

Table 1 has shown the result of maximum *u*-velocity and the locations in the AC with the given results in [2, 6]. A slightly different between the results is illustrated in the table but it is in an acceptable range and may say it is in a great agreement. The quantitative agreement is very encouraging and has enhanced the confidence to the correctness of the obtained solutions. Besides that, Comsol Multiphysics, which is a commercial software applying FEM, was also used to simulate the fluid flow. 8848 elements were generated to mesh the domain and the solutions are in a great agreement with the present result as seen in table 1. Moreover, Comsol Multiphysics was also used to plot the velocity contour for the fluid flow in AC. The results generated by the present computer code (see figure 4(a)) are concurrent to the results determined by the commercial software (see figure 4(b).

Table 1. Comparisons of the maximum *u*-velocity and its location in the AC for the present results with the previous obtained results in the literature.

	Location	maximum *u*-velocity
$\|\textbf{Present}\|\left(\times 10^{-4}\right)$	(0, 0.0005600)	3.71976
$\|\textbf{[2]}\|\left(\times 10^{-4}\right)$	(0, 0.0005775)	3.96000
$\|\textbf{[6]}\|\left(\times 10^{-4}\right)$	(0, 0.0005811)	3.96200
$\|\textbf{Comsol}\|\left(\times 10^{-4}\right)$	(0, 0.0005645)	3.80686

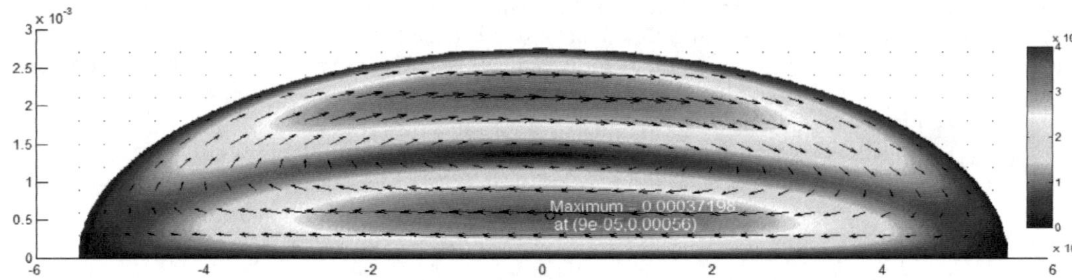

(a)

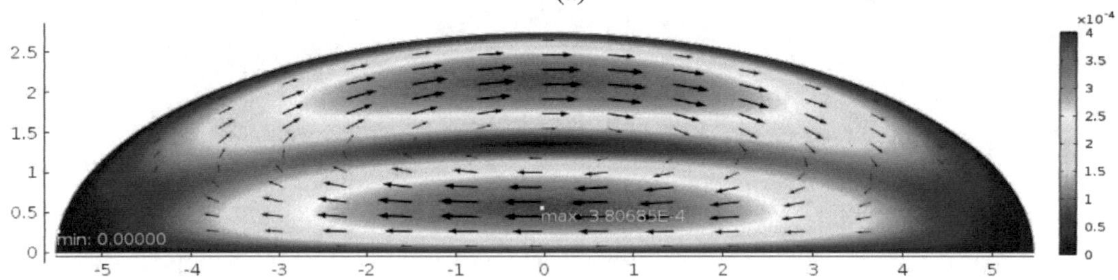

(b)

Figure 4. Velocity contour of the AH flow in the AC (a) Present (b) Comsol.

The effect of the cornea indentation to the behavior of the AH flow in AC are shown in figure 5. When comparing figure 5 with figure 4(a), the indentation decreases the maximum velocity magnitude in the AC. Moreover, the location of the maximum value exist also move upward near to the cornea as seen in Figure 5. Further, when d is increased from 0.3 [mm] to 0.5 [mm], the maximum velocity magnitude is slightly decreased. The location that the maximum velocity magnitude to be found is also changing and is moving further away to the left hand side of the y-axis. The results show that the scale effects which arises from difference in depth of the indentation, do effect the fluid behaviour in the human eye. [6] had stated that the position of the DMD and the fluid flow in AC do induce the spontaneous reattachment of the detached Descemet membrane. They concluded that the upper DMD, which the Descemet membrane is attached to the cornea above the centre of the cornea, are likely to naturally resolve themselves than lower DMD (the Descemet membrane is attached to the cornea below the centre of the cornea). As elaborated in [6], the normal pressure that act form the left hand side of the upper DMD is higher the other side, thus the detached Descemet membrane is pushing closer to the cornea. The indentation of the cornea has decrease the velocity of the AH flow in the AC, therefore, based on the Bernoulli principle the normal pressure the act on the left hand side of the upper DMD may increase. This possible increase the percentage of happening spontaneous attachment of the upper DMD. Further research is needed to fully understand the real mechanism of the AH flow with the DMD and then induce the attachment of the DMD. Besides, in this study only one type of indentation is considered and also the shape of the instrument used to indent the cornea may change the movement of the AH in the AC.

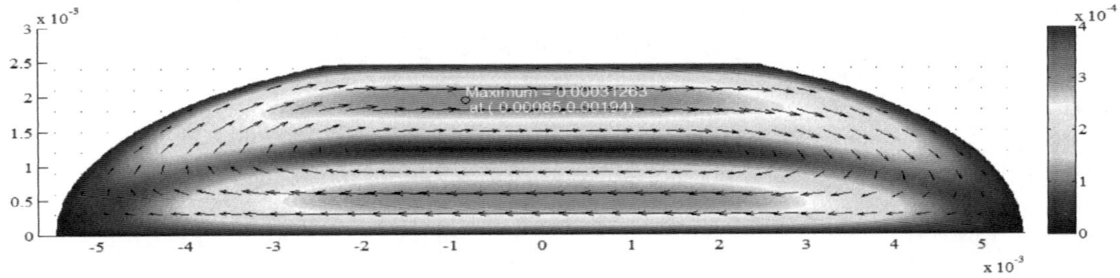

(a)

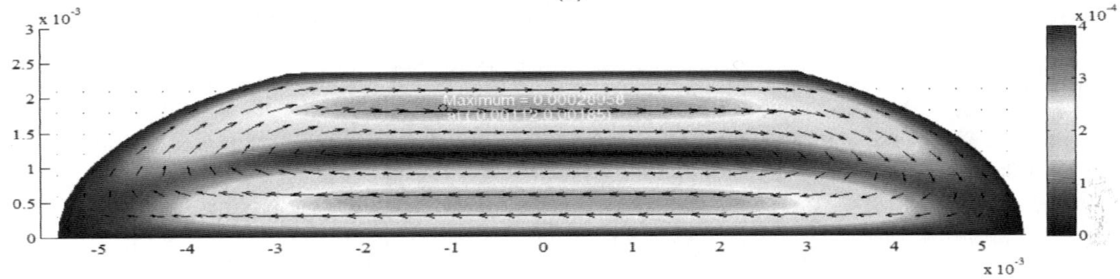

(b)

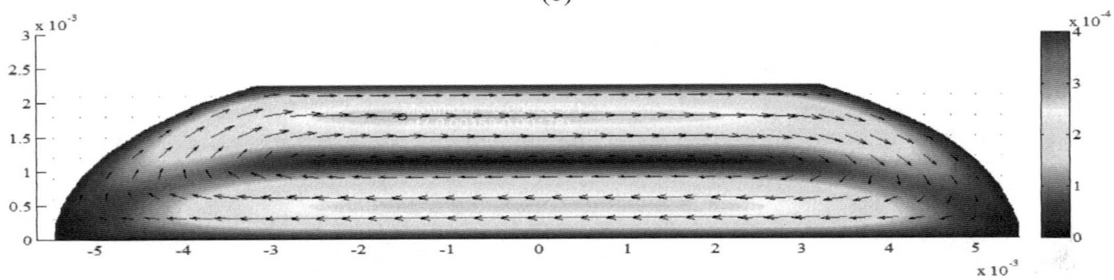

(c)

Figure 5. Velocity contour of the AH flow in the AC for the cornea indentation with depth d= (a) 0.3 [mm], (b) 0.4 [mm] and (c) 0.5 [mm].

4. Conclusions

The behaviour of the AH in the AC under the influence of the cornea indentation have been analyzed. Some figures are plotted to show the behaviour of the fluid under different depth of indentation. Some interesting finding of the study can be concluded as follows:

1. The velocity of the AH is effected by the depth of the indentation.
2. The computation model and the Matlab coding that developed in this study can be used to analyze the fluid flow in human eye with cornea indentation and also suggested to be applied to investigate a more advance research in the future.
3. The indentation of the cornea may increase the possibility of occurring the spontaneous reattachment of the DMD.

As suggestion, the fluid flow in the real shape of the AC and the effect of the shape of the indentation to the cornea should be considered in the future. Furthermore, the influence of the cornea indentation to the phenomena of the spontaneous attachment of DMD need further investigation.

Acknowledgments
The authors would like to acknowledge KPT (MyBrain), MoHE and Research Management Centre – UTM for the financial support through vote numbers 4F632, 4F713, 13H28, 13H74 and 03G53 for this research.

References
[1] Fitt A D and Gonzalez G 2006 *B. Math. Biol.* **68** 53-71
[2] Canning C R, Greaney M J, Dewynne J N and Fitt A D 2002 *IMA J. Math. Appl. Med.* **19** 31-60
[3] Heys J J and Barocas V H 2002 *Ann. Biomed. Eng.* **30** 392-401
[4] Kumar S, Acharya S, Beuerman R and Palkama A 2006 *Ann. Biomed. Eng.* **34** 530-44
[5] Villamarin A, Roy S, Hasballa R, Vardoulis O and Reymond P 2012 *Med. Eng. Phys.* **34** 1462-70
[6] Ismail Z, Fitt A D and Please C P 2013 *Math. Med. Biol.* **30** 1-17
[7] Amini R and Barocas V H 2009 *Invest. Ophth. Visual.* **50** 5288-94
[8] Forbes M 1966 *Arch Ophthalmol-Chic* **76** 488-92
[9] Anderson D R 1979 *Am. J. Ophthalmol.* **88** 1091-3
[10] Matsunaga K, Ito K, Esaki K, Sugimoto K, Sano T, Miura K, Sasoh M and Uji Y 2004 *Am. J. Ophthalmol.* **137** 552-4
[11] Masselos K, Bank A, Francis I C and Stapleton F 2009 *Ophthalmology* **116** 25-9
[12] Amini R and Barocas V 2010 *J. Biomed. Eng.* **132**
[13] Amini R 2010 *Iris Biomechanics in Health and Disease.* (The University Of Minnesota: Faculty of the Graduate School)
[14] Ethier C R, Johnson M and Ruberti J 2004 *Annu. Rev. Biomed. Eng.* **6** 249-73
[15] Scott J A 1988 *Phys. Med. Biol.* **33** 227-41
[16] Beswick J A and McCulloch C 1956 *Br. J. Ophthalmol.* **40** 545-8
[17] Reddy J N 2006 *An Introduction to the Finite Element Method* (Singapore: McGraw-Hill Education)
[18] Jichun L and Yi-Tung C 2008 *Computational Partial Differential Equations Using MATLAB* (Boca Raton: CRC Press)
[19] Zienkiewicz O C, Taylor R L and Zhu J Z 2005 *The Finite Element Method: Its Basis and Fundamentals: Its Basis and Fundamentals* (UK: Elsevier Science)

IOP Conf. Series: Journal of Physics: Conf. Series **822** (2017) 012024 doi:10.1088/1742-6596/822/1/012024

Attributes of oscillatory physiological blood flow through 3-D geometry of single stenosed artery

Khairuzzaman Mamun[1], Mohammad Ali[2], Most. Nasrin Akhter[1]
[1]Department of Mathematics, Dhaka University of Engineering and Technology, Gazipur-1700, Bangladesh.
[2]Department of Mechanical Engineering, Bangladesh University of Engineering and Technology, Dhaka-1000, Bangladesh.

E-mail : nasrin@duet.ac.bd

Abstract. The present study investigates the significant changes of flow behavior for 65 and 85 percentages severities of stenosis by area. The blood is assumed to be incompressible, homogeneous and Newtonian, while artery is assumed to be a rigid wall. The transient analysis is performed using ANSYS-14.5. Oscillatory physiological and parabolic velocity profile has been imposed for inlet boundary condition. The investigation has a Reynolds number range of 96 to 800. Pressure based solver and finite volume method are used for calculations. The flow pattern, wall shear stress (WSS), pressure, velocity, streamline contours, cross sectional and Centre-line velocity distribution are observed at early-systole, peak-systole and diastole for better understanding of arterial disease. Wall Shear Stress distribution shows that as severity increases, shearing of flow also increases for all cases. Thus maximum stress is exerted in throat region at peak systole. Pressure distribution also demonstrates that at all cases 85% stenotic artery creates more force than 65% stenotic artery at their pre-stenotic region. Interestingly, a recirculation region is visible at the post stenotic region in 85% stenotic artery for all cases and recirculation region increases with increasing minimum velocity at the inlet flow. From streamline contours it can be interpreted that 85% stenosis can create vortex at any time phase. So 85% stenosis is very harmful for its high probability of creating vortex.

1. Introduction

Cardiovascular diseases are a class of diseases that involve with heart or blood vessels (arteries and vain). On the other hand, stenosis is a constriction in the blood vessels and a leading cause of stroke and even fatal death in most of the countries. This happens due to our relatively limited understanding about stenosis. Therefore, a study of stenosis can be helpful to understand the biomechanics of vascular diseases. In general the blood flow does not face any obstacles on the way and the flow is regular and smooth. But sometimes due to presence of cholesterol, calcium and other substance in blood, a fatty substance called plaque develops near the inner wall of the artery. As the time passes this plaque grows resulting in hardening the arterial wall and narrowing the blood vessels. This can cause severe diseases such as the development of atherosclerosis. As a result the arterial wall loses its elastic property which limits the area of blood flow. Therefore, the flow turns to be abnormal in the reduced cross sectional area of the artery according to Khader and Shenoy [1]. One of the interesting studies in this area suggested that pulsatile blood flow through artery vessel implicated in several types of hemodynamic forces that could impact in vessel wall structure. These forces are also the cause of development of vascular pathologies and an important factor in atherosclerosis. The regions of high shear stress, which implicates in a direct mechanical harm of the vessel wall, are regions where

IOP Conf. Series: Journal of Physics: Conf. Series **822** (2017) 012024 doi:10.1088/1742-6596/822/1/012024

atherosclerosis occurs. At present, there is no standard procedure to measure the physical severity of the stenosis. Doctors often judge the severity based on a patient's physical symptoms as well as growth rate, constriction size and pressure drop often at high risk to patient's lives as stated by Chua and Shread [3]. But the results of numerical simulation can be better than MR, CT and Ultrasound techniques in analyzing the severity of stenosis. Chua & Shread et al [3] proved that the flow through the constricted tube is characterized by high velocity jet generated in constricted region and flow separation downsream to the stenosis. Khader & Shenoy et al. [1] found the results from numerical simulation that demonstrate that velocity and stenotic jet length increases during increasing the severity of stenosis. Their results also demonstrated that the 3D stenotic CFD model is capable to predict the changes in flow behavior for increased severity of stenosis. Young et al. [4] also studied the wall shear stress and pressure gradient in the stenosis and evaluated the cause of plaque rupture. They found pulsatile blood flow through the stenosis with elastic wall to observe the lumen movement. According to their study the peak WSS occurred just before minimum lumen position. Pinto et al. [5] conducted numerical simulation assuming a physiological pulsatile flow through different models of stenosis. In case of subject-specific anatomically realistic stenosed carotid bifurcation subjected to pulsatile inlet condition, the simulation results demonstrated the rapid fluctuation of velocity and pressure in post-stenotic region by S. Lee et al [7]. Ahmed and Giddens [6] studied both steady and pulsatile flow through 25%, 50% and 75% constriction of a rigid tube where Reynolds number ranges from 500 to 2000. Some of the above mentioned previous experimental and computational studies analyzed the flow aspects in proximal and distal end of stenosis including a detailed observation. Apart from this, an attempt is made in this present study to demonstrate the significant changes of flow behavior for 65 and 85 percentages severities of stenosis. For this a CFD method is used for detailed investigation of flow parameters like velocity, Wall Shear Stress, pressure and streamline contours. The working domain and the boundary conditions of the geometry were defined in pre-processor software ANSYS Workbench. The finite volume analysis was performed using ANSYS Fluent-14.5.

2. Model description:

2.1. Geometry :

This study is considered pulsatile flows of Newtonian fluid in a three dimensional axisymmetric stenotic arteries. For this study, 65% and 85% stenotic artery (by area) are taken. The flow geometry comprises a tube of diameter (D) of 6 mm and a length (L) of 96 mm and divided into three regions, pre stenotic, throat and post stenotic where the length of pre stenotic, throat and post stenotic region are 4D, 2D and 10D respectively. The wall is considered to be rigid. Fig.1(a) shows the schematic diagram of arterial stenosis. The flow field mesh consists of 43657 nodes and 91561 elements for 65% stenotic artery and of 45496 nodes and 95592 elements for 85% stenotic artery. Figure 1(b) shows the mesh in cross sectional inflow plane of a stenotic artery

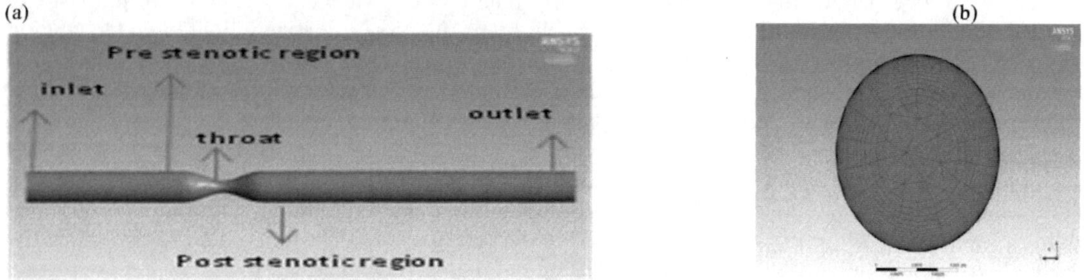

Figure 1. (a) Model of a stenotic artery; (b) mesh in cross sectional inflow plane of a stenotic artery.

2.2 Governing equation and Numerical Method

Blood is taken as fluid where the blood is considered incompressible. The density of the blood is 1050kg/m^3. In a Newtonian model for the blood viscosity, the value of μ is treated as a constant

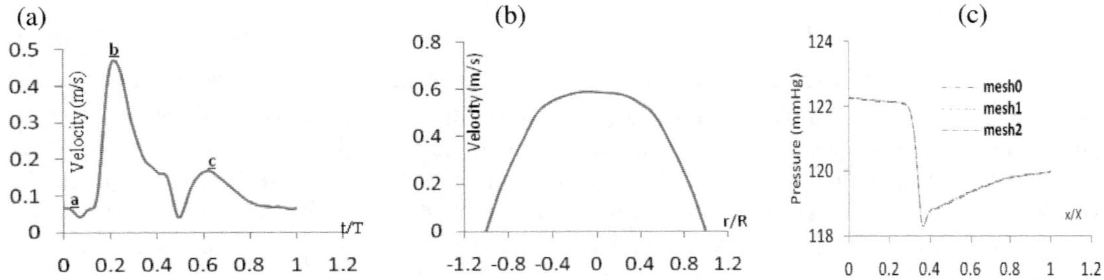

Figure 2. (a) Oscillatory physiological waveform, (b) parabolic inlet velocity profile, and (c) pressure distribution in 65% stenotic artery from different mesh sizes.

usually set to $\mu = 3.45\times10^{-3}\ Pa.s$. Due to constriction blood passes through the throat and post stenotic region with high velocity. Flow velocity at post stenotic region increases but pressure of that decreases. For this region, turbulent flow is expected at post stenotic region. Again, low Re $k-\omega$ model is more acceptable for flow analysis by Shahed et al [8]. Therefore, low Re $k-\omega$ turbulent mode is taken for calculation.

Now, the Navier-Stokes equation can be given by-

$$\frac{\partial u_i}{\partial x_i} = 0 \tag{1}$$

$$\rho\frac{\partial u_i}{\partial t} + \rho\frac{\partial}{\partial x_j}(u_j u_i) = -\frac{\partial p}{\partial x_i} + \frac{\partial}{\partial x_j}(2\mu s_{ij}) \tag{2}$$

where, the strain-rate tensor S_{ij} is given by, $s_{ij} = \frac{1}{2}\left(\frac{\partial u_i}{\partial x_j} + \frac{\partial u_j}{\partial x_i}\right)$ (3)

The k equation: $$\frac{\partial}{\partial t}(\rho k) + \frac{\partial}{\partial x_i}(\rho k u_i) = \frac{\partial}{\partial x_j}\left(\Gamma_k \frac{\partial k}{\partial x_i}\right) + G_k - Y_k + S_k \tag{4}$$

The ω equation: $$\frac{\partial}{\partial t}(\rho\omega) + \frac{\partial}{\partial x_i}(\rho\omega u_i) = \frac{\partial}{\partial x_j}\left(\Gamma_\omega \frac{\partial k}{\partial x_i}\right) + G_\omega - Y_\omega + S_\omega \tag{5}$$

The numerical simulations are performed by well known software ANSYS Fluent 14.5. A pressure based algorithm is chosen as the solver type. This solver is generally selected for an incompressible fluid. An extensive test is carried out with different sizes of mesh such as mesh0 (75511 element), mesh1 (82580 element) and mesh2 (90227 element) respectively. Figure 2(c) shows the pressure distributions for 65% stenosis artery with mentioned mesh sizes. In all cases, the pressure distributions are same. It implies that the solution is grid independent. Before starting of present investigation the numerical simulation is needed to be validated. Validation of the present numerical computation is done by plotting the steady velocity profile at 2.5D downstream from the stenosis throat and comparing it with the velocity profile of Varghese and Frankel [2].

2.3. Boundary condition:

It assumed that the arterial walls are rigid and no-slip condition is imposed at the walls. Since the blood flow through arterial stenosis is an unsteady phenomenon and the blood flow to be fully developed at inlet region, Oscillatory physiological parabolic velocity profile is imposed for inlet boundary condition. For this purpose an user defined function is written in C compiler for the following equation

$$u_x = u\left(1-\frac{y^2+z^2}{radius^2}\right), \text{ where } u=\sum_{n=0}^{16}\left(A_n\cos(\omega t)+B_n\sin(\omega t)\right) \tag{6}$$

Where A_n and B_n are the coefficients which demonstrate the unsteady parabolic nature of velocity at inlet. Sixteen harmonic coefficient are used, which results approximately same physiologically realistic flow wave form shown in figure2(a) and parabolic inlet velocity profile are shown in figure 2(b). In figure 2(a), a, b, and c represent the positions of early systole (0.041sec), peak systole (0.205 sec), and diastole (0.615 sec) respectively. In this study Reynolds number varies from 96 to 800. Since cardiac pulse cycle is 0.82sec, is found from the calculation $\omega=\frac{2\pi}{0.82}=7.66\, rad/\sec.$.

3. Results and discussion:

3.1. Wall shear stress distribution:

WSS is an important factor in determining the severity of arterial stenosis. WSS depends on the viscosity of the fluid and velocity gradient and, is defined as $\tau=\mu\frac{du}{dr}$, where μ is the viscosity and $\frac{du}{dr}$ is the velocity gradient.

(a) (b)

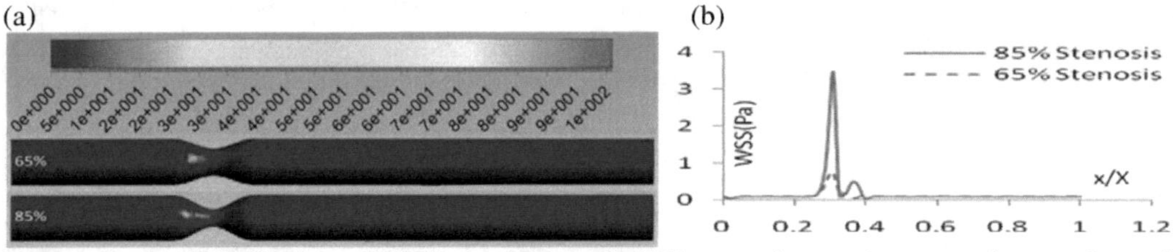

Figure 3. (a) Distribution of WSS in 65% and 85% stenotic arteries at early systole and, (b) comparison of 65% and 85% stenotic arteries for WSS distribution at early-systole.

(a) (b)

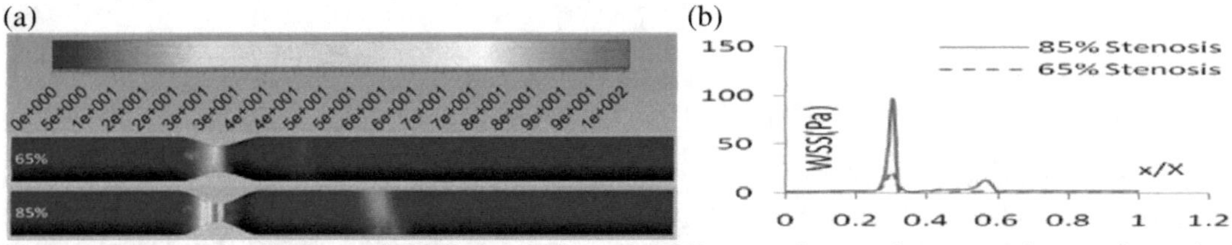

Figure 4. (a) Distribution of WSS in 65% and 85% stenotic arteries at peak systole and, (b) comparison of 65% and 85% stenotic arteries for WSS distribution at peak-systole.

(a) (b)

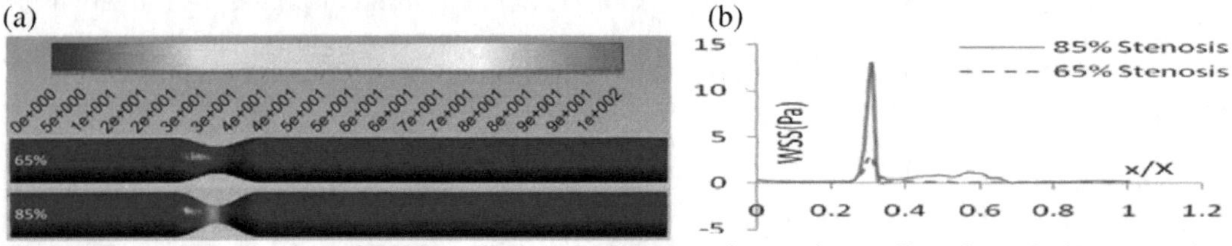

Figure 5. (a) Distribution of WSS in 65% and 85% stenotic arteries at diastole and, (b) comparison of 65% and 85% stenotic arteries for WSS distribution at diastole.

Figure 3(a) shows the distribution of WSS in 65% and 85% stenotic arteries at early-systole. WSS is uniformly distributed throughout the geometries. But in 3(b), a sudden climb to the maximum value of WSS and then sudden decline of the WSS can be noticed in the throat. Fig 3(b) depicts the maximum WSS of 65% and 85% stenotic arteries are 0.76 *Pa* and 3.48 *Pa* respectively. Figure 4 shows the distribution of WSS at peak systole. Figure 4(a) shows that, in both geometries, WSS is uniformly distributed at the pre stenotic region but significant change in WSS is occurred in throat and little variation in WSS is seen at post stenotic region. At peak systole both of the geometries increase their maximum WSS in the stenosis throat. From 4(b), we can notice the maximum WSS of 65% and 85% stenotic arteries are 21.98 *Pa* and, 96.78 *Pa* respectively. Figure 5(a) depicts that, 85% stenotic artery creates little change in WSS at the throat. From 6(b), we can see the maximum WSS of 65% and 85% stenotic arteries are 2.97 *Pa* and, 13.03 *Pa* respectively.

3.2. Pressure distribution:
Pressure distribution in both stenotic arteries at early-systole, peak-systole, and diastole is shown in figure 6.

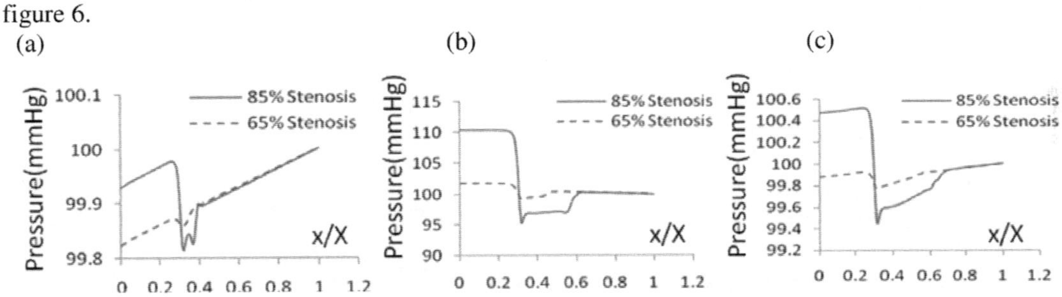

Figure 6. comparison of pressure for 65% and 85% stenotic arteries at (a)early systole (b) pick systole and (c) Diastole

Figure 6(a) shows the distribution of pressure in 65% and 85% stenotic arteries at early-systole. In both geometries the pressure is uniform and varying with a small variation throughout the geometry. No significant change of pressure for 65% stenotic artery but sudden decline of pressure for 85% stenotic artery at the throat region. Figure 6(a) shows the distribution of pressure in 65% and 85% stenotic arteries at peak-systole. High pressure is observed at pre stenotic region but suddenly a large decline of pressure is occurred in the throat region in 85% stenotic artery, and little variation of pressure is noticed in 65% stenotic artery. It describes that the pressure at inlet and throat for 65% stenotic artery are 101.77 *mmHg* and 99.11 *mmHg* and for 85% stenotic artery are 110.42 *mmHg* and 95.27 *mmHg* respectively. Figure 6 (c) shows that the comparison of pressure distribution of 65% and 85% stenotic arteries at diastole. We see that, pressure is uniform and varying with a small variation at the throat. We know that pressure difference may provide some kind of extra force that may drive blood through the stenosis with high velocity. Since The pressure difference in 85% stenotic artery is higher than that of 65% stenotic artery. So 85% stenotic artery will create more force than 65% stenotic artery at their pre-stenotic region.

3.3 velocity distribution:
In Figure 7, inlet velocity in both geometries begins with same magnitude for all cases. Velocities at the throat and post stenotic region for all cases of 85% stenotic artery are relatively higher than that of 65% stenotic artery. Recirculation regions are observed at post stenotic region in 85% stenotic artery for all cases and recirculation region increases with increasing minimum velocity at the inlet flow which have been shown in Figure 8. On the other hand, velocity is laminar in 65% stenotic artery at post stenotic region for all cases. The highest velocity (about 1.97 m/s) is observed in 85% stenotic artery at peak systole and the lowest velocity is observed in 65% stenotic artery at early systole.

4. Conclusion

A numerical study on pulsatile blood flow through 3-D stenotic artery is performed. Pressure and velocity flow field and wall shear stress occurred in the flow field are analysed. Wall shear stress distribution shows that as severity increases, shearing of flow also increases for all cases and maximum stress is exerted in throat region at peak systole. Pressure distribution demonstrate that at every case 85% stenotic artery creates more force than 65% stenotic artery at their pre-stenotic region and 85% arterial stenosis can cause fatal death by blasting the pre stenotic arterial region of the patient of high blood pressure. Numerical simulated results of this study demonstrate that velocity increases in

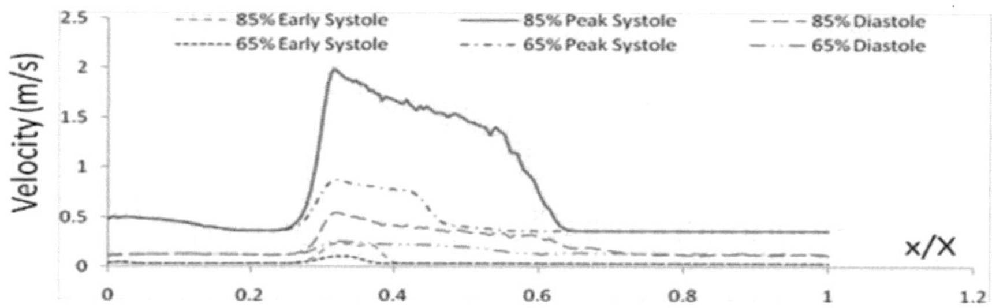

Figure 7. Comparison of velocity of 65% and 85% stenotic arteries

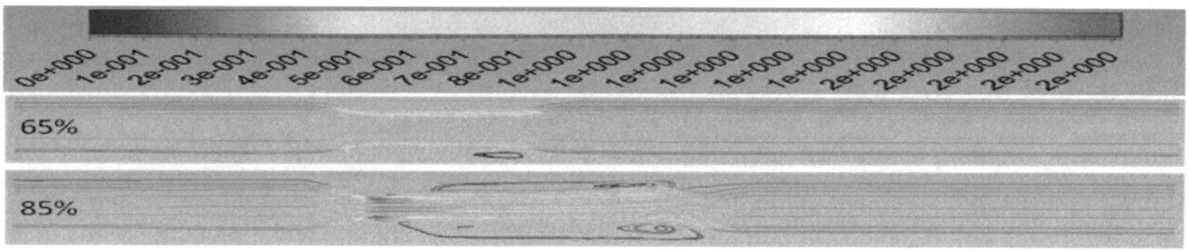

Figure 8. Streamline of velocity at pick systole for 85% stenotic arteries

throat region with the increase in severity for all cases. Due to stenotic constriction, the flow behavior changes abruptly in the downstream of the stenosis. During peak systole the flow increases at the throat region forming a jet and later disrupts suddenly forming eddies due to pressure drop in downstream side. During diastole, due to flow deceleration, the traces of the jet formed prevail for short period and there are no significant changes of flow behavior for both geometries. A recirculation region is visible at the post stenotic region at 85% stenotic artery for all cases and recirculation region increases at minimum velocity of the inlet flow. From this figure it can be interpreted that 85% stenosis can create vortex at any time phase. So 85% stenosis is very harmful for its high probability of creating vortex.

References

[1] S. Khader, and B. Shenoy; Effect of increased severity in patient specific stenosis, World Journal of Modelling and Simulation, Vol. 7 (2011) No. 2, pp. 113-122.

[2] S.S. Varghese and S.H. Frankel, Numerical Modeling of Pulsatile Turbulent Flow in Stenotic Vessels, J. Biomech, Vol. 125, pp. 445-460, 2003.

[3] C. Chua, and G. Shread, Changes in flow and wall stresses through arterial constriction offset from the centre of the vessel, The Anziam Journal, 2009, 50: C744–C759.

[4] V. Young, A. Patterson M. Graves, Z-Y LI, V. Tavani, T. Tang, and J. H. Gillard, The mechanical triggers of rupture: shear vs pressure gradient, The British J. of Radiology, 82 (2009) S39-S45

[5] J. Pinto, K.L. Bessa, D.F. Legendre, and R.H. Mouth, Physiologycal pulsatile waveform through axisymmetric stenosed arteries: Numerical Simulation, ABCM Symposium in Bioengineering, 1 (2006).

[6] S.A. Ahmed, and D.P. Giddens, Pulsatile Poststenotic Flow Studies with Laser Doppler Anemometry, J. Biomech., 17 (1984) 695-705.

[7] S. Lee, and S. Lee, Direct numerical simulation of transitional flow in a stenose carotid bifurcation, Journal of Biomechanics, 2008, 41: 2551–2561.

[8] S.R. Shahed, Md. Ali, S. Saha, and M.N. Akhter, Numerical study on unsteady flow field of arterial stenosis, Presented in the Int. conference on Mechanical Engg. (ICME 2013), 20-21 June, 2014, Dhaka-1000, Bangladesh, and will be published in Procedia Engineering (2014), Elsevier, SciVerse ScienceDirect.

IOP Conf. Series: Journal of Physics: Conf. Series **822** (2017) 012025 doi:10.1088/1742-6596/822/1/012025

Blood-mimicking delivery in polygonal structure of inner quadrupletip microneedle with valveless micro-pump

M D Ibrahim[1], **Y S Yunos**[1], **A R H Rigit**[1], **N A A Mohtadzar**[2], **N Watanabe**[3], **Y Sunami**[4], **M R A Rahman**[1], **L K Wong**[1] **and M Z Mohtar**[1]

[1] Department of Mechanical and Manufacturing, Faculty of Engineering, Universiti Malaysia Sarawak, Malaysia
[2] Department of Electrical and Electronic, Faculty of Engineering, Universiti Malaysia Sarawak, Malaysia
[3] Department of Bioscience and Engineering, Shibaura Institute of Technology, Japan
[4] Department of Mechanical Engineering, Faculty of Engineering, Tokai University, Japan

Email: imdanial@unimas.my

Abstract. This paper presents a titanium quadrupletip micro-needle integrated with a micro-pump with different inner designs, length and diameter of the micro-channels to measure and maximize the velocity flow in the micro-needle as blood delivered into human body. Titanium is used as the material of the micro-needle which are also the common material in manufacturing of micro-needle. The advancement of micro-needle technologies is improved in penetrating human outermost skin, stratum corneum and further to human blood vessels. The micro-needles with channel inner design of circular, square, hexagon, and dodecagon with quadruple tip designs are drawn with inner diameter parameter of 150μm and 100μm with two different channel length which are 10mm and 25mm. The characteristics of blood delivery in geometrically changed inner designs affect the output velocity in microneedle when the micropump is operating. The results showed that, when it is pumped at 0.04m/s, the blood velocity improved by 5.6% than when the pump is increased by 30% of its capacity. This is due to the backflow generated in the micropump.

1. Introduction

Delivering or withdrawing blood techniques are obtained either from arteries or veins mostly by using hypodermic needle to penetrate human skin. For hypodermic needle to reach the arteries or veins are surely uncomfortable and painful. The realization of painless needle from mosquito's mechanism of blood withdrawal, lots of research are done to improve the efficiency of the micro-needles for blood delivery from various sizes of needle and designs. Needles are defined as sharp hypodermic needle that has a function to be painless while completing the task of penetrating the human outermost skin to withdraw blood or injecting fluid into the blood stream [1]~[4]. The advancement of the research leads to the technologies of micro-needles in improving the flow and performance which includes attaching micropump at the microneedle [5]. While keeping microneedle in a longer length, the design parameters of the microchannels differ in order to obtain higher velocity. This is a requirement whenever the control of the intravenous infusion of blood is necessary.

1.1 Micro-needle developments
An early development of microneedles are aimed for the diabetes patients that needs to take insulin regularly. The efforts to bypass the stratum corneum are aimed at altering the medication molecule to make it more lipid soluble which will enhance drug penetration. Micro-needles are inserted at a depth of 1mm within the skin yielded rapid insulin absorption and an adequate reduction of glucose level. Small bore micro-needles can deliver insulin, a relatively small molecule, through the skin in amounts adequate for managing the patient's glucose levels and bypass the stratum corneum easily.

1.2 Expansions of micropump
In recent years, micro-pump has become available for industrial product integration in microfluidic research which is one of special interest of MEMS technology. Micro-pumps were started in middle 1970s. Before 1990s, mechanical pumps were mainly studied. After 1990s, non-mechanical pumps were introduced. In 1975, Thomas and Bessman were presented a documents about designing a micro-pump for implantation for human body. It contain multi piezoelectric disc benders combine with solenoid valve and pumping chamber. In 1984, Smits was patented the silicon micro fabrication technologies and later he published the peristaltic pump which is consists of three piezoelectric discs in 1990. The purpose of the micro-pump was for his insulin delivery systems [6]. Micropumps can be classified into two categories, either mechanical or non-mechanical micropump [7].

1.3 Piezoelectric micropump
In biomedical application, a typical micropump such as MEMS device, the actuation source is placed through a fluid sample (drugs and therapeutic agents) where the fluid is being transferred. A mechanical micro-pump such as piezoelectric micro-pump is widely used in biomedical field. The conversion of mechanical energy to electronic signal (voltage) and vice versa is called piezoelectric effect. A stress applied to such materials will alter the separation between the positive and the negative charges, causing surface net polarization. The piezoelectric effect relates to the coupling between mechanical deformation and electrical polarization. Main advantages of piezoelectric actuators include large actuation force, fast response and simple structure. Their fabrication, however, is complex as is processing of piezoelectric materials [8].

2. Methodology
The analysis of the velocity flow used in the simulation will be using Ansys Fluent software. The micro-needle attached to a micro-pump will be drawn first in 3 dimensional drawing. The actual laminar inflow of fluid will be inserted as the inlet velocity of the micro-needle in order to simulate the velocity flow in the micro-needle. The purpose of this simulation is to show the velocity profile of the blood during delivery in the micro-needle with the direction of the fluid flow.

2.1. Blood as a non-newtonian fluid
Blood is a heterogeneous multi-phase mixture of solid corpuscle contains of red blood cells, white blood cells and platelets [9]. Characteristics of blood are determined by the properties of these components and their interaction with each. Based on properties of blood, 0.004 Pa*s dynamic viscosity and 1060 kg/m^3 density is used in the simulation. External physical condition of the blood rheology is neglected. Although in most cases the effect is not substantial except possibly over short periods of time and normally does not have lasting consequences.

2.2. Design of micropump

The design and drawing of micro-pump is based on one company that sell micro-pump which is suitable for blood delivery. For blood delivery, it requires a micro-pump that handles a high viscosity liquids to pass by the actuator in the micro-pump. The important things in choosing micro-pump is it has to be suitable and has the compatibility in biomedical industry. The dimension of the micropump is 55x35x35 mm. The inner part of the micropump is shown in figure 3.

2.3 Design of micro-needles

For the flow improvement research, the micro-needles with channel inner design of circular, square, hexagon and dodecagon are designed with the parameter of 100μm inner-diameter and 150μm inner-diameter with three different channel length 10mm and 25mm. The figure 1 shows the designated micro-needles that involved in this research.

Micro-needles for tip designs are made up of Titanium 6Al4V, hollow cylinder are previously simulated as they are almost penetrating the rubber, which acts as the outermost layer of human skin [10]. Figure 2 shows the designated tip of the micro-needles that is being used in this research which is quadrupletip micro-needle. The authors previously designed and changed geometries of machineries showed improvements in obtaining better dynamic characteristics [11]~[12].

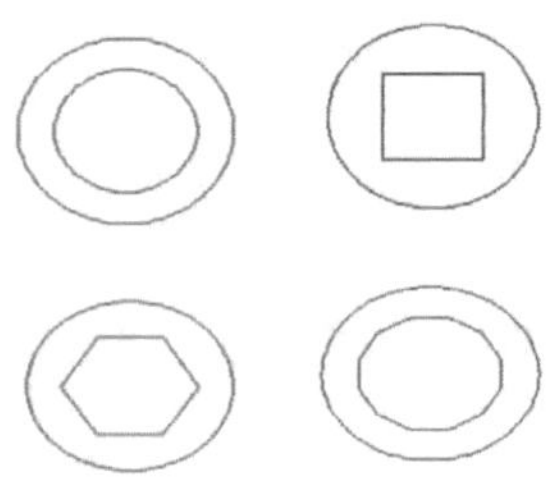

Figure 1. Inner design of micro-needle.

Figure 2. Quadrupletip micro-needle.

2.4 ANSYS Fluent

This research is focused more on the fluid flow velocity in micro-needles respect to the various design, length and diameter of micro-needles during blood delivery. Different translational speed of micropump's diaphragm is applied in the simulation but the properties and inlet velocity remains the same. The translational speed of the diaphragm will be 0.04m/s and 0.08m/s respectively. Wall between the diaphragm and the fluid is applied for these translational speed movement.

3. Modelling

In ANSYS Fluent, microneedle and micropump have to be meshed inside as the delivery of the fluid is flowing from the inlet of micropump to the end of the micro-needle. The meshing of the model helps to receive an accurate, precise result and minimize error occurs during calculation. The finer the mesh, the more accurate result is obtained.

As shown in table 1, meshing for the models are briefly explained in which mesh sizing of the model is used. The meshing for both microneedle and micropump is using fine meshing. Same goes to the fluid flow in the microneedle and micropump, the meshing is set to fine. Table 2 shows the details of number of nodes and number of elements for the whole simulation domain. Figure 4 and 5 below show the meshed model of micropump and microneedle.

IOP Conf. Series: Journal of Physics: Conf. Series **822** (2017) 012025 doi:10.1088/1742-6596/822/1/012025

Table 1. Meshing size.

Type	Details
Relevance center	Fine
Smoothing	Medium
Span angle center	Coarse
Minimum edge length	8.6926×10^{-2} mm

Table 2. Simulation meshed domain.

Parts			No. of nodes	No. of elements
Micro-pump			57832	13956
Micro-needle	Diameter, Ø (mm)	Length (mm)		
Circular	0.10	10	19253	5950
	0.15	10	9551	5567
	0.10	25	12372	7849
	0.15	25	10543	6575
Square	0.10	10	10094	3514
	0.15	10	8339	4526
	0.10	25	7786	4449
	0.15	25	7899	4518
Hexagon	0.10	10	16859	5335
	0.15	10	8382	4909
	0.10	25	10917	6807
	0.15	25	8091	4867
Dodecagon	0.10	10	22245	11722
	0.15	10	18049	11132
	0.10	25	20331	13650
	0.15	25	19310	12702

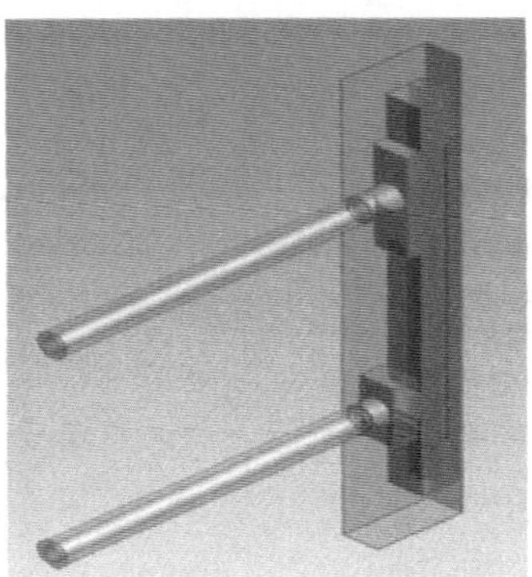

Figure 3. Micro-pump model.

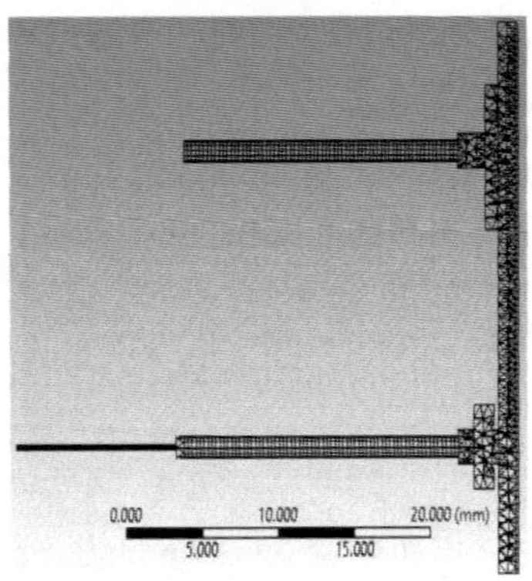

Figure 4. Full model mesh.

Figure 5. Cross-sectional view of meshed micro-pump

4. Results and discussions

The flow simulation of blood delivery in the micro-needle is done by using ANSYS Fluent. The purpose of the simulation is to study the blood flow characteristic during the delivery passing by micropump into the microneedle. Various design parameters of microneedle had been designed and simulated.

4.1. Simulation results

4.1.1. Flow delivery when pumped at 0.04m/s. The microneedles designed as shown in figure 1 and 2 at previous section are used in this simulation attached to the micropump. The micropump is set at 0.04m/s for the diaphragm's translational movement. The results are then tabulated to show the value of the average velocity result of blood delivery in circular, square, hexagon and dodecagon micro-needle penetrating the human skin. The highest velocity using different inner shape of the micro-needle shows the amount that has the best performance of delivering blood. The results are tabulated to compare with other inner design of micro-needle.

Table 3 shows the comparison of different inner design of microneedles between velocities of blood flow in each inner design with their respective details of diameter and length of the microneedles. Figure 6 shows the graph result of the simulation for the blood flow. It shows that having a dodecagon inner design with 0.1mm diameter and 25mm length has the best design for blood delivery when the micropump is being pumped at 0.04m/s. The velocity of the blood flow is 0.8050m/s.

4.1.2. Flow delivery when pumped at 0.08m/s. The microneedles designed as shown at previous section are used in this simulation attached to the micropump. The micropump is set at 0.08m/s for the diaphragm's translational movement. The results are then tabulated to show the value of the average velocity result of blood delivery in circular, square, hexagon and dodecagon micro-needle penetrating the human skin. The highest velocity using different inner shape of the micro-needle shows the amount that has the best performance of delivering blood. The results are tabulated to compare with other inner design of micro-needle.

Table 4 shows the comparison of different inner design of microneedles between velocities of blood flow in each inner design with their respective details of diameter and length of the microneedles. Figure 7 shows the graph result of the simulation for the blood flow. It shows that at 0.08m/s of the diaphragm's translational movement, having a dodecagon design with diameter of 0.1mm and 25mm length has the best blood flow performance. The velocity of the blood flow is 0.7596m/s.

Table 3. Comparison of flow velocity at 0.04m/s.

Dia (mm)	Length (mm)	Circle (m/s)	Square (m/s)	Hexagon (m/s)	Dodecagon (m/s)
0.1	10	0.4334	0.4720	0.5676	0.6622
	25	0.4553	0.4758	0.6566	0.8050
0.15	10	0.1815	0.6798	0.2123	0.2804
	25	0.5416	0.4356	0.2026	0.1948

Table 4. Comparison of flow velocity at 0.08m/s.

Dia (mm)	Length (mm)	Circle (m/s)	Square (m/s)	Hexagon (m/s)	Dodecagon (m/s)
0.1	10	0.430	0.4759	0.5255	0.6293
	25	0.5012	0.4426	0.5736	0.7596
0.15	10	0.1681	0.2151	0.2233	0.2707
	25	0.1250	0.2590	0.2988	0.2393

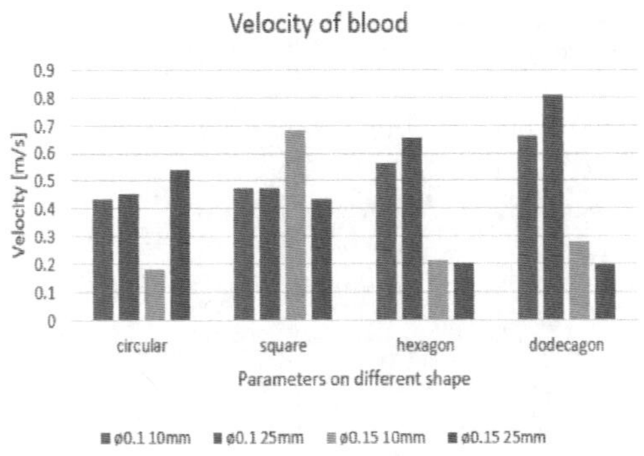

Figure 6. Blood flow performances at 0.04m/s.

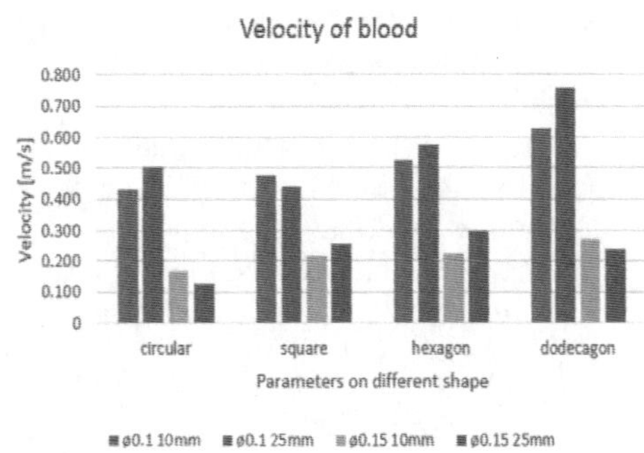

Figure 7. Blood flow performances at 0.08m/s.

5. Conclusions

In this research, it is numerically proven that, the hydraulic diameter and the length of a micro-needle plays an important role in micro-scale technology. Conclusively, dodecagon inner design of the micro-needle using a 0.15mm diameter and a 25mm length has better performance for blood delivery when it is pumped at 0.04m/s and 0.08m/s. The result obtained shows a slight decreased of average velocity. This is due to the backflow and minor turbulence inside the micropump.

This suggests that a micropump with these settings should have a valve equipped with it to ensure no backflow is present during the blood delivery. As blood is a multiphase fluid flow, it behaves time independently. As time passes, the viscosity decreases while shearing force which are applied affects the velocity in a channel. With a higher external applied force, the slip length increase gradually and causes velocity to increase. Slip length is maximum at small channel widths while it seems to reaches its minimum for greater width. In order to increase the volumetric flow rate of fluid in a microchannel, it is required to have a smaller pressure gradient and by controlling the slip length to be at maximum while the fluid is flowing. With this conditions, a higher velocity is achieved, thus, improving the flow performance and volumetric rate of fluid of a micro-needle.

Acknowledgements

The present work was partially funded by Fundamental Research Grant Scheme, Ministry of Education Malaysia, Grant No: FRGS/TK01(01)/1059/2013(05) and UNIMAS Special Short Term Grant, Grant No: F02/SpSTG/1386/16/28. The authors also declares that there is no conflict of interest regarding the publication of this paper.

References

[1] Oki A, Takai M, Ogawa H, Takamura Y, Fukasawa T, Kikuchi J, Ito Y, Ichiki T and Horiike Y 2003 *J. Appl. Phys.* **42** 3722-27

[2] Ogawa H, Nagai M, Kikuchi J and Horiike Y 2003 (741-3 Squaw Valley, CA, USA)

[3] Henry S, McAllister D V, Allen M G and Prausnitz M R 1998 *Proc. IEEE Micro Electro Mechanical Systems, MEMS'98* (Eidelberg, Germany, Jan. 25-29) 494-8

[4] Oka K, Aoyagi S, Arai Y, Isono Y, Hasiguchi G and Fujita H 2002 *Sensor Actuators A* **97-98** 478-85

[5] Khumpuang S, Horade M, Fujioka K and Sugiyama S 2007 *Microsyst. Technol.* **13** 209-14

[6] Nisar A, Afzulpurkur N, Mahaisavariya B and Tuantranont A 2008 *Science Direct, Sensors and Actuators B* **130 (2008)** 917-942

[7] Shiangyu L, Xingguo X, Junyi Z and Prabir P 2014 *ASEE 2014 Zone I Conference (University of Bridegeport, Bridgeport, CT, USA)*

[8] Muhammad Waseem Ashraf, S. T. 2011 *International Journal of Molecular Sciences,*

12(1422- 0067) 3648-3704

[9] Sochi, T 2013 *Non-Newtonian Rheology in Blood Circulation* (University College London, London)

[10] Semait S N 2013 *Minimizing Pain Stresses using Micro-Needles-an Application from Mosquito Syringes* (Undergraduate Student Thesis: UNIMAS)

[11] Sunami Y, Ibrahim M D and Hashimoto H 2013 *International Journal of Rotating Machineries,* **2013** 1-11

[12] Hashimoto H, Ochiai M, Namba T, Ibrahim M D 2009 *Proc. of the Intl. Joint Tribology Conference*

IOP Conf. Series: Journal of Physics: Conf. Series **822** (2017) 012026 doi:10.1088/1742-6596/822/1/012026

Complex fluid flow and heat transfer analysis inside a calandria based reactor using CFD technique

P. S. Kulkarni
Computational Mechanics Laboratory, Department of Aerospace Engineering, Indian Institute of Science, Bangalore, 560012, INDIA

E-mail: psk@aero.iisc.ernet.in, pskdhar@gmail.com

Abstract. Series of numerical experiments have been carried out on a calandria based reactor for optimizing the design to increase the overall heat transfer efficiency by using Computational Fluid Dynamic (CFD) technique. Fluid flow and heat transfer inside the calandria is governed by many geometric and flow parameters like orientation of inlet, inlet mass flow rate, fuel channel configuration (in-line, staggered, etc.,), location of inlet and outlet, etc.,. It was well established that heat transfer is more wherever forced convection dominates but for geometries like calandria it is very difficult to achieve forced convection flow everywhere, intern it strongly depends on the direction of inlet jet. In the present paper the initial design was optimized with respect to inlet jet angle, the optimized design has been numerically tested for different heat load mass flow conditions. To further increase the heat removal capacity of a calandria, further numerical studies has been carried out for different inlet geometry. In all the analysis same overall geometry size and same number of tubes has been considered. The work gives good insight into the fluid flow and heat transfer inside the calandria and offer a guideline for optimizing the design and/or capacity enhancement of a present design.

1. Introduction

In nuclear reactors heavy water is used as the moderator for both energy extraction (primary fluid) and for cooling (secondary fluid). The reactor assembly mainly consist of calandria, Steam Generator, Emergency core cooling system, primary and secondary moderator circulation system, etc. Calandria is a high pressure reactor vessel consisting of stainless steel structure, which holds the hundreds of horizontal fuel channels. In a calandria based reactor fuel channels are arranged in horizontal fashion, which contains fuel material and the primary fluid circulating inside it. Due to fission reaction huge amount of energy is releases in the form of heat, high pressure primary fluid which is circulating inside the fuel channels carries part of this heat to the boiler where it converts ordinary water into steam for energy generation. The secondary fluid which is circulating around the fuel channels is mainly acts as moderator. During loss of primary fluid and the emergency core cooling the moderator inside the calandria must be capable of removing heat to keep a fuel channels within the safe operating conditions.

In order to understand the flow phenomena inside the calandria there have been many experimental studies on scaled test facilities. Koroyannakis et al. [1] first conducted a series of experiments at Sheridan Park Laboratory (SPEL). The dimensions of a scaled model are not exactly mimicking the real calandria geometry, but the scaled model was capable of reproducing important

IOP Conf. Series: Journal of Physics: Conf. Series **822** (2017) 012026 doi:10.1088/1742-6596/822/1/012026

thermohydrulic phenomena. Some experimental studies have been conducted for isothermal [2] and non-isothermal cases [3], [4] to identify the effect of inlet mass flow rate and internal heating on the flow and temperature distribution inside the calandria. It is very difficult to carry out experimental studies on the full scale calandria model and the alternative approach is to carry out numerical analysis. A Computational Fluid Dynamics (CFD) approach has been widely used to study the flow behavior and heat transfer inside the calandria. Initial CFD studies [5] have used simplified geometry representation by approximating porous media for the matrix of the calandria tubes inside the calandria. According to Rhee et al. [5] the flow inside the calandria can be classified into three groups namely a. Momentum dominated flow, b. Buoyancy dominated flow and the c. Mixed type of flow. These flow patterns deepens on the heat load and inlet mass flow rate.

Experimental analysis of fluid flow and heat transfer inside a full scale calandria unit is associated with huge cost and dangerous for handling high heat loads, mass flux conditions makes it practically complex. Conducting experiments on a scaled model or numerical simulations helps for the designer to optimize the design and one can do lot of parametric studies. There are many parameters which govern the heat transfer and intern fuel channel temperature distribution inside the calandria during an event such as loss of coolant accident(LOCA). As to optimize the calandria geometry different geometry configurations have to be studied.

The case study has been done for different injection rates (mass flow rates) of the moderator and varying heat flux (case 1). Inlets are symmetrically located at 3 'o' clock and 9 'o' clock position, outlets are located at 5 'o' clock and 7 'o' clock position. Fuel channels are arranged in staggered fashion. Two sets of configurations have been studied to analyze the flow behavior namely (1) Without considering the fuel channel geometry and (2) with considering fuel channel geometry. For the case with fuel channels, constant heat flux was applied as a boundary condition, in the whole analysis computation is limited to laminar flow only. The inclination of inlet is exactly horizontal (90^0 to the vertical plane). Moderator inlet angel, inlet flow rate, fuel channel configuration, outlet location, etc., are the main geometric and the flow parameters which must be optimized for safe the working of reactor and to increase the overall heat transfer efficiency.

As to improve the heat removal capacity, analysis was carried out to assess the effects of the angle at which the coolant is injected into the calandria on the heat transfer (case 2). The study was carried out for an assigned coolant flow rate and heat dissipation conditions so as to establish the effect of mixing and enhancements in the heat transfer behavior in the reactor on the considered parameter. The simulation is carried out for fixed Reynolds number of $2.5X10^6$ at the inlet, four different heating loads (300MW, 500MW, 800MW and 1200MW) at fuel channels and for different orientation of injection. The aim was to investigate how the orientation of inlet is affecting the heat transfer inside the calandria based reactor and to arrive at the optimal design considerations for safe working of calandria based reactor.

From the study (case 2), where the brute-force approach has been used to arrive at optimum range of inlet angle for constant inlet flow rate, it was found that the inlet angle between 30^0- 60^0 degree (with respect to the horizontal) gives optimum heat removal for all considered heat loads. The combination of mass flow rate and the inlet angle leading to lowest maximum temperatures inside the calendria was found to cause the underutilization of the calendria. Hence the latter part of the analysis has been carried out in the direction of improving the efficiency of the calendria by splitting the mass of flow near the inlet (case 3).

2. Geometry details

The nature of the flow inside the Calandria has significant effect on the safe operation of reactor vessel in PHWR's (Pressurized Heavy Water Reactors). Since the considered geometry is symmetry in nature (X Y Plane), only half portion was considered for computational study and is as shown in figure 1a, which is having diameter of 8 m and in third direction the inlets and outlets are identical throughout its length (Z direction) and hence a section has been considered for computational study. The symmetry

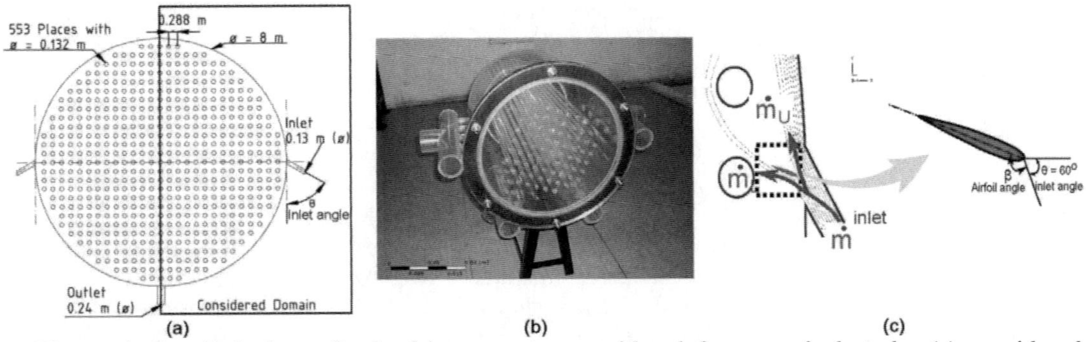

Figure 1. Detailed view of calandria geometry considered for numerical study. (a) considered symmetry model (b) scaled experimental model and (c) Location of airfoil near inlet.

portion of the Calandria contains total 290 fuel channels with diameter of 0.132 m each, one inlet at side and one outlet at the bottom. The diameter of the outlet (0.24 m) is greater than that of the inlet (0.13 m) which is to provide enough pressure difference to flow. The fuel channels are arranged in square form having 0.288 m pitch between them.

A schematic view of the modified design and also a proposed modification near the inlet (placing a splitter) is shown figure 1c. In case of calandria with the splitter, total mass flux at the inlet ($\dot{m}$) splits into two parts, i.e. the mass flux which is coming from below the splitter ($\dot{m}_L$) and from above the splitter ($\dot{m}_U$) as shown in 1c. There are total 553 fuel channels (diameter = 0.132 m) in the full geometry, equally spaced (square pitch = 0.288 m) and are submerged in the moderator. The diameter of the calandria is 8m and has 8m length in Z-direction. Inlets are located diametrically opposite sides and outlet is located at the bottom. Splitter has been added for the case with 1800MW thermal power by placing a NACA0012 airfoil at the center of the inlet and 4m from the calandria center (refer figure 1c). The range of splitter angle (β) studied are from 600 to 1800.

3. CFD model and its validation

3.1. Equations Solved, Boundary Conditions and Mesh details

The steady state flow is simulated by consideration of convection and radiation inside the calandria. Reynolds-Averaged Navier-Stokes equations (RANS) are solved for convection heat transfer to the water, buoyancy term is added to the y-momentum equation using a Boussinesq model. The Boussinesq model is used when density variation is driven by temperature variation. In Boussinesq model a source term is added to the momentum equation (parallel to the gravitational direction) to account for density variation and by keeping constant reference density in all other equations. The buoyancy source term is approximated as

$$\rho - \rho_{ref} = -\rho_{ref}\beta(T - T_r \qquad (1)$$

Where, ρ is the density of the fluid, ρ_{ref} is reference density, β is thermal expansion coefficient, T is temperature of the fluid and T_{ref} is the reference temperature. A standard k-ε turbulence model has been used [6], with default model constants (C_{mu} - 0.09, C1-Epsilon - 1.44, C2-Epsilon - 1.92, TKE Prandtal Number - 1 and TDR Prandtal Number - 1.3). The simulations are carried out for different mass flow rates, different inlet and different splitter angles. Inlet temperature is considered as 44.2°C [7]. The outlet is considered to be at atmospheric pressure. No-slip wall conditions are specified for all the walls.

3.2. Assumption Involved

Because of the geometrical symmetry in the construction of calandria only half portion is considered for numerical simulation and also 0.05 m symmetry section is considered in length wise direction by assuming inlet and outlets are spanning throughout its length in Z-direction. The working fluid (water)

IOP Conf. Series: Journal of Physics: Conf. Series **822** (2017) 012026 doi:10.1088/1742-6596/822/1/012026

is assumed as incompressible single phase flow. The heating of the fuel channels is assumed to be uniform throughout the considered fluid domain. The present numerical model is also validated with the non-isothermal experimental work by Koroyannakis et al. [1].

4. Results and discussion

4.1. Optimization with respect to inlet injection angle

Varying inlet angle step by step is an effective parametric approach for arriving to the optimum inlet injection angle. In this optimization process inlet angle varied from 15^0 (with respect to vertical axis) up-to 90^0. The comparison of fuel channel temperatures reaching boiling point temperature for different injection angles and for different power is shown in figure 2. In figure 2d results for only five cases have been presented (i.e. 90° injection angle is not included). The figure shows the area covered by lines for different injection angles and the area under curve shows the region of calandria where temperature reaching boiling point temperature for different considered thermal power. For all considered thermal power and injection angles the value of fuel channel temperatures are less between injection angles 30° to 60°.

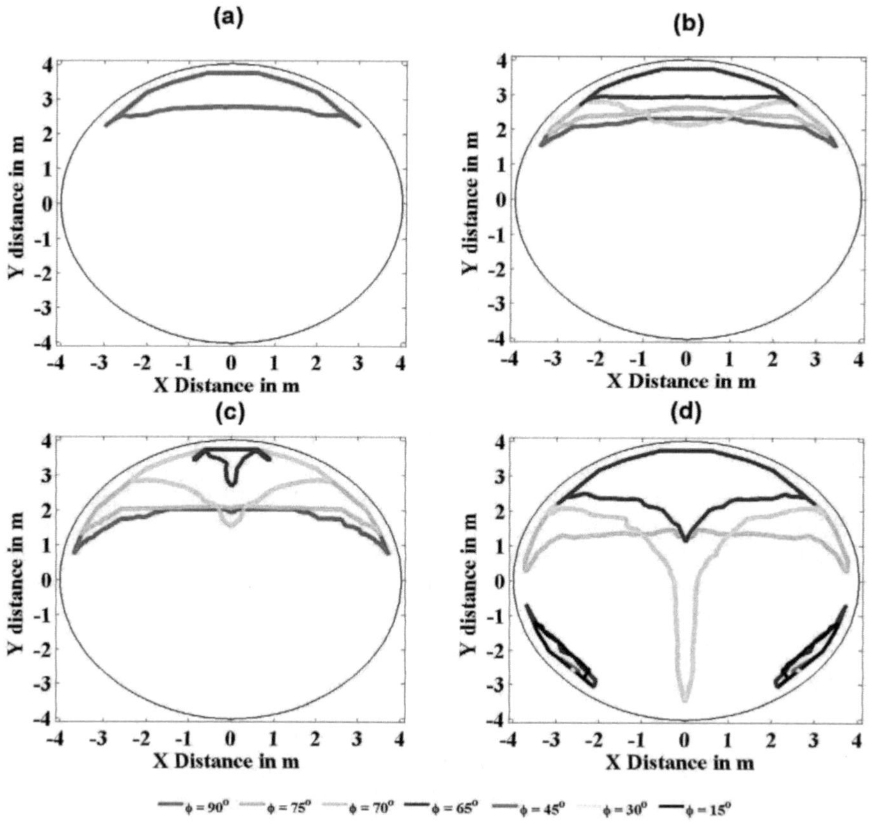

Figure 2. Temperature reaching boiling point temperature for different injection angles (a) 300 MW (b) 500 MW (c) 800 MW and (d) 1200 MW.

Further analysis has been carried out for the case of 1200MW thermal power. The figure 3a shows the variation of temperature for different mass flow rates (dotted lines), the inlet angle is 60^0 and also it shows variation of temperature for different inlet angles. It was observed that any increase in the inlet angle above 40^0 with respect to the horizontal (clockwise) has negligible effects on the resultant temperature. Apart from this, it is also observed that lower mass flow rate cases have higher temperature. But it is important to see the corresponding maximum temperatures of the fuel channel surfaces to prevent any hazardous accidents. The lowest mass flow rates in the present case study (i.e., 60kg/sec and 80kg/sec) leads some of the fuel channels either in the top region or below the inlet close to the calandria wall to attain maximum temperature values greater than 100^0 C. But the lower mass flow rates ensure low pumping power requirements. In the present work, change of phase has not been considered; hence, further analysis has been carried out for 80kg/sec mass flow rate with inlet angle 60^0.

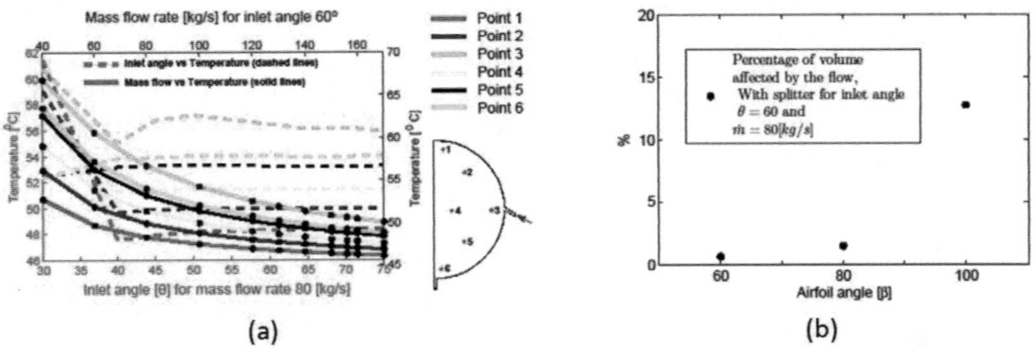

Figure 3. (a) Variation of temperature at few selected locations (points 1 to 6) with respect to inlet injection angle and mass flow rate (b) percentage of volume affected by the flow.

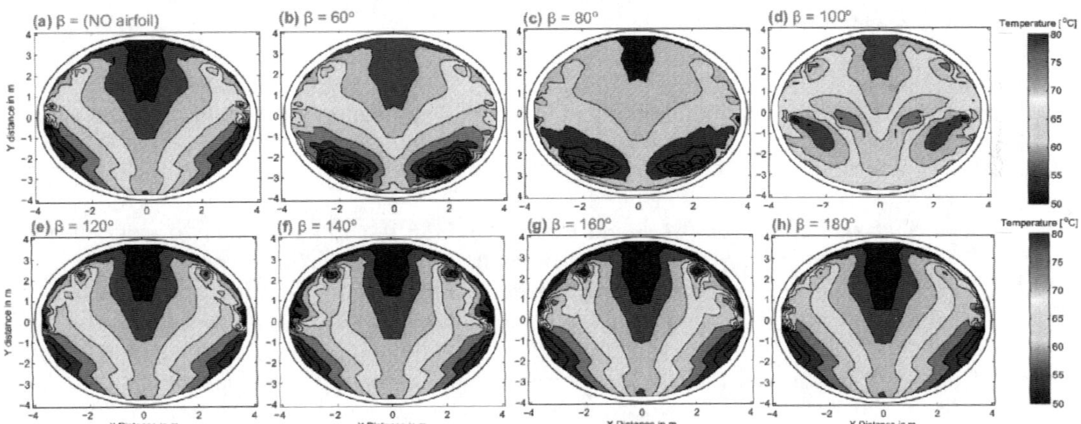

Figure 4. Temperature distribution for different splitter angles.

4.2. Optimizing by introduction of splitter

The idea of using the splitter was to split the mass flow to make the cooler incoming heavy water come in contact with the as many fuel channels as possible, which can ensure enhancement in the heat transfer. The details of splitter and splitter angle studied are explained in the previous section. For splitter angles β = 60° and β = 80° the mass flux from the bottom of the splitter ($\dot{m}_L$) is just flowing along the calandria wall where as for higher β values (i.e. β > 80°) the mass flux ($\dot{m}_L$) is covering some of the fuel channels. For β > 100°, the mass flux $\dot{m}_L$ and the mass flux from the top of the splitter ($\dot{m}_U$) join together and flow towards the upper part of the calandria and the region below the inlet close to the calandria wall is negligibly affected by the flow. For all considered splitter angles the flow diverting towards the bottom part of the calandria can be seen only in splitter angles β = 60°, 80° and

100°. The fuel channel temperature distribution for different splitter angles is as shown in figure 4. The percentage of volume covered by the flow is shown in figure 3b. Even though the percentage of volume covered in the case of $\beta = 60°$, and 80° by the mass flux ($\dot{m}_L$) having some value, but none of the fuel channels come under this volume.

5. Conclusion

Numerical analysis of the fluid flow and heat transfer has been carried out for the calandria based reactor for different flow parameters, heat load conditions and geometric parameters to ensure safe working. The systematic parametric studies made indicate that the heat transfer is largely influenced by the injection angle. The limiting thermal load evaluation is possible to interpret and for each configuration thereby letting a design optimization for the reactor based on the injection angle, as is demonstrated in the temperature plots for different proven levels. The results reveal a good insight to the complete behavior of the calandria based reactors and is considerable of high significance to the preliminary reactor design. It can be brought out that the detail of the CFD study provide initial information of the flow and heat transfer that is not easy in experimental studies or they could be very expensive and also hazardous. The effect of inlet angle on temperature difference is least for angles greater than 40° and lower mass flow rates resulted in higher maximum temperature and temperature difference;

- For all considered thermal power and injection angles the value of fuel channel temperatures are less between injection angles 30° to 60°;
- Splitter orientation considerable affects the streamline and temperature distribution within
- the calandria. Splitter orientation close to 100° with respect to the inlet is found to be optimum.
- In the present analysis, it is found that the mass flux ration required to keep fuel channels within the safe limit is around 50% and it shows that splitter is contributing in an equal amount as that of inlet angle.

Acknowledgment

I thank profoundly Suneel Patil for carrying out numerical simulation and helping me to prepare this paper.

References

[1] Koroyannakis D, Hepworth R D and Hendrie G 1983 Experimental study of combined natural and forced convection flow in a cylindrical tank *Tech. rep. AECL Report*

[2] Ravi S D, Rajan N K S and Kulkarni P S 2008 Computational and experimental studies of fluid flow and heat transfer in a calandria based reactor *Computational Fluid Dynamics*, Springer p 233

[3] Huget R G, Szymanski J K and Midvidy W I 1989 Status of physical and numerical modelling of CANDU moderator circulation *Procedings of 10th Annual Conference of the Canadian Nuclear Society*, Ottawa

[4] Huget R G, Szymanski J K and Midvidy W I 1990 Experimental and numerical modelling of combined forced and free convection in a complex geometry with internal heat generation *Procedings of 9th International Heat Transfer conference* vol **3** p 327

[5] Rhee B W, Yoon C and Min B J 2004 *Journal of the Korean Nuclear Safety* **36** 559–570

[6] Kim M, Yu S O and Kim H J 2006 *Nuclear Engineering and Design* **236** 1155–1164

[7] Bajaj S and Gore A 2006 *Nuclear Engineering and Design* **236** 701–722

Computational aerodynamic study of a supersonic reference projectile

S Zahir[1†] and Waseem Gul[2]

[1,2]Centres of Excellence in Science & Applied Technologies, Islamabad – 44000; PAKISTAN

†e-mail: cfdpak@apollo.net.pk

Abstract. In the present paper results of computational aerodynamic study of a reference projectile in supersonic flow is presented. Pressure distribution along the projectiles nose in longitudinal and circumferential direction was studied using numerical simulations. Static aerodynamic coefficients in the freestream Mach numbers of 3 and 4 at an angle of incidence of -4 to 20 degrees were computed. For validation, the numerical solution was initially compared for the conic part with the experimental results and found to be in good agreement. Nose related aerodynamic features were initially studied in the first part of the study. Further complete projectile's aerodynamic analysis is carried out. The present paper covers the results obtained in the study conducted for the projectile's geometry. Significant flow features were investigated, including correct capturing of bow shock's location, low pressure density/ pressure variations in the low pressure regions along with its influence on the stagnation zone. Further evaluation of peak pressure values associated with the stagnation region on the nose was made. Also, static aerodynamic coefficients were computed to depict projectile's static stability.

1. Introduction

The flow around a projectile under supersonic conditions presents a flow situation comprised of bow-shock formation ahead of the nose, turbulent boundary layer growth on cylindrical mid-section with possible flow separation and formation of a large turbulent wake in the aft of the body [1-2]. In ballistic aerodynamics, prevention or control of the separation of the boundary layer is one of the most important aims commensurate with the design of an appropriate nose section of a projectile [3-4]. Adequate numerical techniques are available in computational fluid dynamics, CFD to simulate the flow situation for steady state and fully turbulent flows. These techniques were applied to determine the static aerodynamic coefficients for reference projectile. With an incoming free stream Mach numbers of 3 and 4 and a combination of angle of attacks, a representative complete flight envelope was investigated in this study. A salient feature of the flowfield that characterizes the supersonic flow interaction of reference projectile geometry is provided by mapping of the axial pressure distribution on the surface. Flowfield maps showing characteristic pressure and Mach number contours are also plotted. All computational fluid dynamics, CFD simulations are made using an in-house NS solver PAK-3D [5], with post-processing for static aerodynamic coefficients is performed using LOOK [6] software.

2. Computational methodology

IOP Conf. Series: Journal of Physics: Conf. Series **822**(2017) 012027 doi:10.1088/1742-6596/822/1/01202

2.1. Projectile model geometry

The geometry of a reference projectile is with a body-wing-tail configuration similar to standard body-tail configuration. It has same nose part geometry as of a reference blunted cone nose, as available in the literature [7].The blunted nose part modeled initially is with a half-cone angle of 10°, a base diameter of 7.65 cm and having a length of 12.0 cm. Apart from blunted conic nose, aft consist of a cylindrical centre-body and a single fin set consisting of four panels. Fins are trapezoidal and with diamond-shaped airfoil cross-section and assumed to be with constant tapering thickness from root to tip. Each fin has a leading-edge sweep. Geometrical details of nose as well complete projectile are, as shown in Table. I.

Table 1. Geometrical details of reference projectile.

Parameter	Nose	Projectile
Reference diameter, cm	7.65	7.65
Reference length, cm	12.0	61.2
No. of fins	-	04
Fin semi-span	-	7.65
Fin root chord	-	10.2
Fin tip chord	-	5.1
Fin Sweep back angle	-	34°

Customary aerodynamic sign convention is used with an in-coming flow is in the positive x-direction and negative angle of attack is with the nose down condition. Axis orientation and geometrical detail (all dimensions are in mm) as a side-view, is also shown in Fig. 1.

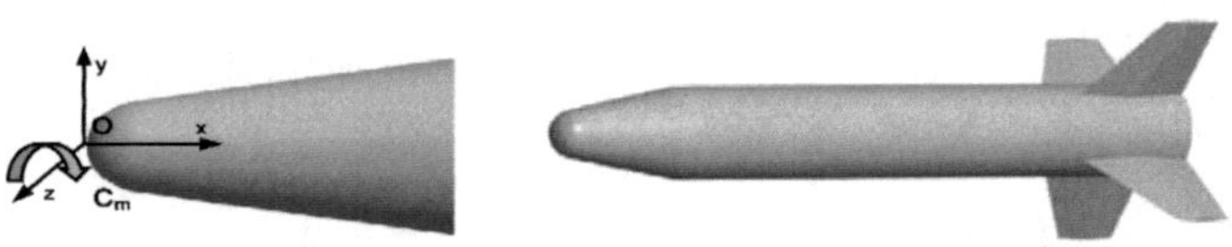

Figure1. Details of nose with complete geometry of the supersonic reference projectile.

2.2. Computational domain and conditions

Quality of computational grid regarding its density and clustering required for capturing flow details in the entire supersonic range of Mach number 3 to 5 is maintained by placing the walls closest cells to lie below a y^+ of 2.0. Satisfactory convergence approaches have been described at more length in literature [8], [9] and [10] for the sake of brevity here it is enough to state that an extensive validation and verification was undertaken to arrive at the present approach. To help convergence rate, a systematic grid generation sequence was followed and it is from course to super-fine grids, ranging from 0.72 to about 1.91 million grid points, using grid generation software PAKGRID [12]. Convergence was deemed to be achieved when the force coefficients ceased to change by less than 0.01% over a subsequent 1000 iterations. Fully-structured multi-block meshes were constructed and tested for grid-independence of results with some local cell adaption for higher-resolution shock-capturing. A grid resolution in nose vicinity having local cell lengths of 0.030, 0.035 and 0.040 cm is used in this region to capture shock stand-off at Mach numbers of 5, 4 and 3 respectively. Grid clustering is also employed to capture bow shock at shock standoff distance. Some details of grids are as shown in Fig.2. Freestream conditions at the inlet supersonic inflow are imposed near and as close as possible to the foreshock location, an axisymmetric flowfield is studied with no-slip adiabatic conditions imposed on cone and pressure outflow condition was employed at the domain exit.

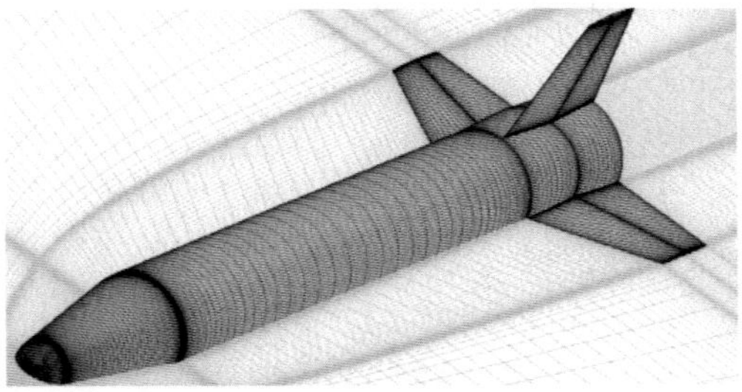

Figure 2. An 180° grid used to compute aerodynamic characteristics of complete projectile.

For the flight of Mach number 3 to 5 in air is considered with a specific heat ratio, $\gamma = 1.4$, Prandtl number, $Pr_t = 0.9$, Gas constant, R = 286.7 J/kg K. Consequently, the freestream flow conditions are with; temperature $T_\infty = 300$ K, pressure $P_\infty = 1.0132 \times 10^5$ Pa and density $\rho_\infty = 1.1767 \times 10^{-4}$ kg/m^3. For Mach number 3 to 5, Reynolds's number per unit reference length, is calculated to be of the order of 8.6×10^6 m^{-1} to 14.3×10^6 m^{-1}.

2.3. Flow computations and solver

Computational fluid dynamics, CFD calculations are made using the Navier-Stokes solver, PAK-3D [5]. PAK-3D is an in-house developed solver and it has been validated against experimental data and other CFD methods [8], [10]. This method is for the block structured meshes and it treats fluid as an ideal gas using perfect gas law, Sutherland's viscosity law and adiabatic heat transfer boundary conditions. Beam and Warming [13] approximation is used to discretize NS equations. Turbulent coefficient of viscosity was calculated by Baldwin-Lomax [14] algebraic turbulence model with modifications based on Reynold's averaging assumptions. Numerical scheme follows approximate factorization formulation for the Navier-Stokes equations and correct differencing in transformation was retained by splitting of inviscid fluxes. Flux vectors got splitted according to the sign of the Eigen values. On attainment of convergence criteria, information of primitive variables at that iteration level got post processed for calculation of non-dimensional axial/circumferential pressure distributions as well as the static aerodynamic coefficients. All simulations were run as steady-state.

3. Results and discussions

3.1. Validation of numerical solutions

This section mainly presents the results based on the numerical simulation of flow investigation made for Mach 3 for the projectile's blunted nose part at a fixed angle of attack. A general aim of comparison is to validate the numerical work for estimation of aerodynamic flow characteristics. One of the characteristic features of the aerodynamic flowfield pertinent to supersonic Mach number range is the computation of axial static pressure distribution on the nose surface. A comparison of computed results and experimental [14] axial static pressure distributions is made. The maximum variation of p/p_∞, at an axial location, x/L from 0.2 to 0.7, is determined to be less than 1%; as shown in Fig.3.In general, overall comparison of the computed results with the experimental data, showed similarity in terms of peak pressure distribution on the leeward side of the cone surface, upstream formation of bow shock and in an aft zone of low pressure.

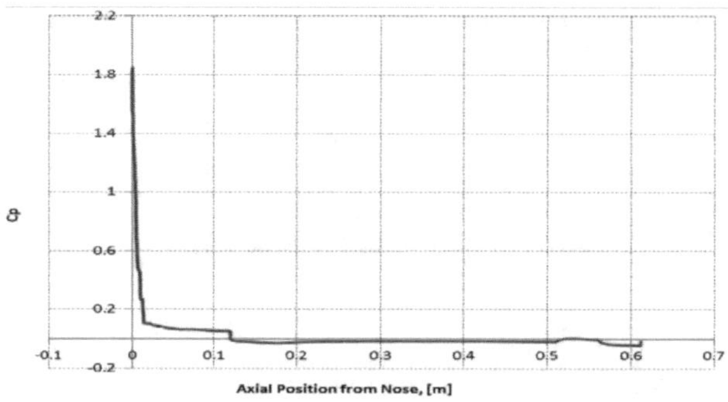

Figure 3. Computed axial C_p distribution of projectile body at leeward side for Mach 4 and $\alpha = 10^o$.

3.2. Computed shock-standoff distance

To understand the behavior of sonic line and bow shock, position of the bow shock at zero angle of attack for different Mach number is computed numerically. The separation distance between the nose and the shock mean thickness is referred as the shock standoff distance. The computed values are given in the Table 1. It is observed that with an increase in Mach number the tendency of the shock to come closer to the nose is evident. With an increase in positive angle of attack, it is noted that the bow shock moves closer to the body on the windward side, while the bow shock moves away from the conic body on the leeward side.

Table 2. Computed shock-standoff distances.

Mach number	Shock-standoff distance [cm]
3.0	0.567
4.0	0.461

3.3. Axial centre of pressure, C_p distribution

C_p distribution was computed for Mach 3 and 4 along fin chord at mid-span locations. The values on leeward and windward sides of fin were calculated at 20 degrees angle of attack for Mach 4, as shown in Fig. 4. There is an increase in C_p value from leading edge to the center of the chord and then a decrease is observed till trailing edge of the fin on windward side. Negative values of C_p are observed at leeward side showing the suction condition, as shown in Fig.4.

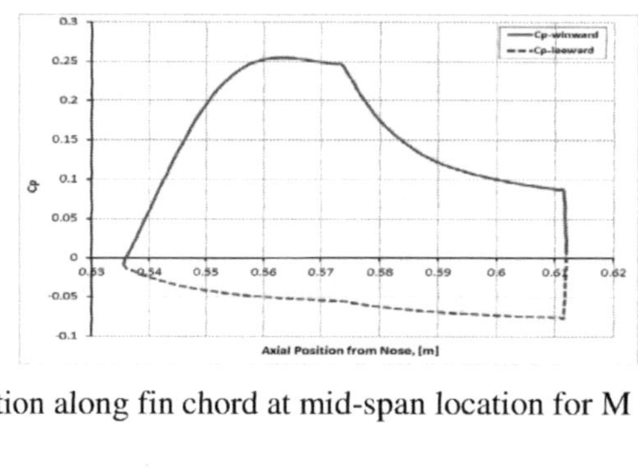

Figure 4. C_p distribution along fin chord at mid-span location for M = 4 and $\alpha = 20$ degrees.

3.4. Circumferential Cp distribution

Numerical simulations were also executed for an angle of attack of 8 degrees, to investigate the dispersion of circumferential C_p distribution on the projectile. Axial locations of these planes are pre-selected at a distance of 7, 35 and 90 mm from the nose. Computed circumferential pressure distribution is shown in Fig. 5. As the bow shock moves away from the leeward side of the nose at an angle of attack of 8 degrees, the low pressure region grows as the axial distance on the nose surface is increased.

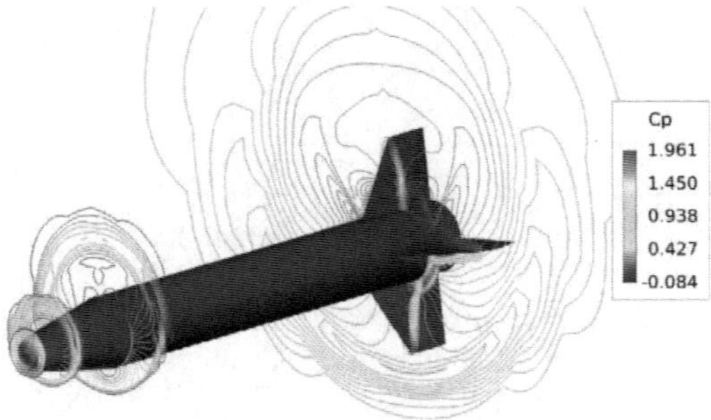

Figure 5. Circumferential pressure distribution at M = 4 and α = 10 deg at axial stations of 0.7, 3.5, 12.0 and 5.61 cm from projectile's nose.

3.5. Static aerodynamic coefficients of projectile

After satisfactory depiction of pressure distribution over the projectile at different Mach numbers and angle of attacks, post processing is done to compute the static aerodynamic coefficients. Normal force, pitching moment, Centre of pressure location and axial drag force coefficients are plotted for Mach 3 and 4. Variation of these coefficients with an angle of attack is shown in Fig. 6 to 8. The normal force coefficient increases with an increase in angle of attack, irrespective of the Mach number, as shown in Fig. 6. An increase in $C_{N\alpha}$ evaluated at an angle of attack of 8 degrees, showed a rise of greater than 10% for both the Mach numbers. Forebody drag coefficient remains almost constant at a value of about 0.26 and 0.29, at angles of attack of 8 to 12 degrees for Mach 3 and 4 respectively. Slight change is observed with a further increase in angle of attack, as shown in Fig.6.

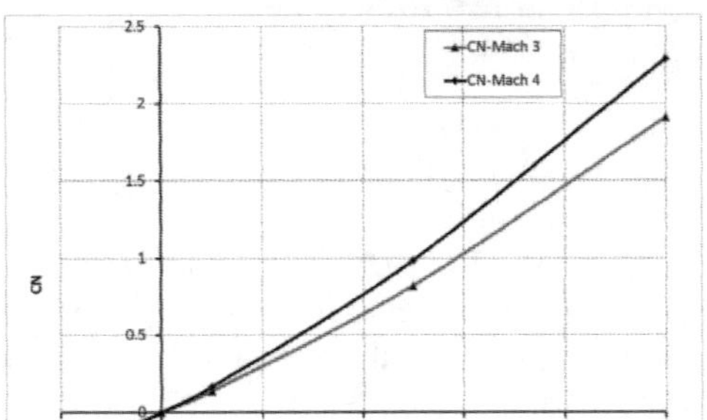

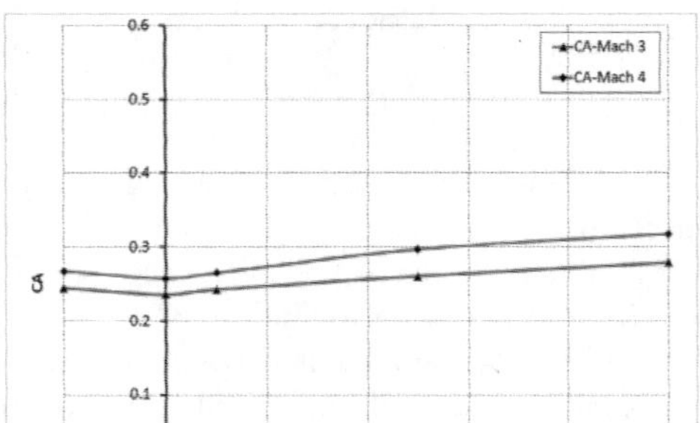

Figure 6. Normal and Axial force coefficients at Mach 3.

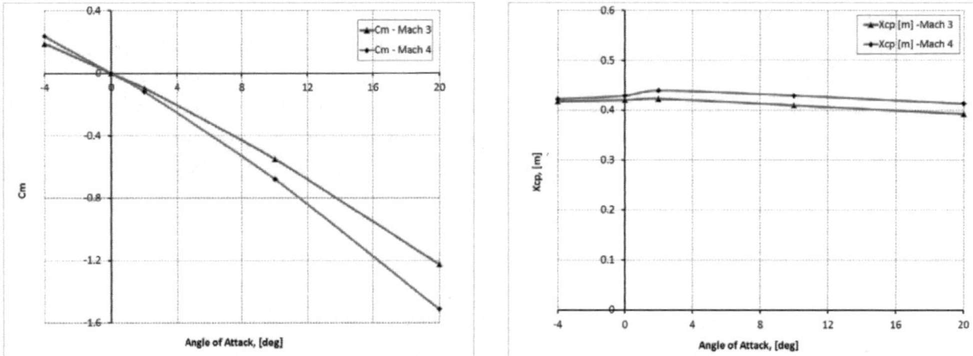

Figure 7. Pitching moment coefficient and Xcp / L at Mach 3.

A similar increasing effect in the pitching moment coefficient is also observed for both the Mach numbers, as shown in Fig. 7. Static stability remained about 0.409m to 0.429m for Mach 3 and 4 respectively, for an angle of attack of about 9 degrees. For a lower value of angle of attacks, slightly less static stability was observed. Generally a blunted nose cone shape exhibit such a trend as its static stability is not much affected by a slight change in α, at small angle of attacks; as shown in Fig.7.

3.6. Qualitative flowfield

Supersonic flowfield around a blunted nose cone consists of a stagnation point, a subsonic patch in the vicinity of the stagnation zone, which is followed by a supersonic conical region. Shock shape for the spherically blunted nose cone of the projectile is computed for Mach 3. Detailed features of the flowfield were obtained for the shock shape and the stagnation areas with a corresponding change in angles of attack. Representative pressure and Mach distribution for Mach 3 are as shown in Fig.10and Fig.11, respectively.

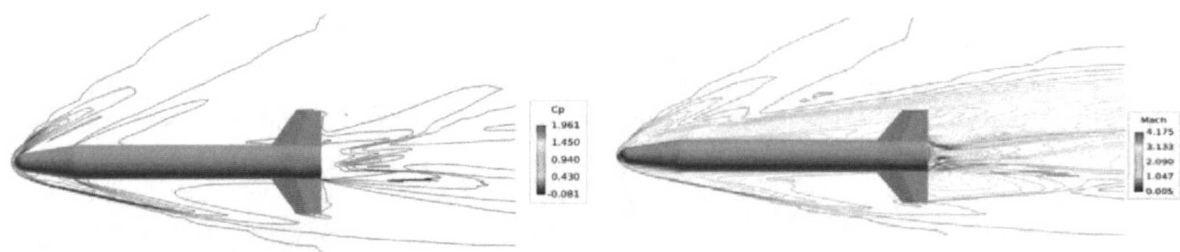

Figure 8. Centre of pressure and Mach contours at M = 4 and α = 10 degrees.

4. Conclusions

A numerical simulation for a reference projectile's nose in supersonic flow was performed to characterize and map the flowfield features. The numerical computation made possible to depict the characteristic features of blunted cone flowfield interactions, typically the formation of bow shock with associated stagnation region on the nose and resultant shock-standoff distance. All computed solutions adequately captured the pressure/ density variations in low pressure region. Further, study was made to perform computational aerodynamic investigations for the nose to determine its static aerodynamic coefficients under supersonic freestream conditions of Mach number 3 to 5 with an increasing angle of incidence from -4 to 20 degrees. Thus at this stage of the study a complete static aerodynamic behavior of the projectile's nose configuration was evaluated under supersonic flow conditions. Though this predicted aerodynamic data needs further validation through windtunnel experimentation.

References

[1] Massey K C McMichael J Warnock T and Hay F 2008 *Development of mechanical guidance actuators for a supersonic projectile* (*The Aeronautical Journal*) Vol. **112** No. **1130**, pp: 181-195.

[2] Marx W J Wessling F C III Peters B R Thies A Strickland B R and Lianos D 2002 *Miniature smart munitions/ guided projectiles for the objective force* (*In 23rd Army Science Conference*)

[3] Spaid F W and Cassel L A 1973 *Aerodynamic interference induced by reaction controls* (McDonnell Douglas Corporation, USA) *AGARDograph* No. 173.

[4] Massey C A Guthrie K B and Silton S I 2005 *Optimized guidance of a supersonic projectile using pin based actuators* AIAA 2005-4966, (*23rd AIAA Applied Aerodynamics Conf*) (Toronto, Canada).

[5] Aerodynamics Division 1998 *A user's guide to PAK-3D* (*Aerodynamics and structural analysis centre*) (Islamabad, Pakistan).

[6] Aerodynamics Division 1997 *User's manual–LOOK* (*Aerodynamics and structural analysis centre*) (Islamabad, Pakistan).

[7] Kurita M Okada T and Nakamura Y 2001 *The effects of attack angle on aerodynamic interaction due to side jet from a blunted body in supersonic flow*. AIAA Paper 01-16173.

[8] Samar R Zahir S, and Khan M A 2010 *Flight control using pin-protuberances for blunted cones* (*The Aeronautical Journal*) Vol. **114**, No. **1154**.

[9] Zahir S Shabbar and Asif M 2005 *Computational Aerodynamic Behaviour of a Plate/Jet–Interaction with a Hypersonic Flowfield for a Biconic Configuration* (*The 8th Inter Symp on Fluid Control, Measurements and Visualization, 8FLUCOM*) (PRC).

[10] Zahir S and Ye Z 2006 *Computational Aerodynamic Interaction of a Short Protuberance /Lateral Plate on Blunted Cone Configurations in Hypersonic Flow* AIAA-2006-3172 (*24th Applied Aerodynamics Conference*) (CA, USA).

[11] Zahir S and Ye Z Hypersonic 2007 *Flow Interaction of Pitched Plates on Blunted Cone at Incidence* (*Inter Journal of Mathematical Models and Methods in Applied Sciences*) Vol.**1**, Issue**3**, pp: 133–136.

[12] Aerodynamics Division 1997 *PAK-GRID User's Manual* (*Aerodynamics and Structural Analysis Centre*) (Islamabad, Pakistan).

[13] Beam R M and Warming R P 1978 *An Implicit factored scheme for the compressible Navier-Stoke equations* (*AIAA Journal*) Vol.**16**, pp: 393-402.

[14] Baldwin B and Lomax H 1978 *Thin-layer Approximation and Algebraic model for separated Turbulent flows*; (*AIAA paper 78-257*).

Acknowledgements

The authors are grateful for the assistance of the Aerodynamics and Structural Analysis Centre, ASAC, Islamabad, Pakistan; for providing computational support.

Computational fluid dynamics simulation of laboratory scale reactor of fast pyrolysis fluidised bed

Seyed Amirmostafa Jourabchi[1*], Hoon Kiat Ng[2], Suyin Gan[2] and Zhong Jian Tan[1]

[1] Faculty of Engineering and Quantity Surveying, INTI International University, Jalan BBN 12/1, Bandar Baru Nilai, 71800 Nilai, Negeri Sembilan, Malaysia
[2] Faculty of Engineering, The University of Nottingham Malaysia Campus, Jalan Broga, 43500 Semenyih, Selangor, Malaysia

E-mail: * amir.jourabchi@newinti.edu.my

Abstract. Euler-Eulerian two-fluid model (EE-TFM), among the computational fluid dynamics (CFD) techniques and module available on the market, have been chosen to study and obtain the operational parameters required for the fluidisation of different materials and different particle diameters of the fluidised bed model. In the present work, the effect of the material, namely stainless steel and sand with the respective diameters of 0.5 and 1 mm have been investigated with the aid of ANSYS FLUENT 15. From the simulation, it has found that the minimum required superficial velocity of the driving gas for fluidisation of steel beads are 70 cm/s and 140 cm/s respectively for diameter of 0.5 and 1.0 mm. On the other hand, the minimum required superficial velocities to fluidise the less dense sand beads are 30 cm/s and 70 cm/s for particle diameter of 0.5 and 1.0 mm respectively. The results show that the minimum fluidisation velocity increases as the density of the particle material increases; while it also increases, when the particle diameter increases. It is concluded that the drag force required to fluidise the specific solid bead material is proportional to both the density and the diameter of the particle chosen.

1. Introduction

In 2010, the global production of biofuel has exceeded over 100 billion litres, comparing to the previous year, it has rose by approximately 17% [1]. The trend has shown that many nations are driven to invest more on the renewable energy, namely biofuel. Glancing through the public concern regarding the human strong dependency on fossil fuels and its insecure supply, biofuel has turned out to be an alternative to reduce the consumption of fossil fuels for transportation need. However, it is crucial to ensure that the demand for biomass does not compete with the regional food supplies, such as corn and sugarcane. Therefore, the biofuels has to be derived from the lignocellulose portion of biomass or in other words, the by-product or waste of agricultural activities to prevent the conflict of food source. Research shows that by supporting the second generation biofuel industry, which is being extracted from cellulosic biomass sources, by 2020; it can be commercially competitive to replace the first-generation ethanol that is generally made from corn [1]. Currently, there are several techniques available in the industry for the conversion of unwanted biomass to combustible, energy-packed fuel; nevertheless, pyrolysis appears to be a relatively promising technique in terms of its simplicity, production yield and economical value.

In recent years, fluidised beds have been extensively used as reactors for pyrolysis as they overcome some of the disadvantages of the conventional pyrolysis reactors such as fixed bed [2]. Fluidised beds have a comparably higher rate of reaction per unit volume due to the highly turbulent flow of the dense solid bed. The turbulence has promoted a better mixing environment for the biomass to have a higher heat transfer rate with the inert particles. One research reported that their designed reactor is able to reach an astounding rate of 1000 °C/s [3]. The turbulent environment has also give rise to the even temperature distribution of the bed, which prevents the formation of local hot spots.

Bubbling fluidised beds have gained its rising popularity in the biochemical industry nowadays due to its even mixing capability and thermal distribution. Large quantities of experimental studies and numerical research have been conducted for different process parameters on pyrolysis; however, the hydrodynamics of the fluidised bed reactors are still not well understood yet due to the complexity of the particle interactions. Therefore, simulations using CFD stands out as a useful tool to study the fluidisation process and ultimately obtain the required parameters for specific operation.

In this study, a CFD model is developed to investigate the complex fluidisation process of particles in a fast pyrolysis fluidised bed reactor, where the required operational parameters are assessed for each condition of particle diameters and density. Later in this assessment feasibility of fluidisation will be evaluated.

2. Simulation Survey

In 2009 a three-dimensional (3D) and transient model was developed to simulate the particle fluid dynamics inside a fluidised bed reactor [4]. In the mentioned simulation, the author used an Eulerian multi-fluid approach to study the interaction between the three phases that consist of solid particles, flowing inert gas and liquid formed throughout the fluidising process. In other word kinetic theory of granular flow (KTGF) was used to describe the particulate phase; which means the solid phase was considered as a granular body that behaves like a fluid, but it is contained of certain volume of solid grains. K-epsilon turbulence model was implemented for the liquid phase to simulate the fluid flow pattern of liquid phase under turbulent condition. The effect of the superficial velocity on the velocity profiles of the other two phases was discussed in the mentioned research. The CFD simulation results have shown good agreement when compared with the experimental data for validation purpose and the authors have devised a conclusion, where the particle interaction and energy flow in three-phase fluidised bed reactors can be predicted via computational simulation [4].

Another research has performed the CFD simulation on the two-phase bubbling fluidised beds using both of the Eulerian and Lagrangian attempts to compare the feasibility and capability in predicting the particle flow behaviour [5]. The authors of this research have found that the Eulerian or two-fluid model (TFM) have shown greater agreement with the experimental results on the bubble possibility compare to Lagrangian or computational particle fluid dynamics (CPFD) model. All bubble possibilities predicted by the CPFD along the bed width are overestimated especially, when it is very close to the sidewall. Their simulations have also demonstrated that the CPFD model is not capable of predicting the bubble formation mechanism during the initialisation of the fluidisation process, whereas TFM have agreed well with the experimental result. However, TFM has its drawback at the computational time consumed, when it is compared with the CPFD model, which is critically important in the simulation of a large-scale industrial fluidised bed [5].

Other research on 3-Dimensional Eulerian granular model to study and validate the behaviour of the biomass and solid grains through fast pyrolysis process in a fluidised bed reactor was developed [3]. The authors have proposed the usage of Eulerian multi-fluid model to predict the velocity profile and volume fraction of the gas phase. The temperature distribution, velocity profile and biomass products were simulated and demonstrated to show the alignment with the experimental measurements. The comparison of experimental measurements and model predictions has shown a satisfying agreement on the accuracy and feasibility of the Eulerian multi-fluid model. The product yields attained from the simulation were found to show good agreement with the bio-oil production yield from experiment [3].

Geldart was the pioneer to categorise the solid flow behaviour after it is fluidised into four different distinctive classes [6]. These classes are characterised and determined by the density difference between the granular particles and the fluidising agent gas and by the mean solid particle size [7]. Although the Geldart's scale is oldest measure of particle class, it is still the most popular and preferable scale used by the researchers for more than 40 years [6]. There are two extreme classes under this system, which are called groups C and D and they occupy each ends of the scale. Group C is very difficult to fluidise because the density difference is too little and the particle size is too small [7]. The group C particles will either flow out of the bed with the fluidising medium or it will just stay as fixed bed or the fluidised part that called grey zone, which is relatively small and difficult to control. On the other hand, group D is a spout able class, where a comparably high velocity surge of gas would force through the granular solid beds and form a void in the middle of the bed and a grain's fountain at the top of the void [7]. Under the spouting condition, some of the particles are transported by the gas surge to the highest point of the bed while other solid grains will forced downward around the spout [8]. Both of the conditions are not preferable as they are hard to control and the heating process will be uneven, causing deterioration in the bio-oil quality.

Another research has investigated the momentum exchange between all the phases involved in a three-phased bubbling fluidised bed [9]. The drag models of Syamlal O'Brien model, Hill Koch Ladd model, Gidaspow model, the representative unit cell (RUC) model and the Richardson Zaki model were all investigated in this work. Preliminary simulations of a two-dimensional (2D) fluidised reactor is conducted by this researcher to investigate and show the effect of using turbulence models in the simulations agreed well with experiments. The comparisons of different drag models were made in their work, which has proven that the entire drag model failed to predict the bubble formation at the centre of the fluidised bed. However, among all the models, the RUC and Gidaspow model have demonstrated a relatively greater matching with the experimental results. Besides, the author has focused on the simulations on the bubble behaviour in both 2D and 3D fluidised bed reactor that was subjected to a uniform inlet gas distribution. The simulations are compared to experiments performed on a fluidised bed, and the simulation results agree well according to bubble frequency [9].

K-epsilon (K-ε) turbulence model has been one of the most common models used in the CFD simulation of the fluidised bed to study the flow characteristics of particles and gas under turbulent flow conditions. However, comparing the use of K-ε and not using it on turbulence model was investigated and the result shows no apparent difference in fluidised bed simulation [10]. Both of their models have provided a satisfying accordance with the experimental results and the trend has shown the volume fraction is concentrated, as it gets closer to the wall, while the centre was relatively low in the fluidised bed. From the results obtained, the authors have concluded that the K-ε turbulence model have virtually no effect on the simulation results in predicting the flow pattern and velocity profile of the solid particles in fluidised beds, hence, turbulence model is not required for future researches [10].

3. Model description

In the present work on fluidising bed, an Euler-Eulerian approach (EE) will be used to define the TFM for a 3D transient simulation as Eulerian simulation consumed less computational power than any Lagrangian approaches. The solid bed materials that will be investigated in the simulation work are silica sand and stainless steel beads of uniform size. These two materials are both inert and able to resist corrosion so that no chemical reactions will be taken place between the biomass and the dense particles. The beads will be situated initially inside the fluidised bed column in static condition and formed a desired height in the bed. The gas phase will then be introduced from the bottom of the bed to initiate the fluidisation of these granular beads.

Nitrogen gas is chosen as the fluidising agent due to its inert properties that prevents any chance of combustion in the fluidised bed. Before entering the fluidised bed column, the nitrogen gas will have to flow through a perforated plate with a sieve size of 250 micrometres to ensure that the gas is distributed evenly for better mixing environment. Table 1 shows the dimensional properties of simulated fluidised bed column.

The designed fluidised bed is compact enough to be placed on a bench, which is convenient for the research purpose of small-scale pyrolysis reactor. Stainless steel was chosen as the dense particle because of its high conductivity that might be effective for boosting the heat transfer rate. AISI 316 is a modified stainless steel that designed for resisting corrosion, which is highly suitable for the pyrolysis fluidised bed application because the solid particles will be constantly under turbulent interaction with the biomass compound such as bio-oil or volatiles that might be corrosive.

Table 1. Physical parameters of the fluidised bed Column

Descriptions	Specifications
Bed type	Bubbling
Shape	Cylindrical
Diameter	50 mm
Column Height	500 mm
Bed Heating Height	100 mm

Silica sand is the most common type of dense solid used by researchers for the study of fluidised bed because it has inert properties and is low cost. In this study, the sand particles will serve as the baseline to evaluate the feasibility of fluidising the stainless steel particles. The minimum fluidisation velocity will be compared in both scenarios and the required process parameters are expected to obtain through the analysis of the simulation results. The stainless steel particles used have a maximum protrusion of 2.5 μm, which is considered to be extremely low comparing to the mean diameter of 0.5 and 1.0 mm. Silica sand has an even smaller value of sphericity than the stainless steel with the same diameter. Therefore, both of the solid particles are considered to be spherical for the ease of calculation. As shown in Table 2, there will be two diameter settings for each type of solids, which add up to 4 kinds of scenarios with different materials and diameters. The objective of testing two different diameters is to investigate the effect of particle size on fluidisation parameters and their corresponding accuracy.

Table 2. Solid materials properties

Descriptions	Specifications	
Beads Type	Silica sand	Stainless Steel
Density (kg/m^3)	2680	7980
Shape	Spherical	Spherical
Diameter (mm)	0.5, 1.0	0.5, 1.0

As this work focuses on fluidisation process of the beads in the fluidised bed while being heated, no biomass grain was simulated in this research. The implementation of valid and reasonable boundary conditions is crucial for the computational domain to compute the governing equations. The inlet boundary condition is a uniform gas flow with constant superficial velocity, while the outlet condition is a pressure boundary with an atmospheric pressure of 1 bar. Gravitational acceleration of 9.81 m/s^2 is applied to demonstrate the fluidisation of solid phase using drag force exerted by flowing gas. The wall boundary condition is set to be free slip for both the solid phase and gas phase because there is no adhesion of gas or solid with the wall. According to the current literature, there will be significant collisions between the solid particles, thus a restitution coefficient with value of 0.95 will be introduced to the solid phase [11]. The solid volume fraction at initial condition is set to be 59.8%

of the static solid bed height. The boundary conditions are recorded in Table 3. Drag model of Gidaspow is used to describe the relationship of Reynolds number or velocity with the drag coefficient because it is reported that the Gidaspow model is the best correlation among the available model such as the Symlal O'Briens and Richardson Zaki [9]. In this research, the Gidaspow model has shown the closest agreement with the experimental results. The transient simulation of fluidisation will be run for a time period of 1 second for each settings of gas inlet velocity ranging from 10 to 150 cm/s to obtain the minimum fluidisation velocity and demonstrate the transformation of fluid behaviour during fluidisation.

Table 3. Process parameters and boundary conditions

Descriptions	Specifications
Gas Type	Nitrogen
Gas Velocity (cm/s)	10 – 150
Coefficient of Restitution	0.95
Drag Model	Gidaspow

Since it is concluded that turbulence model plays an insignificant role in the CFD simulation of fluidisation process, no turbulence model will be implemented to the granular flow simulation in the coming computational work [5], [10]. Initial simulations were validated by mesh refining method. Mesh size refined up to the limitation of research time and available computer speed. After observing of no significant change in the results, the optimum meshing size was chosen to run the rest of the simulation investigation.

4. Results and discussion

Fluidisation will only take place when the minimum fluidisation velocity is achieved or exceeded by the superficial velocity through the solid bed. In other words, the solid particles are fluidised when the nitrogen velocity is equal to the minimum fluidisation velocity. Hence, the minimum fluidisation velocity of a certain material with a specific diameter can be found by studying the changes of solid volume fraction over a period of time. When any drastic bed expansion or bubble formation is spotted from the volume fraction contour, the solid particles can confirmed to be successfully fluidised.

The fluidisation process can be indicated as shown in Figure 1, where the volume fraction of steel beads was monitored with varying velocity of nitrogen. Five different superficial velocities were simulated for the setting of steel particles with diameter of 0.5 mm at the flow time of 0.5 s. In other words, the volume fraction of steel beads at the timing of 0.5 s was captured for each gas velocities using the EE-TFM simulation. As shown in Figure 1, a conspicuous bubble can be seen forming at the middle of the solid bed, when the superficial velocity supplied is 70 cm/s. This is the first velocity that has shown the formation of bubble, when the velocity increased from 60 to 100 cm/s. Therefore, the minimum fluidisation velocity of steel particles with diameter of 0.5 mm can be concluded to be 70 cm/s.

From Figure 1, it can be seen that as the velocity of nitrogen applied increased, the hydrodynamics activity of the solid bed became more violent and turbulent. Before it reached the minimum fluidisation velocity, at the gas velocity of 60 cm/s, not a single bubble can be observed and no significant bed expansion can be seen. The solid particles were only at fixed bed condition, where the nitrogen merely flowed through the empty spaces between each particle without suspending or lifting any solids. When the minimum fluidisation velocity is achieved, a single large bubble can be seen forming at the bottom of the bed that is quickly flowing upward through the bed.

At the superficial velocity of 80 cm/s, after the minimum required velocity was achieved, it can be seen that a flat empty space was formed in the middle of the bed that does not touch the wall, instead of a circular bubble. The flat shaped gas bubble pushes that portion of the bed upward, resembling a piston action, where the particles were raining down from the pushing slug formed and integrated with the remaining bed. A minor bed expansion can be observed from the velocity because of the empty void caused by the relatively large gas holdup as contrasting from the situation at 70 cm/s.

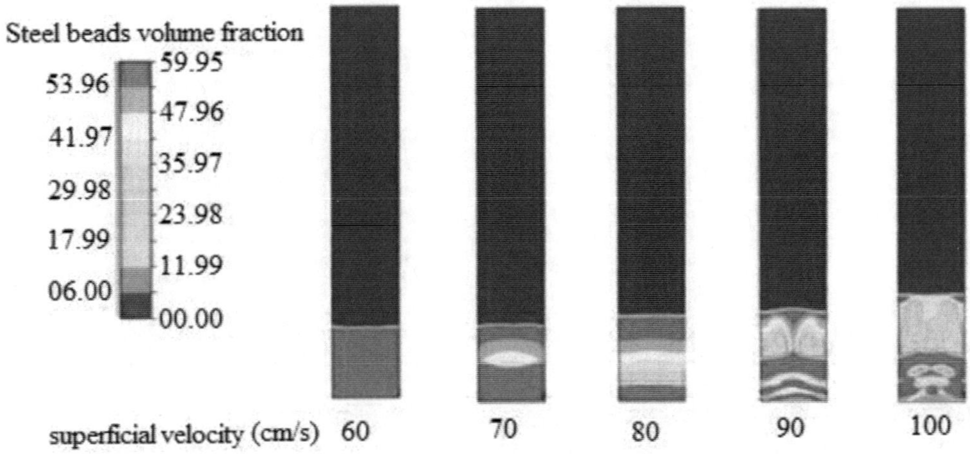

Figure 1. Volume fraction of steel beads when subjected to different gas superficial velocities (Steel, Diameter = 0.5 mm)

As the velocity increased by an extra 10 cm/s, an even greater empty space can be seen from Figure 1. However, instead of forming a flat gas void, an irregular shape of gas bubble was formed near the sides of the wall. It can be seen that the cycle time of bubble formation have been shortened as new bubbles have started to form at the bottom of the bed even before the top bubble escaped from the solid bed. At the highest tested velocity of 100 cm/s, a massive gas void can be spotted as the solid bed was pushed upward by the high pressure gas bubble or slug. The void has caused a comparably significant and large bed expansion.

During the fluidisation process, the velocity of the solid and gas phases change rapidly and drastically in different time and location in the solid bed. The velocity vector pattern for the 0.5 mm steel beads at 0.5 s was shown in Figure 2. It can be clearly seen that the solid flow was in symmetry structure and a two-directional circulation pattern, where it resembles the fluid flow during boiling state.

As shown in Figure 2, the centre area appeared to be a fast bubble flow region where the solid particles were pushed upward by the flow, while the sidewall region, the particles were flowing downward. There is a gas void region, where the steel velocity is zero emerging in the middle of the bed because of the gas bubble formation. The gas bubble was constantly flowing upward, which pushed the steel beads away from the bubble. The bubble movement resulted in the upward motion of steel beads right above the bubble and downward motion at the sides of the gas bubble.

At the bubble-free sidewall, the particles fell in a downward direction after being pushed by the central gas bubble. The bead gained velocity as it moved down the sides, resulting in a relatively high velocity zone that is marked by yellow and red arrows. In the area slightly lower than the high velocity zone, it can be seen that the steel beads were gaining higher velocity as a second bubble was forming at the bottom of the solid bed. It formed a triangular shaped region with upward velocity that is concentrated in the central area. The phenomenon is due to the collision between the falling particles and bubble-caused rising particles near the sidewall. Throughout all the simulation, the flow regime of the fluidisation process has appeared the same for different types of bed materials and varying diameters.

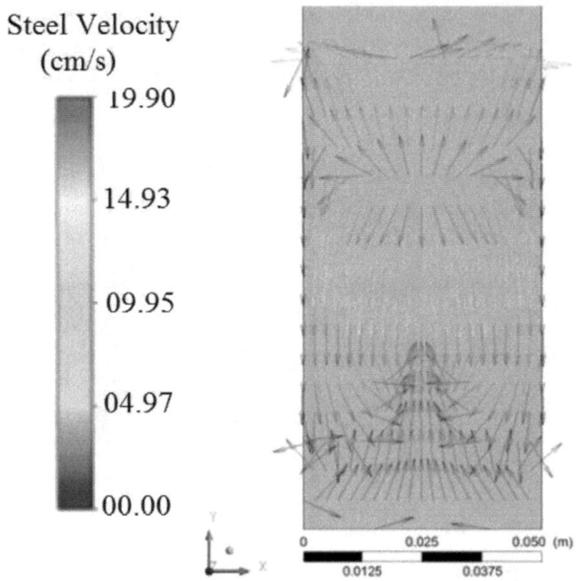

Figure 2. Velocity vector of steel beads during the bubble formation of (Steel, Diameter = 0.5 mm, Superficial Velocity = 0.7 m/s, Time = 0.5 s)

As shown in Figure 3(a), only a single bubble was formed in a consistent cyclic timing, when the minimum fluidisation velocity is achieved. The width of the bubble is almost the same with the diameter of the fluidised bed column, stretching from one end of the wall to the other. This phenomena is called as slugging regime, where it is characterised by the formation of massive gas slugs with the size that is close to the cross-sectional area of the bed column [6].

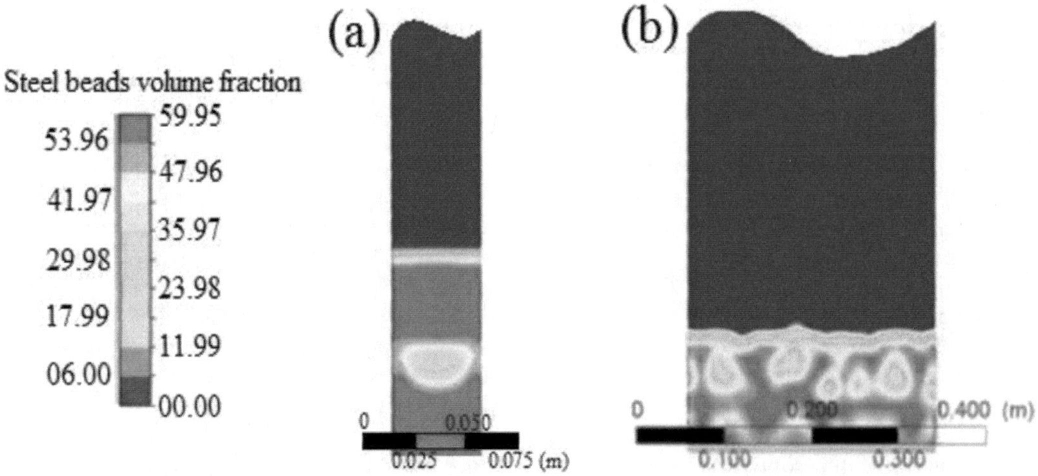

Figure 3. (a) slugging flow regime with a single large bubble formed in 50 mm diameter bed reactor and **(b)** bubbling flow regime in 300 mm diameter bed reactor (in both cases sand diameter is 1 mm and superficial velocity is 0.7 m/s)

As discussed, there are generally two types of slugging phenomena, where type A is a round-nosed gas bubble, while type B is a flat-nosed gas bubble, where the bubble appeared to be flatter at the top [6]. According to this research, the type B slug usually occurs with the bed materials that are more difficult to fluidise such as Group C. In other words, the findings from the simulation has confirmed the theoretical behaviour of group C particles have matched the simulation results, where bubble formed at the minimum fluidisation velocity and flat-nosed slug are more prone to happen with group C solid bed material.

It was also stated that a slugging phenomenon will only take place in fluidised bed that has a ratio of bed height to bed diameter that is larger than or equal to 2 [6]. In the current study, the bed height to bed diameter ratio is exactly 2, which explained why the slugging regime occurred in the designed fluidised bed column. Greater ratio of bed height to diameter provides sufficient time for the newly formed bubbles to fuse and coalesce into a larger bubble as shown in Figure 3(a), hence the bed diameter has to be increased to avoid any slugging regime to take place in the fluidised bed.

In order to test the above theory, a set of simulation was done by modifying the existing design of fluidised bed column. The bed diameter was increased 6 times from 50 mm to 300 mm, while the initial heating section of bed height was kept the same at 100 mm. As shown in Figure 3(b), many gas bubbles were formed instead of a single gas slug that observed in the original narrow column. Hence, from the results in Figure 3(b), it has proven that the slugging regime can be prevented by increasing the bed diameter. Slugging is not recommended or preferable as the flow regime for any fluidised bed application, namely pyrolysis because of the large bubble formation. The phenomena will cause a series of velocity surge in the solid particles as the large bubble pushes a portion of grains upwards with the bubble flow.

As the solid particles carried away by the bubble, it disrupted the heat distribution in the fluidisation system. Local hot spots might form near the sidewall where there is no formation of gas bubble. Hence, when biomass was fed into the fluidised bed, some of the biomass located at the central area will have a lower residence time as it is brought along by the slug; while the other biomass compound will flow slower, if it came near the side wall. It causes a deterioration in the bio-oil quality as portion of the biomass have been heated longer than the remaining compound, which has obsoleted the primary advantage of using a fluidised bed.

The graph of velocity magnitude of sand particles were plotted to monitor the velocity changes and patterns during the fluidisation process. As shown in Figure 4(a), when the sand particles were exposed to the inlet nitrogen velocity of 50 cm/s, a flat peak of velocity magnitude was observed at first 0.2 s. The velocity was first rose to a peak value of 0.025 m/s, and then it dropped slightly and stayed constant before diving to 0.015 m/s. The trend can be explained by the minor bed expansion and compaction, when the bed was introduced to the fluidising medium.

When the nitrogen gas first flowed into the bed, it causes the granular solid particles to move and make way for the gas to escape through the tiny empty spaces. It causes a surge in particle velocity that resulted in the peak shown in Figure 4(a); however, the gas velocity was not sufficient to sustain the weight of the sand beads, which causes the sand particles to slowly falling downward and the velocity to drop slightly.

As the solid particles fell and integrated with the bed, compaction took place, where the bed height returned to the quasi-steady state condition. After the quasi-steady state was achieved, the velocity magnitude of the sand particles became more or less stable and constant at 0.02 m/s. The same situation can be seen, when the superficial velocity was increased to 60 cm/s as shown in Figure 4(b). Nevertheless, a single peak of velocity magnitude was observed from the chart instead of the flat peak in lower superficial velocity.

The difference in the shape of peak can be explained by the shorter time taken by the higher velocity fluidised bed to achieve quasi-steady state condition. At higher superficial velocity, the solid particles reached the steady state in a relatively shorter time compare to lower velocity scenarios, as the drag force created by the gas flow is greater, when the velocity is higher. The drag force caused the sand particles to fall slower into the solid bed, resulting in a relatively lower velocity magnitude value compare to Figure 4(a). It has also lessen the extent of the bed compaction process during the quasi-steady state condition, where the constant velocity magnitude is only 0.004 m/s as compared to the value of 0.020 m/s at the lower velocity in 0.5 s.

At the superficial velocity of 70 cm/s, the velocity magnitude pattern, as shown in Figure 4(c), has changed completely from the previous configurations. A consistent oscillatory pattern was observed with regular peaks of 0.013 m/s and a cyclic or peak to peak period of 0.3 seconds. The pattern indicated that the minimum fluidisation velocity has been achieved by the fluidising agent, where air bubbles were formed in the solid bed. The cyclic pattern of velocity magnitude was caused by the consistent formation of a single, large bubble throughout the fluidising process. As a bubble formed, it forced the particles to move upwards, which causes the velocity magnitude to peak at 0.1 seconds. Then, as the air bubble escaped from the solid bed, the velocity magnitude decreased, when the solid particles slowly rained down onto the top surface of the bed.

Almost immediately after the escape of the first gas bubble, the second bubble was formed at the bottom of the bed at around 0.25 seconds. The second bubble pushed the solid particles upward through the bed again and resulted in an increment of velocity similar to the first peak pattern. The velocity magnitude dropped once again, as the bubble disintegrated and the cycle continues endlessly with the same trend and similar process. A large bubble will form in every 0.3 seconds with a consistent frequency of 3.33 bubbles per second.

According to Geldart's scale, the sand particles with diameter of 1.0 mm can be considered as a Group C particle. The major characteristic of the particle class is that bubbles are formed at the minimum fluidisation velocity, where no intermediate stage of bed expansion happens during the process. The velocity magnitude pattern in Figure 4(c) can confirmed that the theory of Group C flow behaviour is true, where bubbles are formed consistently at the minimum fluidisation velocity itself. Hence, the fluidisation of any solid bed with different properties can be indicated from the corresponding granular velocity magnitude graph in future research. Whenever a cyclic pattern is observed from any Group C particles at a specific superficial velocity, it can be concluded that the solid bed have been fluidised. Besides, the minimum fluidisation velocity can be found without referring to and relying on the transient solid volume fraction contour to visually search for any visible gas bubble present in the solid bed.

Nevertheless, as the velocity increases beyond the minimum fluidisation velocity, a random pattern was exhibited by the solid particle velocity magnitude. As shown in Figure 4(d), when the superficial velocity was adjusted to 80 cm/s, which is above the minimum fluidisation velocity, the cyclic pattern have been distorted to demonstrate many different peaks with several irregular time intervals between each peaks. From the animation video of solid volume fraction with the superficial velocity of 80 cm/s, it has shown that bubbles were formed at a higher rate than the previous case of 70 cm/s. The rate was so high that some of the bubbles actually formed before the previous bubble escaped from the solid bed. Hence, it causes the cyclic pattern to be distorted as shown in Figure 4(d), where some of the cycles were overlapping with each other.

The overlapping conditions resulted in the forming of very high peak such as the highest velocity magnitude at 0.5 seconds, where it has reached 0.025 m/s, boosted almost double comparing to the highest peak of 0.013 m/s at the superficial velocity of 70 cm/s. It was due to the superposition of velocity magnitude that was caused by two bubbles that co-existed in the solid bed at the same time. However, the peaks were lowered as time went on because the formation of bubble has become more irregular. The increased formation rate of bubbles have lowered the possibility or chance of larger bubble to form because the flowing time was too short for the smaller bubbles to coalescence into a larger one before escaping the solid bed. Therefore, since the smaller bubbles were formed at different timings, it has rendered the velocity magnitude pattern to be seemingly more random than the usual case.

Same simulation runs were performed on both materials of choice for beads and two tested sizes with similar output behaviour but in different in range results.

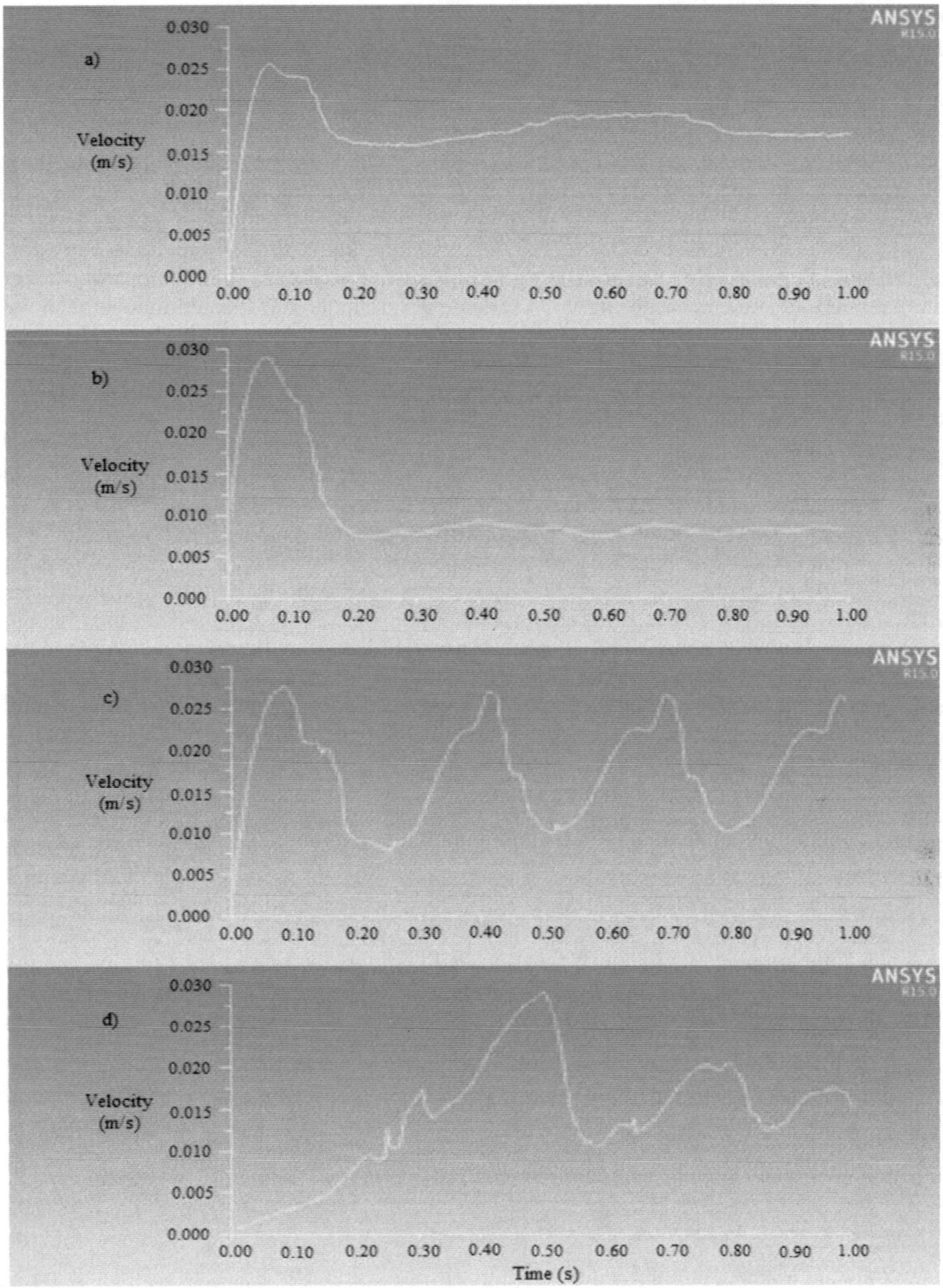

Figure 4. Velocity magnitude of 1.0 mm diameter sand beads against flow time
a) flat peak when superficial velocity is 0.5 m/s
b) single high peak before minimum fluidisation velocity when superficial velocity is 60 cm/s
c) cyclic pattern at the minimum fluidisation velocity when superficial velocity is 70 cm/s
d) random pattern when superficial velocity is 80 cm/s

5. Conclusions

In the present paper, the CFD simulation on the particle hydrodynamics of a two phase fluidised bed have been performed for two different bead materials with two different particle diameters. The Euler-Eulerian two-fluid model was deployed for the numerical simulation to study the fluidisation process and to obtain the minimum fluidisation velocity. Multiple hydrodynamics variables were studied in the ANSYS FLUENT simulation of fluidisation process, which includes the solid volume fraction distribution, the velocity magnitude fluctuation, bubble formation frequency and the velocity profile at different bed heights.

From the simulation, it has been found that the minimum fluidisation of steel beads can be achieved with nitrogen superficial velocities of 70 cm/s and 140 cm/s for diameter of 0.5 mm and 1.0 mm beads respectively. On the other hand, the minimum required superficial velocities to fluidise sand beads are 30 cm/s and 70 cm/s for particle diameter of 0.5 mm and 1.0 mm respectively. It has also discovered that the minimum fluidisation velocity will increase as the density of the particle material increases; while it will also increase when the particle diameter increases. Hence, it can be concluded that the drag force required to fluidise the specific material is proportional to both the density and the diameter of the particle chosen.

Slugging phenomena have been observed from every simulation settings as it often happened when the ratio of initial bed height to the fluidised bed column diameter is more than or equal to 2. Since the height to diameter ratio of the designed column in this research is exactly 2, the simulation results have proven that theory is applicable to the Euler-Eulerian two-fluid model of the fluidisation process. Furthermore, the results have exhibited strong agreement with the theoretical characteristics of the Geldart's scale group C particles, where the fluidisation of the group C particles can be indicated by formation of bubble without any significant bed expansion.

The velocity magnitude fluctuations have been studied, where it has been found that the pattern varies with the superficial velocity of the nitrogen as fluidising agent. Before approaching the minimum fluidisation velocity, the velocity magnitude of the solid particles will appeared to be stagnant after the first peak. However, a relatively uniform cyclic pattern will be observed as soon as the particles are fluidised with a slugging flow regime.

Although the simulation was validated by using mesh refining method, an experiment should be conducted in the future with accordance to the geometrical and operational parameters used in the simulation in order to validate the simulation results. The simulation can be improved by reducing the size of each time step based on time limits and computer's processor's speed. The refinement will yield a higher result resolution, which is essential for the prediction of fluidisation process and its application. Furthermore, the design of the fluidised bed column can be modified by enlarging the reactor diameter to prevent any slugging phenomenon, if the geometry of the practical furnace allows the above modification.

Acknowledgment

The authors would like to acknowledge the Centre for Biomass Research in INTI International University and faculty of Engineering of the University of Nottingham Malaysia Campus for their support towards this work.

References

[1] Stephen J, Mabee W and Saddler J 2011 Will second-generation ethanol be able to compete with first-generation ethanol? Opportunities for cost reduction *Biofuels, Bioproducts and Biorefining* **6(2)** pp 159–76
[2] Trambouze P and Euzen J 2005 Chemical reactors: from design to operation *Choice Reviews Online* **42(05)** p 283
[3] Mellin P, Zhang Q, Kantarelis E and Yang W 2013 An Euler–Euler approach to modeling biomass fast pyrolysis in fluidized-bed reactors – Focusing on the gas phase *Applied*

Thermal Engineering **58(1–2)** pp 344–53

[4] Panneerselvam R, Savithri S and Surender G 2009 CFD simulation of hydrodynamics of gas–liquid–solid fluidised bed reactor *Chemical Engineering Science* **64(6)** pp 1119–35

[5] Liang Y, Zhang Y, Li T and Lu C 2014 A critical validation study on CPFD model in simulating gas–solid bubbling fluidized beds *Powder Technology* **263** pp 121–34

[6] Yang W 2003 *Handbook of fluidization and fluid-particle systems* (New York: Marcel Dekker)

[7] Gupta C and Sathiyamoorthy D 1999 *Fluid bed technology in materials processing* (Boca Raton, Fla.: CRC Press)

[8] Kunii D and Levenspiel O 1991 *Fluidization engineering* (New York: Wiley)

[9] Lundberg J 2008 *CFD study of a bubbling fluidized bed* (Faculty of technology, Telemark University College)

[10] Wang S, Hao Z, Lu H, Yang Y, Xu P and Liu G 2012 Hydrodynamics modeling of particle rotation in bubbling gas fluidized beds *International Journal of Multiphase Flow* **39** pp 159–78

[11] Lohaa C, Chattopadhyayb H, and Chatterjee P K 2014 Effect of coefficient of restitution in Euler–Euler CFD simulation of fluidized-bed hydrodynamics *Particuology* **15** pp 170–7

Effect of mold designs on molten metal behaviour in high-pressure die casting

M D Ibrahim [1], M R A Rahman [1], A A Khan [1], M R Mohamad [2], M S Z M Suffian [1], Y S Yunos[1], L K Wong[1], and M Z Mohtar[1]

[1]Department of Mechanical and Manufacturing Engineering, Universiti Malaysia Sarawak, Kota Samarahan, Malaysia
[2]HICOM Diecasting Sdn. Bhd, Shah Alam, Malaysia

imdanial@unimas.my

Abstract. This paper presents a research study conducted in a local automotive component manufacturer that produces aluminium alloy steering housing local and global markets. This study is to investigate the effect of design modification of mold in die casting as to improve the production rate. Design modification is carried out on the casting shot of the mold. Computer flow simulation was carried out to study the flow of molten metal in the mold with respect to the mold design modification. The design parameters of injection speed, die temperature and clamping force has been included in the study. The result of the simulation showed that modifications of casting shot give significant impact towards the molten flow behaviour in casting process. The capabilities and limitations of die casting process simulation to conduct defect analysis had been optimized. This research will enhance the efficiency of the mass production of the industry of die casting with the understanding of defect analysis, which lies on the modification of the mold design, a way early in its stages of production.

1. Introduction

New technologies, cost pressures, and economic trends have all had a significant impact on the manufacturing industries. This is particularly true in the automotive component market sector, which requires various sizes, complex, tight-tolerance, high performance, and low-cost products. The use of automotive products is expected to escalate because of their promise of decreasing cost and improved efficiency. As a result, the demand for aluminium dies casting processing services and equipment is expected to grow. However, rejected parts especially during the pre-mass production are a major concern for all automotive part manufacturers. In order to improve the production rate, the design and development of a process is enhanced with advanced processing technology so that it will lower its defect rate [1]. The same goes with die casting, mold parameter changes and design modifications should have continuous development and improvement to reduce defect rate as well as improve the production rate.

A research study was undertaken in one automotive components manufacturer which produces automotive aluminium die cast steering housing intended for local and global market using high pressure die casting process. This study is to investigate the effect of design modification of mold in die casting as to improve the production rate. Earlier detection of possible defects that arise based on flow simulation result will help the company make more profits by significantly reducing losses of valuable resources.

2. Methodology

2.1 Mold Design

Designing a mold is the key factor in producing good quality of casting in high pressure die castings. In this paper, the mold designing was initiate with modifying a runner design and air vents system for steering housing. The study mainly focuses on the runner design and air vents in the whole casting system. Runner was designed to support the flow of the molten metal in the cavity. The functional requirement in the mold design is that the design must be able to allow the molten metal to travel along the cavity and filled up the cavity completely. In this paper, two different designs of runner and air vent for the steering housing parts had been modelled using a commercialized 3D mechanical computer-aided design (CAD) program software. There are two designs namely six runner type and four runner type. All of the two different runner design were being illustrated respectively in figure 1 and 2.

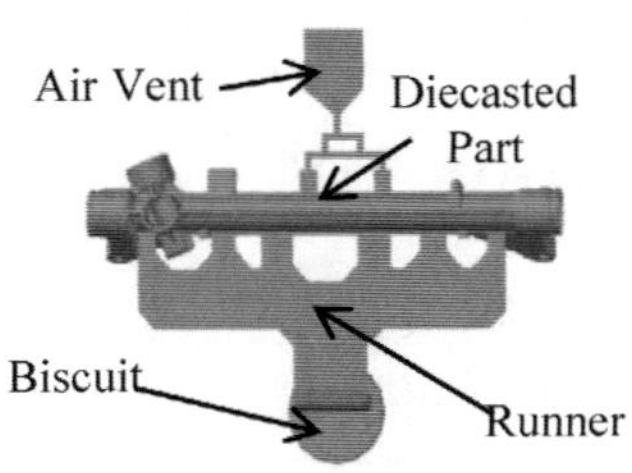

Figure 1. Six Runner Type with One Air Vent

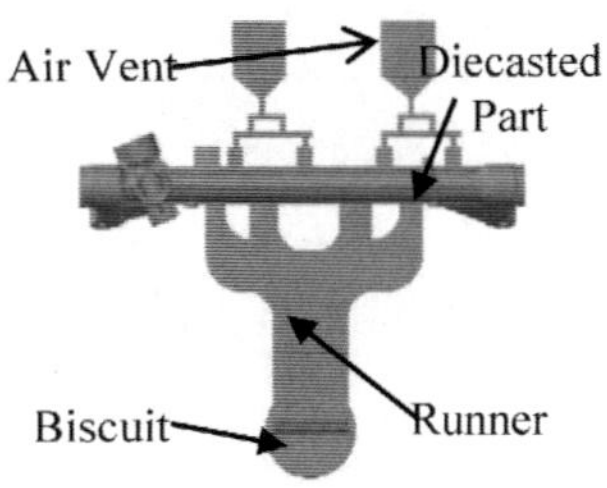

Figure 2. Four Runner Type with Two Air Vent

2.2 Molten Metal Flow Simulation

Simulation is the process of imitating a real process by using a set of mathematical equations implemented in a computer programming. In casting simulation, the mould filling and solidification analysis is done by using an algorithm or program based on finite volume method, to identify the hot spots and hence defects like shrinkage porosities, hot tears, cracks, etc **[2]**. The simulation study is can be regarded as the change of flow pattern, temperature and free surface of liquid metal in filling process. This casting filling process can be described by continuity equation, momentum conservation equation, energy equation and volume of fluid-function equation [3].

2.2.1. Continuity Equation

$$D = \frac{\partial u}{\partial x} + \frac{\partial v}{\partial y} + \frac{\partial w}{\partial z} = 0 \quad (1)$$

Where D is the divergence; u, v, w are components of velocity vector in x, y, x direction of ms^{-1}.

2.2.2. Navier-Stoke Equation

$$\rho\left(\frac{\partial u}{\partial t} + u\frac{\partial u}{\partial x} + v\frac{\partial u}{\partial y} + w\frac{\partial u}{\partial z}\right) = -\frac{\partial p}{\partial x} + \rho g_x + \rho\nabla^2 u \quad (2)$$

$$\rho\left(\frac{\partial v}{\partial t} + u\frac{\partial v}{\partial x} + v\frac{\partial v}{\partial y} + w\frac{\partial v}{\partial z}\right) = -\frac{\partial p}{\partial y} + \rho g_y + \rho\nabla^2 v \quad (3)$$

$$\rho\left(\frac{\partial w}{\partial t} + u\frac{\partial w}{\partial x} + v\frac{\partial w}{\partial y} + w\frac{\partial w}{\partial z}\right) = -\frac{\partial p}{\partial z} + \rho g_z + \rho\nabla^2 w \quad (4)$$

Where *p* is pressure of unit density, Pa; μ is dynamic viscosity, Pas; g is acceleration of gravity, ms^{-2}; ∇^2 is laplacian; ρ is density of fluid; $t$is filling time,s.

2.2.3 Energy Equation

$$\rho c\frac{\partial T}{\partial t} + pcu\frac{\partial T}{\partial x} + pcv\frac{\partial T}{\partial y} + pcw\frac{\partial T}{\partial z} = \frac{\partial}{\partial x}\left(k\frac{\partial T}{\partial x}\right) + \frac{\partial}{\partial y}\left(k\frac{\partial T}{\partial y}\right) + \frac{\partial}{\partial z}\left(k\frac{\partial T}{\partial z}\right) + S \quad (5)$$

Where c is specific heat capacity, J·(kg·K)$^{-1}$; S is heat which resulting from the liquid surface stress applying work for liquid, J; T is temperature of liquid, K; k is thermal conductivity of liquid, w·(m·k)$^{-1}$.

2.2.4 Volume of Fluid-Function Equation

Volume of fluid funcyion equation is chosen as to simulate free surface. A volume of function Fis needed to be define when computational is determined. The state of each grid can be expressed by F value. F is defined as follows:

$$F = \frac{O_V}{A_V} \quad (6)$$

Where O_V is the volume of fluid in one grid, m^3; A_V is the volume of one grid,^{m3}.The value is unity for cells fully occupied by fluid, zero for empty cells, and between 0 and 1 for surface cells. While the volume of fluid-function equation need to be solved:

$$\frac{\partial F}{\partial t} + u\frac{\partial F}{\partial x} + v\frac{\partial F}{\partial y} + w\frac{\partial F}{\partial z} = 0 \quad (7)$$

Where u, v, w are components of velocity vector in x, y, z direction of grid point (x, y, z), ms^{-1}

To support the objectives of computer flow simulation method, there are significant parameters that were undertaken in this project which are (1) injection speed, (2) die temperature, (3) die clamping force. The parameters that are involved in the simulation are stated clearly in table 1 and the thermophysical performance parameters of casting mold material used in this study is shown in table 2.. The analysis of the aluminium molten metal flow behaviour in the casting process is analysed using the flow simulation software. The commercialized 3D mechanical CAD software was embedded with a flow simulation program software where it includes CFD tools that enable us to simulate and efficiently analyse the model of steering housing casting parts.

Table 1. Performance Parameters of Casting

Density	Thermal Conductivity	Heat Capacity	Crystallization Heat	Solidus	Liquidus	Dynamic Viscosity
[kg/m^3]	[W·(m·K)$^{-1}$]	[J·(kg·K)$^{-1}$]	[J·kg]	[°C]	[°C]	[N·s·m^{-2}]
2300	100	1090	394000	635	660	0.001

Table 2. Performance Parameters of Casting Mold

Temperature [°C]	Thermal conductivity [W·(m·k)$^{-1}$]	Heat capacity [J·(kg·k)$^{-1}$]
100	20.1	468.2
200	20.2	525.5
300	22.7	564.0
500	23.4	612.3
700	24.3	685.5

The dies used in die casting are usually made out of hardened tool steels, because cast iron does not able to withstand the high pressures involved, therefore the dies are very expensive, resulting in high start-up costs. Metals that were cast at higher temperatures require dies made from higher alloy steels.

In actual production, the die is always preheated in order to achieved the desired die temperature to allow the process to starts. The most suitable preheating temperature is dependent on the type of casting alloy but normally lies between 420 ~ 660 K (150 ~ 350°C). It is important not to preheat the dies to an excessively high temperature, since the dies may become too hot in the process, causing a tempering back of the die material. The die casting process cycle were typically repeated in seconds. The thermal cycling of a die casting creates stress. This thermal stress will give severe impact to the die surface through heat checking. Heat checking is a phenomenon when surface cracks occur on the die due to a large temperature change on every cycle.

3. Results & Discussion

In this section, there are conditions that will be described in which are very common to the die casting process and usually represent defective castings that must be scrapped. Defects found in castings can categorized into three different classes [4]:

- defects that simply noticed on visual examination or measurement of the casting.
- defects that exists under the surface and revealed by machining, sectioning or radiography.
- material defects discovered by mechanical testing (tensile, bending, impact, etc.) of the casting.

In mass production, the die temperature is kept at an optimum temperature of the respective castings parts of the mold cavity surface. The thermal factors in the die casting process are the temperature of cast alloy, the temperature of pressurized casting chamber and the temperature of the mold. According to Sandeep et al (2014), this temperature depends on the temperature of the material, the quantity of the metal, the method of the cooling of the casting mold, the thermal conductivity of the mold material, and the time during which the casting remains in the mold. The casting of an alloy into a mold with a very insufficient surface temperature results in an early fall of the alloy temperature [5].

In this research, the temperature of the molten metal is set to be 660°C, while the initial die temperature is set to be 220°C. The temperature of the molten metal is always ranging from 650 ~ 670 °C in the actual production. The initial temperature is set to be at 220 °C is because the die is always preheated at an elevated temperature ranging from 150 ~ 350 °C. In mass production, the die is being preheated at an elevated temperature of 1/3 from the casting alloy temperature. Figure 3~4 show the illustration of the surface plot for temperature fluid. It can be observed that the results show a colorful fluid temperature distribution. The color shown in the simulation results indicates the behavior of the molten temperature inside the cavity. This scenario of simulation results is very significant in this study because the color distribution will provide an indicator of possible defects that may arise prior to mass production.

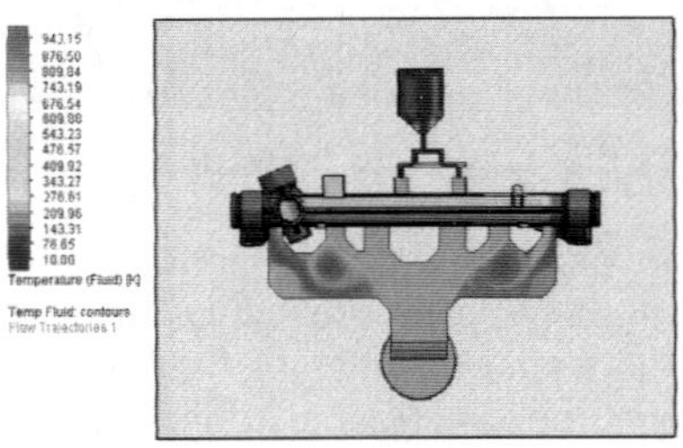

Figure 3. Surface Plot of Fluid Temperature in Six Runner Type Design.

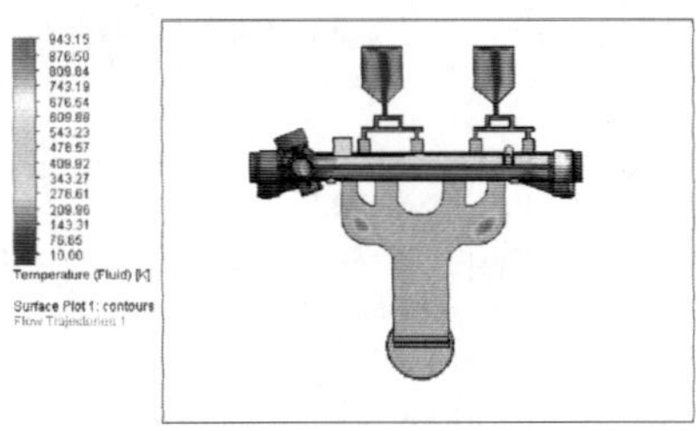

Figure 4. Surface Plot of Fluid Temperature in Four Runner Type.

Defects that could be happening due to thermal factors in die casting process most of the time is surface defects. These surface defects are identified as soldering, cracks, blisters and lack of fill. Soldering happens as the result of the cast metal sticks to die surfaces. Soldering may be caused by excessive metal temperature, incorrect mold temperature, insufficient die release material and incorrect alloy. Generally, soldering starts at the spot facing the injection gate. The presence of high injection velocity and high temperature could cause the protective thin coating on the die surface to worn off and causes the molten metal of aluminium alloy to be in contact directly with the die steel. This scenario is bad as it lets the molten metal stick to the die at the end of the cycle. Eventually, this situation will reduce the die service life [6]. Based on the results illustrated in figure 5, it can be observed that the solid surface plot showing an accumulated high temperature of molten metal at the end of the steering housing. A sample was selected randomly to undergone defects analysis. Based on the analysis, soldering defects on casting defects can be observed as a result of soldering that gives severe defects on die surface as shown in figure 6. In a mass production, if the condition is severe or when other methods technically fail to remove the solder, the molten deposited at the mold surface must be cleaned from the die. It is highly recommended that the die cavity is to be polished and care must be exercised. The amount of die release material applied to the soldering area in order to reduce the soldering must be increased. The die release material is to protect the die steel to be in contact with the aluminium alloy directly. If care is not exercised sufficiently, condition for soldering will be present for the next cycle of the process.

Crack defects also occur due to thermal factors in die casting process. Castings may have cracked due to internal stress or from insufficient pressure during injection. In a real production, cracks are caused by internal stress which comes from insufficient metal or dies temperatures. Even though the die casting process is handled automatically and some are semi-automated, the defects are unpredictable. If this defect continues occurring after adjustments of temperature have been made, it may be necessary to increase the injection velocity.

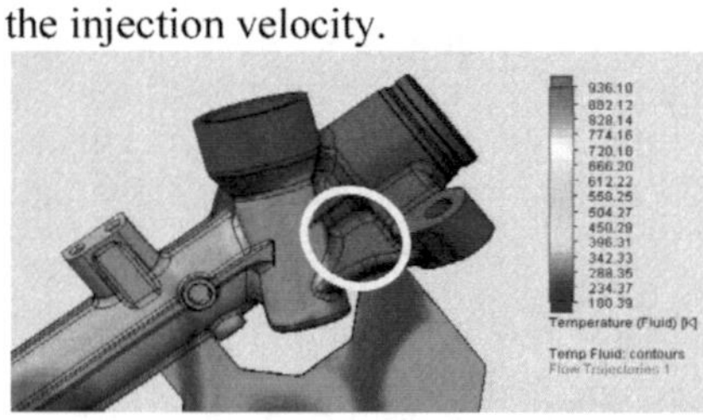

Figure 5. Surface Plot of Temperature of Steering Housing Mold.

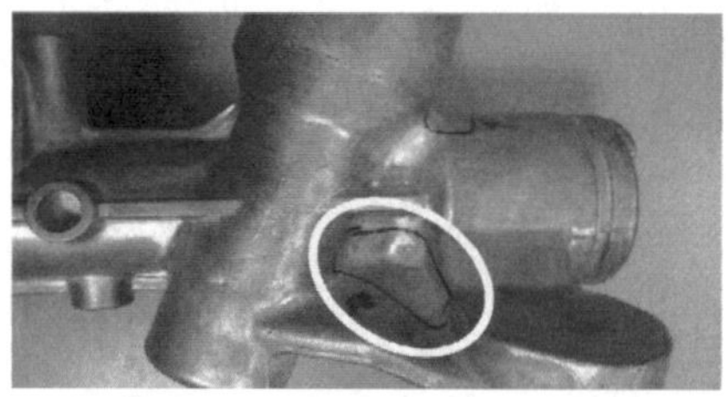

Figure 6. Soldering defects in casting

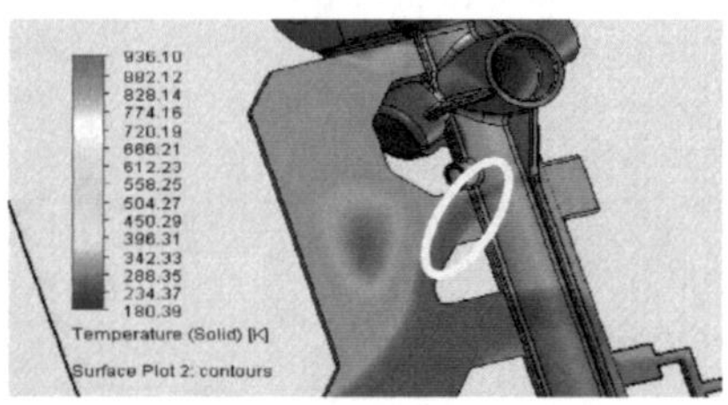

Figure 7. Surface Plot of Temperature of Steering Housing Mold

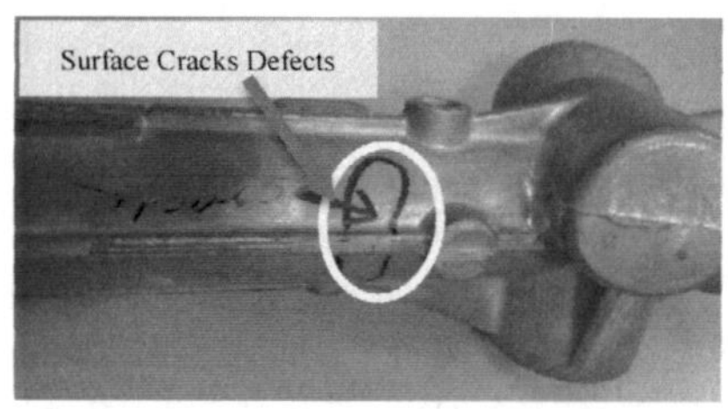

Figure 8. Cracks defects in castings

The cracks in die castings can be classified into two different types, one forms during the solidification known as hot tears and the other forms during the cooling after solidification cold cracks. Hot tears appear at low ductility temperature while the cold cracks appear at a lower temperature in the cooling process based on all simulation done in this study. Cracks defects also may be caused by shrinkage. This happens when the metal is solidified with less volume. Normally, high metal pressures are used at the end of cavity filling to force more metal into the cavity to make up for this shrinkage. This shrinkage occurs at a location that freezes last which leads to cracks or sinks. Cooling this local area will remove the shrinkage from the surface. As we refer to previously illustrated figure 7 and 8, it is can be observed that at the gate of the runner has a slight reduction of temperature which leads to cold crack defects.

4. Conclusions

This paper concludes that four runner type with two air ventilation system is the best design compared to six runner design. With the aid of computer flow simulation the production of the casting process is optimized through defects analysis at its early pre-mass production. This implies that early countermeasures can be carry out to reduce number of defects. Future works can be developed through surface modifications on mold designs in order to enhance the molten flow behavior and reducing defects issues for other parts.

Acknowledgement
This research is partially funded by Grant No: FRGS/TK01(01)/1059/2013(05). Special thanks to the authority of Universiti Malaysia Sarawak and HICOM Diecastings Sdn. Bhd. for the support and facilities to complete this project. the authors claim that there is no conflict of interest regarding the publication of this paper.

References

[1] Ashari M F, Ibrahim M D , A Husaini, A A S and Zulkharnain A. 2014 *Periodical of Key Engineering Materials Vol. 572 in 2014 with the title Advanced Design and Manufacture V Key Engineering Materials Vol.* ***572*** (2014), pp. 304-307, Trans Tech Publications, Switzerland.
[2] Dabade U A, and Bhedasgaonkar R C. 2013 *Procedia CIRP*, **7**, 616-621.
[3] Yuwen X X, Chen L, & Han Y J 2012 *Energy Procedia*, ***17***, 1864-1871.
[4] Gariboldi E, Bonollo F, and Rosso M. 2007 *la metallurgia italiana*, **6**.
[5] Sandeep V C and Rajeev K T 2014. *International Journal of Innovation Research in Science, Engineering and Technology.* **Vol. 3, Issue** 7.
[6] Rahman, M R A. 2015 *Molten Flow Behavior Of Mold Design In High Pressure Die Casting To Improve Production Rate* (Undergraduate Student Thesis: UNIMAS)
[7] Barton H K 1956 *The Diecasting Process.* Long Acre, London: Odhams Press Limited.
[8] Bakemeyer H 2008 *Operating The Die Casting Machine,* Publication **#E-902.** Illinois: North American Die Casting Association.

Implicit gradient reconstruction (IGR) method for compressible flow simulation

Manish K Singh[1,2], N Munikrishna[3], V Ramesh[1], N Balakrishnan[4*]

[1] Scientist, Council of Scientific and Industrial Research-National Aerospace Laboratories, Bangalore, India
[2] Research scholar, Indian Institute of Science, Bangalore, India
[3] Chief Technology Officer, S & I Engineering Solution Pvt. Ltd. Bangalore, India
[4] Associate Professor, Indian Institute of Science, Bangalore, India

*Email: nbalak@aero.iisc.ernet.in

Abstract. The classical CIR scheme is modified by introducing a ϕ parameter which allows an explicit control of dissipation. Also, this permits a seamless integration of the upwind and central difference based schemes. It is demonstrated that the ϕ parameter can be linked to solution reconstruction and thus second order accuracy can be achieved effectively with a first order formula. An objective way of determining ϕ based on solution gradients on the volume interface is established. These gradients in turn are also used for viscous flux computation and therefore come at no additional cost. The efficacy of the proposed methodology is established by solving a number of standard test problems, both in 1D and 2D.

1. Introduction
One of the important developments in the area of upwind schemes is the development of Modified CIR (MCIR) schemes [1]. It simply involves introducing an additional ϕ parameter in the classical CIR scheme, which allows an explicit control of the resulting dissipation. This methodology has been successfully used in conjunction with the framework provided by the kinetic theory of gases for solving many problems of relevance to aerospace industry [2, 3]. The present effort is to further extend the scope of this methodology, by introducing a family of ϕ schemes, which provides a seamless framework for integrating upwind and central difference type schemes. It is demonstrated that the ϕ parameter can be linked to the solution gradients resulting from a reconstruction procedure and therefore the use of appropriate ϕ can be considered as an implicit way to do solution reconstruction. In this work, care is taken to ensure the effort needed for gradient finding for determination of ϕ is no more than what is needed for viscous flux computation. The resulting procedure, which can be used in conjunction with either flux vector splitting or flux difference splitting schemes, is simple, efficient and obviates the need to store solution gradients, leaving significantly less memory foot print as compared to classical linear reconstruction procedure.

2. Methodology
Consider the linear convection equation,

$$u_t + cu_x = 0 \tag{1}$$

where, the property u gets convected with a speed c. The MCIR implementation reads (Figure 1),

$$\frac{u_j^{n+1}}{\Delta t} + \frac{c+\phi|c|}{2}\frac{u_j^n - u_{j-1}^n}{\Delta x} + \frac{c-\phi|c|}{2}\frac{u_{j+1}^n - u_j^n}{\Delta x} = 0 \tag{2}$$

where, ϕ is a user defined parameter. Referring to the Figure 2, representing the convection of a linear profile over a time step Δt, the time averaged flux on the volume interface $j + \frac{1}{2}$ is given by,

$$< f_{j+\frac{1}{2}} > \; = c\left(u_j^n - \frac{(1-\lambda)}{2}\delta u_j^n\right) = cu_j^n + \frac{1}{2}(c - \phi|c|)\delta_+ u_j^n \tag{3}$$

where, the Courant number $\lambda = \frac{|c|\Delta t}{\Delta x}$ and δu_j^n is an undivided difference representing the solution gradient in volume j and $\delta_+ u_j = u_{j+1} - u_j$. For an MCIR scheme, it is easy to show that,

$$\left(\delta u_j^n\right)_{MCIR} = \frac{(1-\phi)}{1-\lambda}\delta_+ u_j^n \tag{4}$$

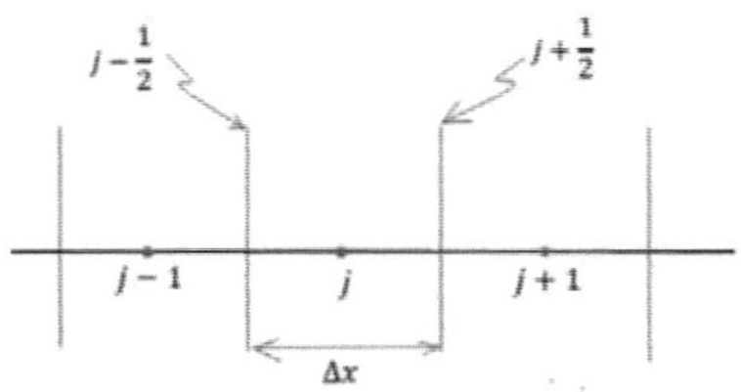

Figure 1. One dimensional stencil

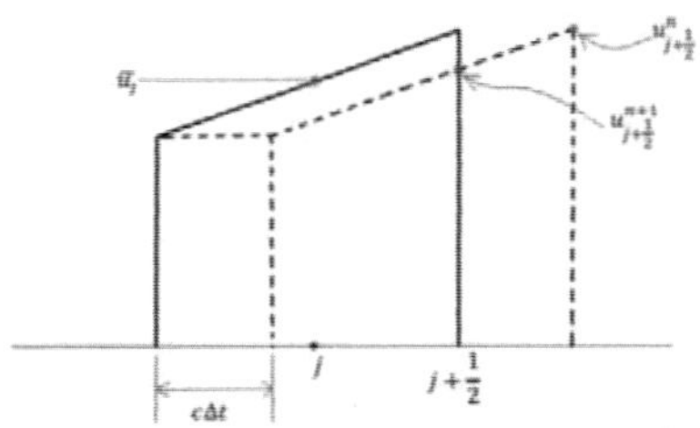

Figure 2. Reconstruction procedure

It is noteworthy to observe that for $\phi = \lambda$, the MCIR gradient reduces to the gradient corresponding to the Lax Wendroff scheme, i.e., $\left(\delta u_j^n\right)_{MCIR} = \delta_+ u_j^n = \left(\delta u_j^n\right)_{LW}$. This observation is very interesting in the sense that a family of ϕ schemes represented by CIR scheme for ϕ = 1.0 on one end and a spatially centered scheme for ϕ = 0.0 on the other end, with varying gradient representation within the finite volume, can be presented in the framework provided by MCIR. It is important to note that this gradient representation is rather implicit as against the classical formulations, where the gradients are computed explicitly in the reconstruction step and therefore the resulting procedure is referred to as Implicit Gradient Reconstruction (IGR) method. At this stage, it is worth remarking that negative values of ϕ would result in an unstable semi-discrete equation and ϕ > 1.0 will accentuate loss in monotonicity. Therefore ϕ is bounded between 0.0 and 1.0. The inference presented above also provides a means to objectively fix the value of ϕ. This is achieved by equating the interfacial flux as obtained using a classical reconstruction scheme and the MCIR scheme:

$$f_{j+\frac{1}{2}} = \frac{c+|c|}{2}\left(u_j + \frac{\delta u_j}{2}\right) + \frac{c-|c|}{2}\left(u_{j+1} - \frac{\delta u_{j+1}}{2}\right) = \frac{c+\phi_{\frac{1}{2}}|c|}{2}u_j + \frac{c-\phi_{\frac{1}{2}}|c|}{2}u_{j+1} \tag{5}$$

This results in the following expression for ϕ associated with the interface $j + \frac{1}{2}$:

$$\phi_{\frac{1}{2}} = 1 - \frac{1}{2}\left(1 + \frac{c}{|c|}\right)\frac{\delta u_j}{\delta_+ u_j} - \frac{1}{2}\left(1 - \frac{c}{|c|}\right)\frac{\delta u_{j+1}}{\delta_+ u_j} \tag{6}$$

It is interesting to note that IGR method chooses the gradient from the upwind cell in the computation of ϕ. The above expression also suggests that if the gradients used in the computation of ϕ are monotonicity preserving then the resulting MCIR scheme will also be monotonicity preserving. However, at this stage it appears as though the projected advantage of IGR method in not requiring an explicit gradient computation is not satisfied as gradients are required even for the determination of ϕ. The above paradox is resolved by imposing an additional constraint that gradient finding associated with the IGR procedure should be limited to that needed for viscous flux computation. This is better explained in the case of 2D, presented in Figure 3, which depicts the determination of ϕ on the Kth interface between cells j and k. The expression for ϕ_K in 2D reduces to

$$\phi_K = 1 - \frac{1}{2}\left(1 + \frac{\vec{c}.\hat{n}_k}{|\vec{c}.\hat{n}_k|}\right)\frac{\delta u_{Kj}}{\delta_+ u_K} - \frac{1}{2}\left(1 - \frac{\vec{c}.\hat{n}_k}{|\vec{c}.\hat{n}_k|}\right)\frac{\delta u_{Kk}}{\delta_+ u_K} \qquad \text{with } \delta_+ u_K = u_k - u_j, \; \overrightarrow{r_{kj}} = \vec{r}_k - \vec{r}_j$$

$$\delta u_{Kj} = (\nabla u_{Kj} . \hat{n}_k)|\overrightarrow{r_{kj}}|, \qquad \delta u_{kk} = (\nabla u_{Kk} . \hat{n}_k)|\overrightarrow{r_{kj}}|$$

where, ∇u_{Kj} and ∇u_{Kk} are gradients computed on the left and right lobes of the covolume constructed around the Kth interface using Green-Gross procedure. $\hat{n}_k$ is the unit vector normal to the Kth interface.

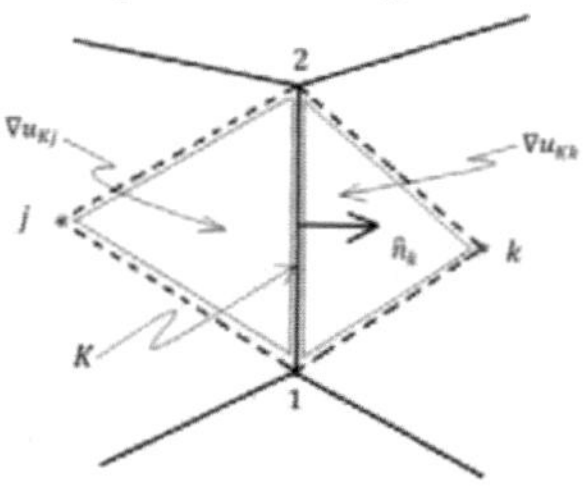

Figure 3. IGR procedure in 2D

It is evident from the above formula; computed gradients for determination of ϕ are essentially as needed for viscous flux computations. Also, all operations associated with the computation of ϕ are limited to the data associated with the corresponding face, which allows us to integrate the determination of ϕ with the face based procedure needed for the computation of both inviscid and viscous fluxes. Extension of the above procedure to Euler equations is rather straight forward, with its application to individual characteristic variables and the associated wave speeds. This would amount to recovering the gradients associated with the characteristic variables W from the gradients of the primitive variables using the following expressions for 1D and 2D, respectively:

$$(DW)_{1D} = \begin{bmatrix} \frac{dp-\rho c du}{2c^2} \\ -\frac{dp-c^2 d\rho}{2c^2} \\ \frac{dp+\rho c du}{2c^2} \end{bmatrix}, \qquad (DW)_{2D} = \begin{bmatrix} \frac{dp-\rho c dq_n}{2c^2} \\ -\frac{dp-c^2 d\rho}{2c^2} \\ \frac{dp+\rho c dq_n}{2c^2} \\ \rho dq_l \end{bmatrix} \quad \text{and} \quad \begin{matrix} q_n = u.n_x + v.n_y \\ q_l = -u.n_y + v.n_x \end{matrix}$$

where, ρ, u, v and p are density, velocity component in x-direction, velocity component in y-direction and pressure, respectively. Once the ϕ values associated with the individual characteristic variables are known, any flux formula like the one by Steger and Warming [4] or by Roe [5], can be easily modified to include the ϕ parameter associated with IGR. The IGR Roe flux formula (a modified Roe flux formula), in 2D is presented below:

$$\left(F_{IGR\_Roe}\right)_I = \frac{F_{\perp L}+F_{\perp R}}{2} - \frac{1}{2}\sum_i\left(\alpha_i\, \phi_i\, |\lambda_i|\, \vec{r}_i\right) \tag{7}$$

where, $F_{\perp L}/F_{\perp R}$ denotes the Euler flux normal to the interface. α_i and λ_i denotes strength and eigenvalue of ith wave, respectively, computed from Roe averaged matrix. $\vec{r}_i$ is Eigenvector corresponding to i^{th} wave.

For IGR Steger and Warming (IGR_SW) flux formula, positive and negative part of the eigenvalues are computed as $\lambda_i^{\pm} = \frac{\lambda_i \pm \phi_i |\lambda_i|}{2}$, and first order Steger and Warming flux formula is used with these eigenvalues. The eigenvalues are defined as $\lambda_1 = u.n_x + v.n_y = v_n$, $\lambda_2 = v_n + c$, $\lambda_3 = v_n - c$ then IGR_SW flux formula is given by

$$(F_{IGR\_SW})^{\pm} = \frac{\rho}{2\gamma}\begin{pmatrix} \alpha \\ \alpha u + c(\lambda_2^{\pm} - \lambda_3^{\pm})n_x \\ \alpha v + c(\lambda_2^{\pm} - \lambda_3^{\pm})n_y \\ \alpha\frac{u^2+v^2}{2} + cv_n(\lambda_2^{\pm} - \lambda_3^{\pm}) + c^2\frac{\lambda_2^{\pm}+\lambda_3^{\pm}}{\gamma-1} \end{pmatrix} \quad \text{where } \alpha = 2(\gamma-1)\lambda_1^{\pm} + \lambda_2^{\pm} + \lambda_3^{\pm}$$

3. Results

3.1 1D test case

The Standard test case of shock tube problem is selected to test the IGR methodology [6]. The diaphragm is located at x_0=0.5 with initial left and right conditions given in table 1. A total of 100 points are used for flow simulation. The Exact solution at time t=0.2 is compared with Roe 2nd order, Steger and Warming (SW) 2nd order, SW 1st order and IGR scheme which modified 1st order SW flux formulation. The density, velocity, pressure and internal energy comparisons are given in Figures 4(a), 4(b), 5(a) and 5(b), respectively. It is evident that IGR scheme improves 1st order SW solution significantly and are comparable to the 2nd order Roe and SW schemes.

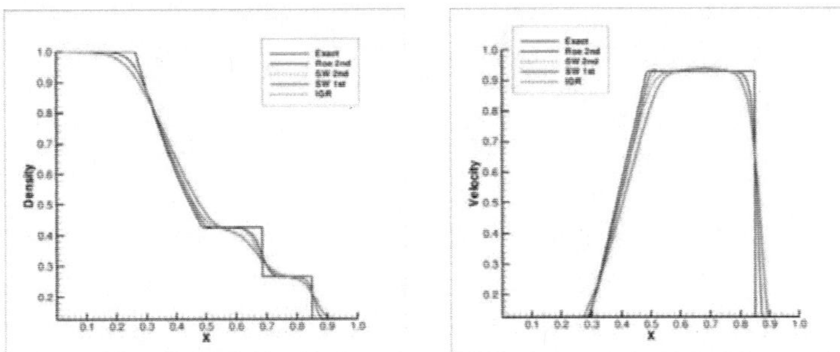

(a) (b)

Figure 4. Shock Tube problem (a) Density plot (b) Velocity plot

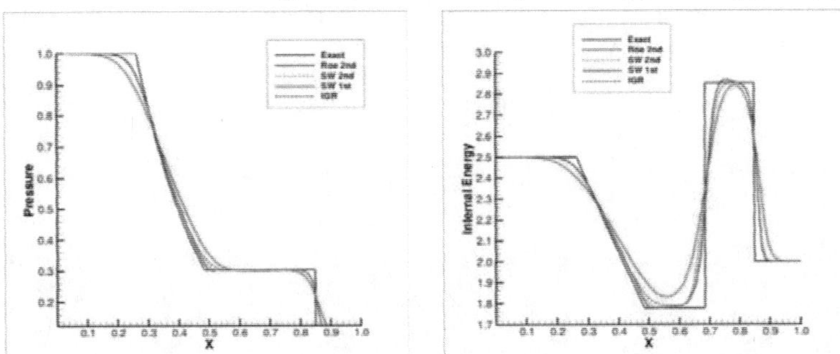

(a) (b)

Figure 5. Shock Tube problem (a) Pressure plot (b) Internal energy plot

Table 1. Shock tube initial condition

Parameter	Left State	Right state
ρ	1.0	0.125
u	0.0	0.0
p	1.0	0.1

3.2 2D Test cases

The IGR methodology have been implemented in HiFUN 2D solver using Roe and SW flux formulation and called as IGR_Roe and IGR_SW, respectively. HiFUN 2D is a finite volume based solver which has been extensively validated for variety of test cases (http://www.cfdcenter.aero.iisc.ernet.in). To demonstrate the capability of IGR methodology, we have selected 3 test cases for flow over NACA 0012 airfoil. The computations from IGR_Roe and IGR_SW are compared with conventional 2nd order Roe and SW schemes which use conventional gradient reconstruction to achieve 2nd order of spatial accuracy. An unstructured grid with 6950 triangular cells is used for all these computations.

3.2.1 Case 1: The first 2-D test case selected is inviscid flow computation over NACA 0012 airfoil at M=0.63 and α=2°. This is a standard test case which has been used in GAMM workshop [7]. The pressure coefficient (Cp) comparison is depicted in Figure 6(a). It is observed that IGR_Roe computation is matching very well with those from 2nd order Roe and SW. The Euler computations from IGR_SW is close to the 2nd order Roe and SW at all the locations except at the suction peak where it's slightly under-predicting it. The density residue ($\log_{10}$) plot is given in Figure 6(b). It is evident that IGR_Roe and Roe 2nd order residue pattern are very close to each other. The Cp contour plots are matching quite well with each other [Figures 7(a)-7(d)]. The comparison between lift coefficient (CL) and drag coefficient (CD) are given in table 2.

Table 2. 2D test case1 (M=0.63, α=2°)

Parameter	Roe 2nd order	IGR_Roe	SW 2nd order	IGR_SW	GAMM (1986)
CL	0.3071	0.3054	0.3048	0.2921	0.3335
CD	0.11E-05	0.29E-02	-0.95E-03	0.91E-02	0.0

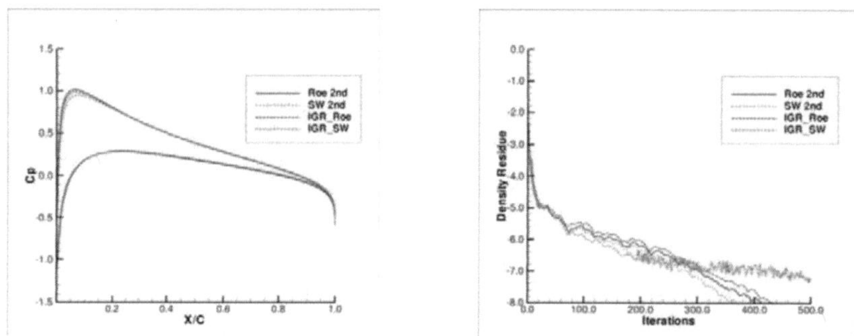

(a) (b)

Figure 6. Comparison between various schemes for Euler solution at M=0.63, α=2° (a) Cp plot (b) Density residue

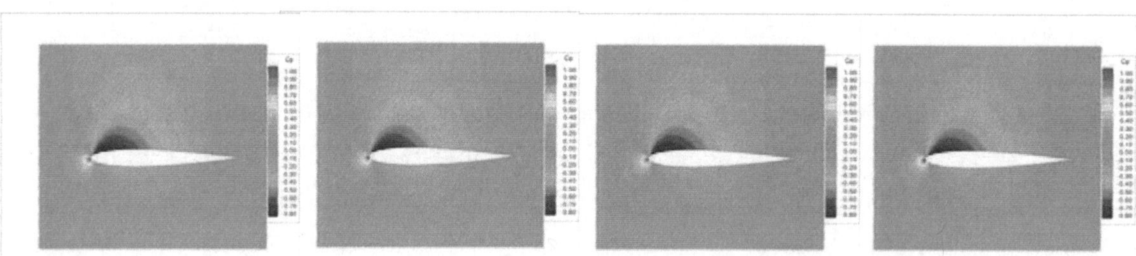

(a) (b) (c) (d)

Figure 7. Comparison of Cp contour for various schemes for Euler Solution at M=0.63, α=2° (a) Roe 2nd order (b) IGR_Roe (c) SW 2nd order (d) IGR_SW

3.2.2 Case 2: The second test case is standard AGARD test case [8] at transonic speed with M=0.85 and α=1°. In this test case, shocks wave appears on both upper and lower surfaces. Figure 8(a) shows the Cp comparison between various schemes. IGR_Roe computations compares quite well with 2nd order Roe and SW results. Figure 8(b) shows convergence plot of density residue. The Cp contour plots are given in figures 9(a)-9(d) for various schemes. Shock wave formation on lower and upper surfaces can be seen clearly in these figures. Table 3 summarizes the CL and CD comparison for this test case.

Table 3. 2D test case2 (M=0.85, α=1°)

Parameter	Roe 2nd order	IGR_Roe	SW 2nd order	IGR_SW	AGARD (1985)
CL	0.325	0.342	0.327	0.293	0.330-0.389
CD	0.0533	0.0564	0.0524	0.0573	0.0464-0.0590

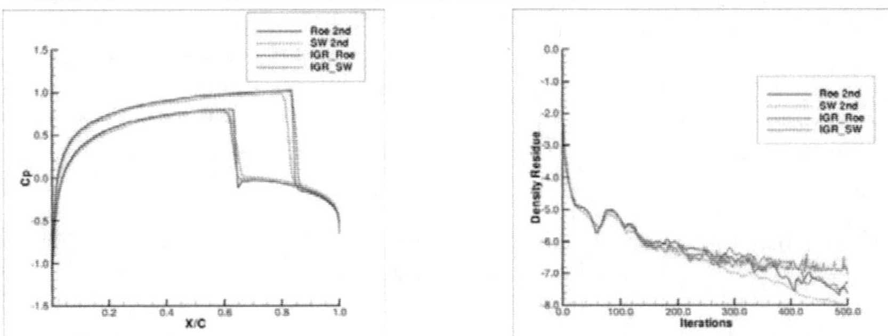

(a) (b)

Figure 8. Comparison between various scheme for Euler solution at M=0.85, α=1° (a) Cp plot (b) Density residue

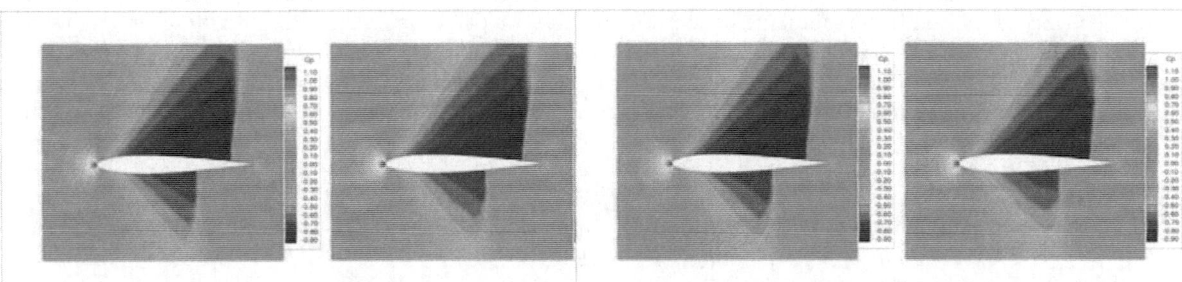

(a) (b) (c) (d)

Figure 9. Comparison of Cp contour for various schemes for Euler Solution at M=0.85, α=1° (a) Roe 2nd order (b) IGR_Roe (c) SW 2nd order (d) IGR_SW

3.2.3 Case 3: The laminar flow over NACA 0012 airfoil at M=0.5, α=3° and Reynolds number (Re) of 5000 is taken as 3rd test case. The Cp and skin friction coefficient obtained from IGR_Roe and Roe 2nd order computations are shown in Figures 10(a) and 10(b), respectively. These computations are compared with reference data available in reference [9]. The IGR_Roe and 2nd order Roe schemes are slightly under-predicting the suction peak. This is due to coarse unstructured triangular grid that we have used. However, there is good agreement between computations and reference data for skin friction coefficient. More details about pressure drag (CD_p), skin friction drag (CD_f), lift coefficient (CL) and separation location is given in table 4.

Table 4. 2D test case3 (M=0.50, α=3°, Re=5000)

	CD_p	CD_f	CD_{total}	CL	X_{sep} (% of chord)
Roe 2nd order	0.0336	0.0343	0.0679	0.0521	46.1
IGR_Roe	0.0347	0.0377	0.0724	0.0404	50.7
Venkatakrishnan(1990)	0.0262	0.0321	0.0589	0.0464	46.4

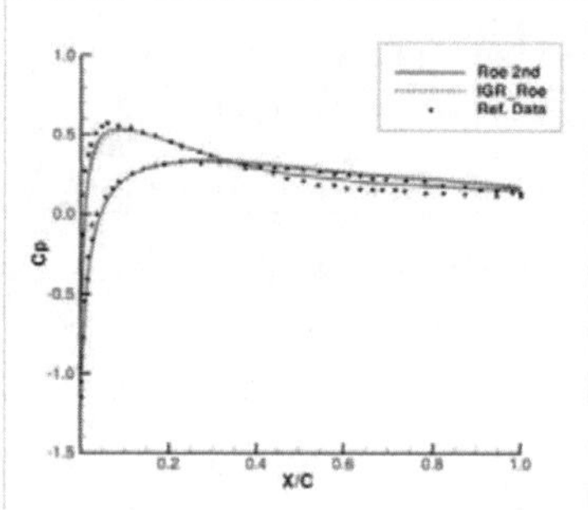

(a)

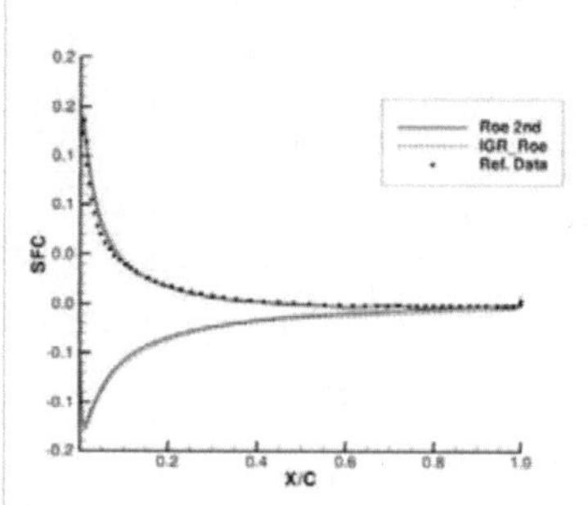

(b)

Figure 10. Comparison between various schemes for laminar flow at M=0.50, α=3°, Re=5000 (a) Cp plot (b) Skin friction coefficient

4. Conclusions

The work establishes an Implicit Gradient Reconstruction procedure modifying the classical CIR scheme using the ϕ parameter. The accuracy of the procedure is established by solving standard test problems in 1D and 2D. As the proposed procedure offers second order accuracy without the need to store solution gradients, the memory foot print is considerably small. This feature is expected to result in significantly enhanced cache utilization for large scale 3D problems on supercomputing platforms with several thousand processor cores, enabling super-optimal utilization of such platforms. In addition, the work also opens out a framework for moderating dissipation associated with individual waves for some of the well established flux formula. The present effort is going on in further fine tuning the ϕ parameter to make the IGR scheme operational for a larger class of problems.

References

[1] Ramesh V and Deshpande SM 2004 *Fluid Mechanics report,* **FM-20**, *Dept of Aerospace Engg., Indian Institute of Science, Bangalore, India.*
[2] Ramesh V and Deshpande SM 2007 *Computers and Fluids* **36** 1592
[3] Singh MK, Ramesh V and Balakrishnan N 2015 *Engineering Applications of Computational Fluid Mechanics* **9** 382
[4] Steger JL and Warming RF 1981 *Journal of Computational Physics* **40** 263
[5] [Roe PL 1981 *Journal of Computational Physics* **43** 357
[6] Sod GA 1978 Journal of Computational Physics **27** 1
[7] GAMM workshop 1986
[8] Viviand H 1985 *AGARD AR-211*
[9] Venkatakrishanan V 1990 *Computers and Fluids* **18** 191

Influence of furnace cross section on the flow and velocity distribution in tangential firing furnace

A F Zainudin, H Hasini and S S A Fadhil

Centre for Fluid Dynamics, Universiti Tenaga Nasional, Jalan IKRAM-UNITEN, 43000 Kajang, Selangor, Malaysia

Abstract. This paper presents an investigation on the influence of different furnace cross sectional configurations on the flow and velocity distribution in a small-scale tangential firing furnace. The flow and combustion of natural gas in three furnaces of different cross section are simulated numerically using CFD. The results show that velocity distribution is non-uniform along the width and depth of the furnace in rectangular furnace while square cross section gives uniform distribution. The effect of furnace cross section on the tangential velocity is more obvious when the length difference of the width and depth of the furnace is larger. The axial velocity is found to be highly influenced by the furnace cross section.

1. Introduction

Tangential firing boiler is the most widely used type of boiler in the world. This is due to its advantages of uniform heat distribution in the furnace and low NO_x emission which imply higher efficiency and less pollution. However, most tangential firing boilers have been experiencing steam temperature imbalance problem which has been long associated with this type of boiler. Earlier investigations on this problem suspect that the root cause is the residual swirl at the upper furnace [1-3]. The residual swirl causes flue gas velocity deviation between the left side and right side of the upper furnace which consequently results in temperature imbalance between the two sides. Excessive steam temperature on one side will shorten the usable life of the heat exchanger tubes and may eventually cause tube failure.

In the light of this issue, there have been many researchers who investigated the occurrence in the boiler utilizing the Computational Fluid Dynamics (CFD) technology [4-12]. Some of the researchers studied the influence of selected parameters on the flow and temperature imbalance occurrence in the boiler. Among the parameters which have been studied are the burner tilt (vertical) angle [7], yaw (horizontal) angle [4, 11], arrangement of the platen superheaters [1, 10], and the structure of the nose region [2]. Up to this point, no literature has been found which study on the effect of the furnace cross section on the flow and velocity distribution in the boiler. Therefore, the main objective of this study is to investigate the influence of different furnace cross section on the flow and velocity distribution in a small-scale tangential firing furnace. CFD simulation was performed on natural gas-fired tangential firing furnaces with different cross sections and the effect on the flow and velocity distribution was observed. The findings from this study may be useful for further investigation in the real full-scale boiler in order to alleviate the temperature imbalance problem.

2. Furnace Description and Case Studied

The furnace studied is a small-scale tangential firing furnace with constant cross section along its height. The typical upper part of a tangential firing boiler, starting from the nose region until the boiler exit is removed for simplicity and to reduce the computational time. In addition, for this study, only the constant-cross-section lower part of the furnace is concerned in order to exclude the renowned effect of the upper furnace structures on the flow. The furnace geometry is shown in figure 1. The geometry is constructed such that its depth, width and height are along x-, y- and z-axis, respectively. The furnace comprises two regions, namely windbox region and furnace region, as illustrated in figure 2. In the windbox region, there are a total of 28 injection ports, 7 ports being at each corner. The 7 ports consist of 3 fuel (F), 2 primary air (PA) and 2 secondary air (SA), arranged as shown in figure 3. The fuel and air are injected tangentially into the furnace through these ports. The reaction between fuel and air produces combustion, forming counter-clockwise swirling fireball which flows upwards towards the furnace exit. As the flue gas rises, heat is transferred to the water walls.

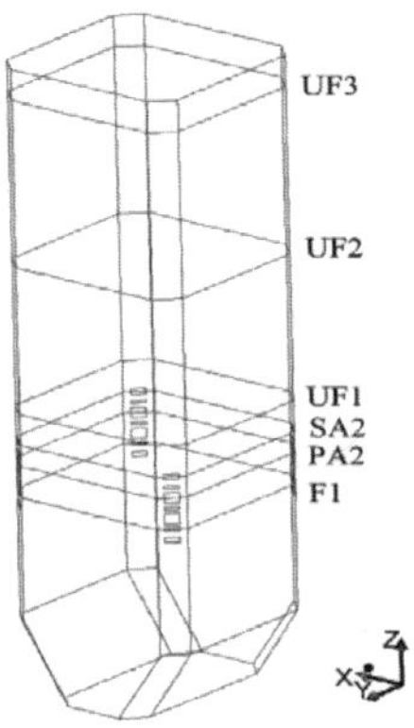

Figure 1. Furnace geometry with the planes selected for analysis.

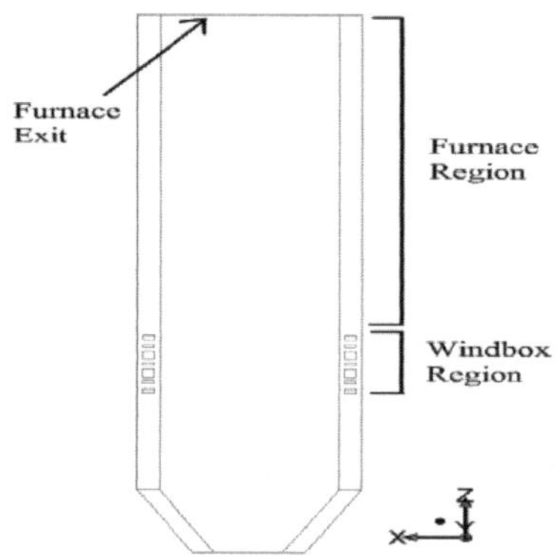

Figure 2. Regions in the furnace.

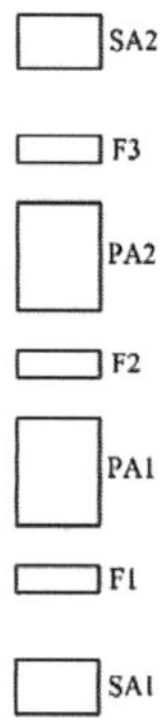

Figure 3. Ports arrangement at each corner in the windbox region.

For the purpose of comparing the influence of furnace cross section on the flow and velocity distribution, three cases were studied. The description of the cases is summarized in Table 1. The furnace with an 8 × 8 cross section is chosen as the base case because it is expected that in a furnace with square cross section (equal width and depth), the flow will be perfectly symmetrical. The depth and height of the furnace are maintained at the length of 8 m and 30 m, respectively for all the cases. To maintain the consistency between all the cases, the injection angles of fuel and air are set such that the imaginary circle formed at the centre of the furnace has the same diameter in all the three cases. The resulting injection angles for all the cases are displayed in figure 4. Six planes at different elevations are selected for the purpose of analysing the results. Three planes are in the windbox region, located at 6 m (F1), 7.5 m (PA2), and 8.5 m (SA2) elevations, respectively. While the other three planes are in the furnace region, located at 10 m (UF1), 17 m (UF2), and 25 m (UF3) elevations, respectively. These planes are shown in figure 1.

Table 1. Summary of Cases Studied.

Case (Furnace)	Cross Section (Width × Depth)
A (Base)	8 × 8
B	7 × 8
C	6 × 8

3. CFD Models

The numerical simulations for this study were carried out using CFD commercial code – FLUENT 14.5. The mesh, models and boundary conditions used are consistent for all the three cases. For each of the cases, the flow domain comprises a total of 1032608 hexahedral grids. The mesh scheme applied on the horizontal cross section throughout the furnace is shown in figure 5. The fuel used is natural gas, modelled as 100% CH_4 by mass. Air consists of 79% nitrogen and 21% oxygen by mole. The air-fuel ratio set for the simulation is 20.64, which corresponds to 20% of excess air. The realizable k-ε model is used to model turbulent flow in the furnace. Turbulent intensity and viscosity ratio are specified at the inlets and outlet, with the value of 5% and 10, respectively. Detail of inlet boundary conditions imposed on the simulation is shown in table 2. The turbulence-chemistry interaction is solved using Eddy-Dissipation model. Heat transfer to the walls is calculated based on fixed wall temperature of 800K whereas radiation heat transfer is solved using P1 gray radiation model. The emissivity of the furnace wall is set to 0.7 [13] whereas the weighted-sum-of-gray-gases model (WSGGM) is used to calculate the emissivity of the flue gas.

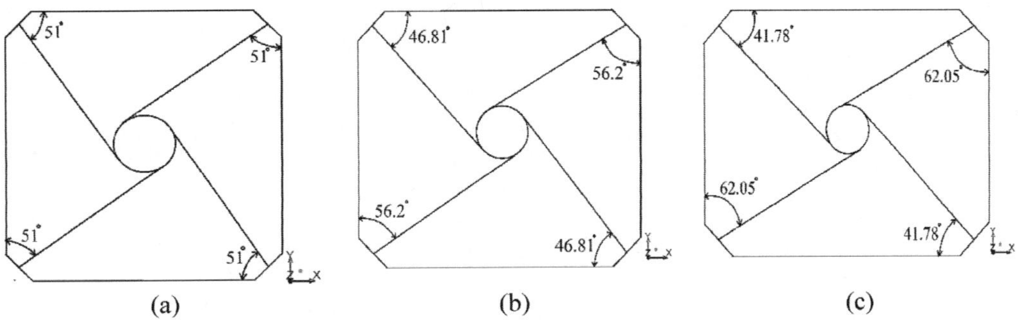

Figure 4. Injection angles of fuel and air for (a) Case A, (b) Case B, (c) Case C.

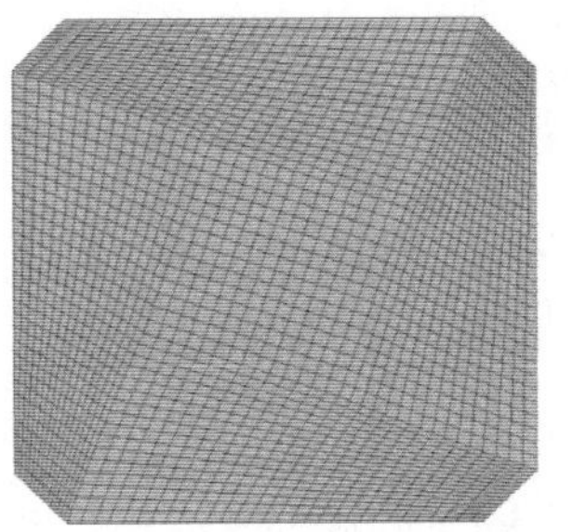

Figure 5. Mesh of the horizontal cross section throughout the furnace.

Table 2. Inlet boundary conditions at each port.

Parameter	Primary Air (PA)	Secondary Air (SA)	Fuel (F)
Mass flow rate (kg/s)	10.32	5.16	0.5
Temperature (K)	500	500	300

4. Results and Discussions

Generally in all the cases, the flow is the most intense in the windbox region where the mixing and reaction of fuel and air take place. Swirling fireball in counter-clockwise direction is formed at the centre of the furnace. In the windbox region and lower part of the furnace region, tangential velocity is dominant whereas at higher elevations in the furnace region, the flow is dominated by the axial velocity (z-component) as the swirl strength has weakened. This is evident from the velocity vectors shown in figure 6. As the flue gas flows upwards, most of the energy is transferred to the walls thus its velocity decreases. At higher elevations, velocity magnitude is higher near to the walls. This implies that as the height increases, the flue gas flows outward towards the walls and this enlarges the swirl diameter. From the velocity vectors, it can also be seen that the swirl remains symmetrical at the centre along the height of the furnace. Since the furnace has constant cross section along its height, the flow is not interrupted.

In case B and C, it was found that the tangential and axial velocity magnitudes along the width are not the same as those along the depth. In case A, the magnitudes are the same along both directions for tangential velocity as well as axial velocity, which can be attributed to the square (equal length of width and depth) cross section of the furnace. Based on these observations, it can be deduced that rectangular (unequal length of width and depth) cross section of the furnace results in non-uniform distribution of tangential and axial velocity between the width and the depth.

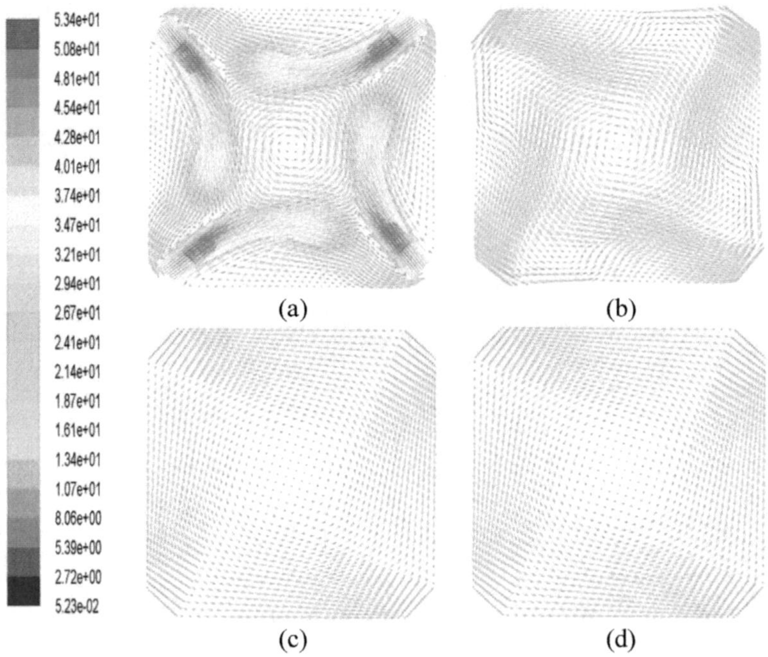

Figure 6. Velocity vectors on planes (a) PA2, (b) UF1, (c) UF2, and (d) UF3 for case A. Generally for case B and C, the pattern is the same, which is elaborated in the discussion.

4.1. Comparison of Tangential Velocity

The tangential velocity of the flue gas is compared between the three cases separately in two directions. The comparison is first made for the tangential velocity along the width followed by the tangential velocity along the depth of the furnace. For the first comparison, the graph in figure 7 is referred. There is not much difference between case A and case B but there is significant difference between case C and the former two cases. The magnitude of tangential velocity in case C is significantly higher than the other two cases. Based on this comparison, it can be expected that if the width of the furnace is made narrower, the tangential velocity along that direction will increase. This suggests that as the difference of length between the width and the depth is larger (narrower rectangle), the effect of furnace cross section on the tangential velocity becomes more significant.

For the tangential velocity along the depth, there is no significant difference between all the cases. The tangential velocity distributions in all the cases are almost identical, since the depth in all the cases are kept constant.

4.2. Comparison of Axial Velocity

Figure 8 shows the plot of axial velocity along the width at selected planes for all the cases. Based on the plot, it can be seen that the axial velocities for case C are obviously higher than those for the other two cases. The axial velocities for case B are also significantly higher than those for case A. Similarly for the axial velocity along the depth, the highest axial velocity is found in case C, followed by case B, and case A, as shown in figure 9. This trend clearly indicates that the smaller cross section produces higher axial velocity magnitude. The reduction in cross-sectional area forces a relatively large portion of the flue gas to flow upwards in the furnace, making the axial velocity component larger. Since the fuel and combustion air are injected horizontally (no tilting), the flue gas will tend to flow mostly in the horizontal plane first before flowing upwards due to the effect of buoyancy. But when the area becomes smaller, some portion of the flue gas will be forced to take the vertical path.

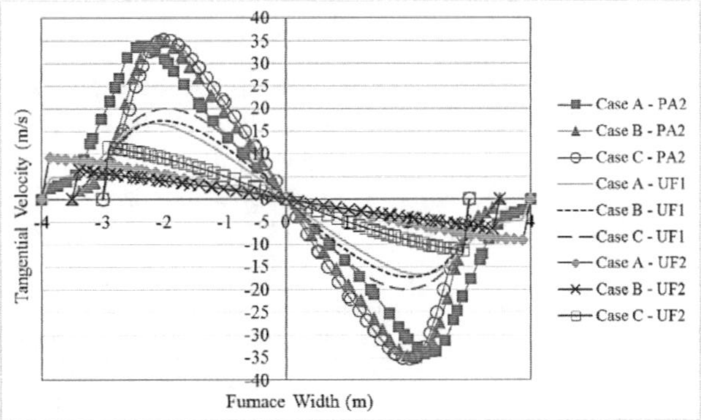

Figure 7. Tangential velocity plot along furnace width for all cases.

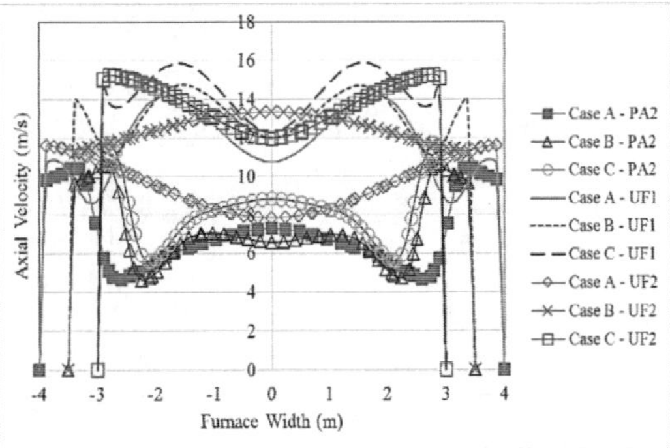

Figure 8. Axial velocity plot along furnace width for all cases.

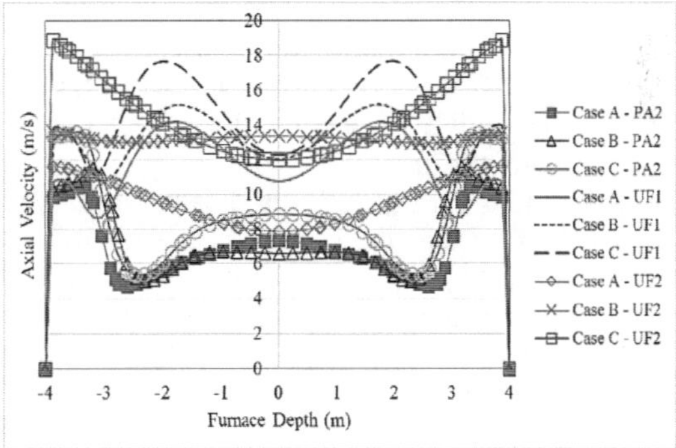

Figure 9. Axial velocity plot along furnace depth for all cases.

5. Conclusion

CFD simulations of flow and velocity distribution in boiler furnace of different cross-sections were carried out. The influence of different furnace cross sections on the flow characteristics in tangential firing furnace were numerically investigated. In general, the flow in the furnace is intense in the windbox region due to swirl which is caused by the tangential firing system. The intensity deteriorates as the combustion gas flows upwards towards the furnace exit. At higher elevations, axial velocity is dominant while the tangential component of velocity weakens. Square furnace cross section produces uniform velocity distribution along the depth and width of the furnace, whereas rectangular cross section causes the tangential and axial velocity components to be non-uniform. The effect of furnace cross section on the tangential velocity is more significant in narrower rectangle, with higher magnitude along the narrower direction. The axial velocity component is highly influenced by the furnace cross section. The smaller the cross-sectional area, the higher the magnitude of axial velocity in the furnace. This can be attributed to the nozzle effect that forces the velocity to increase in smaller cross-sectional flow area.

Acknowledgement
The authors would like to acknowledge the financial support received from the Ministry of Education Malaysia through the Science Fund (06-02-03-SF0273).

References

[1] Yin C, Rosendahl L and Condra T J 2003 Further study of the gas temperature deviation in large-scale tangentially coal-fired boilers *Fuel* **82** 1127–37

[2] Lee Y-P, Xu J-Y, Zhang Q, Gu F and Xu Y-Q 2001 Investigation on residual swirl and flue gas imbalance in large tangentially fired boiler *JSME Int. J.* **44** 378–387

[3] Xu M, Yuan J, Ding S and Cao H 1998 Simulation of the gas temperature deviation in large-scale tangential coal fired utility boilers *Comput. Methods Appl. Mech. Eng.* **155** 369–380

[4] Park H Y, Baek S H, Kim H H, Kim Y J, Kim T H, Lim H S and Kang D S 2016 Reduction of main steam temperature deviation in a tangentially coal-fired, two pass boiler *Fuel* **166** 509–516

[5] Tang G, Wu B, Johnson K, Kirk A, Fu D and Zhou C Q 2016 Numerical study of a tangentially fired boiler for reducing steam tube overheating *Appl. Therm. Eng.* **102** 261–271

[6] Ahmad A, Hasan H, Kannan M, Noor N A W M and Lim M T 2015 Numerical study on the imbalanced flue gas temperature at boiler rear pass area of a tangential-fired coal power plant *J. Mech. Eng. Autom.* **5** 12–18

[7] Tian D, Zhong L, Tan P, Ma L, Fang Q, Zhang C, Zhang D and Chen G 2015 Influence of vertical burner tilt angle on the gas temperature deviation in a 700 MW low NOx tangentially fired pulverised-coal boiler *Fuel Process. Technol.* **138** 616-628

[8] Yang M, Shen Y Y, Xu H T, Zhao M, Shen S W and Huang K 2015 Numerical investigation of the nonlinear flow characteristics in an ultra-supercritical utility boiler furnace *Appl. Therm. Eng.* **88** 237–247

[9] Constenla I, Ferrín J L and Saavedra L 2013 Numerical study of a 350 MWe tangentially fired pulverized coal furnace of the As Pontes Power Plant *Fuel Process. Technol.* **116** 189–200

[10] Zhou Y, Xu T, Hui S and Zhang M 2009 Experimental and numerical study on the flow fields in upper furnace for large scale tangentially fired boilers *Appl. Therm. Eng.* **29** 732–739

[11] Zhou Y, Zhang M, Xu T and Hui S 2009 Effect of opposing tangential primary air jets on the flue gas velocity deviation for large-scale tangentially fired boilers *Energy and Fuels* **23** 5375–82

[12] Belosevic S, Sijercic M, Tucakovic D and Crnomarkovic N 2008 A numerical study of a utility boiler tangentially-fired furnace under different operating conditions *Fuel* **87** 3331–38

[13] Yin C 2013 Refined weighted sum of gray gases model for air-fuel combustion and its impacts *Energy and Fuels* **27** 6287-94

Investigation of flow inside pulverized coal (PC) pipes against coal particle size and air flow rate for a utility boiler

Savithry K. Thrangaraju, Kannan M. Munisamy, Rucgnes Apparao
Universiti Tenaga Nasional, Jalan IKRAM-UNITEN, 43000 Kajang, Selangor, Malaysia

E-mail: savithry@uniten.edu.my

Abstract. This study mainly focuses on investigation on effect coal particle size on flow inside pulverized coal pipes for a utility boiler. The flow that is analyzed here will be the wall shear stress. The objective of the study is to determine the effect of coal particle size and effect of air flow rate on the wall shear stress of pulverized coal pipes. The individual wall shear stress which is computed as area weighted average of the pulverized coal pipes that is studied is compared and analyzed. There are total of 28 pulverized coal pipes in the power plant that is chosen as case study. The study is divided into two parts mainly the effect of coal particle size on the wall shear stress and the other is effect of air flow rate on the wall shear stress. In both configurations the wall shear stress is computed as area weighted average. Prior to simulation the file that is used to study the wall shear stress is modified using Gambit to improve results. The individual wall shear stress of all pipes coming out of a particular mill is observed after contours are developed using CFD tool like Ansys fluent. Parameters like coal flow rate and coal velocity are set in the simulation and results are generated Based on contours and developed graph from the simulation, the effect of both configuration is studied and the range of particle size and range of air flow rate which is suitable for the optimum operation of boiler is suggested.

Keywords: Wall shear stress, pc pipes, coal particle size, air flow rate.

1. Introduction

Major percentage of electrical energy is produced through combustion of coal in many thermal power plants. Tangentially fired coal fired boilers are the mostly used type of boiler in the power generation [1]. The coal combustion process in thermal power plant consists of phenomena such as turbulence, heat transfer by convection and radiation, and chemical reactions. Now we need to see how a typical coal fired power plant operates. First the coal from the stock yard is sent to mill to pulverize into powder. The pulverized coal is carried into boiler through the pulverized coal pipe by the primary air and a blower is used to blow the heated air. The combustion takes place and a fireball forms due to the tangential firing of coal and fuel oil at four different corners. The economizer is the part in which the feed water is reheated by waste heat released from the exhaust gases exiting the boiler. This is a waste heat recovery to increase efficiency. The fireball of burning coal in the boiler will heat up the surrounding water walls that flows up the tubes.

The water flowing in the tubes absorbs heat released by heat flux and radiation from the fireball to the tubes. Partially, the water in the tubes are converted into steam. Then the mixture water and the steam, which is still not fully heated for turbine conditions in the tubes are transferred to the steam drum at top of the furnace. In the boiler drum, separation occurs as the warm feed water flows to the steam drum, whereas the liquid water flows in the down comers on the outer part of the boiler and then to the bottom of the boiler. The steam from the steam drum is transferred to the primary super heater in which the steam temperature is raised to extremely high temperatures and then transferred to the secondary superheaters. At the final superheaters, the hot flue gases flow (product of combustion) on the outer surface of the tubes which causes the steam to be heated to its desired temperature and pressure before it exit from the boiler heading towards turbine. The high pressure and temperature steam contains high amount of enthalpy in which it is converted to mechanical energy by rotating the turbine around 3000rpm. This rotational energy is converted to electrical energy in the generator. The steam from the high pressure turbine will be sent back to reheated region of the furnace to be reheated. This reheated steam will be delivered to intermediate turbine. The main steam that exits the high pressure turbine is reheated and known as reheated steam. The combustion energy from the coal will be transferred as heat energy to the water in the boiler. There will be condensation of steam due to transfer of enthalpy from steam to mechanical energy of rotating turbine, the steam experiences drop in temperature. The condensed steam from the low pressure turbine will be converted to liquid water. This liquid water undergoes the same cycle just like feed water heater at the economizer for the next cycle of steam heating. In order to increase the waste heat recovery, which in turn increase the overall efficiency of the boiler, there are high pressure and low pressure heaters are attached to reclaim heat energy and increase thermal efficiency of the plant. A typical power plant consist of several major parts. One of them is a mill in which the coal is grind into smaller particles and injected into boiler through airflow. In this study there are seven mill included to study effect of coal flow and air flow rate on the flow of each pipe. Each mill has four pipes entering the boiler at four different corners. This is important to enable the tangential firing that creates the fireball. The coal need to be grinded into smaller particles so that the surface area of the coal particles increase and there are more chances of complete combustion. Incomplete combustion can actually cause emissions like NOx and Sox [2, 3, 11] that is harmful the environment. Coal that is grind into powder form is actually easy to be carried by higher flow rate of air into the boiler. Once the coal is injected, it travels along with air along pulverized coal pipes into boiler. The surface area and of the all the pipes differ from each other due to the arrangement of the mill. Once entered the boiler the coal is burn to produce heat energy that heat up the water entering, and converting them into high pressure and temperature steam that rotates the turbine to produce the electrical energy. A typical power plant is divided into many systems namely air system, coal system and ash system. Ash is formed when coal is burned. The ash that is produced is divided into two types, fly ash and bottom ash. Fly ash is trapped in either flue gas desulphurization unit or at times the electrostatic precipitator. The flue gas is the combustion gas that is excreted out through a chimney. Bottom ash is trapped at the bottom of boiler. The correct amount of air is essential as it can cause incomplete combustion. The air system has primary air and secondary air. A research was done to accumulate the available information that is related to the current study. The finding focuses on effect of coal particle size on boiler performance, analysis of wall shear stress and flow inside pipes, thermal analysis of utility boiler and combustion pollution control of utility boiler. An investigation about online monitoring of coal particle size and flow distribution was done in coal-fired power plants. According to finding the fineness of the coal particles and the uniform distribution of diameter of the coal flow sent to the burners are important parameters to attain an effective and also a combustion in coalfield power plants. The authors of published paper studied the effect of a varying centrifugal classifier speed on coal flow distribution. Their finding concludes that the coal flow distribution towards the boiler was significantly improved when the particle size was the smallest. They also managed to find that the maximum deviation from the average flow rate was reduced from 14% at 73 rpm to 9% at 99 rpm [4].

A fundamental study the transport of floating particles was done. It as was about investigation on the transport of fluids composed of water and polypropylene particles. The results are displayed in relation to flow patterns; and the experimental results suggest that the size of the particles has a very small impact on the transport efficiency [5]. Besides, the flow rate and the volumetric concentration have strong and significant impact on transport efficiency, because of their large contribution to the flow pattern. The best operating conditions are close to the limit deposition velocity, when the fully suspended pattern is reached. The solids with high velocity are then well transported with no waste of energy. The optimum concentrations, for a wide range of high efficiency are around 20–25 %vol. The effects of the particle size on combustion characteristics of pulverized coals under O_2/N_2 and O_2/CO_2 atmospheres were studied and compared. The findings summarize four results. Firstly, effect of the particle size on the combustion of pulverized coal is greater than the effect of O2/N2 atmosphere. The effect of the particle size on combustion characteristic differs significantly betweenO2/N2 and O2/CO2 at the low O2 concentration. The changes of combustion characteristic fluctuate with the O2 concentration during O2/CO2 atmosphere. The O2 diffusion rate is the main reason of combustion in the O2/N2 condition, while the product release rate has a significant influence on combustion in the O2/CO2 atmosphere. The fluctuation of combustion characteristic is related to the rate of reaction. A particle which is big in size have lower rate of reaction due to the fact that the exposed surface area is small [6]. Analysis of wall shear stress around a competitive swimmer using 3D Navier–Stokes equations in CFD was done by C.V Popa and H Zaidi [7]. The authors of the paper studied the wall shear stress around a competitive swimmer during underwater glide phases occurring at the beginning and at every turn. The effect of different types of head position studied against wall shear stress. The head positions are namely lifted up, aligned and lowered. Additionally the static pressure distributions is analyzed. The velocity of swimmer are fixed at three different rates which are 1.4, 2.2 and 3.1 m/s. The authors also studied the effect of head configuration on wall shear stress under constant velocity. An important assumption in this study is the flow around the swimmer is turbulent. Their findings show that the wall shear stress increases with the velocity of the swimmer. A study on Computational fluid dynamics (CFD) analysis of single- phase and two phase flow was done. This study was performed in a 90 degree horizontal to vertical elbow with 12.7 mm inside diameter. In order to investigate the effects of different phases, three different combinations consisting of varying velocity of air and water is incorporated. Pressure and velocity profiles at six locations displayed a hike in pressure at the elbow geometry with decreasing pressure as fluid leaves from the elbow. Similar pattern of pressure drop observed for both single-phase and multiphase flows. Comparison of CFD results with available empirical models showed acceptable range of deviation. When the velocity increases the pressure drop increases. This conclusion was made and was applicable for single phase and multiphase [8]. Next X Yan, made a study on Modeling of the Flow within Scaffolds in Perfusion Bioreactors. Although the experimental study over a bioreactor is very complicated and almost impossible, the usage of cfd software enabled him to study the wall shear stress on perfusion bioreactors against different flow rates of fluid. The finding of this study will allow for better control strategies in scaffold fabrication and cell culturing experiments. His finding concludes that the effect of flow rate is significant on wall shear stress on the bioreactor. Flow rate is a controllable parameter in the cell culture process. The higher he flow rate the higher the wall shear stress on bioreactor. The flow rate and walls hear shows a linear relationship. The results also proves that the shear stress value is also affected significantly by the value of the strand diameter and horizontal span [9]. One of the most important elements in electricity generation is the balanced flow of coal and air. This balance need to be achieved at four corners of firing so that a fireball can be created. If the coal and air flow is not in the proper balance, we may face fouls like non uniform combustion in the furnace which is undesired. The coal and air balancing in a power plant is done using CFD. Devices like orifice are used to balance the coal air flow because not all the pipes are same in size and symmetrical. Therefore it is important to place orifice at right point. There was an estimation the pressure drop coefficients with both clean air and coal/air flows in order to select the appropriate size of the orifices. The results implies that the pressure drop is heavily dependent on the piping system geometry [10].

2. Methodology

There were two different configurations namely configuration 1 and configuration 2. Configuration 1 involves the study of wall shear stress against coal particle size under constant air flow rate. The second configuration involves the study of wall shear stress against air flow rate under constant particle size.

IOP Conf. Series: Journal of Physics: Conf. Series **822** (2017) 012032 doi:10.1088/1742-6596/822/1/012032

The current study regarding wall shear stress requires full knowledge on how these pc pipes are arranged. As mentioned above, there are seven mill with each mill contains four pipes heading to boiler at four different corners. This is important to enable tangential firing. All four pipes are not identical, as they have different length and area. There are total of 28 pipes available, but not all pipes are studied. For simplicity only pipes of mill 1, mill 2 and mill 7 is studied. The reason for the choice is that mill 1 and mill 2 are at opposite ends and mill 7 is in the middle. So choosing these three mills is appropriate because an overall conclusion can be made regarding all mill and pc pipes. A few figures describing the pipes and the arrangements of pipes are shown below.

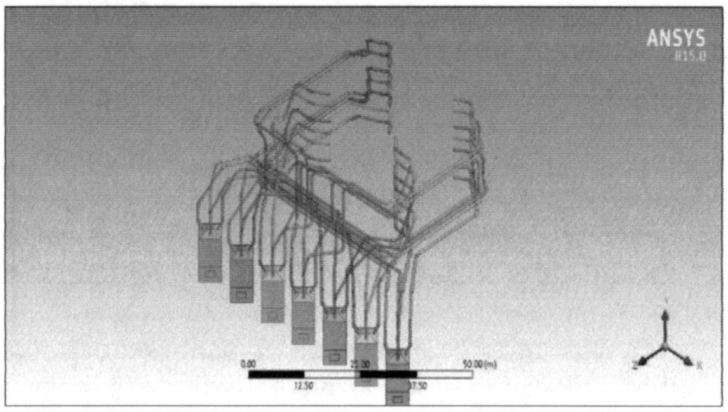

Figure 1. Arrangement of pc pipes leading to boiler

As we compare, the first pipe from mill 1 as shown in figure 3 and the first pipe from mill 2 shown in figure 4 is not same and completely different in geometry and arrangement. Same applies to all other pc pipes. Every pc pipes have different dimensions from each other.

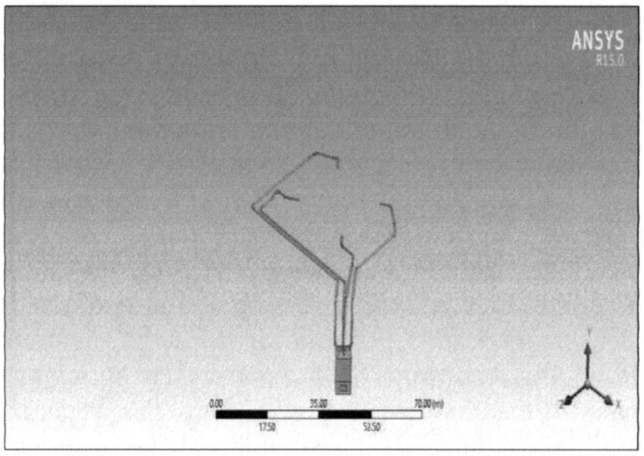

Figure 2. Mill 1 with all four pipes

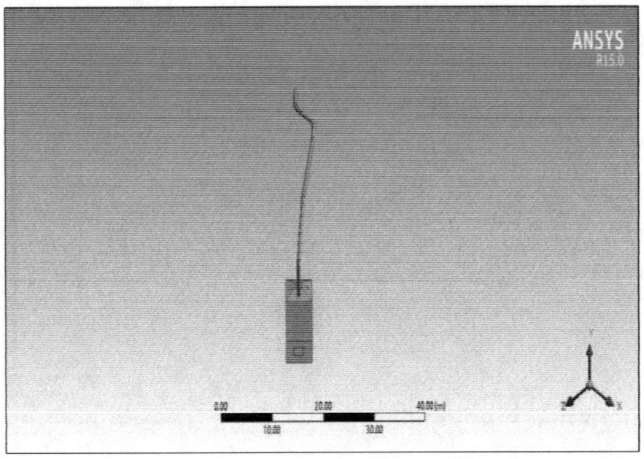

Figure 3. Mill 1 with pipe 1 only

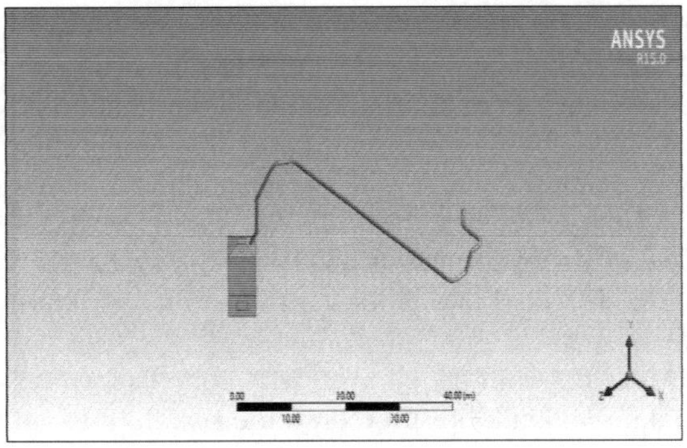

Figure 4. Mill 2 with pipe 1 only

There are total of 12 pipes studies (4 for each mill and 3 mills studies), but only few figures are shown here to illustrate the arrangement and construction of pc pipes of a utility boiler.

After the walls were defined individually, the mesh was imported to Ansys fluent software. The mesh independent study was done to select appropriate mesh configurations against result accuracy. To enable the simulation to be faster, the mills were studied one by one. For instance, in order to study the wall shear stress in mill1 which is named as mill10, the other mills were removed from the file. This will enable the simulation to be faster. Although other mills are removed, it does not affect the accuracy of the results because by default setting, the mills operate independent of other mills. As mentioned earlier there were two configurations were carried out. Both configuration was run in Ansys fluent.

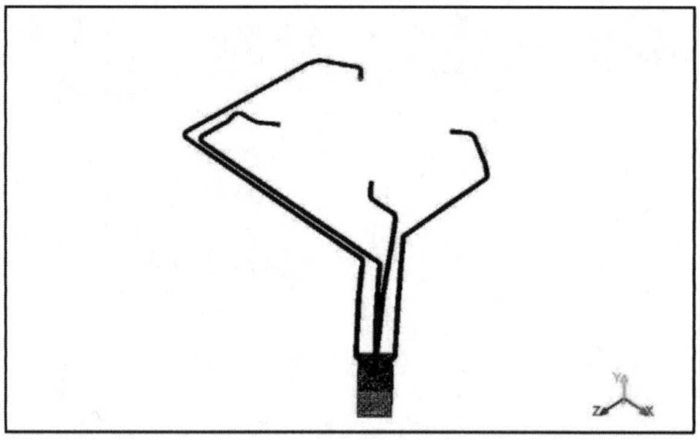

Figure 5. A screenshot of fluent showing mill1 before simulation and all other mills have been removed for faster simulation

2.1 Running configuration 1: Constant air flow rate with varying coal particle size

To achieve the objective which is to find out the relationship between coal particle size and the corresponding wall shear stress, the range of particle size to be examined are from 70 microns to 500 microns. This is the typical range of coal particle size globally. Under injections the coal flow rate is set to 12 kg/s and the coal particle size is set 70 microns. The air flow rate is set at 64 kg/s. The simulation of configuration 1 only involves changing coal particles size. Others parameters are kept constant. A turbulence model of K-Epsilon was used. The wall of pc pipes are set to be stationary wall and the material is aluminum. The coal that is injected is known as coal-mv in fluent software. The velocity of coal entering the mill is set to be 50m/s. Coal is release from surface of coal flow inlet. The outlet of the mill is pressure outlet. After the simulation is complete within acceptable range of error, the CAS file is saved with dat file. Under result report, the surface integral option is selected and the wall shear stress of the pc pipes is computed as area weighted average. The wall shear stress is computed in Pascal (Pa). Then the entire procedure is repeated with 100microns, 300 microns and 500 microns. A graph of wall shear stress against coal particles size is plotted and the trend is observed.

2.2 Running configuration 2: Constant particle size with varying air flow rate

Second configuration is done at coal flow rate of 12kg/s and coal particle size of 70micron. The airflow rate is varied from 64kg/s, 74kg/s, 84kg/s, and 94kg/s. Other parameter and setting were same as configuration 1. For this configuration only mill 1 and mill 2 is studied. Throughout the simulation the coal flow rate is set to 12 kg/s. The reason behind this is that increasing coal flow rate will cause more NOx and SOx emissions. Furthermore increasing coal flow rate increase overall coal consumption. Coal is a non renewable energy, so the usage should be limited. The wall shear stress is computed using the same procedure as configuration 1.

Once both configuration are done, graphs of wall shear stress against coal particle size is plot for configuration 1 for all three mills. Then the graphs of wall shear stress against air flow rate is plotted. Trends of both graphs are analysed.

3. Results and discussion

For configuration 1, the air flow rate is fixed at 64kg/s and coal is injected at various size ranging from 70 microns to 500 microns. The graph below show the coal particle size and the corresponding wall shear stress that is computed as area weighted average.

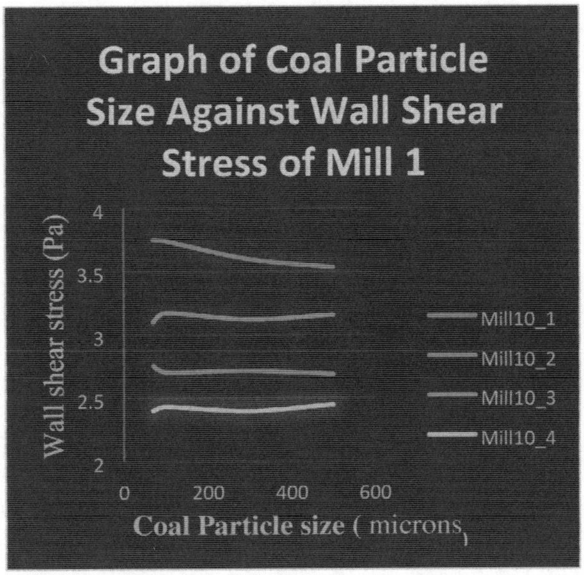

Figure 6. The graph of wall shear stress against coal particle size for mill 1

Based on the graph, the wall shear stress value for all of the four pipes of mill 1 does not show a significant change and almost constant throughout the flow. There is a slight drop can be observed for Mill10_1 which is the pipe 1 of mill 1, in which the wall shear stress reduces as the coal particle size increases. Since the coal are driven by the air flow, the velocity of coal is dependent on air flow rate. For the same amount of force, the smaller object with smaller mass tend to move faster than a bigger object with bigger mass. Same happens within pc pipes with coal particles. Higher the velocity higher the wall shear stress [7,9]. The force here is the air flow. Since air flow is fixed at 64kg/s, smaller particles of coal moves faster with higher velocity compared bigger particles. This is the reason why small coal particles at 70 microns have higher wall shear stress than other particle size like 300 microns and 500microns in mill10_1. For other pipes, for instance mill10_2, the trend line is almost constant. Nevertheless, for mil10_2, the highest wall shear stress is at 70 microns at a value of 2.75 Pascal. Mill10_2 shows same characteristics as mill10_1 for 70 microns. As the coal particle size increase in mill10_2 the wall shear stress is almost uniform. There is no significant change that was seen in the graph. It is not valid to compare mill10_1 with mill10_2 because both are different in geometry. Since the wall shear stress is computed as area weighted average, the area of the pipes play an important role in the average value of wall shear stress. As summarized in table each pipes have different area. From the table it is shown that mill10_4 has the highest surface area compared to all three pc pipes. Thus the wall shear stress in mill10_4 is the lowest in the graph 1. The bigger the area the smaller the wall shear stress. The trend line of mill10_3 also similar to mill10_2, in which highest wall shear stress is observed at small range of particles ranging from 70 to 100 microns. Furthermore the number of bends in the pipe is also an important factor in studying wall shear stress. The bend in a pipe with cause drop in velocity of particles flowing through. In comparison mill10_2 have more bends compared to mill10_1 with is almost a straight pipe. In short, the geometry of the pipe is important parameter in considering the wall shear stress [9].The contours of wall shear stress is also included to display the overall flow behavior.

IOP Conf. Series: Journal of Physics: Conf. Series **822** (2017) 012032 doi:10.1088/1742-6596/822/1/012032

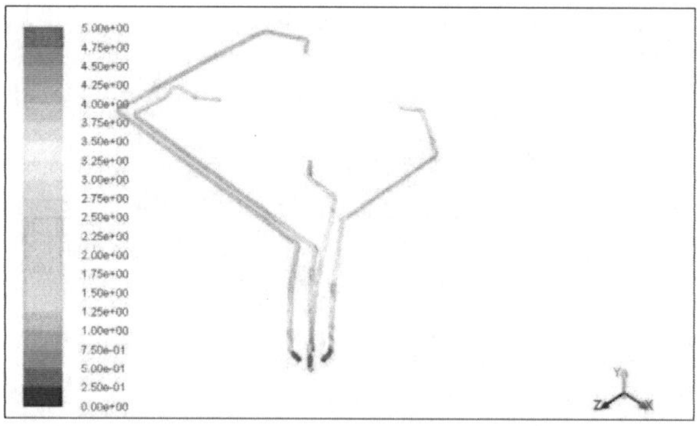

Figure 7. Contours of wall shear stress of mill10 at coal particle size of 70 microns

The figure above shows contour of wall shear stress of mill10 when particles of 70 microns are injected. The highest wall shear stress is observed at the inlet of the pipe. This is because the injected has the highest velocity at the beginning of the pipe and experience a loss in momentum as it travels along the pipe. The drop in velocity in not very significant, thus the wall shear stress also shows only slight decrement. As for mill10_1, the contours have many reddish and yellow region, indicating higher average wall shear stress as shown in graph. The mill10_1 is the shortest pipe, thus the loss of momentum or the drop in velocity is least significant. This is why an almost uniform wall shear stress can be observed. Next is mill10_2, more green contours is observed as compared to mill10_1, this is because the length of the pipe is longer than pc pipe mill10_1. Since it is longer, the drop in velocity is much more significant compared to mill10_1, thus a more greenish region observed. It can also be seen blue contours at the bends of all the pipes, especially mill10_2, mill10_3, and mill10_4. This is due to the fact that bend in pipes causes the drop in velocity of the fluid flowing inside. So lesser the velocity, the lesser the wall shear stress on that bend region. The screenshots of mill10 injected with 100microns, 300 microns, and 500 microns of coal particle size are shown in below.

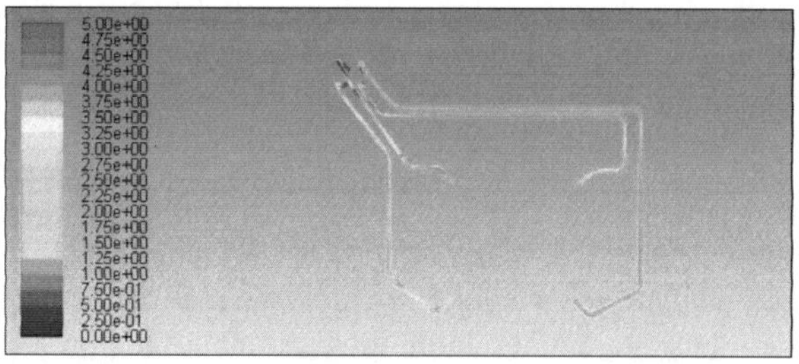

Figure 8. Contours of wall shear stress of mill10 at coal particle size of 100microns

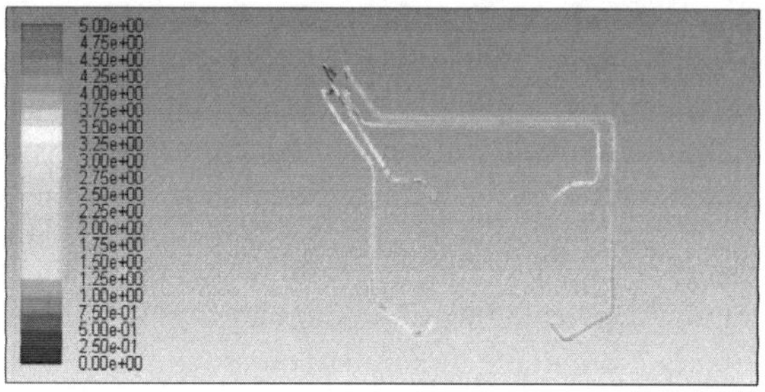

Figure 9. Contours of wall shear stress of mill1 at coal particle size of 300 microns

From the figure above, as the coal particle size is increased, it is observed that greener region is observed throughout pipe 2, 3 and 4, indicating lesser wall shear stress. There is also a slight drop of wall shear stress is pipe 1 which mill10_1 but not so significant. Similarly, the contours of wall shear stress of mill10 that is injected with bigger particle size of 300 microns shows less red regions compared smaller particles of 100 microns and 70 microns. Similar observations for coal particle size of 500 microns.

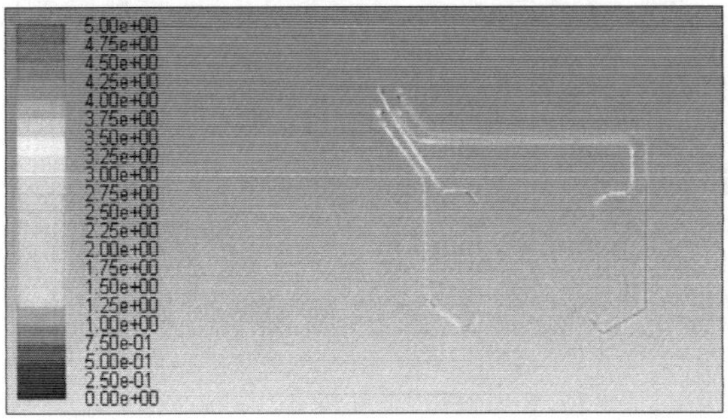

Figure 10. Contours of wall shear stress of mill 1 at coal particle size of 500 microns

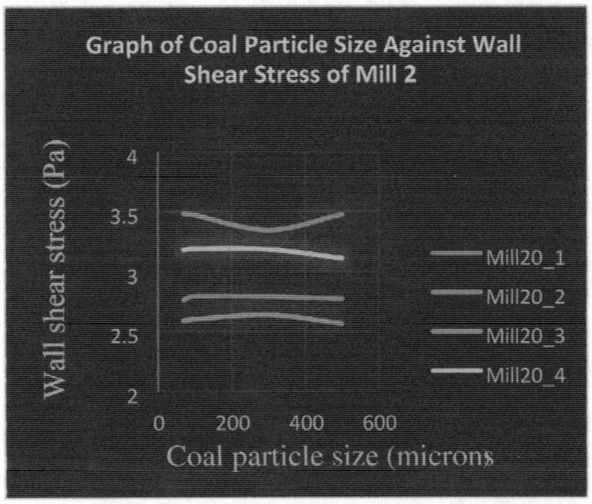

Figure 11. The graph of wall shear stress against coal particle size for mill 2

Similar to mill 1, the wall shear stress of pc pipes of mill 2 did not display significant change and almost uniform. Only for mill20_2, there is a minimum wall shear stress observed at 300 microns and the wall shear stress increases beyond 300 microns of particle size. Although it increases, the percentage of increase is too small. The pipe that experience highest wall shear is mill20_2, in which it has the lowest area as shown in table 1.

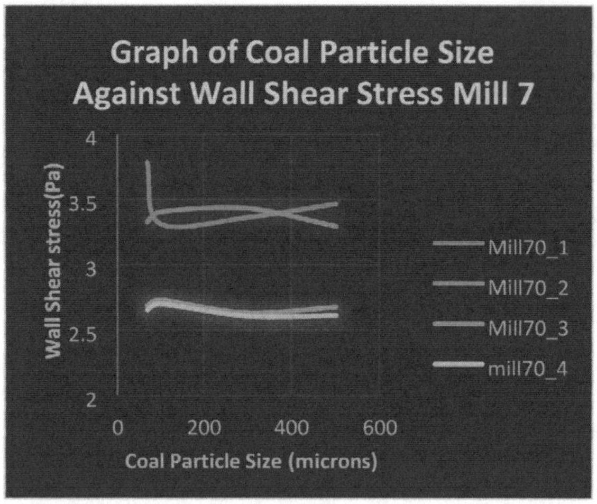

Figure 12. The graph of wall shear stress against coal particle size for mill 7

Unlike mill 1 and mill 2, the graph of mill 7 is little different as there is some intersection between the graphs. However, the pattern is almost linear for all four mills. The highest wall shear stress is observed at mill70_1 which has the smallest area. Mill70_1 and mill70_2 has almost same area, thus both of them have wall shear stress value close to each other. This is why we could see some intersection between the graphs. The graphs of mill70_3 and mill70_4 also very close to each other, thus their value of wall shear stress also very close to each other. From table 2, the area of mill70_3 and mill70_4 only differs a little. By observing figure 1, mill 7 is slightly special when compared to other mills because it is located in the center among all mills, and all four pipe coming out of the mill heading towards the boiler is almost symmetrical. The contours of wall shear stress are shown below.

Figure 13. Contours of wall shear stress of mill 7 at coal particle size of 70 microns

Configuration 2 involves changing the air flow rate and computing the wall shear stress as area weighted average similar to configuration 1. The coal flow rate is fixed at 12kg/s and the coal particle size is fixed at 70 microns. The graph of air flow rate against wall shear stress is shown below. Only mill 1 and mill 2 is studied in this configuration.

The figure 14 shows the behavior of wall shear stress of the pc pipes of mill 1. Unlike the coal particle size, the air flow rate shows a significant increase in wall shear stress. The significant increases is observed in all four pipes. However, the highest wall shear stress is for mill10_1, the shortest pipe with smallest area. For a constant particle, when the air flow rate increases, the velocity of particles in the pipe increases. The force that drives the coal particles is air flow. Higher the flow rate, higher velocity of particles, thus higher value of wall shear stress [9]. In this configuration the range of air flow is from 64kg/s to 94 kg/s only. For instance, with just air flow rate increase of 47 %, the wall shear

IOP Conf. Series: Journal of Physics: Conf. Series **822** (2017) 012032 doi:10.1088/1742-6596/822/1/012032

stress increases by 95 %. Thus, it can be concluded the effect of air flow rate is much more significant than effect coal particle size on wall shear stress.

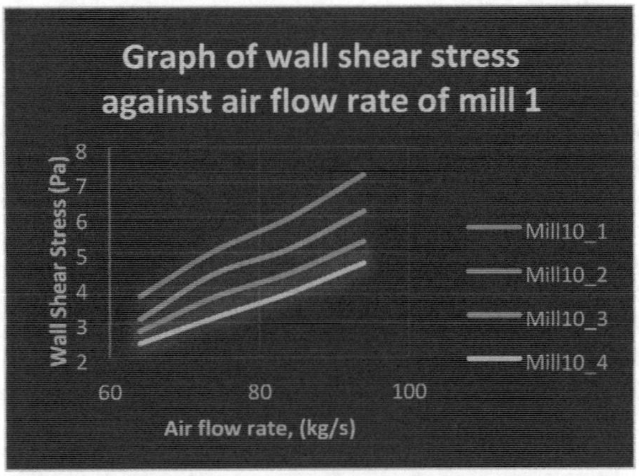

Figure 14. The graph of wall shear stress against air flow rate for mill 1

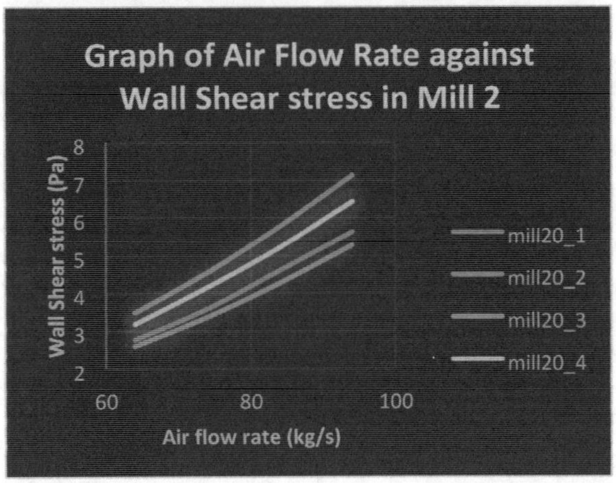

Figure 15. The graph of wall shear stress against air flow rate for mill 2

Similar to mill 1, mill 2 also exhibit same characteristics in which the wall shears stress increases with increasing air flow rate. An almost linear relationship can observed in both graphs. The highest wall shear stress is experienced by mill20_2, which has the smallest area based on table 3.1. Conversely the smallest wall shear stress is experienced in mill20_3 which has the biggest area among all the four pipes. In short, the air flow rate causes a very significant increase in wall shear stress compared to the coal particle size for both mill 1 and mill 2.

4. Conclusion

For configuration 1, the increase in coal particle size does not show significant increase in wall shear stress. However, most of the pipes that has been studied, the wall shear stress is reducing at higher coal particles size. Majority of the pipes displayed minimum value of wall shear stress in coal particle size ranging from 200-300 microns. In order reduce the failure rate of pc pipes due high wall shear stress, it is recommended to use particles of size ranging from 200-300 microns. However, although this range decreases the wall shear stress, on the other hand it can cause the increase of NOx and SOx emissions.

According to J Blondeau, the coal flow distribution towards the boiler was significantly improved when the particle size was the smallest [4]. Increasing coal particle size will cause significant increase in SOX emission [3,12-13]. Using higher particle size with smaller surface area exposed reaction, thus the chances of complete combustion reduces. This can adversely affect the thermal efficiency of the boiler. Furthermore, the grinder in the actual mill practically is not able to grind all coal particles to a single value of coal particle size like 70 microns. Usually the grinding process produces coal at no uniformly distributed particle size within a range. So as per my finding, although the smaller particles cause higher wall shear stress, it is still recommend to inject coal in the small particle range from 70- 100 microns because increase in coal particle size does not show significant drop in wall shear stress but cause significant increase in emissions [3,13]. The problem of uncontrolled emission is way too serious compared failure of pc pipes.

For configuration 2, the air flow rate showed significant increase in wall shear stress unlike coal particle size. So it is concluded that air flow rate ranging from 60-70 kg/ s will be best flow rate to be used in a utility of 700MW. Lower values of air flow rate produces lower wall shear stress. Moreover, higher flow rate will cause the air to fuel ratio of the combustion in the boiler to be affected. In order to balance the air- fuel ratio, more coal need to be injected. This will increase coal consumption. Improper air to fuel ratio increases the chances of incomplete combustion causes SOx and NOx emissions [13]. Contact of the nitrogen from the fuel with the oxygen within the combustion air is reduced there will be notable amount of decrease in the NOx formation.

The objective of the study is achieved. The effect of coal particle size on the wall shear stress is investigated and it is concluded that smaller particles will cause higher wall shear stress. Next objective which is the effect of air flow rate on the wall shear stress. The higher the air flow rate, the higher the wall shear stress. There is a linear relationship between air flow rate and wall shear stress. To enable most optimum operation of a utility boiler, it is best to use coal particle size of range 70-100 microns with causes least emissions and air flow rate of range 60-70 kg/s for an optimum combustion.

References

[1] Energy Information Administration, 2013. International Energy Outlook 2013[R]. Washington, DC: U.S. Energy InformationAdministration. Contract No.:DOE/EIA0484.

[2] Choeng Ryul Choi, Chang Nyung Kim, (2009).Numerical investigation on the flow, combustion and NO_x emission characteristics in a 500 MWe tangentially fired pulverizedcoal boiler. Fuel 88 (2009) 1720-1731.

[3] Ivan Tomanovic, Nenad Crnomarkovic´, Aleksandar Milic´evic,(2016) Numerical study of pulverized coal-fired utility boiler over a wide range of operating conditions for in-furnace SO2/NOx reduction. Applied Thermal Engineering 94 (2016) 657–669

[4] J. Blondeau, R.Kock, J Mertens, AJ Eley, and L Holub, (2016) .Online monitoring of coal particle size and flow distribution in coal- fired power plants: Dynamic effects of a varying mill classifier speed. Applied Thermal Engineering 98(2016) pg 449–454.

[5] Denis Edelin, Pierre-Clement Czujko, Cathy Castelain, Christophe Josset , and Francine Fayolle, (2015).Experimental determination of the energy optimum for the transport of floating particles in pipes. Experimental Thermal and Fluid Science 68 (2015) 634– 643.

[6] Baojun Yi, Liqi Zhang, Zhihui Mao, Fang Huang, and Chuguang Zheng, (2014). Effect of the particle size on combustion characteristics of pulverized coal in an O2/CO2 atmosphere. Fuel Processing Technology 128 (2014) 17–27.

[7] C.V. Popa, H. Zaidi, A. Arfaoui, G. Polidori, R. Taiar, and S. Fohanno,(2011). Analysis of wall shear stress around a competitive swimmer using 3D Navier– Stokes equations in CFD. Act of Bioengineering and Biomechanics, Vol. 13.

[8] Quamrul H. Mazumder, (2012) CFD Analysis of Single and Multiphase Flow Characteristics in Elbow. Engineering, 2012, 4, 210-214.

[9] X. Yan, X. B. Chen and D. J. Bergstrom, (2011). Modeling of the Flow within Scaffolds in Perfusion Bioreactors. American Journal of Biomedical Engineering 2011; 1(2): 72-77

[10] Sowjanya Vijiapurapu, Jie Cui, Sastry Munukutla, (2006). CFD application for coal/air balancing in power plants. Applied Mathematical Modelling 30 (2006) 854–866.

[11] Galen Richards, David Towle, Robert Lewis,Kevin Connolly, Richard Donais, And Todd Hellewell,(200). Development of An Enhanced Combustion Low Nox Pulverized Coal Burner. ALSTOM Power Inc, 2000.

[12] Audai Hussein Al-Abbasa, and Jamal Nasera,(2013). Computational fluid dynamic modelling of a 550 MW tangentially-fired furnace under different operating conditions, Procedia Engineering 56 (2013) 387 – 392.

[13] [Min Kuang, Qunyi Zhu, Zhengqi Li, and Xiang Zhang, (2012). Numerical investigation on combustion and NOx emissions of a down-fired 350 MWe utility boiler with multiple injection and multiple staging: Effect of the air stoichiometric ratio in the primary combustion zone. Fuel Processing Technology 109

Preliminary investigation on the effects of primary airflow to coal particle distribution in coal-fired boilers

N.A.W. Mohd Noor[1], H. Hassan[1], M.F. Hashim[1], H. Hasini[2], K.M. Munisamy[2],

[1] TNB Research Sdn. Bhd., Lorong Ayer Hitam, 43000 Kajang, Selangor, Malaysia.
[2] Department of Mechanical Engineering, Universiti Tenaga Nasional (UNITEN), 43000 Kajang, Selangor, Malaysia.

Email: noorakma@tnbr.com.my

Abstract. This paper presents an investigation on the effects of primary airflow to coal fineness in coal-fired boilers. In coal fired power plant, coal is pulverized in a pulverizer, and it is then transferred to boiler for combustion. Coal need to be ground to its desired size to obtain maximum combustion efficiency. Coarse coal particle size may lead to many performance problems such as formation of clinker. In this study, the effects of primary airflow to coal particles size and coal flow distribution were investigated by using isokinetic coal sampling and computational fluid dynamic (CFD) modelling. Four different primary airflows were tested and the effects to resulting coal fineness were recorded. Results show that the optimum coal fineness distribution is obtained at design primary airflow. Any reduction or increase of air flow rate results in undesirable coal fineness distribution.

Keywords: *coal fineness, pulverizer, primary airflow, CFD simulation*

1. Introduction

Pulverizers system provides pulverized coal to the boiler. Within the pulverizers, the coal is ground and classified to the proper size. Primary air is used to dry and transport the pulverized coal for burning in the boiler [1]. There are various issues related to pulverizer performance such as poor coal fineness and uneven coal flow distribution. The uneven coal flow distribution in furnace could be caused by many reasons. Coarse coal particle entering the furnace is one of the contributing factors [2]. Continuous research in this area has been carried out by many researchers. Richard [3] describes theoretical information on technique in pulverized coal pipe testing and balancing. Shah et al. [4] performed Computational Fluid Dynamics (CFD) simulation to investigate the effect of classifier vane settings on pulverizer performance. Their results show the optimum classifier vane opening can improve the fineness but need to consider other factors such as uniform coal mass flow rate at PC pipes outlets, desired size fraction and classifier efficiency. Sung et al [5] present the influence of coal particle size on Nitrogen Oxide, emission and burnout characteristic. The results show that the burnout performance increases with an increase in the level of coal fineness. The effectiveness of air staging on the NO reduction and burnout performance is significant in the flames with fine pulverized coal particles. Nakamura et al [6] studied the effect of equipment geometry on grinding performance of the pulverizer. Their results show that the stator concave angle affects the particle residence time and total number of the particle stator collision. Blondeau et al [7] present a methodology for the on-line monitoring of both coal particle size and flow distribution and investigate the influence of varying the classifier speed on coal particle size and flow distribution. It was demonstrated that the coal flow distribution between burners was improved when the particle size was the smallest.

Many inefficient operations of boilers and pulverizers were reported in coal-fired power plants in the world. One of the common causes of this inefficiency was due to inefficient operation of pulverizers when coarse coal is allowed into boiler furnace for combustion. These coals are transported into boiler furnace by primary air. The flow rate of primary air is crucial in determining the right coal size that exits the pulverizer mill. In this work, coal fineness and coal flow distribution were measured at different primary air flow rate. The study also involved preliminary CFD modelling of coal particles distribution from pulverizer to burner. The research work aims to improve boiler efficiency and the focus is given to the flow characteristic inside pulverizer mill.

2. Methodology

In this research study, isokinetic coal samplings were carried out at pulverizer to determine coal flow distribution in each pulverized coal (PC) pipes. CFD investigations were carried out focusing on flow characteristics of coal particles inside pulverizer. The CFD model used is based on a real vertical bowl mill at coal-fired power plant in Malaysia.

Coal flow distribution were sampled from the vertical straight PC pipes just after the pulverizer by using Dirty air pitot (DAP) and Pulverized Fuel Sampler (PFS) to measure the flow velocity and extract coal samples [8]. To ensure isokinetic sampling, the sampling gas velocity is maintained at the same level as the gas velocity in the PC pipe. The DAP and PFS are inserted through a dustless connection into the PC pipe. Then, coal fineness was determined by sieving following procedure in ASTM D197-87 [9]. The quantity of coal passing each sieve is used to determine coal particle sizing. The fineness was determined by sieving through ASTM 50, 100 and 200 Meshes and plot into Rosin-Rammler distribution. The sampling activities and fineness test to investigate coal flow distribution are shown in Figure 1-3.

Figure 1: PC pipes velocity measurement by using Dirty Air Pitot

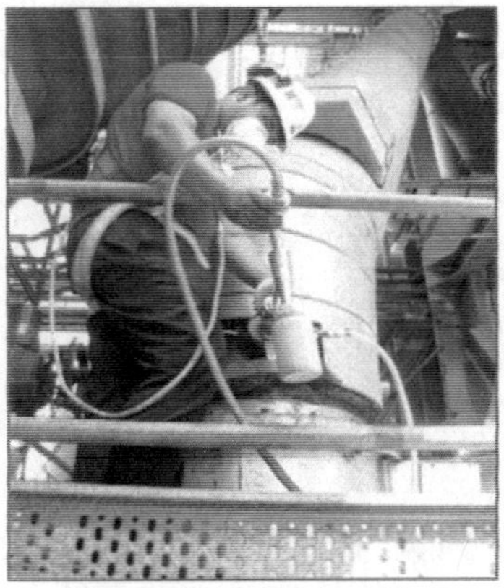

Figure 2: Pulverizer Fuel Sampler - extract coal sample

Figure 3: Coal Fineness test

CFD modelling for pulverizer system was developed to investigate the coal particle distribution at the exit of pulverizer when the amount of primary air varies at 60%, 80%, 100% and 120% of design condition. The recommended range of primary air for power plant for this case study is 80 -100 kNm^3/h. For this project, 90 kNm^3/h is chosen as 100% of the amount of primary air. This calculation is done with the assumption of fixed classifier angle. This system includes all the parts and components that exist in the pulverizer.

Figure 4 shows the schematic diagram of pulverizer model for the simulation of flow and coal particle distribution. Generally, the pulverizer consists of the centre chute where raw coals first enter the system. The raw coal will later grinded on the rotating table with three rollers at the bottom part of the mill. At the same time, hot primary air is injected to carry the grinded coal to the pulverizer exit. Prior to the pulverizer exit, the mixture of grinded coal and hot air will pass through the classifier which is located at the top of the pulverizer. Depending on the classifier angle, fine coal will pass through it and flow downstream to series of complex piping system that leads the mixture to the burner and furnace entrance. The classifier is designed to prevent coarse coal particle to pass through it and these coarse will falls and returned to the bottom of the pulverizer where it will be re-grinded on the rotating table. It is important to carefully take into account the classifier in the model for correct determination of coal fineness leaving the system. In order to model the whole pulverizer system accurately, each components that exist in the mill are taken into account and model separately for simplicity. Figure 5 and Figure 6 show the classifier vane and primary air inlet respectively while Figure 7 shows the cross sectional view of the completed assembly of components that make up the pulverizer.

Figure 4: Classifier vane

Figure 5: Primary air inlet

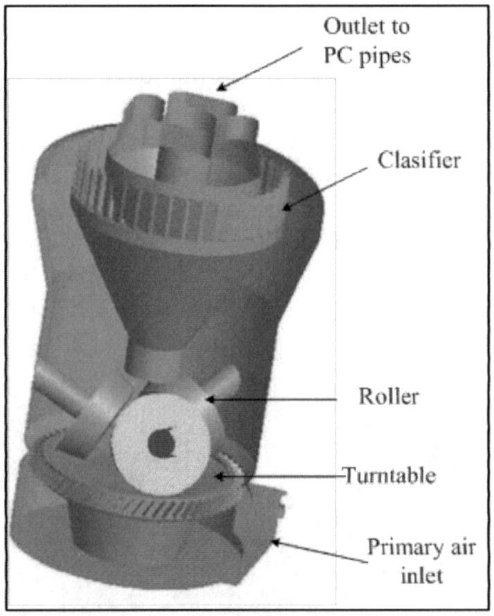

Figure 6: Schematic diagram of pulverizer model

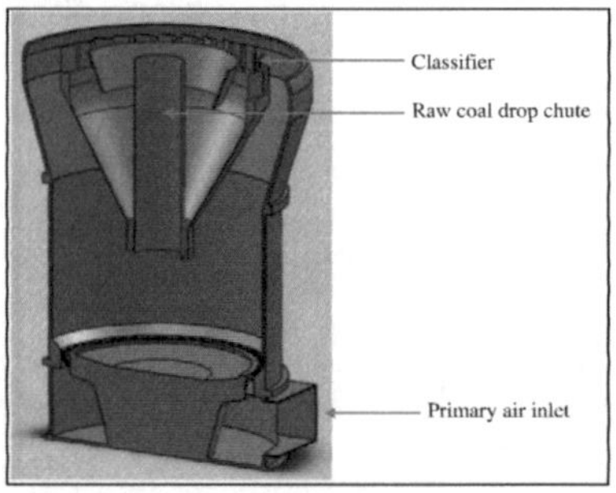

Figure 7: Cross sectional view of pulverizer

3. Results and Discussions

Figure 8 shows the Rosin Rammler chart to represent the sieve test result of coal fineness from the sampled pulverized coal. The x-axis shows the percentage of coal passing through the sieve and the y-axis shows the sieve size in mesh or particle size in μm unit. The ascending line represents the cumulative percentage of coal passing through the sieve. According to plant requirements, more than 70% by weight of the pulverized coal should be smaller than 75μm and less than 1% of the pulverized coal should be larger than 300μm. However, from the coal fineness tests of the sampled pulverized coals, less than 50% were smaller than 75m and more than 10% of the pulverized coal samples were larger than 300μm size. Figure 9 shows the coal flow distribution for the four PC pipes. The figure indicates that there is a significant mal-distribution of coal mass flow in the PC pipes. It shows that the quantity of coal delivered to each burner is not equally distributed. Theoretically, the deviation limit between four PC pipes for each pulverizer is not more than 10% [4].

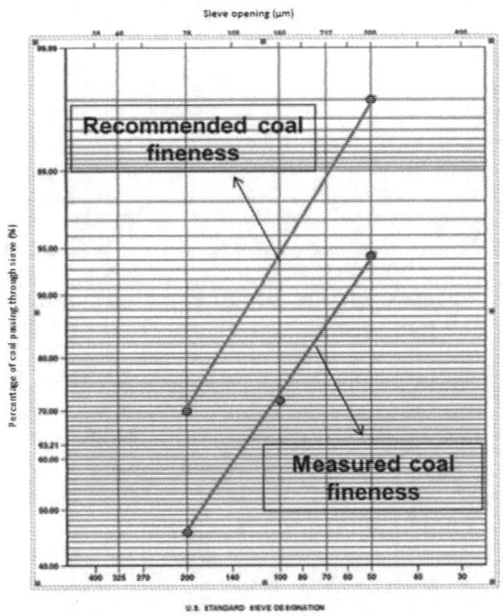

Figure 8: Result of coal fineness

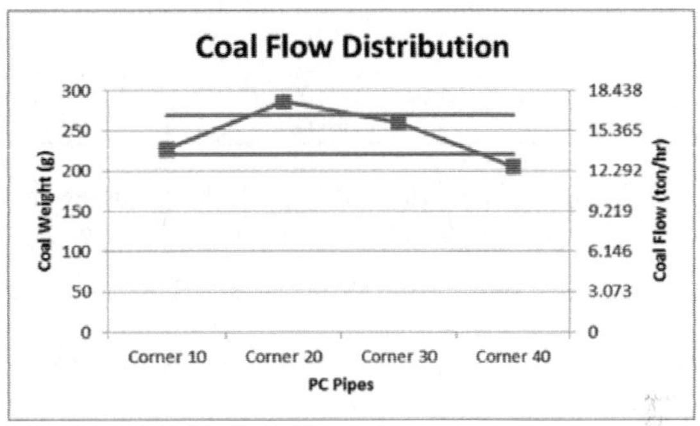

Figure 9: Result of coal flow distribution

The CFD simulation for pulverizer system aims to investigate the coal particle distribution at the exit of pulverizer when the amount of primary air varies at 60%, 80%, 100% and 120% of design condition. This calculation is done with the assumption of fixed classifier angle. Figure 10 shows the prediction of coal particle size distribution at pulverizer exit when the primary air is 60% of the design condition. At this rate, it is predicted that higher percentage of coarse coal passed the classifier and exit the pulverizer. The trend is linear in which fine coal was found to be the lowest. This condition is undesirable in the operation of the pulverizer and explains why it is not practised in actual operation. Similar trend was also found to be true for the case of 80% primary air as illustrated in Figure 11.

However, at 100% primary air the distribution of coal fineness at pulverizer exit changes completely at which most of coal particle group that exits the pulverizer were fine particles. The lowest particle group that escape the classifier was coarse particle as shown in Figure 12. This trend explains why the suggested setting for primary air is at the design condition (100% primary air). Attempt is also made to further increase the amount of primary air to 120% design condition and the resulting coal fineness distribution is shown in Figure 13. It is clear that further increase in primary air is not favourable as the coal fineness distribution falls in the average coal particle size.

To summarize the prediction made on coal fineness distribution at pulverizer exit, comparison is made for all cases and this is shown in Figure 14. It can be concluded that is the supply of primary air is less than the design value, most of the coarse coal particle will escape the pulverizer (with minimum amount of fine coal particle). The optimum setting is given by 100% primary air where it is predicted that with this setting, most of the coal particle that escape the pulverizer is fine particle which is desirable in pulverizer operation. Further increase in primary air does not guarantee continuous supply of fine coal particle as the coal fineness distribution was found to be in a mixture of coarse and fine particle.

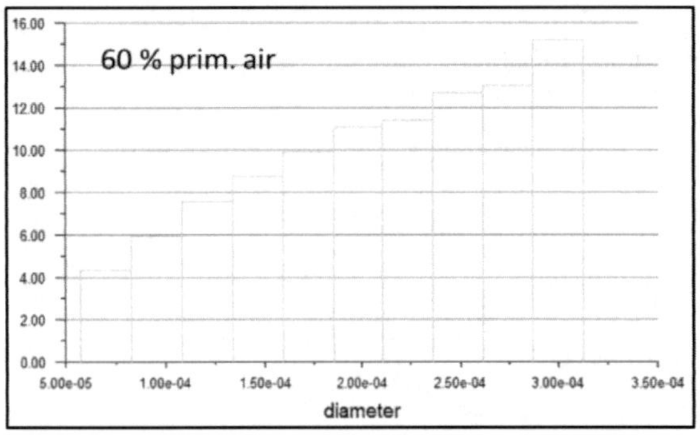

Figure 10: Prediction of particle diameter distribution at pulverizer exit (60% primary air)

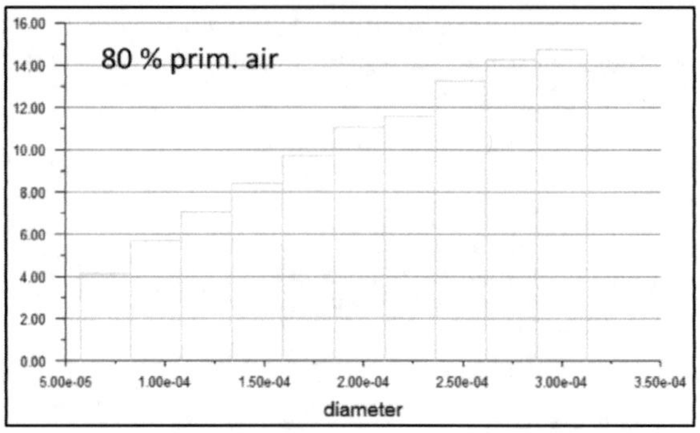

Figure 11: Prediction of particle diameter distribution at pulverizer exit (80% primary air)

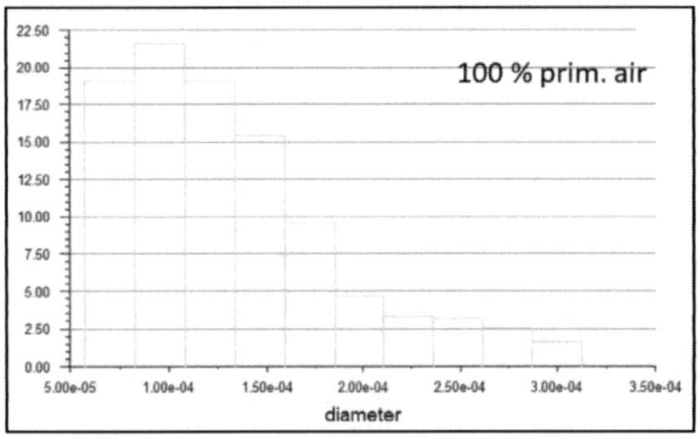

Figure 12: Prediction of particle diameter distribution at pulverizer exit (100% primary air)

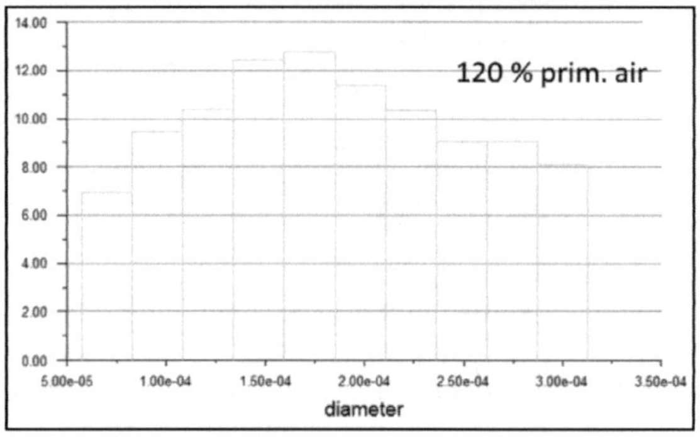

Figure 13: Prediction of particle diameter distribution at pulverizer exit (120% primary air)

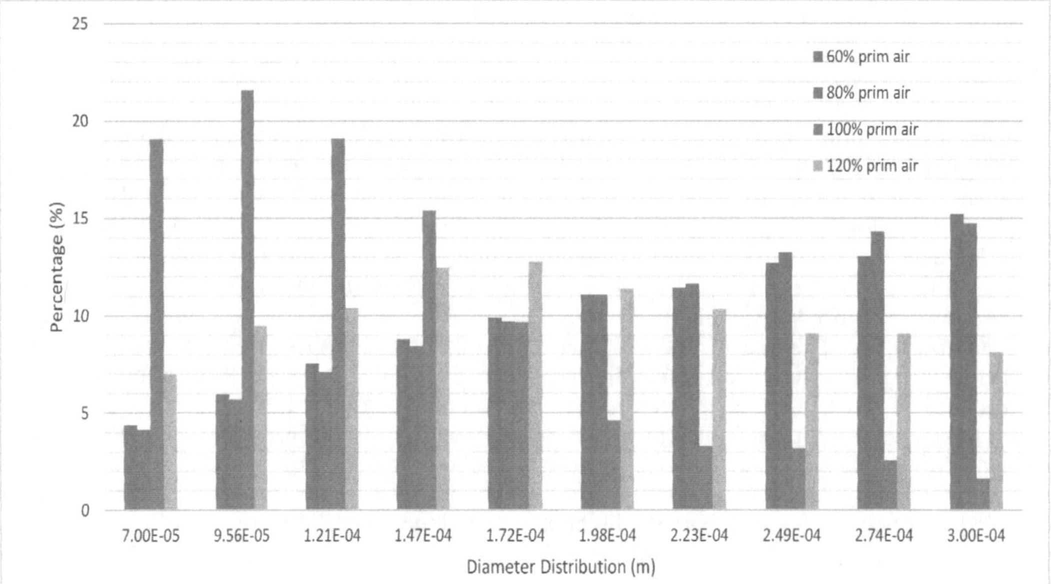

Figure 14: Comparison of particle diameter distribution at pulverizer exit

4. Conclusions

In this work, the effects of different primary air flow on coal flow distribution in coal-fired power plant are simulated using CFD. Isokinetic coal sampling to collect sample for coal fineness analysis were performed. From the measurement, less than 50% pulverized coal were smaller than 75 μm and more than 5% pulverized coal were larger than 300 μm when sampled. To further investigate this problem, CFD simulation was carried out. CFD analysis of coal particle flow and distribution inside pulverizer is carried out. Variation of primary air flow at fixed pulverizer parameters was tested. Based on the simulation, the following conclusions can be withdrawn:

i. Setting of lower primary air (carrier air) results in high percentage of coarse coal particle leaving the pulverizer. Increasing the primary air further will reduce the coarse particle and the trend is valid for air mass less than 100% (design).

ii. At 100% primary air (design), the amount of fine coal particle leaving the pulverizer is the highest.

Based on the prediction made by computational simulation, it is thus recommended that the operation of pulverizer be set at 100% primary air supply. This is to ensure sufficient momentum flux for optimum coal fineness at the pulverizer exit and at the same time avoid pre-mature combustion of coal mixture should the primary air be set lower than design condition.

Acknowledgements

The authors wish to thank Tenaga Nasional Berhad (TNB) for providing the research grant to carry out this project. The financial support received from the Ministry of Education Malaysia via Fundamental Research Grant Scheme (20150210FRGS) is also acknowledged.

References

[1] *Coal Fired Thermal Power Plant (3x700MW).*

[2] "Innovative Combustion Technologies," vol. 35244, no. 205, pp. 1–8, 2007.

[3] B. Richard and F. D. Storm, "Pulverized Coal Pipe Testing and Balancing Four Steps to Fuel Line Balancing Isokinetic Testing," 2010.

[4] K. V. Shah, R. Vuthaluru, and H. B. Vuthaluru, "CFD based investigations into optimization of coal pulveriser performance: Effect of classifier vane settings," *Fuel Process. Technol.*, vol. 90, no. 9, pp. 1135–1141, 2009.

[5] Y. Sung, C. Moon, S. Eom, G. Choi, and D. Kim, "Coal-particle size effects on NO reduction and burnout characteristics with air-staged combustion in a pulverized coal-fired furnace," *Fuel*, vol. 182, no. x, pp. 558–567, 2016.

[6] H. Nakamura, H. Kan, H. Takeuchi, and S. Watano, "Effect of stator geometry of impact pulverizer on its grinding performance," *Chem. Eng. Sci.*, vol. 122, pp. 565–572, 2015.

[7] J. Blondeau, R. Kock, J. Mertens, A. J. Eley, and L. Holub, "Online monitoring of coal particle size and flow distribution in coal-fired power plants: Dynamic effects of a varying mill classifier speed," *Appl. Therm. Eng.*, vol. 98, pp. 449–454, 2016.

[8] "ISO 9931.pdf." .

[9] S. T. Method, "Standard Test Method for Sampling and Fineness Test of Pulverized Coal 1," *Changes*, vol. 87, no. Reapproved, pp. 1–7, 2002.

Nanofluid heat transfer between concentric independently rotating tubes with axial flow

Mohd Rosdzimin Abdul Rahman[1], **Kin Yuen Leong**[1], **Azam Che Idris**[1], **Mohd Rashdan Saad**[1], **Mahmood Anwar**[2]
[1]Department of Mechanical, Faculty of Engineering, Universiti Pertahanan Nasional Malaysia, Kem Sg. Besi, 57000, Kuala Lumpur.
[2]Faculty of Engineering and Science, Curtin University, Sarawak Campus, Malaysia

Email: rosdzimin@upnm.edu.my

Abstract. Nanofluid heat transfer behaviour between concentric independently rotating tubes with an axial flow was numerically investigated. Co-rotating and counter-rotating of the inner and outer tubes was simulated at a fixed rotation number. Flow inside the annulus was turbulent with Reynolds number of 5000 and 10000. *k-epsilon* turbulent model was used and the current work is in agreement with published experimental work. Mixture-multiphase model was used to model the multiphase flows of the TiO_2/water nanofluid. Results show that the Nusselt number increases with the Reynolds number. At a constant Reynolds number, different Nusselt number distribution along the hotted tube for co-rotating and counter-rotating tube was observed. The counter-rotating tube was found to be more efficient in enhancing the heat transfer rate in comparison to the co-rotating tube.

1. Introduction

Flows between rotating tubes is known as Taylor-Couette-Poiseuille flow. This type of flow is imminent in industry due to its significant application in rotating machineries and heat exchangers. To optimize the performance of rotating machineries and heat exchangers, a better knowledge of the convective heat transfer is required. The difficulties in carrying out experiments due to higher inner cylinder rotation and narrow gap between the outer and inner cylinders make numerical simulation to be a powerful tool to understand the knowledge of the heat transfer. Poncet *et al.* [1] simulated the heat transfer in rotating tubes and found that the Nusselt number increases with both the rotation rate of the inner cylinder and the Reynolds number. Recently, a study was conducted on an enclosed rotating tube to investigate the stability of the flow submitted to a radial temperature gradient and found that the heat transfer process along the inner and outer walls were enhanced due to the thermo-convective structure of the flow [2].

All the previous investigations on the topic of heat transfer of rotating tubes were only done using water or air. The heat exchanger requires working fluids that possess high thermal conductivity values for a better heat transfer process. Conventional heat transfer fluid such as water and ethylene glycol exhibit low thermal properties. Countless researches have been done to enhance thermal conductivity of conventional fluids by adding nanoparticles. Choi [3] started to add solid particles to increase the thermal conductivity of fluid. Nanometer diameters of the particles were added into the conventional fluid (water, ethylene glycol and oil), which were eventually called nanofluids, showed significant heat transfer enhancement [3]. Many investigations on nanofluids were later done to study the effect of particle type, size, and volume percentage, based on the fluid type and its application in heat exchanger [4-6]. Ting and Hou [6] numerically discovered that the heat transfer coefficient of 2.5 vol % Al_2O_3-water nanofluid increased by 25.5% compared with pure water in force convection case.

The previous studies of nanofluid heat transfer in rotating tubes were focused on forced, natural and mixed convection of the infinite tube length. In the authors' knowledge, no study has been conducted yet on nanofluid heat transfer in finite length independently rotating tubes. Therefore, the objective of the present work is to investigate numerically the heat transfer of TiO_2/water nanofluid in finite length independently rotating tubes.

2. Numerical Works

2.1 Geometry and boundary conditions

Numerical domain of a two-dimensional axi-symmetry tubes with a length of $L/d_h = 60$ and radius ratio ($k = r_1/r_2$) is shown in Fig. 1. A uniform axial velocity is assigned at the velocity inlet based on the Reynolds number. While, a fully developed condition is assumed at the outlet and all derivatives are taken as zero. The inner and outer wall tubes are assigned as a non-slip moving wall condition with N_i and N_o rotation number. A constant heat flux is fixed at the inner tube. The assumptions made for this numerical analysis are; (1) steady state analysis, (2) effective thermo-physical properties of the nanofluid depends on volume concentration, and (3) fluid is continuum.

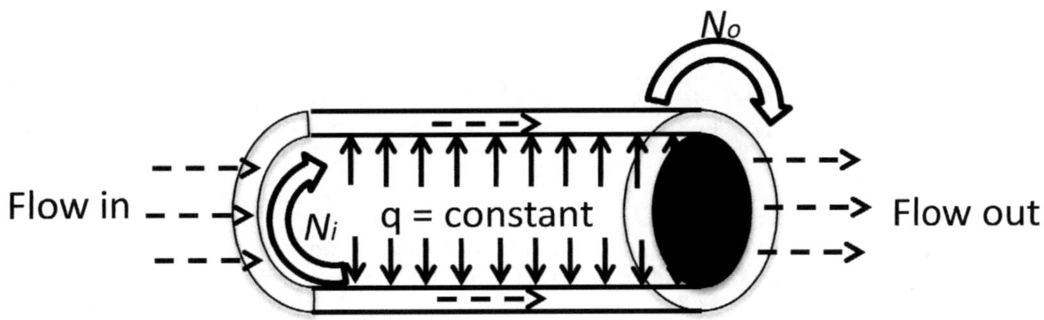

Figure 1. Numerical domain with respective boundary condition.

2.2 Turbulence model

The realizable $\boldsymbol{k-\varepsilon}$ turbulence model was used in this study. Relation of the turbulent viscosity $\boldsymbol{\mu_t = \rho C_\mu k^2/\varepsilon}$ is assigned to combine the turbulent kinetic energy and turbulent dissipation rate.

2.3 Mixture-Multiphase model

The classical mixture theory [7-9] is applied for the Mixture-multiphase model. Generalization principles of continuum mechanics for a single phase to several inter-penetrable continua are employed in this method. Moreover, all phases are assumed to exist at every material point and time in the Mixture model. The Mixture model solves the continuity equation for the mixture, the volume fraction equation for the secondary phase, the momentum equation for the mixture, the energy equation for the mixture and algebraic expressions for the relative velocities.

The continuity equation for the mixture is shown in Eq. 1

$$\boldsymbol{\nabla \cdot (\rho_m \vec{v}_m) = 0} \tag{1}$$

and for the volume fraction of secondary phase is depicted in Eq. 2

$$\nabla \cdot (\alpha_p \rho_p \vec{v}_m) = -\nabla \cdot (\alpha_p \rho_p \vec{v}_{dr,p}) \tag{2}$$

Momentum equation

$$\nabla \cdot (\rho_m \vec{v}_m \vec{v}_m) = -\nabla p + \nabla \cdot [\mu_m (\nabla \vec{v}_m + \nabla \vec{v}_m^T)] \dot{m} + \rho_m \vec{g} + \vec{F} + \nabla \cdot \left(\sum_{k=1}^{2} \alpha_k \rho_k \vec{v}_{dr,k} \vec{v}_{dr,k} \right) \tag{3}$$

where$\vec{v}_{dr,k}$ is the drift velocity for secondary phase *k;*

$$\vec{v}_{dr,k} = \vec{v}_k - \vec{v}_m \tag{4}$$

Energy equation is given as

$$\nabla \cdot \sum_{k=1}^{2} (\alpha_k \vec{v}_k (\rho_k E_k + p)) = \nabla \cdot (k_{eff} \nabla T) + S_E \tag{5}$$

2.4 Numerical procedure

The commercial code Ansys Fluent was employed in the present study. A set of nonlinear partial differential equations was solved by finite volume method. The convergence was set at residual about 10^{-12} for the energy equation.

2.4.1 Validation of the numerical work.

Validation of the Mixture-multiphase model and numerical procedure was done by comparing the axial velocity profile at Reynolds number Re=10000 with the previous experimental work of Rothe and Pfitze [10].

3. Results and discussion

Figure 2 shows a radial distribution of an axial velocity at the outlet of the tubes. It can be seen clearly that the current simulation result is in close agreement with the experimental work by Rothe and Pfitzer [10].

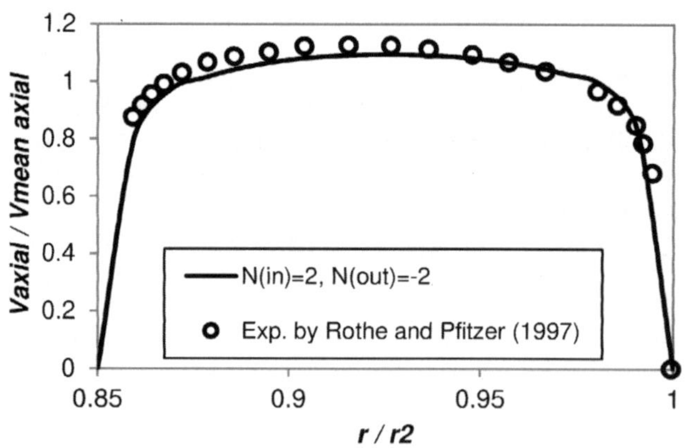

Figure 2. Axial velocity profile for counter-rotating tubes at *Re*=10000.

Figure 3 shows the local Nusselt number along the inner tube wall at Re = 5000 and 10000 for counter-rotating tubes.

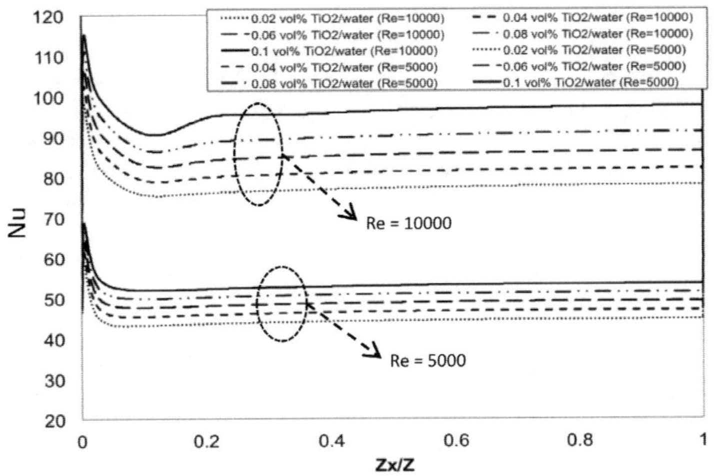

Figure 3. Local Nusselt number along the inner tube wall (N2 N-2) at *Re*= 5000 and 10000.

It can be observed that the Nusselt number is higher at the tube entrance for both cases of Re = 5000 and 10000. The overall recorded Nusselt number for Re=10000 is higher compared to Re = 5000. Moreover, increase in vol% of the TiO_2 nanoparticle contributed to the increase of the Nusselt number. Along the inner tube after the entrance region, the Nusselt numbers recorded were constant for the remaining 80% of the total tube length.

Figure 4 shows the local Nusselt number along the inner tube wall at Re = 5000 and 10000 for co-rotating tubes.

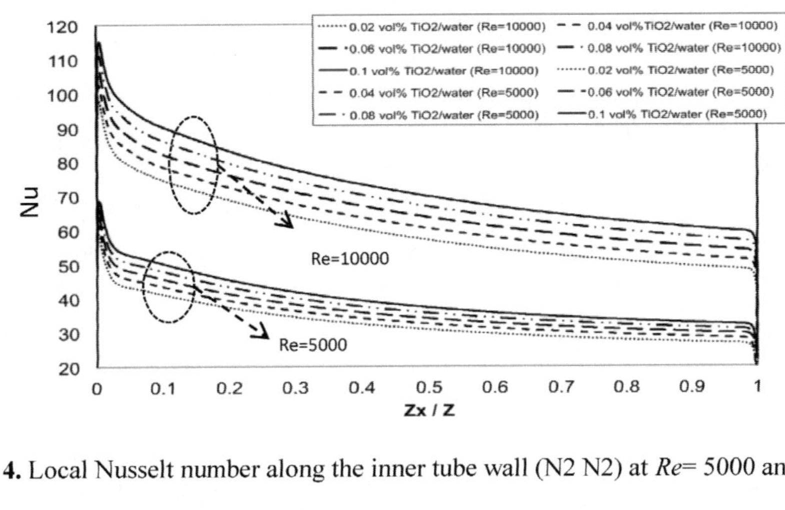

Figure 4. Local Nusselt number along the inner tube wall (N2 N2) at *Re*= 5000 and 10000.

The Nusselt number was found to be higher at the tube entrance for both cases *Re* = 5000 and 10000. The Nusselt number at *Re* = 10000 is higher compared to *Re* = 5000. Moreover, increase in vol% of the TiO_2 nanoparticle also increases the Nusselt number. Along the inner tube, the Nusselt number decreases towards the end of the tube. From the results, it is interesting to discover that the counter-rotating tubes helped significantly to increase the heat transfer compared to co-rotating tubes.

Conclusions

Nanofluid heat transfer behaviour between concentric independently rotating tubes with_axial flow was successfully simulated. The Nusselt number for *Re*=5000 and 10000 cases show similarity at the tube entrances for both cases of co-rotating and counter-rotating. Increase in vol% of the TiO_2 nanoparticle increases the Nusselt number. On the other hand, constant Nusselt number is observed along the inner tube for counter-rotating case. The co-rotating case shows a decrease in the Nusselt number towards the tube end. Moreover, the counter-rotating tubes enhanced the heat transfer more effectively compared to the co-rotating tubes for cases of *Re* = 5000 and 10000.

References

[1] Poncet, S., Haddadi, S., Viazzo, S. 2011 *Int. J. Heat Fluid Flow* **32** 128-144
[2] Viazzo S., Poncet S. 2014 *Computer and Fluids* **101** 15-26.
[3] Choi SUS. 1995 Enhancing thermal conductivity of fluids with nanoparticles development and applications of non-newtonian flows, FED-vol. 231/MD-Vol. 66, ASME, New York, (1995) pp 99-105
[4] Alawi O. A., Sidik N. A. C., Mohammed H. A. and Syahrullail S. 2014 *Int. Commun. Heat Mass Transfer*, **56** 50-62.
[5] Leong, K. Y., Nurfadhillah, M. H., Risby M. S. and Hafizah, A. N. 2016 *Therm. Sci.*, **20** 429-436.
[6] Ting, Hsien-Hung and Hou, Shuhn-Shyurng 2015 *Materials*, **8** 5321-5335.
[7] Bowen, R. M., 1976 Theory of Mixtures, Part I. In: Eringen, A. C. (Ed.), Continuum Physics, Vol. III. New York: Academic Press.
[8] Johnson, G., Massoudi, M. &Rajagopal, K. R. 1991, *Chem. Eng. Sci.* **46** 1713 – 1723.
[9] Joseph, D. D., Lundgren, T. S., Jackson, R. &Saville, D. A. 1990 *Int. J. Multiphase Flow* **16** 35 - 42.
[10] Rothe, T., Pfitzer, H. 1997 *Heat and Mass Transfer* **32** 353-364

IOP Conf. Series: Journal of Physics: Conf. Series **822** (2017) 012035 doi:10.1088/1742-6596/822/1/012035

Novel fully Implicit Collocated grid incompressible flow solver on unstructured meshes

Jobin Joseph Unnupatt[1], N.Munikrishna[2], N.Balakrishnan[3]

1: Research scholar, CAd Lab, Department of Aerospace Engineering, Indian Institute of Science, Bangalore.
2: CTO, S & I Engineering Solutions Pvt. Ltd., Bangalore.
3: Associate Professor, Department of Aerospace Engineering, Indian Institute of Science, Bangalore. Email: nbalak@aero.iisc.ernet.in

Abstract.
This article presents a novel implicit cell center finite volume scheme for solving two-dimensional incompressible steady state flows. A collocated approach has been employed considering its use in conjunction with unstructured meshes. The method inspired by the methodologies for compressible flow solvers is evolved to naturally accommodate several modules from such solvers. The explicit fluxes are computed to second order accuracy using the method of reconstruction. An inconsistent linearization is adopted for the implicit part which employs SGS sub-iterations. At the heart of the implicit procedure is obtaining an auxiliary velocity field, the divergence of which results in a Poisson equation for the pressure field, ensuring a divergence free velocity field resulting from the implicit iteration. The efficacy of the proposed method is demonstrated for several test problems, such as the flow past a Circular Cylinder, NACA 0012 airfoil and lid driven cavity.

1. Introduction
The incompressible flow solvers can be broadly classified as staggered and collocated grid methods, based on the location the pressure and momentum are stored [1,2]. The staggered grid approach, for which the location where the pressure and momentum data are staggered, may not be suited for unstructured data based flow solvers allowing for the use of arbitrary polyhedral volumes. Particularly, when we have access to a compressible flow solver, with its typical modules for solution reconstruction[3], implicit update[4], time accuracy[5] and solution based grid adaption[6], it becomes imperative that most of these modules are adopted for the development of an incompressible flow solver, without resorting to major code changes. Towards this objective, while a preconditioned compressible flow solver can be an interesting option, often, in terms of usage, it gives a lot of flexibility to handle incompressibility in its own right. The present effort in the development of a fully implicit collocated grid incompressible flow solver is inspired by such a requirement.

2. Methodology
Consider the Navier Stokes equations as applied to the incompressible flows given by

$$\frac{\partial U}{\partial t} + \frac{\partial}{\partial x}(f_c + P_1 - f_v) + \frac{\partial}{\partial y}(g_c + P_2 - g_v) = 0 \tag{1}$$

where $U = [\rho u \quad \rho v]^T$ represents the vector of conserved variables, with ρ standing for the fluid density and u and v represent the x and y components of the fluid velocity.

The flux vectors are given as below:

$$f_c = \begin{bmatrix} \rho u^2 \\ \rho uv \end{bmatrix} \quad P_1 = \begin{bmatrix} \mathrm{p} \\ 0 \end{bmatrix} \quad f_v = \begin{bmatrix} \mu\frac{\partial u}{\partial x} \\ \mu\frac{\partial v}{\partial x} \end{bmatrix} \quad g_c = \begin{bmatrix} \rho uv \\ \rho v^2 \end{bmatrix} \quad P_2 = \begin{bmatrix} 0 \\ \mathrm{p} \end{bmatrix} \quad g_v = \begin{bmatrix} \mu\frac{\partial u}{\partial y} \\ \mu\frac{\partial v}{\partial y} \end{bmatrix}$$

Defining $\vec{F_c} = (f_c \quad g_c)$, $\vec{P} = (P_1 \quad P_2)$ and $\vec{F_v} = (f_v \quad g_v)$, Eqn.(1) can be recast as,

$$\frac{\partial U}{\partial t} + \nabla \cdot \left(\vec{F_c} + \vec{P} - \vec{F_v}\right) = 0 \tag{2}$$

For a finite volume i sharing its Jth face with its neighbour j, the state update formula can be written as,

$$\frac{\triangle_t U_i}{\Delta t} + \frac{1}{\Omega_i}\sum_J \left(\vec{F_c} + \vec{P} - \vec{F_v}\right)_J \cdot \hat{n}_J \triangle S_J = 0 \tag{3}$$

where $\triangle_t U_i = U_i^{n+1} - U_i^n$, Ω_i, $\hat{n_J}$ and $\triangle S_J$ represent the volume of the ith cell, unit normal and area (or length in 2D) of the Jth interface respectively. Eqn.(3) can be further simplified as ,

$$\frac{\triangle_t U_i}{\Delta t} + \frac{1}{\Omega_i}\sum_J \left(F_{c\perp} + P_{\perp} - F_{v\perp}\right) \triangle S_J = 0 \tag{4}$$

with, $F_{C\perp} = [\rho u u_{\perp J} \;\; \rho v u_{\perp J}]^T$, $P_{\perp} = [p n_x \;\; p n_y]^T$, $F_{v\perp} = [\mu\frac{\partial u}{\partial n} \;\; \mu\frac{\partial v}{\partial n}]^T$ and $u_{\perp J} = (\vec{u} \cdot \hat{n})_J$. The second term in Eqn.(4), often referred to as residual, a discrete representation for the flux integral, is a function of U and can be determined either at n^{th} time level for an explicit procedure or at $(n+1)^{th}$ time level for an implicit procedure. Though the work primarily deals with the implicit procedure, explicit procedure is described to start with.

2.1. Explicit Procedure

One of the major challenges in the design of incompressible flow solvers is the lack of a governing equation for pressure update. Often the pressure field is recovered by manipulating the mass conservation equation. Hence, we describe an auxiliary velocity based procedure for recovering an equation for pressure update. Towards this, an equation for auxiliary velocity update is written by dropping the pressure term in Eqn.(4) as

$$\frac{\triangle_t \tilde{U}_i}{\Delta t} + \frac{1}{\Omega_i}\sum_J (F_{c\perp} - F_{v\perp})_J^n \triangle S_J = 0 \tag{5}$$

where, $\triangle_t \tilde{U}_i = \tilde{U}_i - U_i^n$. Subtracting Eqn.(5) from Eqn.(4), we have ,

$$\frac{U_i^{n+1} - \tilde{U}_i}{\Delta t} = -\frac{1}{\Omega_i}\sum_J P_{\perp J}^{n+1} \triangle S_J. \tag{6}$$

The RHS of the above equation represents an approximation for the volume integral of the pressure gradient, as prescribed below:

$$\int_{\Omega_i} \nabla p d\Omega = \oint_{\Gamma_i} p d\vec{\Gamma} \cong \sum_J p_J \hat{n}_J \triangle S_J. \tag{7}$$

Realizing that the finite volume state update formula returns cell averaged values, we have from Eqn.(6),

$$\int_{\Omega_i} \left[\frac{U^{n+1} - \tilde{U}}{\Delta t} + \nabla p\right] d\Omega = 0. \tag{8}$$

It follows from Eqn.(8),

$$\Delta t\ \nabla^2 p - \nabla \cdot \tilde{U} = 0 \tag{9}$$

where, the fact that the velocity field in the $(n+1)^{th}$ time is divergence free has been invoked. A more practical way to obtain an iterative solution to Eqn.(8) is introducing a pseudo-time derivative as below:

$$\frac{\partial p}{\partial \tau} = \Delta t\ \nabla^2 p - \nabla \cdot \tilde{U}. \tag{10}$$

Using Eqn.(10), knowing the auxiliary velocity field, the pressure field can be updated. The above pressure equation can be solved implicitly. Eqn.(8) also suggests the auxiliary velocity field can be corrected to obtain the velocity field at the $(n+1)^{th}$ time level as below:

$$U_i^{n+1} = \tilde{U}_i - \Delta t\ \nabla p_i. \tag{11}$$

It is important to note that in Eqn.(5), the convective fluxes are determined using an upwind formulation:

$$F_{c\perp J} = F_{c\perp}^{+}(U_L) + F_{c\perp}^{-}(U_R) = \left[\frac{u_\perp + |u_\perp|}{2}\begin{bmatrix}\rho u\\ \rho v\end{bmatrix}\right]_L + \left[\frac{u_\perp - |u_\perp|}{2}\begin{bmatrix}\rho u\\ \rho v\end{bmatrix}\right]_R. \tag{12}$$

In the above equation, subscripts L and R represent states as projected from left cell i and right cell j using a linear reconstruction procedure. The viscous fluxes in Eqn.(5) are determined using a diamond path reconstruction[7].

2.2. Implicit procedure

As remarked earlier, an implicit update demands that the residual is computed at $(n+1)^{th}$ time level and Eqn.(4) can be recast as below:

$$\frac{\Delta_t U_i}{\Delta t} + \frac{1}{\Omega_i}\sum_J \left[F_{c\perp}^{+}(U_i^{n+1}) + F_{c\perp}^{-}(U_j^{n+1}) + P_\perp^{n+1} - F_{v\perp J}^{n+1}\right]\Delta S_J = 0 \tag{13}$$

An inconsistent linearization of the above equation results in,

$$\frac{\Delta_t U_i}{\Delta t} = -\frac{1}{\Omega_i}\Bigg[\sum_J (F_{C\perp} - F_{V\perp})_J^n \Delta S_J + \sum_J P_{\perp J}^{n+1}\Delta S_J + \sum_J \left(A_i^{+}\Delta_t U_i + A_j^{-}\Delta_t U_j\right)\Delta S_J + \sum_J \nu \frac{\Delta_t U_j - \Delta_t U_i}{|\vec{r_{ji}} \cdot \hat{n}_J|}\Delta S_J\Bigg] \tag{14}$$

The third and fourth terms in the RHS represent the convective and viscous flux linearization respectively, with $A_i^+ = a_i^+ I$, $A_j^- = a_j^- I$ and

$$a_i{}^{+} = \left\{\begin{array}{ll} 0, & \text{for } Pe_i < -1 \\ 0.25d(Pe_i+1)^2, & \text{for } -1 \le Pe_i \le 1 \\ Pe_i \cdot d, & \text{for } 1 < Pe_i \end{array}\right\}$$

$$a_j{}^{-} = \left\{\begin{array}{ll} 0, & \text{for } Pe_j < 1 \\ -0.25d(-Pe_j+1)^2, & \text{for } -1 \le Pe_j \le 1 \\ Pe_j \cdot d, & \text{for } -1 > Pe_j \end{array}\right\}$$

where Pe_i and Pe_j are defined as

$$r = \vec{r_{ji}} \cdot \hat{n} \qquad d = \frac{\nu}{r} \qquad Pe_i = \frac{u_{\perp i} r}{\nu} \qquad Pe_j = \frac{u_{\perp j} r}{\nu}.$$

The above linearization of the convective flux is continuously differentiable and known to significantly enhance the performance of the implicit solver. An update for the auxiliary velocity field may be obtained from Eqn.(14) simply by dropping the pressure terms in the RHS.

$$\frac{\triangle_t \tilde{U}_i}{\triangle t} = -\frac{1}{\Omega_i}\left[\sum_J (F_{C\perp} - F_{V\perp})^n_J \triangle S_J + \sum_J \left(A_i^+ \triangle_t U_i + A_j^- \triangle_t U_j\right) \triangle S_J + \sum_J \nu \frac{\triangle_t U_j - \triangle_t U_i}{|\vec{r_{ji}} \cdot \hat{n}_J|} \triangle S_J\right]. \tag{15}$$

Subtracting Eqn.(15) from Eqn.(14) results in,

$$\frac{\triangle_t U_i - \triangle_t \tilde{U}_i}{\triangle t} = -\frac{1}{\Omega_i} \sum_J P^{n+1}_{\perp J} \triangle S_J \tag{16}$$

Noting that the velocity field at $(n+1)^{th}$ time level is divergence free, the following Poisson equation for pressure may be obtained:

$$\nabla \cdot \tilde{U} = \triangle t \, \nabla^2 p. \tag{17}$$

The above equation is solved in pseudo-time implicitly akin to the explicit procedure. The pressure thus obtained may be substituted in Eqn.(14) for velocity update. Owing to the implicit nature of the equation, it is important to establish a robust iterative procedure to solve Eqn.(14). We employ an SGS iteration as described in the following pseudo code:

- **Step 1** Let $\triangle_t U_i^{(0)} = 0 \quad \forall \ i$
- **Step 2**

$$\frac{\triangle_t \tilde{U}_i}{\triangle t} = -\frac{1}{\Omega_i}\left[\sum_J (F_{C\perp} - F_{V\perp})^n_J \triangle S_J + \sum_J \left(A_i^+ \triangle_t U_i^{(m-1)} + A_j^- \triangle_t U_j^{(m-1)}\right) \triangle S_J + \sum_J \nu \frac{\triangle_t U_j^{(m-1)} - \triangle_t U_i^{(m-1)}}{|\vec{r_{ji}} \cdot \hat{n}_J|} \triangle S_J\right] \tag{18}$$

- **Step 3** For obtaining $p^{(m)}$, solve

$$\frac{\partial p}{\partial \tau} = \nabla \cdot \triangle t \, \nabla p - \nabla \cdot \tilde{U} \tag{19}$$

- **Step 4** Perform a forward and backward sweep of the SGS iteration, with $\triangle_t w_i^{(0)} = \triangle_t U_i^{(m-1)} \ \forall \ i$
  $\Rightarrow$ For forward sweep, $i = 1$ to nc

$$\left[I + \frac{\triangle t}{\Omega_i}\left(\sum_J A_i^+ \triangle S_J + \sum_J \frac{\nu}{|\vec{r_{ji}} \cdot \hat{n}_J|} \triangle S_J\right)\right] \triangle_t w_i^{(k)} = -\frac{\triangle t}{\Omega_i}\left[\sum_J (F_{c\perp} - F_{v\perp})^n_J \triangle S_J + \sum_J P^{(m)}_{\perp J} \triangle S_J + \sum_{\substack{J \\ j<i}} \left(A_j^- + \frac{\nu}{|\vec{r_{ji}} \cdot \hat{n}_J|}\right) \triangle_t w_j^{(k)} \triangle S_J + \sum_{\substack{J \\ j>i}} \left(A_j^- + \frac{\nu}{|\vec{r_{ji}} \cdot \hat{n}_J|}\right) \triangle_t w_j^{(k-1)} \triangle S_J\right] \tag{20}$$

$\Rightarrow$ For backward sweep, i = nc to 1

$$\left[I + \frac{\Delta t}{\Omega_i}\left(\sum_J A_i^+ \Delta S_J + \sum_J \frac{\nu}{|\vec{r_{ji}} \cdot \hat{n}_J|}\Delta S_J\right)\right] \Delta_t w_i^{(k)} = -\frac{\Delta t}{\Omega_i}\left[\sum_J (F_{c\perp} - F_{v\perp})_J^n \Delta S_J + \sum_J P_{\perp J}^{(m)} \Delta S_J + \sum_{\substack{J \\ j>i}} \left(A_j^- + \frac{\nu}{|\vec{r_{ji}} \cdot \hat{n}_J|}\right) \Delta_t w_j^{(k)} \Delta S_J + \sum_{\substack{J \\ j<i}} \left(A_j^- + \frac{\nu}{|\vec{r_{ji}} \cdot \hat{n}_J|}\right) \Delta_t w_j^{(k-1)} \Delta S_J\right] \tag{21}$$

$\Rightarrow \Delta_t U_i^{(m)} = \Delta_t w_i^{(k)} \;\; \forall \;\; i$.

- **Step 5** Repeat the steps till convergence

It is important to note that the above implicit iterations have been made possible primarily because $\Delta_t \tilde{U}_i$ is allowed to lag by one sub iteration. This is of no consequence, assuming a convergence of $\Delta_t U_i$. The number of SGS iterations pertaining to Eqns.[20] and [21] and total number of iterations for the pressure field are left to users choice. In the present work, one SGS iteration corresponding to one forward and one backward sweep and one pressure iteration are found to be optimum. The most important aspect of the present procedure is that velocity resulting from the implicit iterations is divergence free and therefore there is no need for any velocity correction step.

3. Results

In this section, we have presented the validation study of the incompressible solver developed with the proposed methodology. Laminar test cases are considered, the details of the test conditions and grid size are provided in Table.(1). The numerical results are compared with the experiment or standard reference values in Fig.(1) to Fig.(4) and Table(2). The results clearly establish the accuracy of the proposed solution methodology.

Table 1. Test Cases and Grid Details

Sl.no	Geometry	Re	α deg	Grid type	Grid size (no of volumes)
1	Circular cylinder	40	-	O type structured	8000
2	NACA0012	10000	0	hybrid unstructured	12164
3	Lid driven Cavity	10000	0	structured	16641

Table 2. Flow over Cylinder: Comparison of drag coefficient (C_D)

Re	Present work	Lima et al[8]	Park et al [9]	Sucker et al[10]	Triton [11]
10	2.78	2.81	2.78	2.67	
20	2.03	2.04	2.01	2.08	2.22
40	1.53	1.54	1.51	1.73	1.48

4. Conclusions

The preliminary results presented show lot of promise. The proposed method can be seamlessly integrated into any compressible solver. Efforts are on to further extend the method to turbulent and 3D flows.

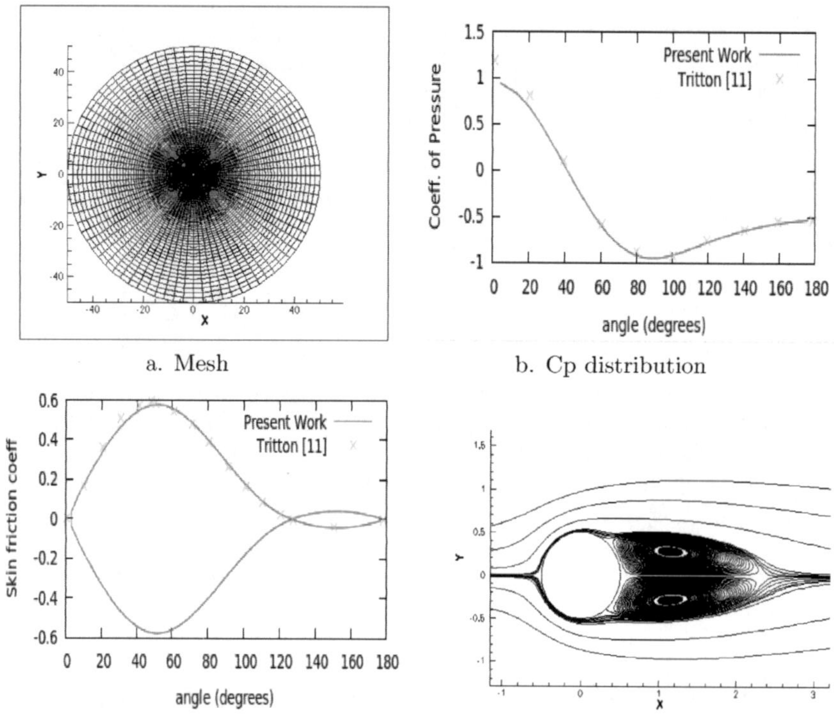

a. Mesh

b. Cp distribution

c. Sfc distribution

d. Separation bubble

Figure 1. Flow over Cylinder at Re = 40

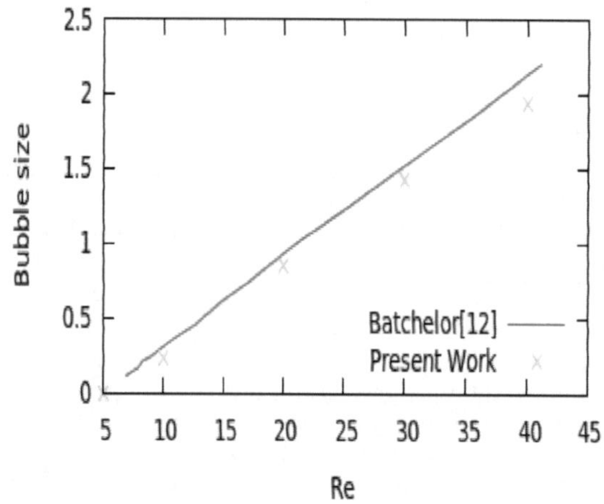

Figure 2. Flow over Cylider: Variation of bubble size with Re

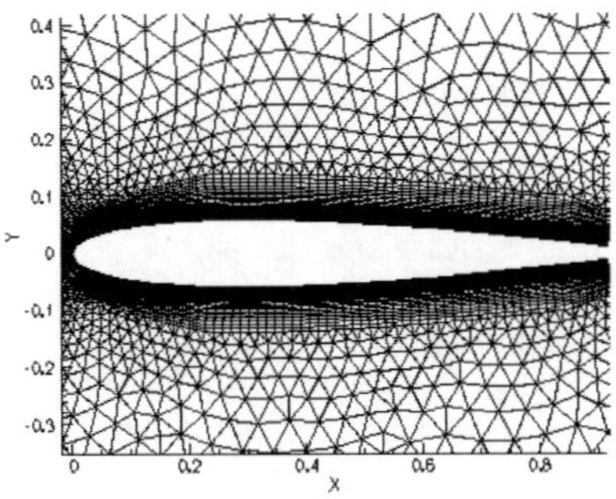

a. Mesh

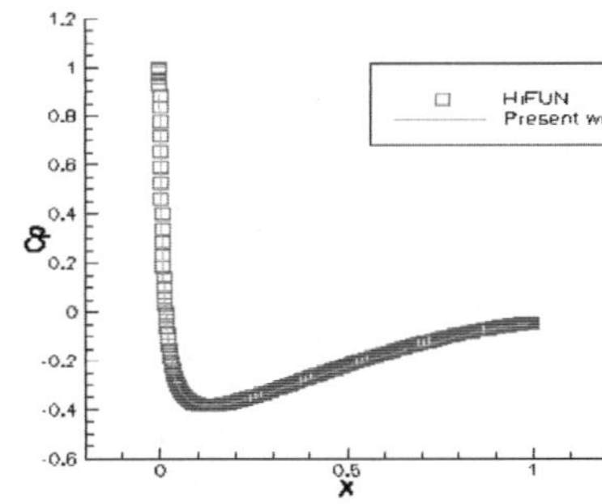

b. Cp distribution

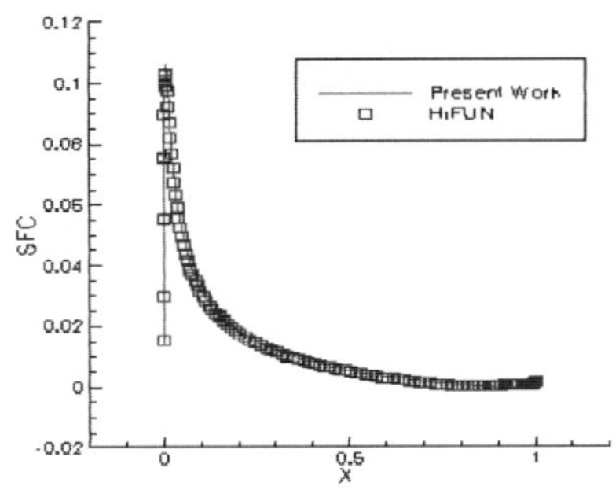

c. Sfc distribution

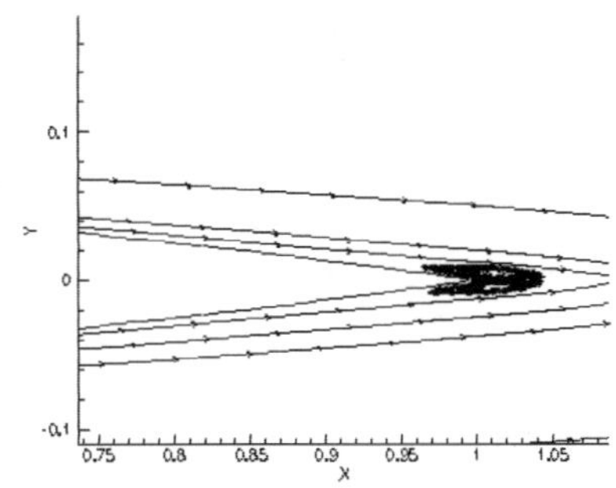

d. Separation bubble

Figure 3. Flow over NACA0012 at Re = 10000

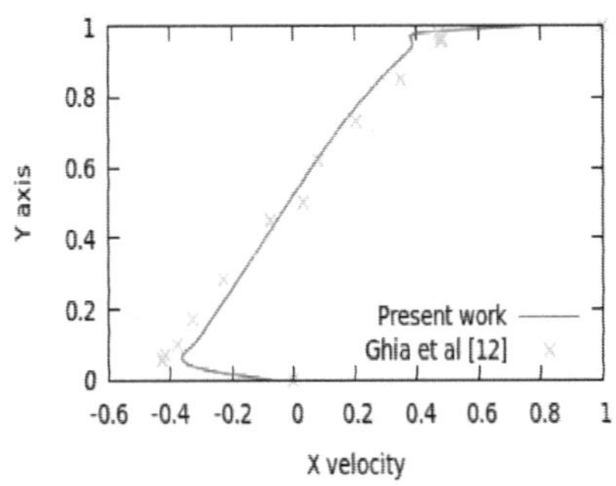

a. X velocity

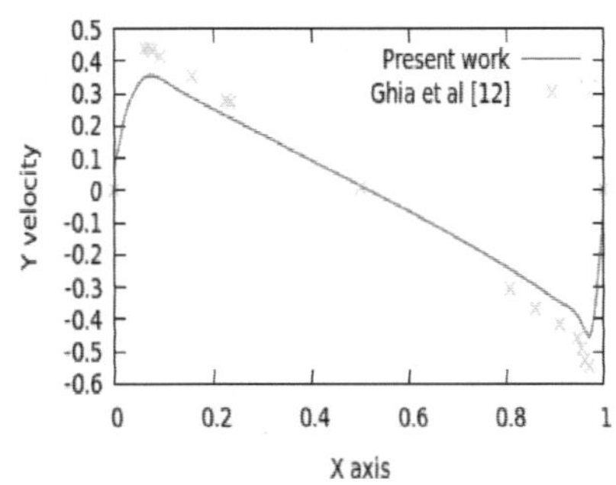

b. Y velocity

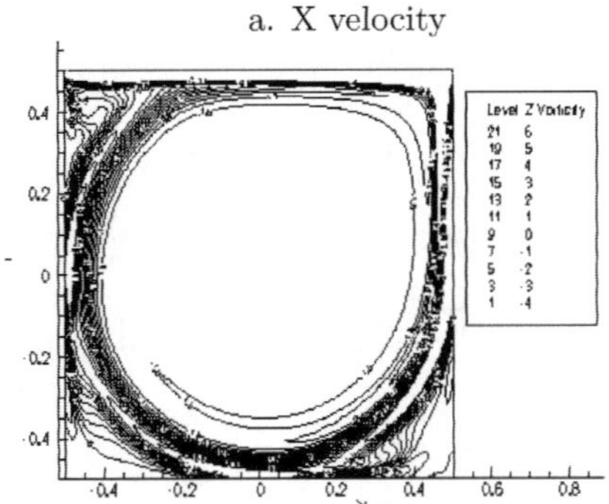

c. Vorticity contour

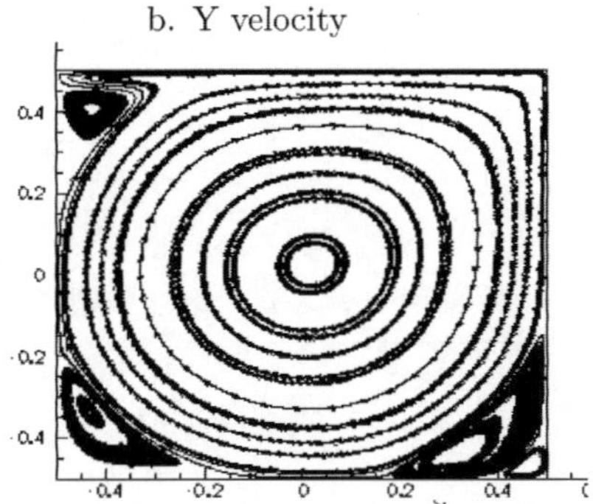

d. Streamlines contour

Figure 4. Lid driven cavity flow at Re = 10000

5. References

[1] Gresho P M, 1991 *Annu. Rev. Fluid Mech.* **23** 413
[2] Suhas V Patankar *Numerical Heat Transfer and Fluid Flow* (Florida: CRC Press)
[3] Barth T J 1992 *Proc. Int. Conf. on Numerical Methods in Fluid Dynamics* **414** 240
[4] Shende N and Balakrishnan N 2004 *AIAA Journal* **42** 1863
[5] Jameson A 1991 *AIAA Paper* 1596
[6] Ganesh V, Nikhil V Shende and Balakrishnan N 2009 *Computers & Fluids* **38** 1799
[7] Jawahar P and Hemanth Kamath 2000 *J. Comp. Physics* **164** 165
[8] Lima E Silva A L F, Silveira-Neto and Damasceno J J R 2003 *J. Comp. Physics* **189** 351
[9] Park J, Kwon K and Choi H 1998 *KSME Int. J.* **12** 1200
[10] Sucker D and Brauer H 1975 *Warme Stoffubertragung* **8** 149
[11] Tritton D J 1959 *J. Fluid Mech.* **6** 547
[12] Batchelor G K *An Introduction to Fluid Dynamics* (Cambridge: Cambridge University Press)
[13] Ghia U, Ghia K N and Shin C T 1982 *J. Comput. Physics* **48** 387
[14] HiFUN *www.sandi.co.in*

IOP Conf. Series: Journal of Physics: Conf. Series **822** (2017) 012036 doi:10.1088/1742-6596/822/1/012036

Numerical simulation of incompressible turbulent flows in presence of laminar to turbulent transition

Satish G[1] , Vashista G A[2] , Sekhar Majumdar[3,4]

Center of Excellence for Computational Fluid Dynamics
Department of Mechanical Engineering
Nitte Meenakshi Institute of Technology, Bangalore 560064, India
[1]Research Associate, [2]Post Graduate student, [3] Professor
[4]Any correspondence to be addressed to the last author

E-mail : essem17@gmail.com

Abstract. Most of the widely used popular mathematical models of turbulence use a judicious combination of intuition, empiricism and the governing equations of instantaneous and mean motion-valid strictly for fully developed turbulence without any laminar region. In reality however, any wall bounded or free shear flow may consist of some laminar flow patches which eventually undergo transition over a finite length to grow into fully turbulent flows. Most of the turbulence models used in commercial CFD codes, are unable to predict the dynamics of turbulent flows with laminar patches. However, accurate prediction of transitional flows is often essential to estimate the pressure losses and/or heat transfer in industrial applications. The present paper implements two different transition models in an existing finite volume URANS-based code RANS3D, developed in house and validated against reliable measurement data for flow past flat plates with different free stream turbulence levels and flow past SD7003 aerofoil at a chord-based Reynolds number of 60,000.

1. Introduction

In the last few decades, various robust CFD codes have been developed to simulate fully turbulent flows in complex configuration, producing results, acceptable for engineering design applications. These codes employ wide range of semi-empirical turbulence models, maintaining a balance between the accuracy requirements, level of information required and the computational resources available. Unfortunately, none of these RANS equation based CFD algorithm, is capable of capturing the important effects of laminar-turbulent transition on the flow field. The underlying physics of the transition process has been explored by several researchers through careful measurements and/or powerful analyses based on complex theories to explain the phenomenon of intermittency. But none of these models are found compatible to be directly coupled to the Unsteady RANS Equation-based CFD codes, used by engineering design community, where turbulent stresses need to be evaluated in both structured and unstructured grid environment.

However, laminar to turbulent transition process is observed to have strong influence specially on the wall shear stress or heat transfer coefficients in wall shear layers for flow past solid surfaces like gas turbine blades. In spite of limited applicability, algebraic correlations based on very careful hot-wire measurements, proposed by Abu Ghanam *et al* [1], have been found to be quite useful for prediction of the inception and the extent of transition during the boundary layer development on a flat plate. Recently a more accurate transition model based on three strongly coupled transport equations *viz.,* laminar kinetic energy (k_L), turbulent kinetic energy (k_T) and specific turbulent dissipation(ω) is proposed by Walters *et al* [2] and validated for a few interesting test cases. In the present work, two different transition models, *viz.*, one algebraic correlation based [1] and another transport equations based model [2] are directly coupled to the flow solver RANS3D, developed by the last author and his research associates at NAL and NMIT Bangalore [3], for prediction of transitional flows in complex configuration. The performance of the models are evaluated through comparison to reliable measurement data reported by ERCOFTAC [4] for flat plates and also for two dimensional flow past a cambered SD7003 aerofoil, operating at a chord-based Reynolds number of 60,000 for which computation data [5,6] on aerodynamic performance are available for validation.

2. Mathematical Modeling

The dynamics of incompressible flow is governed by the continuity equation (Eq.1) and the Reynolds Averaged Navier Stokes (RANS) equations (Eq.2) in non-orthogonal curvilinear coordinates.

$$\textit{Mass:}\quad \frac{\partial}{\partial x_j}\left(\overline{U}_k \beta_k^j\right)=0 \tag{1}$$

$$\textit{Momentum:}\quad \frac{\partial \overline{U}_i}{\partial t}+\frac{1}{J}\frac{\partial}{\partial x_j}\left[\left(\overline{U}_i\overline{U}_k\beta_k^j\right)+\frac{p}{\rho}\beta_i^j\right]=\frac{1}{J}\frac{\partial}{\partial x_j}\left[\frac{\nu_{eff}}{J}\left(\frac{\partial \overline{U}_i}{\partial x_m}B_m^j+\frac{\partial \overline{U}_k}{\partial x_m}\beta_i^m\beta_j^k\right)\right] \tag{2}$$

p & $\overline{U}_i$ are the ensemble-averaged static pressure & cartesian velocity component along i-th direction respectively; ν_{eff} is the effective kinematic eddy viscosity derived from the transition cum turbulence model, J, the Jacobian and β_j^i & B_j^i are metric coefficients of transformation, ν & ρ are kinematic viscosity & density of the fluid.

2.1 Turbulence Model for Fully Turbulent Flow

Standard k-ω turbulence model computes the eddy viscosity, ν_t from the solution of the scalar transport equations (Eqs. 3 & 4) for turbulence energy k and specific dissipation ω respectively, where S_{ij} is the mean strain rate and P_k is the production of turbulence energy.

$$D_t k = P_k - \varepsilon + \nabla.\left[\frac{\nu_{eff}}{\sigma_k}\nabla k\right] \tag{3}$$

$$\textit{and}\quad D_t\omega = \alpha P_k\frac{\omega}{k} - \beta\omega^2 + \nabla.\left[\frac{\nu_{eff}}{\sigma_\omega}\nabla\omega\right] \tag{4}$$

The eddy viscosity is computed as $\nu_t = k/\omega$ and the model constants are $\alpha = 5/9$, $\beta = 3/40$, $\beta^* = 0.09$, $\sigma_\omega = \sigma_k = 2$ and $P_k = 2\nu_t\left|S_{ij}S_{ij}\right|$ However, in case of transitional flows, the updation of the effective eddy viscosity in the above equations is assumed to be a linear combination of the laminar viscosity ν_l and the eddy viscosity ν_t given by the turbulence model for fully turbulent flow The intermittency γ derived from the transition model, described in the next section, is used as the weighting function to compute ν_{eff} as follows

$$\nu_{eff} = \gamma\nu_t + (1-\gamma)\nu_l \tag{5}$$

2.2 Models for Transition Simulation

2.2.1 Algebraic Correlation based Transition Model

This empirical model [1] is based mainly on the transverse velocity profiles measured at different longitudinal stations, using Hot Wire Anemometers for flow over a smooth flat plate in a low speed wind tunnel. The important finding from these detailed measurements are the locations of the start and end of the transition along the plate. The non dimensional transition length parameter is η when $\eta = \frac{\mathrm{Re}_x - \mathrm{Re}_{xs}}{\mathrm{Re}_{xe} - \mathrm{Re}_{xs}}$ and the intermittency $\gamma(x)$ is expressed as $\gamma(x) = 1 - \exp(-5\eta^3)$ where, $\mathrm{Re}_x = \rho x U / \mu$ is the local Reynolds number based on the distance x from the leading edge of the plate and free stream velocity U; Re_{xs} is the Reynolds number based on x_s, the length at the start (onset) of transition, identified as the point where the local momentum thickness based Reynolds number Re_θ exceeds $\mathrm{Re}_{\theta s}$ defined by the following empirical equation derived from measurement data as

$$\mathrm{Re}_{\theta s} = 163 + \exp\left\{ F(\lambda_\theta) - \frac{F(\lambda_\theta)}{6.91} Tu \right\} \tag{6}$$

where the function $F(\lambda_\theta)$ is expressed by Eq. 7, as function of the longitudinal pressure gradient λ_θ imposed on the boundary layer

$$\begin{aligned} F(\lambda_\theta) &= 6.91 + 12.75\lambda_\theta + 63.64\lambda_\theta^2 \quad \text{when } \lambda_\theta < 0 \\ F(\lambda_\theta) &= 6.91 + 2.48\lambda_\theta - 12.27\lambda_\theta^2 \quad \text{when } \lambda_\theta > 0 \end{aligned} \tag{7}$$

where, $\lambda_\theta = \frac{\theta^2}{\nu} \frac{dU_e}{dx}$ is the pressure gradient parameter and U_e is velocity at the free stream edge of the boundary layer. Based on the inception of transition or onset Reynolds number, Re_{xs} and the Reynolds number Re_{xe} at the end of transition location x_e is determined using the following relationship:

$$\mathrm{Re}_{Xe} = \mathrm{Re}_{Xs} + 16.8\,\mathrm{Re}_{Xs}^{0.8} \tag{8}$$

Once the longitudinal variation of the intermittency factor γ is determined from the above set of correlations, one may use the linear formula (Eq. 5) to evaluate the effective eddy viscosity to be used for solving the transport equations for momentum & the relevant turbulence scalars.

2.2.2 Three Equation Eddy Viscosity based Model

The difficulty of correlation-based approach is that one needs to prescribe the location of onset and extent of transition along with some empirical relationship for the variation of intermittency. Specially for complex geometry where non-orthogonal structured or unstructured grid is used, it is difficult to compute the momentum thickness based Reynolds number which decides the onset of transition. In the RANS framework, the major assumption is to relate the effects of turbulence through the local eddy viscosity, represented by transport equations for the velocity and length scales of turbulence. The present three equation model [2] includes a third transport equation for laminar kinetic energy which in the pre-transition zone of laminar flow amplifies external disturbance leading to high amplitude streamwise fluctuations that eventually cause the transition through breakdown. Three scalar transport equations solved along with the mean momentum and continuity equations are as follows :

$$D_t k_T = P_{kT} + R_{BP} + R_{NAT} - \omega k_T - D_T + \nabla . \left[\left[\nu + \frac{\alpha_t}{\sigma_k} \right] \nabla k_T \right] \tag{9}$$

$$D_t k_L = P_{kL} - R_{BP} - R_{NAT} - D_L + \nabla . \left[\nu \nabla k_L \right] \tag{10}$$

$$D_t \omega = C_{\omega 1} \frac{\omega}{k_T} P_{kT} - C_{\omega 2} \omega^2 + C_{\omega 3} f_\omega \alpha_T f_w^2 \frac{\sqrt{k_T}}{d^3} + \nabla . \left[\left[\nu + \frac{\alpha_t}{\sigma_k} \right] \nabla \omega \right] \tag{11}$$

The first source term on the right hand side for all the equations represent the corresponding production term and the next two terms on the right side of the first two equations represent the production terms arising out of transition and the suffices *BP* and *NAT* represent *B*y *P*ass for high free stream turbulence and *NAT*ural transition for low free stream turbulence level respectively. The computation of production and dissipation terms of k_l , k_t and ω and the closure coefficients based on intuition, empiricism and measurement data are discussed in details by Walters *et al* [2].

3. Finite Volume Method

All the computations in the present work use the solver RANS3D for unsteady incompressible flows. The code employs a finite volume procedure [3] in a multiblock structured grid environment, based on a pressure-velocity solution strategy where an implicit predictor-corrector scheme, similar to the SIMPLE algorithm has been used - but modified for cell-centered variable arrangement using Momentum Interpolation technique. Simple stretched Cartesian grids are used for the flat plate, whereas for the aerofoil, an appropriate curvilinear boundary fitted C-Grid topology is generated with geometric stretching near the wall, using a differential-algebraic procedure [4].

4. Results and Discussions

Two different test cases analysed in the present work are – (i) Flow over a smooth Flat Plate with sharp leading edge under zero pressure gradient and (ii) Flow past an isolated SD7003 aerofoil at a chord-based Reynolds number of 60,000 for different angles of attack.

4.1 Flow past a Flat Plate at Zero Pressure Gradient

The first test case considered is transitional flow past a flat plate for two different levels of free stream turbulence marked as T3A and T3B for which detailed measurement data on skin friction coefficient has been reported by ERCOFTAC [5].

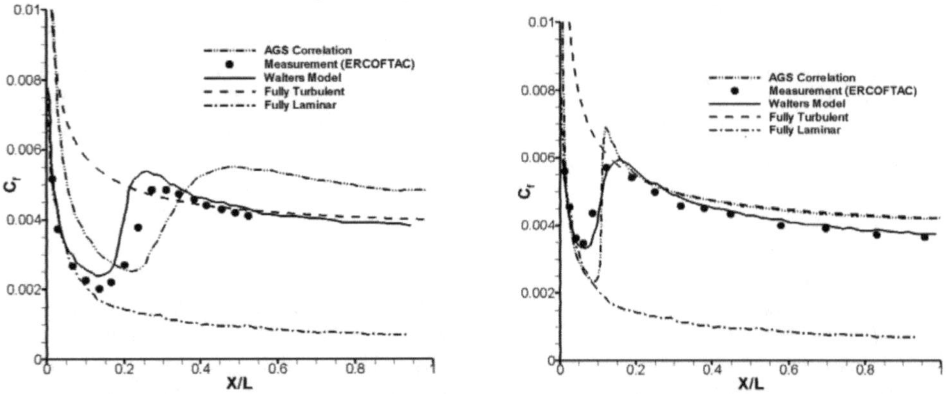

Case T3A : Free Stream Turbulence Level =3.3% **Case T3B : Free Stream Turbulence Level =6.5%**

Figure 1. Variation of Skin Friction coefficient along the flat plate for zero pressure gradient

The present computation uses k-ω turbulence model coupled either to the algebraic correlation model of Abu Ghannam *et al* [1] or the three equation model of Walters *et al* [2] for transition. The variation of C_f along the plate is shown in Fig. 1 for zero pressure gradient. In the laminar zone, reasonable agreement is observed between the present prediction and measurement data. However the skin friction coefficient in both the cases is observed to be overpredicted and this disagreement may be attributed to the gross approximations in linear weighting of the laminar and turbulent viscosity in the transition zone where the dynamics obviously is strongly non-linear. For both the cases of zero pressure gradient flow over a flat plate with a sharp leading edge, the computed skin friction coefficient value over the plate length is in reasonably good agreement with the corresponding measurement data.

4.2 Transitional Flow past an isolated SD -7003 Aerofoil

This test case computes flow past a SD-7003 aerofoil for different angles of attack at a relatively low value of the chord-based Reynolds number (Re=60,000). The farfield is assumed to be at a distance of 10C from the aerofoil trailing edge and a structured C-grid topology (516×103) is used for the computation. A zoomed view of the grid is shown in Fig. 2, and the wall-normal distance is maintained to be 8×10^{-4} times the chord length so that the value of the near wall y^+ is less than unity everywhere in the field. The oncoming flow is assumed to be uniform with constant values of the turbulent kinetic energy (k_T) and the

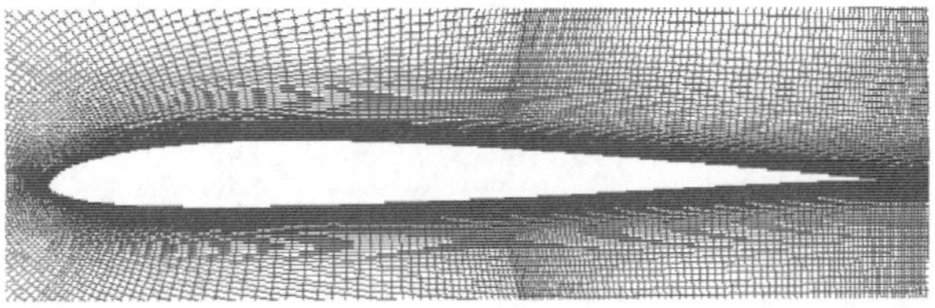

Figure 2. Zoomed view of C Grid for SD 7003 aerofoil

specific dissipation rate (ω), based on the assumption that the mean turbulence level is 0.1% and the eddy viscosity level is ten times the laminar viscosity of the fluid.

4.2.1 Surface Pressure and Skin Friction Coefficient

Fig 3 shows the pressure distribution on the aerofoil surface for two different angles of attack ($\alpha = 2^0$ and $\alpha = 4^0$) at a chord-based Reynolds number of 60000, computed using the transport equation-based transition model [2]. In both the cases, the transition is manifested through a local jump (kink) of Cp at some location on the upper suction surface of the aerofoil. Location

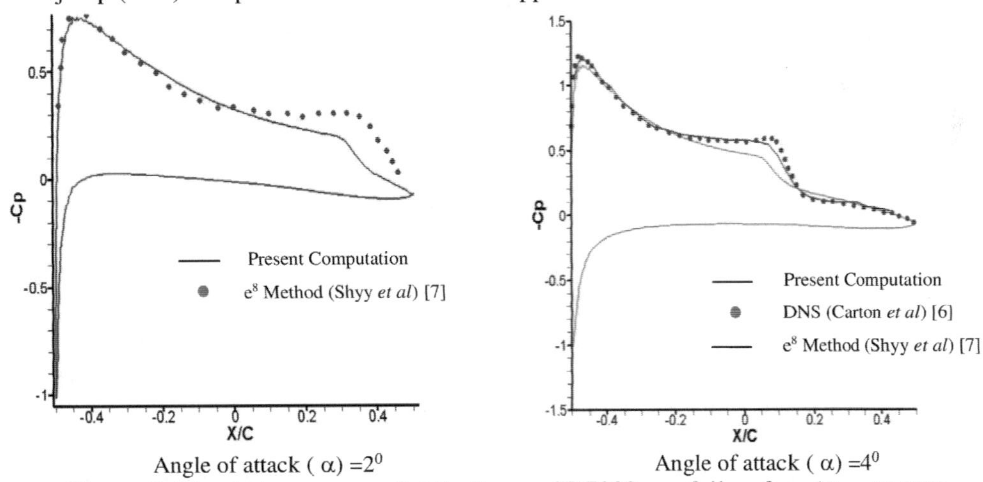

Angle of attack (α) $=2^0$ Angle of attack (α) $=4^0$

Figure 3. Chordwise pressure distribution on SD7003 aerofoil surface (Re = 60,000)

of this kink shifts upstream from x/C=0.3 at $\alpha = 4^0$ to x/C=0.3 at $\alpha = 4^0$. Present prediction is compared to other [6,7] computation results – one is RANS computation [7], where the transition is assumed to occur when the initial disturbance amplitude is magnified by a factor of e^8, whereas the other computation [6] uses DNS for the same flow. It is interesting to observe that the location of the jump in Cp is almost identical in all the three computations. However the discrepancies observed in amplitude of the pressure jump may be improved further by finer local resolution of the grid.

4.3.2 Aerodynamic Coefficients

In order to assess the accuracy level of the present computation, the computed aerodynamic coefficients, have been compared to the corresponding measurement data [9] and other computation results [7,8] in Fig.4. The three-equation transition model is observed to marginally improve the lift coefficient, but considerable improvement is observed specially in the drag coefficient where the present computation is found to be quite close to the measurement data. The present three equation eddy viscosity-based transition-turbulence model is found to be capable of capturing the physics of transition for attached and separated boundary layer flows.

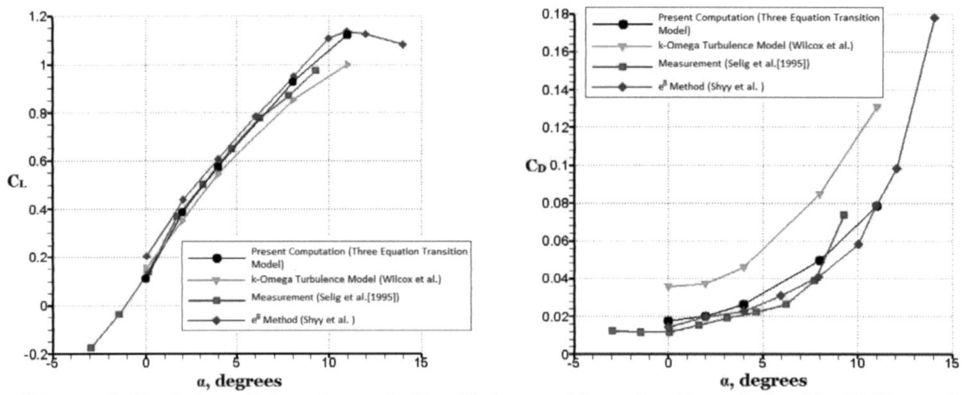

Figure 4. Variation of Aerodynamic Coefficients with angle of attack for SD-7003 aerofoil

5. Conclusion

The present simulation employing appropriate transition-turbulence model in the existing RANS3D code, successfully captures the transitional flow for attached boundary layers on flat plate and also for the separation bubble induced transition on an aerofoil surface. However the present model needs to be further validated for more complex transition problems like the drag crisis for flow around a circular cylinder.

Acknowledgement

The authors thankfully acknowledge the Grant-in Aid research fund provided by the Hydrodynamics Panel, Naval Research Board (DRDO) for the present project. The authors also wish to thank the Director and Principal NMIT, Bangalore for their constant encouragement and permission to present and publish this work.

References

[1] Abu-Ghannam, B. J., and Shaw, R. 1980 Natural transition of boundary layers—the effects of turbulence, pressure gradient & flow history *Journal of Mechanical Engineering Science* **22** 213–228

[2] Walters, D. K and Cokaljat, D. 2008 A three-equation eddy-viscosity model for Reynolds- Averaged Navier-Stokes simulations of transition flow *Journal of Fluids Engineering, ASME* **130**

[3] Majumdar S, Rajani B.N., Kulkarni D.S., and Subrahmanya M.B 2009 CFD simulation of low speed turbulent flow problems using unsteady RANS & large eddy simulation approach *Journal of Aerospace Sciences & Technologies* **Special Issue**

[4] Fathima A., Baldawa N.S., Pal, and S., Majumdar, S.1994 Grid generation for arbitrary 2D configurations using a differential algebraic hybrid method *NAL PD CF9416*

[5] Coupland, J. 1990, ERCOFTAC *Special Interest Group on Laminar to Turbulent Transition and Retransition: T3A and T3B Test Cases*

[6] Carton de Wiart C and Hillewaert K 2012 DNS and LES of transitional flow around

SD7003 using a higher order discontinuous galerkin method, *ICCFD7-3604*
[7] Yongsheng Lian and Wei Shyy 2007 Laminar-turbulent transition of a low Reynolds number rigid or flexible aerofoil *AIAA Journal* **45** 1501-1513
[8] Wilcox D.C.1993 Turbulence Modeling for CFD *DCW Industries Inc.,California*
[9] Selig M.S.,Guglielmo J.J.,Broeren A.P., and Giguere P. 1995, Summary of low speed airfoil data, *University of Illinois at Urbana-Champaign*

Numerical simulation of supersonic separating-reattaching flow through RANS

Rahul Kumar Soni[1], Nitish Arya[1] and Ashoke De[2]

[1]Graduate Student, Department of Aerospace Engineering, IIT Kanpur, Kanpur, 208016, India

[2]Associate Professor, Department of Aerospace Engineering, IIT Kanpur, Kanpur, 208016, India

Email: ashoke@iitk.ac.in

Abstract. The present work reports turbulent supersonic flow subjected to separation-reattachment computed through Reynolds Averaged Navier-Stokes (RANS) based models .The modified density based solver, *rhoCentralFoam* in OpenFOAM framework, is employed for the flow simulation over a backward facing step and a rectangular cavity. .The predictions by the different RANS models show good agreement with the empirical and numerical observations. Qualitatively, mean flow features such as boundary layer separation, re-circulation, and expansion fan and re-attachment shock are predicted with acceptable accuracy; while the quantitative assessment reveals the limitation of different models.

1. Introduction

Separating-reattaching flows have been a subject of various studies in the past due to the complicated flow features as well as due to the wide engineering applications. For example, backward facing steps and rectangular cavities have been used for various applications in the aerospace industry. Backward facing step being a classical problem has been widely studied for its interesting flow physics by various researchers [1-8]. In a compressible flow regime, it includes many features such as the expansion corner, flow separation recirculation zone etc. Hama (1968) in their investigation found that the flow separation occurs slightly below the corner forming a lip shock. The free stream above the corner gives rise to an expansion fan resulting in sharp pressure drop behind the step. The separated boundary layer develops into a free shear layer and on reaching the wall it is compressed and the low velocity part gives rise to a recirculation zone. The free stream layer which was earlier expanded by the corner is turned back to a direction more or less parallel to the original flow through a reattachment shock wave.

Like the backward facing step, cavities have also been known to increase the residence time of the flow inside the combustor. As put forward by various studies [9-15] the striking feature of the flow over a rectangular cavity is the shear layer, which amplifies the flow disturbance and subsequent scattering of these disturbances into acoustic waves at the downstream corner. The acoustic waves generated propagate upstream and create and excite further disturbances in the shear layer generating self-sustained oscillations at various resonant frequencies. The cavity flow may be broadly divided into four types open, closed transitional open and transitional closed as suggested by [12]. The present study only focuses on open cavity (L/D < 13).

A handful of available literature deals with the study regarding the supersonic application of RANS based turbulence models which offer lower computational cost and very much essential for the problems of practical interests and industrial design. However, there is still a need for a systematic investigation of these flows. Therefore, the aim of the present work is to characterize the supersonic flow over the geometries subjected to separation and reattachment and simultaneously assess various RANS based turbulence models. Flow past two geometries is computed namely: backward facing step and rectangular cavity (L/D = 3). The details of these cases can be found in [5] & [11]. Both the cases, as reported widely in the literatures, pose very complicated flow physics involving boundary-layer separation, shock-boundary layer interaction and flow re-attachment. Overall, these cases appear to be ideal candidates for the assessment of the RANS based models in supersonic flow.

2. Numerical Details

2.1 Governing Equations

The Favre-averaged governing equations of fluid motion are discretized and solved using the Finite-Volume method. The equations are expressed as a set of partial differential equations (PDEs) which are derived by the application of the laws of conservation to fluid motion:

Continuity:

$$\frac{\partial \bar{\rho}}{\partial t} + \frac{\partial \bar{\rho}\tilde{u}_i}{\partial x_i} = 0 \tag{1}$$

Momentum:

$$\frac{\partial}{\partial t}(\bar{\rho}\tilde{u}_i) + \frac{\partial}{\partial x_j}[\bar{\rho}\tilde{u}_i\tilde{u}_j + \bar{p}\delta_{ij} + \overline{\rho u_i'' u_j''} - \overline{\tau_{ij}}] = 0 \tag{2}$$

Energy:

$$\frac{\partial}{\partial t}(\bar{\rho}\tilde{E}) + \frac{\partial}{\partial x_j}[\bar{\rho}\tilde{u}_j\tilde{E} + \tilde{u}_j\bar{p} + \overline{u_j'' p} + \overline{\rho u_i'' E''} + \overline{q_j} - \overline{u_i \tau_{ij}}] = 0 \tag{3}$$

Where (~) and (-) in above equation refer to time and Favre averaged quantity, $\overline{\tau_{ij}} = 2\mu S_{ij} - \frac{2}{3}S_{kk}\delta_{ij}$ is considered positive in compression and strain rate tensor is, $S_{ij} = \frac{1}{2}\left(\frac{\partial u_i}{\partial x_j} + \frac{\partial u_j}{\partial x_i}\right)$. Furthermore, total energy is defined as $\breve{E} = \breve{e} + \frac{\tilde{u}_k \tilde{u}_k}{2} + k$; where $\breve{e}$ and k being internal and turbulent kinetic energy respectively. The term $\overline{\rho u_i'' u_j''}$ in momentum equation leads to closure problem which upon utilizing the Boussinesq's hypothesis is treated analogous to viscous stress and defined as $-\overline{\rho u_i'' u_j''} = \mu_t\left(\frac{\partial u_i}{\partial x_j} + \frac{\partial u_j}{\partial x_i}\right) + \frac{2}{3}\rho k\delta_{ij}$ where μ_t is evaluated by complementing the above set of equation with the transport equation of turbulence quantities.

For all the cases, a no-slip boundary condition is enforced at the top and bottom wall along with condition that the normal pressure gradient vanishes at wall. At the outlet, flow variables are extrapolated and non-reflecting boundary condition (NRBC) is imposed to allow outgoing waves to exit flow domain without reflection.

2.2 Numerical Scheme

Numerical simulation of high speed flow involves various challenges due to discontinuties, such as shock waves. Robust solver is required to address these issues in efficient manner. Some of the novel methods are Riemann solvers, Jacobian evaluation and characteristics decomposition. Although these methods are known to offer non-oscillatory and accurate solution but are difficult to implement especially on polyhedral mesh. The density based solver in OpenFOAM framework is based on finite volume discretization utilizing semi-discrete, non-staggered central schemes for collocated variables on polyhedral mesh. The transport equations are solved using operator-splitting approach where initially, convection of conserved variables are solved through explicit predictor equation and then diffusion of primitive variables are solved using implicit corrector equation. This approach offers the quick solution of convection terms through diagonal solver, while diffusion terms are solved implicitly enabling numerical stability. The solver utilizes central schemes proposed by Kurganov & Tadmor [16] which is an alternative approach to Riemann solver offering accurate non-oscillatory solution. In present simulation, second order backward Euler scheme is utilized for the time integration whereas viscid and inviscid fluxes are discretised using central difference and TVD scheme. The parallel processing in OpenFOAM is achieved through the message passing interface (MPI) technique.

3. Results & Discussion

3.2 Backward Facing Step

The numerical results computed through various RANS models are compared against the experimental results. The streamwise velocity, static temperature and pressure profiles at x/h=3 are compared with experimental data, and reported in Fig. 1. Detailed result at other locations (x/h = 1.75 & 6.66) including grid independence can be found in the work of Soni et al. (2015). It is observed that the numerically predicted results match well with the experimental data for almost all the turbulence models, except in the region of y/h < 1.

It is also noticed that the results obtained by k-ε and SA are slightly different than those predicted by SST k-ε and RNG k-ε in the near wall region. The prediction of SST k-ε and RNG k-ε are in better agreement with experimental data. The boundary layer resulting from the formation of reattachment shock could be one reason for the difference observed in the pressure profile especially in case of k-ε, which is known to perform poorly in adverse pressure gradient affected regions.

Streamwise velocity along the wall is presented in Fig. 1 to demonstrate the extent of reattachment length. The prediction of k-ε is lesser as compared to other models which are almost comparable. Since the experimental observation does not offer insight into the reattachment length, comparison with the numerical data is not possible. It is worth noticing that RNG and SST prediction follow each other closely. The minimum velocity prediction by S-A is lowest however other models predict close to each other.

3.3 Rectangular Open Cavity

To perform the grid independence study, two sets of grids are generated namely Grid1and Grid2 with 75500 & 118750 cells, respectively. The normalized pressure distribution along cavity wall is presented in Fig. 2 for both grids. It can be witnessed that the solution for grid 2 appears to follow the experimental observation closely hence all the data reported in the present study corresponds to Grid2.

The velocity contours along with the streamlines for different turbulence models are presented in the Fig. 3. The recirculation region predicted by the k-ε and the RNG k-ε models does not follow the experimental observation accurately only SST k-ε model follows the observation of [11].

The Pressure distribution along the cavity wall predicted by the different turbulence models along with the experimental and numerical findings of Gruber et al. [2001] are presented in Fig. 4. The Pressure remains fairly constant along the front wall and decreases along the cavity floor. The compression of the shear layer at the rear end of the wall results in higher pressure at the rear wall of the cavity. The k-ε and RNG k-ε models show almost the same pressure distribution which is in accordance with the numerical and experimental results.

It is interesting to note that though the SST k-ε model predicts the qualitative data nicely, the pressure distribution near the wall shows small deviations from the empirical and numerical results. The deviations observed near the wall region for the SST k-ε model may be attributed to the poor performance of the wall model which could result from the shifting of the position of the first grid point from the logarithmic region of the boundary layer.

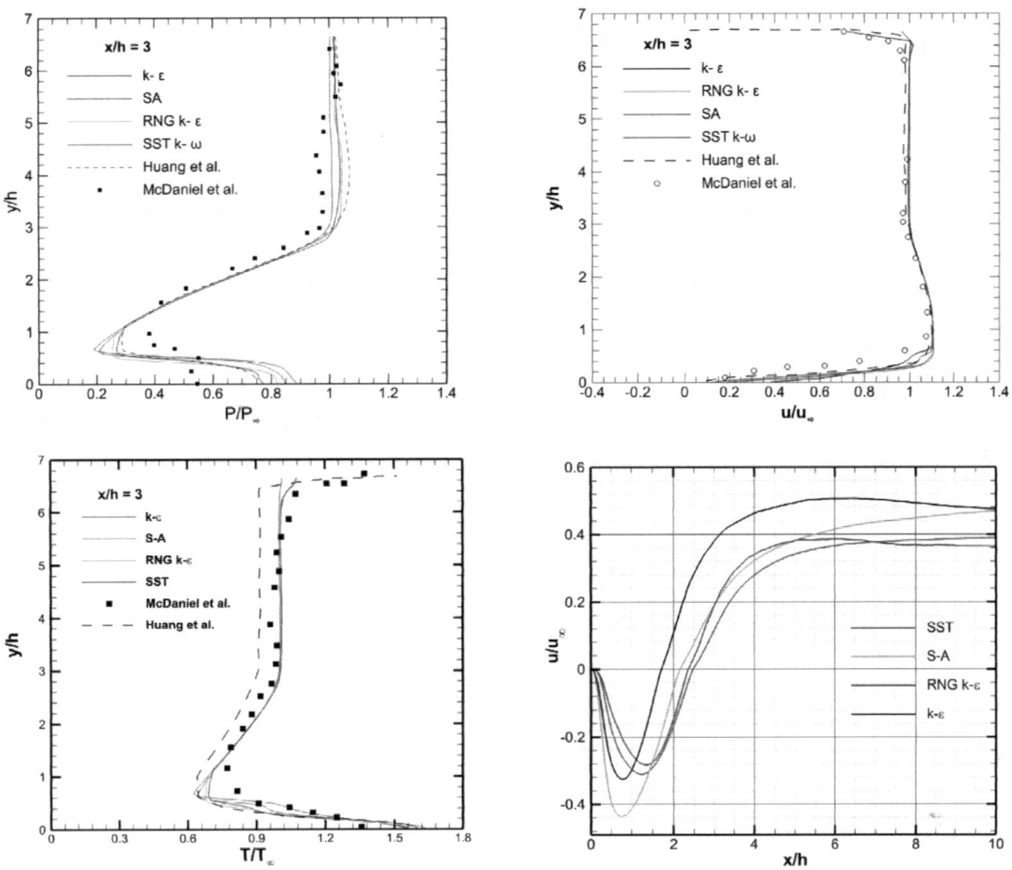

Figure 1: Comparison of numerical results at x/h = 3 and normalized velocity along wall

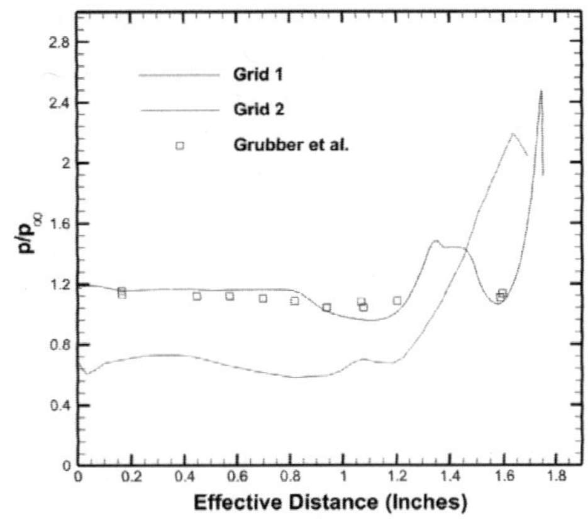

Figure 2: Normalized pressure distribution along cavity wall to demonstrate grid independence

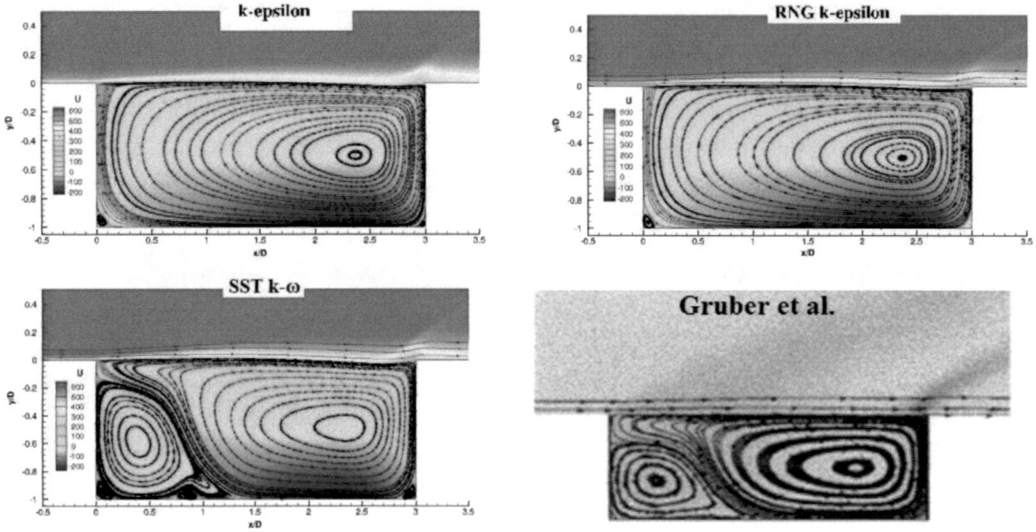

Figure 3: Streamwise Velocity Contours for different turbulence models

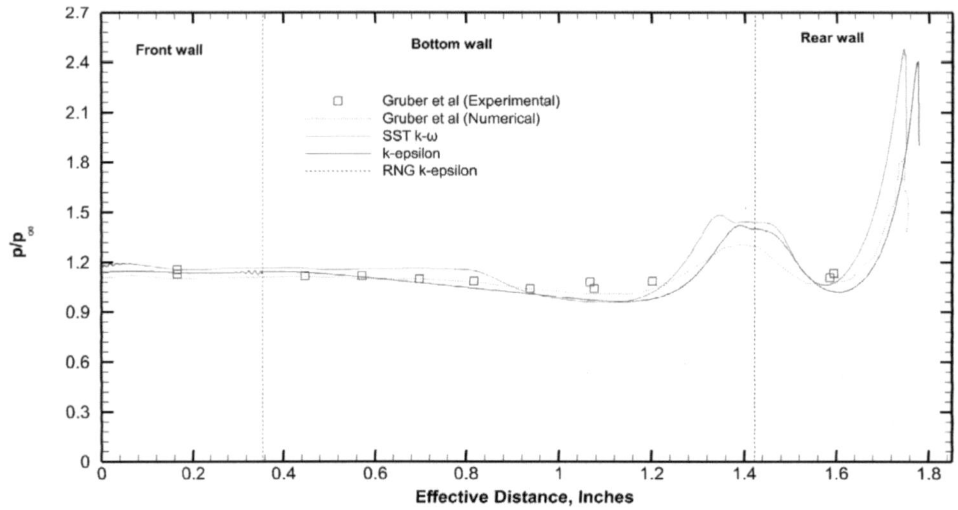

Figure 4: Normalized Pressure Distribution along the cavity wall

4. Conclusion
Systematic assessment of RANS based models is carried out for turbulent supersonic flow over backward facing step and over a rectangular open cavity. For the backward facing step, the numerical and experimental results are in good agreement, except for pressure profile in the region $y/h < 1$, due to poor performance of most of the RANS based turbulence models in the separation region.

For the rectangular cavity, k-epsilon and RNG k-epsilon models under-predict the recirculation region near the front wall. Though SST k-ε model captures essential flow features associated with the cavity, there are discrepancies in the near wall predictions. Overall, SST k-ε appears to perform better in predicting the bulk flow features in the supersonic flow regime considered herein.

References

[1] Chapman D 1951 An analysis of the base pressure at supersonic velocities and comparison with experiment *NACA TR* 1051.

[2] Hama F R 1968 Experimental studies on the lip shock *AIAA Paper* **6** 212-219.

[3] Donaldson I S 1967 On the separation of a supersonic flow at a sharp corner *AIAA Paper* **5** 1086-1088.

[4] Correa S, & Warren R 1989 Supersonic sudden-expansion flow with fluid injection: An experimental and computational study *AIAA Paper* 89-389.

[5] McDaniel J C, Fletcher D G, & Hartfield R J Jr 1992 Staged transverse injection into mach 2 flow behind a rearward-facing-step: a 3D compressible flow test case for hypersonic combustor CFD validation *AIAA Paper* 1992-0827.

[6] Soni R K, Arya N, and De A 2015 Assessment of RANS based models in a supersonic flowSoni *AIP Conference Proceedings* **1648** 030039.

[7] Arya N, Soni R K, and De A, 2015 Identification of coherent structures in a supersonic flow past backward facing step *AIP Conference Proceedings* **1648** 030037.

[8] Arya N, Soni R K, and De A, 2015 Investigation of flow characteristics in supersonic cavity using LES *American Journal of Fluid Dynamics* **5** (3A) 24-32.

[9] Arunajatesan S, and Sinha N 2003 Hybrid RANS-LES modeling for cavity aeroacoustics Predictions *International Journal of Computational Aeroacoustics* **2** 65–93.

[10] Gharib M, and Roshko A 1987 The effect of flow oscillations on cavity drag *J. Fluid Mech.* **177** 501-530.

[11] Gruber M R, Baurle R A, Mathur T, and Hsu K Y 2001 Fundamental studies of cavity-based flameholder concepts for supersonic combustors *J. Propul. Power* **17** 146-153.

[12] Plentovich E B, Stallings R L, Tracy M B 1993 Experimental Cavity Pressure Measurements at subsonic and transonic speeds *NASA Technical Paper* 3358.

[13] Rizzetta D P, Visbal M R 2003 Large-eddy simulation of supersonic cavity flow-fields including flow control *AIAA Journal* **41** 1452–1462.

[14] Stallings R, Wilcox F 1987 Experimental cavity pressure distributions at supersonic speeds *Technical Report NASA–TP–2683* NASA

[15] Zhang X, Edward J A 1988 Computational analysis of unsteady supersonic cavity flows driven by thick shear layers *Aeronautical Journal* 365-374.

[16] Kurganov A, Noelle S, and Petrova G 2000 Semidiscrete central-upwind schemes for hyperbolic conservation laws and hamilton – jacobi equations *J. Comput. Phys* **160** 720–742.

Research in Varying Burner Tilt Angle to Reduce Rear Pass Temperature in Coal Fired Boiler

Savithry K. Thrangaraju[1], Kannan M. Munisamy[1], Saravanan Baskaran[1]

[1]Universiti Tenaga Nasional, Jalan IKRAM-UNITEN, 43000 Kajang, Selangor, Malaysia

Abstract. This research shows the investigation conducted on one of techniques that is used in Manjung 700 MW tangentially fired coal power plant. The investigation conducted in this research is finding out the right tilt angle for the burners in the boiler that causes an efficient temperature distribution and combustion gas flow pattern in the boiler especially at the rear pass section. The main outcome of the project is to determine the right tilt angle for the burner to create an efficient temperature distribution and combustion gas flow pattern that able to increase the efficiency of the boiler. The investigation is carried out by using Computational Fluid Dynamics method to obtain the results by varying the burner tilt angle. The boiler model is drawn by using designing software which is called Solid Works and Fluent from Computational Fluid Dynamics is used to conduct the analysis on the boiler model. The analysis is to imitate the real combustion process in the real Manjung 700 MW boiler. The expected results are to determine the right burner tilt angle with a computational fluid analysis by obtaining the temperature distribution and combustion gas flow pattern for each of the three angles set for the burner tilt angle in FLUENT software. Three burner tilt angles are selected which are burner tilt angle at (0°) as test case 1, burner tilt angle at (+10°) as test case 2 and burner tilt angle at (-10°) as test case 3. These entire three cases were run in CFD software and the results of temperature distribution and velocity vector were obtained to find out the changes on the three cases at the furnace and rear pass section of the boiler. The results are being compared in analysis part by plotting graphs to determine the right tilting angle that reduces the rear pass temperature.

1. Introduction

Generation of electricity through the combustion of fossil fuels in thermal power plants are widely implemented in many countries because it is very efficient in generating large portion of electrical energy to support the needs of the country. Most of the boilers in the thermal power plant are using tangentially fired coal fired boilers in power generating process. Basically coal is preferred to be chosen as the fossil fuel resources globally to run thermal power plant in power generation. A thermal power plant operates on the basic theory of converting heat energy produced through combustion of fossil fuels into mechanical work using working fluid. The mechanical work is further converted into electrical energy using generator. The main function of coal burning system is to convert chemical energy into heat energy by using boiler to convert water into steam which drives the turbine. The combustion element of fuel consists of carbon, hydrogen and small amount of sulphur. The exhaust gases released after combustion contains CO_2, SO_2 and CO [1].

The coal combustion process in thermal power plant consist of phenomena as turbulence, convection and radiation in heat transfer and chemical reaction in combustion of coal. Many factors that are involved in power generation through thermal plant such as furnace scale thickness, air velocity, mass

IOP Conf. Series: Journal of Physics: Conf. Series **822** (2017) 012038 doi:10.1088/1742-6596/822/1/012038

flow rate and temperature of the heated air. The efficiency of the coal fired boiler is dependent to the thermal distribution that is effected by all the listed down factors.

1.1 Brief Description of the Boiler

Figure 1 shows the cross section of a boiler with furnace, cross over pass and rear pass sections in a boiler. The pulverised coal that is mixed with primary air is pumped from the four corners in the furnace section and ignited by injecting liquid fuel oil that starts the combustion. The injection of coal and air follows an imaginary line tangent to a small circle located in the centre of the furnace. The flow provides a rotation motion to the flame and creates additional turbulence for additional mixing of the coal and primary air. The flame spreads in the furnace and produce "fire ball". The system allows fuel particles to have sufficient time for residence in the furnace for the combustion.

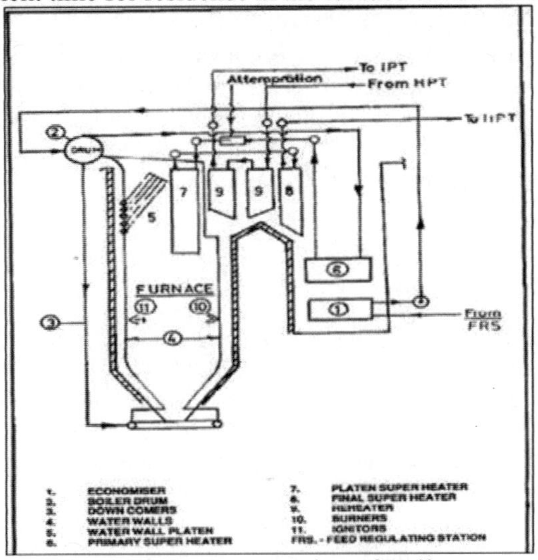

Figure 1: Cross Section of Boiler

Coal is pulverized into fine powder particles. Coal is blown in to the boiler by mixing with primary air in the furnace through a series of burner nozzle that is located at four sides as tangentially in the boiler. Combustion temperature of the furnace can reach up to 1300°C-1500°C.

1.2 Problem Statement and Scope of Study

In the boiler, all the burner nozzles can be tilted at the same time. All the nozzles can be tilted in a vertical direction up to certain angle from the horizontal axis to adjust the reheated steam temperature for the varying fouling conditions of the furnace. The tilting of the burner controls the vertical position of the fireball and also the size of the fireball to produce a uniform temperature distribution to all boiler tubes. The inaccurate tilting angle can lead to high rear pass temperature, imbalance left and right rear pass temperature and hot spots forming on water well that make the tubes to burst. The main aim is find out the right tilting angle for the burner to produce a perfect fireball that creates uniform temperature distribution in the boiler when heat transfers through radiation from the fireball to the tubes. The right tilting angle of the burner also can increases the efficiency of the boiler because the tilting angle can change the temperature of the superheat and reheat temperatures of steam.
The main reason to reduce the rear pass temperature is because to increase the efficiency of the boiler and reduce formation of hotspots on water well that leads to tube burst. Computational Fluid Dynamics has to be used to run the cases by varying the tilting angles at the boundary conditions. The effect of the varying tilting angle of the burners at the rear pass is identified by obtaining flow pattern and temperature distribution at the contours of rear pass. Analysis on the results has to be conducted

and plots have to be drawn to compare all the cases. The right angle tilting angle that reduces the rear temperature will be discussed and determined.

1.3 Summary of Researches Conducted on Boiler Models

The main idea of tilting the burner is to obtain a stable flame that creates an even distribution of thermal flow in the boiler as explained by William Brockman in his research [2]. The stable flame influences the temperature distribution in various sections of boiler as explained by Syed Mujtaba Umair and P.T.Borlepwar in their research on varying the burner tilt as one of the parameters to obtain a stable flame [3]. The research conducted by Masoumi H. and Abroshan H. proves that burner tilting causes changes in different part of the boiler such as upper furnace, crossover pass and rear pass temperature [4]. Other than that, it is proven that tilting angle of the burners influences the slag production in the burner as being explained by Zhou Hao, Cen Kefa and Sun Ping in their research [5]. They prove that slag decreases the efficiency of boiler.

The common method used in many research to determine the results are by running simulations using Computational Fluid Dynamics. The parameters can be tested in actual boiler situations through this software. Other than that, numerical method is also very important in conducting calculation for the data to be used in CFD. Most of the research done identified the temperatures at exit by reading temperature profile graph that obtained through CFD.

2. Methodology

2.1 Development of 700 MW Boiler Geometry

The Computational Fluid Dynamics investigation was done along the furnace path of the boiler and the rear pass of the boiler model which is situated on the other side of the boiler model. Planes and lines are created in the base model to observe the contours on the planes to determine the results such as temperature distribution and flow pattern at different cross section of the boiler. The base model was created according to the full scale furnace that as similar dimensions as the 700 MW coal fired boiler in TNB Janamanjung Power Plant. It has dimensions of 72.58 m height and 35.504 m widths. The base model is consisting of super heaters, burner inlet, furnace outlet, main furnace and bottom ash hopper. The geometry was created using SOLIDWORKS CAD software. The meshing process was done by using GAMBIT software. The meshing done on the boiler model was a fine meshing which enhance a more accurate flow of particles during the simulation in Computational Fluid Dynamics. The model drawn is shown on Figure 2.

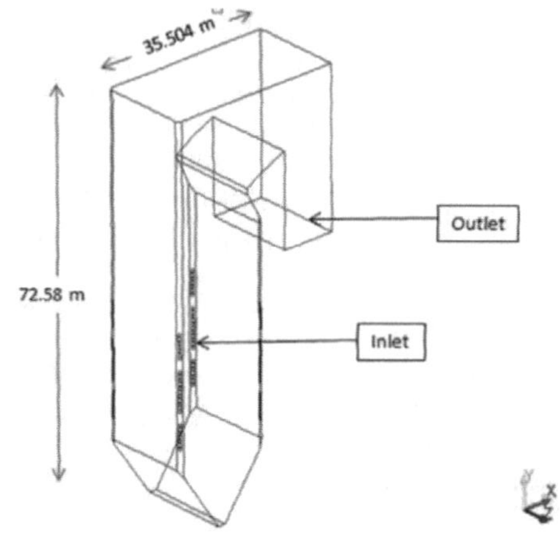

Figure 2: Layout of boiler model

2.2 Planes to Obtain Results

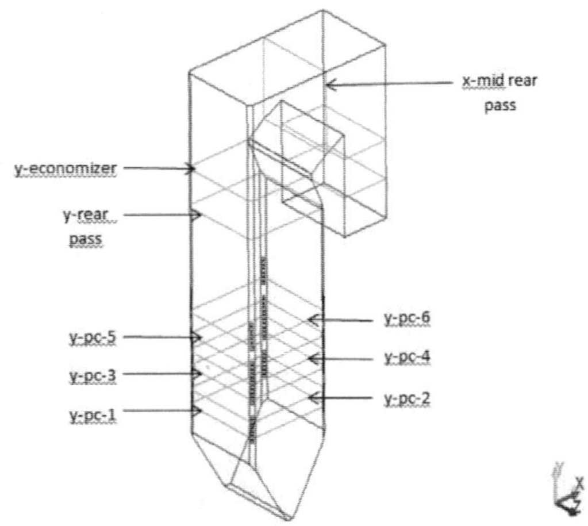

Figure 3: Mesh display of planes

The planes shown on figure 3 were drawn in FLUENT software to measure the changes occurs in the contours as the results when the boundary condition which is the tilting angle of the burner is changed during analysis. The plane y-pc-1, y-pc-2, y-pc-3, y-pc-4, y-pc-5, y-pc-6 are drawn at the inlet part where pulverized coal inlet located. The contours such as temperature distribution, velocity vector of the combustion flow are obtained on these planes in FLUENT. The planes are located at the furnace segment of boiler. These planes are mainly used to obtain contours to observe the fireball formation at the furnace. There are 6 planes drawn because the changes done on the tilting angle in boundary condition will change the position of fireball. The planes 6 planes drawn at the furnace that covers the lower furnace, middle furnace and the upper furnace to observe the formation of fireball for all the three cases tested. The middle rear pass section is used to measure the changes in contours at cross-over segment of boiler for example to observe temperature distribution of the flue gas at the segment. The y- economizer and y-rear pass plane is used to measure the changes in contours in rear pass segment of boiler for all the three cases. The contours on these planes will determine the changes occur at the rear pass from the variation of burner tilt angle.

2.3 Simulation of the Real Cases in Fluent

The analysis were done by selecting three angle for the burner which are case 1(0°), case 2 (+10°) and case 3 (-10°). The burner is at the center for case 1, lifted upwards for case 2 and lifted downwards for case 3. The value that is entered for y-component of flow direction under the burner's boundary condition on FLUENT software is the value obtained from the sin of the angle of the elevation set for the three cases. For example, the value of sin 10 is entered in y-component in all pc inlet and secondary air inlet under boundary condition to run the analysis for the case which that the burner is tilted at (+10°). For the case that the burner is tilted at 0° the value of sin 0 is entered for y-component of flow direction and the value of sin -10 is entered in y-component direction flow the case that the burner is tilted at (-10). The mass flow rate of pulverized coal was set at 32.319 kg/s and it is constant for all the three cases. For all the three cases is then initialized with 0 m/s velocity at x, y, z component and the temperature is set at 300 K for the initialization.

The Rosseland radiation for all three cases is used because this radiation gives out a stable simulation during the calculation running and also quite fast compared to P1. Initially P1 radiation was use when conducting the test run simulation for the test case and find out that the results obtained was not satisfying. The simulation for the real cases were run again using Rosseland radiation type and found out that Rosseland is more stable and takes shorter time for the iterations to run compared with P1 radiation type.

3. Results and Discussion

3.1 Formation of fireball

The formation of fireball on each case is confirmed before analysing the results at the rear pass section of the boiler for all the three cases. Fireball formation can be seen on all at the three cases but the position of the fireball formed varies in each cases. The fireball formation can be observed on different elevation of planes for all the three cases.

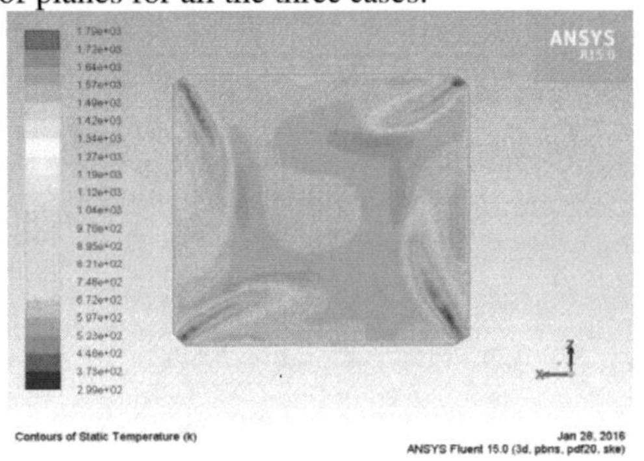

Figure 4: Plane y-pc-3 for (0°)

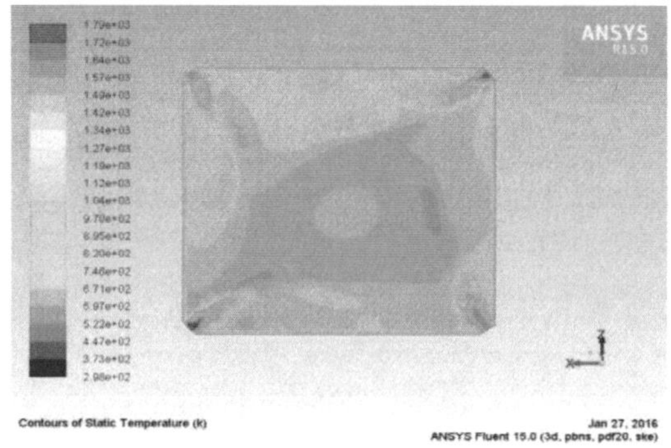

Figure 5: Plane y-pc-6 for (+10°)

The figure 4 shows the fireball formation is about to form on plane y-pc-3 if all the 6 planes at the furnace were being compared for test case 1 which is burner tilt angle at (0°). The plane y-pc-3 figure shows that a fireball is about to form with a surrounding of orange colour that proves higher temperature. Fireball formation occurs at plane y-pc-3 which is considered at the center of the furnace. This is because the burners were set at (0°) tilting angle which is not tilted at all and placed at the center of the furnace.

Fireball formation can be seen clearly on plane y-pc-6 shown in figure 5. The temperature contour proves that fireball formed at the top of the furnace. This is because the burner was tilted upwards for the angle of (+10°) which causes the fireball to be lifted upwards as well.

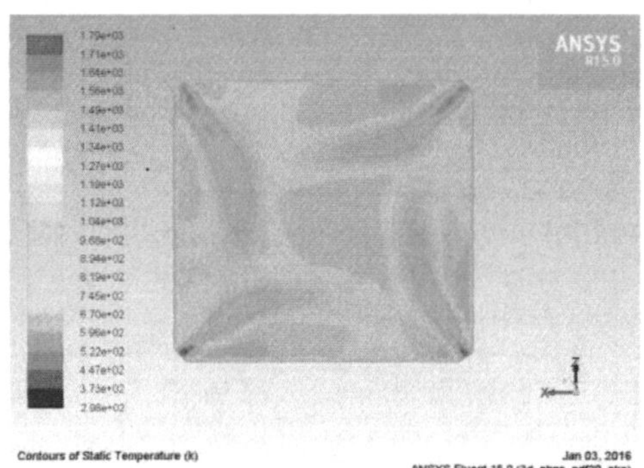

Figure 6: Plane y-pc-1 for (-10°)

The fireball shaped figure can be seen on plane y-pc-1 which is in figure 6. The figure of y-pc-1 which is in figure 6 showing that the fireball formed is not exactly like a ball shape but the shape is sufficient to prove the formation of fireball occurs for (-10°) burner tilt angle case. The other planes can be said that they are showing the flow of flue gas that is rising up from the bottom of furnace to the top part of the furnace on their contours. The result proves that the fireball formed at the lower part of the furnace because the fireball is lifted downwards when the burner is tilted downwards for test case 3 [6].

3.2 Rear Pass Analysis

Two graphs were plotted to show the comparison on temperature distribution at the rear pass section of boiler for all the three cases. The lowest temperature and the highest temperature at the rear pass section of the boiler can be observed on this graph. The distribution on of flue gas from the left hand side rear pass wall and right hand side rear pass also can be analysed through the graphs plotted. The first graph is drawn based on the relationship between temperature distribution at the rear pass and curve length of the rear pass. It is plotted along the line drawn at the mid rear pass of the boiler section. The second graphs show the relationship between temperature distribution at the rear pass and direction vector of the rear pass. The graph plotted on the line along the rear pass is drawn by at the front part of the rear pass section of the boiler following the x position of the mid rear pass line.

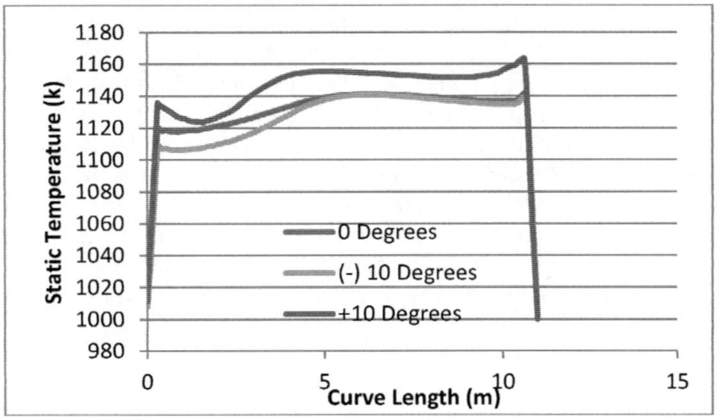

Figure 7: Graph of Rear Pass Temperature with Different Tilt Angle

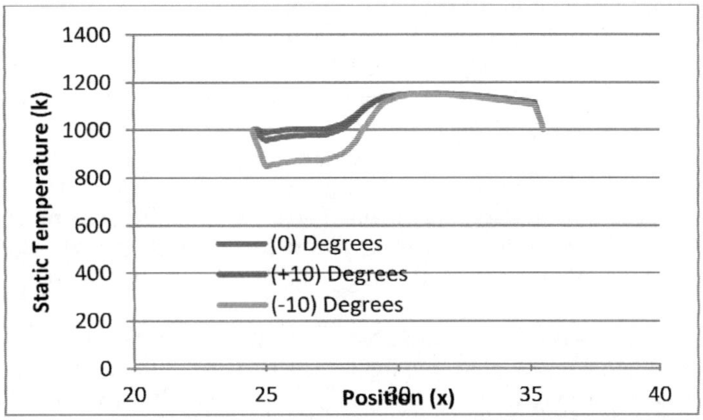

Figure 8: Graph of Rear Pass Temperature with Different Tilt Angle

The graphs on figure 7 and 8 shows that the temperature distributions from the wall of the rear pass at the left hand side until the wall on the right hand side is quite different for all the three cases. The change of temperature distribution from left hand side to the right hand side for test case 2(+10°) shows a lot of difference compared to other two cases. There is also a difference can be seen on the temperature distribution for test case 3 (-10°) when compared with test case 1 (0°). Test case 1 has a stable temperature distribution from left hand side to the right hand side compared to the other two cases. This proves that tilting the burner causes the changes at the rear pass of the boiler. Test case 2 has more difference in the temperature between left hand side and the right hand side of the rear pass section based

on the first graph at figure 7. This shows that as the burner is elevated upwards there is more difference of temperature between left hand side and the right hand side. The temperature distribution on the right hand side of the rear pass is higher compared to the temperature on the left hand side of the rear pass for all the three cases. This is because the temperature distribution from the flue gas that flow at the rear pass are being absorbed more on the left hand side of the rear pass section of the boiler. Other than that, the flow of flue gas is supposed to be at the center because the fireball is produced at the center of the furnace but the flow of the flue gas is disturbed when the flue gas has to rise up until the cross over section from the furnace. This causes the temperature difference between left hand side and right hand side on the graphs. Both graphs prove that the rear pass temperature is the highest for test case 1(+10°) and lowest for the test case 3(-10°). This can be also said that tilting the burner downwards reduces the rear pass temperature and tilting the burner upwards increases the rear pass temperature.
Tilting the burner downwards makes the fireball to be positioned at the lower furnace and tilting the burner upwards raise the fireball to the top of the furnace. The flue gas formed through the fireball produced at the furnace will travel to the cross over section of the boiler where it is located at the top of the boiler and finally travel to the rear pass section. The heat absorption from the flue gas temperature by the furnace wall occurs in that situation. The flue gas takes longer time for it to reach the rear pass when the fireball is located at the bottom of the furnace because the distance for the flue gas increases in that case. Based to that, more heat absorption from the flue gas occurs as the distance of the rear pass and fireball increase in the burner. This causes the temperature of rear pass is at the lowest for test case 3. The distances of the flue gas to travel to the rear pass from the furnace is shorter if the burner is lifted upwards as in test case 1. In this situation, the heat absorption from the flue gas by the furnace wall will be lower and causes the rear pass temperature to increase because all the heat that is not absorbed by the furnace wall.

4. Conclusion and Recommendation

Changes occur at the boiler rear pass when different burner tilt angle were set at the furnace section. This is because the position of fireball changes as the burners tilt angle is changed. The temperature difference between the left hand side and right side at the rear pass also changes as the burners tilt angle is changed. The combustion flow pattern at the furnace also changes when the burner tilt angle is varied and leads to the difference in temperature at the rear pass section of the boiler. The temperature distribution at the rear pass is only analysed as the fireball formation is seen at the temperature and velocity vector contours. The fireball formation can be seen on temperature contour and velocity vector contours on y-pc-3 for test case 1, y-pc-5 and y-pc-6 for test case 2 and y-pc-1 for test case 3. The right tilt angle that reduces the rear pass temperature of the boiler is identified through the graphs and contours. The graphs clearly prove that the rear pass temperature is the lowest when the burner is tilted at angle of (-10°) and highest when the burner is tilted at angle of (+10°). The rear pass temperature is low for test case 3 because more absorption of heat by the furnace wall from the flue gas occurs as the time taken for the flue gas to flow from the fireball at the furnace to the rear pass is longer due to the longer distance.
The recommendations that can be done for this research is more changes can be observed at the boiler such as determine the unburned carbon contain in the boiler when the burner tilt angle is changed. This is because unburned carbon also reduces the efficiency of the boiler because high carbon contained in the flue gas produces more NOx emission that can pollute the environment. Other than that, the ash deposition in the boiler also can be determined based on changes on burner tilt angle. This is because the ash deposition leads to the formation of slag that can actually reduce the efficiency of the boiler. Moreover, the changes of temperature occurs at the rear pass section of boiler can also be investigated by varying the type of coal being used for combustion.

5. References

[1] Blondeau J, Kock R, Mertens J, Eley A J, and Holub L 2016. *Applied Thermal Engineering* **98** p449.

[2] Yan X, Chen X B and Bergstrom D J 2011 *American Journal of Biomedical Engineering* **1(2)** p72.
[3] Umair S M, Birlepwar P T 2015 *International Journal of Engineering Research and General Science* **3(1)**, p2091.
[4] Masoumi H, Abroshan H 2012. *Fuel and Combustion* **5**, p1-13.
[5] Hao Z, Kefa C, Ping S 2002 *Fuel Processing Technology* **79**, p181-195.
[6] Munisamy K M, Yusoff M. Z, Thangaraju S K, Hassan H, Ahmad A 2014 *Materials Science and Engineering* **88**, p1-5.

The effect of nozzle diameter, injection pressure and ambient temperature on spray characteristics in diesel engine

Adiba Rhaodah Andsaler[1, a], Amir Khalid[1, b], Nor Sharifhatul Adila Abdullah[1], Azwan Sapit[1], Norrizam Jaat[1]
[1]Combustion Research Group (CRG), Centre for Energy and Industrial Environment Studies (CEIES),
Faculty of Mechanical Engineering & Manufacturing
Universiti Tun Hussein Onn Malaysia, Batu Pahat, Johor

E-mail: adiba_rhaodah@yahoo.com[a], amirk@uthm.edu.my[b]

Abstract. Mixture formation of the ignition process is a key element in the diesel combustion as it influences the combustion process and exhaust emission. Aim of this study is to elucidate the effects of nozzle diameter, injection pressure and ambient temperature to the formation of spray. This study investigated diesel formation spray using Computational Fluid Dynamics. Multiphase volume of fluid (VOF) behaviour in the chamber are determined by means of transient simulation, Eulerian of two phases is used for implementation of mixing fuel and air. The detail behaviour of spray droplet diameter, spray penetration and spray breakup length was visualised using the ANSYS 16.1. This simulation was done in different nozzle diameter 0.12 mm and 0.2 mm performed at the ambient temperature 500 K and 700 K with different injection pressure 40 MPa, 70 MPa and 140 MPa. Results show that high pressure influence droplet diameter become smaller and the penetration length longer with the high injection pressure apply. Smaller nozzle diameter gives a shorter length of the breakup. It is necessary for nozzle diameter and ambient temperature condition to improve the formation of spray. High injection pressure is most effective in improvement of formation spray under higher ambient temperature and smaller nozzle diameter.

Keywords: CFD, droplet diameter, penetration length, breakup length, nozzle diameter.

1. Introduction

Fuel injected into the chamber is important to the diesel engine because it influence the performance and emission. During combustion, a spray is simply the introduction of liquid into a gaseous environment through a nozzle such that the liquid, through its interaction with the surrounding gas and by its own instability, breaks-up into droplets [1]. Spray characteristics is known to significantly affect the combustion and emission processes in diesel engines. By optimizing spray characteristics, the raw emissions from the diesel engine which are mainly NOx and PM can be minimized [2-4]. Investigations into diesel sprays characteristics have concentrated on the effect of the spray characteristic on engine performance such as the spray tip penetration, break-up length and droplet size and velocity distributions [5].The injection pressure has significant effect on spray liquid penetration [6]. The spray tip penetration gets longer as the injection pressure increases. This result is related to both higher quantity and higher velocity of the droplets at higher injection pressures [7]. Proportional to injection pressure, the spray penetrates faster at higher injection pressures [8]. The break-up length characterizes a point of discontinuity, where the spray changes from a zone of liquid (bulk liquid, or interconnected ligaments and droplets), to a finely atomized regime of droplets [9]. After the disintegration of the liquid column emerging from the nozzle, the generated droplets may further break-up into smaller ones as they move into the surrounding gas [10]. Development of this deformation leads to break-up into smaller droplets. The forces associated with dynamic pressure, surface tension and viscosity control the break-up of a drop [11-12]. However, CFD is a compromise tool being used to predict the critical part that cannot be done in experiment. In this research, the characteristics of spray diesel are investigated focusing on changing ambient temperature, injection pressure with the different nozzle diameter by using Computational Fluid Dynamics. This simulation can capture droplet diameter, spray penetration and breakup length.

2. Simulation set up

The modeling flow chart shows in Figure 1 is the flow of simulation in ANSYS Fluent. The first stage is pre-processing consists of the flow problem in the simulation. Solver stage is complete the calculation of finite element model or solving of governing equation. Post-processing is a process of involving data gathering and visualization.

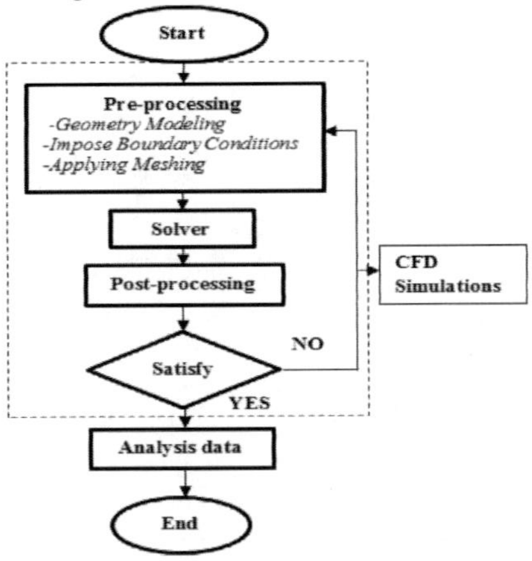

Figure 1. Methodology and modelling flow chart

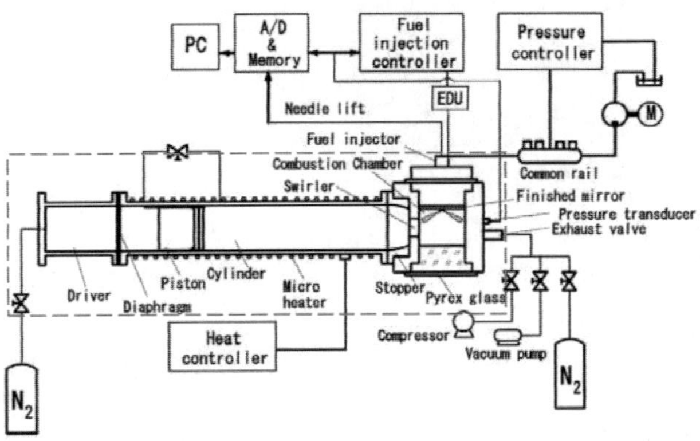

Figure 2.: Experimental setup of RCM.

A rapid compression machine (RCM) was used to generate the actual diesel combustion over a wide range of temperature, pressure and swirl velocities. Figure 2 is a schematic diagram of RCM is shown, together with an outline of the fuel injection system. Figure 3 shows the combustion chamber cross sectional drawing in 3D. The RCM has a disc type combustion chamber with a diameter of 60mm and width of 20mm. The design focuses on injector nozzle and spray chamber.

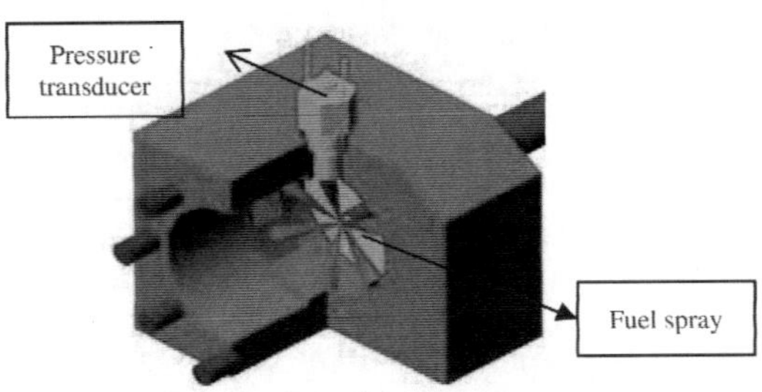

Figure 3. 3D section view of combustion chamber

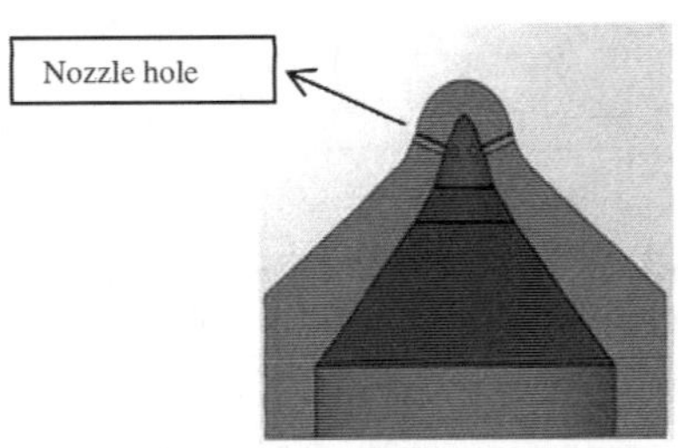

Figure 4. Cross sectional model of spray injector.

The injector was modeled by using Solid Work. In this study the combustion chamber is a space where the spray formation occurs. The model of the injector only considered the tip of the nozzle. Figure 4 shows the 3D model of cross sectional area of the injector tip. Figure 5 shows the sections geometry of the combustion chamber. It is 1/6 section from the overall geometry as there are 6 nozzles in the actual chamber and only 1/6 is adequate and considered sufficient for the simulation analysis. The assigned boundary condition was serving as the initiation and direct to the motion of the flow. Table 1 shows the solver setup of the simulation. The parameters that need to investigate are droplet diameter of particle, penetration length and breakup length.

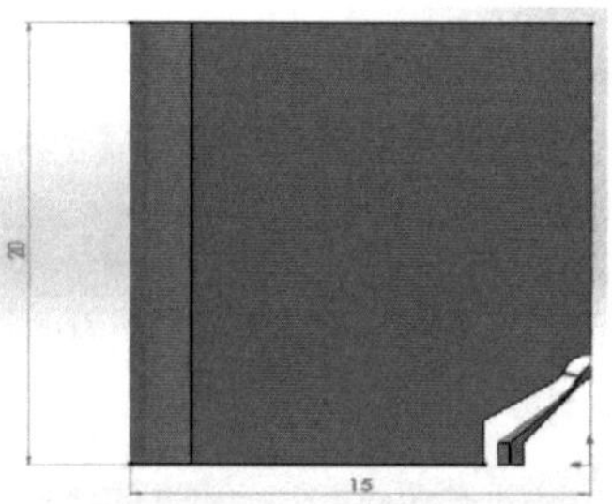

Figure 5. Hollow model for the combustion chamber

Table 1. Solver Setup

Options	Type
Multiphase Model	Volume of Fluid
Time	Transient
Model	Viscous K-Epsilon Model
Material	Diesel Liquid
Boundary Condition	Injection Pressure = 40MPa, 70MPa, 140MPa
	Temperature Inlet = 313Kelvin
	Pressure Outlet = 1MPa
	Temperature Outlet = 500K, 700K

3. Result and discussion

This analysis used two different diameters which is 0.20 mm and 0.12 mm. This simulation using several different ambient temperatures on 500 K and 700 K for every injection pressure simulate with 40 MPa, 70 MPa and 140 MPa. The objective for this simulation is to investigate the different form of spray characteristic that influence by the nozzle diameter, ambient temperature and injection pressure of diesel engine. The analysis of spray characteristic of this simulation is more focus to the droplet diameter, penetration length and breakup length that influence from the parameter use. Grid sensitivity was tested on the model with different grid sizes. The grid size divided into three levels which are coarse, medium and fine grids. Table 2 indicates the number of elements and nodes for each group of different grid sizes. Besides, the number of elements is ranging from 24997 to 114555 elements. The number of elements is obviously directly proportional to the numbers of nodes by ascending the level of grid sizes from coarse to fine with also level of smoothing from low to high.

Table 2. Different grid sizes with its number of elements and nodes

Grid Sizes	Number of Elements	Number of Nodes
Coarse	24997	5175
Medium	46249	9428
Fine	114555	22495

Droplet diameter of the spray was being Figure 6 shows that at 40 MPa droplet diameter at 0.20 mm and 500 K parameter show more droplet size colour red represent to a big size droplet compared to the 0.12 mm 700 K. The droplet distribution size of 0.12 mm and 0.2 mm are shown in the Figure 7 and 8. As illustrated in the graphs, they show that the pattern of the distribution size of the droplet is parallel to each other. The higher of ambient temperature and injection pressure give the smaller droplet diameter. The result describe that slightly fuel injected in the chamber can cause a lower combustion. At the same time droplet diameter depend on the aerodynamic drag in the chamber it can affect the droplet diameter performed smaller when the air collision with the fuel.

IOP Conf. Series: Journal of Physics: Conf. Series **822** (2017) 012039 doi:10.1088/1742-6596/822/1/012039

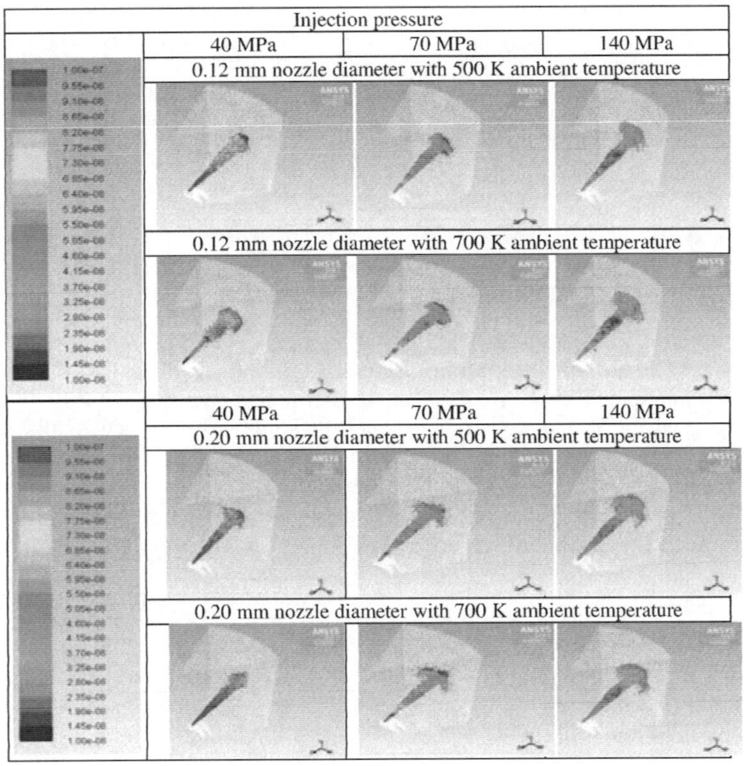

Figure 6. Spray structure of droplet diameter produced by single nozzle, with colour scale represents the droplet diameter.

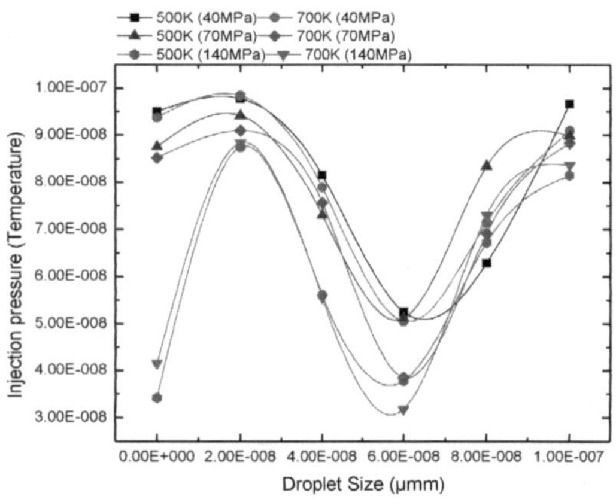

Figure 7. Graph of droplet size distribution of 0.12 mm

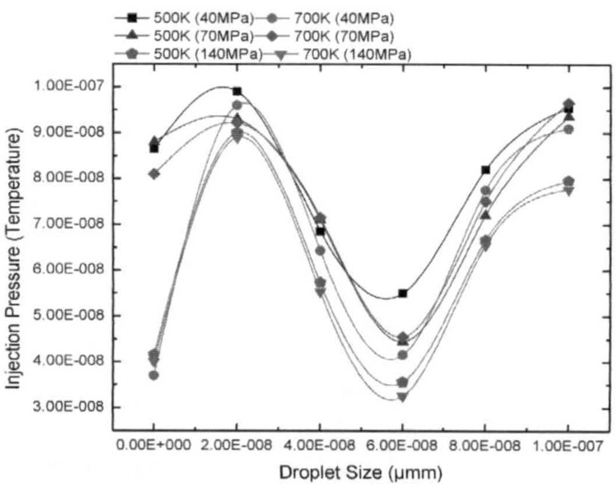

Figure 8. Graph of droplet size distribution of 0.2 mm

As the graph illustrate the penetration length at Figure 9 and Figure 10 using different ambient temperature for 0.12 mm and 0.2 mm nozzle diameter constantly increase with the time taken and both for temperature the penetration length not give much different. This is due to the range of between temperatures not too big. However, the injection pressure influences the spray penetration length, where variant injection pressure which is 40 MPa, 70 MPa and 140 MPa with ambient pressure 1 MPa at atmospheric pressure. As the higher pressure injector, the longer length penetration will influence. This is because the spray was resisted with the ambient pressure.

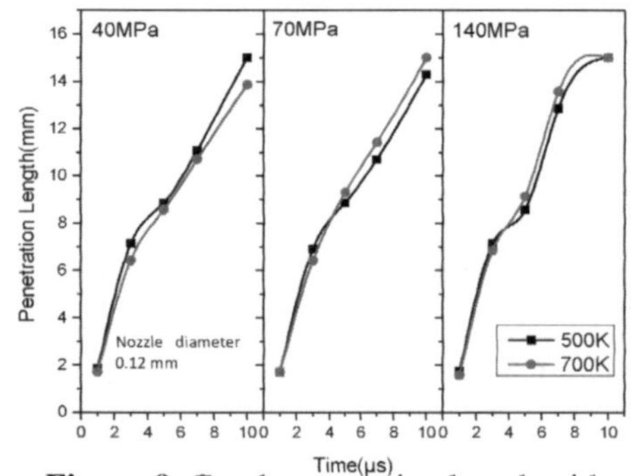

Figure 9. Graph penetration length with nozzle diameter 0.12 mm

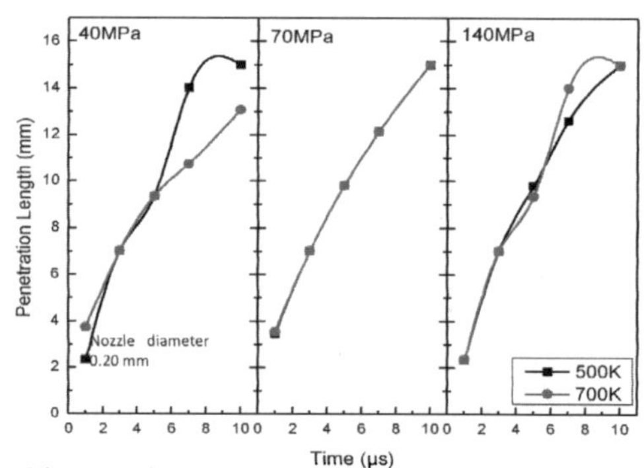

Figure 10. Graph penetration length with nozzle diameter 0.2 mm

Figure 11 and 12 show the graphs of breakup length with 0.12 mm and 0.2 mm nozzle diameter. There is fluctuation result on the temperature 500 K, where injection pressure 140 MPa and nozzle diameter 0.2 mm dramatically increase from the 1 mm stood up to the 2.3 mm. This is maybe due to the droplet did not breakup and its getting longer, this is call as the section of ligament. Compare to the temperature 700 K at the 140 MPa and 0.2 mm breakup length flows steady with the time. Thus the temperature influences to the higher ambient temperature used, the shorter breakup length happen at the core segment.

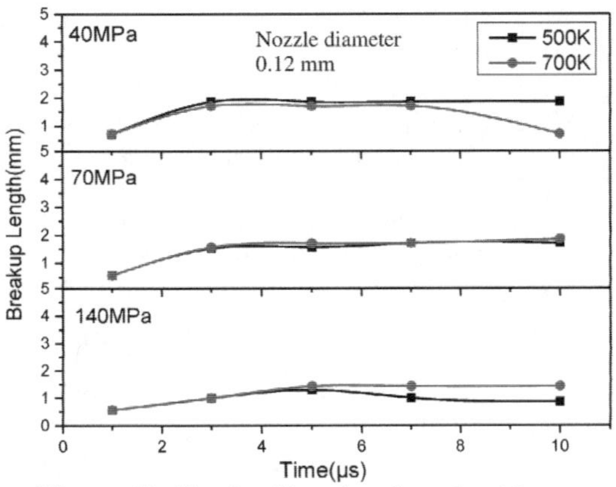

Figure 11. Graph of breakup length with nozzle diameter.

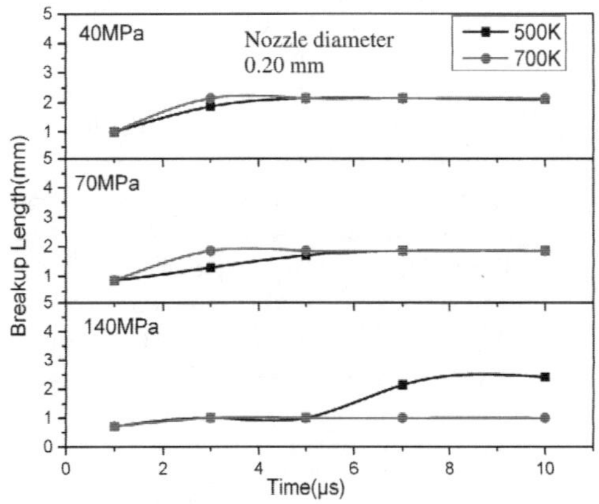

Figure 12. Graph of breakup length with nozzle diameter.

4. Conclusion

This study has shown a simulation flow of the fuel flowing in the nozzle spray before combustion. A low injection pressures in the chamber will disturb the breakdown of the fuel spray to a very fine droplets before start ignition of the fuel. A temperature change in the chamber also affects the length of the spray penetration. For high ambient temperatures, fuel evaporates more easily and ignites the fuel itself. Also, the diameter of the nozzle also affects the breakup length of the spray nozzle as influenced by the higher injection pressure where the breakup length is getting shorter.

Acknowledgement
The authors also would like to thank the Ministry of Higher Education, Malaysia for supporting this research under Exploratory Research Grant Scheme (ERGS) vot. E033 and was partly sponsored by the Centre for Graduate Studies UTHM.

References

[1] Adam, A., Inukai, N., Kidoguchi, Y., Miwa, K. Et Al., 2007,"A Study on Droplets Evaporation at Diesel Spray Boundary during Ignition Delay Period," *SAE Technical Paper* 2007-01-1893.

[2] Khalid, A., Anuar, M.D., Ishak, Y., Manshoor, B., Sapit, A., Leman, M., Zaman, I., 2014" Emissions characteristics of small diesel engine fuelled by waste cooking oil", *MATEC Web of Conferences*, Volume 13, Article number 06006, 2014, DOI: 10.1051/matecconf/20141306006

[3] Khalid, A., Tamaldin, N., Jaat, M., Ali, M.F.M., Manshoor, B., Zaman, I., 2013,"Impacts of Biodiesel Storage Duration on Fuel Properties and Emissions", *Procedia Engineering*, Volume 68, Pages 225-230, Elsevier Ltd, DOI: 10.1016/j.proeng.2013.12.172.

[4] Khalid, A. , Jaat, N., Sapit, A., Razali, A., Manshoor, B., Zaman, I., Abdullah, A.A., 2015,"Performance and emissions characteristics of crude jatropha oil biodiesel blends in a diesel engine" *International Journal of Automotive and Mechanical Engineering*, Volume 11, Issue 1, Pages 2447-2457, DOI: 10.15282/ijame.11.2015.25.0206.

[5] Khalid, A., Hayashi, K., Kidoguchi, Y., Yatsufusa, T., 2011,"Effect of air entrainment and oxygen concentration on endothermic and heat recovery process of diesel ignition", *SAE Technical Papers*, Society of Automotive Engineers of Japan, Inc. and SAE International, DOI: 10.4271/2011-01-1834.

[6] Khalid, A., 2013, "Effect of Ambient Temperature and Oxygen Concentration on Ignition and Combustion Process of Diesel Spray", *Asian Journal of Scientific Research*, Volume 6, Issue 3, Pages 434-444, Asian Network for Scientific Information, DOI: 10.3923/ajsr.2013.434-444.

[7] Camille Hespel, Jean-Bernard Blaisot, Xandra Margot, S. Patouna, Armelle Cessou, Bertrand Lecordier, 2010,"Influence of Nozzle Geometry on Spray Shape, Particle Size, Spray Velocity and Air Entrainment of High Pressure Diesel Spray", *THIESEL*.

[8] I. V. Roisman, L. Araneo, and C. Tropea, 2007,"Effect of Ambient Pressure on Penetration of a Diesel Spray,"*International Journal of Multiphase Flow*, Vol. **33**, No. 8, Pp. 904–920.

[9] Simon Martinez – Martinez, Fausto A. Sanchez – Cruz, Vicente R.Bermudez And Jose M.Riesco-Avila, 2010,"Liquids Spray Characteristic In Diesel Engine", *In: Daniela Siano (Ed.). Fuel Injection. Intech: Sciyo*. 20 – 48 Pages.

[10] Ozg¨Ur O˘guz Tas¸Kiran, 2011,"Experimental Study on Diesel Spray Characteristics and Autoignition Process", *Hindawi Publishing Corporation*.

[11] Naber, J. And Siebers, D., 1996,"Effects of Gas Density and Vaporization on Penetration and Dispersion of Diesel Sprays,"*SAE Technical Paper* 960034.

[12] Andsaler, A.R. , Khalid, A., Jaat, N., Izwan Sukarman, M., " *Effects of ambient density on flow characteristics of biodiesel spray injection using computational fluid dynamics*", ARPN Journal of Engineering and Applied Sciences, Volume 11, Issue 8, Pages 5499-5505, 2016. http://www.arpnjournals.org/jeas/research_papers/rp_2016/jeas_0416_4142.pdf

Vein mechanism simulation study for deep vein thrombosis early diagnosis using cfd

Nabilah Ibrahim[1], Nur Shazilah Abd Aziz[1], and Abreeza Noorlina Abd Manap[2]

[1]Universiti Tun Hussein Onn Malaysia, 86400 Batu Pahat Johor, Malaysia
[2]Universiti Tenaga Nasional, 43000, Kajang, Malaysia

E-mails: nabilah@uthm.edu.my, nurshazilah96@gmail.com, abreeza@uniten.edu.my,

Abstract. Using a Computational Fluid Dynamics (CFD) technique, this work focus on the analysis of pressure, velocity, and vorticity of blood flow along the popliteal vein. Since the study of early stage of Deep Vein Thrombosis (DVT) becomes essential to prevent the pulmonary embolism (PE), those three parameters are analysed to assess the effect of different opening between two valves of a normal popliteal vein. When only one valve is simulated, the result of pressure shows that the highest and lowest velocities are 15.45 cm/s and 0.73 cm/s, respectively. From the visualization of observed data, however, the different size of orifice between the first and second valves influencing the velocity and vorticity of the blood flow. The rotational motion of blood particle at the same region increases the probability of blood accumulating which is associated with the development of thrombus. Thus, a series of experiment has been conducted by changing the size of valve orifice for the first and second valves along the vein distribution. The result of the CFD simulation shows a significant variation in blood flow in terms of velocity and vorticity.

1. Introduction

Coagulation of blood in vessel system could lead to heart attack and other cardiovascular disease such as coronary thrombosis, deep vein thrombosis (DVT), pulmonary embolism (PE) and in the worst case is death [1]. DVT is a condition when accumulation of blood or thrombus is attached on the vessel wall which could prevent the movement of blood along the vessel system. Since the probability of thrombus initiates is along the popliteal vein, thus, this work will focus on the site around the valve area. Veins contain valves to prevent the backflow of blood toward the capillary beds. Dysfunctional of valves will lead to obstruction of blood circulation, thus, this could effects the tissue dysfunctional to whole body. Previous *in vitro* study demonstrates the blood progress of antegrade stream on valve geometry, material properties of the vein, and venous valve [1, 2]. Moreover, it is reported that the dramatic change of the three dimensional velocity profile of blood is due to the shape and size of the orifice [3]. Since there were no studies that investigate the valve aperture orientation in relation to blood flow dynamic, this study is performed to address and visualize the asynchronization between two valves in a popliteal vein.

Conventional advanced computational fluid dynamics (CFD) allows complex numerical reenactment of the cardiovascular framework. CFD investigations have been utilized to assess particular parameters in the body such as wall shear stress, velocity of blood flow in vessel stream and pressure of the blood against the wall of the vessel which have complexity to obtain from the *in vivo* experiment [4]. This work employs the CFD technology, a three-dimensional simulation of blood flow in a venous vessel. The model of the vein structure is then, implemented into ANSYS's CFX software package. Two study cases are performed that control the valves orifice size. The effect of the irregularities in the blood flow will cause valve insufficiency which lead to various disease related to venous vessel such as heart attack, and pulmonary embolism.

2. Methodology

As the objective is to simulate the vein mechanism, the geometrical structure of the popliteal vein needs to be constructed. The image of a popliteal vein is generated from medical imaging equipment model TOSHIBA SSA-580A. The image is extracted from Epiphan DVI2 USB 3.0 which is capable of recording a video at 60 frames per second (fps).

2.1. Valve and Vein Structure

Figure 1 shows the ultrasound B-mode image obtained from the medical imaging equipment. The position of the valve in the popliteal vein could be visualized. By employing the canny-edge detection method, figure 2 shows the clearly edges which could determine the size of the vein and valve [5]. Thus, in the SolidWorks reconstruction image, the length of the vein is set to 10 cm as shown in figure 3. Since this study is focusing on the size of leaflets or valve aperture, therefore the structure of the vein was assumed to be in a cylindrical shape. A pair of leaflets attached at a point of vein wall and the red arrow shows the direction of blood flow. Since valve are reliably situated at particular area in the veins, they are normally present in pairs 3-5 cm separated from one another [3]. Furthermore, it is reported that the leaflets position is fix, thus, *a* represents the size of leaflets opening where in this study it will vary from 30% to 70%. In addition, the maximum leaflets opening is 60-70% of the vein diameter [3].

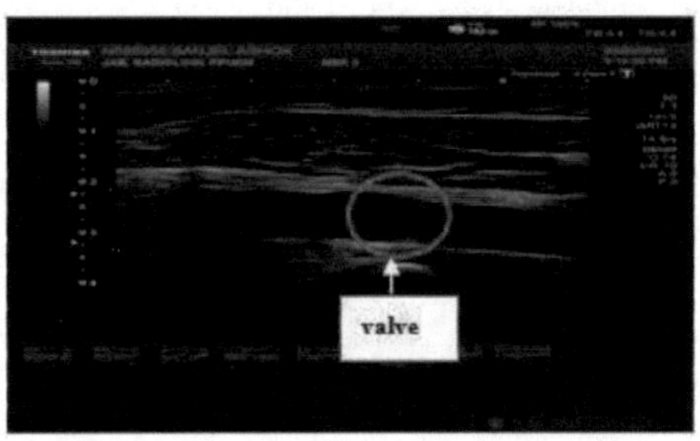

Figure 1. B-mode ultrasound image of popliteal vein

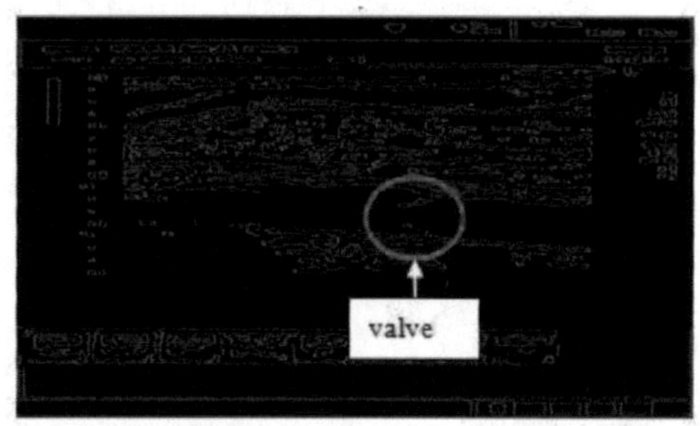

Figure 2. Canny-edge-detection-image of popliteal vein

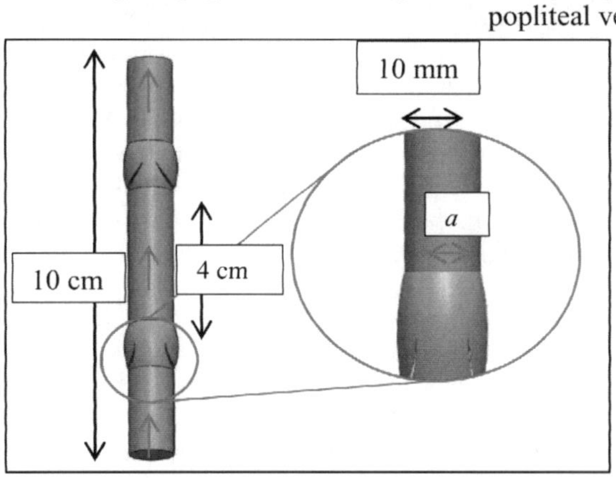

Figure 3. The geometrical structure of popliteal vein

2.2. Designing Popliteal Vein Structure
For the initial step, the boundary condition of the model has to be defined including flow rate, velocity and pressure of the inlet and outlet of the vessel model [6]. Based on the geometrical structure of the vein, a triangle meshes were generated to obtain the optimum grid size for the model. The meshes consist of 65274 nodes and 332503 tetrahedral elements. For the purpose of this study, the blood was assumed as a non-Newtonian fluid flow at isothermal conditions [7], as follows:

$$\begin{cases} \Delta\rho u = 0 \\ \rho\left(\frac{\partial u}{\partial t} + u\nabla u\right) = -\nabla p + \mu\nabla^2 \end{cases} \quad (1)$$

where u is the flow velocity, p is pressure, ρ is density of fluid, and μ is the dynamic viscosity. The parameters used to set up the vein structure are summarized as shown in table 1.

Table 1. Blood parameters

Parameter	Value
Pressure	10 mmHg
Blood Viscosity	0.0035 Pa
Blood Density	1050 kgm^{-3}
Blood Velocity	20 cms^{-1}
Vein Diameter	10 mm
Temperature	37 °C

There are two cases of study in this work. Case A is the set of first valve to 70% orifice size and 50% for the second valve. Meanwhile, for case B, first valve orifice size is set to be 70% and second valve one is 30%. The designated of vein using CFD is shown in figure 4.

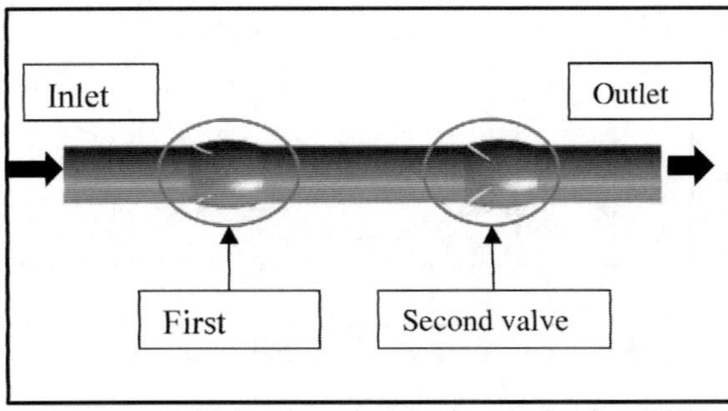

Figure 4. Valve location designated using CFD

2.3. Blood Pressure, velocity, and vorticity
Pressure distribution along the vein is hard to be measured. Therefore, to simulate the initial condition of inlet and outlet, previous in *vivo* experiment has reported the 10 mmHg of pressure and 20 cm/s of velocity [5]. Since the blood clot usually develop at the back of the valve, the region to measure the velocity of vein is focus around the valve along the vein. Thus, three regions of vein is divided to measure the velocity. Vorticity is a condition of rotation fluid elements. The properties of vorticity in blood flow might help in determining the blood particles movement in the body. The observation of vorticity for both cases is conducted to compare the effect of two orifice sizes.

3. Result Analysis

According to previous *in vivo* study, the pressure distribution along the vein is 10 mmHg [8]. However, the pressure before and after valve has different value. Thus, using this simulated vein structure and decided parameters, the visualization of pressure distribution could be clearly visualized, as shown in figure 5. In this case, only one valve is modelled so that the standard based of pressure for the designed vein could be clarified.

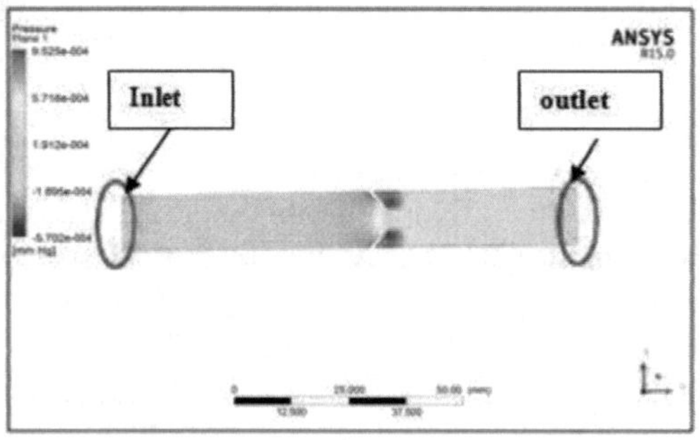

Figure 5. Contour of pressure distribution along the simulated vein

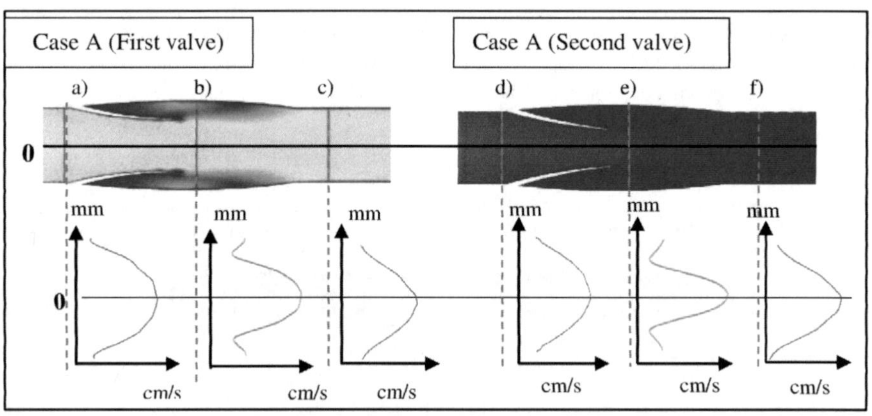

Figure 6. Velocity profile for case A

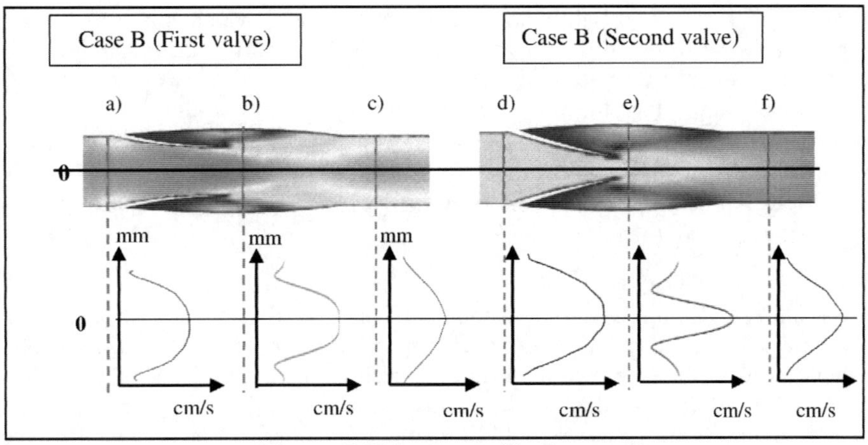

Figure 7. Velocity profile for case B

Figures 6 and 7 show the velocity profile at different locations along the vein distribution, for cases A and B, respectively. The vein was divided into six segments which the three segments for the first valve (a-c) and the other three segments for the second valve (d-f). Every segment was separated by 8 mm. It could be visualized that the red colour represent the regions of high velocity, whereas the blue colour represent the regions of velocity. For case A, obviously, the value of peak velocity decreases after the blood passes through the second valve. Moreover, at the b) segment, the colour contour is green which shows the increasing of velocity as it passing through the leaflets as it following the previous clinical study [7]. Same goes to case B where the pattern of the velocity profile is literally same. However, the velocity range for the both cases is significantly different to one another. This is due to the different second valve orifice size. In addition, the value of the velocity was consistently decreasing after going through the second valve. Since the case B has smaller cross sectional area between the leaflets, therefore, the blood flow could be visualized accelerated in this stenotic area. On the other hand, case A has a large cross sectional area between the leaflets, thus, producing much slower velocity, resulting in reflected stream which against the normal blood flow.

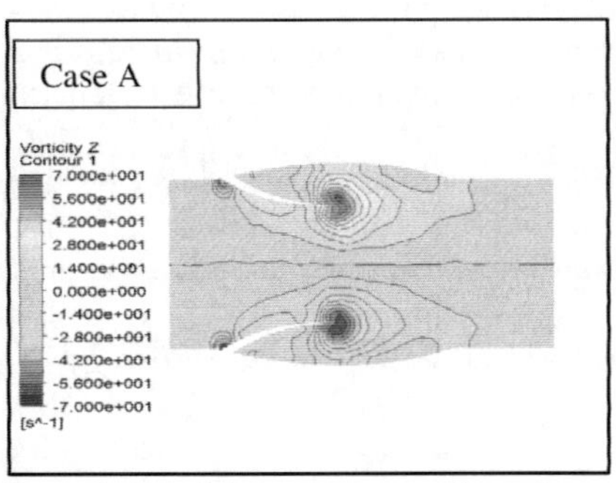

Figure 8. Vorticity contour at the valve area for case A

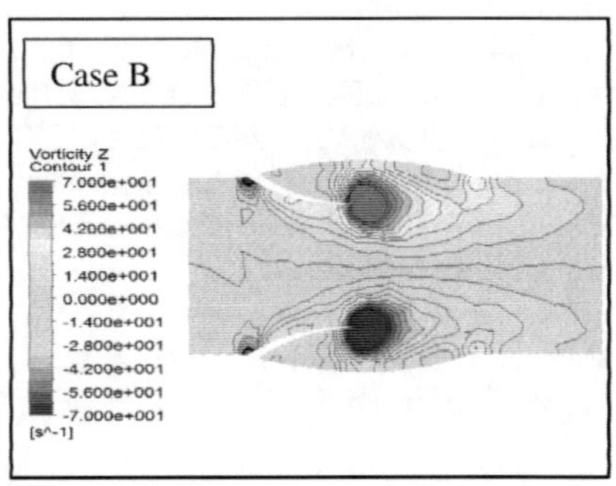

Figure 9. Vorticity contour at the valve area for case B

Figures 8 and 9 indicate the vorticity contour for case A and B, respectively. The blue colour indicate the blood particles which in the clockwise rotation and the red colour indicates the opposite direction. Even though the size of the orifice for the first valve is the same for both case, however the velocity contour shows a significant different in size of the contour from one another. This is due the size of the orifice for the second valve which affecting the blood particle movement at that particular area. The magnitude of vorticity for case B is slightly larger than that of case A, which appear as a larger rotation particles. Furthermore, case B would be one of the factor that affect the accumulating of blood, thus, could lead to the formation of thrombus. In addition, the blood particles can be observed rotating significantly at the valve area since the valve was obstructing the flow of the blood. Hence, once the blood passing through the valve, some of the particles will be entering the valve cusps. That particular particles will rotating in the same area, thus, increasing the contact time of the particles of blood with the vessel wall and the probability of blood clot formation.

4. Conclusion

Two valves were simulated along the designed popliteal vein. The different size of orifice between the first and second valves shows the variation of the velocity and vorticity of the blood flow is signification. For case A which the first valve opened 70% and the second valve opened 50%, the velocity value indicate lower than that of case B, due to the smaller orifice of second valve in case B (30%). In observation of vorticity movement, case B could be seen with a higher vorticity around the second valve which could be assumed has an effect to blood accumulating possibility. In summary, the biomechanics event of the fluid is dependent on the mechanical properties of the valve orifice. In fact, to prevent the reflux of the blood, the valve functions as flow modulator in helping the blood movement in the venous vessel.

Acknowledgement

The authors like to thank Multidisplinary Research Grant MDR (vot:U091) for the sponsorship doing this project. They also like to thank Universiti Tun Hussein Onn Malaysia (UTHM) for having the space of Medical Imaging laboratory and also MiNT SRC laboratory doing the experiment.

References

[1] F. E. Hossler and R. F. West 1988 *The American Journal of Anatomy* **181** 425.
[2] S. Raju, C. A. Hudson, R. Fredericks, P. Neglen, A.B. Greene, and E.F. Meydrech 1999 *Eur. J. Vascular and endovasclualr surgery* **17** 521
[3] Lurie, F., & Kistner, R. L. 2012 *European Journal of Vascular and Endovascular Surgery* **44** 337
[4] Tokuda, Y., Song, M.-H., Ueda, Y., Usui, A., Akita, T., Yoneyama, S. and Maruyama, S. 2008 *European Journal of Cardio-Thoracic Surgery* **33** 164
[5] S. Dummler, S. Eichhorn, C. Tesche, U. Schreiber, B. Voss and M. Deutsch 2011 *Pulsatile ex vivo perfusion of human saphenous vein grafts under controlled pressure conditions increases MMP-2 expression* 1German Heart Center Munich at the Technische Universität München, Department of Cardiovascular Surgery, Lazarettstrasse 36 Munich.
[6] Uus, A., Liatsis, P., Rajani, R. and Man-dic, L. 2014 *The impact of boundary conditions in patient-specific coronary blood flow simulation* 35
[7] Wijeratne, N. S., & Hoo and K. A. 2008 *Numerical studies on the hemodynamics in the human vein and venous valve* 147
[8] H. I. Mat Harun, N. Ibrahim,N. S. Aziz, W. N. S. Wan Zakaria and N. Farhan 2014 *Proc. Int. Conf. on Biomedical Engineering and Sciences (Kuala Lumpur)* 474

IOP Conf. Series: Journal of Physics: Conf. Series **822** (2017) 012041 doi:10.1088/1742-6596/822/1/012041

Vortex stretching in a homogeneous isotropic turbulence

M Hirota, Y Nishio, S Izawa and Y Fukunishi

Mechanical Systems and Design, Graduate School of Engineering, Tohoku University, 6-6-01 Aramaki-Aoba, Aoba-ku, Sendai, Miyagi, 980-8579, Japan

E-mail: masato.hirota.r7@dc.tohoku.ac.jp

Abstract. Stretching vortices whose sizes are in the inertial subrange of a homogeneous isotropic turbulence are picked up, and the geometric relations with the neighboring vortices whose scales are twice larger are studied. Hierarchical vortices are extracted using a Fourier band-pass filter, and each extracted vortex is reconstructed as a set of short cylindrical segments along the vortex axis to discuss the vortex interactions. As a result, it is shown that the directions of larger vortices near the segments of the fast stretching vortices tend to be orthogonal to the direction of the stretching segments, and the locations of the larger vortices that contribute most to the stretching of smaller vortex segments are likely to be found in the direction with the relative angle of 45° from the axes of the stretching vortex segments. And, the vortices with the second highest contributions tend to be in the directions 45° from the stretching segments' axes and orthogonal to the directions of the highest contributing vortices.

1. Introduction

Turbulent flows consist of vortices of various scales interacting with each other. Energy is continuously transported from energy-containing vortices of larger scale to energy-dissipative vortices of smaller scale. This process is well known as the energy cascade of turbulence. Kolmogorov[1] theoretically derived that the energy flux in the inertial subrange was constant, and the slope of the energy spectrum was universally $-5/3$. This $-5/3$ energy spectrum has been confirmed by many experiments[2] and direct numerical simulations (DNS)[3]; however, the physical mechanism behind the theory is less understood.

Statistical properties of turbulent flows such as the energy cascade are believed to be due to the mutual interactions of hierarchical vortices, which occur locally and intermittently. In other words, local vortex behaviors, e.g. vortex formation, stretching and breaking down, should be responsible for the turbulent motions. For example, in free shear flows, primary vortices are firstly generated by the Kelvin-Helmholtz instability, and then, the secondary vortices, named rib vortices, develop between them in the streamwise direction. These secondary vortices are soon stretched owing to the strong shear generated by the primary vortices. As a result, three-dimensional deformation of the primary vortices takes place, and vortical structures become complicated[4]. Similar phenomena can be found in turbulent flows, in a fractal manner. Goto[5, 6] reported that the energy cascade in the inertial subrange is caused by the creation of smaller-scale vortex tubes in straining regions which can be found around the larger-scale vortex tubes. There is little doubt that vortex stretching is an important factor in the energy cascade process of turbulent flows.

IOP Conf. Series: Journal of Physics: Conf. Series **822** (2017) 012041 doi:10.1088/1742-6596/822/1/012041

In our previous studies[7], it has already been found that the vortices of a certain scale in the inertial subrange were likely to be stretched by vortices whose scales were twice larger. Leung *et al.*[8] also reported that vortices of a certain scale were predominantly stretched by the strain field of larger vortices whose scale was larger or comparable. The present study focuses on the stretching mechanism of vortices in a homogeneous isotropic turbulence. The relative positional relationships between a stretching vortex and twice larger vortices around it are investigated in detail.

2. Numerical method

The target is a homogeneous isotropic turbulence in a periodic box of 2π on a side, which is obtained by DNS. The three-dimensional vorticity transport equation is solved by a pseudo-spectral method. Total grid number is 512^3. Aliasing errors are completely removed by the phase shift method. Time integration is performed by a fourth-order classical Runge-Kutta method. In order to keep the flow field in a statistically equilibrium state, a solenoidal external force is randomly given to the low wavenumber region $1 \leq k \leq 2$, where k denotes a wavenumber. Figure 1 shows the energy spectrum of the flow field. The Taylor micro-scale Reynolds number Re_λ of the flow field is $Re_\lambda = 267$.

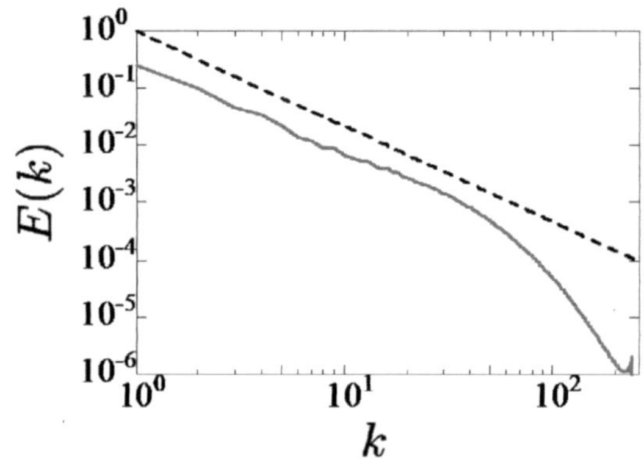

Figure 1. Energy spectrum of turbulent field: ——, present DNS data; - - - -, slope $\propto k^{-5/3}$.

3. Analysis procedures

3.1. Extraction and identification of multi-scale vortices

Vorticity fields of two different scales are extracted by Fourier band-pass filters with the center wavenumbers $k_c = 12\sqrt{2}$ and $6\sqrt{2}$, where the bandwidths are $k_c/\sqrt{2} \leq \boldsymbol{k} < \sqrt{2}k_c$. Vortical structures in the extracted vorticity fields are identified using the Q criterion. Regions where the Q value exceeds 75% of r.m.s. of the enstrophy density $\Omega_{\rm rms}$ are regarded as inside the vortices. Figure 2 shows the vortical structures before and after the filtering. Many tiny vortices are observed in the target turbulent field.

In order to discuss the geometric relation of these extracted vortices, each vortex is reconstructed as a set of cylindrical elements (vortex segments) by the following procedure:

(i) Compute the eigenvalues and the eigenvectors of the Hessian of Q at each computational grid point and determine the vortex axis as a line which connects the local maximum values of Q.

(ii) Place vortex segments which are grid width long along the vortex axes.

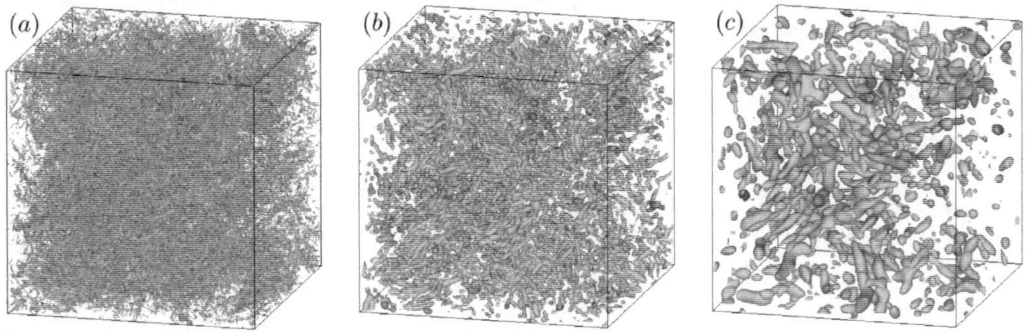

Figure 2. Turbulent vortical structures: (*a*) original vorticity field; (*b*) extracted at target scale $k_1 = 12\sqrt{2}$; (*c*) extracted at the scale twice larger $k_2 = 6\sqrt{2}$. Isosurface of $Q = 0.75\Omega_{\rm rms}$ is shown.

(iii) By applying the wavelet transformation to the vorticity distribution in the plane normal to each axis, estimate the radius of each segment as the most correlated scale of the wavelet coefficient.

(iv) Regard short vortices whose aspect ratios are less than one as noise and eliminate them.

(v) Determine the circulation of each segment so that each reconstructed vortex has the same circulation as that of the vortex region identified by the Q criterion.

By the procedure shown above, the vortical structures of scale $k_1 = 12\sqrt{2}$ and $k_2 = 6\sqrt{2}$ are reconstructed as a set of connected vortex segments. Thus, the velocity at any positions, induced by any vortex, can be estimated by applying the Biot-Savart law.

3.2. Geometric relations between fast-stretching vortex segments and their surrounding vortices

The stretching of a vortex segment of scale k_1 caused by larger vortices of scale k_2 is evaluated by the stretching speed γ, which is,

$$\gamma = \frac{1}{l_s}\frac{dl_s}{dt} = \frac{du_s}{de_s}, \tag{1}$$

where, l_s is the length of the vortex segment, and u_s is the velocity in the vortex axis e_s direction.

Fast stretching vortex segments whose stretching speeds are within top 10% of the whole vortex segments of scale k_1 are picked up, and the geometric relations between these fast stretching segments and their surrounding larger vortices of scale k_2 are investigated using the following procedure:

(i) Choose a fast stretching vortex segment of scale k_1 as the target segment.

(ii) Find vortices of scale k_2 in the vicinity of the target segment within $6R$, where R denotes the radius of the vortex segment of scale k_2.

(iii) Calculate the amount each vortex of scale k_2 stretches the target segment.

(iv) Find the vortex of scale k_2 with the highest contribution to stretching, referred as "No.1", and the vortex of the second highest contribution to stretching, referred as "No.2".

(v) Calculate stretching speeds γ' of the target segment caused by each vortex segment of the No.1 and No.2 vortices.

(vi) Average the radius R_s, the position $\boldsymbol{x}_\mathrm{s}$ and the direction $\boldsymbol{e}_\mathrm{s}$ of the segments of the No.1 vortex, and then the No.2 vortex, using equations below.

$$R_\mathrm{s} = \sum_{i=1}^{n} \gamma'_i R_{\mathrm{s}i} \Big/ \sum_{i=1}^{n} \gamma'_i, \quad (2)$$

$$\boldsymbol{x}_\mathrm{s} = \sum_{i=1}^{n} \gamma'_i \boldsymbol{x}_{\mathrm{s}i} \Big/ \sum_{i=1}^{n} \gamma'_i, \quad (3)$$

$$\boldsymbol{e}_\mathrm{s} = \sum_{i=1}^{n} \gamma'_i \boldsymbol{e}_{\mathrm{s}i} \Big/ \sum_{i=1}^{n} \gamma'_i, \quad (4)$$

where, i denotes the vortex segment number, and n is the total number of segments for each vortex.

(vii) Measure the averaged values of the relative position and the relative angle between the target segment and the No.1 vortex, and then the No.2 vortex.

(viii) Repeat the above procedure for all the fast stretching segments of scale k_1.

4. Results and discussion

Figure 3 shows the probability density function (PDF) of the cosine of an angle between the segment of the fast stretching vortex of scale $k_1 = 12\sqrt{2}$ and the top two vortices of scale $k_2 = 6\sqrt{2}$ that contribute to the stretching. The value of the PDF becomes higher where the cosine value is zero. This result means that the two contribution vortices of the larger scale tend to be orthogonal to a fast stretching vortex segment of the smaller scale. Figure 4 shows the joint probability density function of the relative positions of the No.1 vortices of scale k_2. The centers of the target vortex segments of scale k_1 are placed at the origin. The axes ξ and η correspond to the direction of the target vortex segments and its orthogonal direction. These axes are normalized by the averaged radius R of the No.1 vortices. The joint PDF becomes higher at the region where the relative angle is $\pm 45°$ and where the distance is between R and $2R$. Figure 5 shows the joint PDF of the positions of the No.2 vortices. Again, the origin is the center of the target vortex segment. The axis Ξ is the direction of the No.1 vortices, and the axis H is orthogonal to Ξ. The No.2 vortices tend to appear $1.5R - 2.5R$ away from the target segments and in the direction orthogonal to the direction of the No.1 vortices.

These results indicate that the vortex stretching is likely to be caused under the specific configuration of surrounding vortices, not just randomly distributed vortices.

5. Conclusions

Geometric relations between two different scale vortices were investigated by focusing on the vortex stretching. It was found that the segments of fast stretching vortices tended to be orthogonal to the directions of vortices whose scales were twice larger. Besides, the vortices of the highest contributions to the stretching were likely to be found in the directions with the relative angle of 45° to the axes of the stretching vortex segments, and the vortices of the second highest contributions tended to appear also in directions 45° from the stretching segments and orthogonal to the directions of the vortices of the highest contributions.

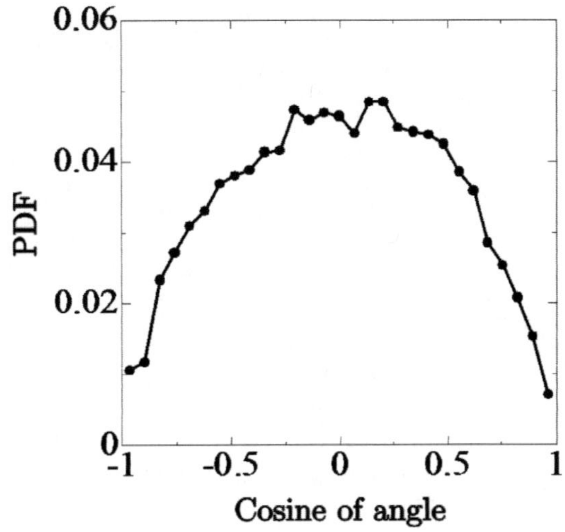

Figure 3. PDF of the cosine of an angle between the segment of the fast stretching vortex of scale $k_1 = 12\sqrt{2}$ and the vortices with the highest contributions to the stretching (No.1) and the second highest contributions (No.2).

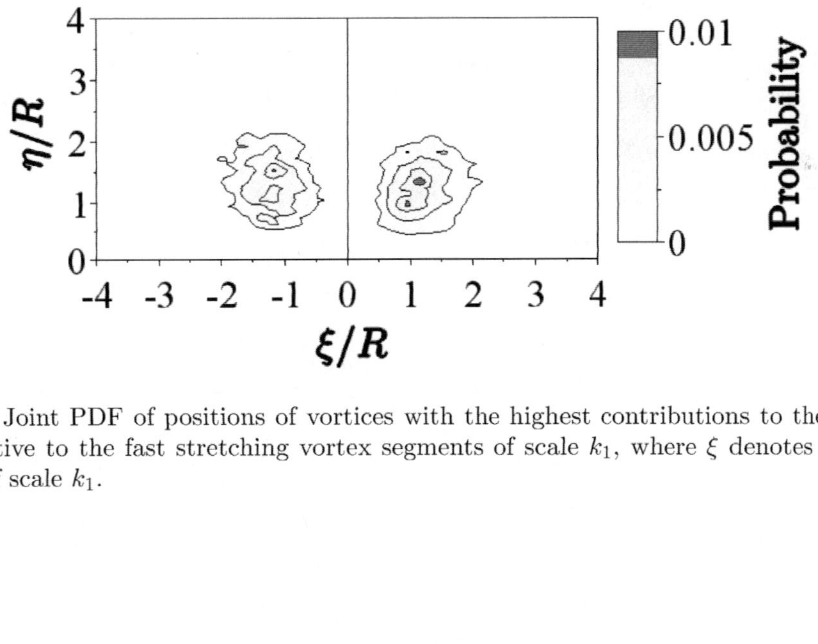

Figure 4. Joint PDF of positions of vortices with the highest contributions to the stretching (No.1) relative to the fast stretching vortex segments of scale k_1, where ξ denotes the axes of segments of scale k_1.

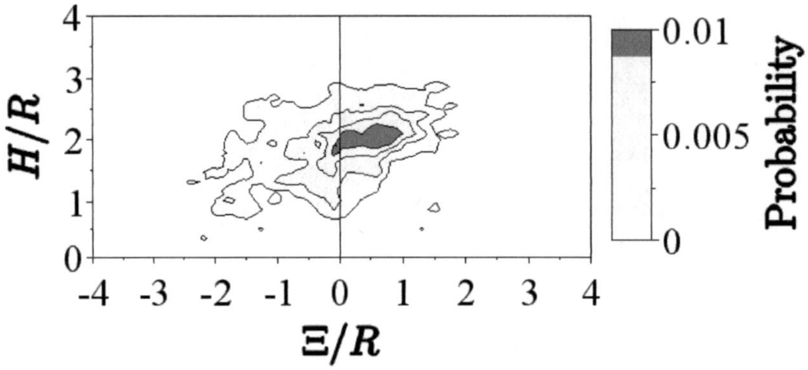

Figure 5. Joint PDF of the positions of vortices with the second highest contributions to the stretching (No.2) relative to the fast stretching segments of scale k_1, where Ξ denotes the direction of the No.1 vortices of scale k_2.

References

[1] Kolmogorov A N 1941 *Dokl. Akad. Nauk SSSR* **30** 301–305
[2] Tsuji Y 2009 *Fluid Dynamics Research* **41** 064003
[3] Ishihara T, Gotoh T and Kaneda Y 2009 *Annu. Rev. Fluid Mech.* **41** 165–180
[4] Lasheras J C and Choi H 1988 *J. Fluid Mech.* **189** 53–86
[5] Goto S 2008 *J. Fluid Mech.* **605** 355–366
[6] Goto S 2012 *Prog. Theor. Phys. Suppl.* **195** 139–156
[7] Hirota M, Shigeta M, Izawa S and Fukunishi Y 2013 *Proc. 27th Symp. Computational Fluid Dynamics* A02-1 (in Japanese)
[8] Leung T, Swaminathan N and Davidson P A 2012 *J. Fluid Mech.* **710** 453–481

A numerical experiment on tidal river simplification in simulation of tide dominated estuaries

X Yin[1], L Jia[2] and L Zhu[1]

[1]School of Civil Engineering and Transportation, South China University of Technology, Guangzhou 510640, China
[2]Center for Coastal Ocean Science & Technology Research, School of Marine Sciences, Guangzhou 510275, China

E-mail: arxlyin@scut.edu.cn

Abstract. In numerical simulation of tide dominated estuaries, introduction of simplified tidal channels into the model for real rivers is one of the strategies to deal with the lack of topographic data. To understand the effects of simplification and their sensitivity to the simplifying parameters, a numerical experiment was conducted to test the parameters such as channel length L, surface width B, bed slope S, bottom elevation ∇_0, bed roughness n and run-off Q_r. The results indicated the values of those parameters which were liable to less tidal prism and greater flood resistance would result in larger simulation errors. For a better simplification the values of parameters for the channel geometry, resistance and upstream inflow needed to be consistent with the average of the natural river as much as possible. The simplification method made the computation stable, fast and saved the storage space and it was adoptable for different time periods and seasons.

1. The Introduction

Estuarine hydrodynamic processes are fundamental to such as aquatic environment, sediment transport and topographical evolution in estuaries. During simulation of them the boundary condition at river end is notably affective to stability and precision of computation [1,2]. In terms of characteristics of hydrodynamic equations and usual discrete procedures of solution, giving flow flux to the landward upstream for inflow while tidal stage to the seaward downstream boundary is commonly favorable to computation stability and convergence [3-5].

In most of tide dominated estuaries there are usually horizontally extensive regions open to sea with funnel shapes. The tidal rivers in the upstream would be quite long or even several times the estuary bay if they have mild slopes, e.g. Delaware Bay, Columbia River estuary and some of sub-estuaries of Pearl River [6-8]. When simulating such estuaries, if the upstream boundaries are situated beyond tidal current range or even tidal limit, the computational region and time consumption will greatly increase and entire topographical data of the long rivers are requested, which sometimes troubled the workers. If we turn to take the head of bay as open boundary, it is not often satisfactory to the computation for significant tidal influences. Therefore, it is a technical problem often met in practice and expecting solutions that how to handle the upstream tidal rivers and their open boundaries economically and conveniently, and in the meantime to ensure the dynamical conditions of the entrance to main body of the estuary.

For this purpose, people used to apply 2-D or 3-D models nested by 1-D models or tried to constitute model of larger and coarser grid system so as to supply boundary conditions for finer simulation of smaller local regions in the estuaries [9,10]. Although some researchers adopted simplified rivers, they didn't discuss the specifications or rules for the simplifications [7]. Now the question is when we take the latter method, what aspects in the simplification are important to the modeling accuracy? This is our motivation to investigate by means of numerical experiment with the case of Huangmaohai estuary of Pearl River and its upstream Yamen waterway and Tan River.

The next part will represent simulation model and study method, then be followed by discussion of the test results and finally the conclusions.

2. Location, model and method

Shown as in figure 1, Huangmaohai estuary bay is located in the middle of the coast of Pearl River delta facing to China South Sea, connecting two of eight sub-inlets of Peal River estuary, Yamen and Hutiaomen.

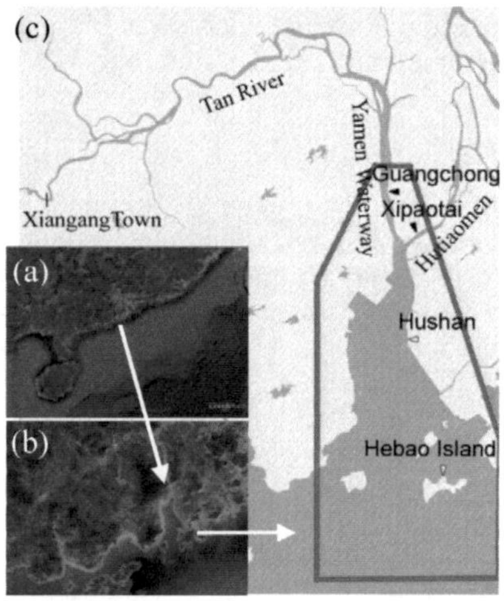

Figure 1. Geo-location of Huangmaohai estuary and the simulation area. The closed triangles in (c) indicate hydrological stations while the open tide stations. The red solid line circles the area with actual terrain in the computation.

Figure 2. Computational Grid. Upstream the broken signs, two simplified channels are added to the zone with actual terrain. Some lengths of the channels are omitted in the figure.

The estuary area in our numerical experiment applied the actual topography in 2003 and was gridded as shown in figure 2. The simplified channels were straight and prismatic, substituting for the real upper tidal rivers. The farthest seaward boundaries reached the 30m isobaths where the tidal level conditions were given by the global ocean tide model, while the upstream boundaries at the simplified river ends were given flow rate series.

The model was operated by the 2-D hydrodynamic module of Delft3D software developed by Deltares and k-ε model was chosen for turbulence calculation. The values of bed roughness were also simplified to be some constants, i.e. 0.012 and 0.014 for estuary bay and open sea respectively but a tested value for the tidal rivers. The simulation covered 10 days dating from 7 to 16 in February, 2001, incorporating spring to neap tides. Owing to the dry season for that time, the upstream run-off flow was stable and thought to be approximately steady. The wind and wave were neglected in the computation due to fine weather during most of the time.

Numerical tests were carried out for the simplified channel length *L*, surface width *B*, bed slope *S*, bottom elevation ∇_0, bed roughness *n* and run-off Q_r one by one. Due to the higher bottom and the smaller cross-section area, for simplicity, the simplified parameters for Hutiaomen waterway were given the averaged values of the actual river all the time. Four major observation points were Guanchong, Xipaotai, Hushan and Hebao Island in figure 1 (c). The simulated hourly levels and flow rates at these points were compared with the on-site observation values.

The root mean square of level or flow rate error were calculated as follows

$$RMS = \left[\frac{1}{k} \sum_{i=1}^{k} (L_p^i - L_m^i)^2 \right]^{1/2}$$

where L_p^i and L_m^i were hourly values in proto observation and model computation respectively, with the hour numbered by *i* in total simulation hours *k*.

For sensitivity analysis, firstly, a reference experiment had been conducted as follows. The upstream boundary flow was given the average of the observation in simulation period, the length and slope of the channel were given the general values based on the real river. The width followed the breadth of the cross-section that connected the un-simplified waterway because the actual width varies mildly. Further, the Manning's roughness was chosen by modulations till the minimum calculation errors at the observation points were reached. On the basis of this reference experiment, taking its simplifying parameters and the corresponding *RMS* values as benchmarks, we conducted comparative tests. Then the percentage change of the tested parameter η_1 and that of the corresponding *RMS* η_2 at Guanchong, the entrance to the main body of the estuary bay, were calculated, respectively.

$$\eta_1 = \frac{P_m - P_0}{P_0} \times 100\% \qquad , \qquad \eta_2 = \frac{RMS_m - RMS_0}{RMS_0} \times 100\%$$

where *P* denoted the tested parameter such as *L*, *B*, or ∇_0 etc. and the subscripts 0 and m denoted the benchmark and the otherwise modeling respectively. Thus, through the variation of η_2 with η_1, the sensitivity of the computed estuarine dynamics to the variation of simplifying parameters could be assessed in some way.

3. Discussions

3.1. The results of reference experiment

The simplifying parameters of the reference experiment was configured as follows, *L*=100km, *B*=1000m, *S*=0.0001, *n*=0.0135 and Q_r=300m^3/s. The computed results are shown in figure 3. It was found to be the best in general among all the tests.

3.2. Sensitivity analysis

The relationship of η_1 for each parameter and the corresponding η_2 is demonstrated in figure 4. On overall, most of the parameters have much greater influences on tidal flow than on tidal level.

By comparing the results in figure 4, we found decreasing of *L* had greater effects on both tidal level and flow than the other changes but increase of it brought very small effects. The influence of decrease of *B* stood the second place to tidal level but increase of it had less effects.

The *RMS* of tidal level was quite sensitive to increasing *S* but not remarkable to decreasing of it. By contrast, the tidal flow was always sensitive to *S* whether it is raised or not. Relative to the average water depth, when bed rising was less than 25%, the tidal level *RMS* was commonly below 3%, otherwise, the error went up quickly. Yet bed lowering brought much mild effects on tide level *RMS*. But for the tidal flow rate, there was an obvious difference with that, the reduction of bottom elevation became the most sensitive parameter while the rising bed are not and slightly bed lifting even diminished the flow *RMS*.

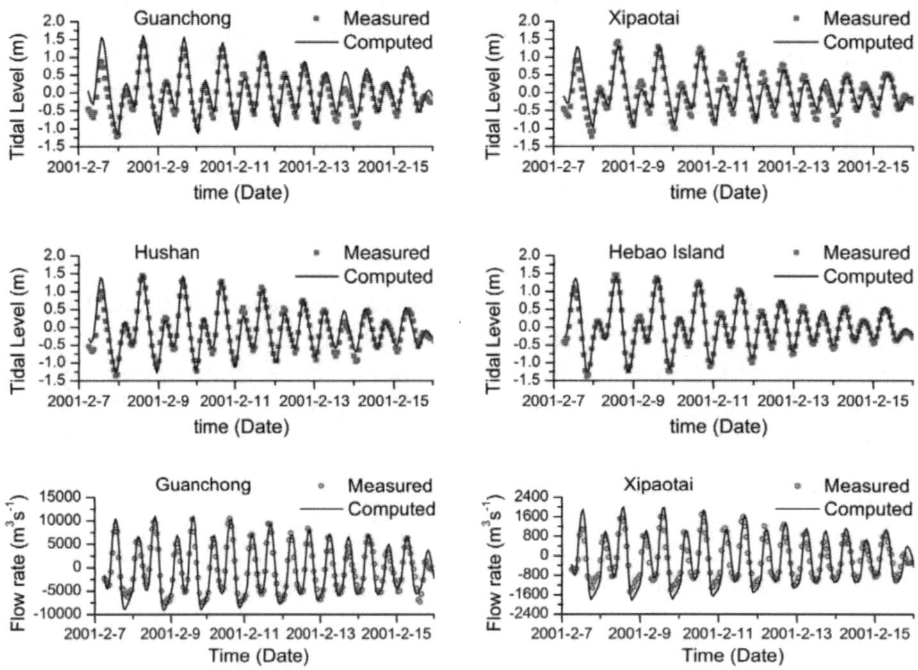

Figure 3. The verification of tide level and flow rate in reference experiment.

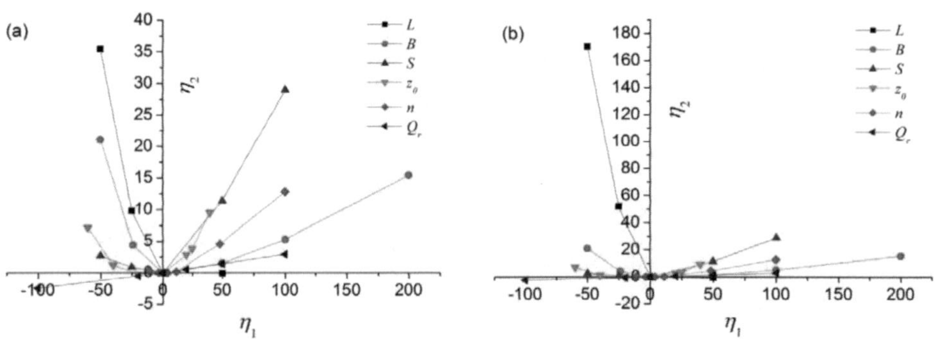

Figure 4. Percentage variations of computational error at Guangchong to simplifying parameters. (a) For tidal level. (b) For flow rate.

As for the bed roughness, it seemed not influencing as imagined whether to the tidal level or to the flow rate. And, the result errors of all runoff tests were not outstanding compared to that of the other parameter tests, especially for the tidal flow. Moreover, both *RMS* varying linearly with the runoff changes was another feature of the parameter.

When we classified the parameters according to the effects on tidal prism or tide storage, except the runoff, we found that for tide level errors, all changes that enhanced the prism such as increase of B or L and decrease of S, V_0 or n affected less. If these parameters were changed inversely, i.e. the prism was reduced, the computation errors showed high sensitive dependence. In addition, decrease of Q_r resulted in negative η_2, which indicated less errors than in the reference experiment. The tidal level *RMS* reached the minimum when Q_r was zero.

3.3. The best simplification

In order to comprehend the best combination of the simplified parameters, we investigated the information of Tan River through relevant documents. In the early 1980s, in the upstream near Xiangang Town a controlling sluice was constructed and then the tidal reach actually has an average bed slope of 0.7×10^{-4}. When the slope in reference experiment was calibrated to that, the flow *RMS* in Guanchong showed decrease by 2.7%. In addition, for channel length, 50% increase of L resulted in a little higher accuracy of computation. This implied the true tidal limit was in the upstream of the sluice project which was the real fact and accorded with the conclusion of other researchers' simulation [11].

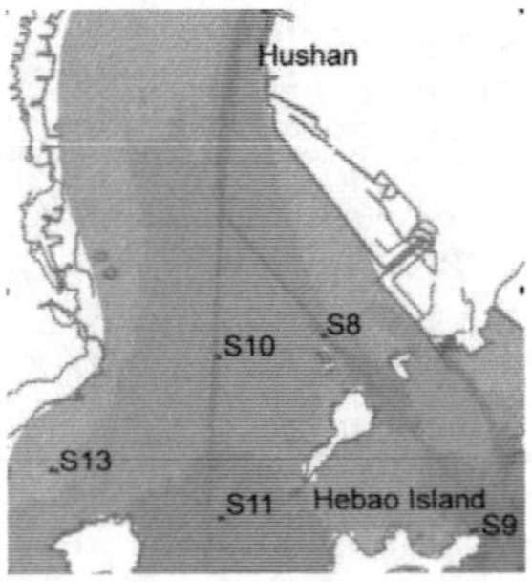

Figure 5. The verification places S8~S13 in downstream of the estuary. The darker line indicates the shipping lanes in planning.

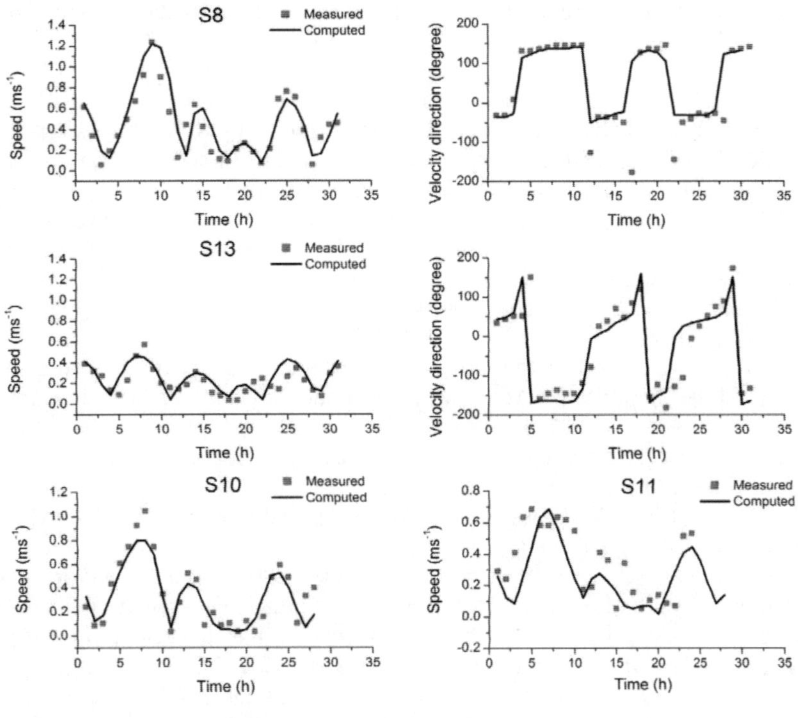

Figure 6. The current verification for S8 and S13 from daytime 20:00 on 1 to 2:00 on 3 in January and S10 and S11 from daytime 9:00 on 27 to 12:00 on 28 in June. True north direction is specified as zero degree. The current direction data in June was missing.

3.4. Flexibility of the simplification

After adopting the optimized length and slope in subsection 3.3, with the other parameters the same as the benchmark except for the runoff, we chose two springs in January of dry season and June of flood season in 2010 to simulate respectively, with the verification points shown in figure 5 and the current results in figure 6. The computed and the measured showed a good accordance in overall, which implied the simplification as well as the parameters can adapt to the different seasons.

4. Conclusions

We conducted numerical experiments by case of Huangmaohai estuary of Pearl River to investigate effectiveness of simplified tidal channels in upstream and significance of the simplifying parameters on hydrodynamics in top of estuary bay. The results show that in simulation of tide dominated estuary if the upstream boundary cannot set to the actual tidal limit due to lack of topographical data or saving computational resources and time, we can choose to simplify the tidal rivers to a straight prismatic channels. During the simplification, the fundamental features of actual tidal prism geometry and resistance characteristics should be maintained. Among the parameters those may diminish the tidal storage and add the resistance to flood tide would result in larger hydraulic calculation errors and they have to be put more cautions and handled more carefully in the simplification.

The simplification method makes the computation stable and fast as well as saving the storage space to some extent. It is adoptable for different time periods and seasons.

Acknowledgments

The work in this paper was supported by the National Natural Science Foundation of China (11572130).

References

[1] Prario B, Dragani W, Mediavilla D and D'Onofrio E 2011 Hydrodynamic numerical simulation at the mouths of the Parana and Uruguay rivers and the upper Rio de la Plata estuary: A realistic boundary condition, *Appl. Math. Model.* **35** 5265-5275

[2] Canestrelli A, Lanzoni S and Fagherazzi S 2014 One-dimensional numerical modeling of the long-term morphodynamic evolution of a tidally-dominated estuary: The Lower Fly River (Papua New Guinea), *Sediment. Geol.* **301** 107-119

[3] Wang B, Fringer O, Giddings S and Fong D 2009 High-resolution simulations of a macrotidal estuary using SUNTANS, *Ocean Model.* **28** 167-192

[4] Yin Z, Xu T, Wang Z, Ma N and Chi W 2015 Dagu Estuary Hydrodynamic Characteristic Research Under the Coupled Action of Runoff and Tide, *J. Ocean Univ. China* **45**(7) 119-124

[5] Xu D, Sun Z, Zhu L and Huang S 2013 Numerical simulation of salinity of Qiantang River estuary, *Ocean. Limnol. Sin.* **44**(4) 829-836

[6] Aristizabal M and Chant R 2014 Mechanisms driving stratification in Delaware Bay estuary, *Ocean Dynam.* **64** 1615-1629

[7] MacCready P, Banas N, Hickey B, Dever E and Liu Y 2009 A model study of tide- and wind-induced mixing in the Columbia River Estuary and plume, *Cont. Shelf Res.* **29** 278-291

[8] Gong W, Shen J and Jia L 2013 Salt intrusion during the dry season in the Huangmaohai Estuary, Pearl River Delta, China, *J. Mar. Syst.* **111-112** 235-252

[9] Zhang W, Yan Y, Zheng J and Zhu Y 2006 1-D and2-D combined numerical model for flow and suspended sediment transport in Pearl River network and Lingding Sea, *J. Sediment Res.* **6** 11-17

[10] Ding P, Hu K, Kong Y and Zhu S 2003 Numerical simulation of total sediment under waves and currents in the Changjiang Estuary, *Acta Oceanol. Sin.* **25**(5) 113-124

[11] Gong W, Liu H, Ren J and Yu H 2012 The study of tidal propagation the in Huangmaohai Estuary and its underlying mechanisms, *Acta Oceanol. Sin.* **34**(3) 41-54

Buoyancy effect on atmospheric surface layer: measurements from the East Coast of Malaysia

Z. Harun[1,2], E. Reda[1] and Rozli Zulkifli[1]

[1]Dept. Mechanical and Materials, Faculty of Engineering and Built Environment, Universiti Kebangsaan Malaysia, UKM Bangi, 43600, Malaysia
[2]Marine Ecosystem Research Centre (EKOMAR), Faculty of Science and Technology, Universiti Kebangsaan Malaysia, UKM Bangi, 43600, Malaysia

E-mail: zambri@ukm.edu.my

Abstract. The nature and evolution of the atmospheric surface layer is still unresolved completely. Many questions regarding the existence of the z-less layer and trend of the cross correlations still remain open. This research analyzes some of the surface layer ambiguities. In this research temperature, velocity and turbulence data were collected from a weather station facility located in the Marine Ecosystem Research Centre (EKOMAR) Mersing on the East Coast of Malaysia. Two high resolution hotwires were utilized at 3 m and 12 m heights above ground. Both gradient Richardson number and Obukhov stability parameter were calculated. Turbulence spectra were plotted at different stability conditions. The results does not show the existence of the z-less layer at deep stable condition. The buoyancy force, under unstable condition, was found responsible for the increase of vertical correlation factor. The fingerprint of the buoyancy force was detected in the spectra at low frequencies.

1. Introduction

The complete understanding of turbulence structure of the atmospheric boundary layer (ABL) is indispensable for the assessment of air pollutant dispersion, natural ventilation through cities and urban heat island intensity. In addition, computational fluid dynamics (CFD) simulations of atmospheric flows rely on experimental data to develop turbulence models and provide realistic boundary conditions and wall functions [1–5]. Much research has been conducted in atmospheric turbulence through the last seven decades since the pioneering work of Monin and Obukhov. Under strong stability conditions all turbulence statistics are expected to be independent of the height z above ground [6, 10]. However, [7] argued that z-less concept holds only for non-dimensional standard deviation of temperature.

The two-point correlation of turbulence in ABL was held by [8] to quantify the structure inclination angle of coherent flow structures. The more unstable the surface layer, the higher the inclination angle in the windward direction with a value of 12° at neutrally stable condition, despite earlier study by [17] who found structure inclination angle to be invariant over three order of the Reynolds number. The two-point vertical correlation was adopted by [9] who found the streamwise and vertical wind components to show little dependence on the turbulence intensity. The spectral analysis of wind components at unstable condition reveals the existence of a spectral region of 2 slope at lower frequencies, indicating the presence of a buoyancy subrange supporting two-dimensional motions [11]. Although the widespread use of ultrasonic anemometers in atmospheric flow measurements, the hot wire anemometry is still a preferred technique for fundamental turbulence studies thanks to its high temporal and spacial resolution [12, 16].

This work aims at enriching the atmospheric turbulence database by reproducing some of the basic correlations using hot wire anemometer measurements at different heights. In this research thermal, mean and turbulent data were collected from Mersing at the East Coast of Malaysia over eleven days (25 March - 5 April, 2016) in the period between the flooding and dry monsoons in Malaysia. A full description of the site and the experimental setup is given in Section 2. The results are illustrated and discussed in Section 3. Finally, the conclusions are listed in Section 4.

2. Method

2.1. Site description

The EKOMAR site (2°34'42.11"N, 103°48'21.05"E) lies on the East Coast of Malaysia, approximately 22 km to the north of Mersing. The pictures in Figure 1 show the entire field facility; it consists of 60 m pole and newly installed 12 m pole for this research. Instrumentation were installed 150 m away from sea line (at high tide) with only a 1 m height site fence and one medium size tree (100 m ahead) in-between. Ground topography is plain and its level is 0.5 m above sea level; the level changes abruptly 100 m away from the instrumentation. Land is covered with 5 cm grass until the sharp level change.

(a) The weather station infrastructure, EKOMAR snapshot taken from the South, showing the South China Sea.

(b) EKOMAR snapshot taken approximately from the West, showing the 12 m pole. The 60 m pole to the right was also utilized, but not discussed here. An intermediary pole in between the two (not clear) was used for the 3 m position.

Figure 1. The EKOMAR weather station, Tangjung Langsat, Endau Johor.

2.2. Instrumentation

Two hotwire probes (5 micron wire diameter) were installed on two adjacent columns (2 m apart) of heights 3 and 12 m. The wires were set vertical such that they read the horizontal magnitude of the wind speed. Samples were acquired at 12 Hz. Each probe was replaced with another calibrated one daily. Calibrations were held in-situ by the Dantec® calibrator. The adverse weather made hotwire measurement more difficult. The authors had to install the entire hotwire etching facility in the EKOMAR laboratory. With this facility, etching the wollaston wire to expose approximately 1.5 mm platinum was possible. A weather station was used to measure temperature, wind speed and direction. It was placed in the middle distance between the two columns at 1.7 m above ground level. The station could captured data continuously. Another temperature sensor was fixed on the 12 m pole. The details of employed devices are listed in Table 1. Data were arranged in continuous sampling groups each of which is 30 minutes long (sampling time) to form a total of 327 groups (163.5 hours of continuous and discrete data).

Table 1. Specs of the utilized instruments.

Sensor	Type and model	Specifications
Hotwire anemometry	5 micron wires Dantec® 54N80 Multichannel Anemometer	Range*: 0.4~46.5 m/s Accuracy*: ±0.2 m/s at 11 m/s
Weather station	Rainwise® MK-III	**Wind speed** Range: 0~67 m/s Accuracy: Greater of 0.45 m/s or 5% of Reading **Wind direction** Range: 0~360° Accuracy: ±11.25° **Temperature** Range: -55~85° C Accuracy: ±0.2°C at 25° C
Temperature sensor	Comet® H6320	Range: -30~60° C Accuracy: ±0.4° C

* Based on in-situ calibration data.

2.3. *Parameters*

2.3.1. Stability parameters Two stability parameters were considered in this study; the Obukhov stability parameter (ζ) and gradient Richardson number (Ri_g) both defined here as following:

$$\zeta = \frac{z}{L} = \frac{z\kappa g\theta*}{u*^2\Theta} \tag{1}$$

$$Ri_g = \frac{\frac{g}{\Theta}\frac{\partial\Theta}{\partial z}}{\frac{\partial U}{\partial z}^2} \tag{2}$$

where L is the Obukhov length, z is the height above ground level, κ is Von Karman constant taken here as 0.41, g is the gravitational acceleration (9.81 m/s^2), U and Θ are the time mean wind speed and potential temperature and $u*$ and $\theta*$ are the friction velocity and friction temperature. In this research, both the friction velocity and temperature were calculated using the Monin-Obukhov similarity theory as follows:

$$u* = \frac{\kappa\Delta U}{ln\left(\frac{z_2}{z_1}\right) - (\psi_{M2} - \psi_{M1})} \tag{3}$$

$$\theta* = \frac{\kappa\Delta\Theta}{ln\left(\frac{z_2}{z_1}\right) - (\psi_{H2} - \psi_{H1})} \tag{4}$$

The two of ψ_M and ψ_H are functions of the stability parameter ζ. The Businger-Dyer relations [13] were adopted to calculate ψ_M and ψ_H as below:

For **stable** surface layer ($\zeta > 0$):

$$\psi_M = \psi_H = -5\zeta \tag{5}$$

For **unstable** surface layer ($\zeta < 0$):

$$\psi_M = 2ln\left(\frac{1+X}{2}\right) + ln\left(\frac{1+X^2}{2}\right) - 2tan^{-1}X + \frac{\pi}{2} \tag{6}$$

$$\psi_H = 2ln\left(\frac{1+X^2}{2}\right) \tag{7}$$

$$X = (1-16\zeta)^{\frac{1}{4}} \tag{8}$$

2.4. Analysis

To calculate the potential temperature, the pressure at the ground level (measured by the weather station) was set as a reference pressure while the barometric pressure at 12 m height was calculated from ideal gas laws. The mean velocities at 1.7 and 12 m were used to calculate Ri_g, $u*$ and $\theta*$. A power spectral density (PSD) analysis was held for the data measured by both hotwires and a Kaiser window [14] was applied to the time series data to remove the undesired noise.

3. Results and discussion

3.1. The weather description

A three-day sample of the temperature and wind speed and direction data gathered by the weather station is shown in Figure 2(a) and Figure 2(b), respectively. The temperature pattern in the sample does not reflect typical Malaysian weather [18] because the field trip was made during a period where El Nino phenomenon was believed to have occurred. The wind speeds have been very steady for the three-day sample at 3 m/s - 5 m/s. The wind direction is very stable where 30 degrees is actually reflecting winds coming from the open seas.

3.2. Turbulence intensities

Figure 3 illustrates the normalized turbulence intensity as a function of Obukhov stability parameter. It is obvious that the intensity fits an asymptotic constant-value line as the surface layer approaches the neutral condition, Figures 3(a) and 3(b). However, at strong stable/unstable conditions ($|\zeta| \gtrsim 0.1$) the turbulence intensities show sharp increase rather than settling at a constant value which does not agree with the existence of the z-less zone. The fitting line tends to a value of 2.1917 from the unstable side and to 2.5772 from the stable side, which is quite around the well-known value of 2.3.

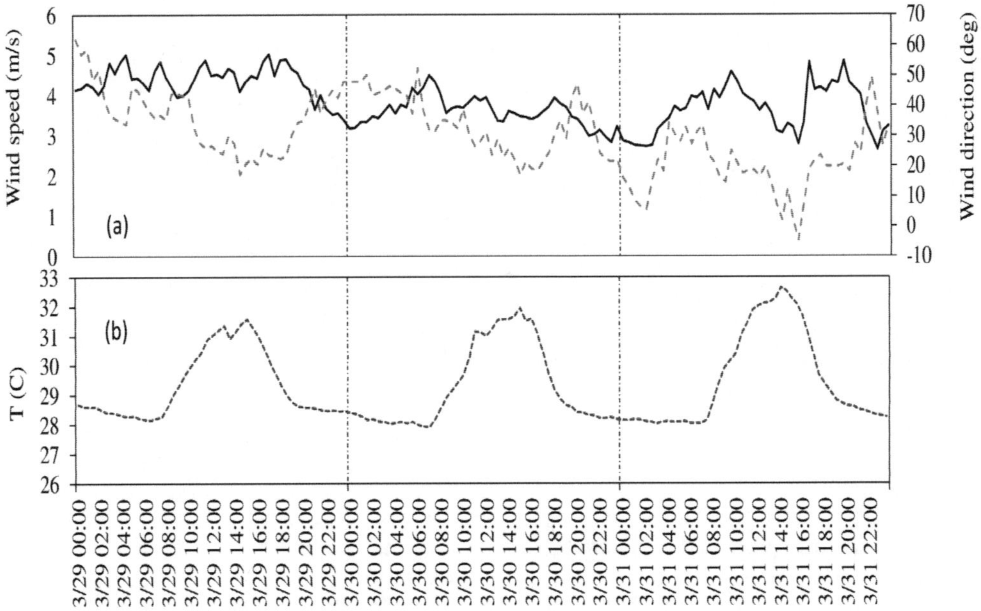

Figure 2. Time series data from the weather station (z = 1.7 m) for the period 29-31 March 2016: (a) wind speed (solid line) and direction (dashed line) (b) temperature.

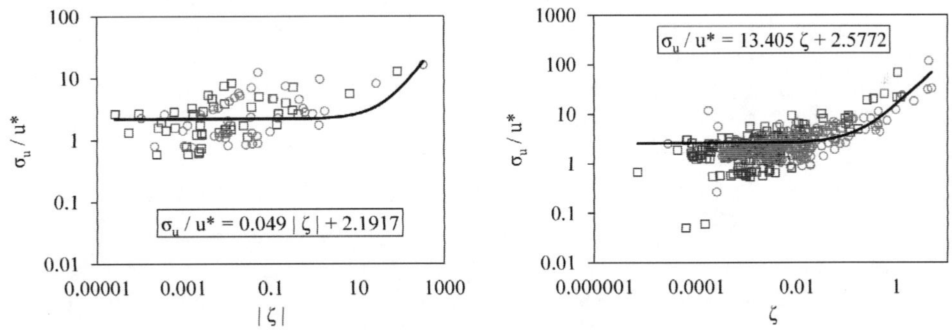

(a) Unstable surface layer.

(b) Stable surface layer.

Figure 3. Turbulence intensity as a function of ζ for unstable surface layer (a) and stable surface layer (b). Profiles of normalized turbulence intensity at z = 3 m (□), at z = 12 m (○) and total data fitting line (solid line).

3.3. Vertical correlations

The variation of the correlation between the two hotwires' velocity components and the gradient Richardson number at the two conditions are illustrated in Figure 4. In unstable condition (Figure 4(a)) the correlation between the two heights increases as the Richardson number decreases (thermal stratification strengthens) [8, 15]. This emphasizes the prominent effects of buoyancy force in mixing the surface layer and breaking down the inclined turbulent flow structures. On the other hand the correlation factor shows almost no dependency on the Ri_g

for stable flow (Figure 4(b)).

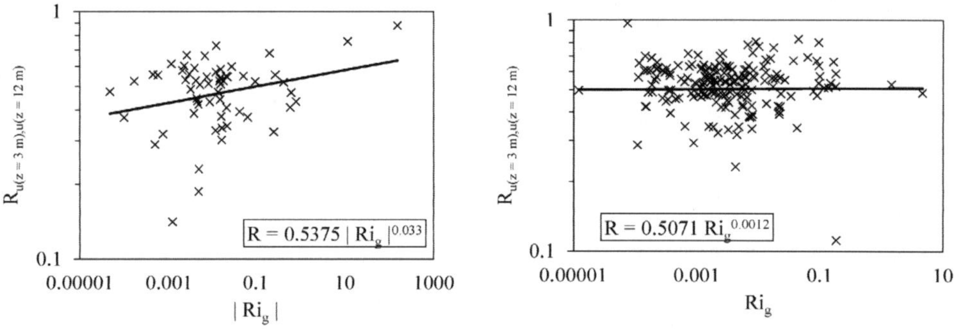

(a) Unstable surface layer. (b) Stable surface layer.

Figure 4. Cross correlation factor between the velocities measures by the two hotwires as a function of Ri_g for unstable surface layer (a) and stable surface layer (b). Profiles of correlation factor (x) and fitting line (solid line).

3.4. Turbulence spectral analysis

Figure 5 shows the spectral analysis of the two hotwires' measurements at unstable (Ri_g = -148.3), neutrally stable ($\text{Ri}_g \approx 0$) and stable conditions (Ri_g = 77.82) where n is the frequency and S_u is the power spectral density function. For both heights, the spectrum of the unstable surface layer experiences a peak at low frequencies, being influenced by buoyant vortical structures. The peak degrades gradually until completely restoring the classic turbulent spectrum profile at profound stable surface layer.

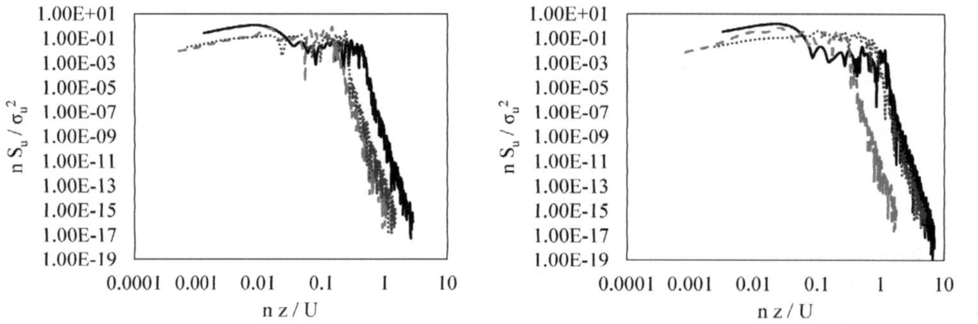

(a) z = 3 m. (b) z = 12 m.

Figure 5. Spectral analysis of velocity components measured by the 3 m level hotwire (a) and the 12 m level hotwire (b). Profiles at unstable condition (Ri_g = -148.3) (solid line), neutral stability condition ($\text{Ri}_g \approx 0$) (dashed line) and stable condition (Ri_g = 77.82). (dotted line).

4. Conclusions

This research sheds light on some of atmospheric layer open questions. 163.5 hours of temperature, velocity and high precision turbulence data were collected from Mersing on the East Coast of Malaysia. The data denies the existence of the z-less zone in very stable surface layer. Also the data showed the effect of buoyancy force in mixing the flow which raises the vertical correlation factor. The buoyant vortical structure was found to alter the turbulence power spectra at low frequencies.

5. Acknowledgement

The author would like to thank the Ministry of Education for providing the Exploratory Research grant. The author would like to express their gratitude for the facility provided at EKOMAR under the Faculty of Science and Technology, UKM.

References

[1] Hargreaves DM Wright N G 2007 *Journal of Wind Engineering and Industrial Aerodynamics* **95** 355–369
[2] Lignarolo L, Gorlé C, Parente A and Benocci C 2011 Large eddy simulation of the atmospheric boundary layer using openFOAM *13th International Conference on Wind Engineering–Amsterdam, The Netherlands*
[3] Defraeye T, Blocken B and Carmeliet J 2011 *Building and Environment* **46** 2130–2141
[4] OSullivan J, Archer R and Flay R 2011 *Journal of Wind Engineering and Industrial Aerodynamics* **99** 65–77
[5] Xie Z T and Castro I P 2009 *Atmospheric Environment* **43** 2174–2185
[6] Wyngaard J and Coté O 1972 *Quarterly Journal of the Royal Meteorological Society* **98** 590–603
[7] Pahlow M, Parlange M B and Porté-Agel F 2001 *Boundary-Layer Meteorology* **99** 225–248
[8] Chauhan K, Hutchins N, Monty J and Marusic I 2013 *Boundary-Layer Meteorology* **147** 41–50
[9] Teichrieb C A, Acevedo O C, Degrazia G A, Moraes O L, Roberti D R, Zimermann H R, Santos D M and Alves R C 2013 *Physica A: Statistical Mechanics and its Applications* **392** 1510–1521
[10] Nieuwstadt F T 1984 *Journal of the Atmospheric Sciences* **41** 2202–2216
[11] Cuxart J, Yagüe C, Morales G, Terradellas E, Orbe J, Calvo J, Fernández A, Soler M, Infante C, Buenestado P *et al.* 2000 *Boundary-Layer Meteorology* **96** 337–370
[12] Lomas C G 2011 *Fundamentals of hot wire anemometry* (Cambridge University Press)
[13] Dyer A 1974 *Boundary-Layer Meteorology* **7** 363–372
[14] Kuo F F and Kaiser J F 1966 *System analysis by digital computer* (Wiley)
[15] Hommema S E and Adrian R J 2003 *Boundary-Layer Meteorology* **106** 147–170
[16] Hutchins N, Nickels T B, Marusic I and Chong M S 2009 *J. Fluid Mech.* **635** 103–136
[17] Marusic I and Heuer W D C 2007 *Phys. Review Letter* **99** 41–45
[18] Latif M T, Dominick D, Ahamad F, Khan M F, Juneng L, Hamzah F M and Nadzir M S M 2014 *Sc. of the Total Environment* **482-483** 336–348

CFD evaluation of erosion rate around a bridge near a sand dune

Wei He, Ning Huang, Hongchao Dun and Wenbo Wang

Key Laboratory of Mechanics on Disaster and Environment in Western China, Lanzhou University
No.222, South Tianshui Road, Lanzhou 730000, Gansu, China

huangn@lzu.edu.cn

Abstract. This study performs a series of simulations through solving the Navier-Stokes equations and the RNG k-ε turbulence model to investigate the wind erosion rates around a bridge in a desert area with sand dunes. The digital elevation model of sand dunes and the bridge model are obtained respectively from hypsographic map and construction drawings. Through combining them into the CFD software of Fluent the simulation zone was formed. The data of wind speed obtained from field observation is fitted into a logarithm format, which was imported into Fluent model as a inlet wind speed condition. Then, the effect of Dun-Go railway on wind-blown sand movement of the neighbouring environment is simulated. The results exhibit that affected by both the sand dune and bridge, the flow field is in a complex condition. It is also shown that the bridge in upstream of the sand dune will not increase the sand transport rate intensively, but change both wind velocity gradient and turbulence kinetic energy over surface of sand dune. On the other hand, when the bridge is built downstream the sand dune, simulation results show that sand deposition rate would be decreased in reference region downstream the pier.

1. Introduction

Wind-blown sand transport over desert is an important subject in the multiphase fluid dynamic. In the December of 2012, a railway was announced to under construction in Northwestern China, between Golmud, Qinghai and Dunhuang, Gansu. Around Shashangou, between Dunhuang and the Altyn-Tagh-Qilian mountain system, the railway will have to cross the eastern edge of the Kumtag Desert. There was a concern that the sand dunes of this area may shift, and even bury the railway. So a 10.7-kilometer-long bridge will be built across a sand valley. The valley stretches itself in a "U" type among complex land forms including compound barchan chain, chain ridges and pyramid dunes. The construction will bring two problems: the effects of the bridge on the wind field and sand transport rate on sand dunes, and the effects of wind-blown sand movement on bridge piers and floor.

This study conducts full scale CFD simulation using the Reynolds averaged Navier-Stokes equations and the RNG k-ε turbulence model to investigate wind field on an area around the bridge. The validity of the simulation is confirmed by comparing the numerical results with the field experiment wind velocity data in same point in the model.

2. Model Description

2.1. Governing equation

To simulate the flow field of the region, the CFD software of fluent 14.0 assumed that a neutral atmosphere boundary layer existed and the 3-D Reynolds averaged N-S equation and the continuity equation should be satisfied. Moreover, the boundary conditions of bridge and sand dunes were set to be non-slip boundary and the flow was considered to be incompressible. The governing equations are listed in the help document of software.

Continuity equation:

$$\frac{\partial\rho}{\partial t}+\frac{\partial}{\partial x_i}(\rho u_i)=0 \tag{1}$$

where ρ means the density of the air and u_i is the fluid velocity.

Momentum equation:

$$\frac{\partial}{\partial t}(\rho u_i)+\frac{\partial}{\partial x_i}(\rho u_i u_j)=-\frac{\partial p}{\partial x_i}+\frac{\partial}{\partial x_j}\left(\mu\frac{\partial u_i}{\partial x_j}\right)+\frac{\partial}{\partial x_j}(-\rho\overline{u_i'u_j'}) \tag{2}$$

where p stands for the static pressure and $-\rho\overline{u_i'u_j'}$ is Reynolds stress.

The turbulence model of k-ε RNG was suggested to be applied when simulating complex flow field [1]. The k-ε RNG turbulence model is given as [2]:

$$\frac{\partial(\rho k)}{\partial t}+u_i\frac{\partial(\rho k)}{\partial x_i}=\frac{\partial}{\partial x_i}\left(\frac{\mu_t}{\sigma_k}\frac{\partial k}{\partial x_i}\right)+G-\rho\varepsilon \tag{3}$$

$$\frac{\partial(\rho\varepsilon)}{\partial t}+u_i\frac{\partial(\rho\varepsilon)}{\partial x_i}=\frac{\partial}{\partial x_i}\left(\frac{\mu_t}{\sigma_\varepsilon}\frac{\partial\varepsilon}{\partial x_i}\right)+C_{1\varepsilon}\frac{\varepsilon}{k}G-C_{2\varepsilon}^*\rho\frac{\varepsilon^2}{k} \tag{4}$$

$$G=2\mu_t S_{ij}S_{ij} \tag{5}$$

$$C_{2\varepsilon}^*=C_{2\varepsilon}+C_{2\varepsilon}' \tag{6}$$

$$C_{2\varepsilon}'=\frac{C_\mu\rho\eta^3(1-\eta\times{\eta_0}^2)}{1+\beta\eta^3} \tag{7}$$

$$\mu_t=\rho C_\mu\frac{k^2}{\varepsilon} \tag{8}$$

$$\eta=S\frac{K}{\varepsilon} \tag{9}$$

$$S=\left(2S_{ij}S_{ij}\right)^{1/2} \tag{10}$$

where k is the turbulence kinetic energy, ε is the kinetic energy dissipation rate, σ_k is the Prandtl number of kinetic dissipation rate, S_{ij} is the shearing rate tensor, C_μ=0.085, σ_k=0.719, σ_ε=0.719, $C_{1\varepsilon}$=1.42, $C_{2\varepsilon}$=1.68, β=0.012 , η_0=4.38.

2.2. Boundary condition

2.2.1. Atmospheric boundary condition. The profile of wind velocity and k-ε were defined as follows [3]:

$$U(z)=\frac{u_*}{\mathrm{K}}\ln\left(\frac{z}{z_0}\right) \tag{11}$$

$$k(z)=\frac{(u_*)^2}{\sqrt{C_\mu}} \tag{12}$$

$$\varepsilon(z)=\frac{(u_*)^3}{\mathrm{K}(z+z_0)} \tag{13}$$

where U stands for velocity, u_* is friction velocity, K is von-Karman number , z is height and z_0 is roughness height (d/30), d is mean particle diameter of sand which is considered to be 250 μm.

2.2.2. Simulating zone. Figure 1 presents a schematic description of the computational domain. At the inlet face, the wind velocity obtained from field observations is imposed. A free developing wind field condition is specified at outlet boundary. On the boundary at the top and the lateral, a zero-gradient for pressure normal to the boundary is supposed. For bridge and ground boundaries, a non-slip condition is set.

2.3. Mesh generation

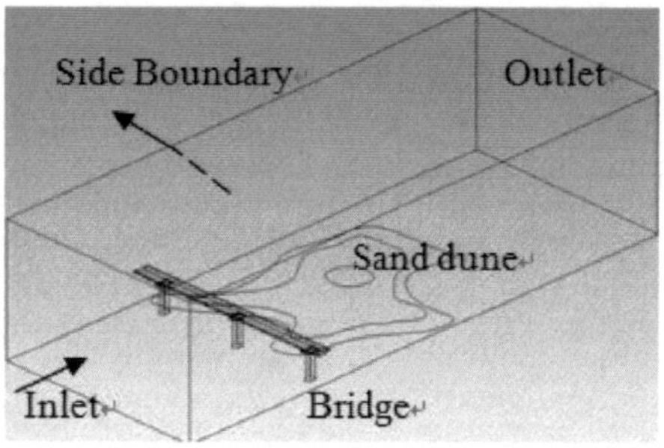

Figure 1. Simulating zone

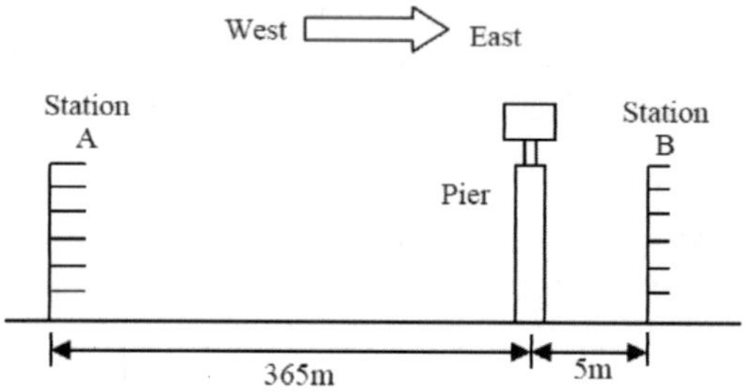

Figure 2. Arrangement of the experiment

The mesh is conducted based on non-uniform grid. All mesh edges were no more than 2-meter-long in domain and no more than 0.2-meter-long near surface of bridge and sand dune. The grid expansion ratio $\gamma < 1.5$ and 5 inflation layers on the solid boundary is set.

2.4. Numerical validation

To validate the numerical method we designed a field observation in this study. Figure 2 shows the layout of experiment. During a process of a 24 h-long sand storm, we obtained wind velocity data from the wind speed profile meter at A station (black points in figure 3(a)), and the data matched into logarithmic type (red line in figure 3(a)). This logarithmic wind speed profile is imposed as inlet wind condition. After a 2000-step iteration until flow field is stable. The simulated wind velocity data at position B can be obtained from CFD-post software and it matches well with wind velocity data obtained from experiment as shown in figure 3.

2.5. Evaluation of wind-blown sand movement

2.5.1. Simulating wind erosion potential on the sand dune. According to Yeh and Tsai [4], wind erosion potential, P , was offered in AP-42 Compilation of Air Pollutant Emission Factors published by U.S. Environmental Protection Agency(EPA):

$$P = 58(u_* - u_{*t})^2 + 25(u_* - u_{*t}) \tag{14}$$

$$P = 0(u_* \leq u_{*t}) \tag{15}$$

where u_{*t} is threshold friction velocity (m s^{-1}).

$$u_* = \sqrt{\tau_w/\rho} \tag{16}$$

In this work, threshold friction velocity is regarded as 0 m/s, which will overstate sand transport rate. Thus, the equation we used to predict wind erosion rate can be written as:

$$P_0 = 58{u_*}^2 + 25u_{*t} \tag{17}$$

2.5.2. Simulating sand depositing rate after the bier. First, to calculate the sand flux blown into controlling zone (figure 5), data of sand transport flux and wind speed is obtained from field observation by setting several sand samplers around pier. The relationship between sand mass concentration C ($kg\ m^{-3}$) and wind speed v ($m\ s^{-1}$) is parameterized (figure 4) and result comes out to be:

$$C = 0.0136 \times e^{-(\frac{v}{1.38894})} - 8.7 \times 10^{-5} \tag{18}$$

analyse every grid on boundary plane a ,b, c, and d , the sand mass flux through each grid can be written as:

$$Q_i = C \times v_i \times A_i \tag{19}$$

where v_i is the wind velocity passing vertically through the grid and A_i stands for the grid area. Sum Q_i of all grids in the boundary planes of a, b, c and d in figure 5, we obtained sand mass flux transported into controlling zone: $Q_{in} = \Sigma Q_i$.

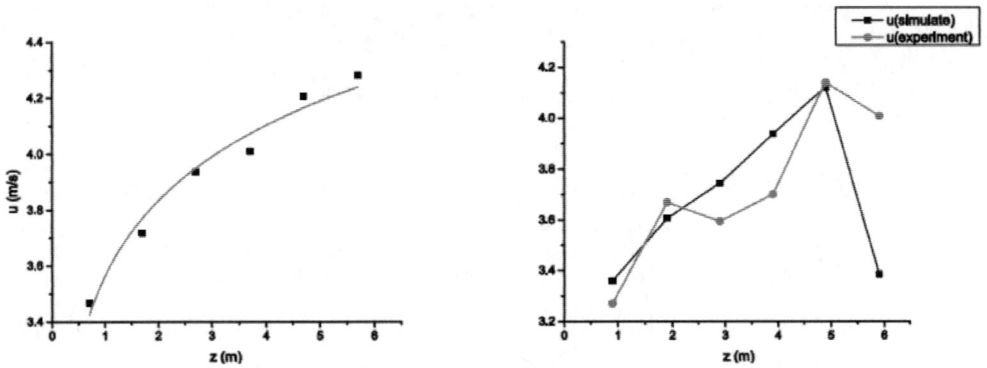

(a) wind velocity data of station A (b) wind velocity data of station B

Figure 3.Wind speed of experiment and simulation

Second, to evaluate the quantity of sand saltating from the land surface and leaving controlling zone, we bring in saltation flux equation of Owen in 1964 [5]:

$$Q_o = \frac{c_0 \rho}{g} u_*^3 (1 - \frac{u_{*t}^2}{u_*^2}) \quad (20)$$

According to simulation result, the value of u_* and u_{*t} of every grid on land surface (shown in figure 5) are already known. Because the mean diameter of sand is considered to be 250 μm, c_0 is suggested to be 1.5. However, the dimension of Q_{in} and Q_0 is disunity, we defined a new saltation flux variable Q_s based on Q_0:

$$Q_s = Q_{oi} \times l_i \times u_{oi} \quad (21)$$

where l_i (m) stands for grid length perpendicular to the ith direction and u_{0i} (m s^{-1}) stands for wind velocity in the ith direction. Sum Q_s of all grids on land surface shown in figure 5, we obtained sand quantity leaving controlling zone, indicate as $Q_{out}=\Sigma Q_s$.

3. Results and Discussion

3.1. The effect of bridge on the sand dune

Figure 6 presents the simulation results of wind speed (figure 6(a)) and wind erosion potential "P_0" (figure 6(b)) over the sand dune without bridge upstream. The results clearly show that the maximum value of "P_0" appears near the top of the sand dune, which indicate particles in this position will leave the ground more easily than other positions. According to the simulation results, the maximum wind erosion potential rate "P_{0max}" has a value of 17.89 and the maximum wind velocity occurs to be 8.2 m s^{-1}. Similar operations have been applied in cases of the bridge is 0 m~20 m from the root of the sand dune.

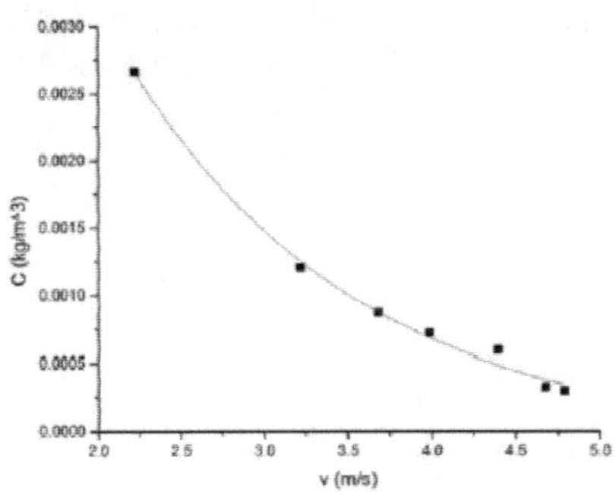

Figure 4. Relationship of C and v

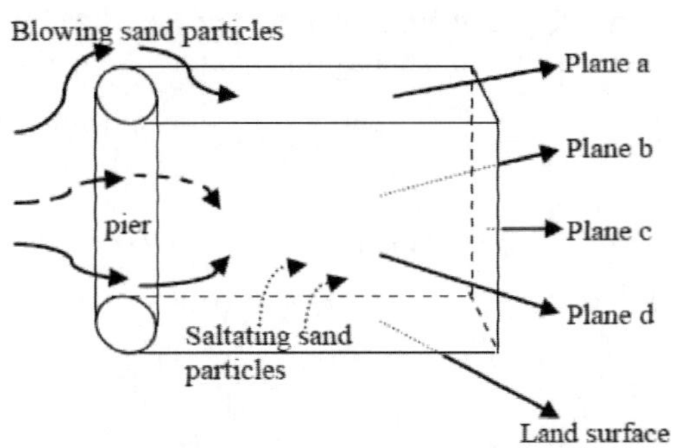

Figure 5.Controlling zone after pier

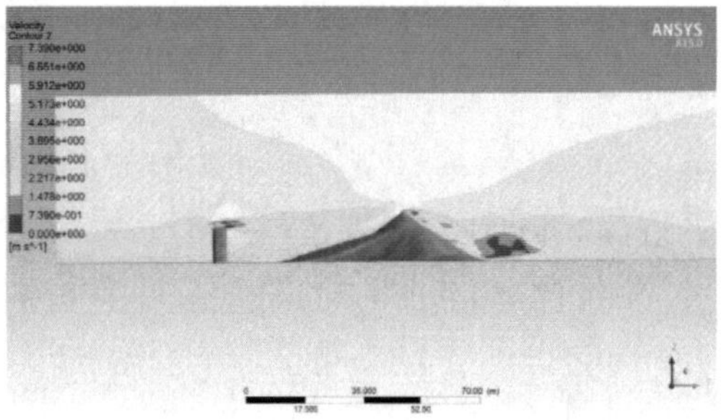

Figure 6(a). The velocity field of air flow

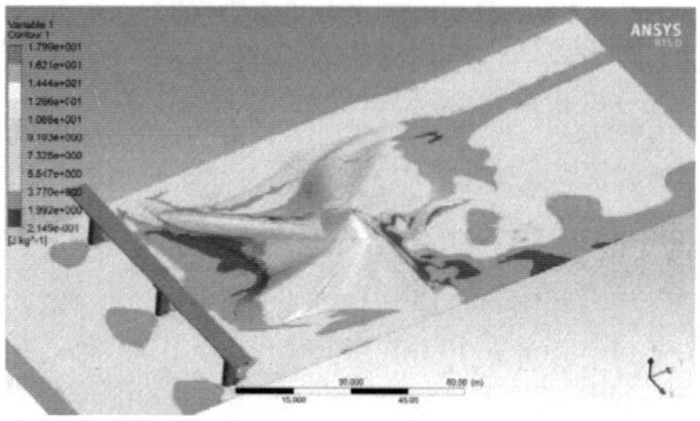

Figure 6(b). The wind erosion potential rate

The values of concerned variables of P_{0max} and velocity u_{max} are listed in table 1. Where S stands for distance between the bridge and the sand dune, P_{0max} is the maximum wind erosion rate on sand dune, $d_u d_h^{-1}$ is velocity gradient and k is turbulence kinetic energy.

Overall, the result shown in table 1 indicates that the difference between wind erosion potential rate with and without bridge is little. However, the wind speed has decreased from 8.2 m s^{-1} to 6.2 m s^{-1} with S increased from 0 m to 20 m. At the same time, wind speed gradient which is related to viscous stress increased from 36.85 to 142.79. Whereas turbulence kinetic energy which is related to Reynolds stress decreased from 0.352 J kg^{-1} to 0.213 J kg^{-1}. We know that shear stress is consists of viscous stress and Reynolds stress and thus, friction velocity u_* is related to wind speed gradient and turbulence [6]. In this study, the reason that P_{0max} did not change observably is determined by the action of velocity gradient and turbulence together.

Table 1.Result of wind erosion potential rate on the top of sand dune

S (m)	u_{max} (m s^{-1})	P_{0max}	$d_u d_h^{-1}$	k (J kg^{-1})
No bridge	8.2	17.89	36.85	0.352
0	8.2	17.77	31.89	0.328
10	6.1	17.99	49.87	0.287
20	6.2	18.03	142.79	0.213

3.2. The effect the sand dune working on the bridge

We have calculated the quantities of sand deposition rate in computation domain at four circumstances: without the sand dune, the distance between the bridge and the sand dune is 1 m, 10 m, and 20 m respectively. Table 2 shows the sand mass flux in every controlling plane. It indicates that the quantity of sand deposition rate decreases when there is a sand dune upstream. With the distance S increasing from 1 m to 20 m, the mass of sand accumulated behind the pier per second decreased about 10%. In another word, the sand deposition quantity around the pier will not be badly affected by the sand dune.

Table 2. Results of sand deposition rate after the pier

Distance (m)	Q_{in} (kg s^{-1})			Q_{out} (kg s^{-1})	Sand deposition rate (kg s^{-1})
	Plane b	Plane d	Plane a and Plane c		
No sand dune	2.889	-0.181	-0.091	-0.033	2.584
1	2.412	-0.261	-0.102	-0.021	2.028
10	1.894	0.117	-0.097	-0.011	1.903
20	2.143	-0.272	-0.022	-0.028	1.821

4. Conclusions

Passing through the desert between Dunhuang and Golmud, the Dun-Go railway is destined to have an effect on neighbouring environment. Although many experiments have been performed to predict how the railway will affect the neighbouring environment, numerical calculation is still needed in consideration of its conveniences. From the consequences of the CFD simulation case in this study, we find out that the upstream bridge will not greatly affect the wind erosion rate on the surface of sand dune for the corporate actions of both wind speed gradient and turbulence make the friction velocity changes little. On the other hand, the sand deposition rate after the bier decreases because of the existence of the upstream sand dune. In a summery, the bridge will not affect wind erosion on this sand dune, and the sand dune will decrease sand deposition rate of the downstream pier.

Acknowledgments

This work is supported by the Major projects of national forestry public welfare industry research (201404306).

References

[1] Choudhury D 1993 Introduction to the renormalization group method and turbulence modeling *Technical Memorandum* TM 107
[2] ANSYS Fluent 6.3.26 Fluent Inc 2006. Lebanon, NH, USA
[3] Richards P J and Hoxey R P 1993 Appropriate boundary conditions for computational wind engineering models using the k–e turbulence model *J. Wind Eng. Ind.***46&47**,145–53
[4] Cheng P Y, Chien H T and Ruey J Y 2010 An investigation into the sheltering performance of porous windbreaks under various wind directions *J. Wind Eng. Ind.* Aerodyn. **98 (2010)** 520–32
[5] Owen R P 1964 Saltation of uniform grains in air *J. Fluid Mech* **20** 225–42
[6] Yaping S 2008 *Physics and Modelling of Wind Erosion*, ed Lawrence A. Mysak and Kevin Hamilton (Germany: University of Cologne) p 68

Drifting snow and its sublimation in turbulent boundary layer

Guang Li[1], Ning Huang[1,*], Zhengshi Wang[1]

[1]Key Laboratory of Mechanics on Disaster and Environment in Western China, Lanzhou University, Lanzhou 730000, Gansu, China
[*]Correspondence to: huangn@lzu.edu.cn

Abstract. Drifting snow is a special process of mass-energy transport in hydrological cycle especially in alpine region. It can not only change the snow distribution, but also result in phase change of ice crystal into water vapour, which is so called drifting snow sublimation. Thus drifting snow is of glaciological and hydrological importance in cold regions. In this paper, recent research on drifting snow and its sublimation is reviewed, and some new progresses by our research team in Lanzhou University are also introduced.

1. Introduction

Snow distribution is an important part of eco-hydrology in alpine region, which can be influenced by many factors. Wind blowing snow acting with terrain is the major reason for snow redistribution [1]. Wind blowing snow is a specific hydrological phenomenon, which reflected in two aspects. One is change the distribution of precipitation by transporting the snow from one side to another [2-4]. The other is letting snow particles jump and suspend with high speed, to promote their sublimation, which can lead to mass loss even larger than surface sublimation [5-7]. And these two processes have great influence on other hydrology processes in alpine region. So it is of great significance on wind blowing snow study.

Research on wind blowing snow can be traced back to the early last century, when only observational or descriptive works on specific disasters were studied. Because of the great effects of drifting snow on global environment, human's production and life protection, in 1980s, more and more researchers focused on the forms and trajectory of snow particle motion by wind-tunnel, theoretical analysis and numerical simulation. However, idealized hypotheses are accepted by most previous researchers, for example snow particle is regarded as a sphere with equivalent volume, and wind speed is time-averaged. With the development of computer science, complex factors like turbulence and rugged terrain have been taken into consideration on modelling drifting snow. Then many drifting snow models were developed in the early 21th century [8-10]. But most of these models didn't take into consideration of the coupling action between snow particles and wind field, as well as effects of turbulence on particle motion. In this paper, wind-blown snow movement in turbulent boundary layer is simulated, and some new ideas then are proposed.

2. Simulation methods

Based on ARPS (The Advanced Regional Prediction System), a non-hydrostatic regional prediction system that developed in Oklahoma University [11, 12], codes of two-phase flow considering coupling of wind and particles, grain-bed collision, air entrainment of snow particles, and drifting snow sublimation, are added to simulate the drifting snow and its sublimation in turbulent boundary layer. The filtered Navier-Stokes equations are as following:

$$\frac{\partial\bar{\rho}\tilde{u}_i}{\partial t}+\frac{\partial\bar{\rho}\tilde{u}_i\tilde{u}_j}{\partial x_j}=-\frac{\partial\tilde{p}^*}{\partial x_i}-\frac{\partial\tau_{ij}}{\partial x_j}+B\delta_{i3}+F_i \quad (1)$$

Where ρ is air density, u is wind velocity, p is pressure, τ_{ij} is sub-grid stress, B is buoyancy, F_i is the force that particles act on wind.

The particle motion equations are:

$$\frac{du_{pi}}{dt}=\frac{3\rho\nu(u_i-u_{pi})}{4\rho_p {d_p}^2}C_D Re_p+g_i(1-\frac{\rho}{\rho_p}) \quad (2)$$

Where u_{pi} is snow particle velocity, ν is kinematic viscosity of air, ρ_p is snow particle density, d_p is diameter of snow particle, C_D is drag coefficient, $Re_p=\frac{(u_i-u_{pi})d_p}{\nu}$ is particle Reynolds number, g is gravitational acceleration.

Splash function developed by Kok and Renno [13] is used to simulate grain bed collision, and semi-empirical formula proposed by Doorschot and Lehning [14] is used to model the air entrainment.

Thorpe and Mason [15] gave an expression of mass loss of an ice crystal by blowing sublimation:

$$\frac{dm}{dt}=\frac{2\pi r(RH-1)}{\frac{L_s}{KNuT_a}\left(\frac{L_s}{R_v T_a}-1\right)+\frac{R_v T_a}{ShDe_s}} \quad (3)$$

Where m is the mass of ice crystal, $RH$is relative humidity, T_a is air temperature, L_s is the latent heat of sublimation ($2.84\times10^6 Jkg^{-1}$), K is the thermal conductivity of air, R_v is gas constant of water vapor ($461.5kg^{-1}K^{-1}$), D is rate of molecular diffusion of water vapor, Nu and Sh are non-dimensional numbers relative to wind velocity, $e_s=610.78\exp[21.87(T_a-273.16)/(T_a-7.66)]$ is saturated vapor pressure relative to ice surface.

In this paper, every snow particle is tracked to calculate its mass loss during its saltation, and then we get the sublimation rate by summing the mass lose of all saltating particles in the simulation domain.

Considering the negative feedback of sublimation process, we take the water heat balance equation into account [16]:

$$\frac{\partial\theta}{\partial t}=\frac{\partial}{\partial y}\left(K_\theta\frac{\partial\theta}{\partial y}\right)-\frac{L_s S}{\rho C} \quad (4)$$

$$\frac{\partial q}{\partial t}=\frac{\partial}{\partial y}\left(K_q\frac{\partial q}{\partial y}\right)-\frac{S}{\rho}-Q \quad (5)$$

Where$\theta=T_a(\frac{1000}{P})^{0.286}$ is the potential temperature, $q=0.622\cdot\frac{e_s}{p-e_s}\cdot RH$ is the specific humidity, $K_\theta=\kappa u_* y+K_T$ and $K_q=\kappa u_* y+K_v$ are diffusion coefficient of water vapor and heat [17]. $S=\frac{1}{V}\sum\frac{dm_i}{dt}$ is the sublimation rate, C is the specific heat of air, $Q=u\nabla q$ is convection flux.

3. Results and discussions

3.1 Snow stripes

Uniform horizontal grids and refined vertical grids are used to simulate the wind of an 10m × 10m × 4m area, and grain size distribution obtained from natural snow by Omiya [18] is used to simulate drifting snow. Our simulated wind profile agrees well with the wind-tunnel experiment [19], shown as figure 1.

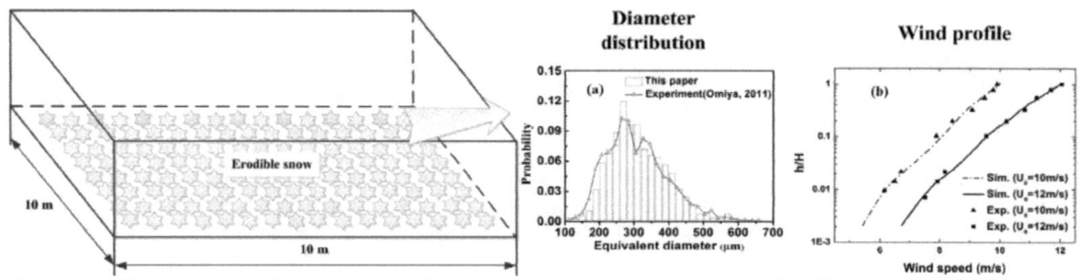

Figure 1. Computational domain. (a) Snow particle size distribution (b) Horizontal wind velocity profile in clear wind field.

Figure 2 shows that simulation results have good consistency with experiment data [20-22], while figure 3 (a) shows that mean particle spanwise horizontal velocity decreases with height for low wind friction velocity, and evolves with opposite phases for high wind friction velocity, which is firstly be detected; (b) shows that streamwise horizontal velocity of particles will concentrate with increment of wind velocity, while the peak value is nearly invariable.

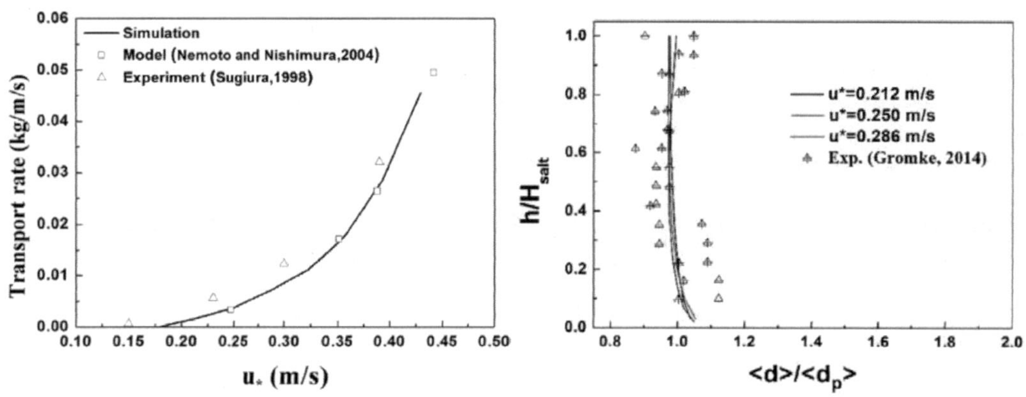

Figure 2. (a) Transport rate with wind friction velocity (b) Particle size distribution with height

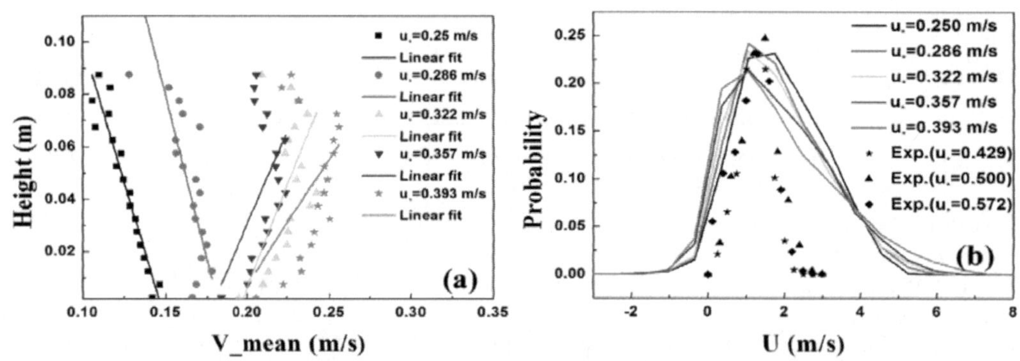

Figure 3. (a) Mean particle spanwise horizontal velocity distribution with height (b) Probability density distribution of particle streamwise horizontal velocity

Snow streamers are first reproduced in drifting snow simulation in this paper, and spectral analysis is used to explain the relationship between snow streamers and turbulent flow. Figure 4 shows similar spectrogram of snow streamer width and wind streamer width, which indicates that intermittency of turbulent flow may be the reason of formation of snow streamers.

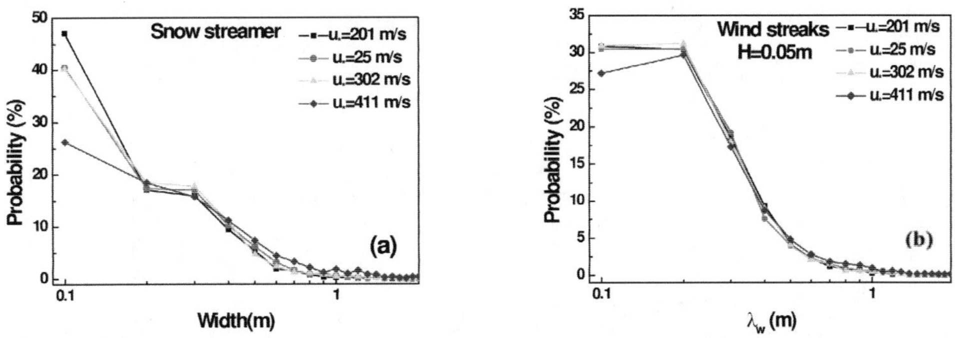

Figure 4. (a) Spectrogram of snow streamer width. (b) Spectrogram of wind streamer width.

3.2 Drfting snow sublimation

Figure 5(a) shows the sublimation rate versus friction velocity on situation one. It indicates that sublimation rate increases with friction velocity and air temperature, and decreases with relative humidity. Figure 5(b) shows that sublimation rate in saltation layer can be higher than that in suspension layer in low friction velocity, which indicates that sublimation rate in saltation layer cannot be ignored.

Figure 6 show the sublimation rate that sublimation rate will quickly reach a peak value and then decreased to zero at the circumstance without moisture transport, which is called a negative feedback effect. However, moisture and heat diffusion and advection will weaken the negative feedback effect, which leads to continuous sublimation events in saltation layer. So it is inadvisable to easily see no blowing snow sublimation in saltation layer.

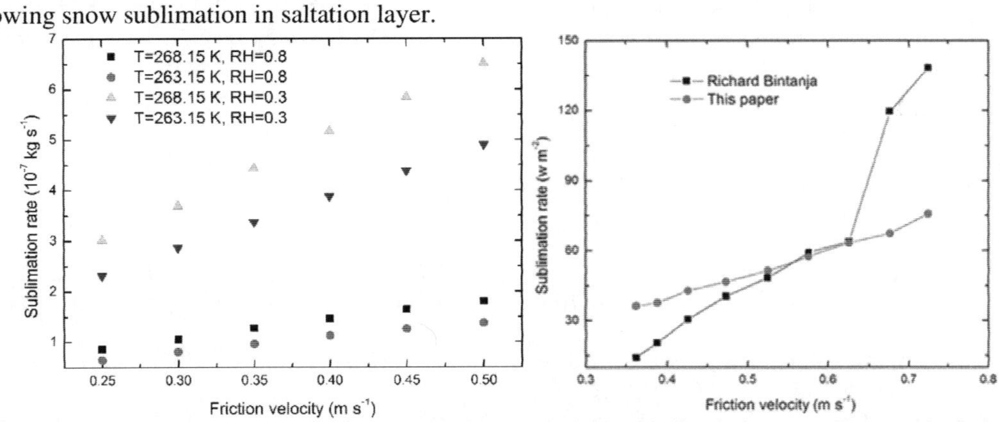

Figure 5. (a) Sublimation rate change with friction velocity by fixed air temperature and relative humidity. (b) Sublimation rate change with friction velocity in different layers

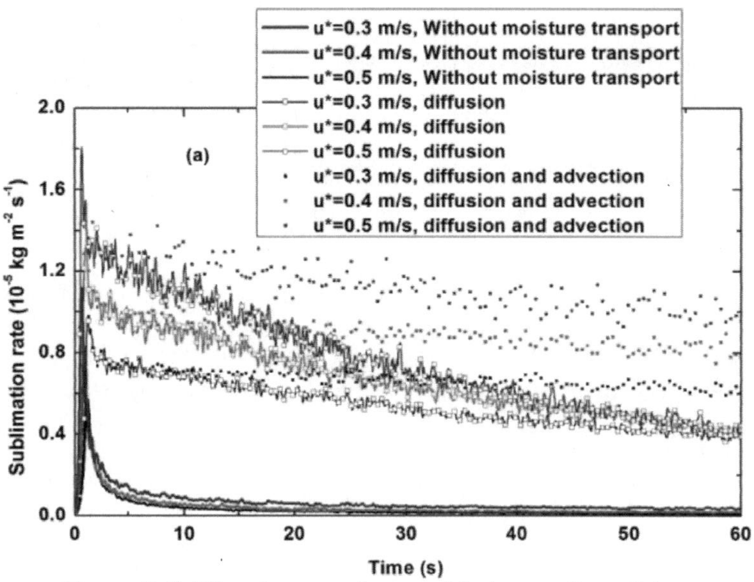

Figure 6. Sublimation rate change with time on 3 conditions

4. Conclusions

In this paper, research on drifting snow and its sublimation in turbulent boundary layer is introduced, Snow streamer is reproduced and its formation mechanism is analysed. The mean particle spanwise horizontal velocity distribution with height is presented. Drifting snow sublimation in saltation layer is simulated, and the results show it is an indispensable part of snow sublimation, which is usually ignored before.

Acknowledgements

This work is supported by the Major projects of national forestry public welfare industry research (201404306).

References

[1] Dadic R, Mott R, Lehning M and Burlando P 2010 Wind influence on snow depth distribution and accumulation over glaciers *Journal of Geophysical Research Atmospheres* **115**(F1) pp 460-72

[2] Pomeroy J W, Marsh P and Gray D M 1997 Application of a distributed blowing snow model to the arctic *Hydrological Process* **11**(11) pp 1451–64

[3] Déry S J, Crow W T, Stieglitz M and Wood E F 2009 Modeling snow-cover heterogeneity over complex arctic terrain for regional and global climate models *Journal of Hydrometeorology* **5**(1) pp 33-48

[4] Gallée H, Trouvilliez A, Agosta C, Genthon C, Favier V and Naaimbouvet F 2013 Transport of snow by the wind: a comparison between observations in adélie land, antarctica, and simulations made with the regional climate model *Boundary-Layer Meteorology* **146**(146) pp133-47

[5] Pomeroy J W and Essery R L H 1999 Turbulent fluxes during blowing snow: field tests of model sublimation predictions *Hydrological Processes,* **13**(18) pp 2963-75

[6] Fang X and Pomeroy J W 2009 Modelling blowing snow redistribution to prairie wetlands *Hydrological Processes* **23**(18) pp 2557-69

[7] Macdonald M K, Pomeroy J W and Pietroniro A 2009 Parameterizing redistribution and sublimation of blowing snow for hydrological models: tests in a mountainous subarctic

catchment *Hydrological Processes* **23**(18) pp 2570–83

[8] Liston G E and Elder K 2006 A distributed snow-evolution modeling system (SnowModel) *Journal of Hydrometeorology* **7**(6) pp 1259-76

[9] Lehning M, Völksch I, Gustafsson D, Nguyen T A, Stähli M and Zappa M 2006 ALPINE3D: a detailed model of mountain surface processes and its application to snow hydrology *Hydrological Processes* **20**(10) pp 2111-28

[10] Pomeroy J W, Gray D M, Brown T, Hedstrom N R, Quinton W L, Granger R J and Carey S K 2007 The cold regions hydrological model: a platform for basing process representation and model structure on physical evidence *Hydrological Process* **21**(19) pp 2650–67

[11] Xue M, Droegemeier K K and Wong V 2000 The advanced regional prediction system (ARPS)-A multiscale nonhydrostatic atmospheric simulation and prediction model. Part I: model dynamics and verification *Meteorol. Atmos. Phys.* **75** pp 161–93

[12] Xue M, Droegemeier K K, Wong V, Shapiro A, Brewster K, Carr F, Weber D, Liu Y and Wang D 2001 The Advanced regional prediction system (ARPS)-A multi-scale nonhydrostatic atmospheric simulation and prediction tool. Part II: model physics and applications. *Meteorol Atmos. Phys.* **76** pp 143–65

[13] Kok J F and Renno N O 2009 A comprehensive numerical model of steady state saltation (COMSALT) *Journal of Geophysical Research: Atmospheres* **114**(D17)

[14] Doorschot J J and Lehning M 2002 Equilibrium saltation: mass fluxes, aerodynamic entrainment, and dependence on grain properties *Boundary-Layer Meteorology* **104**(1) pp 111-30

[15] Thorpe A D and Mason B J 1966 The evaporation of ice spheres and ice crystals *British Journal of Applied Physics* **17**(4) pp 541-8.

[16] Déry S J and Yau M K 1999 A Bulk Blowing Snow Model *Boundary-Layer Meteorology* **93**(2) pp 237-51

[17] Bintanja R 2000 Snowdrift suspension and atmospheric turbulence. Part I: theoretical background and model description *Boundary-Layer Meteorology* **95** pp 343–68

[18] Omiya S, Sato A, Kosugi K and Mochizuki S 2011 Estimation of the electrostatic charge of individual blowing-snow particles by wind tunnel experiment *Annals of Glaciology* **52**(58), pp 148-52

[19] Lv X H, Huang N and Tong D 2012 Wind tunnel experiments on natural snow drift *Science China Technological Sciences* **55**(4) pp 927-38

[20] Nemoto M and Nishimura K 2004 Numerical simulation of snow saltation and suspension in a turbulent boundary layer *Journal of Geophysical Research: Atmospheres* **109**(D18)

[21] Sugiura K, Nishimura K, Maeno N and Kimura T 1998 Measurements of snow mass flux and transport rate at different particle diameters in drifting snow *Cold Regions Science and Technology* **27**(2) pp 83-9

[22] Gromke C, Horender S, Walter B and Lehning M 2014 Snow particle characteristics in the saltation layer *Journal of Glaciology* **60**(221) pp 431-9

IOP Conf. Series: Journal of Physics: Conf. Series **822** (2017) 012046 doi:10.1088/1742-6596/822/1/012046

Effects of fractal grid on spray characteristics and flame development in burner combustion

S H Amirnordin[1], A Khalid[1], M Suardi[1], B Manshoor and M F Hushim
[1] Combustion Research Group (CRG), Centre for Energy and Industrial Environment Studies (CEIES), Faculty of Mechanical and Manufacturing Engineering, Universiti Tun Hussein Onn Malaysia, 86400 Parit Raja, Batu Pahat, Johor, Malaysia.

E-mail: amirk@uthm.edu.my

Abstract. Turbulence generator plays an important role in enhancing turbulence in combustion and determining the flame characteristics in burner combustion. This research demonstrated the effect of a fractal grid on the spray and flame characteristics in burner combustion. Three geometrical configurations of fractal grid were investigated with different equivalence ratios of 0.5–1.0. The images were captured using direct photographic method. The characteristics of the spray and flame were studied, including the length, angle, and area. The results from this fractal grid were compared with the swirl. The results showed that the fractal geometry and ratio of air-to-fuel mixture affected the performance of the burner. The correct combination of fractal geometry and air-to-fuel ratio resulted in complete combustion and improved the overall performance of the burner.

1. Introduction

Diesel is a natural resource that is consumed heavily in the world because of the increasing demand for energy. It can be used in numerous applications, such as fuel in automotive and burner systems. Therefore, the continuous improvement of controlled diesel combustion is urgently required [1].

The generation of turbulence from fractal grids has attracted many researchers [2-6]. Krogstad [5] investigated turbulence from fractal grids generated by wind tunnels. Measurements using hot wire and laser Doppler anemometry revealed the strong dependency of turbulence on the grid geometry. The advantages of fractal grid were further studied by Soulopolous *et al.* [6]. The fractal grid was used to generate turbulence in the premixed flame. The parameters of the grid were blockage ratio, effective mesh size, and perimeter. Results illustrated the advantages of fractal grids. Under the same heat release and downstream distance, fractal grids created intense turbulence and changed the combustion regime. Compared with regular grids, fractal grids exhibited intense turbulence at a given downstream in which the flame angle and turbulent flame speed increased by 50%. Physically, the curvature and flame brush thickness presented large corrugations and intense burning.

In this research, the effects of fractal grids of different geometry were investigated, focusing on the behaviour of the flame. A closed chamber system (Figure 1) was used with direct photographic methods to study the angle, length, and area of the flame.

2. Experimental Setup

A schematic of the experiment is presented in Figure 1 [7]. The main component was an atomizer with eight nozzle holes of 1 mm in diameter. The atomizer [8] was equipped with an air compressor for primary air, a blower to supply excess air, and a fuel pump. During the operation, the air from the compressor at 1 bar (0.1 MPa) was directed to the bottom inlet of the atomizer, whereas the side inlet was for incoming fuel. Excess air was supplied to assist in combustion in the closed chamber. The spray and flame video images were captured in color using Sony FDA-EV1S. The images were analyzed using GOM Player and ImageJ software.

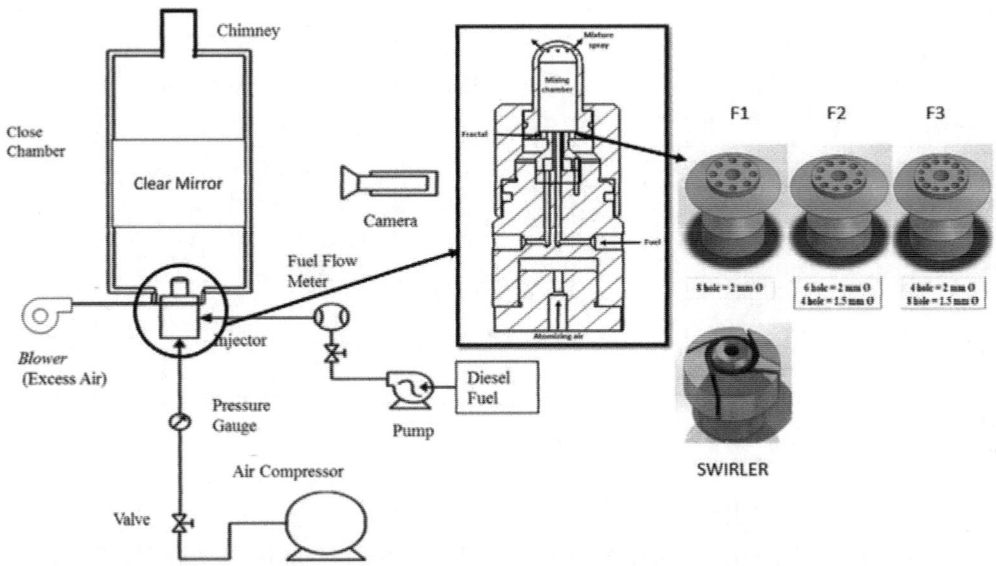

Figure 1. Schematic diagram of experimental setup

Table 1 shows the properties of diesel used in this experiment . The density of the diesel fuel is 0.83374 g/cm^3. The details of the experimental equipment and testing conditions are presented in Table 2. The primary air pressure P_i was constant at 0.1 MPa, which corresponded with the primary air flow rate of 40 L/min. The adjustable parameters were fuel mass flow rate and secondary air mass flow rate. Diesel was sprayed via the 8-hole atomizer with a diameter of 1 mm using three different fractal geometries. The following fractal grids were fabricated and tested in this system: Fractal 1 with eight holes of 2 mm, Fractal 2 with six holes of 2 mm and four holes of 1.5 mm; and Fractal 3 with four holes of 2 mm and eight holes of 1.5 mm.

3. Result and Discussion

Figure 2 shows the spray formation of Fractal 1 at the equivalence ratio of 0.6 to 0.9 (lean) and 1.0 (stoichiometry). The diesel fuel was sprayed upward as the mixtures were directed into the injector. As time increased by 0.03 s for each image, spray penetration rose. Analysis of spray penetration (at 0.12 s when the penetration peaked) showed that the penetration also increased as the equivalence ratio increased (Table 3). The highest penetration was achieved at stoichiometric conditions (equivalence ratio of 1.0) of 587 mm.

Figure 3 shows the flame images of Fractal 1 developed from 0 s to 0.12 s at different equivalence ratios. An increase in the air–fuel ratio affected flame stability and burner combustion. At a high equivalence ratio, the flame area increased because of the increment of the air–fuel ratio. Table 3 indicates that the highest flame area was 317,585 mm^2 under stoichiometric conditions. The image of the flame area obtained from ImageJ is highlighted in Figure 4.

Table 1. Properties of diesel fuel

Properties	Value
Density (g/cm^3)	0.83374
Kinematic viscosity (Cp)	3
Flash point (°c)	80
Water content (%)	0.00796
Acid value (mgKOH/g)	0.423

Table 2. Equipment and experimental condition [7]

Air Compressor	Model	QUASA HDC-D3050
	Capacity, L/min	200
	Pressure, kg/cm^2	8
Water Pump	Model	SFDP1-014-080-22-Seaflo
	Voltage, V	12
	Flow rate, L/min	5.1
Fuel Pump	Model	CNY-3805
	Pressure, bar	3
	Flow rate, L/hr	100
DC Voltage Regulator	Model	Teletron TC-1206A
	Current, A	64 (max)
Operating condition	Air Pressure, MPa	0.1
	Air Density, kg/m^3	1.16
	Ambient Temperature, K	300
	Equivalence Ratio	0.5 - 1.0

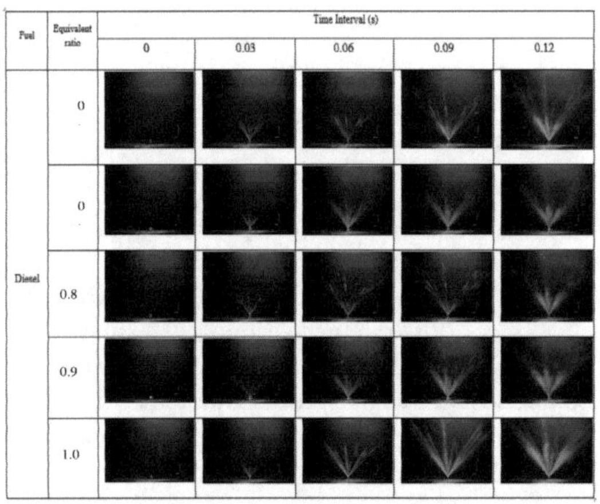

Figure 2. Spray development using Fractal 1 (8 holes of Ø2 mm)

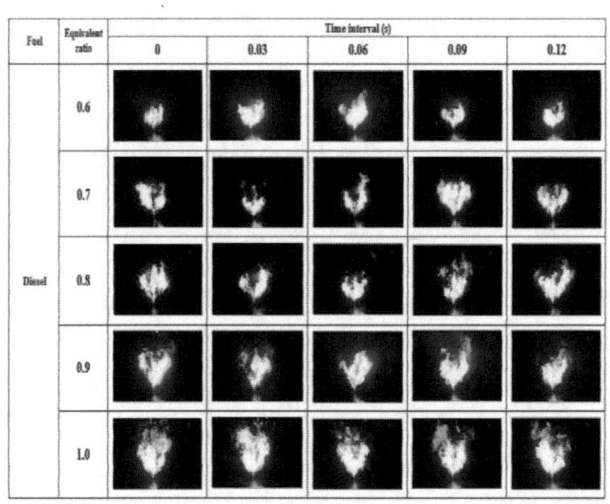

Figure 3. Flame development using Fractal 1 (8 holes of Ø2 mm)

Table 3. Spray penetration and flame area of Fractal 1

Equivalence ratio	0.6	0.7	0.8	0.9	1.0
Spray penetration (mm)	310	407	502	547	587
Flame area (mm^2)	118,669	170,264	183,295	260,237	317,585

Figure 4. Flame area (indicated by the white area) of an image from Fractal 1 at equivalence ratio 1.0

The flame characteristics consisted of the flame length, flame angle, and flame area. Figure 5 exhibits the characteristics of the three fractals. In general, the flame length, angle, and area showed an increasing trend with the equivalence ratio for each type of fractal. This trend indicated the strong relation between the air–fuel ratio and burner combustion.

For Fractal 1, the flame length was higher compared with the swirl. However, the swirl showed a larger flame area and angle as the equivalence ratio approached stoichiometric conditions (0.8, 0.9, and 1.0). For Fractal 2, the swirl showed a higher flame length and flame area. Fractal 3 exhibited higher flame length, flame angle, and flame area compared with the swirl. Therefore, Fractal 3 displayed potential as a turbulence generator in the burner system with better flame characteristics compared with the swirl.

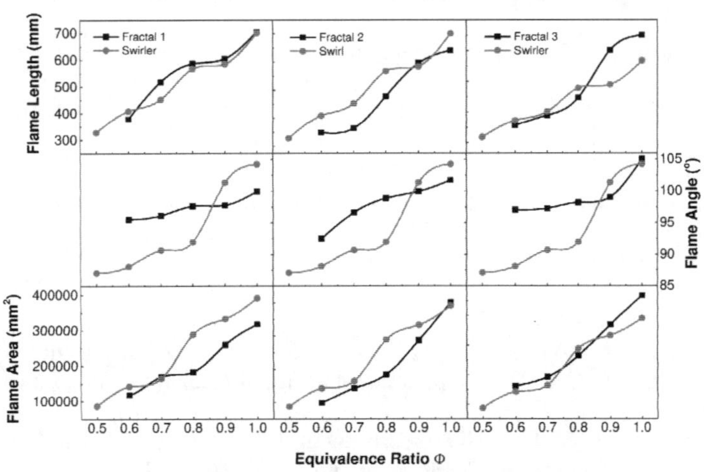

Figure 5. Flame characteristics of each Fractal

Figure 6 shows the influence of fractal geometry on the flame characteristics in the burner. The changes in the flame length, angle, and area of the flame were also influenced by the equivalence ratio. The equivalence ratio of 1.0 produced the largest flame angle, highest flame length, and highest flame area. The figure also indicated that Fractal 3 demonstrated the highest flame length, flame angle, and flame area, especially at the equivalence ratios of 0.8, 0.9, and 1.0. Therefore, Fractal 3 at the equivalence ratio of 1.0 resulted in the highest combustion performance.

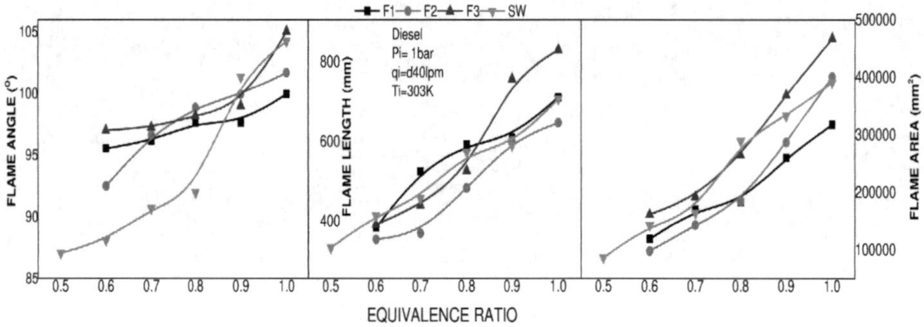

Figure 6. Flame characteristics of all Fractals

4. Conclusion

A fundamental study on the effect of fractal grid in the rapid mixing injector using diesel fuel in the burner system was conducted. Discussion focused on the relation between the equivalence ratio and flame development during the burning process. Results are summarized as follows:

1. A high equivalence ratio will result in a long flame length, wide flame angle, and large flame area. The flame length will affect the flame area. As the equivalence ratio increased, the intensity of the flame also increased.
2. The results of the flame length were dependent on the fractal geometry. The fractal grid significantly influenced combustion. Fractal 3 was found to improve combustion compared with the other fractals and swirl.

Acknowledgements

The authors would like to thank the Ministry of Higher Education Malaysia for supporting this research under Exploratory Research Grant Scheme (ERGS) Vot E032 and also partly sponsored by the Centre for Graduate Studies UTHM.

References

[1] Mishra D 2008 *Fundamentals of Combustion* (New Delhi: Prentice-Hall of India Private Ltd)
[2] Seoud RE and Vassilicos JC 2007 Dissipation and decay of fractal-generated turbulence *Physics of Fluids* **19**(10):105108
[3] Verbeek AA, Bouten TW, Stoffels GG, Geurts BJ and van der Meer TH 2015 Fractal turbulence enhancing low-swirl combustion *Combustion and Flame* **162**(1):129-43
[4] Othman MF, Manshoor B and Khalid A 2013 Circle grid fractal plate as a turbulent generator for premixed flame: an overview In *IOP Conference Series: Materials Science and Engineering* (Vol. **50**, No. 1, p. 012050) IOP Publishing
[5] Krogstad P 2012 Turbulent decay in the near field of multi-scale and conventional grids *International Journal of Heat and Fluid Flow* **35** 102-8
[6] Soulopoulos N, Kerl J, Sponfeldner T, Beyrau F, Hardalupas Y, Taylor AM and Vassilicos JC 2013 Turbulent premixed flames on fractal-grid-generated turbulence *Fluid Dynamics Research* **45**(6):061404
[7] Khalid A, Amirnordin SH, Lambosi L, Manshoor B, Sies MF and Salleh H 2014 Spray characteristic of diesel-water injector for burner system *Advanced Materials Research* **845** pp 66-70
[8] Yatsufusa T, Kidoguchi Y and Nakagawa D 2014 Improvement of emissions and burning limits in burner combustion using an injector on the concept of fuel-water internally rapid mixing *Journal of Energy and Power Engineering* **8**(1) 11-9

Investigation on terminal velocity and drag coefficient of particles with different shapes

Bin Xu, Ning Huang, Wei He, Youxing Chen

Key Laboratory of Mechanics on Disaster and Environment in Western China, Lanzhou University
No.222, South Tianshui Road, Lanzhou 730000, Gansu, China

huangn@lzu.edu.cn

Abstract. Multiphase flow are involved in many fields in natural phenomenon as well as industrial applications. Drag coefficient is an important parameter in the modelling and simulation of multiphase flow. Non-sphere particles are usually simplified to sphere particles in multiphase flow simulation. However , non-sphere particles are more common in practice. Hence, it is significant to investigate shape effects on the drag coefficient of non-sphere particles. This article selects several typical shapes of particles as research objects. 3-D printers are used to build particle models. And high speed cameras are used in record the whole process of particle settling. Detailed information of particle path and speed are gotten through image processing of high speed cameras.

1. Introduction

Fluid-solid system exists widely in natural world and human's life such as sand storm, haze, sediment transportation etc.[1] Sand storm and haze are increasing in recent years. And particle transportation and catalytic reaction propose higher requirement for the research of multiphase flow in industrial application. Because of force analysis is the base of multiphase flow predictions, many scientists have done a lot of researches on frag force of particles in the fluid.

The earliest research on drag coefficient may trace back to 19th century. In 1851, Stokes[2] solved N-S equation in creeping flow (Re<<1) lose sight of inertia part. He gave the formula for calculating drag of smooth sphere particle as:

$$F = 3\pi\mu du \tag{1}$$

Where μ is dynamic coefficient of viscosity, d is diameter of sphere, u is the relative speed of particle and fluid. According to historical document, drag coefficient are defined as equation.(2):

$$C_D = \frac{\mathbf{F}}{\frac{1}{2}\rho u^2 A} \tag{2}$$

And can be formed as equation.(3):

$$C_D = \frac{24}{Re} \tag{3}$$

Where Re is Reynolds number defined as $\mathrm{Re} = \rho ud/\mu$. This formula is appropriate when Re<<1; as Re grows, equation.(3) shows larger deviation compared to experiment.

Oseen[3] reserved the linear part of the inertia part, and he get an approximate solution:

$$C_D = \frac{24}{Re}\left(1 + \frac{3}{16}Re\right) \quad (4)$$

As Re grows , the inertia part turned non-ignorable, and we can barely get the analytical solution. Nevertheless , it is still necessary to use experiment data to get fitted curve. The existing literature gave several formulas based on mass experiment data of free fall experiment and fluidized bed.

Schiller and Naumann[4] gave an formula based on experiment:

$$C_D = \frac{24}{Re}\left(1 + 0.150Re^{0.687}\right) \qquad \text{for } Re<800 \quad (5)$$

White[5] has given a similarity fitted curve formula :

$$C_D = \frac{24}{Re} + \frac{6}{1+\sqrt{Re}} + 0.4 \qquad \text{for } Re<2*10^5 \quad (6)$$

Clift and Gauvin[6] gave a more accurate formula beneath the critical Reynold's number:

$$C_D = \frac{24}{Re}\left(1 + 0.15Re^{0.687}\right) + \frac{0.42}{1+\frac{42500}{Re^{1.16}}} \qquad \text{for } Re<3*10^5 \quad (7)$$

Concerning non-sphere particle, many researchers proposed their shape factor to amend the formula for non-sphere[7][8][9][10][11]. Some frequently used are showed as follow:

Ganser:
$$\frac{C_D}{K_2} = \frac{24}{Re K_1 K_2}\left(1+0.1118\left(Re K_1 K_2\right)^{0.6567}\right) + \frac{0.43056}{1+\frac{3305}{Re K_1 K_2}} \quad (8)$$

Chien:
$$C_D = \frac{30}{Re} + 67.289 exp(-5.03\psi) \quad (9)$$

Tran-Cong:
$$C_D = \frac{24}{Re}\frac{d_A}{d_n}\left[1+\frac{0.15}{\sqrt{c}}\left(\frac{d_A}{d_n}Re\right)^{0.687}\right] + 0.42\left(\frac{d_A}{d_n}\right)^2 / \sqrt{c}\left(1+42500\left(\frac{d_A}{d_n}Re\right)^{-1.16}\right) \quad (10)$$

Haider:
$$C_D = \frac{24}{Re}\left[1 + exp\left(2.3288 - 6.4581\psi + 2.4486\psi^2\right)\right]Re^{(0.0964+0.5565\psi)} + \frac{Re \times exp(4.905 - 13.8944\psi + 18.4222\psi^2 - 10.2599\psi^3)}{Re + exp\left(1.4681 + 12.2584\psi - 20.7322\psi^2 + 15.8855\psi^3\right)} \quad (11)$$

All these formula have improved the departure of drag coefficient for sphere particles when it is used to calculate drag coefficient of non-sphere particles. But the accuracy of these formulas are relatively low because of the limited experimental technique. Due to the development of experiment technique, some new equipment could be used in the experiment to get some more detailed information, and here in this paper we bring in high speed camera and 3-D printer techniques to do some investigations on the movement of non-sphere particles.

2. Experiment and method

We use five different shapes of rigid particles as research objects, which are sphere, ellipsoid, cylinder, hexahedron and octahedron. All particle were printed by 3-D printer, and the material is resin, the density is 1.16E3Kg/m^2, all the surfaces of particles are smooth. Each shape of non-sphere particles contains various sphericities.

The method used in this article is particle settling, which is widely uesd in literatures[9][12][13][14].ie. let the particle fall from the top surface of a water tank, when the particle reaches its fall velocity, we consider it attains balanced state. Under this circumstance, the value of drag force and gravity of particle are equal, and the drag coefficient can be calculated;

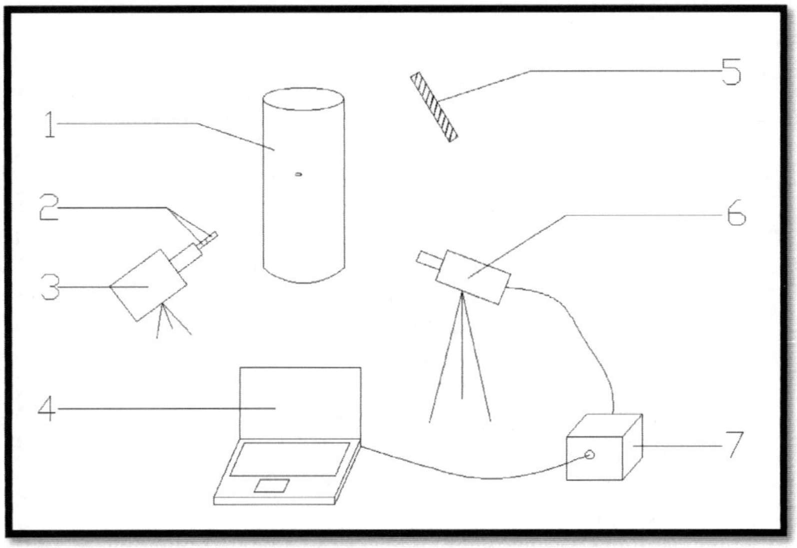

Figure 1. experiment setup

1. water tank 2.cylindrical lenses 3.laser device 4.computer 5.mirror 6.high speed camera 7.flash memory

The setup of experiment is showed in figure1, which consist of 3 parts, lighting system, settling particles and image capture system. Lighting system are formed by laser device, cylindrical lenses, mirror. Laser device emits a dot laser light source. Then a volume light is formed after the dot laser light source passes through two orthogonal column prisms. The reflected light of volume light by the mirror can supply fully light for water tank and particles. The high speed camera is placed vertical to the lighting direction of laser device. The high speed camera is opened as a particle releases in water tank. The liquid in this experiment is water, which is standing for at least 24 hours in room temperature. It ensure the gas dissolving in water discharge sufficiently so that it won't attach on the particle and affect the experiment result. The frame rate are 250fps and 500fps. Any case that the particle fall too close to wall will be deleted.

2.1 Data processing

High-speed camera get digital image through CCD sensor, which is actually a two-dimensional array[15]. In this experiment, a series of gray level images with 1024*1024 pixel are obtained. The particle can be separated from background by picture processing. The velocity of a particle can be obtained through counting the distance of the particle positions in two adjoin images and time interval of the two adjoin images.

2.2 Settling behaviour and result

The experimental results of drag coefficients of 5 shapes of rigid particles are shown in figure 2, it is obvious that the experiment data of sphere coincide well to the well-known standard curve. As regard to drag coefficient of non-sphere particles, the experiment data show great difference, they all turn bigger than that of sphere.

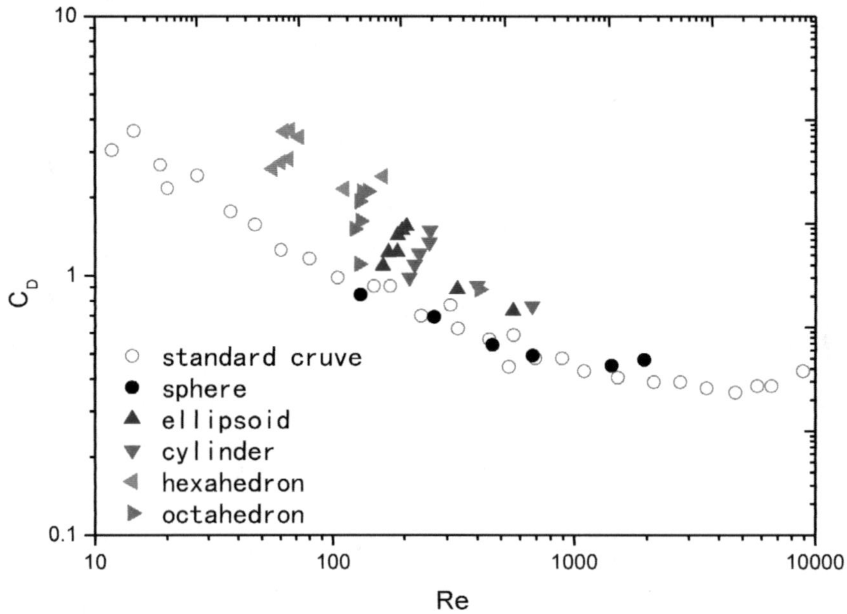

Figure 2. The experimental results of drag coefficients of 5 shapes of rigid particles

To test the accuracy of the modified formulas for non-sphere particle, we the drag coefficients of these formulas are compare the experiment data. The result are shown in figure 3:

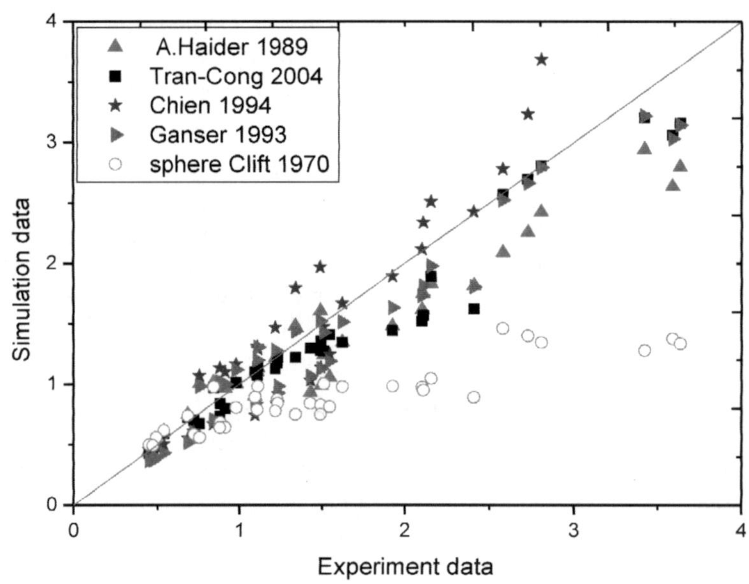

Figure 3. Comparison of modified formulas

As we can see from figure 3, each formula improves the accuracy to a certain extent; The mean errors are listed in table 1. It can be seen that the formula proposed by Tran-Cong in 2004 is the most accurate one.

Table 1. mean error of each formula

author	A.Haider	Tran-Cong	Chien	Ganser
Mean error	16.72%	10.26%	18.69%	15.11%

Through high speed camera, our experimental results successfully record the oscillation of particles when they fall in the tank. The images shows that non-sphere particles do oscillate regularly at high Reynolds number, which approve the phenomenon in previous literatures that sometimes particles will not fall straightly.

To dig detail information, the horizontal velocity and vertical velocity of hexahedron and Octahedron particles, are calculated. And the line graph of velocity-time are drawn as figure 4. The Reynold number of the particles are 55.4 (figure 4-1) and 139.6 (figure 4-2) respectively. The figures show that velocities in both horizontal and vertical directions are affected by the unsteady motion of particles. The horizontal motion of particles shows a regular forth and back movement with periodic velocity. The vertical velocities show fluctuate characteristics with periodicity in some extent. The velocity-time line graph shows that the fluctuation of velocities in two directions has some connections, it may result from the periodic drop of wake vortices from the particles. As we can see from figure 4, the extreme value of horizontal velocity can be up to 30%-40% higher than that of the mean vertical velocity, which means the oscillation motion may have sufficient influence on particle velocity, posture, as well as the shape of projective plane. Further it may lead to the wrong judgement of particle orientation, projected area and finally affect the calculation of drag coefficient.

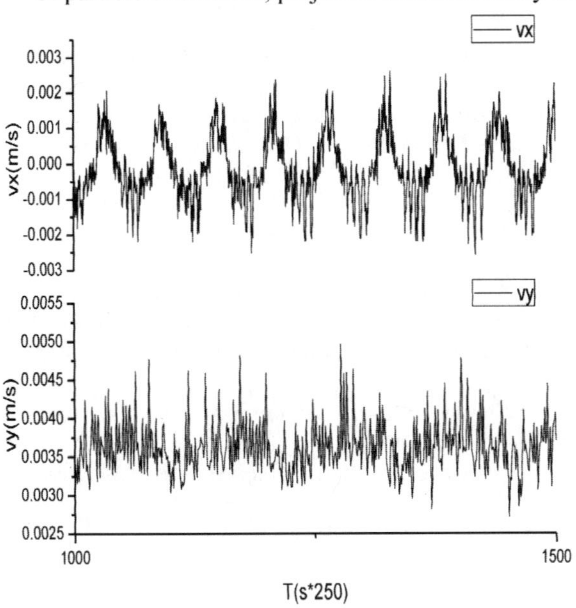

Figure 4-1. the line graph of velocity-time for Re 55.4

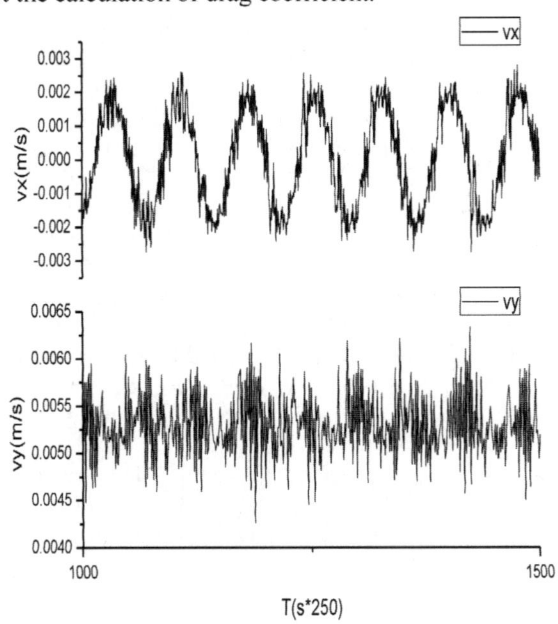

Figure 4-2. the line graph of velocity-time for Re 139.6

3. Conclusion

This study use 3-D printer and high-speed camera techniques reproduce the particle settling experiment, and drag coefficients for 5 shapes of particles are calculated. The results of sphere fit well with exist formula. A comparison between our experiment data of non-sphere particles that of the modified formulas shows that each formula improves the accuracy to a certain extent. Among these formulas, formula proposed by Tran-Cong is the most accurate one.

As for particle oscillation, we get the image of particle oscillation throughout its settling process. The results indicate that non-sphere particles do oscillate regularly at high Reynolds number, which may lead to the wrong judgement of stress state, particle orientation and projected area, and finally affect the calculation of drag coefficient.

Acknowledgments

This work is supported by the Major projects of national forestry public welfare industry research (201404306).

References

[1] Roland Clift 1978 *Bubbles, Drops, and Particles*
[2] Stokes G G 1851 *On the effect of the internal friction of fluids on the motion of pendulums* vol 9 (Pitt Press)
[3] Oseen C W 1910 Ueber die stokes' sche formel, und iiber eine venvandt. e aufgabe in der hydrodynamik *Ark. Math. Astron. Fys* **6**
[4] Schiller L and Naumann A 1933 Über die grundlegenden Berechnungen bei der Schwerkraftaufbereitung *Z. Ver. Dtsch. Ing* **77** 318–20
[5] White F 2010 Fluid Mechanics 862
[6] Clift R and Gauvin W H 1971 Motion of entrained particles in gas streams *Can. J. Chem. Eng.* **49** 439–48
[7] Gary H. Ganser 1993 A rational approach to drag prediction nonspherical particles **77** 143–52
[8] Chien S-F 1994 Settling Velocity of Irregularly Shaped Particles *SPE Drill. Complet.* **9** 281–9
[9] Tran-Cong S, Gay M and Michaelides E E 2004 Drag coefficients of irregularly shaped particles *Powder Technol.* **139** 21–32
[10] Haider A and Levenspiel O 1989 Drag coefficient and terminal velocity of spherical and nonspherical particles *Powder Technol.* **58** 63–70
[11] Loth E 2008 Drag of non-spherical solid particles of regular and irregular shape *Powder Technol.* **182** 342–53
[12] Göğüş M, İpekçi˙ O and Kökpinar M 2001 Effect of Particle Shape on Fall Velocity of Angular Particles *J. Hydraul. Eng.* **127** 860–9
[13] Willmarth W W, Hawk N E and Harvey R L 1964 Steady and unsteady motions and wakes of freely falling disks
[14] Survey G and Paper P 1969 The Behavior of Large Particles Falling in Quiescent Liquids The Behavior of Large Particles Falling in Quiescent Liquids
[15] Gonzalez R C 2009 *Digital image processing* (Pearson Education India)

Propagation of landslide induced impulse wave in channel type reservoirs

An Yi[1], Liu Qingquan[1,2*], Shi Chuanqi[1] and Yang Jiaxiu[3]

[1] Key Laboratory for Mechanics in Fluid Solid Coupling Systems, Institute of Mechanics, Chinese Academy of Sciences, Beijing, P.R. China
[2] Department of Mechanics, School of Aerospace Engineering, Beijing Institute of Technology, Beijing 100081, P.R. China
[3] Guiyang Engineering Corporation, Power Construction Corporation of China, Guiyang, P.R. China

E-mail: * qqliu@imech.ac.cn

Abstract. The propagation characteristics and parameter sensitivity of the landslide induced impulse wave in practical channel type reservoirs are investigated in this study. The two-dimensional Saint-Venant equation and dry-wet boundary treatment method are used to simulate the wave generation and propagation processes in real reservoirs. In order to better reproduce the initial wave, a matching technique of initial wave between nearfield and far field simulation is implemented, while the nearfield wave generation processes are simulated based on full Naiver-Stokes equation. A real landslide induced impulse wave event is simulated with this technique and the propagation characteristics and parameter sensitivity are investigated. The simulation results and comparison between different cases indicate that the longitudinal shape of the channel-type reservoir is a crucial factor for propagation. Classical engineering methods which ignore this factor might produce improper estimation of the max wave height.

1. Introduction

The landslide induced impulse wave (LIIW) is a common but destructive disaster which happens generally in reservoirs and sea coast. In the mountainous area, many steep slopes locate at the bank of reservoir. Numerous events could trigger slope failure of these slopes, thereby massive landslide body might run into the water with high speed and thus cause destructive huge wave which will propagate to the upper and lower reaches in the reservoir. Comparing with LIIW in the basin-type reservoir, the LIIW in the channel-type reservoirs might be more danger as wave energy density of the latter is larger than the former at the same far field site.

Heller and Hager (2010) [1] developed two models for both 2D LIIW and half-plane 3D LIIW. Their estimation is directly based on impact parameters and the distance between observation site and the impact site without explicit discrimination of the generation and propagation processes. In fact the generation process and the propagation process of LIIW are controlled by different physical mechanism and thus can be very different. The kernel mechanism of the wave generation is the energy transport and transform between landslide body and the water flow. Different energy magnitude and distribution lead to different wave form, while the energy dissipation in the inner generation zone is so strong that wave cannot sustains. The Naiver-Stokes equation instead of any depth averaged equation should be used to represent this process, as the velocity in the height direction could not be ignored and significant energy is consumed in the wave generation processes. As for the propagation process, the key mechanism is the wave dissipation and reflection in reservoirs with complex boundaries, while the energy dissipation in this process is much smaller than it of the generation process. Thus, the depth averaged theory such as 2D Saint-Venant equation could be applied in this problem. As the spatial size of interested area for the wave propagation calculation is at least one order larger than it of wave generation calculation, a united method of N-S equation based simulation for wave generation and S-V equation based simulation for wave propagation is the best choose for the whole problem except some special events.

2. Numerical Model

A united model of 3D N-S equation based SPH model and 2D S-V equation based model is developed in this study. The generation process is modeled with the 3D SPH method while the wave propagation is modeled with 2D S-V equation based model. The latter is developed on the open source project Basilisk by Popinet [2]. The two models are connected by the initial wave height at a specific position where the waves become steady and most wave energy could propagate. In this study, this position is about two times of the maximum water depth away from the rim of the landslide body.

2.1. *Simulate wave generation with SPH method*

Using the weak compressible SPH method, the Naiver-Stokes equation could be transform to the following discrete form represented on particles:

$$\frac{d\rho_a}{dt} = \sum_b m_b\left(\boldsymbol{v}_a - \boldsymbol{v}_b\right)\cdot\nabla_a W_{ab} \tag{1}$$

$$\frac{d\boldsymbol{v}_a}{dt} = -\sum_b m_b\left(\frac{P_b}{\rho_b^2} + \frac{P_a}{\rho_a^2} + \Pi_{ab}\right)\nabla_a W_{ab} + \boldsymbol{g} \tag{2}$$

where all information (density ρ, pressure P, velocity v and mass m) of particle "a" could be calculated by integral of nearby particles. The W_{ab} is kernel function act as weighting function, and the Π_{ab} is artificial viscosity term representing fluid viscosity. Detailed information of SPH method and its application on LIIW problem could be found in Shi et al.'s papers [3,4].

2.2. *Simulate wave propagation with Basilisk*

Two-dimensional Saint-Venant equation could be written in the following conservative form:

$$\frac{\partial q}{\partial t} + \frac{\partial f(q)}{\partial x} = S \tag{3}$$

$$q = \begin{pmatrix} h \\ hu \\ hv \end{pmatrix},\quad f(q) = \begin{pmatrix} hu & hv \\ hu^2 + \frac{1}{2}gh^2 & huv \\ huv & hv^2 + \frac{1}{2}gh^2 \end{pmatrix},\quad S = \begin{pmatrix} -\partial Z_b/\partial t \\ gh(S_{bx} - S_{fx}) + hv_t\Delta u \\ gh(S_{by} - S_{fy}) + hv_t\Delta v \end{pmatrix} \tag{4}$$

where h is the water depth, z is the elevation of reservoir bottom, $\eta = z + h$ is the elevation of water surface, (u, v) is the horizontal averaged velocity of flow, g is gravity, the first row of q and $f(q)$ are from continuity equation, the second and third row are from moment conservation equation, the drag from the bottom is consider as source term of equation (3), and take the following form:

$$S = -\frac{n^2\left\|\vec{U}\right\|^2}{h^{4/3}}\vec{U} \tag{5}$$

where n is the Manning coefficient, and $\vec{U}$ is the velocity vector.

The adaptive octree mesh refine technique of Basilisk project is employed in this study. Mesh in any region which needs accurate wave simulation (such as shallow area, steep wave and wave run-up) will be refined to achieve good accuracy. While mesh in those regions that do not need detailed simulation will be coarsened to cut down the computational cost and reduce numerical dissipation.

Compared with classical method, this adaptive mesh refine method could achieve the balance between accuracy and computational cost.

3. Case study

Two LIIW cases with real reservoir configuration are studied. The first one is the Huangtian LIIW event in which the longitudinal shape of the channel is curvy and furcate. The second one is a presumed LIIW event happens in a real straight channel-type reservoir, which is a good comparison for the former.

3.1. Huangtian LIIW event

Huangtian LIIW is a real event happened in 20th Jul, 2009. Huangtian landslide which located in Xiaowan reservoir is about 7.5 Km away from the Xiaowan dam. Stormed by a very heavy rain, the Huangtian slope failed and about 1 million cubic meters slid into the reservoir, induced 30 meters high huge wave. A wave height about 2m is recorded at the survey site located just in front the dam. The DEM of the calculation area is shown in Fig. 1.

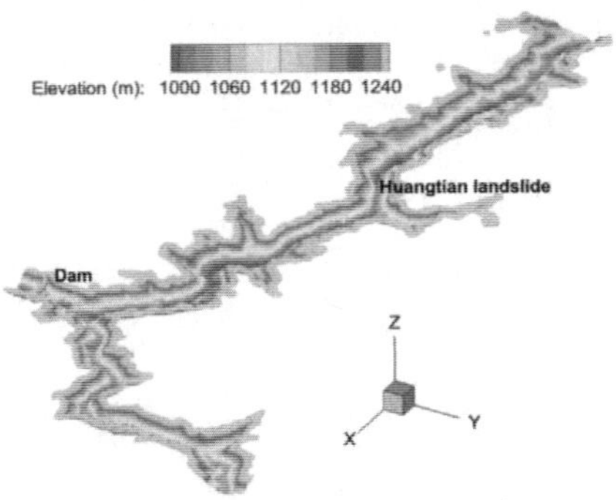

Figure 1. DEM model of Huangtian LIIW simulation

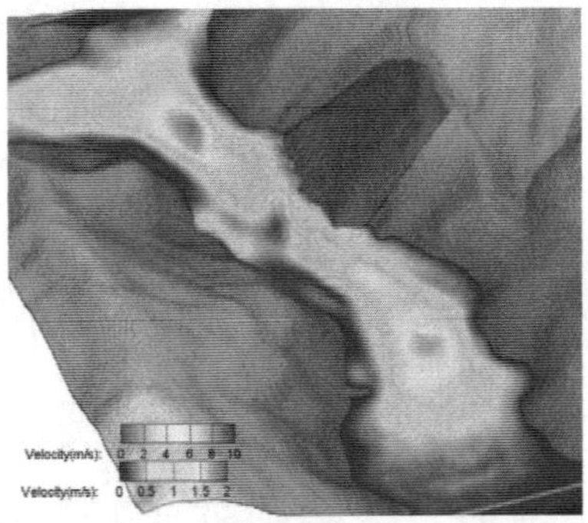

Figure 2. Snapshot of wave generation process colored by velocity

3.1.1. Wave generation simulation. Shi et al. (2015) [3] had simulated the generation process of this LIIW event with SPH method. The length of the calculation area is about 14 times of water depth, and the wave hydrograph at the 2 times of water depth away from the rim of landslide is recorded. The recorded wave parameters such as wave height and wave period are used for wave propagation calculation. Due to the limit of space, here we just show one snapshot of the simulation (Fig. 2). More details on wave generation could be found in Shi et al.'s paper.

3.1.2. Wave propagation simulation. The simulation zone (19262m × 8030m^2) is as large as 3 times of the distance between the Huangtian landslide and the dam. Thus the whole LIIW propagation processes could be simulated while no wave reflection at the upper and lower boundaries will influence the interested area. We use the digital elevation model data from ASTER v2, of which resolution is 30m, to represent the simulation area. The most coarsen mesh is about 75m and the finest mesh is about 4.7m. The maximum total elements in the simulation is about 2 million. The simulation start from the generation of the initial wave and last for ten minutes while the time step is 0.02s. Each case cost about 12 CPU hours in the super computer of "Yuan", Chinese Academy of Sciences.

All physical parameters are set according to the real situation, while two main empirical parameters to be determined are Manning drag coefficient and vortex viscosity. Considering the shape of channel cross section is like "V", the contribution of the side wall on the wet perimeter is significant. Also, the numerous junction points and large scale roughness will result larger drag comparing with normal channels. Thus the Manning coefficient is set to 0.06 which is generally suit for gravel rivers. The determination of vortex viscosity is very empirical as different studies often use very different values. Actually, this parameter is related with the mesh size. Il Won Seo et al. (2014) [5]

choose the value of 5e-3 by calibrating with small scale laboratory flume experiment. If we use the Smagorinsky 2D vortex viscosity relationship to scale this value:

$$\upsilon_t = (C_s \Delta_g)^2 \sqrt{(\frac{\partial u}{\partial x})^2 + (\frac{\partial v}{\partial y})^2 + \frac{1}{2}(\frac{\partial u}{\partial y} + \frac{\partial v}{\partial x})^2} \tag{6}$$

we can obtain the value of 94 for vortex viscosity, which is a proper choice as it is close to empirical values that the previous study has used.

A typical simulation snapshot is shown in Fig. 3 coloured by water elevation. In Fig.3, we can see that the wave is propagating to the upper and lower reaches, while the wave height at the lower reach is significantly smaller than it of upper reach. It is clear that a part of wave energy is transport to a branch gully instead of going to the lower reach. This phenomena could also be seen in Fig.4 which shows the time evolution of water elevation at specific points. The wave height of leading wave is dramatically decreasing along with the channel, drop to 1.7m at the crossroad, and then increase by swelling effect to 2.0m in front of the dam, which is consistent with the observation data. Also the decrease rates at the branch section and the turning section are obviously larger comparing with other sections.

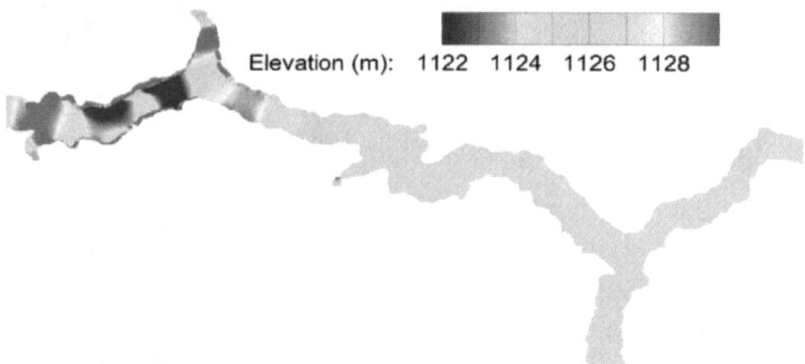

Figure 3. Snapshot of wave propagation process coloured by water elevation

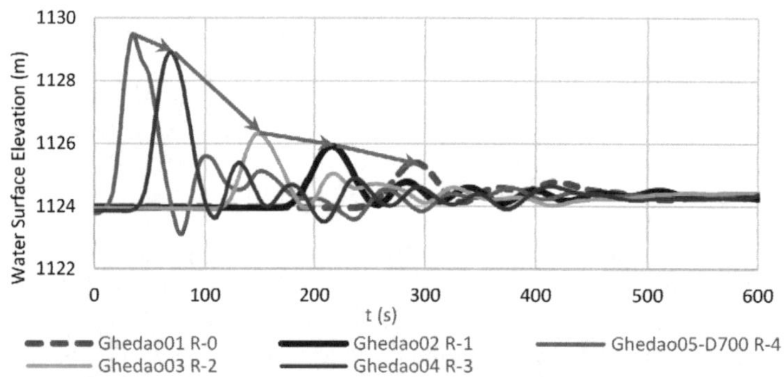

Figure 4. The water elevation evolution at specific points

3.2. A straight channel-type reservoir case

To study the influence of the turning and the branch on the propagation of the LIIW. We found a real straight channel-type reservoir which is similar to the Xiaowan reservoir in many aspects except the longitudinal shape. In this case, the influence of tuning and the branch is very small and the dam axis direction is perpendicular to the wave incidence direction.

Two snapshot on water elevation of the initial and medium stages of wave propagation are shown in Fig. 5, respectively. It is clear that the wave energy is mainly dissipated in the generation zone while the wave energy which could propagate to the downstream is relatively small comparing with the initial wave. Also, as the channel is rather straight, the wave height does not decrease very much in the channel.

IOP Conf. Series: Journal of Physics: Conf. Series **822** (2017) 012048 doi:10.1088/1742-6596/822/1/012048

Besides, as the dam axis is perpendicular to the longitudinal direction of the channel, the wave will run-up at the dam slope and all wave energy will accumulate which might result very huge wave at the dam site. We could also see this phenomena in the quantitative hydrograph of wave height at different location in the reservoir as shown in Fig.6.

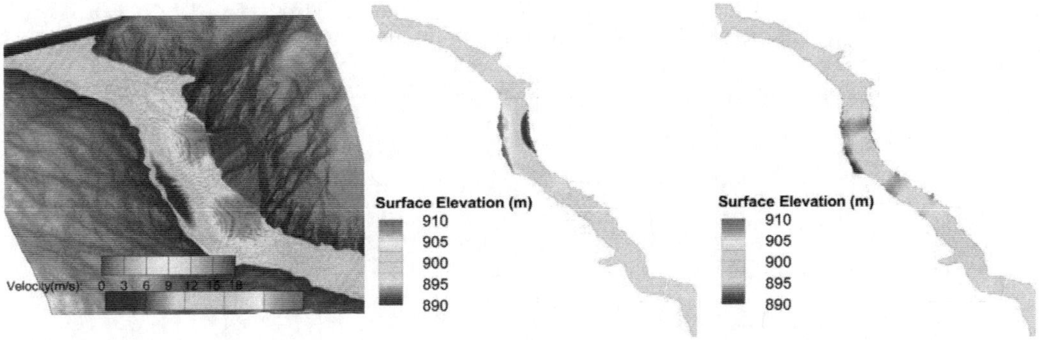

Figure 5. Snapshots of the simulated wave generation and propagation processes

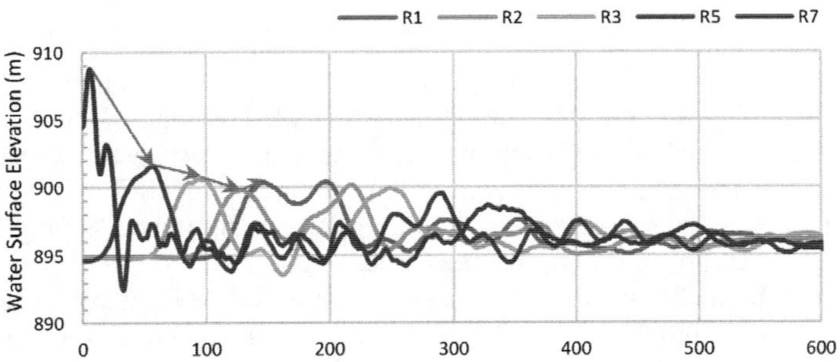

Figure 6. Water elevation at the specific locations in the reservoir

4. Discussion on parameters sensitivity

The wave propagation calculation based on two empirical parameters: the Manning roughness and the vortex viscosity. Here we discuss the sensitivity of those two parameters based on the latter case study.

To discuss the sensitivity of the Manning roughness, four values of Manning roughness (0.01, 0.0.3, 0.06 and 0.09) are chosen for comparison. We can see in Fig.7 that the result is not sensitive to the value of the Manning roughness.

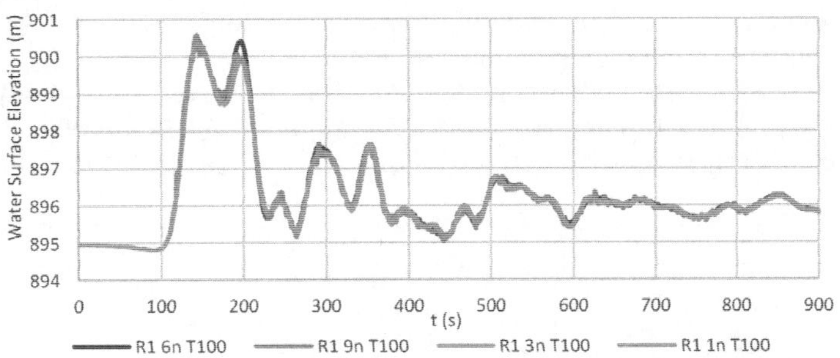

Figure 7. Parameter sensitivity analysis of the Manning roughness

To discuss the sensitivity of the vortex viscosity, three values of vortex viscosity (70,100 and 130) are chosen for comparison. The value of 100 is the same as the case study and the other two is ±30% of the parameter value. Figure 8 shows the hydrograph at the R1 point which locates at the front of the dam. We can see that the sensitivity of the result to the parameter value is medium. A change of ±30% of the value will result about 5% variation of wave height. That is because the vortex viscosity mainly represents the two dimensional moment transport and dissipation which definitely influence the wave height, however as the water depth in this case is very large and the large velocity zone in real 3D situation is always located at the surface, the depth averaged moment would not be too sensitive to the vortex viscosity.

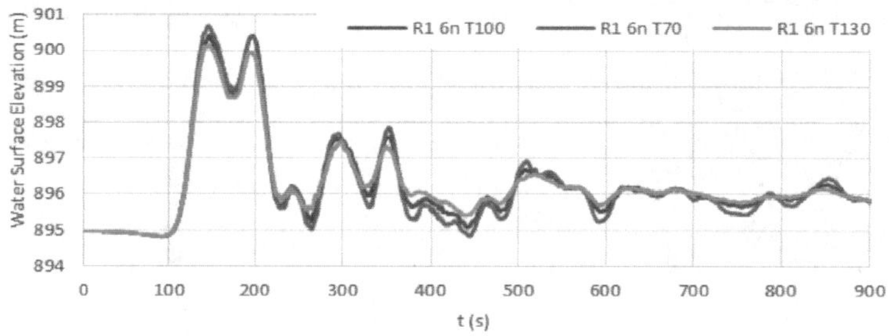

Figure 8. Parameter sensitivity analysis of the vortex viscosity

5. Conclusion

The propagation characteristics and parameter sensitivity of the LIIW in practical channel type reservoirs are investigated in this study. It is found that the united method of 3D SPH simulation for wave generation and 2D S-V equation based simulation, along with a matching technique of initial wave between nearfield and far field simulation could provide better result on LIIW simulation. A real LIIW event and a presumed LIIW event based on real configuration are simulated with this technique. The propagation characteristics for both LIIW events are discussed.

We found that the longitudinal shape of the channel-type reservoir, such as turning and branch gully, is crucial factor controlling the propagation of LIIW. Classical engineering methods which ignore this factor might produce improper estimation of the max wave height. The parameter sensitivity of the Manning roughness is weak as the water depth is generally very large. While the parameter sensitivity of the vortex viscosity is medium and should be carefully treated.

Acknowledgements

This work was financially supported by the Natural Science Foundation of China (No. 11372326, No.11432015) and the National Program on Key Basic Research Project of China (973 Program) (No.2014CB04680202).

References

[1] Heller V and Hager W 2010 *J. Waterw. Port C.-ASCE* **136** 145-55
[2] Popinet S 2003 *J. Comput. Phys.* **190** 572-600
[3] Shi CQ, An Y and Yang JX 2015 *Sci. Sin. Phys. Mech. Astron.* **45** 104706 (in Chinese)
[4] Shi CQ, An Y, Wu Q, Liu QQ and Cao ZX 2016 *Adv. Water Resour.* **92** 130-41
[5] Seo IW, Kim YD and Song CG 2014 *Scientific World J.* **2014** 197539

Study of a possible wake vortex encounter of an aircraft over the Arabian Sea

T N Venkatesh, C Arshad Shameem and J Saravanakumar

Aerospace Weather Research Group, CTFD Division, CSIR-National Aerospace Laboratories, Bengaluru, 560017

E-mail: tnv@nal.res.in

Abstract. A Jet airways aircraft encountered severe turbulence over the Arabian sea. We have performed high-resolution numerical simulations of the weather conditions at the time of the incident to ascertain the possible cause of the incident: atmospheric clear-air turbulence or a wake vortex encounter. Our analysis of the turbulence indices and parameters related to vortex propagation/decay at the flight altitude, suggests that wake-vortex encounter was the more likely cause of the incident.

1. Introduction

Atmospheric turbulence is one of the factors which can affect passenger comfort and flight safety. Other than pilot reports, clear-air turbulence (CAT) indices estimated from synoptic weather charts have been the traditional method of warning pilots. The evolution of high-resolution numerical weather models coupled with extensive observation networks has led to improvements in the prediction skill of these models. This has led to the feasibility of using high-resolution forecasts for not only turbulence forecasts but also optimization of air routes (See for example, study by Sharman et al [1]). Another possibility is of distinguishing the source or nature of the encounter. It is well known that CAT is not a single phenomena, but can be caused by different mechanisms, like Kelvin-Helmholtz instability, mountain waves, inertia-gravity waves etc. At times, even other phenomena like wake-vortex encounters at cruise level are reported as CAT in the absence of accurate first-hand information. We examine a case of a flight over the Arabian sea which was initially reported as a turbulence event by external sources, but diagnosed by its pilots [2] to be an encounter with the wake vortices of another aircraft. By performing an analysis of the results of an NWP model, we assess the possibility of which of the explanations is more likely.

2. Turbulence encounter: a case study

It was reported that a Boeing 737-800W flight of Jet Airways (9W565) from Dammam to Thiruvananthapuram encountered turbulence along its route on 4th October 2014. According to available information in the public domain [3], the estimate of the incident time was 01 UTC and location, was close to the Indian coast. The possibility of turbulence at this location was estimated using numerical models and some of the initial findings were reported in [4]. Later, we learnt of the exact location and time through a private communication with the captain of the aircraft [2].

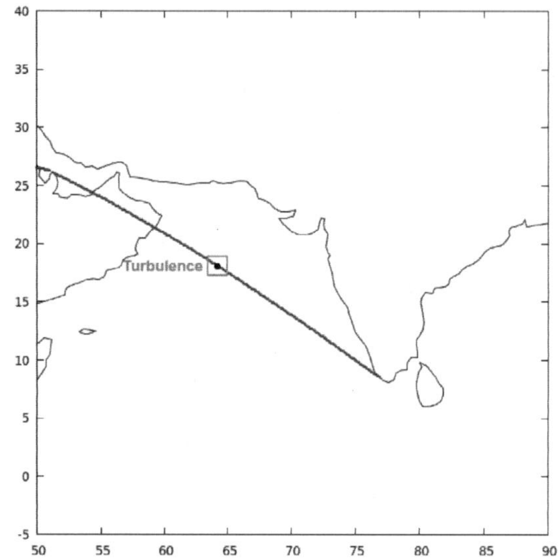

Figure 1. Flight path (source: [5]) of 9W565 with incident location marked in a red square

The encounter happened at 23:11 UTC when it was cruising at an altitude of 37000 feet. The exact incident location was at 63.99E and 17.56N. Prior to the incident, the pilots noticed an Airbus A380 passing in the opposite direction, 1000 feet above. The aircraft underwent a "strong rolling downward motion towards the left" and lost altitude of around 500 feet. There were further "violent excursions to the left and right" [2]. The loss of control and reduction in airspeed which followed was typical of an encounter with a strong wake vortex.

Wake vortex encounters during takeoff and landing have been well studied and there are operational procedures to avoid/minimize such hazards. In recent times, there have been a few reports by the commercial airliners in which wake-upset during *cruise* was suggested as a possible cause [6]. This could possibly be due to the increase in large aircraft such as Airbus A380. It is therefore important to study reported cases so that such encounters can be avoided/minimized in the future.

The typical flight path, with a mark of incident location, is shown in Figure 1. The infrared images [7] from Kalpana satellite are shown Figure 2. From Figure 2, we can observe that there is not much of convective activity in the flight route except near Cochin and Thiruvananthapuram. Also, at the location of the incident, the skies were clear.

3. Numerical simulations of atmospheric conditions

The atmospheric conditions during the turbulence incident were simulated using two numerical weather prediction models in a coupled way. The first is a hydrostatic, global model, Varsha GCM, which solves the governing equations using spectral techniques [8]. This model was integrated at a coarser resolution (about 80km) and the simulated fields were fed to a limited area model (LAM) as lateral boundary conditions. The second model used was the public domain, high resolution, non-hydrostatic model [9, 10], WRF, developed by National Center for Atmospheric Research (NCAR). This LAM is highly scalable and has multiple options to solve various sub-grid scale physical processes. Figure 3 shows the schematic of the coupling method, where the GCM is initiated using global analysis and then the output is used to initiate the

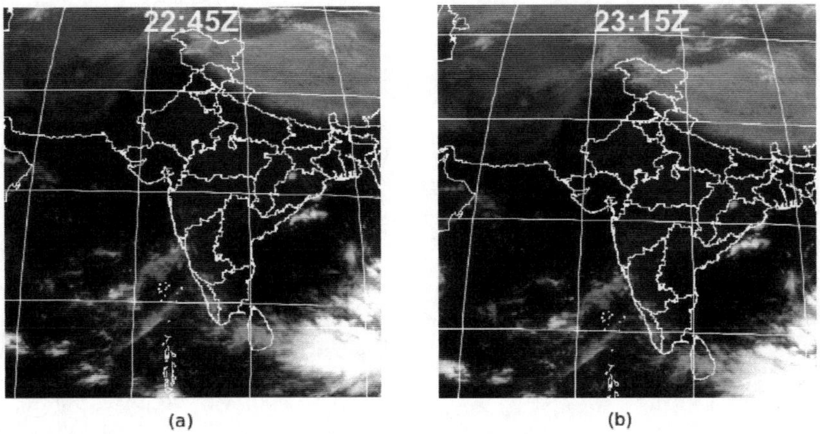

Figure 2. Infrared satellite images from the Kalpana satellite around the time of the incident: (a) 22:45UTC; (b) 23:15UTC .

LAM.

We conducted two simulation experiments with different horizontal and vertical resolutions for the LAM, the first one is with a single computational domain having horizontal resolution of 20km and 28 levels in vertical and the second one is with four nested domains of resolutions 27, 9, 3 and 1km in horizontal and 50 levels in vertical. In both the cases, we have used the analysis data from NCEP-FNL [11] as the input condition. The results presented in this paper are based on the first simulation experiment.

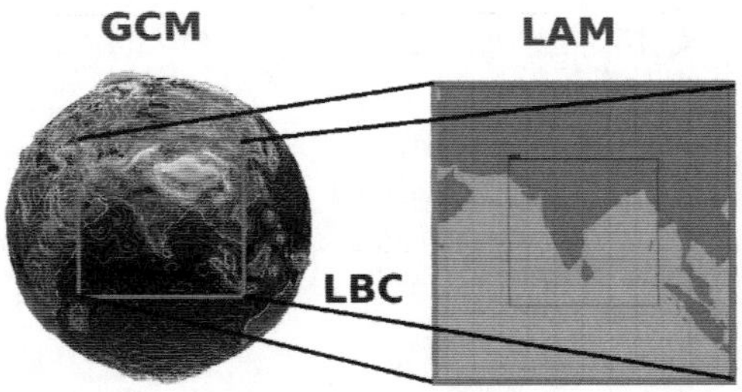

Figure 3. Schematic diagram showing coupling of the Varsha GCM and the WRF limited area model

The large scale upper level (at 250hPa) wind fields, from the model simulations, at 22UTC and 23UTC are shown in Figure 4. One can see that at the cruise altitude, winds were around 5 m/s and mainly from west to east. Also, there was a region of horizontal wind shear close to the location of the incident.

From the WRF simulations, we analyze the possibilities of CAT due to atmospheric conditions and suitability of conditions for a wake vortex encounter. It is clear that given the resolution of

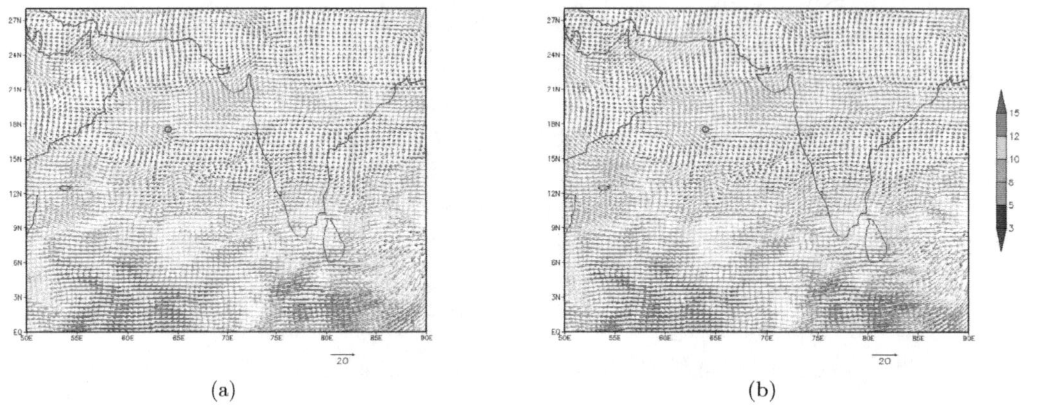

Figure 4. Large scale wind field from WRF on 4th Oct 2014 at 250hpa, Wind speed in m/s: (a) 22UTC; (b) 23UTC.

the WRF model, CAT which occurs at the aircraft scale cannot be directly resolved. However, since the energy associated with turbulence on this scale cascades down from the larger scales of atmospheric motion, the attempt is to infer the presence of CAT from the large-scale fields which are resolved in the computation.

3.1. Turbulence indices

There are many theories in literature explaining the phenomenon of clear air turbulence (CAT) and some of them have been successful in predicting various classes of turbulence (i.e.,. light, moderate and severe). But, the forecasting skill of severe CAT has only improved recently [12]. There are many turbulence forecast indices, which are estimated from NWP model outputs, reported in literature [13, 14, 15]. Most of them were developed to predict the CAT from the large scale atmospheric fields. Algorithmic details of such indices are summarized in [16]. The formula used for the two such indices are given below.

Brown's index

$$\Phi = \sqrt{0.3{\zeta_a}^2 + {D_{SH}}^2 + {D_{ST}}^2} \tag{1}$$

where, ζ_a is the absolute vorticity, $D_{SH} = \frac{\partial v}{\partial x} + \frac{\partial u}{\partial y}$ is the shearing deformation and $D_{ST} = \frac{\partial u}{\partial x} - \frac{\partial v}{\partial y}$ is the stretching deformation.

Ellrod index

$$E_1 = S_v * \sqrt{\left({D_{ST}}^2 + {D_{SH}}^2\right)} \tag{2}$$

where $S_v = |\frac{\partial V}{\partial z}| = \sqrt{\left(\frac{\partial u}{\partial z}\right)^2 + \left(\frac{\partial v}{\partial z}\right)^2}$ is the vertical wind shear.

As the first step, we estimated these indices, from the model simulations, to examine the possibility of CAT near the incident location. The estimated values of Brown's and Ellrod indices are shown in Figure 5 as an example. It can be seen that the indices does not show any indication of CAT in that region. There are regions where the indices are high, but these are far away from the region where the incident took place.

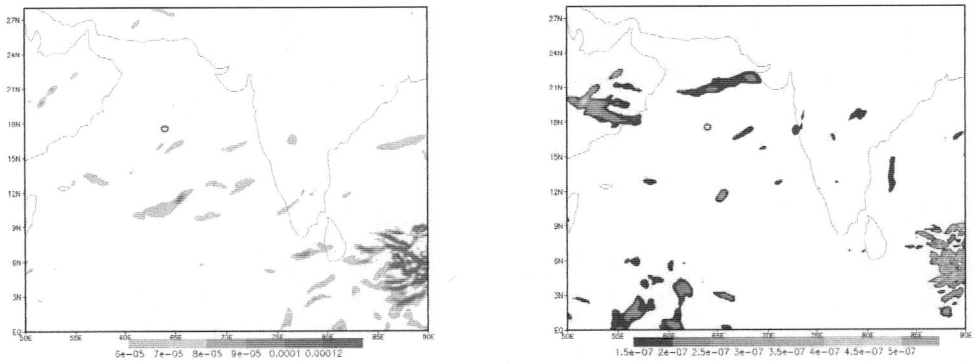

(a) (b)

Figure 5. Clear air turbulence indices at 200hPa on 4th Oct 2014: (a) Brown Index; (b) Ellrod Index.

3.2. Wake vortex encounter

Next we examine the quantities which govern the strength, propagation and decay of wake vortices at the location and altitude of interest. Basic aircraft parameters, are shown in Table 1.

Table 1. Aircraft basic characteristics

Type of Aircraft	Boeing 737-800	Airbus A380
Maximum Takeoff Weight (kg)	79015	575000
Wing span b_L (m)	35.79	79.8
Cruise Speed $V_{m,c}$ ($m\ s^{-1}$)	231.40	242.98

Transportation of aircraft trailing vortices is complex phenomena. Environmental condition plays a substantial role in the transportation of vortices. Contours of crosswind and vertical velocity in a plane perpendicular to the flight path at the location and time of the incident are shown in Figure 6. One can see that there is considerable cross-wind at the flight altitudes (corresponding to both the aircraft). The vertical velocities are however small.

From the basic aircraft parameters, the estimate of vortex strength (Γ_0) for the A380 is $977.5m^2/s$, initial vortex Spacing, $b' = b_L\pi/4$ is $62.67m$, descent velocity based on vortex induction w_0 is $2.48m/s$, Brunt Vaisala Frequency is $0.01182s^{-1}$, and a time scale for decay is $t' = b'/w_0$ is $25.2s$.

Since the atmospheric vertical velocities are small (fig 6), an estimate for the descent time to the altitude of the 737 aircraft is $t_1 = 300/w_0 \sim 120$ seconds. Non-dimensionalizing with t', we get $t_1^* = 4.8$.

Since there is decay of vortex strength with time and lateral spread of the vortices, the relevant questions are i) is the vortex of sufficient strength at this time to affect the second aircraft? ii) is one of the vortices within the span of the second aircraft?

Proctor et al [17], have studied the problem of vortex decay away from ground in stratified and turbulent environments. Using their results, we estimate the non-dimensional vortex strength (Γ/Γ_0) at t_1^* is 0.22. This implies a vortex strength of 215 m^2/s which is sufficiently strong.

Proctor and Ahmad [18] studied the cross wind effects on aircraft trailing vortices. They concluded that if the gradient of the crosswind shear is negative $(\partial^2 U_c/\partial z^2 < 0)$ then, clockwise rotating vortex will descend at a rate faster than counter-clockwise rotating vortex. Also, if the vertical gradient of crosswind is positive $(\partial U_c/\partial z > 0)$ then, the lateral separation between the vortices increases. This increase in lateral and vertical separation between the vortices increases their linking time and decay rate. Thus one of the vortices will have comparatively more life span than the other.

From our numerical simulation, we estimated that $\partial^2 U_c/\partial z^2 = -4.36 \times 10^{-8}$ which is negative and $\partial U_c/\partial z = 1.08 \times 10^{-4}$ which is positive. The estimated values from our simulation was found in accordance with the criteria given by the Proctor and Ahmad. This shows that one of the vortices can have comparatively more life span than the other.

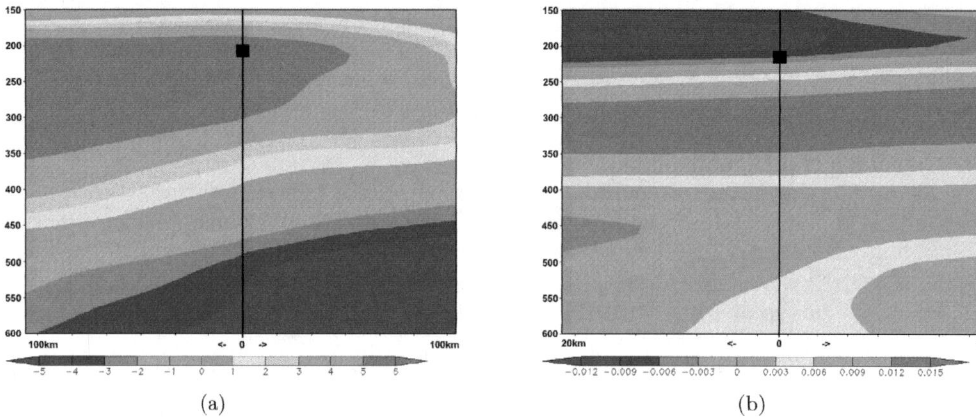

Figure 6. Velocity field in a plane perpendicular to the aircraft path at 23UTC: (a) cross wind velocity; (b) vertical velocity.

Also, the magnitude of the cross-wind (around 6 m/s) is sufficient for one of the vortices to be blown back in to the path of the second aircraft.

4. Summary

We have performed high-resolution numerical simulations of the atmospheric conditions corresponding to the time and region where an aircraft encountered severe turbulence. From our estimates of both CAT indices and parameters related to vortex descent and decay, we find that the encounter was more likely to be due to the wake vortex from an A380 which passed overhead and less likely due to atmospheric conditions alone. While more analysis may be required to confirm these findings, we have demonstrated the utility of using such numerical simulations to understand aviation weather phenomena and also use for purpose of advance warnings.

Acknowledgments

This work was funded by Science Engineering and Research Board of Govt. of India. We thank NCAR for making WRF available in the public domain and NCEP for providing the FNL datasets.

References

[1] Sharman R D, Trier S B, Lane T P and Doyle J D 2012 *GRL.* **39** L12803
[2] Pilot report about incident on VTJBU 2015 (Private communication)
[3] The Times of India 2010 http://timesofindia.indiatimes.com/city/mumbai/Aircraft-encountered-turbulence-over-Bangalore/articleshow/5857479.cms
[4] Shameem C A and Venkatesh T N 2015 *Proc. 17th AeSI Annual CFD Symp* 160
[5] FlightAware 2016 http://uk.flightaware.com/
[6] Nelson R C 2004 *AIAA 2004-5171*
[7] IMD 2014 http://www.imd.gov.in/section/satmet/dynamic/
[8] Sinha U N, Venkatesh T N, Mudkavi V Y, Murthy A S V, Nanjundiah R S, Sarasamma V R, S R, Bhagyalakshmi K, Verma A K, Sreelekha C K and Resmi K L 2005 *NAL PD FS 0514*
[9] Michalakes J, Dudhia J, Gill D, Henderson T, Klemp J B, Skamarock W C and Wang W 2005 *Proc. 11th ECMWF Workshop on the Use of High Performance Computing in Meteorology*
[10] Klemp J B, Skamarock W C and Dudhia J 2007 *Monthly Weather Review* **135** 2897
[11] NCEP 2014 http://rda.ucar.edu/datasets/ds083.2/
[12] Venkatesh T N, Mathew J and Nanjundiah R S 2014 *Meteorology and Atmospheric Physics* **126** 139
[13] Brown R 1973 *Meteor. Mag.* **102** 347
[14] Ellrod G P and Knapp D I 1992 *Wea. Forecasting* **7** 150
[15] Sharman R, Tebaldi C, Wiener G and Wolff J 2006 *Weather Forecast.* **21** 268
[16] Femin K J, Asha V, Shameem C A and Venkatesh T N 2014 *NAL PD-FSU/2014/1003*
[17] Proctor F H, Ahmad N N, Switzer G S and Duparcmeur F M L 2010 *AIAA 2010-7991*
[18] Proctor F H and Ahmad N N 2011 *AIAA 2011-3038*

The effect of mountain wind on the falling snow deposition

Zhengshi Wang[1,2] and Ning Huang[1,2]

[1]Key Laboratory of Mechanics on Disaster and Environment in Western China (Lanzhou University), the Ministry of Education of China, 730000

[2]Department of Mechanics, School of Civil Engineering and Mechanics, Lanzhou University, Lanzhou, 730000, China

Corresponding to: wangzhsh2013@lzu.edu.cn

Abstract. The air-flow over morphologic prominence is of vital importance for the temporal and spatial variations of snow distribution, which may further affect the hydrological cycle, climate system, ecological evolution as well as other natural processes. Most previous studies have neglected the effect of non-uniform mountain wind on the trajectory of falling snow particle and thus the uneven snow distribution over complex terrain is little known. Here in this paper, we carry out a numerical study on the deposition process of mixed grain size snow particles over mountain terrains. A three-dimensional large eddy simulation (LES) code is used to predict the wind field and the fluid-solid coupling effect is considered, the Lagrangian particle tracking method is employed to track the movement of each falling snow particle. Our results suggest the boundary layer flow over steep slopes or complex terrains exhibits obvious uneven characteristics. The mountain wind has obvious effect on the motion of snow particles and finally results in a non-uniform distribution of snow. This research is of significance to understand the evolution of high resolution snow distribution in cold highland areas.

1. Introduction

Snow is widely distributed on Earth with significant interannual and seasonal variations. Subject to the action of wind, snow always shows a nonuniform distribution especially in mountainous areas with complex terrain. On the one hand, the distribution unevenness in the mountainous areas may induce and aggravate snow floods, avalanche, mudslides, landslide and other natural disasters [1], which not only results in huge direct and indirect economic losses, but also may cause casualties. On the other hand, some characteristics of snow cover, such as the snow cover area, the snow distribution and snow depth, are important input parameters of global energy balance, climate, hydrological and ecological models. Thus, to understand the evolution of high resolution snow distribution is essential.

At this stage, snow distributions are mainly examined from field measurement or pattern prediction. For the observation, remote sensing retrieval is the most commonly used method, such as the multispectral imaging AVHRR (Advanced Very High Resolution Radiometer) [2] and moderate resolution spectral image MODIS (Moderate Resolution Imaging Spectroradiometer) [3, 4]. The temporal and spatial resolution of remote sensing retrieval is poor for snow cover changes fast along with snowfalls, and the accuracy of remote sensing retrieval could not cope due to the interference of various nature factors. In addition, almost all the inversion algorithms for the remote sensing have a strong geographical restriction [5]. For example, the AVHRR does not apply to the complex terrain. More importantly, the snow depth cannot be acquired by remote sensing retrieval. On the other hand, large amount of studies on the snow distribution over complex terrain have been conducted by many researchers. Such as, the seeder-feeder mechanism [6-9] is described to be the main process to generate the enhancement of precipitation over terrains. And Houze [10] went on a study about the fundamental formation process of precipitation cloud for different terrains and conditions. On a much smaller scale that near the topography surface, the variation of advecting and vertical winds contributes greatly to the falling trajectories of snow grains and results in the locally enhanced deposition [8, 11-13]. However, most models or patterns pay little attention to the real motion processes of the falling snow particles. In other words, the behaviors of falling particles in the complex turbulent boundary layer on snow distribution are neglected.

In this paper, the Lagrangian particle tracking model was introduced into ARPS, and the interaction between particles and wind flow is taken into account. An imaginary snowstorm over rugged terrain is performed through tracking the trajectory of each falling snow particle. Then the snow deposition condition around the terrain is simulated. Finally, the movement feature of snow particles in the air and the distribution of snow on the ground are analyzed in detail.

2. Model and methods

2.1. Wind field prediction and particle motion

The Large Eddy Simulation (LES) method is used to simulate the wind field and the source code is obtained from ARPS. ARPS is developed by the University of Oklahoma that incorporates numerical simulation, data assimilation and data processing [14, 15].

The filtered three-dimensional, nonhydrostatic and fully compressible Navier-Stokes equations can be written as:

$$\frac{\partial \bar{\rho}\tilde{u}_i}{\partial t} + \frac{\partial \bar{\rho} u_i \tilde{u}_j}{\partial x_j} = -\frac{\partial \tilde{p}^*}{\partial x_i} - \frac{\partial \tau_{ij}}{\partial x_j} + B\delta_{i3} + F_i \tag{1}$$

where the tilde symbol indicates the filtered variables and line symbol represents grid volume-averaged variables. x_i ($i = 1,2,3$) indicate the coordinates of streamwise, spanwise, and vertical directions, respectively; u_i is the instantaneous velocity component, $\bar{\rho}$ is the grid volume-averaged air density and t is time, τ_{ij} is the subgrid stress tensor and δ_{ij} is the Kronecker delta; $p^* = p' - \alpha\nabla(\bar{\rho}\mathbf{u})$ is the logogram of the pressure deviations and a divergence damping terms designed to attenuate acoustic waves, in which the Δ represents the divergence and α is the damping coefficient. $B = -g\,\rho'/\bar{\rho}$ is the buoyancy term comes from the density fluctuation ρ', g is the gravitational acceleration.

To take into consideration of the reaction force of snow particles in the air on the wind field, a volume force F_i is added in the right side of Equation (1), which can be written as [16]:

$$F_i = -N\frac{1}{\rho V_{grid}}\sum_{s=1}^{N_p} m_p \frac{u_i(x_p(t),t) - u_{pi}(t)}{T_p} f(Re_p) \tag{2}$$

in which, N_p and V_{grid} are respectively the particle number and the volume of the grid cell. m_p is the mass of snow particle, $u_{pi}(t)$ and $u_i(x_p(t),t)$ are the velocity of snow particle and the wind velocity at the location of grain at time t, respectively; $T_p = \rho_p d_p^2 / 18\rho\nu$ represents the particle response time and N means that one particle represents N snow particles for the purpose of reducing computation.

$f(Re_p)$ can be express as [17]:

$$\begin{aligned} &f(Re_p) = 1 && (Re_p < 1) \\ &f(Re_p) = 1 + 0.15\,\mathrm{Re}_p^{0.687} && (Re_p \geq 1) \end{aligned} \tag{3}$$

where $Re_p = d_p |u - u_p| / \nu$ is the particle Reynolds number and ν is the kinematic viscosity of air.

The snow particles are assumed to be only subject to gravity and drag force in the air, and the governing equation of particle motion can be expressed as [18-20]:

$$\frac{du_{pi}}{dt} = \frac{3\rho\nu(u_i - u_{pi})}{4\rho_p(d_p)^2} C_D Re_p + g_i(1 - \frac{\rho}{\rho_p}) \tag{4}$$

where $C_D Re_p$ can obtain from Wallis' empirical formula:

$$C_D Re_p = \begin{cases} 24 + 3.6(Re_p)^{0.687}, (Re_p \le 1000) \\ 0.44 Re_p, (Re_p > 1000) \end{cases} \tag{5}$$

2.2. *Wind field prediction and particle motion*

The study area is in the Qilian Mountains in northwestern China ($E100^{\circ}14'15", N38^{\circ}00'50.4"$), as shown in Fig. 1, and the topographic data for this study are derived from a DEM database with 30 m resolution. The initial base state wind is set as horizontally homogeneous logarithmic profile based on the field measurement. The atmosphere boundary layer is assumed to be stably stratified. The humid atmosphere is simulated with the constant relative humidity of 90% based on the measurement in the snowfall processes. The roughness length is 0.01 m and the boundary layer initial depth is set as 500m.

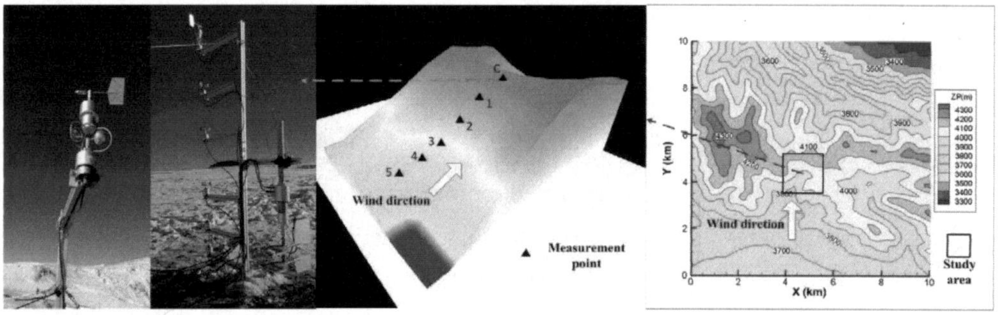

Figure 1. The study area and measurement point schematic diagram

The rigid ground boundary condition is applied at the ground and a Rayleigh damping layer is used at the top of the domain (1000 m in depth). Along the lateral boundaries, the open radiation boundary conditions are used. The snow particle size is gamma distributed and the periodic boundary conditions are used for the snow grains. The periodic boundary cannot be used in the flow field because the terrain is unsymmetrical, this may lead to slightly deviation near the boundary. In order to avoid this influence, a much bigger computational domain is selected. In our simulation, the dimensions of the computational domain are set at 10 km × 10 km × 4 km and the study area is 2 km × 2 km × 4 km. The grid has a uniform horizontal size of 50 m and the nested grid size is 5 m. In the vertical direction, the grid is stretched with a cubic function for the purpose of acquiring more detailed information on the surface layer. The mean vertical grid space is 50 m and the smallest grid is 2 m. This grid resolution is necessary to reproduce the small-scale flow features of complex terrains [21].

3. Result and discussion

The snow deposition around the terrain under the condition of uniform snowfall is simulated based on the coupled LES and particle tracking pattern. The total simulation time is 2 hours. First of all, the wind field without particle is analyzed. Figure 2 is the ground surface (2 m height) wind vector map of the study area at T=4200 s. As can be seen from the figure, the wind is accelerated significantly on the windward slope and there formed some clear recirculation zones and vortexes on the leeward slope. In addition, the flow around the topography is quite clear. To further analyze the wind speed information, the wind speed-up ratio on the windward slope is calculated and the result is compared with the observation results, in which the speed-up ratio is a mean value of 60 seconds and equal to the specific value of local wind speed and reference wind speed at measurement point C. It can be seen that the simulated wind speed-up ratio is consistent with the experimental results. In this condition, the particle is introduced into the calculation. In our simulation, the snow particles fall uniformly from the height of 5000m (the reference ground height is 3000m) with a release rate of 10000 particles per second, and the initial velocity of each snow particle is zero.

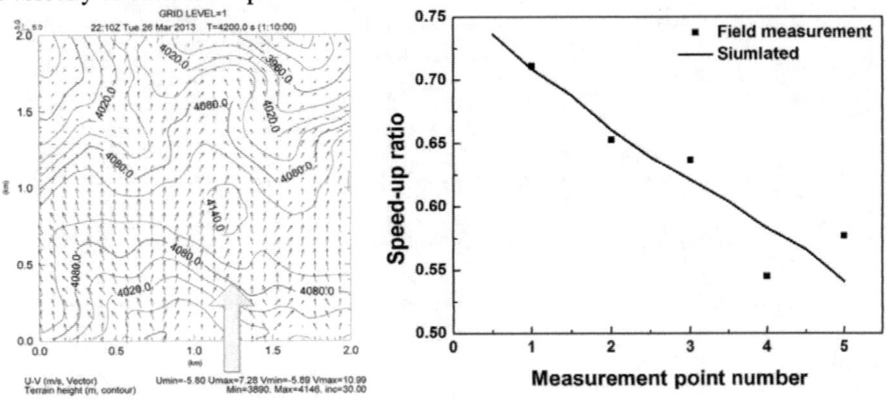

Figure 2. The wind vector illustration of study area and the wind speed-up ratio of the measurement point.

The probability distributions of one hour averaged vertical wind speed (10m in height) and the particle falling speed (into the boundary layer) are displayed in Figure 3, in which the positive values mean the up direction and the red lines represent that the coupling effect between particles and wind flow is included. It can be seen from this figure that the proportion of larger vertical wind speed increases when the coupling effect is considered. The percentage of absolute value of vertical wind speed increases from 19.2% to 32.7%. At the same time, the falling velocity of snow particles is obviously changed if the coupling effect is included. As shown in the figure, another peak value appeared with the coupling effect. The percentage of smaller terminal velocity is obviously increased because the increase of the upward-moving particles.

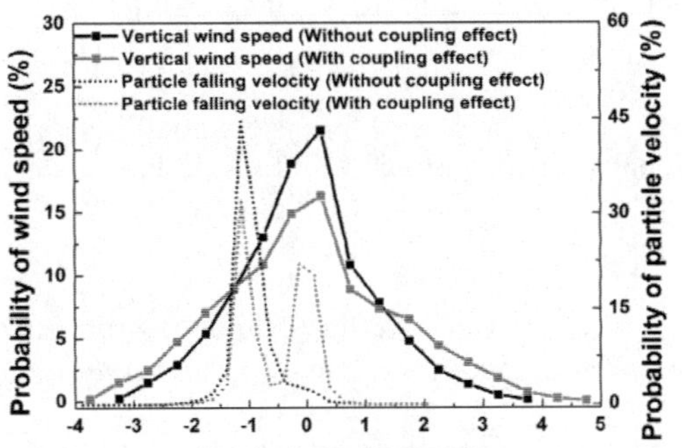

Figure 3. The probability distributions of vertical wind speed and particle falling velocity.

Figure 4 shows the evolution of snow depth distribution of the study area at different times. The circle on the map is at the windward slope position and the dotted circle represents the leeward slope position. We select 10 points in the circle to obtain the mean snow depth. As shown in the figure, there is significant difference in snow depth at different locations of complex terrains. In general, snow particles are easier to accumulate on the windward slope than on the leeward slope, and the difference between windward slope and leeward slope intends to become larger with time. For example, the average snow depth of windward slope (solid circle) and leeward slope (dotted circle) is almost equal at t=30 min. However, at t=60 min and t=90 min, the mean snow depth on the windward slope are respectively 0.545 mm and 0.987 mm, and the average snow depth on the leeward slope are 0.364 mm and 0.503 mm, respectively. The average snow depths of windward slope are 1.497 times and 1.962 times of leeward slope at t=60 min and t=90 min, respectively. This is mainly because the updraft on the windward slope has an inverse influence on the particle fall velocity and thus snow particles will convergence on the windward.

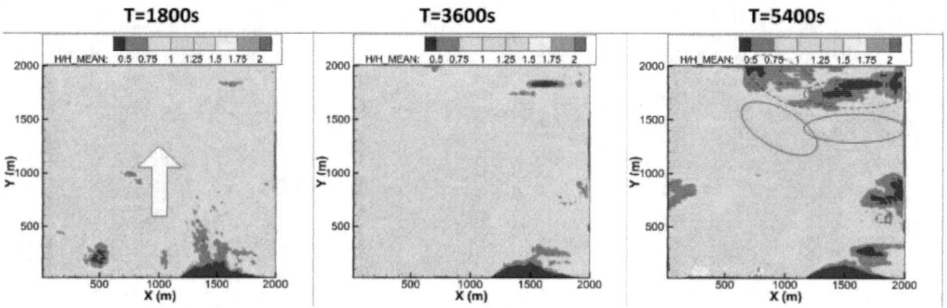

Figure 4. Snow deposition mass (kg/m2) of study area at different time.

4. Conclusions

The Lagrangian particle tracking model is introduced into ARPS Numerical Simulation System and the snowfall deposition process over rugged terrain is simulated. The coupling effect between snow particles and air flow is considered. The snow deposition mass and velocity distributions are obtained through statistical method. The main conclusions are as follows:

1) The wind field is non-uniform over rugged terrain. The wind velocity and direction vary greatly at different positions. In general, the wind is accelerated in the windward slope; there are large amount of recirculation zones and vortexes in the leeward slope.

2) The updraft has obvious influence on the fall velocity of snow particles in the air. The percentage of smaller terminal velocity or even upward-moving particles is obviously increased under the action of updraft.

3) The non-uniform mountain wind is an important mechanism to enhance the uneven distribution of the orographic precipitation and snow particles are easier accumulated on the windward slope than on the leeward slope.

Acknowledgments

This work is supported by the Major projects of national forestry public welfare industry research (201404306).

References

[1] Michaux J L, Naaim-Bouvet F and Naaim M 2001 Drifting-snow studies over an instrumented mountainous site: II. Measurements and numerical model at small scale *Annals of Glaciology* **32**(1) pp 175-81

[2] Zheng Z J, Liu Y J and Zhang B C 2003 Improved remote sense monitoring on snow cover of china in winter *Journal of Applied meteorological science* **15**(Suppl.) pp 75-84

[3] Wang J Chen Z D, Li W J, Tan H, Lu A X and Li S 2000 Reasearch on spectral reflectance porperties of snow using moderate resolution imaging spectroradiometer data *Journal of Glaciology and Geocryology* **22**(2) pp 165-70

[4] Hao X H, Wang J, and Li H Y 2008 Evaluation of the NDSI threshold value in mapping snow cover of MODIS-a case study of snow in the middle qilian mountains *Journal of Glaciology and Geocryology* **30**(1) pp 132-38

[5] Nolin A W 2010 Recent advances in remote sensing of seasonal snow *J. Glaciol.* **56** pp 1141-50

[6] Bergeron T 1965 On the low-level redistribution of atmospheric water caused by orography *International Conference on Cloud Physics* Tokyo

[7] Choularton T W and Perry S J 1986 A model of the orographic enhancement of snowfall by the seeder-feeder mechanism *Q. J. R. Meteorol. Soc.* **112** pp 335–45

[8] Zängl G 2008 The temperature dependence of small-scale orographic precipitation enhancement *Q. J. R. Meteorol. Soc.* **134** pp 1167–81.

[9] Stoelinga M T, Stewart R E, Thompson G and Theriault J M 2013 Microphysical processes within winter orographic cloud and precipitation systems *Mountain Weather Research and Forecasting* pp 345–408

[10] Houze R A 2012 Orographic effects on precipitating clouds *Rev. Geophys.* **50** RG1001

[11] Colle B 2004 Sensitivity of orographic precipitation to changing ambient conditions and terrain geometries: An idealized modeling perspective *J. Atmos. Sci.* **61** pp 588–606

[12] Dore A J and Choularton T W 1992 Orographic enhancement of snowfall *Environ. Pollut.* **75** pp 175–79

[13] Lehning M, Löwe H, Ryser M and Raderschall N 2008 Inhomogeneous precipitation distribution and snow transport in steep terrain *Water Resour. Res.* **44** W07404

[14] Xue M, Droegemeier K K and Wong V 2000 The advanced regional prediction system (ARPS)-A multiscale nonhydrostatic atmospheric simulation and prediction model. Part I: model dynamics and verification *Meteorol. Atmos. Phys.* **75** pp 161–93

[15] Xue M, Droegemeier K K, Wong V, Shapiro A, Brewster K, Carr F, Weber D, Liu Y and Wang D 2001 The Advanced regional prediction system (ARPS)-A multi-scale nonhydrostatic atmospheric simulation and prediction tool. Part II: model physics and applications. Meteorol *Atmos. Phys.* **76** pp 143–65

[16] Yamamoto Y, Potthoff M, Tanaka T, Kajishima T and Tsuji Y 2001 Large-eddy simulation of turbulent gas-particle flow in a vertical channel: effect of considering inter-particle collisions *Journal of Fluid Mechanics* **442** pp 303-34

[17] Clift R 1978 Bubbles, *Drops and Particles* (New York: Academic)

[18] Anderson R S and Haff P K 1988 Simulation of eolian saltation *Science* **241** pp 820–23

[19] Shao Y and Li A 1999 Numerical modelling of saltation in the atmospheric surface layer *Bound-Lay. Meteorol.* **91** pp 199–225

[20] Lopes A M G, Oliveira L A, Ferreira A D and Pinto J P 2012 Numerical simulation of sand dune erosion *Environmental Fluid Mechanics* **13**(2) pp 145-68

[21] Raderschall N, Lehning M and Schar C 2008 Fine-scale modeling of the boundary layer wind field over steep topography *Water Resources Research* **44** W09425

Basic Coandă MAV Fluid Dynamics and Flight Mechanics

H Djojodihardjo[1]and RI Ahmed[2]
[1]The Institute for the Advancement of Aerospace Science and Technology, Jakarta, Indonesia;
[2]Graduate Student, Aerospace Engineering Department, Universiti Putra Malaysia;
[1]harijono.djojodihardjo@yahoo.com; [2]engriyadh64@gmail.com

Abstract. Capitalizing on the basic fundamental principles, the Fluid Dynamics and Flight Mechanics of a semi-spherical Coandă MAV configurations are revisited and analyzed as a baseline. A mathematical model for a spherical Coandă MAV in hover and translatory motion is developed and analyzed from first physical principles. To gain further insight into the prevailing flow field around a Coandă MAV, as well as to verify the theoretical prediction presented in the work, a computational fluid dynamic CFD simulations for a Coandă MAV generic model are elaborated. The mathematical model and derived performance measures are shown to be capable in describing the physical phenomena of the flow field of the semi-spherical Coandă MAV. The relationships between the relevant parameters of the mathematical model of the Coandă MAV to the forces acting on it are elaborated subsequently.

Keywords: Circulation control, Coandă effect, Coandă MAV Flight Dynamics, Coandă MAV Fluid Dynamics, Micro air vehicle (MAV), Unmanned Air Vehicle (UAV)

1. Introduction

Coandă effect and Coandă jet have been found and utilized for flow control in many applications in engineering and health science, among others, such as in aircraft and vehicle technology [1-5], wind-turbines [6-9], and upper respiratory system [10]. The science of flow control can be traced back to Prandtl in his seminal paper in 1904 with his breakthrough in the science of fluid mechanics by introducing the boundary-layer theory and elucidated the physics of the separation phenomena and the control of the boundary layers. Various Coandă MAV configurations have been proposed and developed. In the design and operation of a Coandă MAV which could meet the desired mission and design requirements, the basic working relationships among various relevant variables and parameters governing the aerodynamics forces should be established. To assist the analysis, design and developments of Coandă MAV's, several tools can be resorted to. The first is the analytical tool, which capitalizes on the basic fundamental principles. The second is the utilization of Computational Fluid Dynamics (CFD) which has the advantage of providing visualization for significant insight and identification to the problem. In addition, CFD visualization can be utilized in gaining better insight and identifying specific details. Experimental tools can benefit from the insight gained by analysis, CFD computational and visualization studies, in the preliminary study stage by designing specific experiments as well as in the conceptual and prototype design stages. The analysis carried out here is associated with hovering and cruising conditions, which should give further insight on and can be further elaborated into Coandă MAV maneuvering performance.

The present analysis is carried out on Coandă MAV spherical configuration as a baseline, which utilizes some results from the authors' previous analysis. The equations can be easily modified for other configurations and can be compared to other investigators' results for further assessments. CFD visualization studies will also be utilized to gain further understanding in developing the physical model in the analysis.

In the model development and analysis of a semi-spherical Coandă MAV, a propulsion system is utilized to introduce a Coandă jet blanket which is deflected downward due to the curved upper surface of the vehicle, to provide both hover as well as cruise capabilities. Thus lift will be generated, first to hover and later for cruise propulsion. Such Coandă jet has been utilized for circulation enhancement for fixed wing aircraft in forward flight and turbine blades movement [3-9]. The principle of Coandă MAV lift generation as well as the equation of motion for its translatory motion will be derived and elaborated.

Some basic results drawn from the physical and mathematical model has been obtained to describe the physical phenomena of the flow field related to the relevant surfaces influenced by the Coandă effect jet sheets, and rewritten in a more convenient form as necessary to gain better insight on the relationship among the relevant parameters.

2. Fluid Dynamic Analysis for hovering Semi-spherical Coandă MAV as a Baseline Configuration

The mathematical model of Coandă MAV elaborated here follows closely that of Ahmed et al [10-11] and Djojodihardjo and Ahmed [12-13], which was based on first principles and are summarized and reproduced here as a baseline reference for successive development. For Coandă MAV with the configuration depicted in Figure 1, an actuator rotor can be visualized to be located at the center of the body. For conceptual development, the dimension of the rotor can be set to be small as necessary. Part of the flow being drawn by the actuator can be utilized for lift (like in a helicopter), and part of it will be utilized for introducing radial flow on the surface of the body as a Coandă jet blanket. Momentum analysis is carried out in the analysis of Coandă effect related to finding the relevant aerodynamic forces and defining the Coandă MAV performance parameters. The latter is developed for a baseline and simplified semi-spherical Coandă MAV as depicted in Figure 3(b).

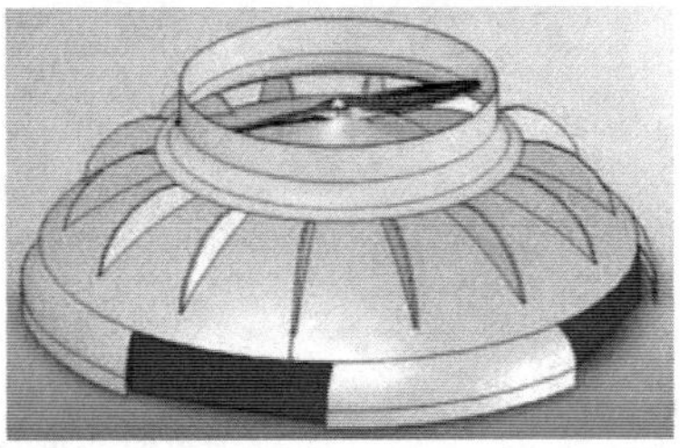

Figure 1: An impression of semi-spherical Coandă MAV configuration utilizing a propeller to develop Coandă jet over its upper surface

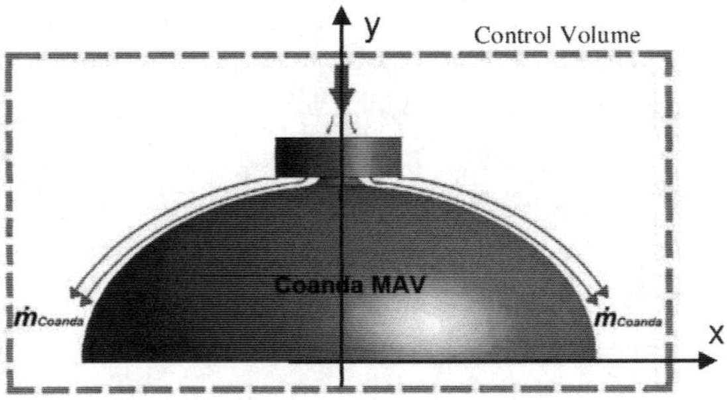

Figure 2: An equivalent semi-spherical Coandă MAV for generic analysis.

In the following, a rigorous analysis is carried out to elaborate how Coandă effect contributes to the generation of lift capability of Coandă MAV. First, the analysis will address the Coandă effect for lift by applying the fundamental conservation analysis on the control volume CV (red dashed rectangle in Figure 2).

Applying the continuity equation on the control volume, which reduces to the Coandă Blanket as depicted in Figure 3, there is obtained

$$\rho \cdot V_{j\text{-}R} \cdot 2\pi Rh = \rho \cdot V_{j\text{-}in} \cdot 2\pi R_i h_i = \rho \cdot V_{j\text{-}out} \cdot 2\pi R_o h_o \tag{1}$$

and

$$V_{j\text{-}R} = \frac{R_i h_i}{Rh_R} V_{j\text{-}in} = \frac{\dot{m}}{2\pi\rho Rh_R}$$
$$V_{j\text{-}out} = \frac{R_i h_i}{R_o h_o} V_{j\text{-}in} = \frac{\dot{m}}{2\pi\rho R_o h_o} \tag{2}$$

where $V_{j\text{-}R}$ is the jet flow velocity, R is the vehicle body radius, h is the jet slot thickness and $\dot{m}$ is the jet mass flow rate. The momentum equation applied to the control volume in the y (vertical) direction can be differentiated into the contribution due to the Coandă Blanket momentum and the pressure difference on the body due to Coandă Blanket:

$$\begin{pmatrix} \textit{Total Lift force} \\ \textit{due to Coandă Blanket} \end{pmatrix} = \begin{pmatrix} \textit{Vertical component of momentum} \\ \textit{balance due to Coandă Blanket} \end{pmatrix} + \begin{pmatrix} \textit{Pressure difference on the body} \\ \textit{of MAV subject to Coandă Blanket} \end{pmatrix} \tag{3}$$

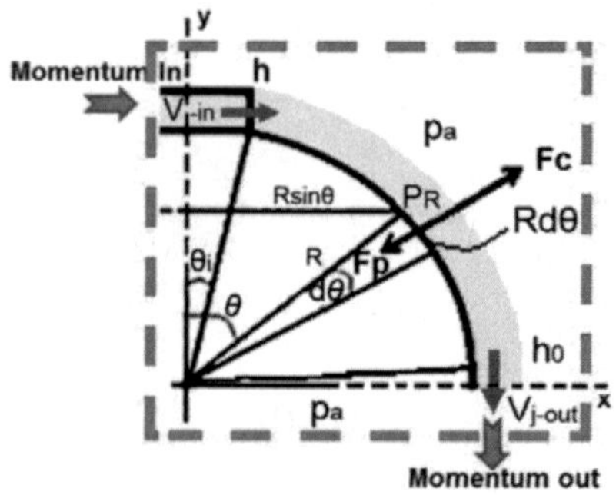

Figure 3: Schematic of Coandă jet blanket for a Spherical MAV momentum balance.

The contribution of lift from the momentum flux through the control volume CV in the y (vertical) direction is given by:

$$\begin{pmatrix} \textit{Force in the} \\ \textit{y-direction} \end{pmatrix} = \begin{pmatrix} \textit{Momentum Out} \\ \textit{in the y-direction} \end{pmatrix} - \begin{pmatrix} \textit{Momentum In} \\ \textit{in the y-direction} \end{pmatrix} \tag{4}$$

Since the Momentum in the radial direction does not contribute to lift, then the momentum equation in the y-direction for the control volume CV depicted in Figure 4 is:

$$F_{\text{Coanda jet Blanket}} = \dot{m} V_{j\text{-}out} = 2\pi R_i h_i \, \rho \cdot V_{j\text{-}in} \cdot V_{j\text{-}out} \tag{5}$$

The contribution of lift from the pressure difference between the upper part (curved MAV body covered by Coandă jet Blanket) and the lower part of the MAV body can be found by considering various fundamental physical approach. The detail of the elaboration can be found in Ahmed and Djojodihardjo [10-11] and yields:

$$F_{Induced\ pressure\ difference} = F_{lower-surface} - F_{upper-surface}$$
$$= \pi\left(R_0^2 - R_i^2\right)p_a - \int_{\theta_i}^{\theta_0}\left(p_a - \frac{h}{R}\rho\left(\frac{\dot{m}}{2\pi\rho Rh}\right)^2_{j-R}\right)\pi R^2 \sin 2\theta d\theta \quad (6)$$

where θ_i and θ_0 are the turning angles at which the jet flow injected and separated from the Coandă surface respectively, as depicted in Figure 3. The jet flow assumed to be uniform outflow separating at the sharp edge of the Coandă MAV curved surface. Then the contribution for lift due to the pressure difference across the surfaces of the MAV body, for the latter approach is given by (as elaborated by Djojodihardjo and Ahmed [12-13]), for the significant value of θ_i):

$$F_{Induced\ pressure\ difference} = \left(\frac{\dot{m}^2}{4\pi\rho h_R R}\right)_{j-R} \quad (7)$$

Hence the total lift due to Coandă jet blanket momentum and Coandă jet blanket induced pressure difference is given by

$$Lift_{\substack{Spherical-Coanda\ jet\ Blanket+\\ induced\ pressure\ difference}} = 2\pi R_i h_i\ \rho V_{j\text{-}in} \cdot V_{j\text{-}out} + \left(\frac{\dot{m}^2}{4\pi\rho Rh}\right)_{j-out} = \dot{m}\cdot V_{j\text{-}out} + \frac{1}{2}\dot{m}V_{j-out} \quad (8)$$

Therefore, due to the presence of the Coandă blanket, the Coandă MAV has additional lift given in (8). It should be noted that a more exact approach requires additional information, i.e. the energy conservation equation, which is then utilized as the fourth equation. This is elaborated subsequently.

Applying the energy conservation equation on the same control volume, which has been redrawn for convenience in Figure 4, assuming uniform properties across the sectional areas at the input and output of the Coandă jet blanket. Ignoring the entrainment energy exchange between the ambient air and the Coandă jet blanket, the energy equation can be written as:

$$\left(\frac{p}{\rho}\right)_{in} + \frac{1}{2}\left(V_{j\text{-}in}{}^2\right) = \left(\frac{p}{\rho}\right)_{out} + \frac{1}{2}\left(V_{j\text{-}out}{}^2\right) \quad (9)$$

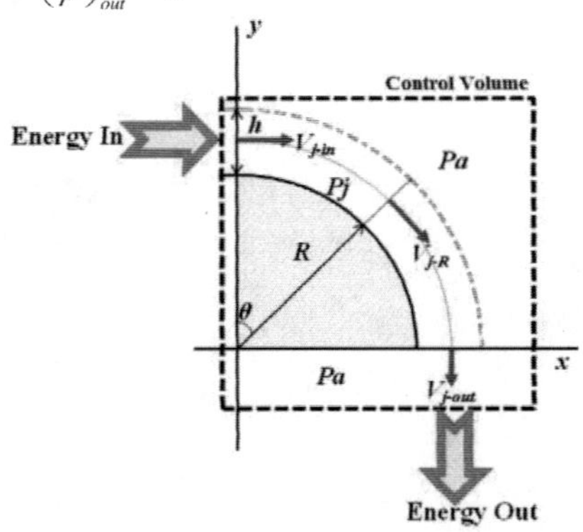

Figure 4: Schematic of the control volume around the MAV for the application of the energy conservation equation

which is essentially the Bernoulli equation along any streamline between the inlet and the outlet sections. The contribution of the entrainment energy exchange along the Coandă jet blanket can later be incorporated, such as by adopting certain assumptions as a heuristic approach or, applying the complete viscous equation by numerical simulation or CFD approach. The latter should be carried out for more meticulous approach, using the present simplified analytical approach as guidelines. At the same time, the CFD results, by appropriate considerations of the relevant parameters utilized there can be used in developing the heuristic approach as well as for validating and assessing the merit of the analytical approach. Noting that from the outset, the flow is considered to be incompressible, equation (9) reduces to:

$$\frac{1}{2}\rho V_{j\text{-in}}^{2} = \frac{1}{2}\rho V_{j\text{-out}}^{2} \tag{10}$$

Hence, by the application of the conservation energy principle for incompressible flow, and assuming ambient pressure at the inlet and outlet of the Coandă jet blanket and neglecting the energy exchange between the ambient air and the Coandă jet blanket, the following velocity relationship between the inlet and outlet sections of the Coandă jet blanket is given by

$$V_{j\text{-}out} = V_{j\text{-}in} \tag{11}$$

This latter equation can eliminate the use of the earlier assumption taken to relate the inlet and outlet velocity by a plausible heuristic assumption (which can be established by using CFD results), and replace it by equation (11). This should simplify the solution given in equation (8) and reduce the use of idealization or the number of assumption since fewer unknowns are included in the equations. Combining equations (8) and (11)

$$Lift_{\substack{Coanda-jet\ Blanket+ \\ induced\ pressure\ difference}} = \frac{3}{2}\dot{m}V_{j-in} \tag{12}$$

The results will be assessed a posteriori by using more structured CFD simulations, although the latter may also contain uncertainties and inaccuracies.

3. Flight Dynamic Analysis for Semi-spherical Coandă MAV in Translatory Motion

3.1. System of Coordinates

For the derivation of the equation of motion of the Coandă MAV, the system coordinates that are required for setting the equation will be defined. Following the convention in flight dynamics, four main reference frames are identified, namely the inertial coordinate system, the local horizon reference frame, the body reference frame, and the wind reference frame, as required in the analysis, and are depicted in Figure 5a.

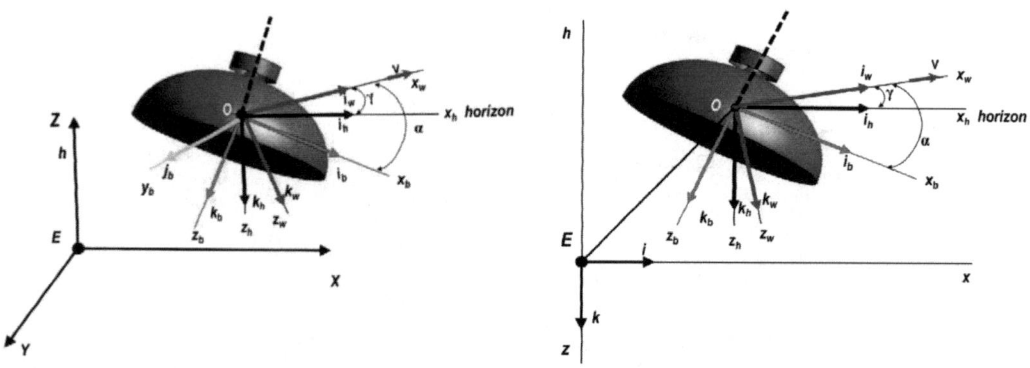

Figure 5: (a) The details of the four coordinate systems for Coandă MAV flight vehicle; (b) Axes systems in the vertical plane perpendicular to the Earth's Surface

Without loss of generality, for simplicity and instructiveness, only two-dimensional coordinate systems in the plane of symmetry of the Coandă MAV will be elaborated. These four coordinates are:

1. ***Inertial coordinate system***, which is used for defining and the application of the Newton's law of motion. In two dimensions, this coordinate system is depicted by the $O_E$xyz.

2. ***The local horizon coordinate system $Ox_hy_hz_h$***, which is fixed at the center of mass of the vehicle, and is parallel with the earth Inertial Coordinate system.
3. ***The body coordinate system $Ox_by_bz_b$***, which is fixed to the vehicle and follows a conveniently chosen axis of the vehicle.
4. ***The wind axes system $Ox_wy_wz_w$***, which moves with the vehicle and the x_w axis is coincident with the velocity vector (flight path of the vehicle).

In the figure, the wind axes are oriented to the flight path angle γ relative the local horizon axes and by the angle of attack α relative to the body axes. The two dimensional configuration in the vertical plane perpendicular to the earth surface is shown in Figure 5b.

3.2. Flight Dynamic Analysis

In the development of the flight dynamics of Coandă MAV in translatory motion, without loss of generalities, the equation of motion is developed in the vertical plane perpendicular to the earth motion, rendering two-dimensional planar motion. The hovering state, as depicted in Figure 6, will be utilized as a reference.

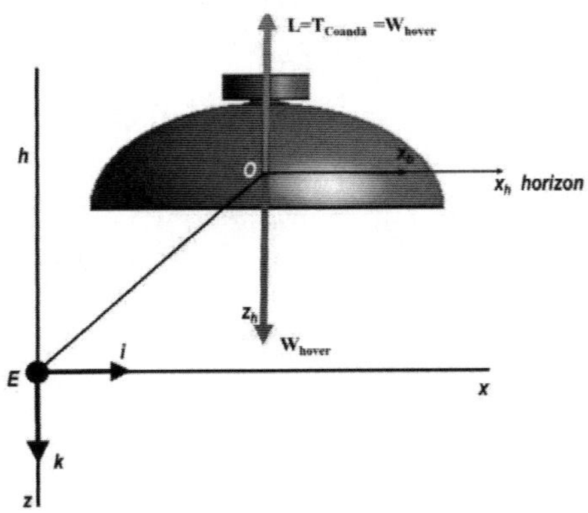

Figure 6: Coordinate systems for Coandă MAV in hover, in two-dimensional configuration (at the vertical plane of symmetry and motion)

Balance of forces in the free-body diagram exhibited in Figure 6 and the use of the Coandă lift from equation (12) leads to:

$$L_{Hover} \equiv T_{Coand\check{a}-MAV} = \frac{3}{2}\dot{m} \cdot V_{j\text{-}in} = 3\pi\rho R_i h_i V_{j\text{-}in}^2 = W_{Coand\check{a}-MAV-Take-off} \quad (13)$$

In the development of the equation of motion during translatory motion, a further simplifying assumptions will be made, which could be refined at later stage to incorporate more realistic ones.
These are:

1. The thrust of the Coandă MAV will not be affected by the attitude of the Coandă MAV configuration during hover. Such simplifying assumption can be later modified using more elaborate analytical model, assisted by CFD simulation.
2. During the translatory motion, the thrust of the Coandă MAV is assumed to be:

$$T_{Coand\check{a}-MAV} = \frac{3}{2}\dot{m} \cdot V_{j\text{-}in}\Big|_{hover-equivalent} = 3\pi\rho R_i h_i V_{j\text{-}in}^2\Big|_{hover-equivalent} \quad (14)$$

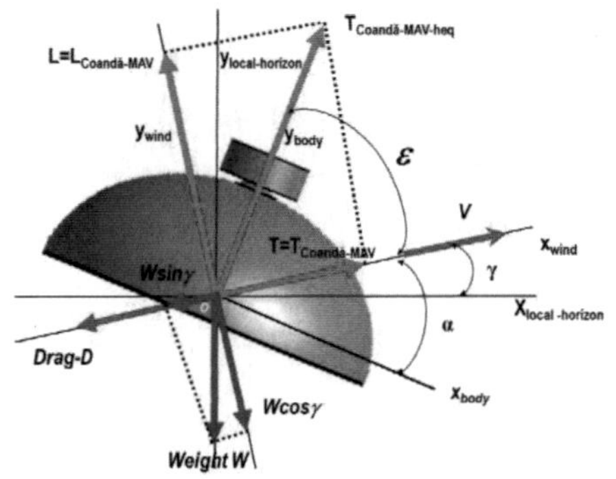

Figure 7: Schematic of Coandă MAV in two-dimensional tranlatory motion in the vertical plane (at the vertical plane of symmetry and motion).

The equation of motion of the Coandă MAV in translational motion in the two-dimensional vertical plane can be written by referring to Figure 7 using the basic flight dynamic equation for flight vehicle:
Kinematic:

$$\dot{x} = V\cos\gamma \qquad \left[LT^{-1}\right] \tag{15}$$

$$\dot{h} \equiv \dot{y} = V\sin\gamma \qquad \left[LT^{-1}\right] \tag{16}$$

$$\varepsilon = tan^{-1}\frac{\ddot{y}_{wind}}{\dot{V}} = tan^{-1}\frac{g\left(L - W\cos\gamma\right)}{W\dot{V}} \qquad [-] \tag{17}$$

$$\dot{\varepsilon} = tan^{-1}\frac{\ddot{y}_{wind}}{V} = tan^{-1}\frac{g\left(L - W\cos\gamma\right)}{WV} \qquad \left[T^{-1}\right] \tag{18}$$

Coandă MAV Fluid Dynamics:

$$T_{Coandă-MAV-heq} \equiv T_{Coandă-MAV-hover-equivalent} = \frac{3}{2}\dot{m}\cdot V_{j\text{-}in} = 3\pi\rho R_i h_i V_{j\text{-}in}^2 \qquad \left[MLT^{-2}\right] \tag{19}$$

$$T_{Coandă-MAV-heq} \equiv T_{Coandă-MAV-hover-equivalent} = \frac{3}{2}\dot{m}\cdot V_{j\text{-}in} = \frac{3}{2}\dot{m}\cdot\frac{\dot{m}}{2\pi\rho R_i h_i} = \frac{3}{4}\frac{\dot{m}^2}{\pi\rho R_i h_i} \qquad \left[MLT^{-2}\right] \tag{20}$$

Coandă MAV Flight Dynamics:

$$\dot{V} = \left(g/W\right)\left(T_{Coandă\ MAV-heq}\cos\varepsilon - D - W\sin\gamma\right) \qquad \left[LT^{-2}\right] \tag{21}$$

$$\ddot{y} = \left(g/W\right)\left(T_{Coandă\ MAV-heq}\sin\left(\varepsilon+\gamma\right) - W - D\sin\gamma\right) \qquad \left[LT^{-2}\right] \tag{22}$$

Weight-Fuel Consumption:

$$\dot{W} = -CT \qquad \left[MT^{-1}\right] \tag{23}$$

It should be noted, that the above relationships are derived for balance of forces and Newton's equation for semi-spherical Coandă MAV treated as a point mass moving in a plane perpendicular to the planar Earth.

4. Performance Measure

4.1. Performance Measure during hover

The feasibility of using Coandă techniques to enhance aerodynamic performance of Coandă MAV can be outlined through some non-dimensional quantities of performance measure. The most logical measure that has commonly been utilized is the Coandă jet momentum coefficient, C_μ (Mamou and Khalid [14], 2007; Djojodihardjo [9]), which are defined as:

$$C_\mu = \frac{\dot{m}V_{Coandă-jet}}{\frac{1}{2}\rho_j V_\infty^2 A_{ref}} \tag{24}$$

While (24) may be appropriate for any Coandă activated vehicle (i.e. Coandă MAV), particularly during hover and lift-off, one can define another performance measure to evaluate the aerodynamic performance for spherical Coandă MAV considered here, by assessing the output lift as compared to the input momentum rate, following consideration discussed by Djojodihardjo and Ahmed [12], and is given by:

$$PM1_{\substack{Spherical\,Coandă-\\MAV}} = \frac{Lift_{Coanda-jet\,Blanket+induced\,pressure\,difference}}{Coanda\ Jet\ Momentum\ In} = \frac{3}{2} \tag{25a}$$

Alternatively, this obvious dimensionless performance measure may be defined based on the rate of the energy of the Coandă jet ejected by Coandă MAV at its ideal inlet velocity $V_{j\text{-}out}$ at its peripheral outlet compared to the rate of momentum influx of the Coandă jet times its inlet velocity $V_{j\text{-}in}$ (non-dimensional):

$$\begin{aligned} PM3_{\substack{Spherical\\Coandă-MAV}} &= \frac{Lift\ produced\ by\ Coandă\ MAV\ times\ V_{j-out}\ at\ its\ peripheral\ outlet}{the\ rate\ of\ momentum\ influx\ of\ the\ Coandă\ jet\ times\ its\ inlet\ velocity\ V_{j-in}} \\ &= \frac{Lift.V_{j-out}}{2\pi R_0 h_0 \rho V_{j\text{-}in}^{\ 2}\cdot V_{j\text{-}in}} = \frac{\frac{3}{2}\dot{m}V_{j\text{-}in}^{\ 2}}{2\pi R_0 h_0 \rho V_{j\text{-}in}^{\ 3}} = \frac{3\pi R_0 h_0 \rho\cdot V_{j\text{-}in}^{\ 3}}{2\pi R_0 h_0 \rho\cdot V_{j\text{-}in}^{\ 3}} = \frac{3}{2} \\ &= 3\rho A_{MAV\_base}\left(\frac{h_0}{R_0}\right) \end{aligned} \tag{25b}$$

It is realized at this point, that more information is required to insure uniqueness. Resorting to the present heuristic approach, further realistic assumption can be drawn from CFD visualization, which also added further insight to the problem. Similar expression can be defined for Cylindrically Shaped Coandă MAV. Alternatively, it may be noted that Mirkov and Rasuo [15-16] considered the inlet velocity to thrust ratio for a given inlet size could be the most important output value resulting either from experiment on UAV's or by numerical simulation. In view of that, another performance measure could be defined to be the lift to inlet velocity, which is a dimensional quantity and substituting equation (13) for the lift, it gives;

$$PM2_{\substack{Spherical\\Coandă-MAV}} = \frac{Lift_{Coanda-jet\,Blanket+induced\,pressure\,difference}}{Coandă\ jet\,inlet\,Velocity} = \frac{\dot{m}\cdot V_{j\text{-}in} + \frac{1}{2}\dot{m}V_{j-in}}{V_{j\text{-}in}} = \frac{3}{2}\dot{m} \tag{26}$$

A third performance measure may be defined based on the Coandă jet rate of momentum influx per unit Coandă jet kinetic energy input (dimensional):

$$PM3_{\substack{Spherical\\Coandă-MAV}} = \frac{Lift_{Coanda-jet\,Blanket+induced\,pressure\,difference}}{Jet\ Inlet\ velocity^2} = \frac{2\pi R_0 h_0 \rho \cdot V_{j\text{-}in}{}^2 + \pi R_0 h_0 \rho V_{j-in}{}^2}{V_{j\text{-}in}{}^2} = 3\rho A_{MAV\_base}\left(\frac{h_0}{R_0}\right) \tag{27}$$

where

$$A_{MAV\_base} = \pi R_0{}^2 \tag{28}$$

4.2. Performance Measure during translation

As a baseline, performance measure during translation can be defined for two cases:

For level flight

For climbing

For level flight (by referring to Figure 7):

$$PM_{\substack{Coandă-MAV\\LevelFlight}} = \frac{Power\ required\ to\ overcome\ Drag}{Power\ Input\ by\ Coandă\ jet\ action} = \frac{Drag * Velocity}{Coandă\ Jet\ Momentum\ Gain * Coandă\ Jet\ Velocity\ In} = \frac{D \cdot V\cos\gamma^2}{\dot{m}V_{j\text{-}in}{}^2} \quad \left[\frac{ML^2T^{-2}}{ML^2T^{-2}}\right] \tag{29}$$

For purely level flight, as depicted in Figure 8, γ=0, and Equation (29) reduces to:

$$PM_{\substack{Coandă-MAV\\LevelFlight}} \approx \frac{1}{4} C_{D\to x} \frac{\pi\rho R_o{}^2 \sin\alpha}{\dot{m}_i} \frac{V^3}{V_{j\text{-}in}{}^2} \tag{30}$$

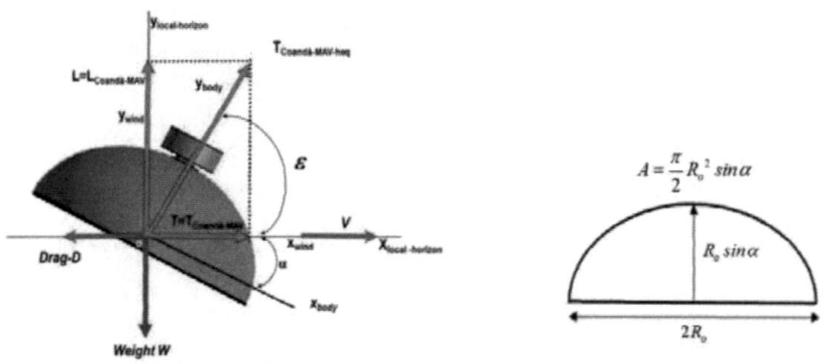

Figure 8: Schematic of Coandă MAV in two-dimensional cruising motion in the vertical plane of symmetry and motion. Inset: Cross sectional area perpendicular to direction of motion.

For climbing:

$$PM_{\substack{Coandă\\-MAV\\Climbing}} = \frac{Power\ required\ to\ Climb}{Power\ Input\ by\ Coandă\ jet\ action} = \frac{\mathbf{T}_{Coandă-MAV-heq} \bullet \dot{\mathbf{y}} - \mathbf{D} \bullet \dot{\mathbf{y}} - \mathbf{W} \bullet \dot{\mathbf{y}}}{Coandă\ Jet\ \text{Momentum Gain} * Coandă\ Jet\ \text{Velocity In}} = \frac{3}{2}\frac{\cos(\alpha-\gamma)V\sin\gamma}{V_{j\text{-}in}} - \frac{(D\sin\gamma + W)V\sin\gamma}{\dot{m}V_{j\text{-}in}{}^2} \quad \left[\frac{ML^2T^{-2}}{ML^2T^{-2}}\right] \tag{31}$$

For purely climbing flight, γ=π/2, and Equation (29) reduces to

$$PM_{\substack{\text{Coandă}-MAV \\ Climbing}} = \frac{\frac{3}{2}\dot{m}\cdot V_{j\text{-}in}V - \left(\frac{1}{2}\rho C_{D\rightarrow y}\pi R_o^{\,2}V^2 + W\right)V}{\dot{m}V_{j\text{-}in}^{\;2}} = \frac{\frac{3}{2}\dot{m}\cdot V_{j\text{-}in}V - \left(\frac{1}{2}\rho C_{D\rightarrow y}\pi R_o^{\,2}V^2 + \left(\frac{3}{2}\dot{m}V_{j-in}\right)_{hover}\right)V}{\dot{m}V_{j\text{-}in}^{\;2}} \tag{32}$$

Which has been further developed from that given in [16]. The Performance Measure $PM_{\text{Coandă-MAV Level Flight}}$ can be expressed as a function of C_D, R_o, $\dot{m}$ h_i, α, V, $V_{j\text{-}in}$. These are the basic relationships that can further be modified and elaborated for other flight conditions.

5. Some Results and Examples

5.1. Results and examples for hover

To assess the merit and plausibility of the theoretical analysis using first principles carried out here, a comparison of the theoretical prediction for the performance measures PM1, PM2 and PM3 given by (25) - (27) are compared with CFD simulation carried out using ANSYS FLUENT® for Spherically Shaped Coandă MAV. The results are depicted in Figures 9 and 10.

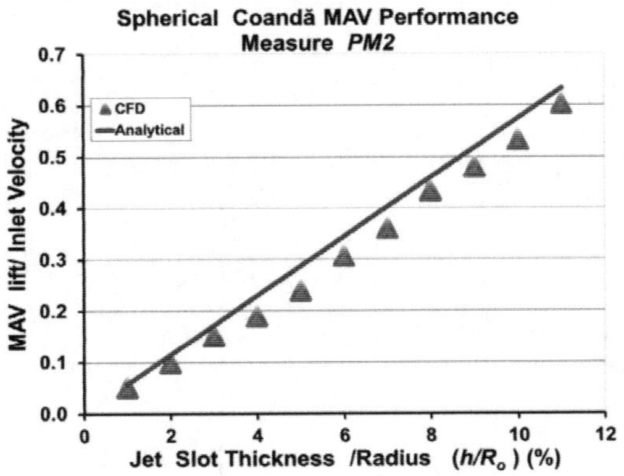

Figure 9: Performance measure for spherical Coandă MAV for various Jet slot thickness at different jet velocity.

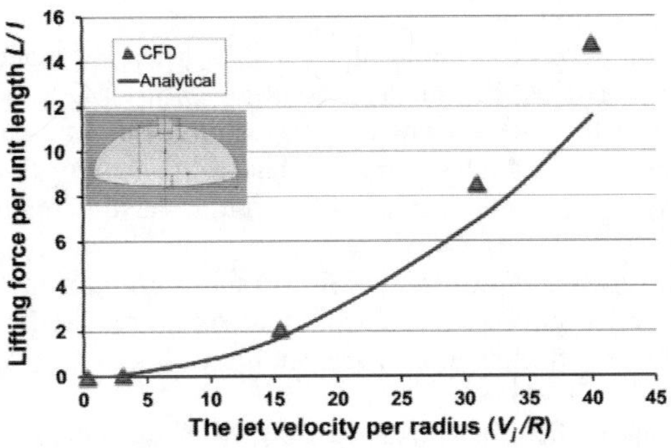

Figure 10: Analytical and CFD models comparison for jet inlet velocity influence on performance of Spherical Coandă MAV at constant jet thickness (h=50mm).

The CFD simulation was performed using steady RANS with two equations k-ω Shear Stress Transport (SST) turbulence model. Simple pressure-velocity coupling scheme with the least squares cell base as discretization gradient was applied in the solution method together with second order upwind for the momentum equation and the turbulent kinetic energy.

As depicted in Figure 9, the influence of the ratio of the jet slot thickness h to the reference radius R_o which influence the jet inlet velocity on the performance measure PM2 is clearly noticeable. In addition to the above performance measures, the influence of the jet inlet velocity on the Coandă MAV lift may also be of interest from designer's point of view which enables them to befittingly select the right propulsion system for such MAV. The jet inlet velocity influence on performance of Spherical Coandă MAV at constant jet thickness (h=50mm) is depicted in Figure 10.

For Cylindrical Coandă MAV the corresponding theoretical prediction for PM2 is given by [11-12]

$$PM2_{\substack{Cylinderical\\ Coandă-MAV}} = \frac{Lift_{Coanda-jet\ Blanket+induced\ pressure\ difference}}{Coandă\ jet\ inlet\ Velocity} = \frac{\rho V_{j-in}^2 l\left(\frac{h^2}{t}\right)+\frac{1}{2}\rho\left(\frac{h}{t}V_{j-in}\right)^2 Rl}{V_{j-in}} \tag{33}$$

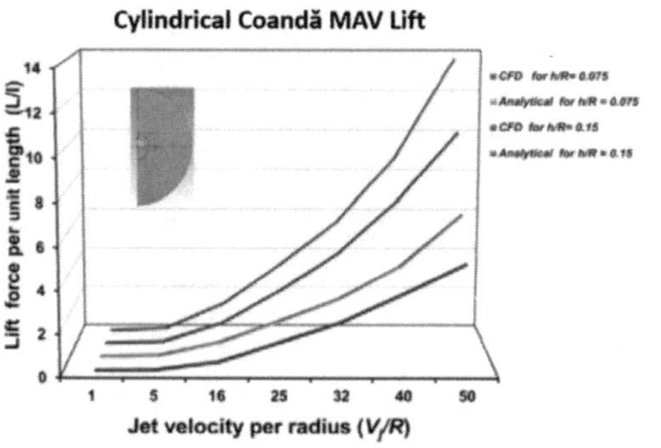

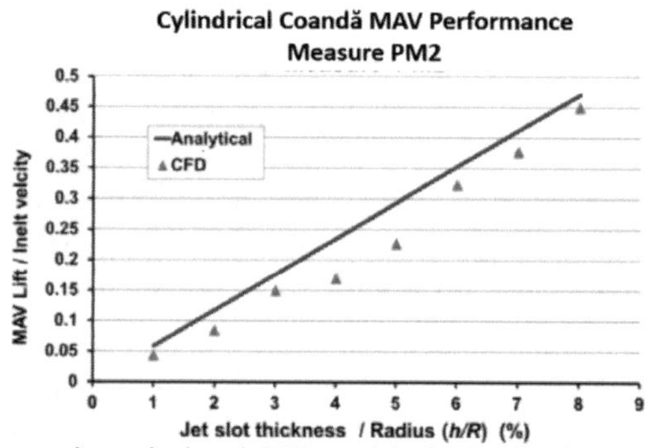

Figure 11: Performance measure comparison between mathematical and CFD models for Cylindrical Coandă MAV for various (a) Inlet velocity and (b) Jet slot thickness

For comparative and validation purposes, Figure 11 (a) and (b) are produced and compared to CFD simulation results to assess the present theoretical prediction with CFD simulation results a cylindrical Coandă MAV. Relatively good qualitative agreements between the theoretical results and CFD simulations shown there lend support to the present analysis, noting that in the analytical study carried out here many simplification have been introduced, while in the CFD simulation, the full Navier-Stokes equation for incompressible fluid was used.

The effect of viscosity that can be revealed by CFD simulation is exemplified by Figure 12 for two different locations of the Coandă jet introduction. The CFD images exhibited in Figure 12 (a) and (b) show the velocity magnitude contours for the given configuration and inlet conditions for cases considered by Mirkov and Rasuo [17-18] and Ghozali [19].

Figure 13 shows structured numerical mesh used in the flow domain. Through meticulous attempts and grid sensitivity studies the size of the mesh cells was generated small to enable visualizations of the flow around the whole body with best details. The numerical study was performed with a moderately small Reynolds number, Re= 68458, based on mean velocity and jet inlet height. The velocity is uniform ($V_{j\text{-}in}$ = 20 m/s) across the inlet which has a thickness of 0.05 m. The number of the mesh cells was 52830, and the mesh quality was found to be acceptable. The orthogonality quality was 4.96943e-01, which should be acceptable from the ANSYS orthogonality quality requirements. As elaborated by Ahmed and Djojodihardjo [10-11] and Djojodihardjo and Ahmed [12-13], the present generation of mesh is considered to be acceptable.

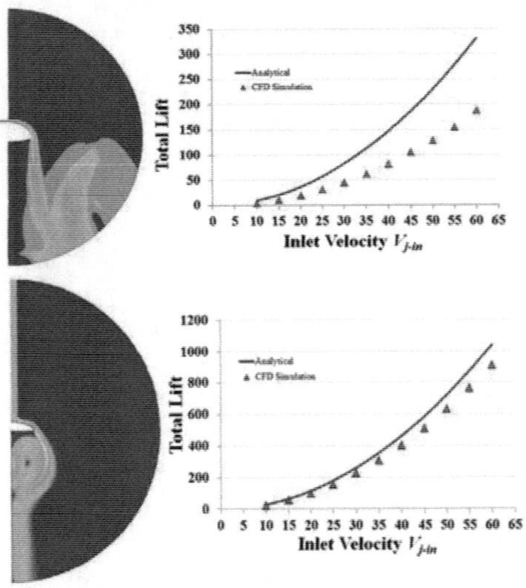

a)

b)

Figure 12: Lift versus input jet velocity for similar configuration considered in (a) Mirkov and Rasuo [17], (b) Ghozali [19]

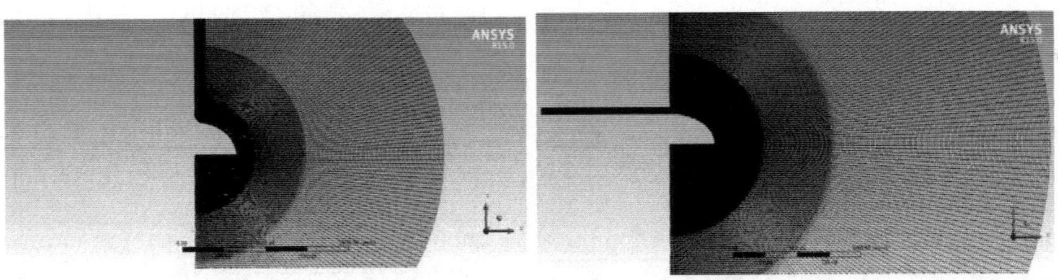

Figure 13: Mesh details for similar configuration considered in (a) Mirkov and Rasuo [18], (b) Ghozali [19]

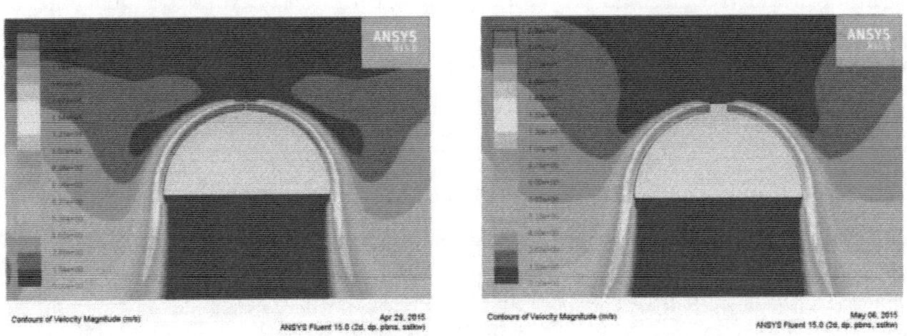

Figure 14: Symmetrical Coandă effect - velocity magnitude contour with different inlet radius a) R_i=5 mm, b) R_i=50mm

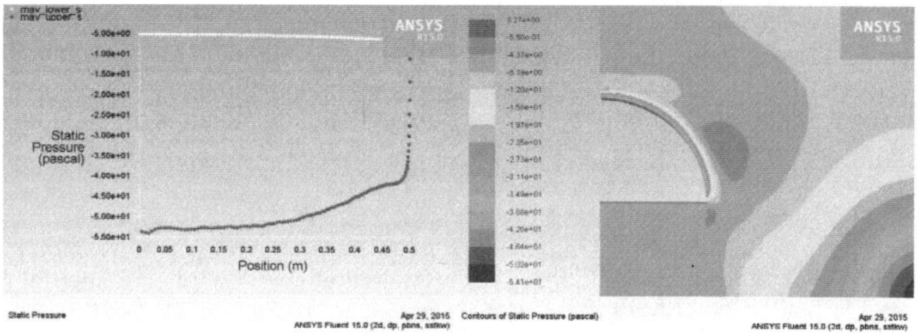

Figure 15: Comparison between the CFD simulation and plot for pressure contours along semi-spherical Coandă MAV surface.

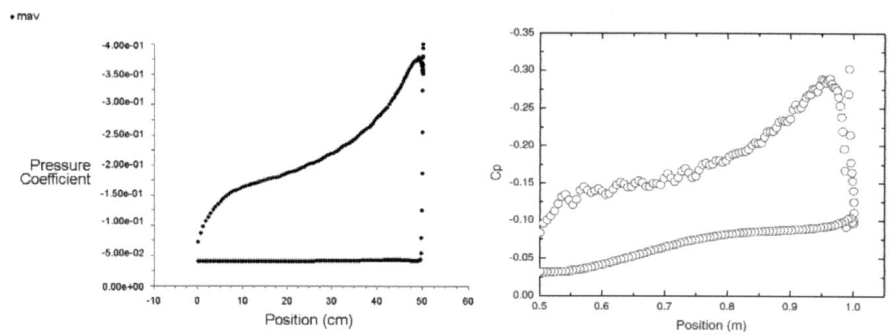

Figure 16: Comparison of the coefficient of pressure on the Coandă MAV surface jet blanket; a. present CFD Simulation, b. simulation by Mirkov and Rasuo [18].

The influence of the inlet jet radius on the air vehicle performance (lift) is also investigated using the CFD simulation, The results are presented as velocity magnitude contours, which were carried out for two inlet jet radius, R_i=5 mm and R_i=50mm and as shown in Figure 14. For the semi-spherical Coandă MAV, the CFD simulation and visualization results are exhibited in Figure 15, which exhibit the pressure contours along semi-spherical Coandă MAV surface obtained by CFD simulation results (left), and the associated static pressure color-coded contour (right).

The CFD numerical study presented here was performed with a moderately small Reynolds number, Re= 68458, based on mean velocity and jet inlet height. The velocity across the inlet, which has a thickness of 0.05 m, is uniform ($Vj\text{-}in$ = 20 m/s). The number of the mesh cells was 52830, and the mesh quality was found to be acceptable. The orthogonality quality was 4.96943e-01, which is acceptable from the ANSYS orthogonality quality requirements.

5.2. Results and examples for translation

For translatory flight, some examples of Performance Measures simulation studies for Coandă MAV in purely level and climbing flights are exhibited by Figs. 17 and 18, obtained using equations (30) and (32) for level and climbing flights, respectively. Fig. 17 shows the influence of Coandă jet slot thickness on the cruise performance for various Coandă jet inlet velocity, while Figure 18 shows pure Coandă MAV climbing performance for various Coandă jet slot thickness and inlet velocity. The latter figure shows the threshold of these parameters that will provide Coandă MAV climbing capability.

IOP Conf. Series: Journal of Physics: Conf. Series **822** (2017) 012051 doi:10.1088/1742-6596/822/1/012051

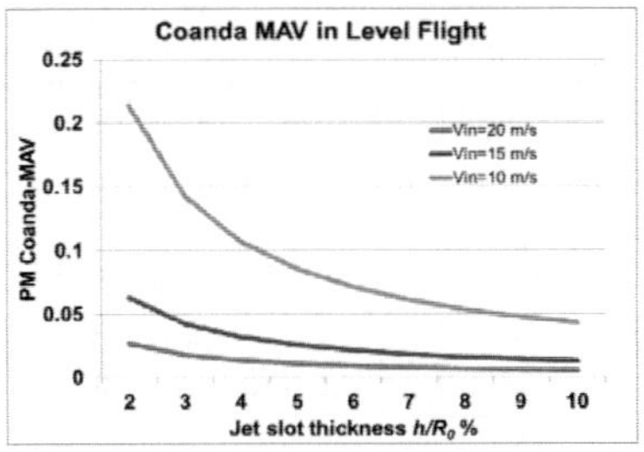

Figure17: Coandă MAV Performance Measure as a function of jet slot thickness for various cruising velocity, calculated using Eq. (29) and CFD simulation for a baseline Coandă MAV configuration.

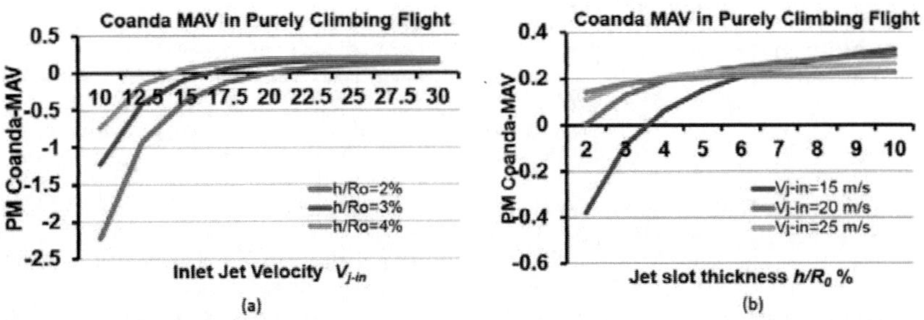

Figure 18: Coandă MAV Performance Measure as a function of jet slot thickness for various climbing rate (velocity), calculated using Eq. (31) and CFD simulation for a baseline Coandă MAV configuration; (a) Climbing Performance as function of Coandă jet velocity at various jet slot thickness, (b) Climbing Performance as function of Coandă jet slot thickness at various jet velocity.

Parametric study results exhibited by these figures can be utilized to give insight on the performance of the semi-spherical Coandă MAV considered as well as for preliminary sizing purposes.

6. Conclusions

A comprehensive effort has been made to analyze and describe the governing equations applicable to Coandă MAV in hover and translatory motion utilizing first principles as articulated in fluid dynamics, using conservation principles for control volume analysis, and flight mechanics, using free-body diagram analysis for the application of Newton's law of motion. The basic working relationships among various relevant variables and parameters governing the aerodynamics forces and performance measures of Coandă MAV in hover and translatory motion are then established.

CFD simulations for the Coandă MAV generic models considered are carried out with two objectives, to gain insight into the flow situation for establishing the analytical model, and to assess and validate the analytical model subject to the theoretical assumptions.

Various CFD studies have also indicated that there are still significant discrepancies between CFD and experimental studies [6], which necessitate the existence of specific baseline configurations for validation purposes.

From the theoretical analysis, performance measures are derived as further elaboration of the significance of Coandă jet momentum coefficient, which are shown to be capable in assessing the effectiveness of the Coandă jet as lift generating mechanism of the and Coandă MAV, as well as its propulsion mechanism in the Coandă MAV flight.

Comparison of the theoretical analysis using first principles on the Coandă MAV generic configurations with Reynolds averaged Navier Stokes CFD simulation using appropriate turbulence model serves also to assess the uncertainties and accuracy of the theoretical approach. In addition, the CFD computational and visualization studies, can provide further insight in revealing other characteristics of the flow field, and can assist further in-depth analysis, as well as in the design of experiments to that end. For example, to find out the effect of viscosity, Coandă jet input variables and Coandă MAV jet blanket on the Coandă jet stability and attachment. CFD visualization has certainly assist the analytical work in assessing the application of fundamental conservation principle in continuum mechanics as well as elaborating the equation of motion for the flight dynamics of the Coandă MAV.

References

[1] Gad-el-Hak M 2000 Flow Control - Passive, Active and Reactive Flow Management (Cambridge University Press, London)

[2] Liu Y, Sankar L N, Englar R J, Ahuja K K 2001 Numerical Simulations of the Steady and Unsteady Aerodynamic Characteristics of a Circulation Control Wing Airfoil *AIAA Paper* **2001-0704**

[3] Kweder J, Panther C, Smith J E 2011 Applications of circulation control, yesterday and today *International Journal of Engineering, (IJE)* **195 vol.4 (5)**, pp. 411-429

[4] Djojodihardjo H and Thangarajah N 2014 Research, development and recent patents on aerodynamic surface circulation control-A critical review *Recent Patents on Mechanical Engineering* **vol. 7**, pp. 1-37.

[5] Djojodihardjo H 2013 Progress and development of Coandă jet and vortex cell for aerodynamic surface circulation control – An overview *The SIJ Transactions on Advances in Space Research & Earth Exploration (ASREE)*, **Vol. 1, No. 1**, September-October.

[6] Rumsey C L and Nishino T 2011 Numerical study comparing RANS and LES approaches on a circulation control airfoil International Journal of Heat and Fluid Flow **32** 847-864.

[7] Tongchitpakdee C, Benjanirat S and Sankar L N 2006 Numerical Studies of the Effects of Active and Passive Circulation Enhancement Concepts on Wind Turbine Performance *J.Sol.Energy Eng* **128(4)** 31023089

[8] Abdul-Hamid, M F, Djojodihardjo H, Suzuki S and Mustapha F 2010 Numerical Assessment of Coanda Effect as Airfoil Lift Enhancer in Wind-Turbine Configuration, *Regional Conference on Mechanical and Aerospace Technology* Bali, February 9 – 10

[9] Djojodihardjo H, Abdul-Hamid M F, Jaafar A, Basri S, Romli F I, Mustapha F, Mohd-Rafie A S and Abdul-Majid D L A 2013 Computational Study on the Aerodynamic Performance of Wind Turbine Airfoil Fitted with Coandă Jet *Journal of Renewable Energy* **2013**, 839319

[10] Djojodihardjo H, Ahmed R I 2016 CFD Simulation of Coandă Effect on the Upper Respiratory System *Journal of Medical Imaging and Health Informatic* in print

[11] Ahmed R I, Djojodihardjo H, Abu-Talib A R, Abd-Hamid M F 2014 Application of Coandă jet for generating lift for micro air vehicles – Preliminary design considerations, *Applied Mechanics and Materials* **629** 139-144

[12] Ahmed R I, Djojodihardjo H, Abu-Talib AR, Mohd-Rafie A S 2016 First Principle Analysis of Coandă Micro Air Vehicle Aerodynamic Forces for Preliminary Sizing *Aircraft Engineering and Aerospace Technology* DOI: 10.1108/AEAT-03-2015-0080.R2 in print,

[13] Djojodihardjo H and Ahmed R I 2014 An Analysis on the Lift Generation for Coandă Micro Air Vehicles *IEEE, ICARES conference proceeding*, Yogyakarta.

[14] Djojodihardjo H, Ahmed R I, Abu-Thalib A R and Mohd-Rafie A S 2015 Analytical and CFD visualization studies of Coandă MAV *Proceedings The 13th Asian Symposium on Visualization* Novosibirsk

[15] Mamou M and Khalid M 2007 Steady and unsteady flow simulation of a combined jet flap and Coanda jet effects on a 2D airfoil aerodynamic performance *Revue des Energies Renouvelables* **CER'07 Oujda** 55 – 60

[16] Djojodihardjo H and Ahmed R I 2016 Analytical, Computational Fluid Dynamics and Flight Dynamics of Coandă MAV, *conference paper ICAS2016_0114, 30th International Congress of the International Council of the Aeronautical Sciences*, Daejon.

[17] Mirkov N and Rasuo B 2019 Numerical Simulation of Air Jet Attachment to Convex Walls and Applications *ICAS 2010-621*

[18] Mirkov N and Rasuo B 2012 Maneuverability of a UAV with Coanda Effect Based Lift Production *ICAS 2012-617*

[19] Ghozali D 2013 Analysis of Coandă Effect Using Computational-Fluid-Dynamic, *Thesis, Universitas Gajah Mada,* Indonesia

Acknowledgement

The authors would like to thank Universiti Putra Malaysia (UPM) for granting Research University Grant Scheme (RUGS) No.9378200, and the ministry of higher education ERGS: 5527088 ; FRGS:5524250 under which parts of the present research is carried out.

Blowing momentum and duty cycle effect on aerodynamic performance of flap by pulsed blowing

Ping Zhou[1], Yankui Wang[2], Jinjun Wang[3], Yongxiang Sha[4]

[1] Ministry-of-Education Key Laboratory of Fluid Mechanics, Beihang University, China
[2] Ministry-of-Education Key Laboratory of Fluid Mechanics, Beihang University, China
[3] Ministry-of-Education Key Laboratory of Fluid Mechanics, Beihang University, China
[4] Ministry-of-Education Key Laboratory of Fluid Mechanics, Beihang University, China

[1] E-mail: shakazp@163.com
[2] E-mail: wangyankui@buaa.edu.cn

Abstract: Control surface, which is often located in the trailing edge of wings, is important in the attitude control of an aircraft. However, the efficiency of the control surface declines severely under the high deflect angle of the control surface because of the flow separation. To improve the efficiency of control surface, this study discusses a flow-control technique aimed at suppressing the flow separation by pulsed blowing at the leading edge of the control surface. Results indicated that flow separation over the control surface can be suppressed by pulsed blowing, and the maximum average lift coefficient of the control surface can be 95% times higher than that of without blowing when average blowing momentum coefficient is 0.03 relative to that of without blowing. Finally, this study shows that the average blowing momentum coefficient and non-dimensional frequency of pulsed blowing are two of the key parameters of the pulsed blowing control technique. Otherwise, duty cycle also has influence on the effect of pulsed blowing. Numerical simulation is used in this study.

Keywords: Flow Control, Pulsed Blowing, Duty cycle

Nomenclature

C_μ	=	Average blowing momentum coefficient
V_∞	=	Velocity of the freestream (m/s)
Re	=	Reynolds number based on chord length of model
c_0	=	Chord length of the model (m)
m_j	=	Mass flow rate of blowing (kg/s)
V_j	=	Jet velocity from the blowing slot (m/s)
S_e	=	Reference area of the flap (m^2)
S_j	=	Area of the blowing slot (m^2)
f	=	Frequency of pulsed blowing (Hz)
Str	=	Non-dimensional frequency of pulsed blowing
h_j	=	Width of the blowing slot (mm)
C_p	=	Pressure coefficient
C_L	=	Average lift coefficient of the flap
c	=	Chord length of flap (m)
ρ_∞	=	Density of free stream (kg/m^3)
α	=	attack angle of main wing (°)

δ_e = deflect angle of flap (°)
t = time (s)

1. Introduction

In modern aircraft designing, the attitude control of an aircraft is still relies on traditional control surface such as flap, which is often located in the trailing edge of wings. However, the efficiency of the control surface declines severely under the high deflect angle of the control surface because of the flow separation. This condition leads to the penalty of attitude control and limited aerodynamic performance of an aircraft. In order to improve the efficiency of flap, flow control is necessary. In the past decades, scientists have exerted considerable effort to develop flow control techniques. Moreover, various flow control techniques [1] has been used to suppress flow separation such as moving surface control technique [2-4], plasma flow control technique [5-8] and co-flow jet control technique [9-12]. Although flow separation on the control surface can be suppressed substantially by these flow control technique, the application of these techniques in engineering are limited because of complicated devices or other reasons. Relatively, the flow control technique by blowing at the leading edge of flap catch people's eyes because of its simple device, rapidly remarkable lift increment and low gas consumption. In order to reduce the gas consumption further, pulsed blowing technique is taken into consideration. Study shows that lift coefficient of flap can be higher with pulsed blowing compared with that of continuous blowing.

This study introduces an flow control technique by pulsed blowing near the leading edge of the flap to suppress the flow separation over the flap that is located on the tailing edge of an airfoil. First, the effect of pulsed blowing on the aerodynamic performance of the flap is investigated thoroughly. Thereafter, the effect of cycle duty of pulsed blowing are discussed. All the results of this study is completed based on the following conditions: the angle of attack of the main wing is 0° and the deflect angle of the flap is 20°.

2. Numerical Model and Data Process

2.1 Numerical Model

The airfoil model used in the study is based on NACA0025 airfoil with a 0.6 m chord length. This model can be divided into two parts (see Figure 1): the main wing and a flap that is as long as 0.206m.

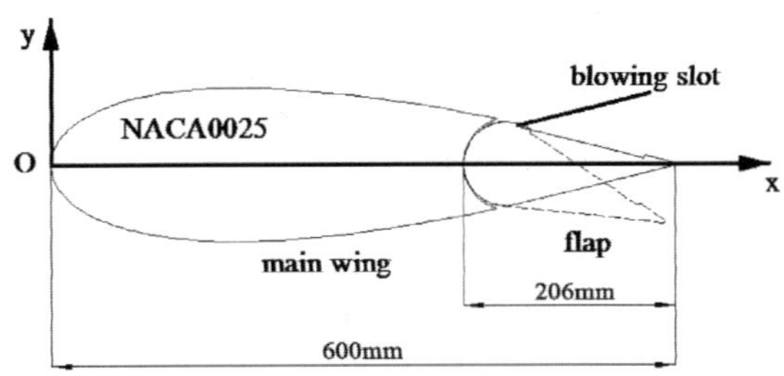

Figure 1. Sketch of the Model

A spanwise blowing slot with a 0.5mm height is located near the leading edge of the flap which is selected based on the flow separation point on the upper surface of the flap (see Figure 2). Compressed air flows downstream along the tangential direction of the upper surface of the flap from the slot.

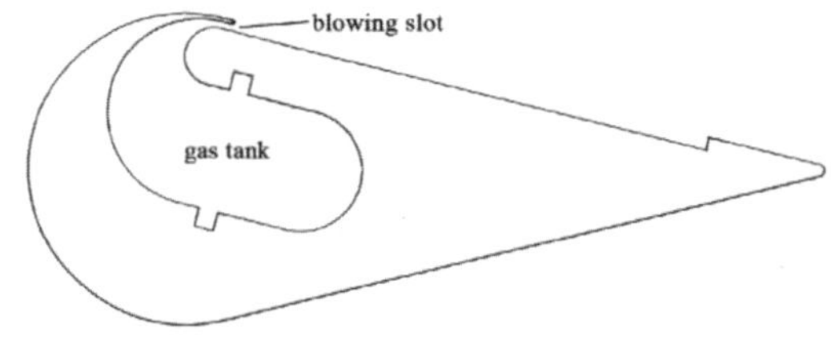

Figure 2. Sketch of the Flap

The 2D mesh of the model used in numerical simulation is as shown in. The minimum gap between the main wing and the flap is 0.5 mm, and the width of blowing slot is also 0.5 mm. An O-mesh with circle far-field, which is 20 chord length away from the model, is used in the simulation. In addition, the k-omega SST turbulence model is used. The boundary condition of the mass flow rate is used on the blowing slot. The height of the first layer of the grid is 0.02 mm, and the number of cells is 79757.

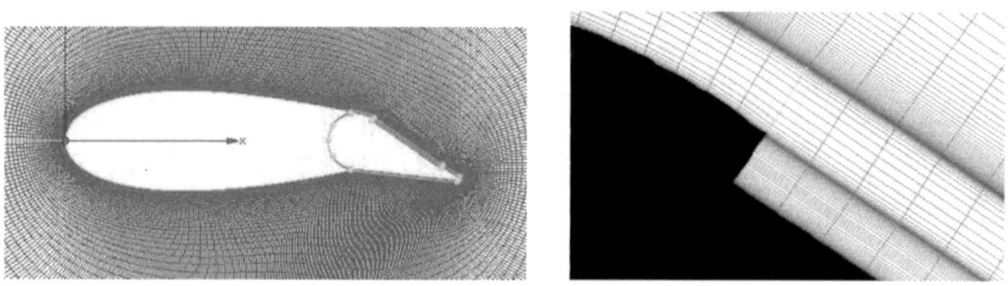

(a)Mesh of Full Model (b) Mesh near Blowing Slot

Figure 3. Mesh Used in Numerical Simulation

The pulsed blowing wave is as shown in Figure 4. The wave is a standard square wave with a duty cycle of 0.5.

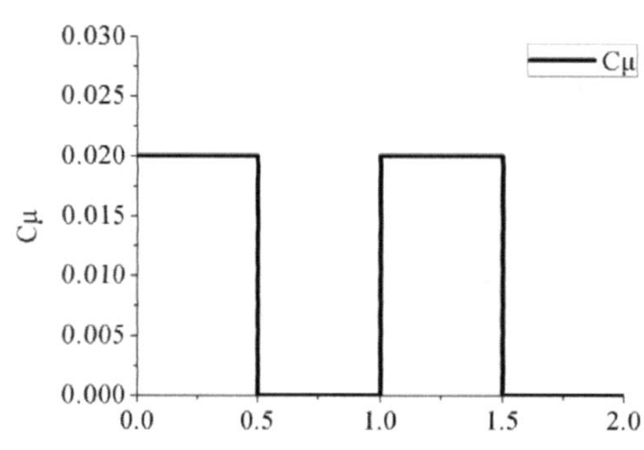

Figure 4. Sketch of Wave Form

2.2 Data Process

Average blowing momentum coefficient is used to represent the gas consumption. The equation is as follows:

$$C_\mu = \frac{1}{T}\sum_T \frac{m_j V_j}{\frac{1}{2}\rho_\infty V_\infty^2 S_e}\Delta t \quad (1)$$

where C_μ represent average blowing momentum coefficient, T is the cycle time, *mj* is the mass flow rate of the blowing jet, *Vj* is the jet velocity from the blowing slot, ρ_∞ is the density of freestream, S_e is the reference area of the flap, V_∞ is the velocity of freestream and Δt is the interval time of two adjacent time step.

The Strouhal number *Str* is the parameter that represents pulsed frequency. The equation of the Strouhal number is as follows:

$$Str = \frac{fc}{V_\infty} \quad (2)$$

where Str is the Strouhal number. *f* is the pulsed frequency and *c* is the chord length of the flap.

Duty cycle *dr* is defined as the proportion of blowing time in one single cycle time.

2.3 Validation of the Numerical Simulation Results

The validation experiment is conducted in the D-4 low-speed wind tunnel of Beihang University. Figure 5 gives out the model used in validation experiment. The model has 1 aspect ratio and 0.6 m the spanwise length. The proposed model is equipped with 44 pressure taps at the longitudinal symmetric section along the upper and bottom flap. Two fiberglass plates are set on both sides of the model spanwise to simulate a two-dimensional flow in the test. DTC initium electronic scanivalve is used to measure the pressure distribution of the model with an accuracy of 0.05% and the highest sampling frequency is 650Hz.

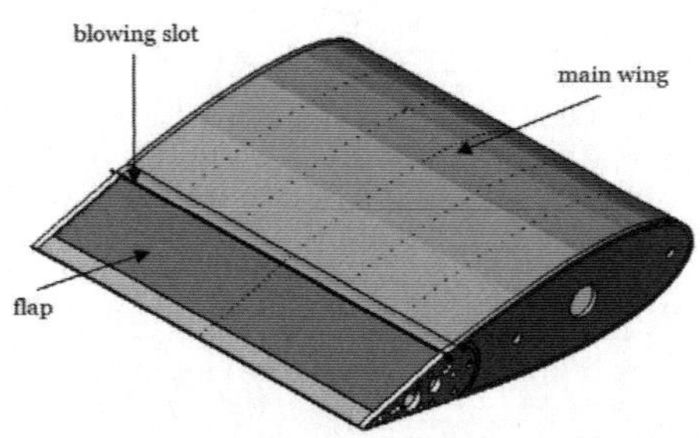

Figure 5. The Sketch of Model Used in Experiment

Figure 6 gives out the pressure distribution on flap with a blowing momentum coefficient of 0.055 under continuous blowing. The dash line stands for experimental result and the solid line stands for numerical simulation result. The angle of attack of the main wing is 0° and deflection angle of flap is 20° with a Reynolds number of 0.8×10^6. It can be found that the pressure distribution is well coincided except the fore part of flap which is covered by main wing. In fact, the width deviation of gap between main wing and flap is the main reason of this difference between numerical and experimental result. Thus, the numerical result can be considered of appropriate and credible.

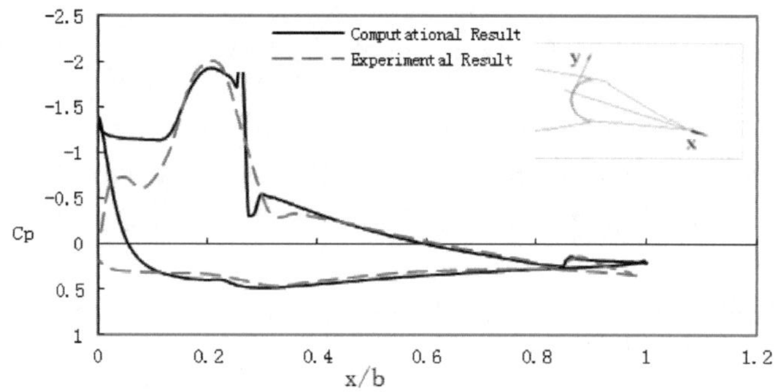

Figure 6. Comparison of Pressure Distribution between the Experimental and Numerical Results

3. Results and Discussion

3.1 Effect of Pulsed Blowing on Aerodynamic Performance of Flap

Pulsed blowing at the leading edge of flap can cut the consumption of gas greatly while higher lift increment is get. Figure 7 shows the pulsed blowing effect on the average lift coefficient of the flap changing with the Strouhal number under different average blowing momentum coefficient as well as the comparison with that of the non-blowing under the following conditions: angle of attack α=0°, deflection angle of the flap δ_e=20°, V_∞=20m/s, and Re=0.8×10^6, duty cycle dr=0.5. The solid line with square mark represents $C\mu$=0.01, solid line with circle mark represent $C\mu$=0.02, solid line with triangle mark represent $C\mu$=0.04 and dash line represent $C\mu$=0.

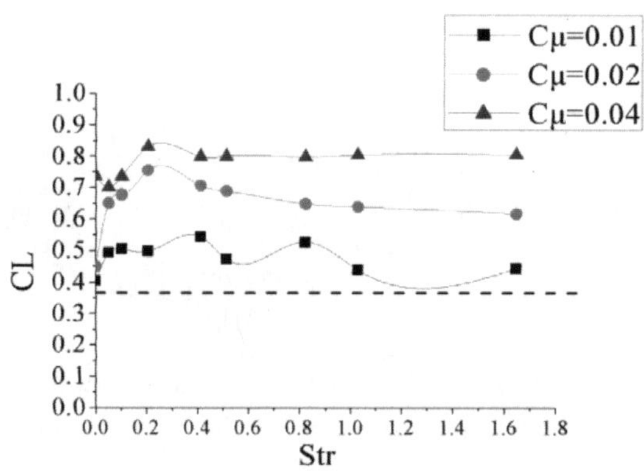

Figure 7. Effect of Str on Average Lift Coefficient of the Flap

When Str=0, the intersection of y axis and three solid line, respectively, represent the lift coefficient of the flap under continuous blowing under different $C\mu$. With the increase of $C\mu$, the lift coefficient of the flap increase. When $C\mu$ is 0.04, lift increment of the flap is 100% compared with the lift coefficient of the flap without blowing. When pulsed blowing is used, further increment of lift coefficient of the flap occurs. If $C\mu$ is low (such as $C\mu$=0.01), the flow over flap is a separated flow and pulsed blowing cannot master the flow over the flap. The evolution of lift coefficient of the flap with different Strouhal number is irregular and the lift increment is not evident enough. When $C\mu$ is high enough (such as $C\mu$=0.02), the lift coefficient of the flap increases rapidly with the increase of

Strouhal number. *CL* reaches the maximum value when *Str*=0.206. With the continuously increase of *Str*, *CL* decrease slightly. However, the lift coefficient of pulsed blowing is constantly higher than that of continuous blowing under the same $C\mu$. When *Str*=0.206, *CL* of pulsed blowing is 67% higher than that of continuous blowing under the same $C\mu$.

3.2 Effect of Duty Cycle on Average Lift Coefficient of the Flap

When $C\mu$ keep constant, the duty cycle also has effect on *CL*. Figure 8 shows the *CL* changing with *Str* under different duty cycle and the following conditions: angle of attack α=0°, deflection angle of the flap δ_e=20°, V_∞=20m/s, and Re=0.8×10^6 and $C\mu$=0.02. The solid line with square mark represents *dr*=0.25, solid line with circle mark represent *dr*=0.5 and solid line with triangle mark represent *dr*=0.75.

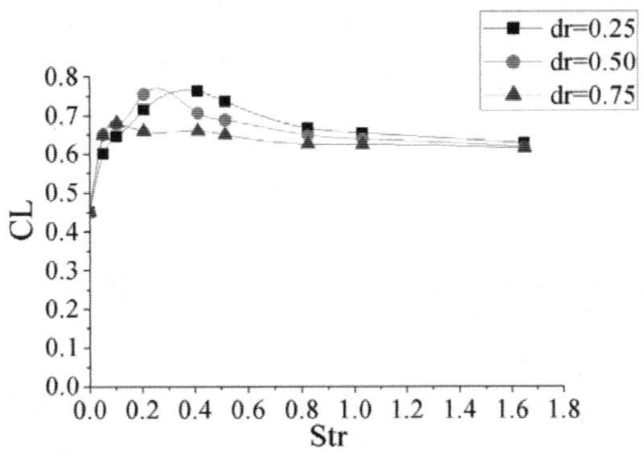

Figure 8. Effect of Duty Cycle on Average Lift Coefficient of the Flap

The tendency of evolution of *CL* under different duty cycle is similar: increase rapidly with the increase of *Str* and reach the maximum value, then decrease slightly. But the difference reflect in the maximum value and the corresponding *Str*. If define the critical *Str* as the *Str* when *CL* reach the maximum value, then the maximum value of *CL* increase with the increase of duty cycle. The critical *Str* also increase with the increase of duty cycle.

Based on the study before, the mechanism of improving the lift of the flap by pulsed blowing can be divided into two components. First, blowing jet can be treated as an injector. The flow upstream of the blowing slot is injected by the blowing jet and induces a suction peak, which is the same mechanism as continuous blowing. Second, the vortex generated by the switch of blowing and non-blowing over the blowing slot induces a suction peak. Lift increment is generated by these two components of suction peak. Both the inject effect and vortex need sufficient time to develop. When *Str* is low, the lift increment supplied by the inject effect can reach its maximum value and stability. The lift increment supplied by the vortex reaches the maximum value and decreases thereafter. A critical value of *Str* balances the both components of the lift increment and leads their sum into maximum value. When *Str* is higher than the critical value, the vortex generated by the switch from blowing and non-blowing cannot obtain sufficient time to develop, thereby leading to a lift increment penalty in the non-blowing period. By contrast, the blowing jet cannot induce a high speed-up in the upstream of blowing slot, thereby leading to a lift increment penalty in the blowing period. When $C\mu$ keep instant, the higher the duty cycle is, the lower the blowing momentum coefficient in blowing-period is. Thereby, the lift increment generated by inject effect decreases with the increase of duty cycle. Meanwhile, the time of non-blowing period is decrease with the increase of duty cycle. This leads to a lower lift increment of vortex.

4. Conclusions

Based the above discussion, the following conclusions can be made:

(1) Pulsed blowing can increase the lift coefficient of the flap evidently and improve the control efficiency of the flap.

(2) The lift coefficient of the flap increase with the increase of the Strouhal number rapidly and decrease slightly if Strouhal number continuously increase. When Strouhal number is high enough, the lift coefficient of the flap will keep instant.

(3) The maximum value of average lift coefficient of the flap and the critical Strouhal number increases with the increase of duty cycle when the average blowing momentum coefficient keeps constant.

Acknowledgements

The research work is supported by the National Natural Science Foundation of China (11272035) and the Aeronautical Science Foundation of China (2011ZA51003).

References

[1] Sellers, W. L., III, Singer, B. A., and Leavitt, L. D. 2003 Aerodynamics for revolutionary air vehicles *21st AIAA Applied Aerodynamics Conference* 2003-3785

[2] Modi VJ, Munshi SR, Bandyopadhyay G, and Yokomizo T. 1998 High-performance airfoil with moving surface boundary-layer control *J. Aircraft* **35(4)** 544-553

[3] Asrokin A, Ramly MR, and Ahmad AH. 2013 Rotating cylinder design as a lifting generator *2nd International Conference on Mechanical Engineering Research*

[4] Zhang YY, Huang DG, Sun XJ, and Wu GQ 2010 Exploration in Optimal Design of an Airfoil with a Leading Edge Rotating Cylinder 2010 *J. Thermal Science* **19(4)** 318-325

[5] Salmasi A, Shadaram A, and Taleghani AS 2013 Effect of plasma actuator placement on the airfoil efficiency at poststall angles of attack *IEEE Trans. Plasma Science* **41(10)** 3079-3085

[6] Wang JJ, Choi KS, Feng LH, Jukes TN, and Whalley RD 2013 Recent developments in DBD plasma flow control *Prog. Aerospace Sciences* **62(1)** 52-78

[7] Corke TC, Enloe CL, and Wilkinson, SP. 2010 Dielectric barrier discharge plasma actuators for flow control *Annual Rev. Fluid Mechanics* **42(1)** 505-529

[8] Houser NM, Gimeno L, Hanson RE, Goldhawk T, Simpson T, and Lavoie P. 2013 Microfabrication of dielectric barrier discharge plasma actuators for flow control *Sensors and Actuators A-Phys.* **201(1)** 101-104

[9] Zha GC, Gao W, Paxton CD. 2007 Jet effects on coflow jet airfoil performance *J. AIAA* **45(6)** 1222-1231

[10] Im HS, Zha GC, and Dano BPE 2014 Large eddy simulation of coflow jet airfoil at high angle of attack *J. Fluids Eng. Trans. Asme* **136(2)** 1101-1111

[11] Wang BY, and Zha GC 2011 Detached-eddy simulation of a coflow jet airfoil at high angle of

attack *J. Aircraft* **48(5)** 1495-1502

[12] Gan WB, Zhou Z, Xu XP, and Wang R 2014 Delayed detached-eddy simulation and application of a coflow jet airfoil at high angle of attack *Adv. Computational Modeling and Simulation* **444(1)** 270-277

[13] Wu Peng, Deng Xue Ying, and Wang Yan Kui 2012 Application of pulsed blowing technique in high-lift control surface design *Adv. Materials Research* **482(1)** 121-125

Effect of nose perturbation on asymmetric vortices over a blunt-nose body at high angle of attack

Sha Y.X.[1], Wang Y.K.[2] and Qi Z.Y.[3]
[1] Master Student, Ministry-of-Education Key Laboratory of Fluid Mechanics, Beihang University, Beijing, 100191, China
[2] Professor, Ministry-of-Education Key Laboratory of Fluid Mechanics, Beihang University, Beijing, 100191, China
[3] Ph.D Candidate, Ministry-of-Education Key Laboratory of Fluid Mechanics, Beihang University, Beijing, 100191, China

[1] E-mail: yx_sha@163.com
[2] E-mail: wangyankui@buaa.edu.cn

Abstract. To improve the maneuverability, the modern missiles of blunt-nose slender body configuration are required to flight at high angle of attack where many research show complex asymmetric vortices flow will develop over the slender body. The effect of artificial perturbation located on the nose of a blunt slender body at high angle of attack(α=50°) on asymmetric vortices has been investigated with low speed wind tunnel test. The Reynolds number(Re_D) of the experiment was 1.48×10^5. The experimental results are shown to be the following: Asymmetric flow over blunt-nose body was extremely sensitive to the machining tolerances of the nose and can be governed by the artificial perturbation; with the circumferential angle of artificial perturbation varied a period, asymmetric vortices present a behavior of single-period.

Keywords: Blunt-nose slender body; high angle of attack; asymmetric vortices; low speed; perturbation

Nomenclature

α	=	Angle of attack (degree)
β	=	Angle of sideslip (degree)
φ	=	Rolling angle of model (degree)
θ	=	Azimuthal angle around model (degree)
γ	=	Circumferential angle of artificial perturbation (degree)
ω	=	Meridian angle of artificial perturbation (degree)
ρ_∞	=	Density of freestream (kg/m³)
V_∞	=	Velocity of freestream (m/s)
q_∞	=	Dynamic pressure ($q_\infty = \rho_\infty V_\infty^2 / 2$)

IOP Conf. Series: Journal of Physics: Conf. Series **822** (2017) 012053 doi:10.1088/1742-6596/822/1/012053

Re_D = Reynolds number based on model diameter
C_p = Pressure coefficient ($C_p = (P-P_0)/ q_\infty$)
C_y = Sectional side force coefficient
D = Reference diameter of the slender body (m)
x/D = Non-dimensional pressure station

1. Introduction

With the development of combat aircraft in performance of mobility and agility, the key to winning in modern air combat often depends on air-to-air missiles (AAM). Air combat simulation results show that [1] about 40% of that possibly are close combat. America's most advanced combat aircraft F-22 has a variety of weapons, but each of them has at least two AAM, which fully illustrate the important role of AAM in the future air combat. In order to adapt to the requirement of combination of air, space and land battle [2], high maneuverability is not only one of the most significant factor to short-range combat missiles, but also the focus and difficulty in researching which all countries in the world are made unremitting efforts to. Currently, strict performances of infrared guidance [3] and requirements of thermal protection are put forward for the close combat AAM. As a result, a typical configuration of blunt-nose and slender body has been selected for this kind of high mobility missiles generally. Furthermore, they usually are required to flight at high angle of attack to improve the maneuverability as much as possible [4].

Nevertheless, many research [5-20] shows that complex asymmetric vortex flow will develop over the slender body at high angle of attack. Because of the uncertainty and unpredictability of asymmetric vortex flow [5-7], the large side force induced by it is also uncertain. Since the flow phenomena of asymmetric vortex over slender body at high angle of attack was discovered [8] about half a century, a great number of studies on finding the flow mechanism of the uncertain asymmetric vortex flow and how to control it has been carried out. The research of wind tunnel turbulence [9-12], vibration model [12-13], and roughness of model processing [14-17] indicated that asymmetric vortex is extremely sensitive to the processing roughness near the nose of the model. Since 2002, Deng X.Y. and Wang Y.K. [18-19] also have found even the model machining tolerances of the nose can result in different asymmetric flow over two models with same geometry. Since the model machining tolerances were random in fact, it made the research on the asymmetric vortex flow lagged. However, a novel idea was promoted based on the previous studies conducted by Deng X.Y. and Wang Y.K. [18-19], in which a small artificial micro-perturbation with diameter 0.2mm located close to the model tip is introduced. It was proved that asymmetric vortices flow over slender pointed-nose body was mainly determined by the artificial micro-perturbation and the asymmetric flow over two models with same geometry was repeatable. On this basis, to the question of how to control the asymmetric vortex so as to control side force, in 2004, Defence Research and Development Canada [20] conducted a series of detailed studies on pointed-nose missile flight control using micro-actuated flow effectors by means of wind tunnel experiments and numerical simulations. The results show that the setting of micro-actuated flow effectors could provide considerable side force to the missile flight at high angle of attack, and different side force could be obtained when the micro-actuated flow effectors located in different position.

However, at current, it is not enough for the understanding of asymmetric vortex flow over blunt-nose slender body. For example, if the artificial micro-perturbation could play the same role for blunt-nose slender body? What the effect of nose artificial perturbation on the asymmetric vortices flow over blunt-nose slender body at high angle of attack? Can the side force induced by this asymmetric flow be utilized to control the missile flight? In order to answer these questions, this paper investigates the effect of nose artificial perturbation on asymmetric flow over a blunt-nose slender body at high angle of attack and make better understanding of it through wind tunnel experiments.

2. Experimental Setup

2.1 Wind Tunnel

The wind-tunnel experiments were performed in the closed-loop, low-speed wind tunnel with a test section 1.5×1.5m at Beihang University. The test section was equipped with a C-strut model support fixed on a rotating platform to adjust the model attitude. It is equipped with a servo motor at the end of the model to change the roll angle φ. The maximum freestream velocity of test section is 60m/s and the turbulence level is 0.08%. The Reynolds number for the experiment was $Re_D=1.48\times10^5$ based on diameter of afterbody (D=66mm).

2.2 Model

The model used, sketched in Figure 1, is a blunt-nose slender body with the overall length of L=1060mm and the afterbody diameter of D=66mm. The fineness ratio of model is L/D=16.1. The radius of blunt nose is r=27.2mm. Between the nose and cylindrical afterbody, there is a conical part with the apex angle of 18 deg. The pressure taps are located at 7 stations along the body axis as Figure 1 and each station has 24 equal spaced taps. The wind tunnel test model is shown in Figure 2.

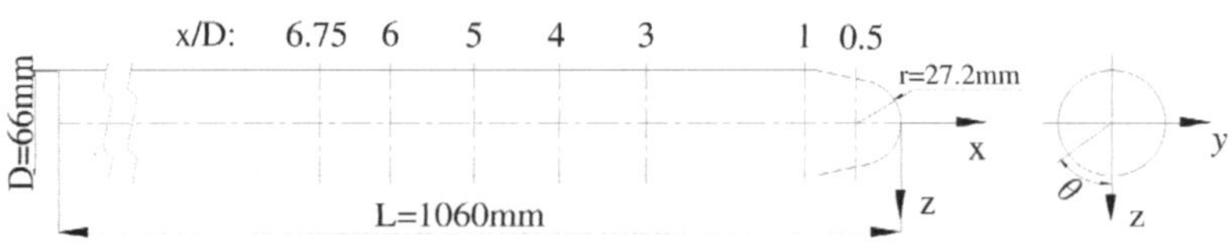

Figure 1. Sketch of test model

Figure 2. Wind tunnel test model

3. Results and Discussion

3.1 Asymmetric vortices flow over blunt-nose slender body at high angle of attack

The side force is usually used to quantify the overall flow asymmetry. In this paper, sectional side forces (C_y) is calculated by the integral of pressure distribution. Figure 3 shows the sectional side force coefficient changing with roll Angle of model at α=50°, β=0°, V_∞=35m/s.

As can be seen from Figure 3 (a), the sectional side force coefficient present irregular variation, not only in the magnitude but also the direction, even in the same cross-section (*x/D*=3) since the model is slender body in fact. Figure 3 (b) also shows the sectional side force distributions along slender body model at various roll angles. From the figure the C_y-*x/D* curves do not coincide with another. Because the side force is induced by asymmetric vortex, it can be seen that asymmetric vortices changing with roll Angle of model present unrepeatable and non-deterministic.

The results from the studies of pointed-nose slender body model [18-19] showed that asymmetric vortex flowfield over slender body was extremely sensitive to the natural disturbances which were produced by geometric imperfect machining tolerances of the nose. As a result, when the model was rolled at a different angle, it would result in a different response of asymmetric vortices as shown in Figure 3.

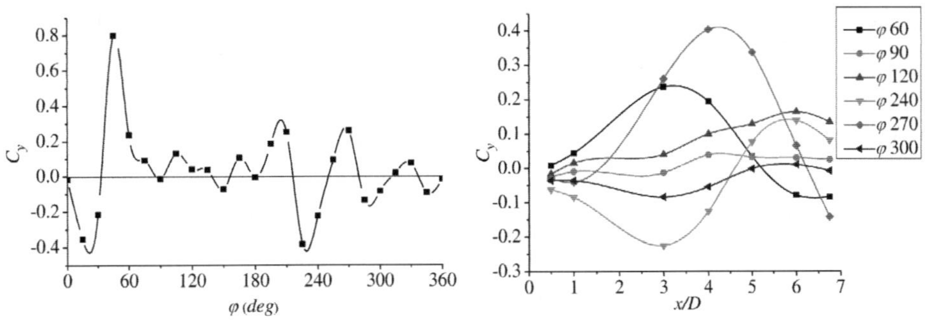

(a) *x/D*=3 (b) C_y-*x/D*

Figure 3. Sectional side force coefficient changing with roll Angle of model (α=50°, β=0°, V_∞=35m/s)

3.2 Effect of artificial perturbation on the asymmetric vortices flow

In order to investigate the effect of artificial perturbation on the asymmetric vortices flow over blunt-nose slender body model, a small artificial perturbation located on the nose of the model is introduced.

In this paper, a small wood cube with length of L_p=2mm, width of W_p=1mm and height of H_p=1mm is selected, which is shown in Figure 4. The location of artificial perturbation is present in Figure 5, in which ω represents the meridian angle and γ is the circumferential angle.

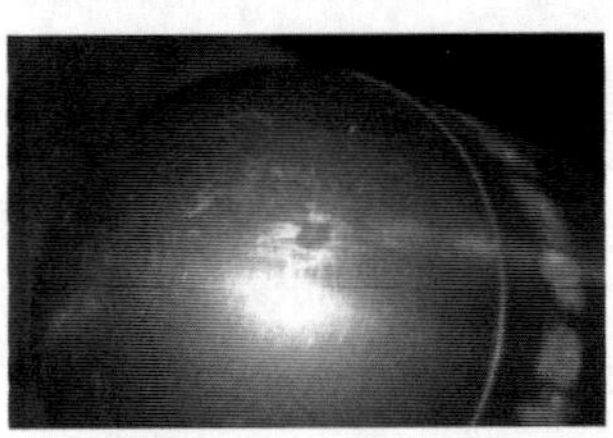

Figure 4. Artificial perturbation model

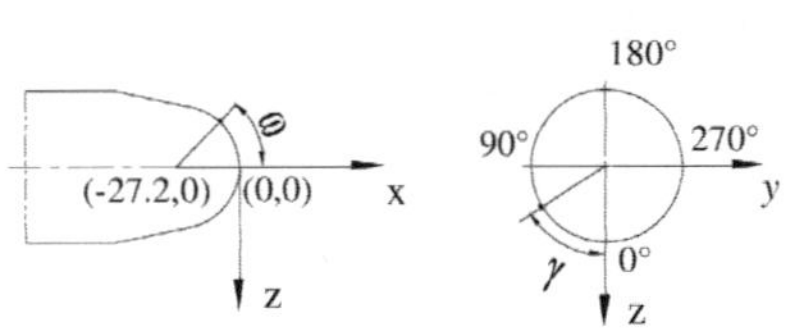

Figure 5. Location of artificial perturbation

Figure 6 presents the characteristic of sectional side force coefficient changing with circumferential angle of artificial perturbation for the conditions of α=50°, β=0°, V_∞=35m/s. Compared with Figure 3, it can be proved that the behaviors of asymmetric vortices relate to the location of artificial perturbation, and it means, compare to natural disturbances, the artificial perturbation on the nose of model has the main control to the asymmetric vortices flow. Figure 6 also shows that with the circumferential angle of artificial perturbation varies a period, asymmetric vortices present a behavior of single-period.

Figure 7 and Figure 8 give the pressure distributions of section x/D=3 and sectional side force along slender body at 6 different circumferential angle of artificial perturbation, respectively. When the circumferential angle (γ) is between 0° to 180°, the asymmetric vortices flowfield over the model exhibit Left Vortex [19] and there is a strong suction peak near leeside of θ=165°. When γ is between 180° to 360°, the asymmetric vortices exhibit Right Vortex [19] and the suction peak appear in leeside of θ=195°.

Therefore, it is feasible that the asymmetric vortices flowfield over the blunt-nose slender body can be controlled by adjusting the circumferential location of artificial perturbation on the nose so that a novel side force control technique can be promoted by adjusting artificial perturbation.

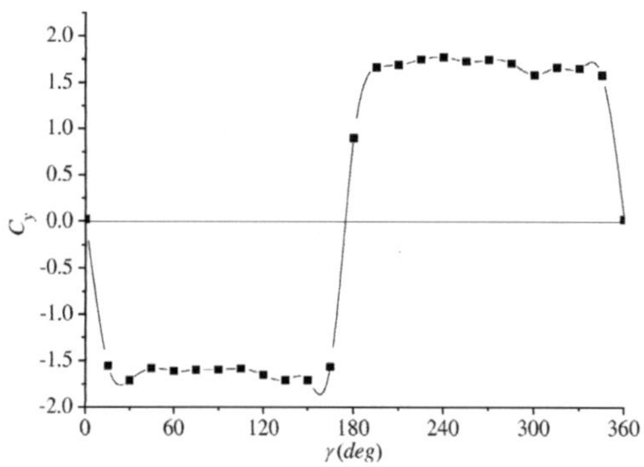

Figure 6. Sectional side force coefficient vs circumferential angle of artificial perturbation (α=50°, β=0°, V_∞=35m/s, ω=40°, x/D=3)

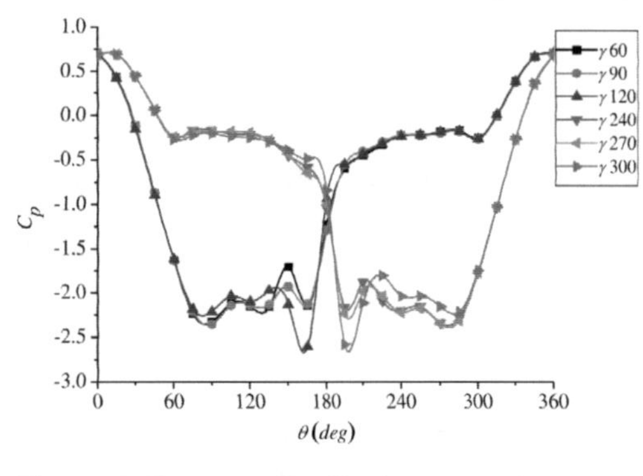

Figure 7. Pressure distributions at various circumferential angle of artificial perturbation (α=50°, β=0°, V_∞=35m/s, ω=40°, x/D=3)

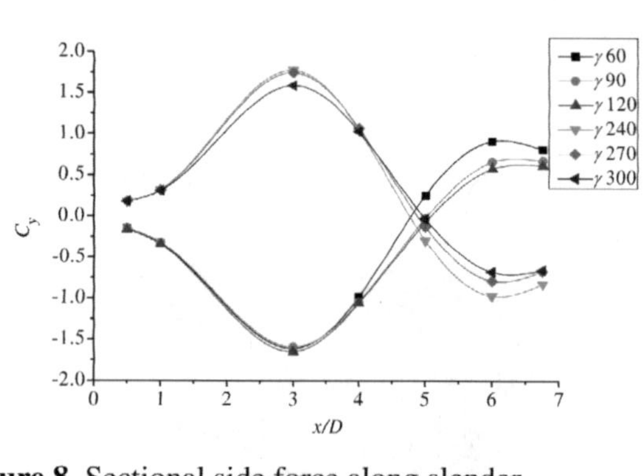

Figure 8. Sectional side force along slender body at various circumferential angle of artificial perturbation (α=50°, β=0°, V_∞=35m/s, ω=40°)

4. Conclusions

Based the above discussion, the following conclusions can be made:

(1) Asymmetric vortex flowfield over blunt-nose slender body was extremely sensitive to the natural disturbances which were produced by geometric imperfect machining tolerances of the nose.

(2) Compare to natural disturbances, the artificial perturbation on the nose of model has the main control to the asymmetric vortices flow.

(3) With the circumferential angle of artificial perturbation varied a period, asymmetric vortices present a behavior of single-period. As a result, a novel side force control technique can be promoted by adjusting artificial perturbation.

Acknowledgements

The project is supported by the National Natural Science Foundation of China (11472028), the Open fund from State Key Laboratory of Aerodynamics and Shanghai Aerospace Innovation Fund (SAST201443).

References

[1] Fan H T, Liu D J. Development trends of short-range dogfight IR air to air missile[J]. I. and L. Engineering, 10, 2005
[2] Zhang B J, Yang X and Ma M. The foreign development of air-to-air missile[J]. China E and Trade H, 73-75, 9, 2014
[3] Fan H T, Liu D J. The development prospects of close combat about infrared air-to-air missile[J]. I. and L. Engineering, 5, 2005
[4] Zhu B L. Air-to-air missile and the modern air combat[B]. Air-to-air missile Issue
[5] Deng X Y, Wang Y K. Asymmetric vortices flow over slender body and its active control at high angle of attack. ACTA Mechanica Sinica, 20(6): 567-579, 2004.
[6] Chen X R, Deng X Y, Wang Y K. Influence of nose perturbations on behaviors of asymmetric vortices over slender body. ACTA Mechanica Sinica, 18(6): 581-593, 2002.
[7] Deng X Y, Tian W, Ma B F, et al. Recent progress on the study of asymmetric vortex flow over slender bodies. ACTA Mechanica Sinica, 24(5): 475-487, 2008.
[8] Allen, H.J. and Perkins, E.W., Characteristics of Flow over Inclined Bodies of Revolution, NACA RM A50L07, 1951.
[9] Hunt B. L. Asymmetric Vortex Forces and Wakes on Slender Bodies[R]. AIAA-82-1336
[10] Ericsson L. E. Vortex Unsteadiness on Slender Bodies at High Incidence[J]. J. Aircraft, 1987, V24(4): 319-326
[11] Dexter P. C. A Study of Asymmetric Flow over Slenr Bodies at High Angles of Attack in a Low Turbulence Environment[R]. AIAA-84-0505
[12] Zilliac G. G., Degani D., Tobak M. Asymmetric Vortices on a Slender Body of Revolution[J]. AIAA Journal, 1991, V29(5): 667-675
[13] Ericsson L. E., Reding J. P. Aerodynamic Effects of Asymmetric Vortex Shedding from Slender Bodies[R]. AIAA-85-1797
[14] Levy Y., Hesselink L., Degani D. A Systematic Study of the Correlation Between Geometrical Disturbances and Flow Asymmetries[R]. AIAA-95-0365

[15] Bridges D. H., Hornung H. G. Elliptic Tip Effects on the Vortex Wake of an Axisymmetric Body at Incidence[J]. AIAA Journal, 1994, V32(7): 1437-1445

[16] Ward K. C. Boundary Layer Separation and the Vortex Structures around an Inclined Body of Revolution[R]. AIAA-87-2276

[17] Moskovitz C. Experimetal Investigation of a New Device to Control the Asymmetric Flowfield on Forbodies at Large Angles of Attack[R]. AIAA-90-0069

[18] Deng Xueying, Chen Xuerui, Wang Yankui, et.al., Influence of Nose Perturbations on Behaviors of Asymmetric Vortices over Slender Body, AIAA paper 2002-4710, 2002.

[19] Deng Xueying, Wang Yankui et.al., Deterministic Flow Field and Flow structure Model of Asymmetric Vortices over Slender Body, AIAA paper 2003-5475, 2003.

[20] Wong F.C. Missile Flight Control Using Micro-actuated Flow Effectors, Review of fiscal year 2004/2005 progress. DRDC Valcartier TN 2005-282

IOP Conf. Series: Journal of Physics: Conf. Series **822** (2017) 012054 doi:10.1088/1742-6596/822/1/012054

Effect of splitter plate on fluid flow characteristics past a triangular cylinder

Sunakraneni Soumya[1] and K Arul Prakash[2]

[1,2] Fluid Mechanics Laboratory, Department of Applied Mechanics,Indian Institute of Technology Madras, Chennai 600 036, India

[2]corresponding author, Associate professor

E-mail: `arulk@iitm.ac.in`[2]

Abstract. In the present numerical investigation, the effect of splitter plate length on fluid flow characteristics past an equilateral triangular cylinder apex facing the flow configuration is studied. Simulations are carried out by using in-house solver based on Streamline Upwind/Petrov-Galerkin (SUPG) finite element method in laminar regime. The range of Reynolds numbers (Re) based on side (b) of the triangular cylinder is varied from 50 to 200 and the splitter plate length (L) is varied from $L/b = 0$ to $= 6$. The effect of splitter plate length on parameters like lift and drag are studied. It is observed that drag decreases with the attachment of splitter plate. In addition, vortex shedding suppression is observed for $Re = 50 - 150$ whereas, for $Re = 200$ vortex shedding suppression is not observed even for plate length of $L/b = 6$. The obtained results indicate that for a specific value of Re drag is minimized at a particular splitter plate length. The drag is minimized approximately from 9% to 57% based on Re by attaching a splitter plate to the triangular cylinder.

1. Introduction

Bluff bodies offer more resistance to the fluid flow and suffer oscillatory forces, and the fluid flow over bluff bodies is very complex. Moreover, the phenomenon of vortex shedding is observed in most of the fluid flows over bluff bodies. As a result, the forces acting on the body, and the transport mechanisms around the body are influenced by vortex shedding. The fluctuating periodic forces generated on the bluff body due to the vortex shedding leads to damage to the structure thereby shortening the life of the structure. As a consequence, techniques to suppress vortex shedding and reducing drag has remained a potential area for investigation. In one of the foremost experimental study by [1] it was found that a splitter plate length $5D$ attached to the circular cylinder (D is cylinder diameter) suppressed the vortex shedding and reduced the pressure drag to approximately 63% of the the value for the plain cylinder. In addition, the effect of gap between the cylinder and splitter plate was also investigated by [1]. [2]

IOP Conf. Series: Journal of Physics: Conf. Series **822** (2017) 012054 doi:10.1088/1742-6596/822/1/012054

conducted experiments on different shapes of bluff bodies attached to the splitter plate to examine the vortex shedding characteristics and found that vortex shedding from bluff bodies with extended splitter plates is characterized by impinging-shear-layer instability. Numerical study done by [3] for fluid flow past circular cylinder attached with splitter plate found that for each *Re* there exists a critical length of the splitter plate where the complete suppression of vortex shedding is obtained. Drag is decreased with splitter plate attachment and there exists an optimum length for each *Re* where minimum drag is experienced. It is observed from the literature that the majority of studies carried out on the effect of splitter plate on laminar wakes of canonical body shapes such as circular and square cylinders ([3], [4]). In recent times fluid flow past non canonical shapes are getting more attention due to its novel flow features. The novelty of the present study is to find the critical splitter plate length where the vortex shedding disappears and minimum drag is obtained for each *Re*. To the best of authors knowledge, studies are not available on triangular cylinder attached with a splitter plate in low *Re* regime.

2. Problem definition

The problem under investigation is to study the fluid flow characteristics over an unconfined equilateral triangular cylinder attached with a thin rigid splitter plate of lengths (L/b) varying from 0.0 to 6.0 (in steps of 0.5) with thickness (t_h/b) = 0.1. The thickness of the splitter plate is constant throughout the study. Fig.1 shows the computational domain with boundary conditions of present study. In the present simulations side of triangular cylinder (b) is taken as characteristic dimension and free stream velocity U_∞ is the characteristic velocity. The simulations are carried out for *Re* in the range of 50-200. The fluid flow is considered as two-dimensional, laminar and incompressible.

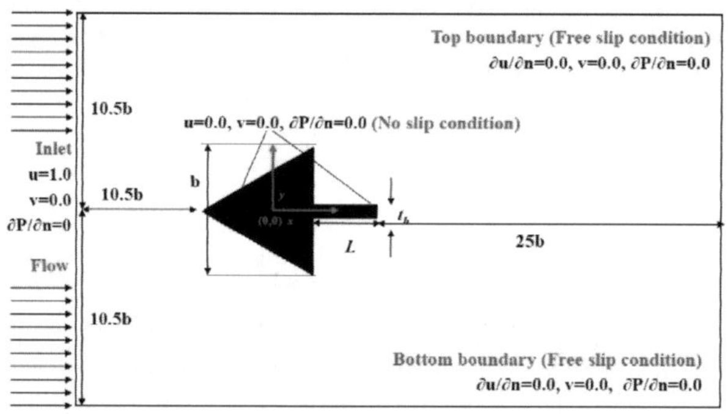

Figure 1: Computational domain with boundary conditions

2.1. Governing equations

The governing equations for unsteady, incompressible, laminar and two-dimensional flow with constant viscosity fluid in non-dimensional form are as follows.

Conservation of mass :

$$\frac{\partial u}{\partial x} + \frac{\partial v}{\partial y} = 0 \tag{1}$$

Conservation of momentum in x direction :

$$\frac{\partial u}{\partial \tau} + u\frac{\partial u}{\partial x} + v\frac{\partial u}{\partial y} = -\frac{\partial p}{\partial x} + \frac{1}{\mathrm{Re}}\left(\frac{\partial^2 u}{\partial x^2} + \frac{\partial^2 u}{\partial y^2}\right) \tag{2}$$

Conservation of momentum in y direction :

$$\frac{\partial v}{\partial \tau} + u\frac{\partial v}{\partial x} + v\frac{\partial v}{\partial y} = -\frac{\partial p}{\partial y} + \frac{1}{\mathrm{Re}}\left(\frac{\partial^2 v}{\partial x^2} + \frac{\partial^2 v}{\partial y^2}\right) \tag{3}$$

In the above equations, geometrical parameters (x, y) are non-dimensionalized using side of triangular cylinder b, velocity is non-dimensionalized using free-stream velocity U_∞, pressure is non-dimensionalized by $\rho \mathrm{U}^2{}_\infty$, and time is non-dimensionalized by $\frac{b}{\mathrm{U}_\infty}$. Here, ρ is density of the fluid, and μ is the dynamic viscosity of the fluid. As depicted in Fig.1, the origin of the domain is located at the center of the cylinder. The outlet boundary is always maintained at a distance of $25b$ from the trailing edge of the splitter plate. Hence, the downstream distance is adjusted in such a way that it varies with respect to the splitter plate length.

3. Numerical procedure and validation

Eulerian-Velocity correction approach is used to solve the governing equations iteratively and discretization is done by using SUPG based finite element method. To avoid node to node oscillations in the results when standard Galerkin method is applied to convection dominated flows, [5] proposed a method in which discontinuous weighting functions are used and the method is known as SUPG based finite element method. The details of the numerical formulation are given in [6, 7]. The present algorithm has been validated well with several experimental and numerical studies by [6].

Table 1: Comparison of various fluid flow parameters with available literature values for flow past a square cylinder at $Re = 100$

Reference	Cd_{p_avg}	Cd_{avg}	Cl_{rms}
[8]	1.441	1.4878	0.188
[9]	1.4436	1.4936	0.1922
[10]	-	1.64	-
[11]	-	1.61	-
[12]	-	1.8	-
[13]	-	1.47	0.156
[14]	-	1.44	0.152
[15]	-	1.47	0.159
[16]	-	1.5287	0.1928
[17]	-	1.542	0.1915
Present	1.464	1.534	0.189

Validation study of present scheme is done for fluid flow past a square cylinder. It is evident from Table 1 that the drag and lift values obtained from present numerical algorithm are in good agreement with the literature values.

4. Results and discussion

The performance of a splitter plate as a drag reduction device as well as a vortex shedding suppression device significantly depends on Reynolds number (Re), cylinder shape and length of the splitter (L/b). The combined effect of these parameters on fluid flow characteristics of an equilateral triangular cylinder is elucidated with the aid of instantaneous vorticity contours. Fig.2 depicts instantaneous vorticity contours.

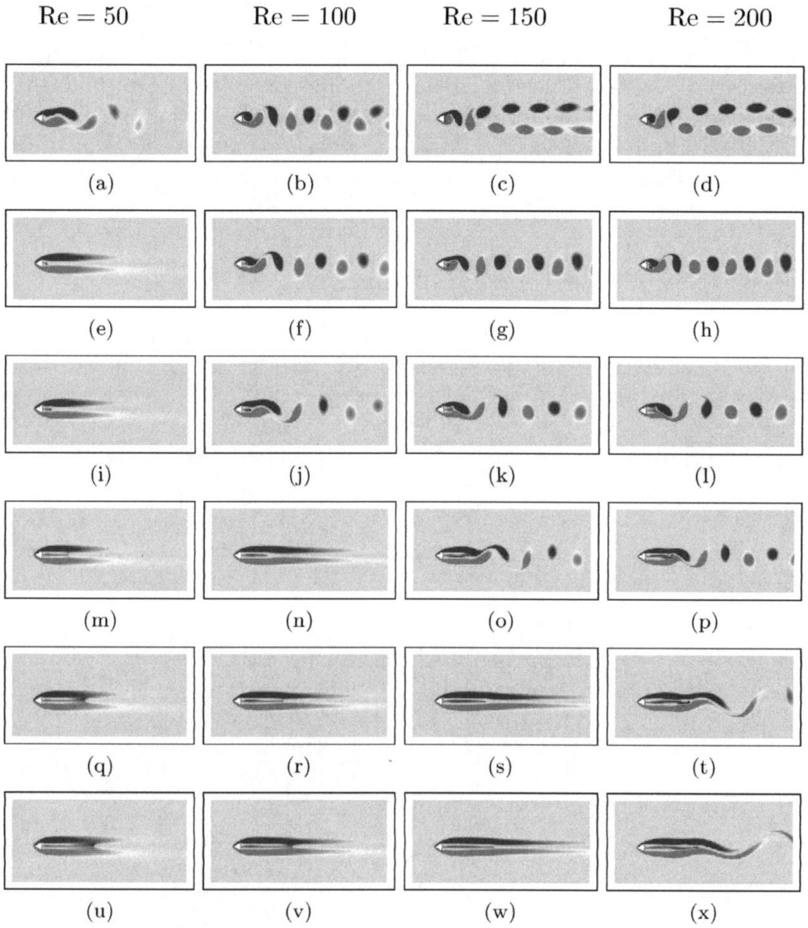

Figure 2: Instantaneous vorticity contours for an equilateral triangular cylinder attached with splitter plate of various lengths.(a-d) $L = 0.0b$, (e-h) $L = 0.5b$, (i-l) $L = 1.0b$, (m-p) $L = 3.0b$, (q-t) $L = 5.0b$, (u-x) $L = 6.0b$

IOP Conf. Series: Journal of Physics: Conf. Series **822** (2017) 012054 doi:10.1088/1742-6596/822/1/012054

Wakes of all bluff bodies are similar in structure ([1]) as far as instabilities are concerned. From Fig.2 it is to be observed that when a small plate of length $L = 0.5b$ is attached to the cylinder the separated shear layers are elongated and their point of interaction is located downstream in the wake region. A small-scale vortex is observed at the trailing edge of the plate. With the further increment of plate length the separated shear layers are convected more downstream and shedding occurs until $L < L1_{cri}$, and at $L = L1_{cri}$ (example: Fig.2n) vortex shedding is suppressed. $L1_{cri}$ is defined as the minimum plate length for which vortex shedding suppression is attained. From Fig.2 it is worth noticing that, for higher Re lengthier plates are required to obtain vortex shedding suppression. Further it is also observed that for Re = 200 even with a plate length of $L = 6.0b$ shedding is not suppressed but shear layers are elongated which results in reduction of vortex shedding frequency. The fluid flow is unsteady in nature for plate length of $L < L1_{cri}$ whereas for $L \geq L1_{cri}$ flow is steady in nature. These various flow patterns can be clearly identified from Fig. 3, which gives parametric space graph of various regimes depending on plate length (L/b) and Reynolds number (Re). From Fig.2 it is inferred that due to splitter plate attachment vortex formation process is altered, wake region is narrowed, vortex formation region is elongated.

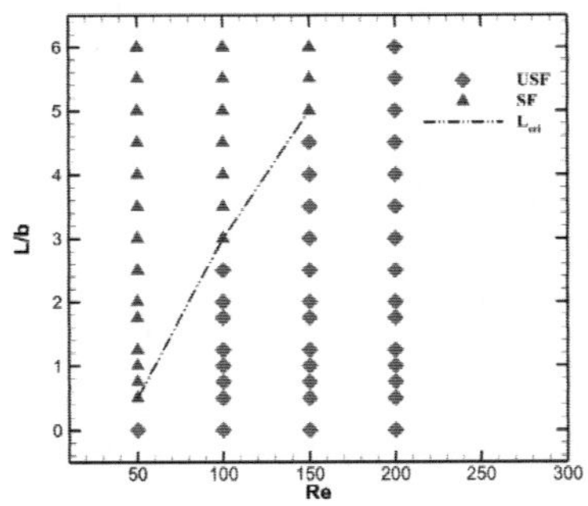

Figure 3: Parametric space graph of various fluid flow regimes

Figure 4 shows variation of viscous drag, pressure drag and total drag with splitter plate length. From Fig.4a it can be observed that skin friction drag decreases as the plate length increases upto certain length of the plate, and then with further increment of plate length skin friction drag increases, due to increased plate length causes additional frictional drag. From Fig.4b it is observed that pressure drag decreases almost monotonically as plate length increases upto $L = L1_{cri}$, beyond $L1_{cri}$ pressure drag remains constant. Total drag is summation of skin friction drag and pressure drag, hence it shows the combined result of both. Fig.4c shows total drag variation with plate length. From Fig. 4 it is observed that for higher Re, significant drag reduction is observed when compared to lower Re. It is found that with the attachment of splitter plate reduction in total drag $\simeq 9\%$ to $\simeq 57\%$ is attained from cylinder without splitter plate.

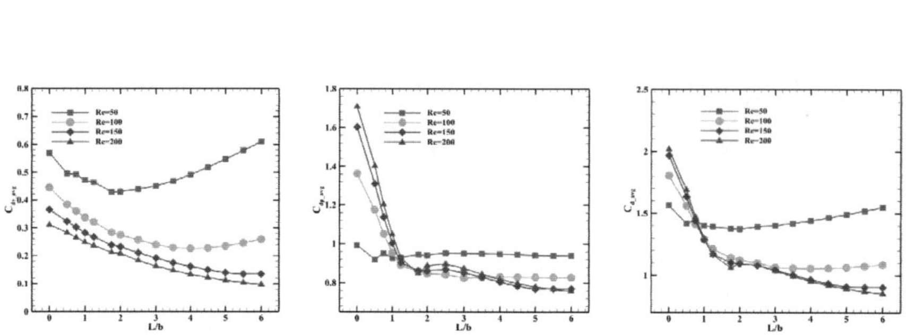

(a) Cd_{v_avg} (b) Cd_{p_avg} (c) Cd_{avg}

Figure 4: Effect of splitter plate length on drag

5. Conclusions

In the present numerical investigation, study has been carried out to notice the effect of splitter plate on equilateral triangular cylinder on fluid flow characteristics for low Re. From this study it is observed that vortex shedding suppression is observed for $Re = 50 - 150$. Decrease in drag is observed with splitter plate attachment.

References

[1] Roshko, A., 1954. Tech. Rep. 3169, NACA.
[2] Nakamura, Y., 1996. *Journal of Fluids and Structures* , **10**, pp. 147–158.
[3] Kwon, K., and Choi, H., 1996. *Physics of Fluids*, **8**, pp. 479–486.
[4] Mat Ali, M. S., Doolan, C. J., and Wheatley, V., 2011. *Physics of Fluids*, **23** (3), pp. 033602–12.
[5] Brooks, A. N., and Hughes, T. J., 1982. *Computer Methods in Applied Mechanics and Engineering*, **32**, pp. 199 – 259.
[6] Prakash, K. A., Biswas, G., and Rathish Kumar, B., 2006. *International journal of heat and mass transfer*, **49**(23), pp. 4633–4652.
[7] Prakash, K. A., Biswas, G., and Rathish Kumar, B., 2012. *International journal of Numerical Analysis and Modelling,series B*, **3**, pp. 259–269.
[8] Sahu, A. K., Chhabra, R., and Eswaran, V., 2009. *International Journal of Heat and Mass Transfer*, **52**(3), pp. 839–850.
[9] Sharma, A., and Eswaran, V., 2004. *Numerical Heat Transfer, Part A: Applications*, **45**(3), pp. 247–269.
[10] Davis, R., and Moore, E., 1982. *Journal of Fluid Mechanics*, **116**, pp. 475–506.
[11] Franke, R., Rodi, W., and Schönung, B., 1990. *Journal of Wind Engineering and Industrial Aerodynamics*, **35**, pp. 237–257.
[12] Kelkar, K. M., and Patankar, S. V., 1992. *International Journal for Numerical Methods in Fluids*, **14**(3), pp. 327–341.
[13] Sohankar, A., Norberg, C., and Davidson, L., 1998. *International journal for numerical methods in fluids*, **26**(1), pp. 39–56.
[14] Cheng, M., Whyte, D., and Lou, J., 2007. *Journal of Fluids and Structures*, **23**(2), pp. 207–226.
[15] Lankadasu, A., and Vengadesan, S., 2008. *International Journal of Heat and Fluid Flow*, **29**(4), pp. 1054–1059.
[16] Sen, S., Mittal, S., and Biswas, G., 2011. *International Journal for Numerical Methods in Fluids*, **67**(9), pp. 1160–1174.
[17] Dey, P., and Das, A. K., 2015. *Engineering Science and Technology, an International Journal*.

IOP Conf. Series: Journal of Physics: Conf. Series **822** (2017) 012055 doi:10.1088/1742-6596/822/1/012055

Error propagation in the procedure of pressure reconstruction based on PIV data

Zhongyi Wang[1], Qi Gao[1*], Runjie Wei[2] and Jinjun Wang[1]

[1] Key Laboratory of Fluid Mechanics, Ministry of Education, Beijing University of Aeronautics and Astronautics, Beijing, China
[2] MicroVec., Inc, Beijing, China

[*]Email: qigao@buaa.edu.cn

Abstract. Pressure reconstruction based on particle image velocimetry (PIV) has become a popular technique in experimental fluid mechanics. Errors in the raw velocity field significantly affect the accuracy of pressure gradient field and further reduce the quality of reconstructed pressure. Thus, error propagation deserves a serious concern. The format and magnitude of errors are investigated using a probability density function (PDF) based method. Theoretical derivation and numerical validation are operated in both Eulerian and Lagrangian descriptions. The influence of spatial and temporal resolutions on error propagation is discussed. A criterion of parameter selection for error suppression in the Lagrangian method is proposed. The results show that error propagation in the Lagrangian method has definite format and magnitude which makes this method more suitable for error control through specific treatment. Time interval during pressure reconstruction should be carefully determined since a large uncertainty appears when the time interval is employed improperly.

1. Introduction

Particle image velocimetry (PIV) based pressure reconstruction has become an important technique in modern experimental fluid mechanics [1]. This technique is realized by solving the Navier-Stokes (N-S) equations. Different methods have been developed for pressure reconstruction which can be categorized into two types: the direct path integration method [2] and the Poisson solver method [3, 4]. For every method, the calculation of pressure gradient is unavoidable. Since the viscous term is generally neglected for at least two-order smaller than the other terms in the N-S equations, the main task for pressure integration is to calculate the material acceleration [1]. Noise propagation from velocity to material acceleration has been investigated by several researchers [5, 6] which mainly focused on the aspect of uncertainty. The format of the errors during noise propagation and how the relevant parameters, such as the temporal and spatial resolutions, affect the error propagation process have not been well investigated.

Therefore, error propagation in the procedure of pressure reconstruction is discussed using a statistical tool of probability density function (PDF) in this work, through which both the magnitude and format in error propagation can simultaneously be estimated. Different error propagation mechanisms are found between the Eulerian and Lagrangian methods. Assessments are conducted using a 2D synthetic solid body rotation flow. Effect of the spatial and temporal resolutions on error propagation is investigated and a criterion for determining the measurement parameters is provided.

2. Methodology

2.1. Basic theory for pressure reconstruction
The standard N-S equation for incompressible flow is [1]

$$\nabla p = -\rho \frac{\mathrm{D}\mathbf{u}}{\mathrm{D}t} + \mu \nabla^2 \mathbf{u}, \tag{1}$$

where the viscous term is neglected in the following discussion for its small magnitude. Both the Eulerian and Lagrangian methods are used to calculate the pressure gradient field (the material acceleration term) in this paper which can be expressed respectively as [7]

$$\frac{\mathrm{D}\mathbf{u}}{\mathrm{D}t} = \frac{\partial \mathbf{u}}{\partial t} + (\mathbf{u}\cdot\nabla)\mathbf{u}, \tag{2}$$

$$\frac{\mathrm{D}\mathbf{u}}{\mathrm{D}t} = \frac{\mathrm{d}\mathbf{u}_p(t)}{\mathrm{d}t} = \frac{\mathrm{d}\mathbf{u}(\mathbf{x}_p(t),t)}{\mathrm{d}t}. \tag{3}$$

After obtaining pressure gradient, pressure is directly integrated using the following equation

$$p(\mathbf{s}_{\mathrm{end}}) = p(\mathbf{s}_{\mathrm{start}}) + \int_{\mathbf{s}_{\mathrm{start}}}^{\mathbf{s}_{\mathrm{end}}} \nabla p \cdot \mathrm{d}\mathbf{s}, \tag{4}$$

In this paper, the Eulerian method uses a central difference scheme to calculate both the local and convective acceleration [8]. The Lagrangian method applies a pseudo-tracing method [2]. Finally, pressure is integrated through a widely used omni-directional integral algorithm [2, 9].

2.2. Error propagation derivation
Two reasonable assumptions about the PIV determined velocity field are proposed:

- The value of error on every grid is independent from its velocity.
- The value of error on every grid is spatially and temporally independent from each other.

It should be reminded that the locally correlated errors due to overlap in PIV correlation would not affect the spatial independency of PIV errors in global statistics analysis. Velocity error ε_u with a Gaussian distribution as follows is studied [10]

$$f(\varepsilon_u) = \frac{1}{\sqrt{2\pi}\sigma}\exp(-\frac{(x-\mu)^2}{2\sigma^2}), \tag{5}$$

Equation (5) is denoted as $\varepsilon_u \sim N(\mu,\, \sigma^2)$. Here, μ and σ^2 are the mathematical expectation and variance, respectively. μ is generally set as 0.

2.2.1. Error propagation to pressure gradient. Error propagation to pressure gradient is derived in both the Eulerian and Lagrangian systems.

(1) Eulerian perspective. With a central difference scheme, the errors of local acceleration ε_u^{el} along x-direction (only one direction is considered for easy explanation) can be derived as

$$\varepsilon_a^{el} = \frac{\partial(u|_{\mathbf{x},t} + \varepsilon_u^e|_t)}{\partial t} - \left.\frac{\partial u}{\partial t}\right|_{\mathbf{x},t} \approx \frac{\varepsilon_u^e|_{t+\Delta t} - \varepsilon_u^e|_{t-\Delta t}}{2\Delta t}. \tag{6}$$

Here, $\varepsilon_u^e|_t$ and $\varepsilon_u^e|_{t\pm\Delta t}$ are random errors of $u|_{\mathbf{x},t}$ and $u|_{\mathbf{x},t\pm\Delta t}$, respectively. Since the random errors satisfy the PDF of $N(\mu,\, \sigma^2)$, the PDF of ε_a^{el} could be obtained as

$$\varepsilon_a^{el} \sim N(\mu\Delta t^{-1}, \sigma^2\Delta t^{-2}/2). \tag{7}$$

Equation (7) indicates that errors propagating from Gaussian velocity errors to local acceleration don't change the format. Definite PDF could be calculated for ε_a^{el} (ignoring the truncation errors). This result could also be obtained using the central limit theorem [11].

However, the nonlinear convective acceleration in the Eulerian system produces unclear error distribution. In the same way as the derivation of ε_a^{e1}, errors of the x-component of the convective acceleration could be obtained as

$$\varepsilon_a^{e2} \approx \left.\frac{\partial(u\varepsilon_u^e)}{\partial x}\right|_{x,t} \approx \frac{(u\varepsilon_u^e)\big|_{x+\Delta x,t} - (u\varepsilon_u^e)\big|_{x-\Delta x,t}}{2\Delta x}. \quad (8)$$

In equation (8), a high-order smaller term has been truncated. It indicates that the distribution of ε_a^{e2} depends on both u and ε_u^e. As u always changes with different flows, the PDF of ε_a^{e2} is hard to be determined in advance. Therefore, the PDF of the full material acceleration noise by the Eulerian method is also indefinable which makes it a difficult task to design a targeted error suppression treatment.

(2) Lagrangian perspective. The Lagrangian method calculates acceleration by tracking virtual particles [2, 9]. Random noises propagating from velocity to the Lagrangian acceleration are derived as

$$\varepsilon_a^l = \frac{D(\mathbf{u}_p|_t + \varepsilon_{\mathbf{u}}^l)}{Dt} - \frac{D\mathbf{u}_p|_t}{Dt} \approx \frac{\varepsilon_{\mathbf{u}_p}^l\big|_{t+\Delta t} - \varepsilon_{\mathbf{u}_p}^l\big|_{t-\Delta t}}{2\Delta t}. \quad (9)$$

Equation (9) has the similar form as the local acceleration error which results in

$$\varepsilon_a^l \sim N(\mu\Delta t^{-1}, \sigma^2\Delta t^{-2}/2). \quad (10)$$

2.2.2. Error propagation to pressure. The pressure error ε_p is related to the pressure gradient error $\varepsilon_{\nabla p}$ by a linear equation. In the Lagrangian method, $\varepsilon_{\nabla p} = \varepsilon_a^l$. Dividing the integration field into n intervals with a space interval of Δx and denoting the averaged value of $\varepsilon_{\nabla p}$ in the ith interval as $(\varepsilon_{\nabla p})_i$, we can get

$$\varepsilon_p\big|_{\mathbf{x}_n} = \lim_{n\to\infty}\sum_{i=1}^{n} \varepsilon_{\nabla p}\big|_{\mathbf{x}_i}\Delta x. \quad (11)$$

Finally, the PDF of the pressure errors is derived as

$$\varepsilon_p \sim N(n\mu\Delta x\Delta t^{-1}, n\sigma^2\Delta x^2\Delta t^{-2}/2), \quad (12)$$

where n denotes the integration length. For a measurement domain with a grid size of $n_x \times n_y$, the largest variance for the pressure errors from the expectation should be less than $\sqrt{n_x^2 + n_y^2}\,\sigma^2\Delta x^2\Delta t^{-2}/2$. The Eulerian method is not included in this pressure error analysis for its unclear error distribution.

In a summary, errors of the pressure gradient and pressure using the Lagrangian approach has a certain distribution which is independent from the flow velocity. However, the Eulerian approach suffers from unclear acceleration errors due to the convective term. This special characteristic of Lagrangian approach makes it a more preferable method in practice since targeted treatment could be designed for noise reduction. Therefore, the Lagrangian method is specifically investigated later.

3. Results and discussion

Assessment is conducted using a synthetic solid body rotation flow [2] since quantitative evaluation of errors is impossible in experiment with unknown noise value. Two main tests are shown: one is the qualitative investigation about error sensitivity of different methods; the other one is a quantitative validation about the PDF of errors as derived in section 2, as well as a criterion for measurement parameter selection. Synthetic velocity fields with a size of 101×101 grids are produced. The time and space intervals are set as dt = 0.01 s and h = 1 in both directions. Rotation angular velocity is ω = 4 rad/s. Gaussian errors as equation (5) are added into accurate velocity fields and 3σ value is used to represent the maximum of velocity errors. As the flow is axisymmetric, only the x-component of a vector is discussed in the following description. Parameters of the Gaussian distribution are chosen as $\mu = 0$ and $3\sigma_u = 0.02\sqrt{\overline{u_0^2}}$. $\overline{u_0^2}$ represents the average of the accurate velocity component u_0^2 .

3.1. Error propagation in both the Eulerian and Lagrangian system

Contours of the theoretical and calculated total material accelerations ($\|\mathbf{Du}/\mathrm{D}t\|$) are given in figure 1. As can be seen, the Eulerian material acceleration suffers from more fluctuation than the Lagrangian acceleration which indicates that the Eulerian method is more prone to errors [5, 7]. Correlation coefficients between each of the calculated acceleration and the theoretical acceleration are 0.86 and 0.95, respectively. Although the Lagrangian method has better accuracy than the Eulerian method, obvious fluctuations caused by errors are still visible which indicates that an error suppression treatment is necessary.

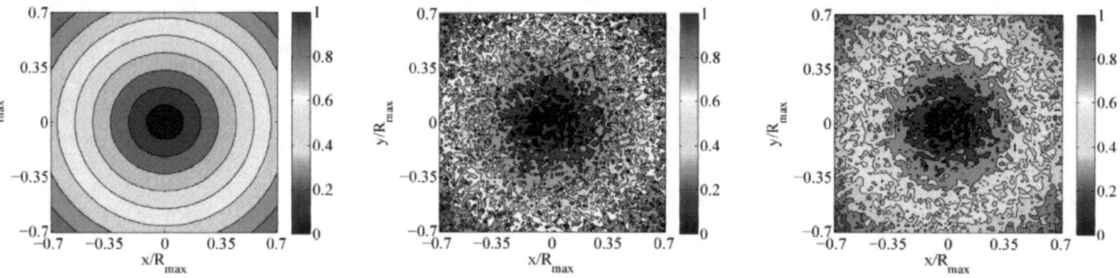

(a) Theoretical acceleration (b) Eulerian acceleration (c) Lagrangian acceleration

Figure 1. Comparison between total material acceleration. $\|\mathbf{Du}/\mathrm{D}t\|$ is non-dimensionalized by $\omega^2 R_{\max}$.

The calculated and theoretical pressure values have much smaller difference than that of the total acceleration which is undetectable directly from the pressure contours. To show the difference clearly, the contours of the pressure errors are shown in figure 2. The result is similar to what figure 1 shows that the Eulerian pressure has larger errors.

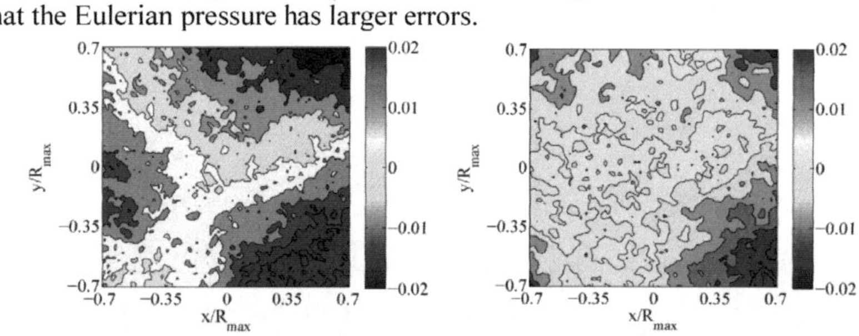

(a) Eulerian pressure error (b) Lagrangian pressure error

Figure 2. Comparison between pressure errors. Pressure errors are non-dimensionalized by $0.5\rho\omega^2 R_{\max}{}^2$.

3.2. Validation and control of the error propagation

Considering the better accuracy and certain error distribution by Lagrangian method, the error propagation to the Lagrangian acceleration has the highest priority for investigation. Figure 3 gives the PDF of the Lagrangian acceleration errors with different temporal and spatial resolutions. In figure 3(a), the space interval is fixed to 1. As can be seen, the Lagrangian acceleration errors have a Gaussian distribution. The maximum acceleration error is inversely proportional to dt [12], which is consistent with equation (10). In figure 3(b), dt is specially set as 0.001 s so that the maximum displacement of virtual particles in each direction is about 0.2. As a result, PDF of the errors at $h \le 0.2$ coincide with each other, which suggests that these two spatial resolutions (h = 0.1 and 0.2) are high enough for an accurate particle tracking when dt = 0.001 s. On the contrary, space interval of h larger than 0.2 would cause increasing errors as h increases.

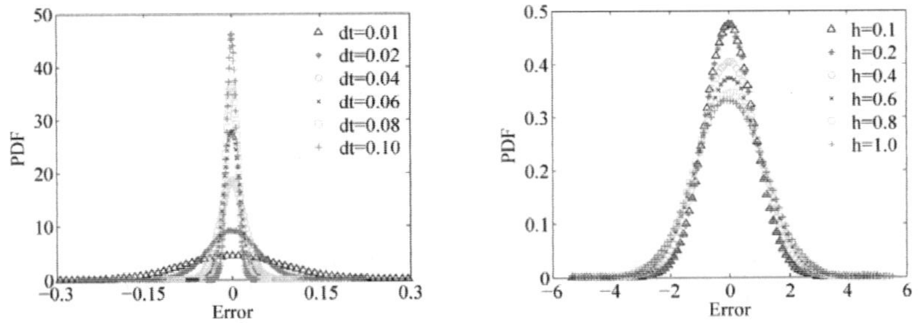

(a) $h = 1$ (b) $dt = 0.001$ s

Figure 3. PDF of the Lagrangian acceleration errors at different temporal and spatial resolutions.

PDFs of the acceleration errors are fitted using the *normfit* tool of MATLAB and result is shown in table 1. A noise enlarged multiple k is defined as $k = \sigma_a/(\sigma_u/dt)$ to describe how the error propagates from velocity to acceleration. According to equation (10), $k \approx 0.71$. Statistically, a value of $k \approx 0.87$ is obtained for properly selected time and space intervals. The larger value of k is caused by tracking errors, truncation errors and Gaussian fitting errors. Noting that the maximum displacement of virtual particles in each direction is about 0.2 when $dt = 0.001$ s, 1 when $dt = 0.005$ s, and 2 when $dt = 0.01$ s. It can be summarized from the table that:

(1) For fixed space interval of 1, k increases as dt decreases when $dt < 0.005$ s and increases as dt increases when $dt > 0.005$ s. Therefore, if the spatial resolution is considered to be high enough for an accurate particle tracking, two conclusions could be obtained as

- When $dt < h/u_{max}$, uncertainty increases dramatically as dt decreases, which results in dramatically increased Lagrangian acceleration errors.
- When $dt > h/u_{max}$, tracking errors and truncation errors increase slowly as dt increases, which results in slowly increased Lagrangian acceleration errors.

(2) For fixed time interval of 0.001 s, k increases as space interval increases when $h > 0.2$ and does not change when $h \leq 0.2$. For fixed time interval of 0.01 s, k does not change when $h \leq 2$. Therefore, if the time interval is considered to be high enough for an accurate pseudo-tracking, another two conclusions could be obtained as

- When $h > u_{max}dt$, tracking errors and truncation errors increase as h increases, which results in increased Lagrangian acceleration errors.
- When $h < u_{max}dt$, Lagrangian acceleration errors are independent from space interval. The reason that decreased h would not cause increased uncertainty like dt lies in the different roles of h and dt during particle tracking. As equation (9) shows, dt is directly used as a denominator for Lagrangian acceleration determination, while h is mainly related to the interpolation during particle tracking. Therefore, h would not affect the Lagrangian acceleration accuracy as long as the spatial resolution is high enough for an accurate interpolation.

Table 1. Gaussian fitting parameter and its relationship with the maximum of velocity errors. (a) Influence of dt on k ($h = 1$); (b) Influence of h on k under two time intervals.

(a)

dt (s)	0.001	0.002	0.004	0.006	0.008	0.01	0.02	0.04	0.06	0.08	0.10
σ_a	1060.7	438.97	182.86	106.07	83.93	67.85	33.94	17.34	11.57	8.72	7.05
σ_u/dt	777.53	388.75	194.38	129.58	97.19	77.75	38.88	19.44	12.96	9.72	7.75
k	1.36	1.13	0.94	0.82	0.86	0.87	0.87	0.89	0.89	0.90	0.91

(b)

	$dt = 0.001$ s						$dt = 0.01$ s					
h	0.1	0.2	0.4	0.6	0.8	1.0	0.2	0.4	0.6	0.8	1.0	2.0
σ_a	700.13	703.66	778.21	883.30	954.67	961.67	67.18	67.39	67.82	67.10	67.59	67.39
σ_u/dt	777.53	777.53	777.53	777.53	777.53	777.53	77.75	77.75	77.75	77.75	77.75	77.75
k	0.90	0.90	1.00	1.14	1.23	1.24	0.87	0.87	0.86	0.87	0.87	0.87

A proper selection of the temporal and spatial intervals helps suppress the error enlargement [5, 7]. A simple criterion is proposed as $\mathrm{d}t \geqslant h/u_{\mathrm{ref}}$, where u_{ref} is the characteristic velocity of a structure in flow. It suggests that dt is better to be larger than smaller around the optimal dt since uncertainty increases dramatically when dt is very small.

4. Conclusions

The paper proposed a probability density function (PDF) based method for error propagation analysis through which the magnitude and format of the errors could be simultaneously estimated. Error propagation in pressure reconstruction is theoretically and statistically investigated in both the Eulerian and Lagrangian perspectives. Influence of the spatial and temporal resolutions on the error propagation is discussed and assessed using a 2D synthetic solid body rotation flow. A criterion is developed for the parameter determination. The main conclusions are summarized as follows. (1) Acceleration errors determined by the Lagrangian method have a certain PDF which is independent from the flow velocity, while the Eulerian material acceleration errors depend on the velocity fields.(2) The Lagrangian acceleration errors is related to the velocity errors by a noise enlarged multiple which is about 0.71 in an ideal condition and is about 0.87 in statistical result. (3) The spatial and temporal resolutions have an important effect on error propagation. A criterion is developed as $\mathrm{d}t \geqslant h/u_{\mathrm{ref}}$ that helps suppressing the error enlargement. Small dt should be carefully used in case of high uncertainty.

Acknowledgements

This work is supported by the National Natural Science Foundation of China (11472030, 11327202, 11490552).

References

[1] van Oudheusden B W 2013 PIV-based pressure measurement *Meas Sci Technol* **24**(3)
[2] Liu X and Katz J 2006 Instantaneous pressure and material acceleration measurements using a four-exposure PIV system *Exp Fluids* **41**(2) 227-40
[3] Gresho P M and Sani R L 1987 On pressure boundary conditions for the incompressible Navier - Stokes equations *Int J Numer Meth Fl* **7**(10) 1111-45
[4] Gurka R, Liberzon A, Hefetz D, Rubinstein D and Shavit U 1999 Computation of pressure distribution using PIV velocity data Workshop on particle image velocimetry
[5] Violato D, Moore P and Scarano F 2011 Lagrangian and Eulerian pressure field evaluation of rod-airfoil flow from time-resolved tomographic PIV *Exp Fluids* **50**(4) 1057-70
[6] de Kat R and Ganapathisubramani B 2013 Pressure from particle image velocimetry for convective flows: a Taylor's hypothesis approach *Meas Sci Technol* **24**(2) 024002
[7] de Kat R and van Oudheusden B 2012 Instantaneous planar pressure determination from PIV in turbulent flow *Exp Fluids* **52**(5) 1089-106
[8] Christensen K and Adrian R 2002 Measurement of instantaneous Eulerian acceleration fields by particle image accelerometry: method and accuracy *Exp Fluids* **33**(6) 759-69
[9] Liu X and Katz J 2013 Vortex-corner interactions in a cavity shear layer elucidated by time-resolved measurements of the pressure field *J Fluid Mech* **728** 417-57
[10] Foucaut J-M, Carlier J and Stanislas M 2004 PIV optimization for the study of turbulent flow using spectral analysis *Meas Sci Technol* **15**(6) 1046
[11] JCGM 2008 Evaluation of measurement data - guide to the expression of uncertainty in measurement *available online: http://wwwbipmorg/en/publications/guides/gumhtml* Measurement science and technology joint commitee for guides in metrology
[12] Jensen A and Pedersen G K 2004 Optimization of acceleration measurements using PIV *Meas Sci Technol* **15**(11) 2275

Flow morphing by coaxial type plasma actuator

S Toyoizumi[1], H Aono[2] and H Ishikawa[2]

[1] Graduate School of Engineering, Tokyo University of Science, Tokyo, JAPAN
[2] Department of Mechanical Engineering, Tokyo University of Science, Tokyo, JAPAN,

E-mail: 4515635@ed.tus.ac.jp

Abstract. The purpose of study is to achieve the fluid drag reduction of a circular disk by Dielectric Barrier Discharge Plasma Actuator (DBD-PA). We here introduced "Flow Morphing" concept that flow around the body was changed by DBD-PA jet, such as the body shape morphing. Coaxial type DBD-PA injected axisymmetric jet, generating the vortex region on the pressure side of the circular disk. The vortex generated by axisymmetric plasma jet and flow around circular disk were visualized by tracer particles method. The fluid drag was measured by compression type load cell. In addition streamwise velocity was measured by an X-type hot wire probe. The extent of fluid drag reduction by coaxial type DBD-PA jet was influenced by the volume of vortex region and the diameter of plasma electrode.

1. Introduction

There is a problem of energy resources depletion in recent years and the necessity of saving energy is increasing in various fields. Certainly, energy saving devices of fluid machinery are also required. The fluid drag reduction is important; however it is an unsolved problem. The fluid drag reduction of a circular disk, as a typical example of three-dimensional bluff bodies, has been researched by many researchers. Igarashi et al. have proposed two methods of changing the circular disk shape to control the flow field around disk. One was applying a spherical surface to side wall of disk, the other was cut to a stepwise shape with smaller diameter[1][2]. They reported the stepwise shape disk can reduce the fluid drag by about 75 percent. Other active flow active researches, such as an electromagnetic actuator[3] , a synthetic jet[4] and laser-pulse[5] were reported. Bigger et al. have proposed that a symmetrical actuation by the electromagnetic actuators in disk produced effective reduction in the recirculation region length of 10 percent. The problems with the electromagnetic actuators are complication of mechanism and noise.

A Dielectric Barrier Discharge Plasma Actuator (DBD-PA) has actively studied as a promising flow control device[6]. Figure 1 shows the schematic diagram of typical DBD-PA. DBD-PA consists of two electrodes and a dielectric material. When a high alternating current voltage is applied between two electrodes, DBD occurs on the surface of a dielectric material. This induces the plasma jet along the wall surface. In this study, the fluid control around the body to use an axisymmetric jet by the coaxial type DBD-PA was investigated. Figure 2 shows the schematic of structure of the coaxial type DBD-PA. The advantages of DBD-PA are easy installation due to simple structure, excellent responsiveness, active controllable and instantaneous controllable due to electrically control.

We here proposed "Flow Morphing" concept[7] that flow field around the body was changed by DBD-PA jet, just as the shape morphing technology. The shape continuously change by morphing is one of

the control method. However, several problems of shape morphing are devices to be heavier and complication of mechanism.
In this study, we visualizing the vortex changed by axisymmetric jet and flow around circular disk by tracer particles method. The fluid drag was measured by compression type load cell. In addition streamwise velocity was measured by the X-type hot wire probe.

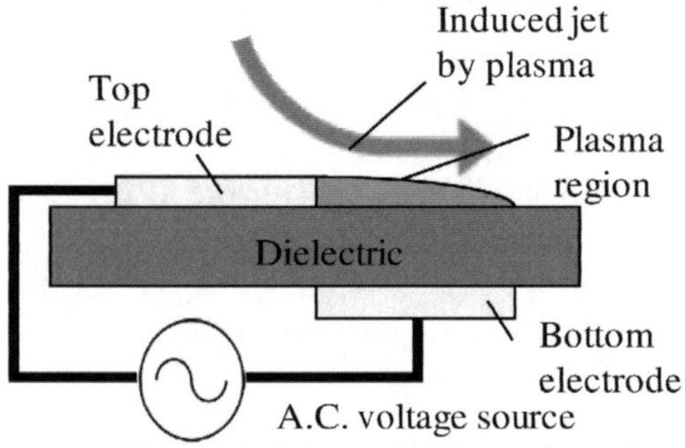

Figure 1. Schematic illustration of plasma actuator

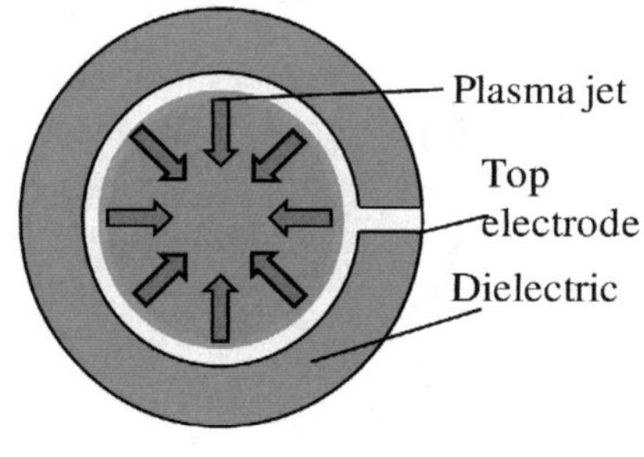

Figure 2. Schematic of coaxial type plasma actuator

2. Experimental Setup and Methods

In present study, experiments were carried out in the suction-type wind tunnel that has cylindrical measuring section. Circular disk (Diameter: D=50 [mm], Thickness: t=5 [mm]) was supported by a strut (d=3 [mm]) from downstream side. The strut was supported vertically on linear bush with low friction. The circular disk was fixed vertically to mainflow. Considering the general blockage ratio, less than 5% seems reasonable. The circular disk (D=20 [mm], t=5 [mm]) was chosen because the blockage ratio is 3%. However this disk has three problems, the influence of the electrical noise form PA, the scattering of measurements and PA shape pattern was restricted by the small surface area of the disk. Therefore the circular disk (D=50 [mm], t=5 [mm]) was chosen. The fluid drag are influenced of the blockage effect. Figure 3 shows the different diameters of coaxial type DBD-PA. The coaxial type DBD-PA was chosen in order to induce the axisymmetric jet. We used 4 different diameters of PA from generating plasma diameters (D_p=10, 20, 30 and 40 [mm]). The dielectric of PA is polyimide film that has a thickness of 125 [μm]. The both electrodes of PA are copper foils that have a thickness of 12 [μm]. Top and bottom electrodes are connected to a high voltage and frequency power source. In this study, the applied voltage is 5.0 [$kV_{p\text{-}p}$] (subscript p-p means the value between a peak to peak) with the frequency of 6.0 [kHz]. The Reynolds number (Re) based on the mainflow velocity and disk diameter was calculated to be 5,000 and 10,000. The PA in static air generated plasma jet velocity of 0.4~0.7 [m/s].

(a) D_p= 10 [mm] (a) D_p= 20 [mm] (a) D_p= 30 [mm] (a) D_p= 40 [mm]

Figure 3. Coaxial type plasma actuators

2.1. Flow visualization

Figure 4 shows the schematic of the flow visualizing experiment. The flow was visualized by the tracer particle method. The photograph was taken using the high-speed camera (PHOTRON, FASTCAM-SA3) and the YAG laser as the illumination. The flame rate was 1,000[fps]. The main flow velocity U_∞ was 1.7 and 3.3 [m/s].

2.2. Fluid drag measurement

Figure 5 shows the fluid drag measurement setup. The drag is measured by compression type load cell (KYOUWA, LMB-A-10N-P). The load cell was installed to the linear bushing and can only measure the force to the mainflow direction. The sampling frequency was 1.0 [kHz], the number of sampling points are 16,384, low pass filter is 10 [Hz]. The measurements were carried out 20 times for each main flow velocity. The measuring time was about 16 seconds. The PA power source is turned on in 8 seconds. Considering the effect of hysteresis, interval time needs 30 seconds.

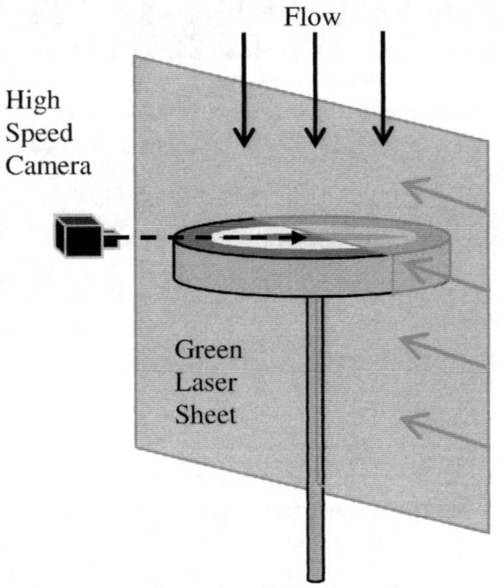

Figure 4. Flow visualizing experiment

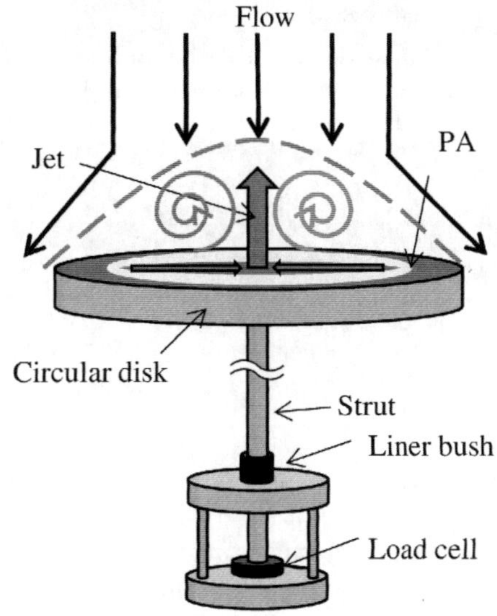

Figure 5. Fluid drag measurement setup

2.3. Streamwise velocity measurement

The streamwise velocity behind the disk was measured by the X-type hot wire probe. The hot wire probe was connected to constant temperature anemometer (KANOMAX, Model 1010) and the linearizer (Model 1013). The velocity profiles depending of plasma edge diameters were obtained by the X-type hot wire probe at x/D=3, r/D=0.16~0.9 (at 0.04 space).

3. Results and Discussion

3.1. Flow visualization

Figure 6 shows flow structure of induced jet from the coaxial type DBD-PA. Large vortex structure was generated on the pressure side of circular disk. The yellow symbol indicates ejected position of induced plasma jet (i.e, the edge of plasma electrode). For no control case (Figure 6 (a)), in-coming flow passed though along the surface of disk. With the induced plasma jet case (Figure 6 (b)~(e)) it showed the vortex structure growth. The shape of generated vortex was observed in conical shape when D_p of 10, 20 [mm] case. The generated vortex has conical shape when D_p were 10 and 20 [mm]. On the other hand, shape of vortex generated region in D_p = 30 and 40 [mm] case were observed as ring structures. From these instantaneous images, the size of vortex generated region was not proportion to D_p. The vortex generated region showed the maximum size when D_p was 20 [mm]. As the D_p decreased, plasma jet collided near the centre of circular disk. As the D_p was increased, the ring type vortex region generated. In the case of D_p=20 [mm] the vortex yielded largely changes of flow structure than the case of D_p=10 [mm] because induced jet velocity was enough large. In that there was no steady vortex region, repeating vortex generating and collapsing at 5 [Hz] intervals. Figure 8 suggested that flow morphing effects were weak when this unsteady process was occurred.

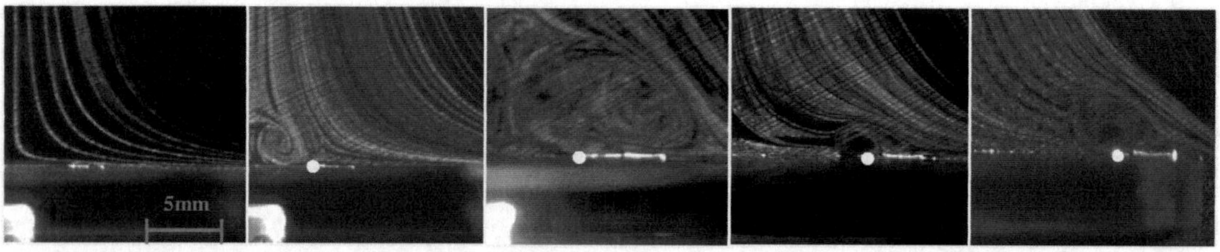

(a) Without PA jet (b) D_p = 10 [mm] (c) D_p = 20 [mm] (d) D_p = 30 [mm] (e) D_p = 40 [mm]

Figure 6. Flow visualization of vortex region on the pressure side of circular disk

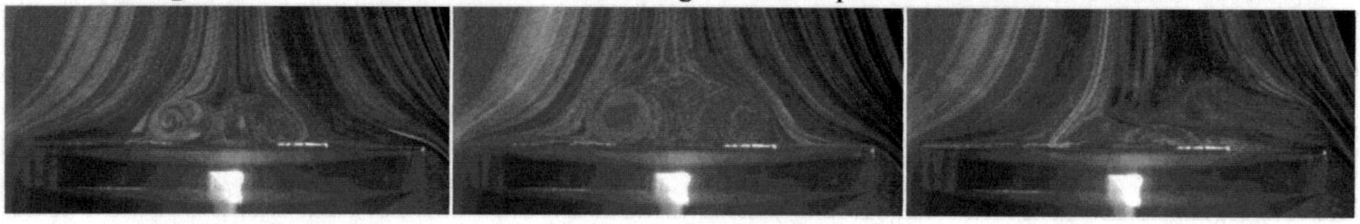

(a) Vortex generation (b) Vortex development (c) Vortex collapse

Figure 7. Vortex growth process

3.2. Fluid drag measurement

Figure 8 shows the drag coefficient in all D_p cases. The error bar shows standard error that measured by 20 times (Standard error is standard deviation devided by the square root of sample size). The drag coefficient was calculated by formula (1). The reducing effect by plasma jet cases was shown in Table 1. Due to inducing plasma jet, the fluid drag reduction is about 6~11% lower than that of without PA case. The case of D_p=20 [mm] is lowest of all D_p cases of drag reduction.

$$C_D = \frac{2D}{\rho S V_\infty^2} \ (D: Drag, \rho: flow\ density, S: Representative\ area) \quad (1)$$

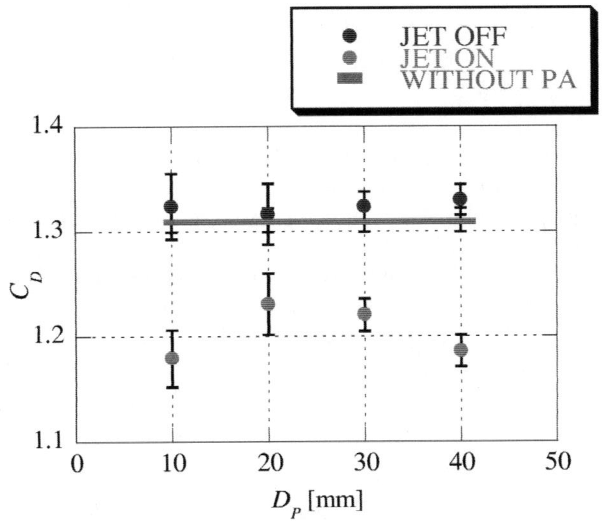

Figure 8. Relationship of C_D and PA edge diameter

Table 1. Drag reduction rate

	D_p=10 [mm]	D_p=20 [mm]	D_p=30[mm]	D_p=40[mm]
Re = 5,000	10.9 [%]	6.5 [%]	7.8 [%]	10.6 [%]

3.3. Streamwise velocity measurement

Figure 9 shows the streamwise velocity distributions in Reynolds number of 5,000. Increment of streamwise velocity in the wake behind the circular disk without PA jet was confirmed in all D_p cases. Above all, the velocity increasing effect in the center of disk wake was greater than that of the outside region of disk. Streamwise velocity defect behind the circular disk became smaller by inducing plasma jet. This suggested that the reverse flow region and the drag acting on the circular disk were decreased.

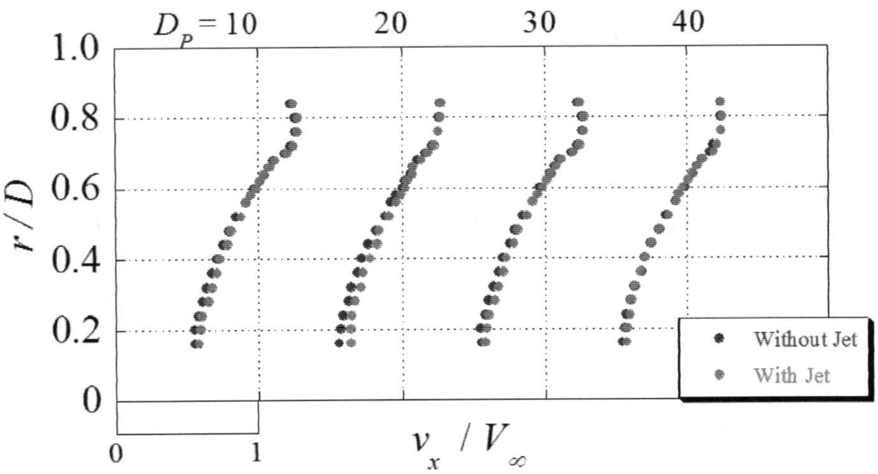

Figure 9. Time mean velocity distributions with plasma jet

4. Conclusions

We have performed wind tunnel experiments in order to clarify the drag reduction mechanism by flow morphing with coaxial type DBD plasma actuator. Flow morphing by the coaxial type DBD-PA achieved about 10% drag reduction compared with no PA in this study. Flow visualization showed the growth process of vortex by induced plasma jet on the pressure side of circular disk. The hot-wire probe measurement showed that the streamwise velocity in the wake behind the disk was accelerated by generated vortex induced by the coaxial type DBD-PA. In the case of the disk diameter of $D_p = 20$ [mm], the vortex generation, development and collapse process were repeated with low frequency. This unsteadiness affected the effects of drag reduction and streamwise velocity distribution. This suggested that time independency of flow structure played a role in favorable effect of flow morphing.

Acknowledge

The authors are grateful for the fruitful discussions and cooperation with Mr. Eguchi, M, a graduate student of Tokyo University of Science.

References

[1] Igarashi T, Muranaka M, and Nakamura N 2006 Drag reduction of a circular disk normal to air-stream applying turbulent transition *JSME Transaction* **72** 719 83-90(in japanese).
[2] Igarashi T, Muranaka M, and Nakamura N 2007 Drag reduction of a circular disk with small cutout at its edges normal to air-stream *Bulletin of JSME* **73** 725 147-153(in japanese).
[3] Bigger R P, Higuchi H and Hall J W 2009 Open-loop control of disk wakes *AIAA Journal* **47** 5
[4] Feng L H and Wang J J 2010 Circular cylinder vortex-synchronization control with a synthetic jet positioned at the rear stagnation point *J.Fluid Mech* **662** 232-259.
[5] Kim J H, Matsuda A, Sakai T and Sasoh A 2011 Wave drag reduction with acting spike induced by laser-pulse energy depositions *AIAA Journal* **49** 9.
[6] Corke C C, Enloe C L and Wilkinson S P 2010 Dielectric barrier discharge plasma actuator for flow control *Annu. Rev. Fluid Mech*.**42**.505-529.
[7] Wlezien R W et al, 1998 The aircraft morphing program *AIAA98* 1927.

Solving Wave Flow Energy Propagated by Twin-Hull Ships (Catamarans)

Md Salim Kamil[1], Mazlan Muslim[2] and Ishak Ali[3]

[1,2,3] Universiti Kuala Lumpur Malaysian Institute of Marin Engineering Technology, Dataran Industri Teknologi Kejuruteraan Marin, Bandar Teknologi Maritim, Jalan Pantai Remis, 32200 Lumut, Perak, Malaysia

[1]mdsalim@unikl.edu.my, [2]mazlanm@unikl.edu.my, [3]ishak@unikl.edu.my

Abstract. The main objective of this paper is to put forward a method of solution of wave flow energy propagated by twin-hull ships utilizing ship wave resistance integral equation by Tuck-Lazauskas [8]. The underlying principle of the governing equation is based on the linear thin ship theory pioneered by Michell. The ship wave resistance integral equation by Tuck-Lazauskas is improvised in which the upper limit of the mathematical integration of the governing equation is set to the final root instead of $\pi/2$ as in the original equation to eliminate the problem of solving such a divergent improper integral. The method of solution and the theoretical results are validated against works of others experimentally and theoretically. The results obtained by the proposed method are found to be very consistent and more closely fit to the experimental results in comparison with those obtained by other theoretical method as presented in this paper.

1. Introduction

Twin-hull ships or catamarans are more advantageous than mono-hull ships of equal displacement for their higher operational speed, significantly larger deck area, more stable platform, shallower draft, less susceptible to rolling, less wave resistance to a certain extent of Froude numbers due to their hull slenderness and invariably less total resistance. Nevertheless, the disadvantages of catamarans over mono-hull ships amongst others are more vulnerable to pitch and heave responses, structurally more complex and complicated waves flow between the hulls. Continuous studies had been made on predicting ship wave resistance of multi-hill ships for more accurate solutions since the past centuries until today and this reflects for its importance.

Many fundamentals and modern methods evolved through these periods of times. The search for better methods and more accurate solutions has driven for deeper studies and motivates to bring together the fundamentals of the past and current works and the relevant theories. The objective of the study is to search for an effective and reliable method of solution of the wave resistance or wave energy utilizing the wave resistance integral by Tuck and Lazauskas (1998) based on the underlying linear thin ship theory by Michell (1898) and the subsequent related works by Wigley (1926 - 1942) and Newman (1977). The ship wave resistance in this study is in fact equivalent to the wave energy propagated or generated by the catamaran while underway or in motion cutting through the water.

2. Governing Equation

The study made to a specific reference to the governing equation of the ship wave resistance of multi-hull ships by Tuck and Lazauskas suited for twin-hull configuration. This equation is connected to the Michell-Wigley's and Newman's wave amplitude functions of the demi hull ships. The waves system in a global coordinates (x,y,z) and origin local coordinates (x_{j0}, y_{j0}, z_{j0}) of a multi-hull ships is expressed as;

$$\zeta_j(x,y) = Re \int A_j(\theta)\ exp[(-ig/U^2cos^2\theta)(x_{j0}cos\theta + y_{j0}sin\theta)].exp[(-ig/U^2cos^2\theta)(xcos\theta + ysin\theta)]d\theta;$$
$$-\pi/2 < \theta < \pi/2 \quad (1)$$

The above expression is the real part of the complex equation defining the waves profile or system generated by a multi-hull ship with $j = 1, 2...n$ and suited for a twin-hull ship or catamaran with $j = 1, 2$. Where $A_j(\theta)$ is the wave amplitude function for j numbers of ship hull, U is the speed, g is the gravitational acceleration, x_{j0} and y_{j0} are the axial and transverse positions referred to the origin of the local coordinate system for j numbers of ship hull. Further explanation on the above expression are given below.

The wave amplitude function is generalized as;

$$A(\theta) = \sum A_j(\theta)\ exp[(-ig/U^2cos^2\theta)(xcos\theta + ysin\theta)];\ j = 1, 2, ...n \quad (2)$$

Taking the real part;

$$A(\theta) = \sum A_j(\theta)\ cos[(-g/U^2cos^2\theta)(xcos\theta + ysin\theta)];\ j = 1, 2, ...n \quad (3)$$

The centre plane wave amplitude function would be for y = 0;

$$A(\theta)_{CP} = \sum A_j(\theta)\ cos[(-g/U^2cos\theta)x];\ j = 1, 2\ ...n \quad (4)$$

The wave resistance of a twin-hull ship R_w is much dependent on the square of the speed U^2, square of the wave amplitude function $|A(\theta)|^2$, the transverse hull interference factor $HIF_T(\theta)$ and the cube of the cosine of the wave propagation angle θ (i.e. $cos^3\theta$) associated with the waves system. The mathematical expression of the ship wave resistance for a twin-hull ship due to Tuck-Lazauskas can be derived from the ship wave resistance expression of a multi-hull ship for j = 1 to 2, non-staggered hull configuration with x = 0 and wave amplitude function of the demi hull by Newman may finally be written in the following forms;

$$R_w = (\pi\rho U^2/2) \int |A(\theta)_{CP}|^2\ HIF_T(\theta)cos^3\theta d\theta;\ -\pi/2 < \theta < \pi/2 \quad (5)$$

As for the demi hull wave amplitude function due to Michell-Wigley, the mathematical expression of the ship wave resistance for a twin-hull ship due to Tuck-Lazauskas is of the form;

$$R_w = (4\rho g^2/\pi U^2)\int (I^2 + J^2)\ HIF_T(\theta)sec^3\theta d\theta;\ 0 \leq \theta \leq \pi/2 \quad (6)$$

$$J = bd \iint (\delta\eta/\delta\xi)exp(-dg\zeta sec^2\theta/U^2)sin(\ell g\xi sec\theta/U^2)\delta\xi\delta\zeta;\ -1 \leq \xi \leq +1,\ 0 \leq \zeta \leq 1 \quad (7)$$

$$I = bd \iint (\delta\eta/\delta\xi)exp(-dg\zeta sec^2\theta/U^2)cos(\ell g\xi sec\theta/U^2)\delta\xi\delta\zeta;\ -1 \leq \xi \leq +1,\ 0 \leq \zeta \leq 1 \quad (8)$$

The transverse hull interference factor $HIF_T(\theta)$ is expressed as;

$$HIF_T(\theta) = 4cos^2[0.5(2p/L)\ sin\theta/(Fn^2 cos^2\theta)] \tag{9}$$

With reference to Michell-Wigley for wave amplitude function of a thin ship is given as;

$$\iint_{CP} (\partial\eta/\partial x)exp[\upsilon\ sec^2\theta(z + ixcos\theta)]dzdx \tag{10}$$

Where;

Subscript CP means the central plane, $\upsilon = g/U^2$, $\zeta = z/d$, $\xi = x/\ell$, $\ell = L/2$, $\eta_x = \partial y/\partial x$, L – Length of ship, d – Draft of ship

3. Problem Formulation

Hypothetically the magnitude of wave resistance or the wave energy propagated by a moving ship in water is also dependent on a precise maximum resultant propagation angle of the combined divergent and transverse waves. In this study the ship wave resistance equation by Tuck and Lazauskas is improvised to eliminate the problem of solving such a divergent improper integral that the denominator tends to zero (i.e. improper integral of type 2). The ship wave resistance is solved by first determining the final root of solution of the ship wave resistance integral which sets the upper limit of the mathematical integration. This unique angle is termed as the maximum wave propagation angle abbreviated by θ_{max} replaces the maximum limit of integration $\pi/2$ of the original expressions. θ_{max} dictates the absolute solution of the ship wave resistance problem and the method is herein called Final Root Method. The theoretical results are validated utilizing a twin-hull ship model of National Physical Laboratory (NPL) round bilge hull form.

Mathematically, the integral of the ship wave resistance R_w contains infinite roots of solution. This complex integral is not easily differentiable directly to determine the roots of solution. Nonetheless, practically one may solve for the root iteratively from the wave amplitude function or the ship wave integral for zero values. The study zoomed on the final root of minima which exists in the range of θ approaching the value of $\pi/2$. The so-called precise maximum resultant wave propagation angle θ_{max} would be the final root of the ship wave resistance integral function and that the values of the integral I, J or R_w are equal to zero on the x-axis, beyond this point there should distinctively no other root exists.

The improvised ship wave resistance integral or equation for a twin-hull ship is hence written in the form as shown below in which the upper limit of the mathematical integration is reassigned with the precise final root of solution θ_{max};

$$R_w = (\pi\rho U^2)\int|A(\theta)_{CP}|^2\ HIF_T(\theta)cos^3\theta d\theta;\ 0 < \theta < \theta_{max} \tag{11}$$

$$A(\theta)_{CP} = (2\upsilon/\pi)sec^3\theta\iint(\partial\eta/\partial x)exp[\upsilon\ sec^2\theta(z + ixcos\theta)]dzdx$$

$$= 2\upsilon/\pi)sec^3\theta\iint \eta_x exp[(gz/U^2)sec^2\theta]cos[(gx/U_2)sec\theta)]dzdx;\quad 0 < z < d,\ -\ell < x < +\ell \tag{12}$$

4. Computational Technique

In implementing the calculation technique the model of principal particulars as given below are divided into frames or stations of equal interval or spacing longitudinally along the length of the ship and divided into waterlines of equal intervals vertically down from the free surface to the keel bottom. The mathematical calculation technique makes use of Simpson's Rules of Integration as appropriate.

The integration performed in strip-wise sequence, first in the η-ζ-direction at each cross-sectional plane of the hull followed by the integration in the ξ-direction along the length of the ship hull to compute the integral I with the origin at amidships. The coordinate systems for the twin-hull models are shown in Figures 1 and 2. The integration is then continued in the angular θ-direction from zero up to the limiting resultant propagation angle i.e. θ_{max} to finally determine R_w. The integration in the angular θ-direction is done at every interval and finer intervals near the final wave propagation angle θ_{max}. The values of integrals I and J are read from calculation of separate subroutines for the demi hull. The effect of transverse hull interference factor $HIF_T(\theta)$ or the interacted wave amplitude function is necessary to be calculated following the given expression.

5. Theoretical Analysis

The theoretical analysis was performed according to the improvised ship wave resistance equation described in Paragraph 3. The values of integrals I and J for the selected interval of angular spacing θ are first calculated separately. Next are to calculate the transverse hull interference factor $HIF_T(\theta)$ for each interval of θ and the ship wave resistance R_w for a chosen speed. The calculation of R_w is then necessary to be iterated until the final root of solution θ_{max} is found. The calculations for R_w are then continued for different speeds and varying ratios of transverse hull spacing to the ship length 2p/L. Several validations of method and results are also made against works done by M. Insel, A.F. Molland and J.F. Wellicome [13] using an NPL hull form model with the model principal particulars are give in Table 1. The coordinate system is given Figure 1 and Figure 2 below.

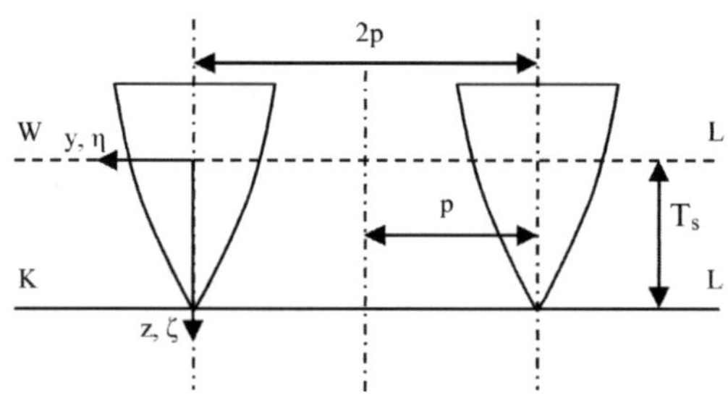

Figure 1. Twin-Hull Configuration and the Coordinate System (Cross-sectional View)

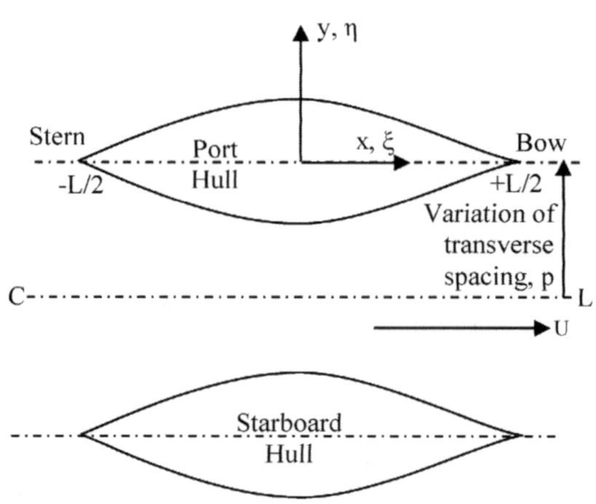

Figure 2. Twin-Hull Configuration and the Coordinate System (Plan View)

Table 1 - Model Principal Particulars (NPL Hull):

Length Between Perpendiculars, L_{BP}	- 1.600 m
Length Waterline, L_{WL}	- 1.600 m
Draft, T	- 0.089 m
Beam Waterline, B_{WL}	- 0.178 m
Wetted Surface Area, S at draft T	- 0.338 m^2

6. The Results

The validation for the twin-hull ships are carried out utilizing the NPL Hull for the ratio of transverse hull spacing to the ship length of 2p/L = 0.3 and 0.5 and for Froude numbers, F_n between 0.3 to 1.0. The coefficients of ship wave resistance C_w were evaluated ($C_w = R_w/0.5\rho SV^2$, Rw – ship wave resistance, ρ – water density, S – hull wetted surface area, V – ship speed), for the range of the Froude numbers. The calculation results are shown in Tables 2 and 3 and plotted as shown in Figures 2, 3, 4 and 5.

Table 2. Theoretical and Experimental Coefficient of Ship Wave Resistance C_w, 2p/L = 0.3

V (m/s)	F_n	Insel-Molland-Wellicome (IMW)		Newman (final root)
		C_w x 10^3 (Experimental)	C_w x 10^3 (Theoretical)	C_w x 10^3 (Theoretical)
1.19	0.30	2.200	3.000	2.132
1.39	0.35	3.000	1.000	2.909
1.59	0.40	4.000	3.000	3.875
1.82	0.46	6.600	5.900	6.398
1.98	0.50	6.000	5.600	5.812
2.38	0.60	4.000	3.600	3.873
2.77	0.70	2.700	2.600	2.616
3.17	0.80	1.800	1.700	1.744
3.57	0.90	1.500	1.400	1.453
3.96	1.00	1.400	1.300	1.356

Table 3. Theoretical and Experimental Coefficient of Ship Wave Resistance C_w, 2p/L = 0.5

V (m/s)	F_n	Insel-Molland-Wellicome (IMW)		Michell-Wigley (final root)
		C_w x 10^3 (Experimental)	C_w x 10^3 (Theoretical)	C_w x 10^3 (Theoretical)
1.19	0.30	2.200	2.400	2.132
1.39	0.35	2.400	1.200	2.325
1.58	0.40	3.800	2.400	3.875
1.82	0.46	5.200	4.300	5.038
1.98	0.50	4.800	4.200	4.651
2.38	0.60	3.400	3.200	3.294
2.77	0.70	2.700	2.600	2.617
3.17	0.80	2.100	2.000	2.035
3.57	0.90	1.700	1.600	1.647
3.96	1.00	1.500	1.400	1.453

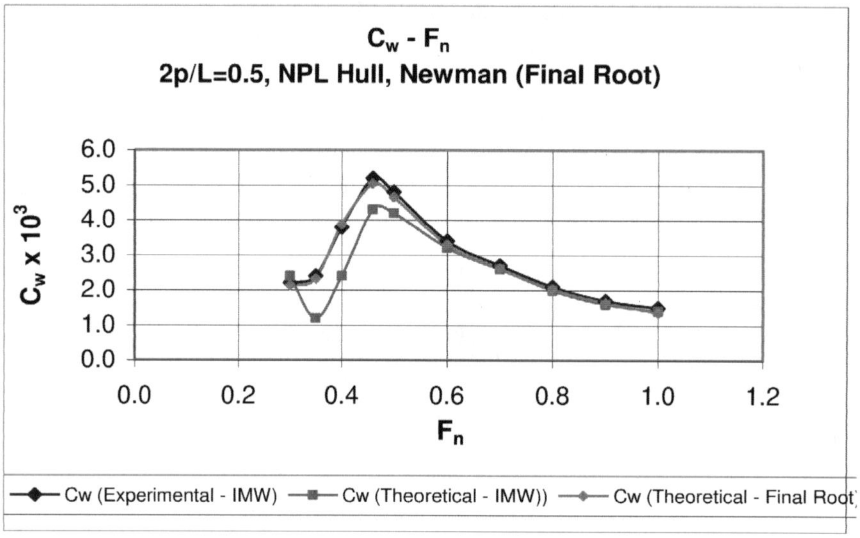

Figure 2. Graphs of Experimental and Theoretical (Newman – Final Root) C_w Versus F_n, 2P/L = 0.3

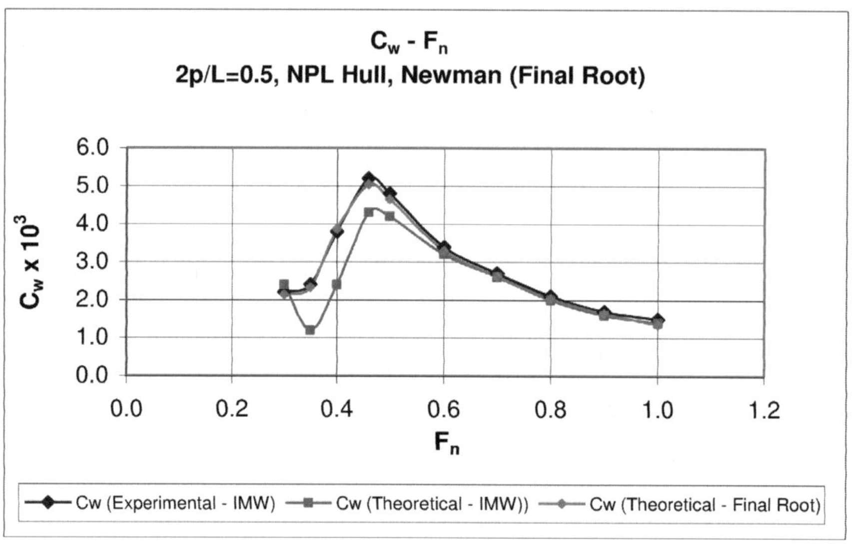

Figure 3. Graphs of Experimental and Theoretical (Newman - Final Root) C_w Versus F_n, 2p/L = 0.5

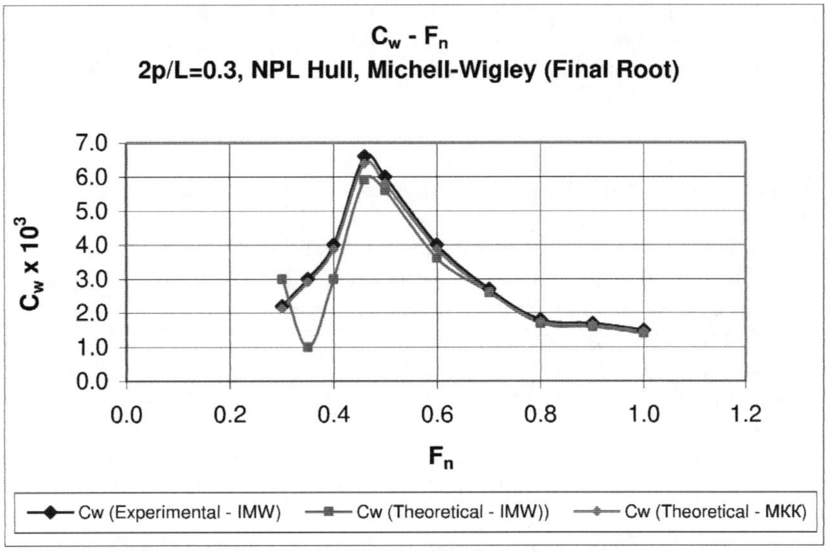

Figure 4. Graphs of Experimental and Theoretical (Michell-Wigley - Final Root) C_w Versus F_n, 2P/L = 0.3

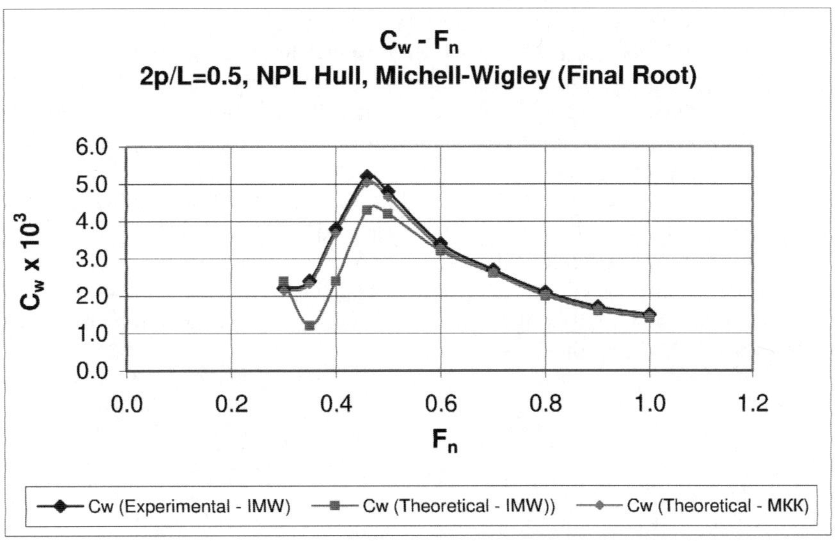

Figure 5. Graphs of Experimental and Theoretical (Michell-Wigley - Final Root) C_w Versus F_n, 2P/L = 0.5

7. Conclusion

The results calculated theoretically in comparison with those experimentally indicated that the method of solution employed in this study is indeed useful, consistent and worthwhile. The accuracy of the calculated results are well within 3% to 5%. Obviously, this new method of solution could provide an option as alternative method for solving such a divergent improper integral equation of ship wave resistance visa-vis the flows of wave energy propagated by a moving twin-hull ship (catamaran). The method could also be used for the study of hull optimization and hydrodynamics performances of twin-hull ships particularly in terms of the wave resistance component during the ship design activities with the ultimate objective is for predicting the powering requirement and the selection of the appropriate propulsion plant.

References

[1] J.H. Michell (1898), Wave–Resistance of a Ship, Philosophical Magazine, Vol. 45, Ser.5, pp106–123.
[2] W.C.S. Wigley (1926), Ship Wave Resistance, A Comparison of Mathematical Theory with Experimental Results, 67th-Session of the Spring Meetings INA.
[3] W.C.S. Wigley (1934), A Comparison of Experiment and Calculated Waves Profiles and Wave Resistance for a form Having Parabolic Waterlines, Proceedings of the Royal Society, London.
[4] W.C.S. Wigley (1942), Calculated and Measured Wave Resistances for a Series of Forms Defined Algebraically, the Prismatic Coefficient and Angle of Entrance Being Varied Independently, Trans. INA, Vol. 84.
[5] J. N. Newman (1977), Marine Hydrodynamics, Massachusetts, MIT Press, pp260-284.
[6] M. Insel, A.F. Molland and J.F. Wellicome (1994), Wave Resistance Prediction of a Catamaran by Linearised Theory, Proceedings of CAMO 94, 59-67M.
[7] P.R. Couser, A.F. Molland, N.A. Armstrong, I.K.A.P. Utama (1997), Calm Water Powering Predictions For High-Speed Catamarans, FAST '97, Sydney, Australia.
[8] E. O. Tuck and L. Lazauskas (1998), Optimum Hull Spacing of a Family of Multihulls, Applied Mathematics Department, University of Adelaide, Australia.
[9] E. O. Tuck, D. C. Scullen and L. Lazauskas (2002), Wave Patterns and Minimum Wave Resistance for High-Speed Vessels 24th Symposium on Naval Hydrodynamics Fukuoka, JAPAN.
[10] O. M. Faltinsen (2005), Hydrodynamics of High-Speed Marine Vehicles, Norwegian University of Science and Technology, Cambridge, Cambridge University Press, pp115-135.
[11] P. K Sahoo, L. J. Doctors and L. Pretlove (2006), CFD Prediction of the Wave Resistance of a Catamaran With Staggered Demihulls, International Conference on Marine Hydrodynamics, Visakhapatnam, India.
[12] M. S. Kamil, M. P. Abdul Ghani, A. M. Abdul Malik, K.V. Rozhdestvenskiy (2011), Revitalizing the Theoretical Approach of Predicting Ship Wave Resistance, Marine Science and Technology Conference, Kuala Lumpur, Malaysia.
[13] M.S. Kamil (2014), Solving Michell Ship Wave Resistance Integral of Mono-Hull Ships by Final Root Method. Journal of Intellectual Marine Technology (Russian Publication), № 3 (25) T.1 2014 ISSN 2073-7173, pp83-90.

Comparision on dynamic behavior of diesel spray and rapeseed oil spray in diesel engine

Azwan Sapit[1, a], **Mohd Azahari Razali**[1], **Mohd Faisal Hushim**[1], **Norrizam Jaat**[1], **Akmal Nizam Mohammad**[1], **Amir Khalid**[1, b]
[1]Centre for Energy and Industrial Environment Studies (CEIES), Universiti Tun Hussein Onn Malaysia, Parit Raja, Batu Pahat, 86400 Johor, Malaysia.

Corresponding authors: azwans@uthm.edu.my[a,] amirk@uthm.edu.my[b]

Abstract. Fuel-air mixing is important process in diesel combustion. It significantly affects the combustion and emission of diesel engine. Biomass fuel has high viscosity and high distillation temperature and may negatively affect the fuel-air mixing process. Thus, study on the spray development and atomization of this type of fuel is important. This study investigates the atomization characteristics and droplet dynamic behaviors of diesel engine spray fuelled by rapeseed oil (RO) and comparison to diesel fuel (GO). Optical observation of RO spray was carried out using shadowgraph photography technique. Single nano-spark photography technique was used to study the characteristics of the spray while dual nano-spark shadowgraph technique was used to study the spray droplet behavior. Using in-house image processing algorithm, the images were processed and the boundary condition of each spray was also studied. The results show that RO has very poor atomization due to the high viscosity nature of the fuel when compared to GO. This is in agreement with the results from spray droplet dynamic behavior studies that shows due to the high viscosity, the RO spray droplets are large in size and travel downward, with very little influence of entrainment effect due to its large kinematic energy.
Keywords: Diesel spray, Atomization, Biomass fuel, Rapeseed oil

1. Introduction

Fuel that derived from biomass such as a vegetable oil fuel is a viable alternative to current fossil fuel and can be use readily to current day diesel engine [1-2]. However studies show that, when diesel engines use neat biomass fuel, NOx emissions increase [3]. Particulate emissions decreases at heavy load due to oxygenated nature of biomass fuel; in contrast, SOF (Soluble organic fraction) emissions increases at light load operation [4]. These kinds of emission formation are greatly affected by spray atomization. Precise optimization of the fuel injection, spray atomization and fuel-air mixing are among the fundamental factors that control the combustion and emission formation [5-6]. Application of high-pressure injection and high ambient temperature could promote better spray atomization and fuel-air mixing, and this should be applicable to biomass fuel too. There are also studies which suggest the geometry of piston cavity design can improve mixture formation and emission level. Generally, biomass fuel is subjected to transesterification process to improve biomass fuel combustion and prevent carbon buildup during engine combustion [7]. However, it is still desirable to use neat biomass fuel without transesterification, especially when the technology to process biomass fuel to biodiesel is not accessible (rural area) and when considering cost. Thus, it is important to firstly understand the spray characteristics of biomass fuel, which is the objective of this study by the means of optical technique. In addition, by using the same technique fitted with dual nano-spark shadowgraph optical arrangement, rapeseed oil spray droplet behavior can be studied.

2. Design of Experiment

Figure 1 shows the experimental setup of shadowgraph photography. The experimental system is the same as previous work [8-9]. It is composed of spray chamber, rapid compression machine (RCM), fuel injection device and optical system. A rapid compression machine was used to create diesel atmosphere in the spray chamber. The spray chamber has two quartz windows in diameter of 60mm for the access of nano-spark light [8] to a still camera. The spray chamber was filled with inert gas mixture of nitrogen and argon to prevent spray ignition. When ambient temperature reaches designated temperature after termination of rapid compression, fuel was injected into the spray chamber and spray behavior was taken by the optical system. Figure 2 shows the dual nano-spark optical arrangement [8] which can take simultaneously two pictures with little time interval of the same spray. All the images taken by films were scanned and changed to digital pictures. Then using in house algorithm system developed by Yatsufusa[8], the image is analyze and spray fuel droplet can be verified.

Fuels used were rapeseed oil (RO) and diesel fuel which is also known as gas oil (GO). The kinematic viscosity of RO is 47.78mm2/st@303K (density 911kg/m3) and GO 3.30 mm2/st@303K (827kg/m3). The 50% distillation temperatures of RO and GO are about 880K and 570K, respectively.

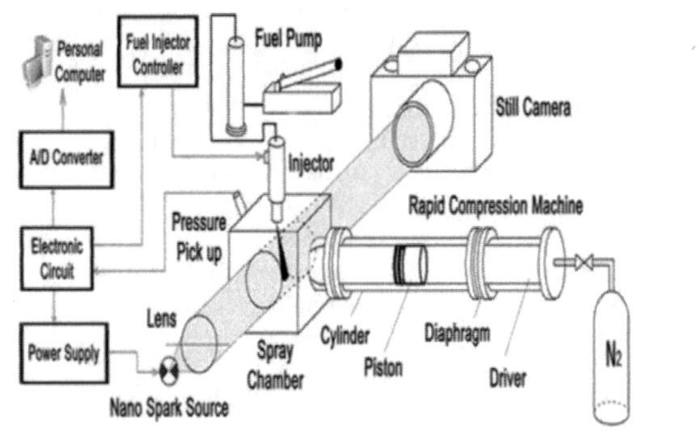

Figure 1. Spray observation system and optical arrangement for single nano-spark shadowgraph photography method.

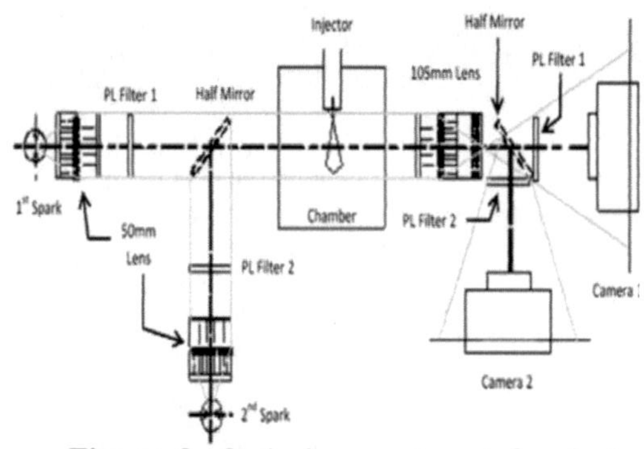

Figure 2. Optical arrangement for dual nano-spark shadowgraph photography technique

3. Result and discussion

Figure 3 shows GO spray image and its analysis at T_i=298K, P_{inj}=70MPa, t=0.50ms aSOI (after start of injection), taken by dual nano-spark shadowgraphy and time interval between nano-spark 1 to 2 is Δt=25μs. Figure 3 (a) shows the original images, (b) shows the droplet vector calculated by comparing the image from camera 1 to camera 2, (c) shows the averaged droplet flying velocity and (d) shows the averaged droplet flying angle. Figure 3 (b) y axis shows the spray travel distance with Z_{sp}=0 equal to nozzle tip. The analysis first divides spray length into 4 sections, 0-25% (upper), 25-50% (middle), 50-75% (lower) and 75-100% (tip) of spray length for notation purpose. In fig.3(c) and (d), for Z_{sp}=0 to 60mm, each 10mm is divided into individual sections, with the averaged value is calculated for each section. The flying angle is considered to be 0° along the x axis, with positive value clock-wise. Error bar shows the standard deviation.

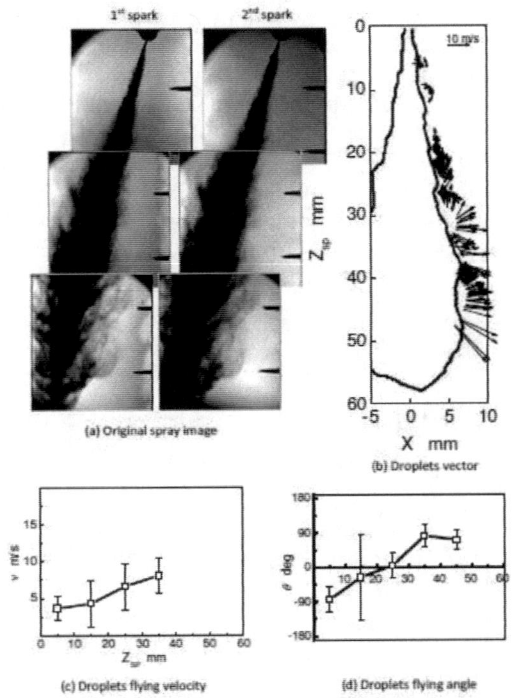

Figure 3. Gas oil spray images and droplets behavior (T_i=298K, P_{inj}=70MPa, Δt=25μs , t=0.50ms)

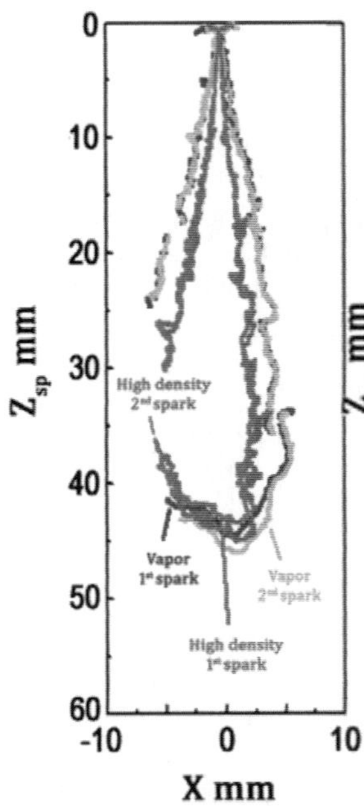

Figure 4. Progression of high density region and vapor boundary in gas oil spray.(T_i=500K, P_{inj}=70MPa, Δt=25μs, t=0.50ms)

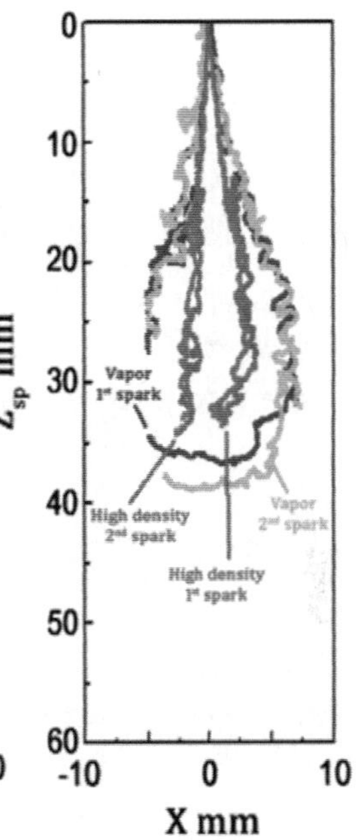

Figure 5. Progression of high density region and vapor boundary in gas oil spray (T_i=700K, P_{inj}=70MPa, Δt=25μs, t=0.50ms)

From the vector line shown in Figure.3(b), it can be seen that the droplets in the middle and lower section of the spray moving outward from the spray boundary, meanwhile the droplets in the upper section is moving inside toward the spray body. This can be confirmed by observing the droplet flying angle graph in Figure.3(d), as the section of Z_{sp}=0~20mm shows negative value. Droplets of the upper section, which subjected to atomization for a longer duration, has smaller diameter and easier to be influence by air entrainment due to small kinetic energy compared to droplets of the medium and lower section. This made the upper section droplets seem to be sucked into, or entrained into the spray body. In comparison, the middle and lower section photography at T_i=500K, P_{inj}=70MPa and t=0.50ms (aSOI).

The 4 lines shows high density liquid phase and vapor phase with the former dictated by red line and the later by blue line. Furthermore, the darker color represent analysis image from camera 1 (1st spark) and the lighter color for camera 2 (2nd spark). Y axis is the spray length and X axis is the spray cross section thickness. From the figure, it can be observe that the vapor phase become large at spray tip section. This characteristic cannot be seen in the upper and middle spray section. It is believed that compared to the upper section, larger spray droplets at the spray tip section with relatively high velocity that is continuously formed and then vaporized, made the vapor phase in this section become larger. Figure 5 shows the condition at T_i=700K, P_{inj}=70MPa and t=0.50ms (aSOI), similar to Figure 4 with different on the temperature. When compared to T_i=500K, it can be observed that the vapor phase on the spray droplets formed after the formation of spray body thus having a bigger kinetic energy fly outward. Figure 4 shows image analysis of high density liquid phase and vapor phase taken by dual nano-spark photography at T_i=500K, P_{inj}=70MPa and t=0.50ms (aSOI). The 4 lines shows high density liquid phase and vapor phase with the former dictated by red line and the later by blue line. Furthermore, the darker color represent analysis image from camera 1 (1st spark) and the lighter color for camera 2 (2nd spark). Y axis is the spray length and X axis is the spray cross section thickness. From the figure, it can be observe that the vapor phase become large at spray tip section. This characteristic cannot be seen in the upper and middle spray section. It is believed that compared to the upper section, larger spray droplets at the spray tip section with relatively high velocity that is continuously formed and then vaporized, made the vapor phase in this section become larger.

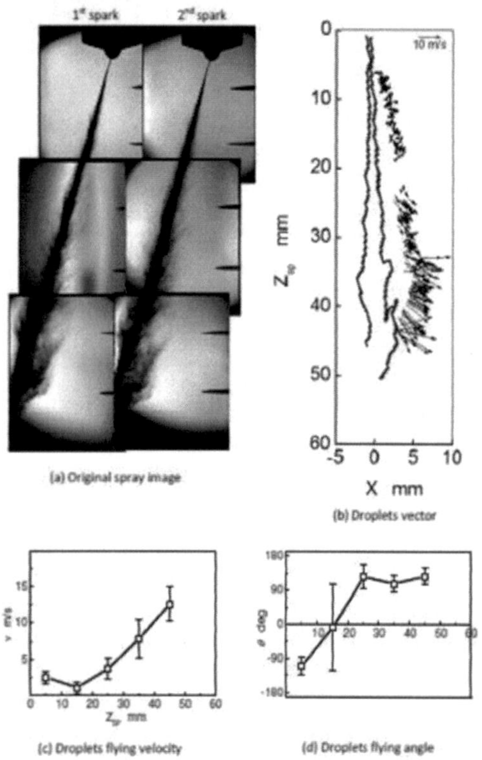

Figure 6. Rapeseed oil spray images and droplets behavior (T_i=298K, P_{inj}=70MPa, t=0.50ms, Δt=25μs)

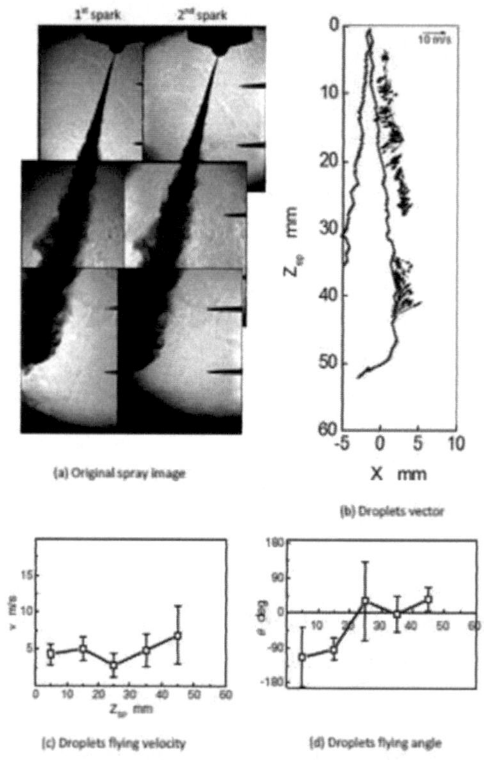

Figure 7. Rapeseed oil spray images and droplets behavior (T_i=700K, P_{inj}=70MPa, t=0.50ms, Δt=25μs)

Figure 5 shows the condition at T_i=700K, P_{inj}=70MPa and t=0.50ms (aSOI), similar to Figure 4 with different on the temperature. When compared to T_i=500K, it can be observed that the vapor phase on the spray tip region become more large. In addition, the high density liquid phase that existed in the spray tip region at T_i=500K can be seen quickly vaporized, making the spray penetration shorter. Figure 6 shows RO spray image and its analysis at T_i=298K, P_{inj}=70MPa, t=0.50ms aSOI (after start of injection) taken by dual nano-spark shadowgraphy and time interval between the color and have different sizes of sphere for red color. The prototype color detection assistive device, for experimental purposes only detects 4 base

IOP Conf. Series: Journal of Physics: Conf. Series **822** (2017) 012058 doi:10.1088/1742-6596/822/1/012058

colours and HSV within its range. Besides the HSV range, the result will display unknown or not detected. nano-spark 1 to 2 is Δt=25μs, similar to Figure.3. It can be observed from Figure 6(b), droplets vector in the upper section especially at Z_{sp}=0~20mm is small, similar to that of the GO. Air entrainment influence also can be seen as the droplets are moving toward the spray body. By observing the section of Z_{sp}=0~20mm of Fig.4(c) and (d), which has similar value to GO, it is understood that the RO spray upper section droplets dynamic is similar to GO.

In contrast, at Z_{sp}=20~40mm, the droplets move along the spray axis, with smaller entrainment effect. The same tendency can be observed when Z_{sp}=40mm and larger. This flying angle tendency of RO spray droplets at middle and lower section is different from GO. Difference in droplet size, with the larger droplets of RO have bigger kinetic energy and the not easily affected by air entrainment. This made the droplets move along the spray axis with little deviation and believed to the one of the reason that made RO spray cone angle smaller when compared to GO. Fuel droplets on the spray tip is continuously under surrounding gas aerodynamic forces, decreasing its velocity, and at the same time, the droplets that came after that pushes forward and replace the existing droplets on the spray tip. The droplets with low kinetic energy are pushed aside and form the outer spray region. This mechanism can be seen on the GO and RO spray. RO spray has high viscosity which made the droplets (quantity) that were pushed aside due to aerodynamic forces was small in numbers. This in turn made the spray cone angle smaller than of the GO and also the branching structure was not well developed. The final result is poor atomization for RO spray. In contrast, as the RO spray droplets kinetic energy is greater than of the GO, the spray penetration length is longer.

Figure 7 shows RO spray image and its analysis at T_i=700K, P_{inj}=70MPa, t=0.50ms aSOI and time interval between nano-spark 1 to 2 is Δt=25μs, similar to Figure 6. The high ambient temperature decrease RO spray viscosity and this can be confirmed with the increase of spray cone angle. In addition when compared to Figure 5(b) at T_i=298K where air entrainment effect can be clearly seen especially at the middle and lower section, it can be observed that droplets fly inward to the spray body exhibiting the influence of air entrainment. Figure 7(c) and (d) shows that the droplets velocity vector becomes small and at the lower section can be seen to fly outward of spray body. Decrease in the droplets size due to lower viscosity of RO spray is believed to be the reasons of these results. Based on these findings, it is clear that high ambient temperature improves atomization of RO spray.

Figure 8 show image of RO spray taken at T_i=850K, P_{inj}=70MPa, t=0.50ms, with a magnified section located on the right side. The magnified image represents area from the full spray image marked with red boundary. From the magnified image it can be seen that at T_i=850K, the formation of vapor phase at spray boundary can be observed. This is very interesting as it is known from previous work, even at T_i=700K, vapor phase is difficult to be observed. This result shows that in the condition of very high ambient temperature, RO spray (middle and lower section) undergoes atomization which eventually leads to vaporization. Figure 9 shows the high density liquid and vapor phase analysis of the image by using in-house image processing algorithm. The high density liquid dictated by red line and the vapor phase by blue line. Y axis is the spray length and X axis is the spray cross section thickness. The result shows that large area of vapor phase in the middle section. In contrast, high density liquid phase can be observed advancing at the spray tip. It is believed that in order to promote faster vaporization at the spray tip section, higher ambient temperature is required.

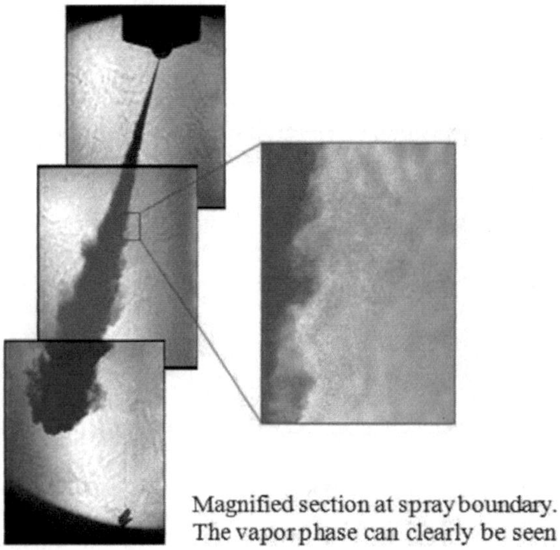

Figure 8. Vaporization of rapeseed oil spray (T_i=850K, P_{inj}=70MPa, t=0.50ms)

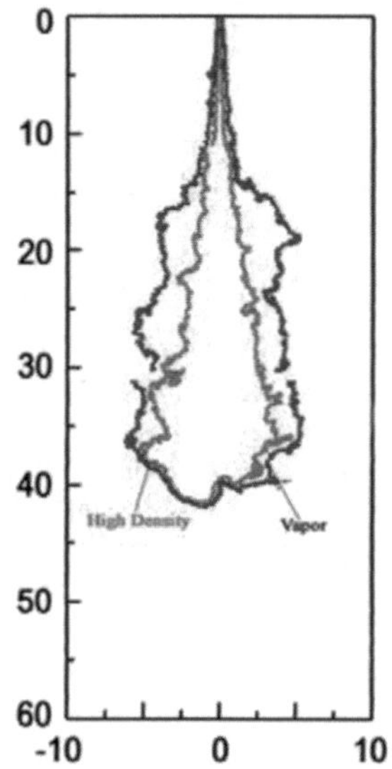

Figure 9. Boundary analysis of vaporization of rapeseed oil spray (T_i=850K, P_{inj}=70MPa, t=0.50ms)

4. Conclusion

Rapeseed oil (RO) has very poor atomization when compared to gas oil (GO) spray. This is in agreement with the results from spray droplet dynamic behavior studies that show due to the high viscosity, RO has larger droplets size with bigger kinematic energy and not easily influence by the air entrainment, especially at the lower section of the spray. High ambient temperature promotes RO spray to expands further from the center line and improve atomization. However, it should also be noted that even at T_i=700K, evaporation of RO spray can little be seen, and vapor region almost cannot be detected. This characteristic poses a problem to engine with low ambient temperature and running at low load, as it can be the cause of SOF increase due to poor atomization of the RO fuel. Very high ambient temperature is needed to improve the atomization of RO as shown in the result at T_i=850K. This also indicate that other methods beside of very high ambient temperature; which difficult to attain when running at low speed, are needed to improve the atomization process is a must when using RO.

Acknowledgement

The authors also would like to thank the Ministry of Higher Education, Malaysia for supporting this research under Exploratory Research Grant Scheme (ERGS) vot.E033. The author would like to express his gratitude to Prof. Dr.Yoshiyuki Kidoguchi from Power Laboratory, The University of Tokushima, Japan for his advice and guidance in this research.

References

[1] R. D. Misra and M. S. Murthy 2010 Straight vegetable oils usage in a compression ignition engine—A review *Renewable and Sustainable Energy Reviews* **14** 3005-3013

[2] Khalid, A. 2013 Effect of ambient temperature and oxygen concentration on iIgnition and combustion process of diesel spray *Asian Journal of Scientific Research* **6(3)** 434-444

[3] Nabi M. N., Shahadat M. M. Z., Rahman M. S. and Beg M. R. A. 2004 Behavior of diesel combustion and exhaust emission with neat diesel fuel and diesel- biodiesel blends. *SAE Paper* **01** 3034

[4] W. Jazair, T. Harada, S. Kubo and Y. Kidoguchi 2007 Improvement of emissions in a DI diesel engine fuelled by bio-diesel fuel and waste cooking oil *SAE Paper* **01** 2029

[5] A.Sapit et al 2013 Effect of wall configuration on atomization of rapeseed oil diesel spray impinging on the wall *Applied Mechanics and Materials* **315** 320-324

[6] Khalid, A., Hayashi, K., Kidoguchi, Y., and Yatsufusa, T 2011 Effect of air entrainment and oxygen concentration on endothermic and heat recovery process of diesel ignition *SAE Technical Papers* **01** 1834

[7] Khalid, A., Tamaldin, N., Jaat, M., Ali, M.F.M., and Manshoor, B., Zaman, I 2013 Impacts of biodiesel storage duration on fuel properties and emissions *Procedia Engineering* **68** 225-230

[8] A. Sapit, S. Nagayasu, Y. Tsuboi, Y. Nada, and Y. Kidoguchi 2011 A study on improvement of diesel spray characteristics fueled by rape-seed oil *SAE Paper* **32** 0561

[9] Yatsufusa T., Kidoguchi Y., Adbullah A., and Gomi T 2009 Measurement of particle diameter in diesel spray by long-distance microscopic shadowgraph photography *Atomization* **18(63)** 2-9

Flow visualization around an apple with and without bagging

H Matsumoto[1], Y Kubota[2], M Ohishi[3] and O Mochizuki[4]

[1] Graduate school of Biomedical Engineering, Toyo University, 2100 Kujirai, Kawagoe-shi, Saitama, 350-8585, Japan
[2] Faculty of Science and Engineering, Toyo University, 2100 Kujirai, Kawagoe-shi, Saitama, 350-8585, Japan
[3] Daikirika Kogyo Inc., 212-8 Akagidai, kounosu-shi, Saitama, 365-0001, Japan
[4] Department of Biomedical Engineering, Toyo University, 2100 Kujirai, Kawagoe-shi, Saitama, 350-8585, Japan

Emails: [1] s36b01510026@toyo.jp, [2] kubota548@toyo.jp, [3] ohishi@daiki.co.jp, [4] mochizuki@toyo.jp

Abstract. The typhoon often causes the vast damage to drop the apple before harvest. Many apples fall from trees by the strong wind. These apples are usually bagged to protect them from insects and control sun light for the apples colouring while they are ripening on the tree. We directly measured the drag force acting on an apple with and without bagging experimentally to bare the influence of the bagging on the dropping mechanism. There are two interesting results through the experiment: the drag coefficient of a naked apple is smaller than a sphere, and the bagging is a cause of increasing drag coefficient. To know the reason of these results, we visualized flow around the apple with and without bagging by using the hydrogen bubbles method in an open water channel in this study. We found two facts as follows: the hollow on the top of an apple plays reduction of width of the wake of an apple and reason of increasing the wake width is the flow separation from peripheral edge of the bagging.

1. Introduction

News tells us that apples drop by strong wind of a typhoon every year. These apples are often bagged to protect them from insects and control sun light for the apples colouring while they are ripening on the tree. When we estimate the drag force on the apple by the wind, the drag coefficient of the apple is necessary. The drag coefficient C_D of a sphere is well known as 0.47 [1], but there are no data about C_D of the apple and much more several shapes of body. Therefore, we have experimentally measured C_D of the apple with and without the bagging, to investigate the influence of the bagging on the drag force of the apple [2]. As a result, two interesting facts become clear through our experiments: the drag coefficient of a naked apple is smaller than a sphere, and the bagging is a cause of increasing drag coefficient. To know the reason of these results, we visualized flow around the apple with and without bagging by using the hydrogen bubbles method in an open water channel in this study. We found two facts as follows: the hollow on the top of an apple plays reduction of width of the wake of an apple and reason of increasing the wake width is the flow separation from peripheral edges of the bagging.

2. Experimental setup

Figure 1 shows an open water channel with 0.18 m width × 0.22 m height × 1.0 m length of a test section. The free stream velocity is 0.18 m/s. The apple model showing in figure 1 was modelled on an apple, *Jonathan*. This model was made with the silicon (Shin-Etu silicon KE-12) to realize the hollow of a real apple and unique curve of the profile of the apple. The dimension of this apple model based on the average of that of Jonathan is as follows: diameter is 76 mm, height is 75 mm. The hardness of this model is 40 durometer angstrom based on the real apple. The Reynolds number of the flow around the apple is 1.4×10^4. The blockage effect of the apple in the test section is 11%.

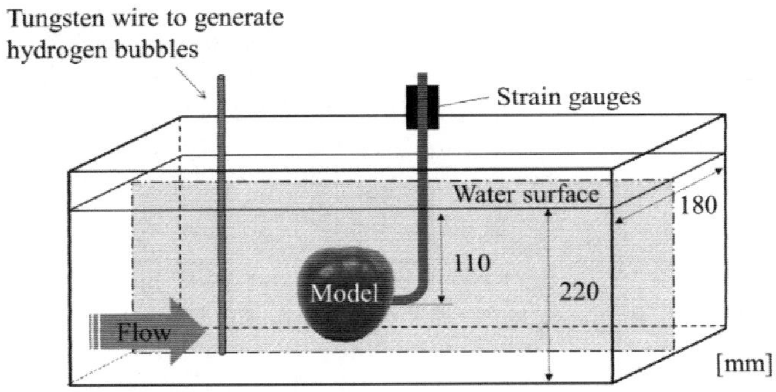

Figure 1. Experimental set up. Apple model in the test section of an open water channel.

The apple is wrapped by the paper bagging (height: 165mm, width: 137mm) that is popularly adopted by famers in Aomori prefecture in Japan. Its surface is waterproofed. The bag has a long slid near the opening of the bag for putting in a stem. Some short slides are on the bottom part for some reasons: preventing a temperature rising in a bag, draining out rain water and allowing air to pass through easily. A wire embedded along one edge is used to close a bag by twisting. Opening a bag along perforation on the middle of the bag makes bright color of apple by exposing to sunlight before just harvest. Figure 2 shows the covered models in top, front and side views. Measured angle is defined in Figure 2(d). Measurement of the drag force acting on the covered model and visualizing the flow around an apple model is carried out with the different angle, since C_D and wake depends on the geometry.

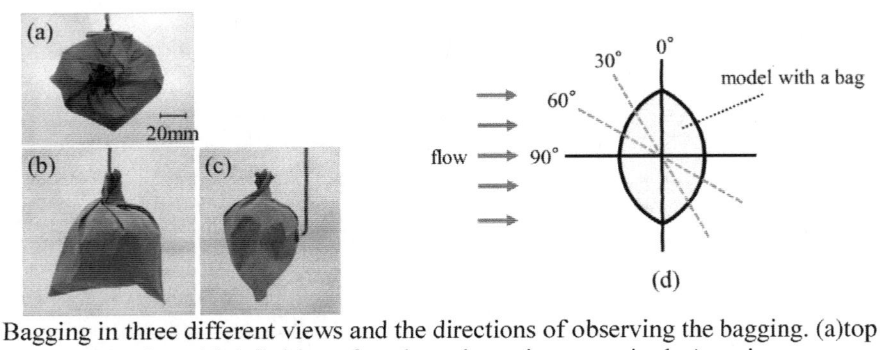

Figure 2. Bagging in three different views and the directions of observing the bagging. (a)top view, (b) front view, (c)side view, (d)definition of angle to the main stream in the top view

Flow around the apple is visualized by a hydrogen bubble method to study the bagging on the drag and the influence of the hollows at top and bottom part of the apple on the drag. Hydrogen bubbles emanate from a tungsten wire connected to a positive electrode to that DC voltage applied continuously. The diameter of wire is 0.05 mm that is placed 30 mm above from the model upright in water tank. A copper plate connected to negative electrode is placed in the downstream of water tank. A light sheet illuminates a vertical plane of the flow around the apple. The measurement of half width is carried out by interval of 10 second in every 5 times due to wavelength of vortex.

A digital camera is used for recording the flow patterns by means of hydrogen bubbles. Behaviour of hydrogen bubble was recorded as flow around the apple model 2 minutes by digital camera. The half width of the wake is measured to correlate with the magnitude of the drag from the flow patterns of the wake. The shape of the covered model has a different projected area with different angle against flow as shown in figure 2.

3. Result

The visualized flow patterns, drag coefficients, half widths are shown in table 1. For comparison, the data of a sphere with similar dimension with the apple obtained in this experiment is presented in the table. We maintain the condition of a model during the measurement of C_D in Table 1 to investigate the effect of the bagging to drag force acting on an apple. Drag coefficient is calculated from the definition. The value of C_D on a sphere is $C_D = 0.45$. This warrants adequacy of our experiment. The drag coefficient of the naked apple with hollows is $C_D = 0.36$. This is the minimum value in our experiment. Furthermore, this is less than the known C_D of a sphere. The result of the bagging for C_D, when the model was set 0, 30, 60 and 90degs, is 0.80, 0.69, 0.61, and 0.55 respectively. This means that the drag force acting on an apple becomes larger due to the bagging. Among the results of the C_D acting on the bagging, the maximum value appears at 0 deg. The minimum value is at 90 deg., this is due to similar shape to a streamlined body. The C_D of the bagging is about twice as larger than the naked apple. The bagging shows a possibility of the cause of dropping apple from the tree easily. This means that if we can find the low influenced shape or surface material of the bagging, the number of dropping apples due to the strong wind of typhoon may be decreased.

To understand the reason why the C_D of the naked apple is smaller than the sphere, we show the flow around the apple, C_D and half width of the wake as shown in Table 1. We can see that the half width of the naked apple is smaller than that of the sphere, and that of the filled hollows. The drag coefficient C_D and half width of the apple filled hollows are a little bit larger that the naked one. This means that the hollow plays an important role to reduce the drag. We think that small eddy in the hollow retards the separation as similar to sucking from the surface of the airfoil.

The C_D and half width of the bagging obtained in each direction show larger values than those of the naked apple. The flow along the bagging apple separates from the peripheral edge of the bagging because of singularity. Thus, the half width of the wake depends on the frontal shape of the bagging. The half width increases with increasing the angle of orientation. The minimum value appears at 90 deg. and the maximum 0 deg. This is similar to change of the C_D.

4. Conclusion

We visualized flow around the apple with and without bagging by using the hydrogen bubbles method in an open water channel in this study. We found two facts as follows: the hollow on the top of an apple plays reduction of width of the wake of an apple and reason of increasing the wake width is the flow separation from peripheral edges of the bagging. Thus, we need to review the shape of paper bagging to decrease the drag force for preventing the apple dropping by the strong wind.

References

[1] Handbook for mechanical engineering Basic edition α4 2006 *The Japan society of mechanical engineers* 80–93

[2] Matsumoto H, Kubota Y, Ohishi M and Mochizuki O 2016 Relation between bagging of an apple and strong wind *Transactions of the JSME* **82** 1–8 in Japanese

[3] Choi J, Jeon W P, and Choi H 2006 Mechanism of drag reduction by dimples on a sphere *Phys. Fluids* **18** 041702

IOP Conf. Series: Journal of Physics: Conf. Series **822** (2017) 012060 doi:10.1088/1742-6596/822/1/012060

Table 1. Flow pattern around a model, C_D and half width of wake

Model	Set-up	Flow visualization using hydrogen bobbles	C_D	Half width
Sphere	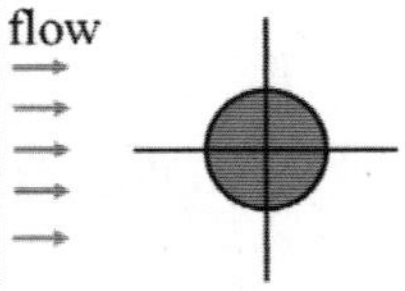	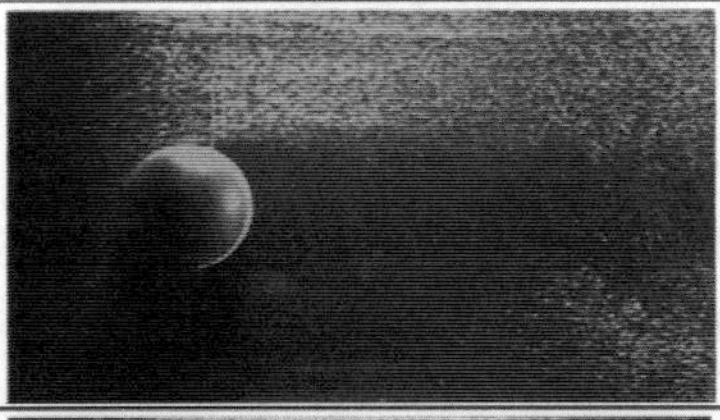	0.45	1.1
Apple	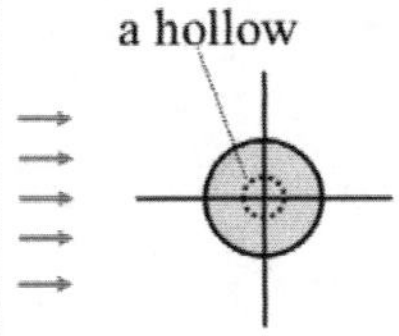	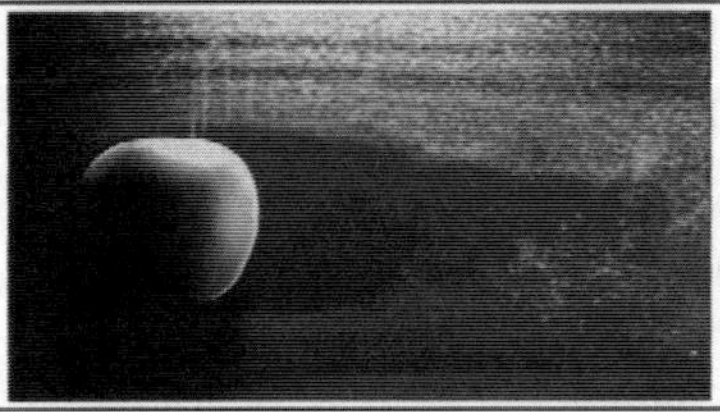	0.36	0.26
	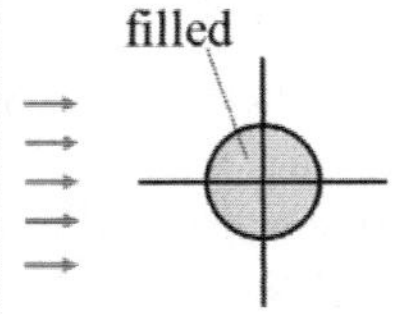	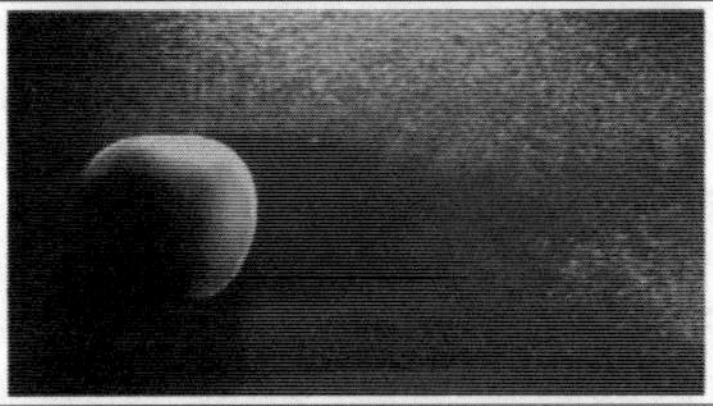	0.39	0.37
Covered apple	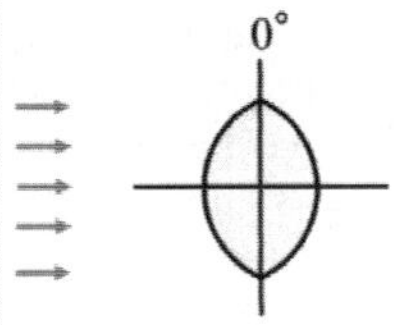		0.80	1.7
	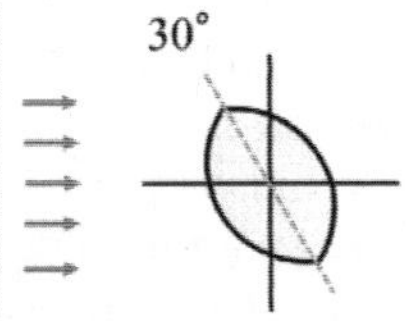	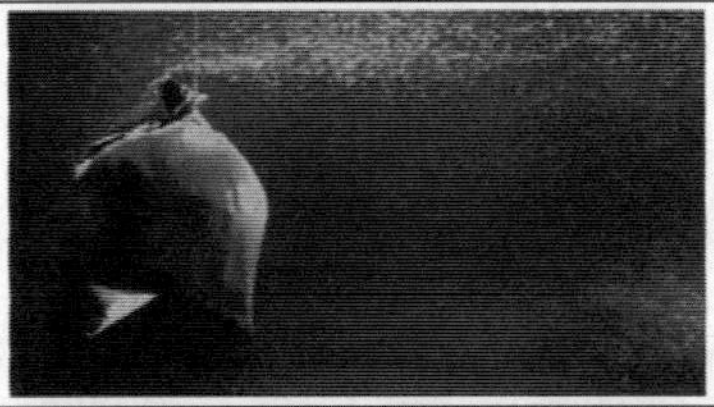	0.69	1.1
	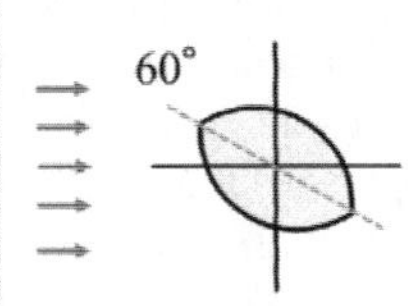	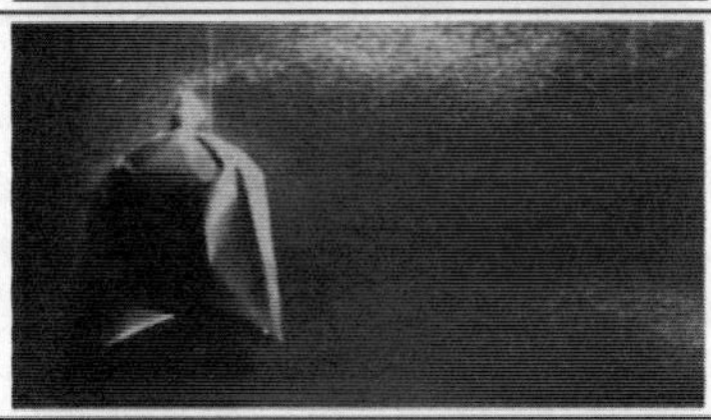	0.61	0.95
	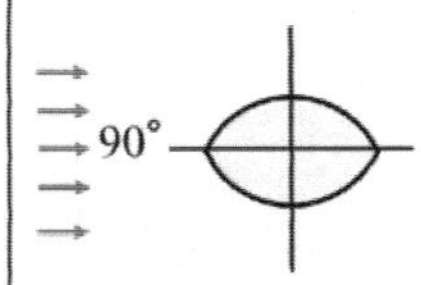		0.55	0.62

IOP Conf. Series: Journal of Physics: Conf. Series **822** (2017) 012061 doi:10.1088/1742-6596/822/1/012061

Characteristics of a wind-actuated aerodynamic braking device for high-speed trains

H Takami[1] and H Maekawa[2]

[1] Brake Control, Vehicle Control Technology Division, Railway Technical Research Institute, Tokyo, Japan
[2] Department of Mechanical Engineering and Intelligent Systems, University of Electro-Communications, Tokyo, Japan

E-mail: takami@rtri.or.jp

Abstract. To shorten the stopping distance of the high speed trains in case of emergency, we developed a small-sized aerodynamic braking unit without use of the friction between a rail and a wheel. The developed device could actuate a pair of two drag panels with a travelling wind. However, after the drag panel fully opened, vibrational movements of the drag panel characterized by its slight flutter were repeated. In this study, to stabilize the opened panel, matters pertaining to the angle of attack with respect to the drag panel and pertaining to the arrangement of the two panels were examined by a wind tunnel experiment using a scale model. As a result, to stabilize the opened panel and to keep the good performance of the braking device, it is found out that an angle of attack of 75 to 80 degrees is suitable provided that the interval of the two panels is narrow enough.

1. Introduction

There is a strong demand for improving the brake performance of high-speed trains in Japan [1] to shorten the stopping distance in case of emergency such as an earthquake. The emergency brake for the conventional train is ensured by a wheel disk brake system. However, its performance is affected by rail surface conditions (dry or wet condition) especially in a high-speed region because of its dependency on the adhesion force between a wheel and a rail. Therefore, an additional brake system which is unaffected by the adhesion force is required, and it should be lightweight and compact.

As s braking device satisfying to this objective, we developed a small-sized aerodynamic braking device for trains [2]. This device can get air drag force directly for braking by a panel set on the train roof. An aircraft is generally equipped with a similar device, but trains have never been equipped with it. One of the originality of the developed device is actuating a pair of two drag panels with a traveling wind (a head wind). In this study, we report matters pertaining to the drag and stability of the aerodynamic braking device which were examined in a wind tunnel experiment using a scale model.

2. Overview of the developed aerodynamic braking device

2.1. Prototype device

The author attempted to miniaturize the device in order for the device to be installed flexibly on the train roof, whereby many devices with small-sized drag panels were appropriately arranged throughout the train roof to obtain higher drag force.

Figure 1 shows the prototype of the aerodynamic braking device. Figure 1(a) shows the drag panel folding position and figure 1(b) shows the opening position (braking position). To raise the drag coefficient of the panel, the drag panel was a rectangular thin plate. The maximum design load by the drag was set to 3,000 N/unit on the assumption that the train runs at 400 km/h in a tunnel section.

A pair of drag panels rotating around a horizontal axis are connected by a pair of spur gears called a 'torque balancer' so as to make each of them rotate in the opposite directions. When the panels open slightly by a spring, the rotational force by the drag acts on the panels so as to open or close the panels; it depends on the train traveling direction and the panel opening direction, as shown in figure 2.

At this time, due to the difference between the drag coefficient of one panel and that of the other, the rotational force acting so as to open the panel becomes greater than that acting so as to close the panel. The difference of drag coefficient is caused by difference in attack angle. Therefore, the pair of drag panels can be actuated by the traveling wind without external power. Besides, the prototype has a mechanical panel locking mechanism actuated by a pneumatic cylinder so that the drag panel does not lift unintentionally by vibration or disturbance when the train running normally.

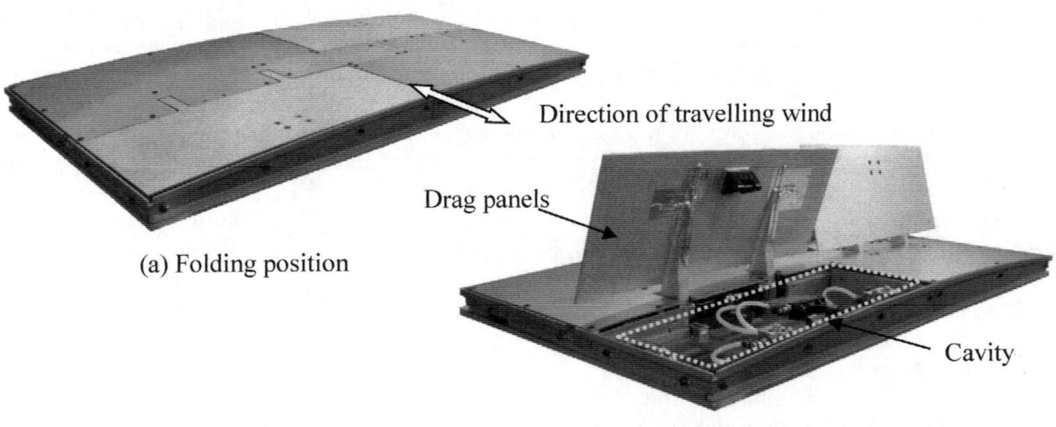

Figure 1. Prototype of aerodynamic braking device.

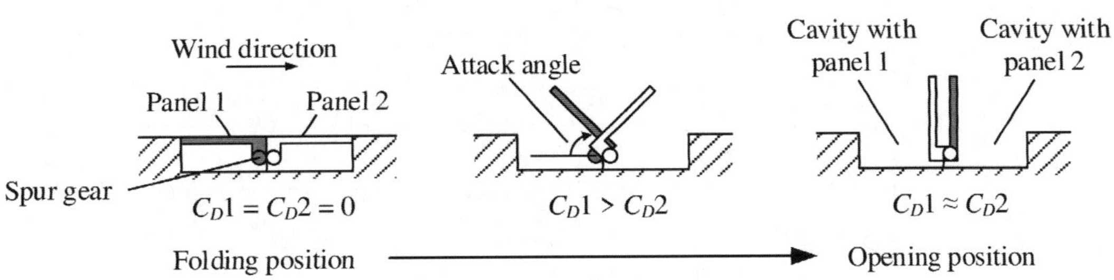

Figure 2. Motion of two-panels.

2.2. *Stability of the prototype device*

The prototype of the aerodynamic braking device was tested in large-scale wind tunnel facilities at a flow speed of 400 km/h (111 m/s). Drag panels immediately opened and produced the drag force, after an operation command was outputted to the device. The process time of the motion from the folding position to the opening position required for getting drag force by the wind was only 0.4 seconds. This rapid motion is a great advantage for emergency braking.

However, after the drag panel fully opened, vibrational movements of the drag panel characterized by its slight flutter were repeated. We considered that this motion was caused by the instability of the rotational force by the drag acting on the pair of the two panels. The stability of the opened panel is as follows:

- Stable: $C_D1 > C_D2$ *and* $C_T1 > C_T2$
- Unstable: $C_T1 \leq C_T2$

Torque coefficient C_T is obtained by formula $C_T = 2T/hA\rho U^2$, where T represents rotational force of the panel; h is the panel height; and A is the front projection area of the panel.

The magnitude of the drag force and that of the torque vary with the following factors:

i) attack angle with respect to a panel,
ii) arrangement of two panels,
iii) location of a cavity for folding a panel.

Here, the cavity that is a box-formed space of around 50 mm in depth for folding a panel is located at the upstream or the downstream side of the panel. According to scale model experiments and numerical analysis with respect to a cavity [3], the drag force of a panel with a cavity at the upstream side was approximately 15% larger than that of a panel with a cavity at the downstream side because of differences in stagnation pressure. For the mechanical reason, the cavity of the panel 1 is always located at the front of the panel, and the cavity of the panel 2 is always located at the back of the panel (cf. figure 2). In this case, the relation of the drag force of the two panels are stable in $C_D1 > C_D2$.

Therefore, to stabilize the opened panel, i) the attack angle with respect to the panels and ii) the arrangement of the two panels were examined by a wind tunnel experiment using a scale model.

3. Drag and stability of the two panels

3.1. Experimental setup

Figure 3 shows an experimental set up of the wind tunnel. The experiments were carried out in small-scale wind tunnel facilities of Railway Technical Research Institute. The test section has a cross section

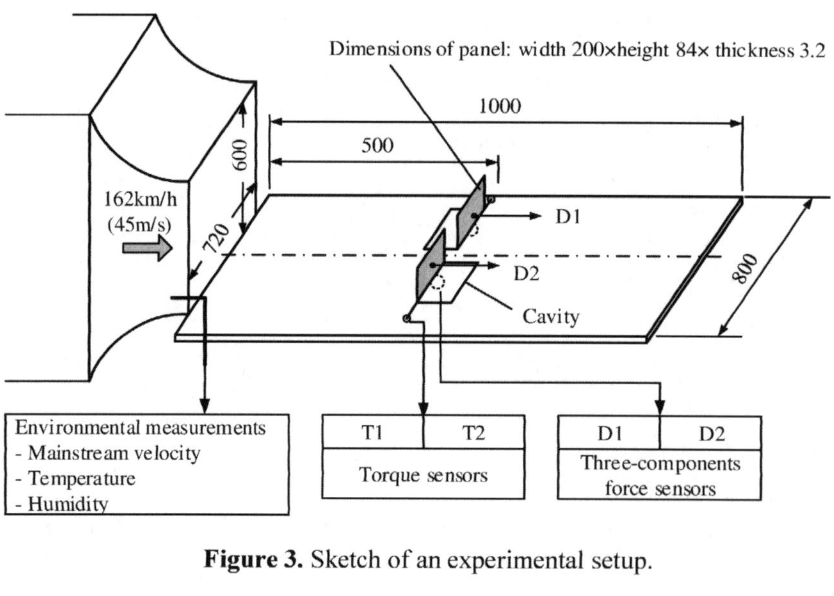

Figure 3. Sketch of an experimental setup.

of 0.6 m in height and 0.72 m in width, and is 1 m in length, and the maximum test velocity is 162 km/h (45 m/s). The model scale is 1/2.5. The Reynolds number based on the width of the drag panel is Re = 6.0×10^5. Dynamic similarity with the full-scale prototype is not completely ensured, because the Reynolds number of the full-scale prototype is about 6 times as large as the scale model experiment. Hence, the purpose of the scale model experiments is to study the basic effect of matters pertaining to the arrangement of the two panels and pertaining to the attack angles with respect to the panel.

To measure the drag force, each panel mounted on three-component force sensor (*NISSHO-ELECTRIC-WORKS*, LMC-3891, full scale ±100 N) with a gap around the fixed parts. In addition, to measure the rotational force of the panel, each axle of the panel is equipped with a torque sensor (specialized sensor consisting of strain gauges). The scale model experiments are examined with static angle (nonrotation); consequently, the measured rotational force is unaffected by a kinetic friction of the spur gear. Furthermore, the torque sensor is calibrated to cancel the rotational force of the panel weight.

3.2. Arrangement of the two panels

Figure 4 shows a schematic of the arrangement of the two panels. One of the streamwise intervals between the two panels S1 is 27 mm based on the prototype length, and the others of the intervals are 10 mm and 6 mm based on the practical dimensions of the rotating axis and the pair of spur gears. All interval conditions are S1 > 0; namely the panel 1 is always located in the downstream of the panel 2.

The spanwise interval between the two panels S2 is unlimited except that S2 < 0. However, it is desirable to make it short as much as possible in order to reduce the device size. Hence, one of the intervals between the two panels S2 is 20 mm based on the prototype length, and the others of the intervals are 10 mm and 1 mm based on the practical dimensions.

Figure 5 shows drag force and torque on each panel vertically opened; here, the plus sign of the torque coefficient is defined as the direction in which it should act to lift each panel. In the case of the streamwise interval S1 = 27 mm, the relation of the drag force of the two panels are unstable in $C_D1 < C_D2$, and the torque coefficient C_T at the 'torque balancer' is negative. At this time, the main stream acts on the panel 2 first located in the upstream, and then a separated flow acts on the panel 1 located in the downstream. In addition, a jet between the panels flows over the cavity at the back of the panel. In the case of the streamwise interval S1 = 6 mm that is the arrangement of the two panels approximately in parallel, the relation of the drag force of the two panels changes to a stable condition in $C_D1 > C_D2$ because C_D1 with the upstream cavity is larger than C_D2 with the downstream cavity in Ref. [3]. However, the torque coefficient C_T is still negative. It could be attributed to the lift force caused by a secondary flow around the panels.

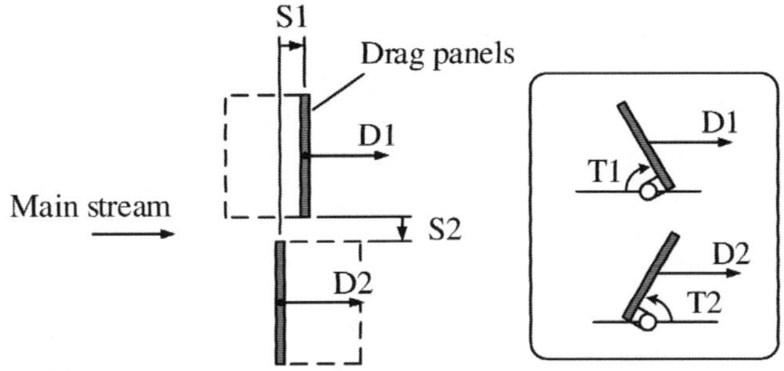

Figure 4. Schematic of arrangement of the two panels. The streamwise intervals S1 are 27 mm, 10 mm and 6 mm. The spanwise intervals S2 are 20 mm, 10 mm and 1 mm.

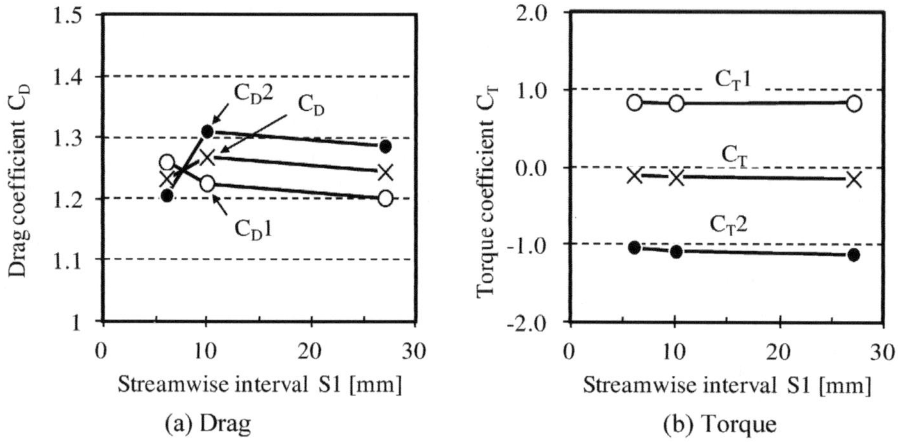

Figure 5. Relationship of the streamwise intervals S1 on each panel vertically opened; spanwise interval S2 is 20 mm.

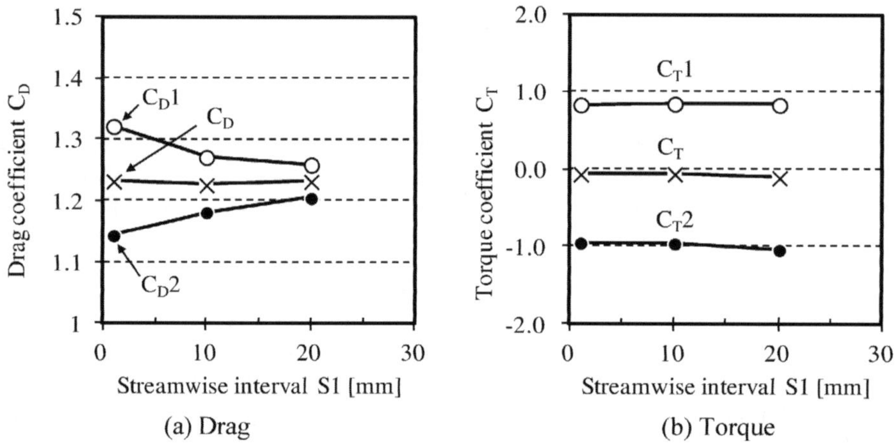

Figure 6. Relationship of spanwise intervals S2 on each panel vertically opened; streamwise interval S1 is 6 mm.

Table 1. C_D and C_T on experimental conditions.

		Streamwise interval S1		
Spanwise interval S2		27 mm	10 mm	6 mm
20 mm	C_D	1.24	1.27	1.23
	C_T	-0.15	-0.13	-0.10
10 mm				1.23
				-0.06
1 mm	C_D	0.98	0.98	1.23
	C_T	-0.13	-0.06	-0.06

Furthermore, the narrow spanwise interval of S2 tends to stabilize the relation of the drag force of the two panels, according to figure 6 under S1 = 6 mm conditions. It is found that both narrower S1 and

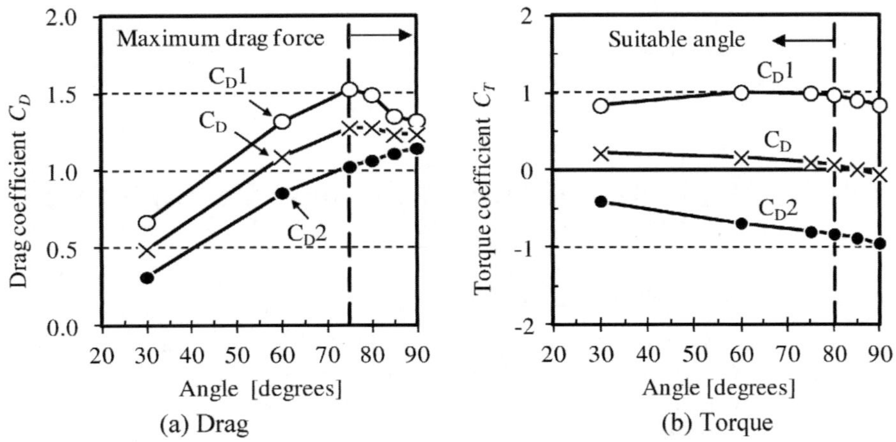

Figure 7. Relationship of the attack angles; streamwise interval S1 is 6 mm, and spanwise interval S2 is 1 mm.

narrower S2 would tend to stabilize the opened panels. Nevertheless, it was impossible to improve the C_T condition by changing the arrangement of the two panels.

3.3. Attack angle with respect to the panel

In this section, to stabilize the opened panel, the attack angle with respect to the fully opened panel is reduced from 90 degrees.

Figure 7 shows the drag and torque according to the attack angle with respect to a panel, provided that the streamwise interval and the spanwise interval are S1 = 6 mm, S2 = 1 mm. It is pointed out that the C_T is in a positive state when the attack angle is 80 degrees or less. On the other hand, the drag coefficient C_D falls off when the attack angle is less than 75 degrees, and the performance of the aerodynamic braking device goes down.

Therefore, to stabilize the opened panel and to keep the good performance of the braking device, it is found out that an attack angle of 75 to 80 degrees is suitable on this arrangement of the two panels.

4. Conclusion

A wind-actuated aerodynamic braking device consisted of a pair of two drag panels. However, after the drag panel fully opened, vibrational movements of the drag panel characterized by its slight flutter were repeated.

In this study, to stabilize the opened panel, matters pertaining to the angle of attack with respect to the drag panel and pertaining to the arrangement of the two panels were examined by a wind tunnel experiment using a scale model. As a result, to stabilize the opened panel and to keep the good performance of the braking device, it is found out that an angle of attack of 75 to 80 degrees is suitable provided that the interval of the two panels is narrow enough.

References

[1] Arai H, Kanno S, Fujino K, Kato H and Asano K 2010 *JR EAST Technical Review* **31** 17–21
[2] Takami H 2013 *Transactions of the Japan Society of Mechanical Engineers Series* B **79**-803 1254-1263
[3] Inoue Y, Suzuki M, Maekawa H and Takami H 2015 *Proc. of ASME-JSME-KSME joint Fluids Eng. Conference*

Fluids vertical transfer utilizing VFD based centrifugal pumps

Vinesh Thiruchelvam[1], Lim Min Hong[2]

[1]Faculty of Computing, Engineering & Technology
Asia Pacific University of Technology & Innovation
Kuala Lumpur, Malaysia
[2]Institute of Engineers Malaysia
Bangunan Ingeniur
Petaling Jaya, Selangor
Malaysia
[1]dr.vinesh@apu.edu.my
[2]blitz.1992@hotmail.com

Abstract. Efficiency of centrifugal pumps has been consistently improving with technology enhancements. Pump manufacturers have increased the variety of pump dimension. Pump selection is based on application selection primarily on energy conversation. In a conventional pumping system, the pump will be driven at the same speed even though the output usage is below the actual demand. As a result, energy is wasted and the pump could have been driven at a lower speed for the same operational function. One of the ideal solutions is to vary the pump speed based on real-time demand. The paper illustrates the system simulation for a high rise building water transfer and the relevant design calculations such as the volumetric flow rate and net positive suction head. Supporting this operational function, the variable frequency device (VFD) has also been designed and simulated by using the Proteus software.

1. Introduction

Pump is an electro-mechanical system which was invented with the purpose of transferring fluids based on variety, velocity and volumetric figures. Water pumps allows for transfer of water to a higher height or a determined length. In dry climatic countries, the water pump has played a vital role water supply across borders from other countries. The most popular application is via centrifugal pumps which have lower maintenance cost and has high flexibility in terms of implementation. This philosophy is centrifugal application is relevant to changes in the pump speed or diameter of the impeller and this is known as the Affinity Law.

2. Research Problem

Pump for a conventional pumping system is selected based on the demand of the application and the maximum output usage will be the data to be referred. This ensures that the pump will be able to provide sufficient water in any kind of circumstances or physical arrangement. Wasted energy will be consumed when the output flow rate is lower compared to the maximum usage. On the other hand, relevant calculations such as minor losses and Net Positive Suction Head (NPSH) have been normally neglected during the design stage. This will affect pump performance and lead to the risk of cavitation. The flow rate of the application is normally controlled by the mechanical valve. This current developed solution will effectively control the flow rate of the system. However, the drawback is the heat dissipation of the pump that will increase since the flow will be restricted. This will shorten the life span of the pump as the pressure on the pump will significantly increase as well.

3. System Design

For the project a scaled down prototype has been designed and built to analyse the performance of the pump selected. The Proteus software has been used to design the VFD circuit. Arduino Uno has been chosen as the micro-controller to vary the frequency range of the VFD. To increase the precision of the system, all mechanical hardware has been designed using the SolidWorks software ensuring zero dimension errors. Fixed dimensions were the base to carrying out the design calculations to determine the pump capacity. Engineering calculations were based on meeting criteria are listed as below. Studies were done on all five criteria meeting the objective of ascertaining energy efficient pumping conditions.

- Volumetric Flow Rate
- Required Net Head
- Net Positive Suction Head
- Affinity Law
- Minor Loss

3.1 Volumetric Flow Rate

Understanding the characteristic of a system is important when selecting the size of a pump. According to the H – Q curve which is provided by the manufacturer, the information required when selecting the pump are the volumetric flow rates and the required net head. For this research project, the energy is conserved by changing the speed of the pump based on the output flow rate. The pump speed reduced according to the real time demand. Therefore, the volumetric flow rate of the output pipe was determined. To calculate the volumetric flow rate of the system, the Bernoulli principle was applied as shown in Equation 1;

$$\frac{P}{\rho} + \frac{V^2{}_1}{2} + gz_1 = \frac{P}{\rho} + \frac{V^2{}_2}{2} + gz_2 \quad (1)$$

Where,
P = pressure
V= velocity of the water
g = gravitational force
z = height
ρ = density

3.2 Required Net Head

The required Net Head is essential information when selecting the pump capacity. To determine the net head required from the pump, the equation in Equation 2 was used.

$$H_{required} = \frac{P_2 - P_1}{\rho g} + \frac{\alpha_2 V_2{}^2 - \alpha_1 V_1{}^2}{2g} + (Z_2 - Z_1) + h_{L,total} \quad (2)$$

In the equation, P_1 and P_2 are the pressures at point 1 and 2 while V_1 and V_2 are the velocities at point 1 and 2 respectively. For the α1 and α2, these are the kinetic energy correction factors. In most cases, these two terms are often ignored since the flow of the system is turbulent. Z1 represents the height of the inlet water level while the Z2 represents the highest point of the outlet pipe and both are measured from the ground level.

3.3 Net Positive Suction Head

Net Positive Suction Head (NPSH), is important term when it comes to the pump selection. Based on the Equation 3, it is found that the NPSH is the value of the suction head based on the consideration of external factors such as location above sea levels. In a pump application, there are two types of NPSH which are known as NPSH required and NPSH available. NPSH required is the value of suction head required by the pump and it is provided by the manufacturer [4]. On the other hand, NPSH available is the suction head which is available from the pipeline system. These are values in which the user should apply design calculations according to the layout and operating conditions of the system. As a rule of thumb, the NPSH available from the system should be always greater than the NPSH required by the pump. The consideration of the NPSH value during the design stage will enhance the reduction for cavitation to occur [6].

$$NPSH_a = \pm h_s - h_L + h_A - h_v \tag{3}$$

Where,
$NPSH_a$ = net positive suction head available, ft
h_s = static suction head (if positive value) or static suction
h_v = vapour pressure of liquid at the operating temperature
h_A = absolute pressure at liquid surface
h_L = head loss due to the friction in the suction line of the pump

3.4 Affinity law

Affinity Law is a principle which is specifically used to determine the performance of the pump when either the speed of the pump or the size of the impeller had been changed. According to this law, the torque created by the pump is proportional to the square of the speed of the pump. In other words, the torque of the pump will be increased by a factor of four when the original speed of the pump has been increased twice. As an engineering principle, the power of the device is the multiplication product of the speed and torque. Therefore, the power produced by the pump will be triple of the speed [7].

$$\left(\frac{N_1}{N_2}\right) = \frac{V_1}{V_2}$$

$$\left(\frac{N_1}{N_2}\right)^2 = \frac{H_1}{H_2}$$

$$\left(\frac{N_1}{N_2}\right)^3 = \frac{P_1}{P_2}$$

Where,
N = speed of the pump's motor
V = volumetric flow rate
H = dynamic head
P = power of the pump

3.5 Minor Losses

In fluid mechanics, the minor loss can be defined as the friction loss caused by the roughness of the internal surface of the pipeline. According to the calculations performed, it also showed flow restriction caused by the joint and valve. The actual flow rate will be significantly affected by friction losses [1]. To determine the losses, the equations utilized include the Reynolds Number, Volumetric Flow Rate, Colebrook Equation, and the Darcy Weisbach. The equations shown in at below have been used for the performance monitoring on the research project.

Reynolds number;

$$Re = \frac{V_{avg} D}{v}$$

Volumetric flow rate;

$$V = \frac{\dot{v}}{A}$$

Colebrook equation;

$$\frac{1}{\sqrt{f}} = -2 \log_{10}\left(\frac{\varepsilon/D}{3.7} + \frac{2.51}{Re\sqrt{f}}\right)$$

Darcy Weisbach equation;

$$h_L = f \frac{L}{D} \frac{V^2}{2g}$$

To apply equations, the Engineering Equation Solver (EES) software was required since there are more than two unknowns found in the equations.

4. Prototype Implementation

The prototype for this research project was constructed based on the engineering drawing development using SolidWorks as shown in Figure 1. As for the media of transferring water, a PVC pipe with the diameter of 1 inch was been chosen and used throughout whole system. Meanwhile, the VFD has been built based on the required electronics and component rating.

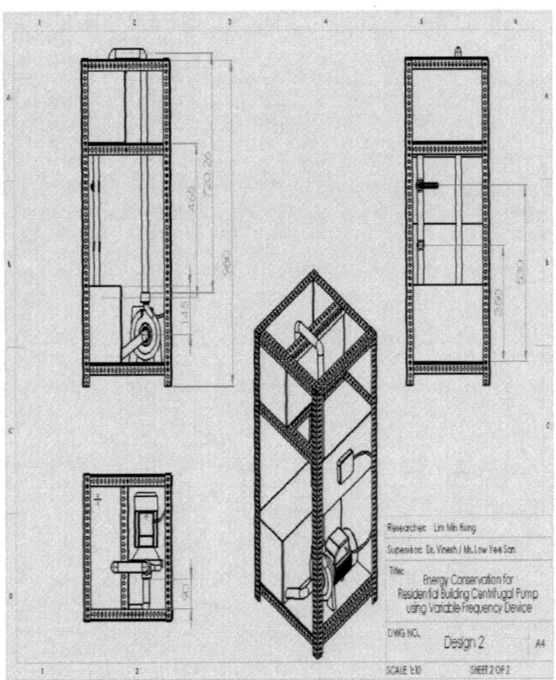

Figure 1. Prototype Design in SolidWorks

5. Variable Frequency Drive

In a Variable Frequency Drive (VFD), there are three main sections which operate in different type of functions to optimize the system and generate the pulse width modulation (PWM). These sections included the rectifier, DC Bus, and inverter [3]. The main function of the rectifier is to rectify the AC input so that the output does not have negative voltage. It was constructed by diodes with a full wave bridge layout. The DC bus operates as the stabilizer and smoothen the rectified voltage. To stabilize the voltage from its source, the capacitor with a large capacitance is connected to the rectifier. To design and simulate the circuit, the VFD was constructed using Proteus. The VFD circuit diagram is illustrated on Figure 2. To isolate the power switch and microcontroller, the gate driver was connected between the gate of the MOSFET and the Arduino. The system will be operated by using the frequency of 50Hz. To calculate the capacitance value, the inductance value has been assumed as 100mH.

$$f = \frac{1}{2\pi\sqrt{LC}}$$

$$50 = \frac{1}{2\pi\sqrt{(100m)C}}$$

$$C \approx 100\mu C$$

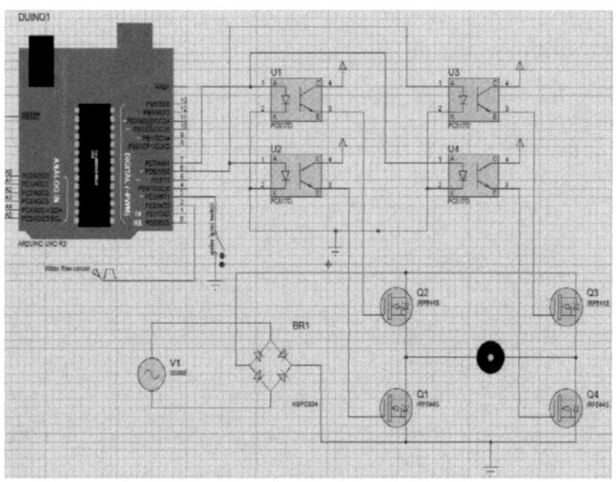

Figure 2. VFD Circuit Design Using Proteus

The inverter was constructed with power switches and is controlled by an external controller. The amount of power switches connected was dependant on the number of input phases of the application. At this point, the DC source from the DC bus converted into pulse width modulated waves before it supplied current to the motor. In order to control the speed of the motor, the power switches were consistently switched on and off according to the desired frequency and duty cycle [5]. For the filtering part of the circuit, the resonance circuit or known as the RL circuit was been used. With the RL circuit, the square wave AC output will be smoother to achieve a sine wave AC output. The diode arrange in the H Bridge configuration for the VFD is shown in Figure 3.

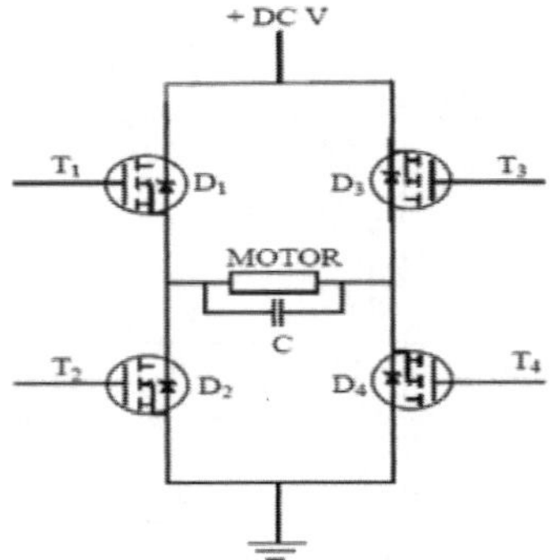

Figure 3. The Inverter of the VFD in H bridge configuration

6. AC Motor Speed Control

To determine the speed of the induction motor used for this research, there were two common terms which were considered, synchronous speed and the rated speed. The rated speed is the actual speed of the motor while the theoretical speed is represented by the synchronous speed. The rated speed was less than the synchronous speed due the external factors. As per the synchronous speed equation determinant used, it was found that the main factors which affected the synchronous speed were the frequency and the number of poles. The frequency is directly proportional to the synchronous speed while the number of poles is reverse proportional to the speed [2]. Controlling an AC induction motor requires an inverter circuit which can produce a forward and reverse voltage. Therefore the inverter circuit was constructed with the H bridge layout and the pulse was sent to the gate of the MOSFET. In each of the half cycle, there were only two MOSFETs triggered by the microcontroller. This design resulted in six different operating conditions if any two of the MOSFETs are triggered. However, it was found that only two of the conditions can be used to power the load. During operation, the MOSFET T1 and T4 worked as a pair to supply forward voltage to the motor while the reverse voltage was supplied when T2 and T3 are triggered. The Figure 4 below shows the timing diagram for each of the power switches.

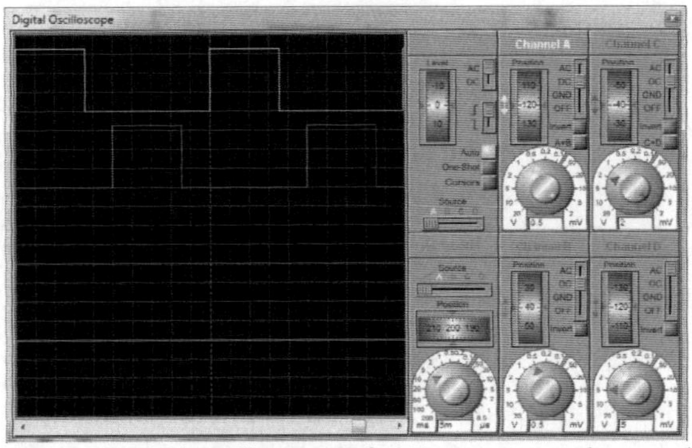

Figure 4. Timing diagram of AC motor speed control

7. Testing & Analysis

During testing, the first test was based on the speed of the motor that was changed by feeding different frequency ranges to the VFD. The frequency range was set between 20 Hz to 50 Hz considering safety precautions. The flow rates of the pump was recorded when the speed of pump was changed. This is shown in Table 1.

Table 1. Flow Rates of the Pump at Different Speeds

Frequency (Hz)	Speed of the motor (rpm)	Flow rate (litre/min)
20	1200	36.34
30	1800	55.67
40	2400	74.83
50	2850	90.95

The second test carried out was on the measurement of power consumption at different pump speeds. In this test, the wattage meter was used to measure the power consumption of the pump. Similar to the first test, the pump speed was varied between 1200 rpm to 2850 rpm and the power consumption was recorded at each speed difference as shown in Table 2.

Table 2. Pump Power Consumption at Different Speeds

Pump Speed	Wattage of the pump (kW)	
	Theoretical	Practical
1200	27.62	28.23
1800	93.21	95.46
2400	220.95	226.24
2850	370.00	372.30

Based on the data recorded during the two tests conducted, it was found that the flow rate generated by the pump is directly proportional to the pump speed.

8. Testing & Analysis

In conclusion, the research has proven that the power consumption of the pump can be reduced by applying the philosophy of the Affinity Law. To achieve the objective of energy reductions, the VFD is accepted based on the test results as the most suitable solution since it will able to reduce the pump speed without sacrificing on the pump lifespan. With the combination of feedback system, energy conservation can be achieved. For instance, a flow rate sensor can be implemented at the output of the tank to examine the usage of the water consumption. By using the data of a particular sensor, the pump speed can be varied based on the current status of the water demand via the water circuit monitoring.

References

[1] Yunus, A. Cengel & John, M. Cimbala. (2010). *Fluid Mechanics: Fundamentals and Applications*. 2nd Edition. Europe: Mcgraw HIll Education.

[2] Atul M. Gajare & Nitin R. Bhasme, (2012). A Review on Speed Control Techniques of Single Phase Induction Motor. *International Journal of Computer Technology and Electronic Engineering*. 2(5). p. 33 – 39.

[3] Jigar N. Mistry (2012). Variable Frequency Drive. *Research Expo International Multidisciplinary Research Journal*. 3(2). p. 252 – 256.

[4] Maher Abdul Ameer & Audai Hussein Al-Abbas (2013). Experimental and Numerical Investigations of the Effects of Net Positive Suction Head on Water Hammer in Pipeline Systems. *International Journal of Engineering and Advanced Technology*. 3(1). p. 86 – 91.

[5] Neetha John, Mohandas R., Suja C Rajappan (2013). Energy Saving Mechanisms Using Variable Frequency Devices. *International Journal of Emerging Technology and Advanced Engineering*. 3(3). p. 784 – 790.

[6] Mohammed Khudhair Abbas (2010). Cavitations in Centrifugal Pump. *First Engineering Scientific Conference.* Baghdad, Iraq, 22 - 23 December 2010. Baghdad: Diyala University. Pp. 170 – 180

[7] Walter V. Jones (2011). Motor Selection for Centrifugal Pump Applications Make Easy. *Pulp and Paper Industry Technical Conference*. Nashville, TN, 19 – 23 June 2011. Nashville: IEEE. pp. 140 – 150.

IOP Conf. Series: Journal of Physics: Conf. Series **822** (2017) 012063 doi:10.1088/1742-6596/822/1/012063

Kinetic modeling of liquefied petroleum gas (LPG) reduction of titania in MATLAB

Tan Wei Yin[1], Sivakumar Ramakrishnan[1], Sheikh Abdul Rezan[1*], Ahmad Fauzi Mohd Noor[1], Noor Izah Shoparwe[1], Reza Alizadeh[2] and Parham Roohi[2]

[1]School of Materials and Mineral Resources Engineering, Universiti Sains Malaysia,14300, Nibong Tebal, Penang ,Malaysia
[2]Sahand University of Technology, Iran

E-mail:*srsheikh@usm.my

Abstract. In the present study, reduction of Titania (TiO_2) by liquefied petroleum gas (LPG)-hydrogen-argon gas mixture was investigated by experimental and kinetic modelling in MATLAB. The reduction experiments were carried out in the temperature range of 1100-1200°C with a reduction time from 1-3 hours and 10-20 minutes of LPG flowing time. A shrinking core model (SCM) was employed for the kinetic modelling in order to determine the rate and extent of reduction. The highest experimental extent of reduction of 38% occurred at a temperature of 1200°C with 3 hours reduction time and 20 minutes of LPG flowing time. The SCM gave a predicted extent of reduction of 82.1% due to assumptions made in the model. The deviation between SCM and experimental data was attributed to porosity, thermodynamic properties and minute thermal fluctuations within the sample. In general, the reduction rates increased with increasing reduction temperature and LPG flowing time.

Keywords: Titania, LPG, Kinetic Modelling, MATLAB, Shrinking Core Model, Titanium Carbide

1. Introduction

Titanium-containing compounds are very important in the advance technologies especially in the field of aerospace and biomedical. Commercially, titania was reduced into metallic titanium and titanium based compounds such as titanium carbide and nitride which have applications in catalyst, paint pigment or manufacture of composite materials [1]. The current technology used for production of metallic titanium through the Kroll process is by chlorination of titania at high temperature at about 800°C-1100°C [2]. According to Adipuri et al. [3] ,chlorination of titanium oxycarbide (TiO_xC_{1-x}) occurs at much lower temperature, around 400-500°C than TiO_2 in the production of titanium tetrachloride through the Kroll process. Synthesis of TiO_xC_{1-x} can be introduced through different technologies. It is mainly produced by reduction of titania with carbon in the temperature range of 1700 to 2000°C [4]. Studies by Dewan et al. [5] have shown that carbothermal reduction of titania in different gas atmosphere was significant and it was found that hydrogen reduced titania and reacted with carbon to form methane gas. This methane gas promoted the formation of titanium sub oxides which was reduced into TiO_xC_{1-x}. For this paper, synthesis of TiO_xC_{1-x} by liquefied petroleum gas (LPG) could have few advantages such as the replacement of the coal as the reducing agent. Besides this, a low CO_2 emission can be achieved compared to conventional reduction and being environmentally friendly by reusing natural gas which is abundant in Malaysia [6].

2. Experimental Setup

The experimental setup is shown in Figure 1. Raw material was titania powder with an average particle size of about 3.71 μm supplied by Alfa Aesar, USA. The sample mass was 1.0 gram for each reduction experiment. The reduction experiments were carried out in a reactor tube furnace which was purged with a mixture of commercial liquefied petroleum gas (LPG), hydrogen and high-purity argon as gaseous reductants. The gas mixtures was 10/45/45 vol. % of LPG-H_2-Ar gas. Reduction was carried out between 1100-1200°C for 1-3 hours. LPG gas analysis by gas chromatography have shown its composition to be composed of 69.9 vol. % propane (C_3H_8) and 28.7 vol. % butane (C_4H_{10}) [7]. At high temperature, all of the C_3H_8 and C_4H_{10} will be converted to CH_4 due to presence of TiO_2 as a catalyst and thermodynamic stability of CH_4 at the reduction temperature used [8]. All gases were supplied by Goldenair Industrial Supply Sdn. Berhad, Malaysia. The total gas flow rate was maintained at 1.5 Liter/min.

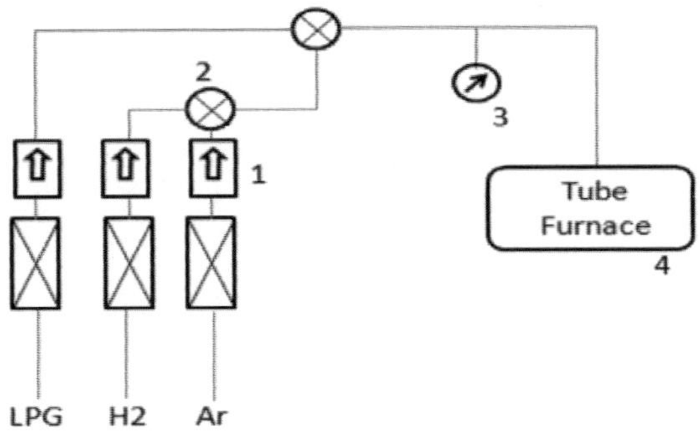

Figure 1. Schematic flow sheet of gas system (1) Regulator with flow meter, (2) gas splitter, (3) gas flow rate gauge (LPM), (4) reactor tube furnace

3. Experimental

TiO_2 reduction data for the kinetic modelling work has been published by Yin et al. [9]. Besides this, experimental data by Sheikh Abdul Rezan et al. [11] was also used in the kinetic model. The reduction mechanism of TiO_2 by H_2 follows a shrinking core model (SCM) as proven by the latter studies. For methane (CH_4) reduction of TiO_2, studies by Guangqing Zhang et al. [12, 13] have shown the SCM model gives a good representation of the kinetics.

4. Kinetic Modelling Methodology

The kinetic model was coded using MATLAB R2011b software based on the SCM. Details on the MATLAB model have been reported elsewhere [10, 14]. In order to solve the complicated equations regarding CH_4 reduction of TiO_2 in the kinetic model, step by step modelling process was done based on its mass and thermal transport properties [15]. The kinetic model developed by MATLAB in this study has been successfully applied previously on isothermal and non-isothermal reduction of TiO_2 and Fe_2O_3 in H_2 gas atmosphere [10, 14, 22]. A flowchart on the computational methodology in the SCM is given in Figure 2.

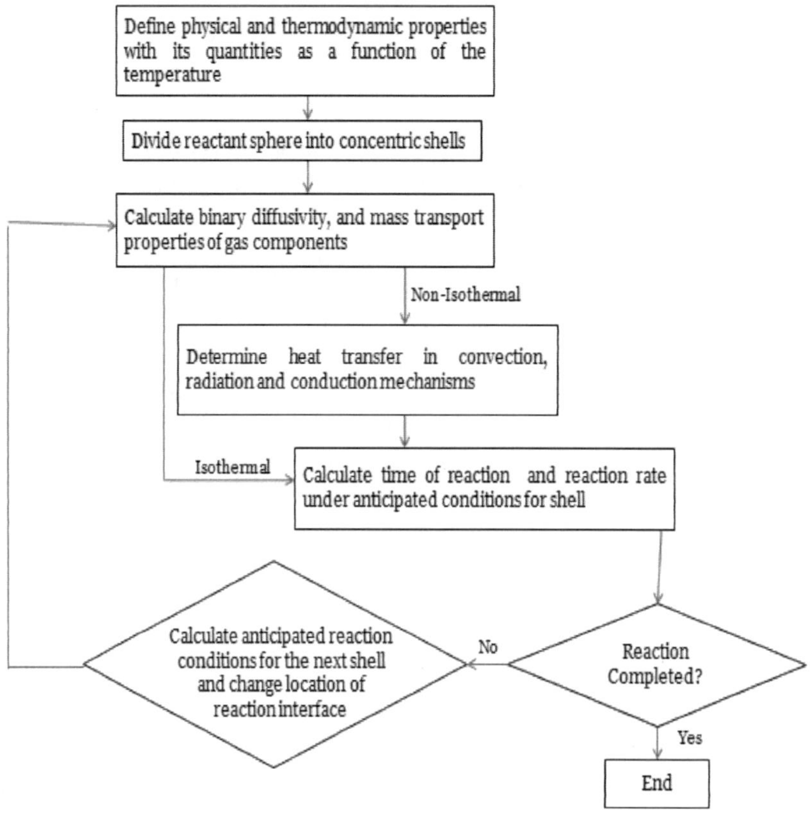

Figure 2. Flowchart on the computational methodology

4.1 Determination Rate of Transfer

Consider the chemical reaction in Eq. 1, it was assume TiO_2 is a solid spherical reactant metallic oxide compound, CH_4 is the reactant gas, TiC is the solid reaction product and CO with H_2 is the evolved gaseous product. The evolved gas product was mainly H_2 since CO content was less than 1 vol. % [9]. The overall reaction can be described by Eq. 2 with the symbol D designated as the product gas mixture of CO and H_2. Since LPG gas concentration was 10 vol.% whereas H_2 gas was 45 vol. %, the CO gas concentration could be ignored as it was less than 1 vol. % due to the small TiO_2 pellet mass. If gaseous transport away from the pellet was defined to be positive magnitude in nature, the rates of mass transport for n_A (CH_4) and n_B (H_2) for ideal gas behavior was represented by Eq. 3-4.

$$TiO_2 + 3CH_{4(g)} = TiC + 2CO_{(g)} + 6H_{2(g)} \quad (1)$$
$$\Delta G° = -0.157*T + 145.3 \text{ (kcal/mol)}$$

$$TiO_2 + 3CH_{4(g)} = TiC + 2CO_{(g)} + 6H_{2(g)}, \; A_{(s)} + B_{(g)} = C_{(s)} + D_{(g)} \quad (2)$$
where $A = TiO_{2(s)}$, $B = CH_{4(g)}$, $C = TiC_{(s)}$ and $D = CO_{(g)}$ & $H_{2(g)}$ combined

$$n_A = \frac{4\pi Ro^2}{R} K_m \left(\frac{PCH_4(s)}{Ts} - \frac{PCH_4(o)}{Tg}\right) \quad (3)$$

$$n_B = \frac{4\pi Ro^2}{R} K_m \left(\frac{P(H_2)_{(s)}}{Ts} - \frac{P(H_2)_{(o)}}{Tg}\right) \quad (4)$$

where K_m = Mass transfer coefficient (cm/sec), R = Gas constant of 82.05 atm. cm^3/g-mole-°K, $P_A(a)$ = Pressure of reactant gas in atmosphere, $P_B(a)$ = Pressure of product gas in atmosphere and a= surface (s) or bulk(o) gas.

4.2 Prediction of Convective Mass Transfer Coefficients
The convective mass transfer coefficient between a gas and a particle with spherical shape was calculated from Eq. 5-6,

$$\mathbf{Sh} = \mathbf{2.0} + (\mathbf{Re})^{\frac{1}{2}}(\mathbf{Sc})^{\frac{1}{3}} \quad (5)$$

$$Sh = \frac{K_m \times d_o}{D_{AB}} \quad (6)$$

where Sherwood number, *Sh* is a function of the mass Reynolds number (Re), Schmidt number (Sc), and binary diffusivity (D_{AB}). AB was the reductant and product gas defined as CH_4-H_2.

4.3 Prediction of Binary Diffusivity
The method was based on kinetic theory and gives predictions with an average deviation of 4 pct. (%) with a maximum deviation of 16% [16]. The binary diffusivity $\boldsymbol{\beta_{AB}}$ (cm^2/sec)prediction was made using Eq. 7.

$$\beta_{AB} = \frac{BT^{3/2}\sqrt{(\frac{1}{M_1} + \frac{1}{M_2})}}{\mathrm{P} * \mathrm{r}\,(r_{AB})^2 \times ID} \quad (7)$$

where r_{AB} = collision diameter (Å), taken as the mean of the collision diameters of components A (CH_4) and B (H_2), M_1, M_2 = molecular weights of CH_4 and H_2, B = [10.7 – $2.46\sqrt{(1/M_A + 1/M_B)}$] x 10^{-4}, ID = Collision integral of diffusion (unitless), P = absolute pressure (atm) and T = Temperature (K). The symbol A & B was designated as the main reductant-product gas.

4.4 Prediction of Convective Heat Transfer Coefficients
Equations 8-9 below enables accurate prediction of the convective heat transfer coefficient between a sphere and flowing gas using dimensionless parameter called Nusselt (Nu) number.

$$Nu = 2.0 + 0.6(Re)^{\frac{1}{2}}(Pr)^{\frac{1}{3}} \quad (8)$$

$$Nu = \frac{h_c \times d_o}{K_f} \quad (9)$$

where Re =Reynolds number, Pr = Prandtl number, h_c = convective heat transfer coefficient (cal/cm^2-sec-°K) and K_f = thermal conductivity of the gas film (cal/sec-cm-°K).

4.5 Determination Rate of Reaction

The location of the reaction interface on the spherical particle determines the reaction points E and F in Eq. 10. These points enable the calculation of R' (% Reduction) or extent of reduction. This, along with the bulk gas composition, evaluation of transport mass coefficients, and the temperatures of the reaction interface and gas stream enables calculations of the instantaneous reaction rate (n_C) at any given time during the reaction. The extent of conversion of TiO_2 to TiC was proportional to the magnitude of n_C.

$$n_C = \frac{(P_B(o) - K_{eq} * P_A(o))}{T_g * (K_{eq}\left(\frac{E}{D_{A,eff}} + \frac{F}{K_{mA}}\right) + \left(\frac{E}{D_{B,eff}} + \frac{F}{K_{mB}}\right)} \tag{10}$$

where $E = \frac{R*(R_o - r_i)}{4\pi(ri \times Ro)}$ and $F = \frac{R}{4\pi R_o^2}$ are location points of reaction interface on the sphere, R= final radius, mass transfer coefficient of molecules A and B for gases (K_{mA}, K_{mB}), $D_{A,eff}$ and $D_{B,eff}$ are the effective Knudsen diffusion for molecules A and B.

4.6 Determination of Spherical Pellet Temperature

Prediction of the variation of spherical pellet temperature change during reaction was more difficult than the determination of the gaseous mass transport rate. The reasoning behind this argument was that heat transfer at the reactant core consisting of TiO_2 occurred simultaneously with heat transfer through the accumulated solid reaction product (TiC) and the bulk gas. To simplify the heat transfer calculations in MATLAB, a few assumptions was made. The first assumption, the accumulation of heat in reaction zone was negligible. The 2nd assumption, heat transfer was 1-dimensional and by conduction, convection and radiation. The 3rd assumption, the average pellet core temperature of T_C was defined to account for the overall heat transfer away the pellet. This allowed the simplification of the temperature calculations by Eq. 11-14.

$$T_C = \frac{\left[\begin{array}{c}\Delta t\left(n_B * C_{pH2}(T_C)*(T_C - T_o) + n_A C_{pA}(T_p)*(T_p - T_o) - \frac{4\pi K_p r_i R_o}{(R_o - r_i)} * T_s\right) + T_{C_{z-1}} \\ * \frac{4}{3}\pi ri^3 * C_{pC}(T_C) * \rho_C + 4\pi r_i^2 \Delta r * \rho_C * (-\Delta H(T_C))\end{array}\right]}{\left[{}^4/_3 \pi r_i^3 * C_{pC}(T_C) - \frac{4\pi K_p R_o r_i}{(R_o - r_i)} * \Delta t\right]} \tag{11}$$

$$T_p = T_C - \frac{R_o}{(r_i + R_o)} \times (T_C - T_S) \tag{12}$$

$$T_S = \frac{\left[\begin{array}{c} n_A[C_{pA}(T_P) * (T_P - T_o) - C_{pA}(T_g) * (T_P - T_o)] \\ + n_B[C_{pB}(T_C) * (T_C - T_o) - C_{pB}(T_P) * (T_P - T_o)] \\ + 4\pi R_o^2 \varepsilon\alpha T_o^4 - \frac{4}{3}\frac{\pi(R_o^3 - r_i^3)C_{PP}*(T_P)}{\Delta t}\rho_P * (T_P - T_{P_{z-1})} \\ + \frac{4\pi K_p r_i R_o}{(R_o - r_i)} * T_C + 4\pi R_o^2 h_c * T_g \end{array}\right]}{\left[\frac{4\pi K_p r_i R_o}{(R_o - r_i)} + 4\pi R_o^2 hc + (4\pi R_o^2 \varepsilon\alpha * T_{S_{z-1}}^3)\right]} \tag{13}$$

$$T_g = \frac{\left[\begin{array}{c} n_A C_{pA}(T_g) * (T_g - T_o) + G C_{pg}(T_o) * T_o + (4\pi R_o^2 h_c * T_S) \\ + 4\pi R_o^2 \alpha\varepsilon(T_S^4 - T_o^4) + n_B C_{pB}(T_P) * (T_P - T_o) \end{array}\right]}{[G C_{pg}(T_o) + 4\pi R_o^2 h_c]} \tag{14}$$

where T_p= Average temperature of the particle (°K) C_{pp}= heat capacity of solid product (calorie/mole-°K), C_{pC} = heat capacity of solid reactant core (calorie/mole-°K), C_{pg}= heat capacity of reducing gas (calorie/mole-°K).

The overall chemical reaction for the system was represented by Eq. 2 from SCM [17] and the reaction rate can be predicted by the simultaneous solution of Eq.10-15 done in MATLAB. Lastly, after the reaction rate (n_c) and time of reaction (t) have been determined, the R' can be calculated based on Eq. 15-16:

$$\Delta t = -4\pi r_i \rho_c \frac{\Delta r}{n_c} \quad (15)$$

$$R_o \rho_c = [1 - (1 - R')^{\frac{1}{3}}] = kt \quad (16)$$

where R' = % Reduction (100*instantaneous weight of pellet/ initial weight of pellet sphere), R' = (M_o-$M_{(t)}$)/ (M_o-M_f), ρ_c = molar density, M_o = initial weight of pellet (TiO_2), $M_{(t)}$ = weight of pellet at time t (Ti_2O_3 + TiC), M_f= final weight of pellet (TiC) and k = kinetic constant (s^{-1})

5. Results and discussion

The experimental data for extent of reduction (R') versus time at different temperatures is shown in Figure 3. Table 1 tabulates the final R' calculated from the weight loss of samples. The R' was increasing with reduction temperature and time because a higher amount of TiO_2 was reduced into Ti_2O_3 and TiC. As shown in Figure 3, there was significant difference in R' between higher and lower reduction temperature. At 1200°C and 180 min of reduction time with 20 min LPG flow time, the R' was the highest at 37.9%. The titania sample reduced at latter conditions contained 3.68 wt.% carbon when analyzed by elemental carbon analysis.

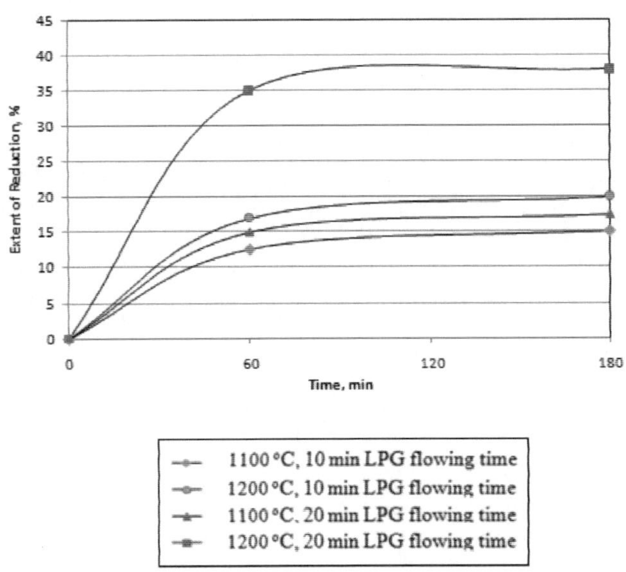

Figure 3. Reduction of titania by LPG-hydrogen-argon mixture (10 vol. % LPG -45 vol.% H_2-45 vol. % Argon) at different temperatures and LPG flowing times

Table 1. Extents of titania reduction at different temperatures with 20 min of LPG flowing time

LPG Flowing Time (min)	10				20			
Temperature, °C	1100		1200		1100		1200	
Reduction Time, min	60	180	60	180	60	180	60	180
Extent of Reduction, %	12.48	14.98	16.98	19.97	14.98	17.48	34.95	37.96
Carbon, wt.%	0.97	1.07	2.05	2.11	1.51	2.40	2.43	3.68

During the reduction process, thermal cracking of propane and butane took place, which is the major component in LPG; propane was cracked into ethane, ethylene, methane, and hydrogen [18, 19]. The final stable state was methane (CH_4). Deposition of solid carbon occurred at the end of the cracking process, as shown in Eq. (17). The presence of carbon monoxide (CO) gas was due to the chemical reaction in Eq. (1).

$$CH_4\,(g) \rightarrow C\,(s) + 2H_2\,(g) \tag{17}$$

The cracking process enhances the rate of reduction as more carbon is available for reduction. Furthermore, similar work done by Itao, G.B. and N.M. Anacleto [6] showed the thermodynamic activity of C_3H_8 and C_4H_{10} was much higher than CO and H_2. This higher activity allowed a greater degree of cracking into solid carbon. The longer flowing time of LPG was able to produce more free carbon and thus increase the reduction rate. Free solid carbon can reduce and carburize titania into TiC but the reduction was slow due to the low reduction temperature studied in this work.

5.1 Predicted Reduction Kinetics of Titania

The reduction of titania has been one of the most important and complex gas-solid reactions in kinetic study by various researchers [9, 12, 13, 22]. The complexity was due to the formation of titania sub-oxides such as Ti_4O_9, Ti_3O_5 and Ti_2O_3. These sub-oxides affects the reaction rate and differ greatly in terms of its physical and material properties. A shrinking core model (SCM) developed by Szekely et al. [20] was employed to determine the reaction rate. From the mass and heat transport equations listed above, the transport-limited reaction rate for reduction of titania was calculated in MATLAB. Figures 4 show the results of isothermal reduction at 1100°C and 1200°C. From the figure, both rate and extent of reduction increased with temperature from 1100°C to 1200°C.

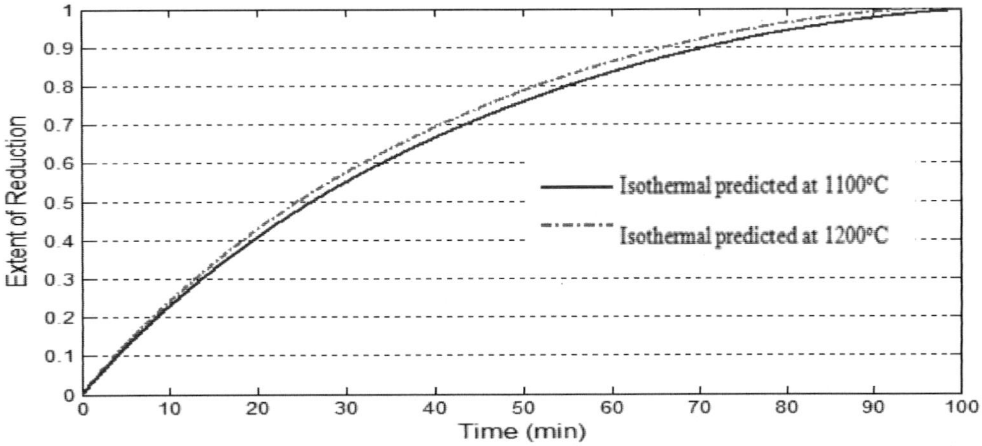

Figure 4: Extent of Reduction Predicted in Isothermal Conditions at 1100°C and 1200°C

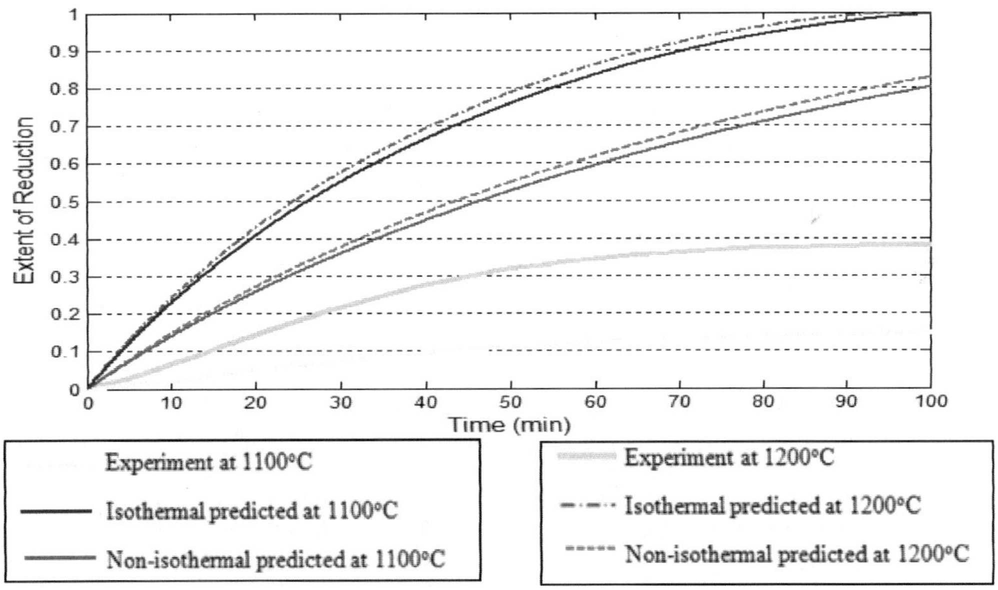

Figure 5: Comparison of predicted and experimental extent of reduction at different temperatures

Figures 5 shows the comparison between the predicted and experimental R' at 1100°C and 1200°C in LPG-H_2-Ar gas atmosphere. From the figure, the deviation between the predicted and experimental results are significant. All the simulated plots are similar in which the predicted non-isothermal R' data was closer to the experimental results compared with isothermal conditions. Even though the deviation was significant, it can be observed that the R' increased with temperature for both simulated and experimental. The source of the deviation will be discussed below.

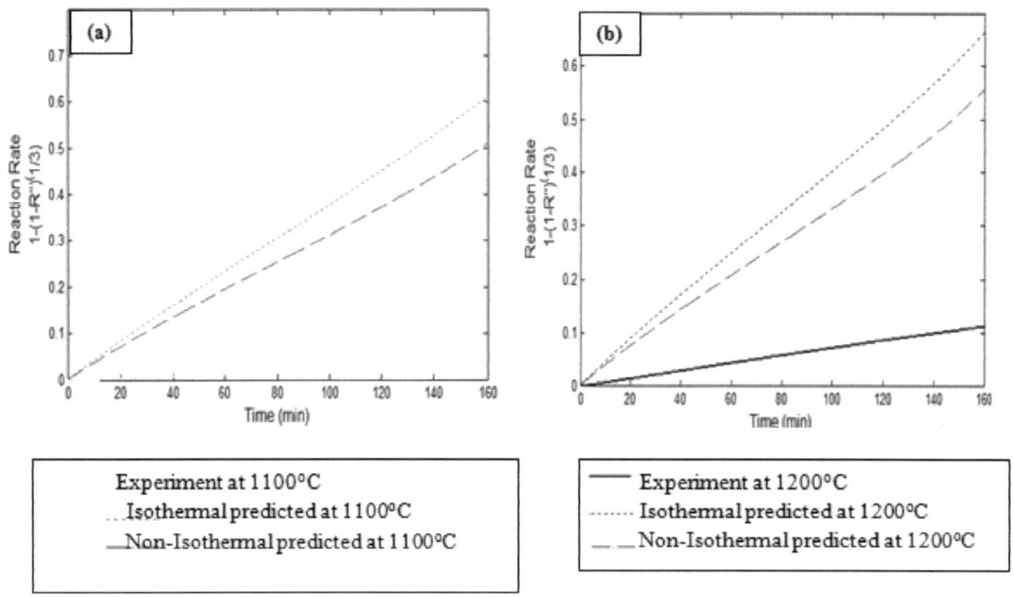

Figure 6. Comparison of non-isothermal predicted and isothermal predicted with experimental reaction rate for reduction of titania at different temperatures (a) 1100°C (b) 1200°C

The comparison of temperature variation with respect to reaction rate was shown in Figure 6. From the figure, it shows the deviation between predicted and experimental measured results. On the other hand, the experimental reaction rate changed linearly with reduction time, which showed the reaction was chemically controlled. In this case, temperature was a key parameter in determining the overall reaction rate. The graphical trend of reaction rate for experimental and predicted results should be similar. An explanation for the deviations observed was due to the SCM model assumptions that were made upon developing the kinetic model. Those assumptions were that the porosity and pellet size of sample was constant during the reduction.

Firstly, the sample was assumed to be porous without close pores for gas transport in and out during the reaction time. However, Ronald Charles Gower [15] have suggested that this situation was ideal to take place because of the presence of open and closed pores. Moreover, the porosity throughout the sample also varied and was not constant. These pores formed within the product and restricted the movement of the CH_4 gas. Therefore, until the product layer cracks to allow free movement of CH_4, the diffusion of oxygen and carbon atoms in the solid phase or gaseous reactants can only contribute to the reaction kinetics. Delay in this cracking of intermediate product layers of Ti_2O_3 also slows down the reaction. Furthermore, Ti_2O_3 does not convert directly to TiC but as another intermediate phase of TiO_xC_{1-x}. Studies by Dewan et al. and Guangqing Zhang et al. [5, 13] have shown that the end product to be TiO_xC_{1-x} instead of TiC at the reduction temperature used. The presence of these intermediates phases also was difficult to model by SCM. The slow reaction for intermediate product to transform into TiO_xC_{1-x} caused the build up in non-porous Ti_2O_3. As a result, mass transfer for carbon and CH_4 to penetrate the Ti_2O_3 layer was difficult.

Pure TiC was employed in developing the kinetic model instead of using the actual end product of the reaction, which was a solid solution titanium oxycarbide (TiO_xC_{1-x}). There was insufficient thermodynamic and physical properties for TiO_xC_{1-x} phases to be used for the kinetic model compared with TiC. Added to this, it was impossible to reduce titania into pure TiC at this low temperatures. Instead of producing pure TiC, a solid solution of TiO_xC_{1-x} with similar cubic crystalline structure as

TiC was produced during reduction [21]. Due to these supporting evidence, the predicted and experimental results deviation was observed. Even though there were deviations between the predicted and experimental results, the model still shows the kinetic trends that was expected from SCM. The trend indicated the increasing in rate and extent of reduction when temperature increases when governed by chemically controlled reaction rate.

6. Conclusions

Titania was reduced to TiO_xC_{1-x} in a temperature range of 1100 - 1200°C by liquefied petroleum gas (LPG)-hydrogen-argon gas mixtures. The variation in the LPG flowing time and reduction time showed a significant effect on the extent of reduction. The results obtained confirmed that the reduction rate was accelerated by the longer flow time of LPG. The highest experimental extent of reduction of about 37.9% occurred at a temperature of 1200°C with 3 hours reduction time and 20 min of LPG flowing time. The reduction rates increased with increasing reduction temperature and LPG flowing time. The SCM was implemented in MATLAB that modelled isothermal and nonisothermal reduction. For the nonisothermal reduction at 1200°C at 3 hours with 20 min of LPG flowing time, the R' calculated by the kinetic model was about 82%. There were differences between the predicted and experimental results, but the overall kinetic trend of reduction rate was similar. The differences was due to limitation of SCM to predict changes in porosity, thermodynamic properties of TiO_xC_{1-x} and minute changes in temperature in the concentric shells during reduction. Future models will address this limitations.

Acknowledgement

The author wishes to thank Universiti Sains Malaysia (USM) and Ministry of Higher Education (MOHE) of Malaysia for supporting this work with three research grants. This research was supported by two grants from MOHE Fundamental Research Grant Scheme (FRGS) with grant number of 203/PBAHAN/6071230 and 203/PBAHAN/6071261. The third grant was from USM Research University Individual (RUI) grant of 1001/PBAHAN/814273. Special thanks to USM technicians, Mr. Shahrul Ami and Mr. Mohamad Syafiq for supporting the author's experimental work.

Nomenclature

A	reactant solid
B	reactant gas
C	product solid
Cp_A	Heat capacity of gas reactant (calorie/mole.°K)
Cp_B	Heat capacity of gas product (calorie/ mole.°K)
D	product gas
Eq.	Equation
G	mass flow rate (gram/cm^2.sec)
K_{eq}	Equilibrium constant
R'	Percent of reduction
R_o	Initial radius (cm)
R	Final radius (cm)
T_g	Gas temperature (°K)
T_o	Initial temperature (°K)
T_s	Surface pellet temperature
atm	Atmosphere
d_o	Sphere diameter
r_i	Position of reaction interface (cm)
t	time

References

[1] Rezan, S.A., G. Zhang, and O. Ostrovski, *Carbothermal Reduction and Nitridation of Titanium Dioxide in a H_2-N_2 Gas Mixture*. Journal of the American Ceramic Society, 2011. **94**(11): p. 3804-3811.

[2] Gerasimov, I.A.I., A.N. Krestovnikov, and A.S. Shakhov, *Chemical Thermodynamics in Nonferrous Metallurgy: (Khimicheskaya Termodinamika V Tsvetnoi Metallurgii) Vol. III: Thermodynamics of Tungsten, Molybdenum, Titanium, Zirconium, Niobium, and Tantalum and of Their More Important Compounds; a Reference Book*. 1965: Israel Program for Scientific Translations.

[3] Adipuri, A., Y. Li, G. Zhang, and O. Ostrovski, *Chlorination of reduced ilmenite concentrates and synthetic rutile*. International Journal of Mineral Processing, 2011. **100**(3): p. 166-171

[4] Sharifah, A.S.S., H.S. Saidin, N. Baharun, S.A. Rezan, and H. Hashim, *Microstructural Study of Reduced Malaysian Ilmenite by Carbothermal Reduction and Nitridation in Nitrogen Atmosphere*. Advanced Materials Research, 2014. **858**: p. 265-271.

[5] Dewan, M.A.R., G. Zhang, and O. Ostrovski, *Carbothermal reduction of titania in different gas atmospheres*. Metallurgical and Materials Transactions B, 2009. **40**(1): p. 62-69.

[6] Itao, G. B. and N. M. Anacleto (2011). *Reduction of Nickeliferous Laterite Ores by Liquefied Petroleum Gas (LPG)*. The Mindanao Forum.

[7] YIN, T.W., *Kinetic Modelling Of Reduction Of Titania With Liquefied Petroleum Gas (Lpg) Using Matlab*, in *School Of Materials And Mineral Resources Engineering*. 2014, Bachelor in Engineering Thesis, Universiti Sains Malaysia: Nibong Tebal, Penang. p. 119.

[8] Koc, R. and J.S. Folmer, *Carbothermal synthesis of titanium carbide using ultrafine titania powders*. Journal of Materials Science, 1997. **32**(12): p. 3101-3111.

[9] Yin, T.W., S. Ramakrishnan, Sheikh Abdul Rezan Sheikh Abdul Hamid, A. Fauzi, and N.I. Shoparwe. *Liquefied Petroleum Gas (LPG) Assisted Reduction of Titania*. in *5th International Conference on Recent Advances in Materials, Minerals and Environment (RAMM) & 2nd International Postgraduate Conference on Materials, Mineral and Polymer (MAMIP)*. 2015. Vistana Hotel, Penang, Malaysia: Universiti Sains Malaysia.

[10] Nurul Syazwina Binti Che Ibrahim, S. Ramakrishnan, N.b. Baharun, S.A. Rezan, R. Alizadeh, and P. Roohi, *Nonisothermal Kinetic Modelling For Hydrogen Reduction of Ferric Oxide Using Matlab*. Materials Science Forum 2015. **819**: p. 6.

[11] Rezan, S., G. Zhang, and O. Ostrovski, *Carbothermal Reduction and Nitridation of Titanium Dioxide in a H_2–N_2 Gas Mixture*. J. Am. Ceram. Soc., 2011. **1-8**.

[12] Zhang, G. and O. Ostrovski, *Kinetic modeling of titania reduction by a methane-hydrogen-argon gas mixture*. Metallurgical and Materials Transactions B, 2001. **32**(3): p. 465-473.

[13] Zhang, G. and O. Ostrovski, *Reduction of titania by methane-hydrogen-argon gas mixture*. Metallurgical and Materials Transactions B, 2000. **31**(1): p. 129-139.

[14] Giap, T.W., S. Ramakrishnan, S.A. Rezan, A. Hamid, R. Alizadeh, and P. Roohi, *Kinetic Modelling for Hydrogen Reduction of Ferric Oxide Using MATLAB*, in *Proceedings of the International Conference on Science, Technology and Social Sciences (ICSTSS) 2012*, A. Kasim, Editors. 2014, Springer Singapore: Singapore. p. 441-447.

[15] Gower, R.C., *A Determination of the Transport-limited Reaction Rate for a Gas-Solid Reaction Forming a Porous Reaction Product*. 1971, Phd Thesis, Lehigh University.

[16] Nurul Syazwina, S. Ramakrishnan, N.b. Baharun, S.A. Rezan, R. Alizadeh, and P. Roohi, *Nonisothermal Kinetic Modelling For Hydrogen Reduction of Ferric Oxide Using Matlab*. Materials Science Forum 2015. **819**: p. 6.

[17] Homma, S., S. Ogata, J. Koga, and S. Matsumoto, *Gas–solid reaction model for a shrinking spherical particle with unreacted shrinking core*. Chemical Engineering Science, 2005. **60**(18): p. 4971-4980.

[18] Shah, N., Y. Wang, D. Panjala, and G.P. Huffman, *Production of Hydrogen and Carbon Nanostructures by Non-oxidative Catalytic Dehydrogenation of Ethane and Propane.* Energy & Fuels, 2004. **18**(3): p. 727-735.

[19] Wang, Y., N. Shah, and G.P. Huffman, *Simultaneous production of hydrogen and carbon nanostructures by decomposition of propane and cyclohexane over alumina supported binary catalysts.* Catalysis Today, 2005. **99**(3): p. 359-364.

[20] Szekely, J., J. W. Evans and H. Y. U. Sohn (1976). Gas-solid reactions, Academic Press.

[21] Lyubimov, V.D., T. A. Timoshchuk, and M. V. Kalacheva, *Mechanism Of Formation Of Titanium Carbide During Carbothermal Reduction Of Titanium-Dioxide.* 1992.

[22] Chien, L. C., S. Ramakrishnan, S. A. Rezan, R. Alizadeh and P. Roohi (2016). *Mathematical Modeling of Hydrogen Reduction of Titania Using MATLAB. 2nd International Science, Technology and Engineering Conference (2nd ISTEC).* Abdul Mutalib Md Jani and Ali H. Jawad (Editors). Equatorial Hotel, Penang, Malaysia., Universiti Teknologi MARA (Perlis).

Numerical study on passive convective mass transfer enhancement

G P Aravind, K M Muhammed Rafi, and M Deepu
Department of Aerospace Engineering, Indian Institute of Space Science and Technology, Trivandrum, 695547, India. Email: anu.aravind007@gmail.com

Abstract. Passive mixing mechanisms are widely used for heat and mass transfer enhancement. Vortices generated in flowfield lead to gradients that favour convective mass transfer. Computations on enhancement of convective mass transfer of sublimating solid fuel by baroclinic torque generated vortices in the wake of a swept ramp placed in high speed flow is presented here. Advection Upstream Splitting Method (AUSM) based computational scheme employed in the present study, to solve compressible turbulent flow field involving species transport, could capture the complex flow features resulted by vortex boundary layer and shock boundary layer interactions. Convective mass transfer is found to get improved in regions near boundary layer by horseshoe vortex and further transported to other regions by counter rotating vortex pair. Vortices resulted by flow expansion near aft wall of wedge and recompression wave-boundary layer interactions also promotes convective mass transport. Extensive computations have been carried out to reveal the role of vortices dominance at various lateral sweep angles in promotion of convective mass transfer in turbulent boundary layer.

1. Introduction

Passive mixing devices are widely used in conjugate heat and mass transfer applications such as chemical processing systems, hybrid rocket propulsions systems, refrigeration systems, and thermal protection systems. Promotion of mixing leads to improvement of performance and operational efficiencies of such systems; thereby energy and environmental issues associated with such systems can be brought down considerably. Passive mixing devices are often simple to implement in practical applications that do not require any power or external interference. Vortices generated by solid boundary features such as winglets sharp edges surface protuberance features [1, 2] have been widely investigated by many researchers over the last few decades for the possible improvement in heat/mass transfer effects.

Compressible turbulent flows in passive mixing devices involves complex flow features such as shock/expansion waves and their interaction with boundary layers associated with solid surfaces. It generates sharp gradients in flowfield. Misaligned gradients of the flow variables such as pressure and density and presence of viscous stress in density gradient generate baroclinic torque induced vortices [3]. Vorticity generated by this effect will be perpendicular to density is given by

$$\frac{D\boldsymbol{\omega}}{Dt}=\frac{\boldsymbol{\omega}}{\rho}\cdot\nabla\mathbf{u}+\frac{1}{\rho^{3}}\nabla\rho\times\nabla p \tag{1}$$

Baroclinic torque induced vortex generated in a high speed flow past a swept ramp and its role promotion of convective mass transfer is analysed in the present study. A schematic of major features in this flowfield is given in figure 1. Flow expands as it spills over the ramp and thereby produces low pressure region near to the wall. This results in the baroclinic torque induced vortex generation. Large scale streamwise vortex develops when flow spills over the ramp which advects more fluid in the free stream to the centre. Horseshoe vortices generated by flow separation behind the ramp moves downstream in boundary layer region. Counter-rotating vortex pairs have been observed in both vortex systems. These vortex systems die out due to the viscous effects in developing turbulent boundary layer. Species evolved by the vaporisation of solid fuel gets transported to the mixing zone initially by the horseshoe vortices and later carried by streamwise counter-rotating vortex pair. In this process fuel gets mixed up well with high speed oxidiser stream as the vortex system provides low velocity region to promote molecular diffusion.

Vortex system in the flow field depends on sweep angle of the wedge. Role of sweep angle in enhancing mass transfer from a mass evolving boundary is analysed in the present computational study.

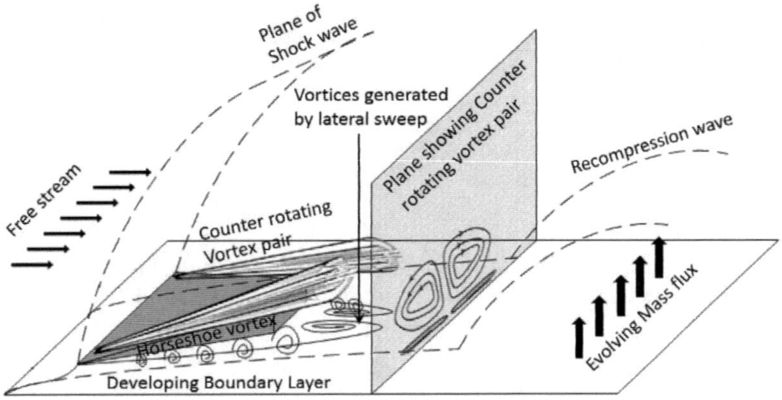

Figure 1. A schematic of flow features

2. Background

Literature review focus on previous work reported in the field of convective heat/mass transfer enhancement using passive methods based on typical surface features. A summary of major contributions has been presented in table 1.

Table 1. Summary of salient literature in convective heat/mass transfer enhancement

Authors	Experimental Methodology	Results
Gentry et al [4,5]	Used delta wing Vortex generators(VG) on flat plate for Heat transfer enhancement and also to modify a developing channel flow	Noticed a 50 – 60 % enhancement in flat plate and a 300% increase of local heat transfer coefficient in developing channel flow
Bruce et al [6]	Experimental investigation of the development and structure of stream wise vortices embedded in a turbulent boundary layer	Observed an enhanced circulation decay in arrays of closely packed vortices that may be due to the merging of neighbouring counter rotating vortex cores
Promvonge et al [7]	Investigated the effects of combined ribs and winglet type VGs	Inline ribs shows highest increase in Nusselt number and staggered ribs shows better thermal performance.
Debora et al [8]	Studied the vortex produced by a simple round wall jet with a 2D turbulent boundary layer	Max vorticity levels strongly dependent on skew angle and jet velocity and obtained skew angle is between 45 – 90deg

Liou et al [9]	Investigated heat transfer and fluid flow in a square duct with 12 different shaped vortex generators	Direction and strength of secondary flow with respect to heat transfer wall is the most important factor which promotes heat transfer enhancement.
Petersen et al [10]	Investigated baroclinic vortex production	They found that local azimuthal temperature gradients introduces vorticity.
Storey et al [11]	Effect of vortices on the frost growth in developing channel flow	Noticed a 7% increase in the frost growth with the induced stream wise vortices
Henze et al[12]	Investigated longitudinal tetrahedral VGs	Found that ratio between height of VG and hydrodynamic boundary layer thickness will affect the heat transfer enhancement
Azize [13]	Delta wing type VGs	Experimentally studied the flow structure in horizontal ducts having double rows of half delta wing VGs

3. Governing Equations

Present problem involving mass transport enhancement in Baroclinic induced vortex mass transfer enhancement can be modelled using conservative form of Navier-Stokes equations which govern a three dimensional turbulent compressible flow incorporating fuel and oxidizer species as follows.

$$\frac{\partial U}{\partial t}+\frac{\partial F}{\partial x}+\frac{\partial G}{\partial y}+\frac{\partial H}{\partial z}=S \tag{2}$$

Where U is the vector of all conservation variables. F and G are fluxes in x and y directions, and S is the vector of source terms.

The turbulence model used in the present study is SST κ- ω two equation model [14].

4. Numerical Modelling

In the present study, AUSM [15] scheme available in a Finite Volume Method (FVM) based commercial package (Ansys Fluent), has been made use for the inviscid flux computations. Key aspect of the present problem is implemented in conjunction with boundary conditions given on fuel layer represented in bottom of the computational domain. The schematic of mass and energy exchange process occurring in the convective mass transfer boundary layer is shown in in figure 2.

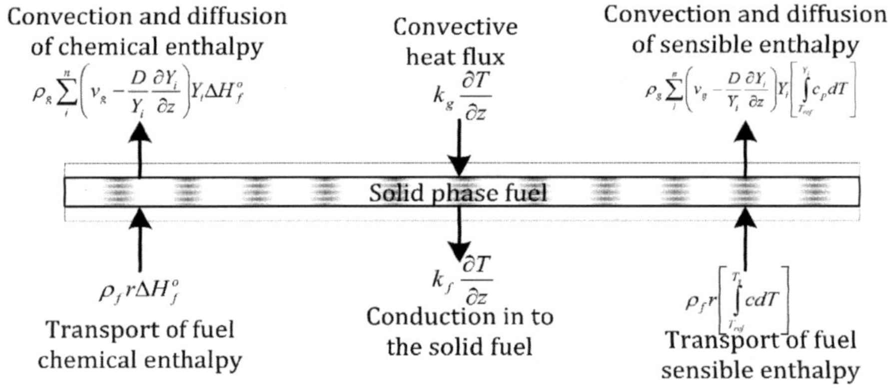

Figure 2. Modelling of Mass and Energy balance

Simulations have been performed on 3D computational domain to compare the enhancement due to baroclinically induced vortex interaction. The Computational domain consists of a flat hydrocarbon fuel layer undergoing mass transport in simple convective boundary layer as shown in figure 3. A simple flow manipulator (swept ramp) is placed on the flat hydrocarbon fuel layer to create vortices so as to bring in effect of the vortex boundary layer interaction in flat fuel layer. Geometry and dimensions of the ramp are as given in figure 3. The computational domain were initially discretized with 8×10^5 control volumes and later refined to 1.6×10^{6} level for which convergence of species profile has been attained. The entire computational domain was initialized with oxidizer fluid inflow conditions. Grid is refined near all no-slip walls and near to the swept ramp, by giving biasing in direction normal to no-slip walls, maintaining a y+ of 1. For the present density based solver using explicit time integration, convergence of the order of 10^{-5} for all residuals has attained after 125000 iterations with a CFL of 0.5.

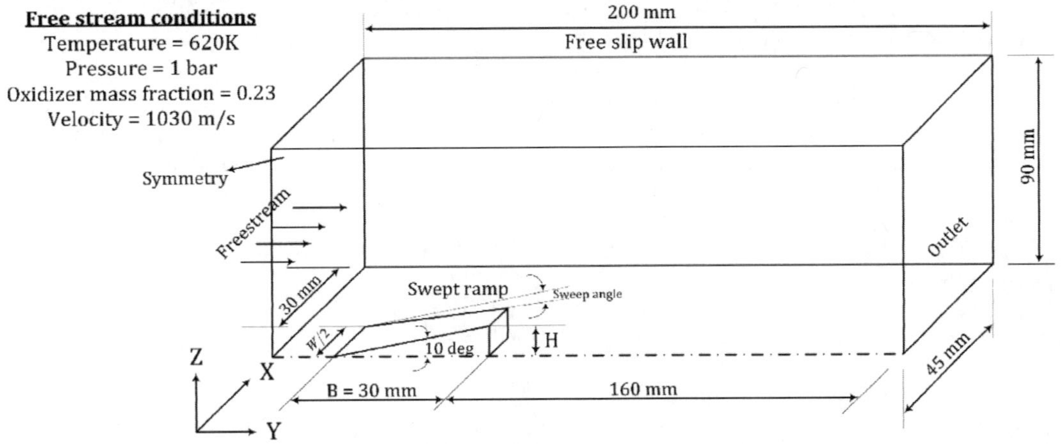

Figure 3. Computational Domain

5. Results and Discussion

The computational procedure was validated using experimentally measured centreline wall pressures for a similar ramp placed in supersonic flow [16]. Streamline plot of the computed results for the 3D domain is given in figure 4. All salient flow features, described before (figure 1), has been captured. Velocity magnitudes associated with regions of vortex boundary layer interaction as well as counter rotating vortex pair are observed to be relatively smaller compared to other regions. This enables the formation of gradients favouring enhancement of mass transfer.

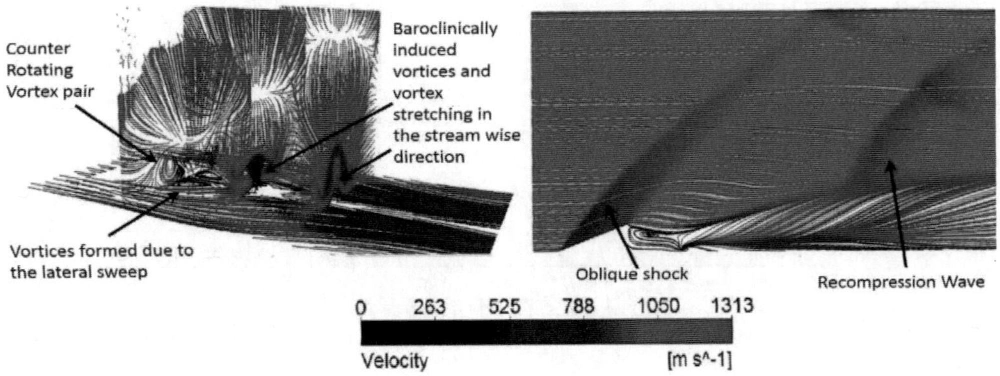

Figure 4. Streamline plot of computations

5.1. Flow features

A plot of computed streamlines in the flow field is given in figure 4. A boundary layer develops on fuel surface when approaching oxidizer flow interacts with it, due to no-slip boundary condition given on the surface. A strong oblique shock is developed at the leading edge of swept ramp. Boundary layer separates in the region of interaction of expansion wave with boundary layer. Reattachment compression wave is also generated from the boundary layer after the ramp. A high temperature gradient region also formed upstream of expansion wave reattachment. A low velocity recirculation region is formed in the wake region of ramp. In short, the present flow field of shock interaction with developing boundary layer on a convective mass transfer boundary layer involves many zones of sharp gradients.

5.2. Effect of Swept ramp on mass transfer

Three swept ramp configurations are used in the present study viz. No sweep, 5^0 sweep and a 7^0 sweep. Field plot showing the lift of species from the plate is shown in figure 5. Two longitudinal counter rotating vortex pair are generated along the slanting surface of the ramp along with lateral vortices due to the sweep given in the ramp. Enhancement in mass transfer coefficient is observed in two regions along the axial direction as shown in figure 6. First enhancement is observed at the wake region of the ramp and the latter at a region where the baroclinic torque induced vortices are predominant. This enhancement in mass transfer improve further with increase in sweep angle of wedge. This is due to the dominance of vortex with given amount of sweep. Spanwise variation of mass transfer coefficient at wake region of the swept ramp and baroclinic torque induced regions are plotted in figure 7 and 8. Aforementioned enhancement are observed in spanwise direction as well.

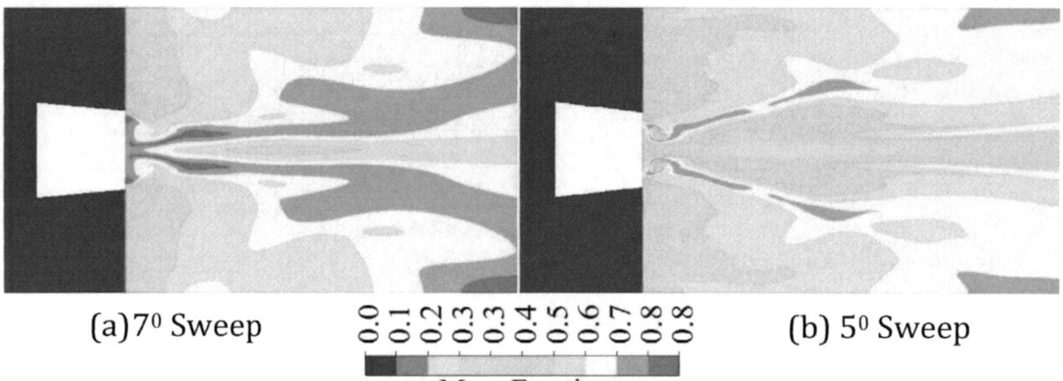

Figure 5. Field plot showing the lift of species from the plate for 7^0 Sweep (a) and 5^0 Sweep (b)

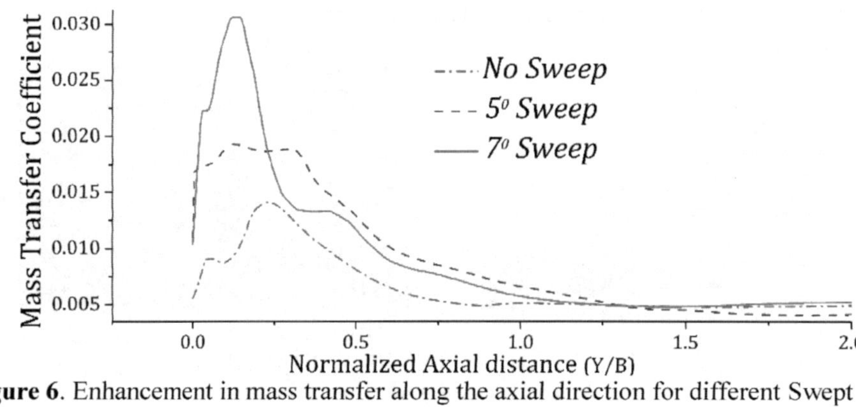

Figure 6. Enhancement in mass transfer along the axial direction for different Swept angles

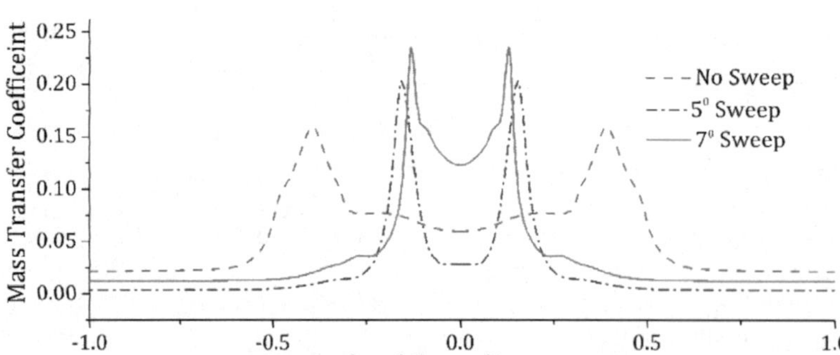

Figure 7. Enhancement in mass transfer for spanwise direction in the wake region

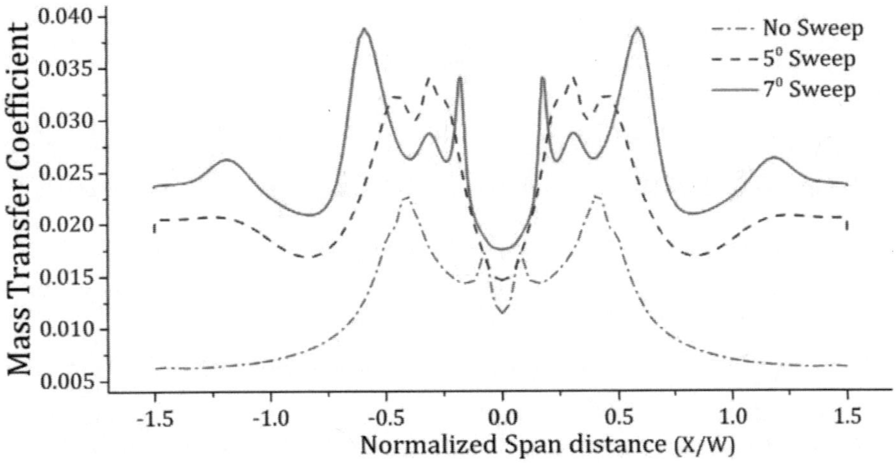

Figure 8. Enhancement of mass transfer for spanwise direction (baroclinic torque induced region)

6. Conclusions

Three dimensional compressible turbulent flow field with mass transport of sublimating fuel has been simulated using AUSM scheme available in a FVM based solver. Computational procedure has been validated using experimental results and scheme was found to be robust in negotiating sharp gradients resulting due to shock/vortex interactions with boundary layer. Extensive computations have been performed to reveal the role of lateral sweep of vortex generator for convective mass transfer enhancement in high speed flows. Misalignment of isopycnals and isobars becomes more dominant in flow field with higher sweep angles, which leads to production of baroclinic torque induced vortices. Enhancement of mass transfer is observed to be prominent in two regions of the flow field viz. in the wake region of the swept ramp and baroclinic torque induced vortex region. Enhancement of convective mass transfer in spanwise direction, resulted by mass transport in counter rotating vortex pair, is also improved with sweep angle. This establishes the sweep angle factor of wedge-based vortex generators in passive mixing systems.

References

[1] Webb, R. L. and Kim, N. H., 1994. *Taylor Francis: New York, NY, USA*.

[2] Jacobi, A. M. and Shah, R. K., 1995. *Experimental Thermal and Fluid Science*, **11**, p 295-309.

[3] Waitz, I., Marble, F., and Zukoski, E., 1991. *AIAA Journal*, **31**, p 3140.

[4] Gentry, M. C. and Jacobi, A. M., 1997. *Experimental Thermal and Fluid Science*, **14**, p 231-42.

[5] Gentry, M. C. and Jacobi, A. M., 2002. *Journal of heat transfer*, **124**, p 1158-168.

[6] Wendt, B. J., Greber, I., and Hingst, W. R., 1993. *AIAA journal*, **31**, p 319-25.

[7] Promvonge, P., Chompookham, T., Kwankaomeng, S., and Thianpong, C., 2010. *Energy Conversion and Management*, **51**, p 1242-249.

[8] Compton, D. A. and Johnston, J. P., 1992. *AIAA journal*, **30**, p 640-47.

[9] Liou, T. M., Chen, C. C., and Tsai, T. W., 2000. *Journal of Heat Transfer*, **122**, p 327-35.

[10] Petersen, M. R., Julien, K., and Stewart, G. R., 2007. *The Astrophysical Journal*, **658**, p 1236.

[11] Storey, B. D. and Jacobi, A. M., 1999. *International Journal of Heat and Mass Transfer*, **42**, p 3787-802.

[12] Henze, M. and Von Wolfersdorf, J., 2011. *International Journal of Heat and Mass Transfer*, **54**, p 279-87.

[13] Akcayoglu, A., 2011. *Experimental Thermal and Fluid Science*, **35**, p 112-20.

[14] Menter, F. R. 1994, *AIAA journal*. 32, p 1598-605.

[15] Liou, M. S. and Steffen, C. J., 1993. *Journal of Computational physics*, **107**, p 23-39.

[16] Donohue, J. M. and cDaniel, J. C., 1996. *AIAA journal*, **34**, p 455-62.

Thermal cut-off response modelling of universal motors

Kashveen Thangaveloo[1] and Yung Shin Chin[2]

Technology Analysis Team, Dyson Manufacturing Sdn Bhd, Jalan Cyber 14, Senai Industrial Estate IV, 81400 Senai, Johor, Malaysia

E-mail: [1]kashveen.thangaveloo@dyson.com, [2]chin.yungshin@dyson.com

Abstract. This paper presents a model to predict the thermal cut-off (TCO) response behaviour in universal motors. The mathematical model includes the calculations of heat loss in the universal motor and the flow characteristics around the TCO component which together are the main parameters for TCO response prediction. In order to accurately predict the TCO component temperature, factors like the TCO component resistance, the effect of ambient, and the flow conditions through the motor are taken into account to improve the prediction accuracy of the model.

Keywords: Thermal model, Universal motors, Thermal cut-off, Heat and mass transfer.

1. Introduction

Thermal cut-off unit, or more commonly known as TCO, is a mechanical device that functions through an increase in resistance when there is rise in temperature [1]. When the temperature reaches a designated calibration temperature, the bimetallic strip snaps open and breaks the circuit [2]. The bimetallic strip will return to its original position after it is allowed to be cooled down, as shown in figure 1.

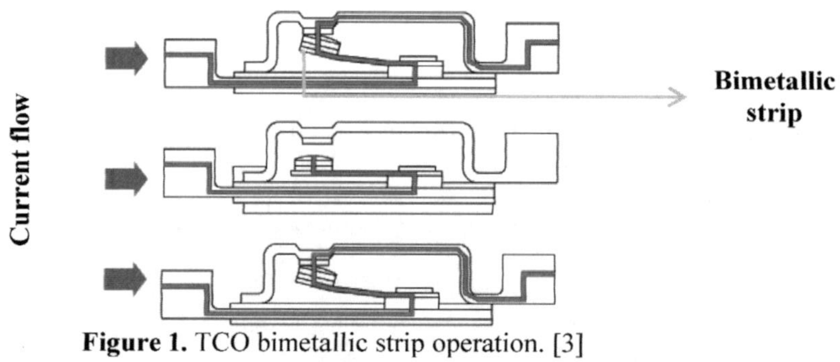

Figure 1. TCO bimetallic strip operation. [3]

A typical TCO usually consists of a bimetallic strip, a metal casing, and a protective glass cloth as shown in figure 2.

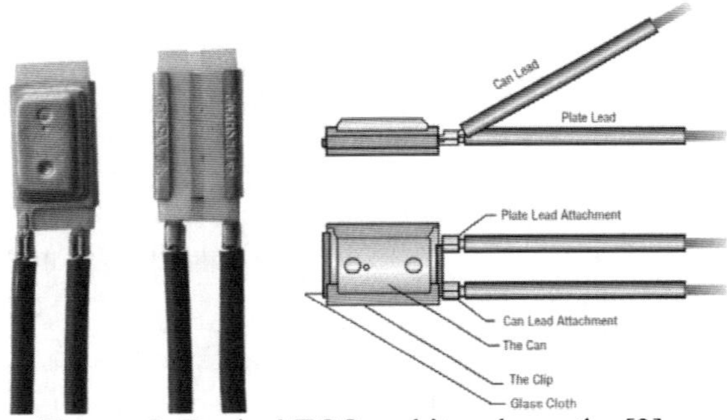

Figure 2. Typical TCO and its schematic. [3]

TCO's are used widely in many small motors that are present in electrical appliances. In the case of universal motors, the waste heat generated by windings and bearings is an example of unwanted heat. The TCO acts as a protection for the motor when the motor overheats/current overloads during abnormal operation. A motor will overheat when it is overloaded for either an extended period of time or on a frequency basis [4]. Placement of the TCO on the motor is important since the accuracy of the TCO can be affected significantly by load variations and flow characteristics within the motor [5]. With this in mind, the TCO is usually placed at the hottest practical location inside the motor. For example, a TCO is placed at near electric motor's winding so that TCO can activate in time if the motor winding overheats, as shown in figure 3.

Figure 3. TCO inside a universal motor.

Choosing a correct bimetal type for a particular application is crucial to ensure the TCO functions as required. According to Kanthal [6], the factors affecting bimetal operation are operational temperature and deflection, mechanical stress, electrical resistivity, thermal conductivity, corrosion resistance, and machinability. The operating condition of a motor is crucial when selecting a proper TCO to be used inside a motor. For a universal motor, amongst the main parameters that affect TCO selection are the electrical current, the motor winding insulation class, and the motor minimum flow rate. Apart from these parameters, the overall construction of motor, position of TCO and its internal resistance will also affect the working characteristic of a TCO. The TCO used in a motor has to achieve a balance between protecting the motor from overheated condition and avoiding nuisance tripping (early cut-off).

In order to avoid undesirable TCO cut-offs, it is crucial to understand the TCO response behaviour and the operational window in any electrical appliance that uses universal motor. This operational window is the margin between which the TCO activates before motor failure and nuisance tripping. In this study, the TCO response model is being developed to help understand TCO's response towards ambient temperature, operating flow rate, electrical current, and TCO resistance. By doing this, the thermal risks of any TCO dependent appliance can then be evaluated to provide sufficient margin between the TCO tripping at motor overheat condition/current overload and nuisance tripping at normal use.

The studies [5] [9] done by other researchers so far are focused on the motor thermal protection model mainly through the use of resistance temperature detectors (RTD) or positive temperature coefficient (PTC) thermistors. TCO's that are widely used in many small universal motors for its inexpensive design has not been studied much in detail. The present paper aims to make a motor thermal protection model which is based on TCO wherein it will also quantify its variations from various factors.

2. Mathematical Model

The mathematical model for TCO response prediction includes formulating a thermal model based on heat transfer from motor windings to the TCO unit. This model can be divided into two: first by modelling heat transfer from windings to the exhaust air and secondly the heat transfer from the exhaust air to the TCO component.

The thermal model for the heat transfer between the windings and the exhaust air is made by comparing the resistive heating of motor windings, thermal resistance of the system and thereby calculating the temperature rise of the exhaust air based on its thermal capacity, as shown in equation (1).

$$I^2(R_f+R_a) - \frac{\Delta T}{R} = \dot{m}C_p\Delta T \tag{1}$$

In equation (1), I is the current passing through the motor circuit, R_a is the armature coil resistance, R_f is the field coil resistance, R is the thermal resistance, $\dot{m}$ is the mass flow rate, ΔT is the temperature difference between the exhaust and TCO surface *and* C_p is the specific heat capacity of air.

Based on equation (1), the thermal model terms are rearranged into an iterative transient model [7] as shown in equation (2).

$$T_n = \left[\frac{I^2}{\tau}\,dt + T_{n-1}\left(1 - \frac{dt}{\tau}\right)\right] \tag{2}$$

where τ is the time constant,

T_n is the exhaust air temperature at time n (expressed in units of I^2)

T_{n-1} is the exhaust air temperature at time n-1 (expressed in units of I^2)

The final exhaust temperature is multiplied by $(R_f + R_a)R$ to get the temperature in unit of degree Celsius.

The incremental form of equation (2) is ideal in calculating the real time temperature response. The transient nature of this model would also make it easier to calculate the winding resistance when it changes as a function of temperature.

The thermal model for the heat transfer between the exhaust air and the TCO surface is a steady state temperature calculation compared to equation (2) which is time dependent. The steady state model is made with the assumption that the convective heat transfer is the major contribution to the temperature rise compared to the other heat transfer modes. This is a reasonable assumption to make since radiation is not going to be significant at such low temperature range of 20°C to 150°C as compared to forced convection. As for conduction, most motor designs have TCO units primarily held in the direction of airflow by negligible contact to solid parts. Even in cases where they are primarily attached to solid parts, the part they are attached to is a plastic surface which has poor thermal conductivity. These assumptions help to reduce the complexity of the model.

The steady state model can be derived for a thermal condition where the TCO surface and the motor exhaust air have achieved thermal equilibrium. At this state, the heat produced by the TCO is equal to the heat dissipated through the motor exhaust air. This is shown in equation (3).

$$I^2 R_{TCO} = hA(T_{TCO} - T_{exhaust}) \tag{3}$$

In equation (1), R_{TCO} is the TCO component resistance, A is the flow facing surface area of TCO, T_{TCO} is the temperature of the TCO component and $T_{exhaust}$ is the temperature of the motor exhaust air.

By rearranging the terms, we would arrive at equation (4) which can be used to calculate the steady state TCO surface temperature.

$$T_{TCO} = \frac{I^2 R_{TCO}}{hA} + T_{exhaust} \tag{4}$$

The heat transfer coefficient h value in equation (4) is derived from equation (5).

$$h = \frac{\mathrm{Nu}k}{l} \tag{5}$$

where Nu is the Nusselt number that is calculated using $\mathrm{Re_D}$ and Pr which are the respective Reynolds and Prandtl number. The below equation is derived based on the condition of flow around rectangular solids [8].

$$\mathrm{Nu} = 0.158\mathrm{Re_D}^{0.638}\mathrm{Pr}^{1/3} \tag{6}$$

The flow characteristics around the TCO would be different for different TCO design configurations. Based on the relative location of TCO in the motor, certain parameters like the velocity of air over TCO, the characteristic length for Reynolds number, and the heat transfer coefficient around TCO would be changed in the model.

To make this predictive model without the need of additional testing, the input data for the proposed thermal model is taken from the supplier datasheets of universal motor and TCO component. Sample data are shown in figure 4 and figure 5.

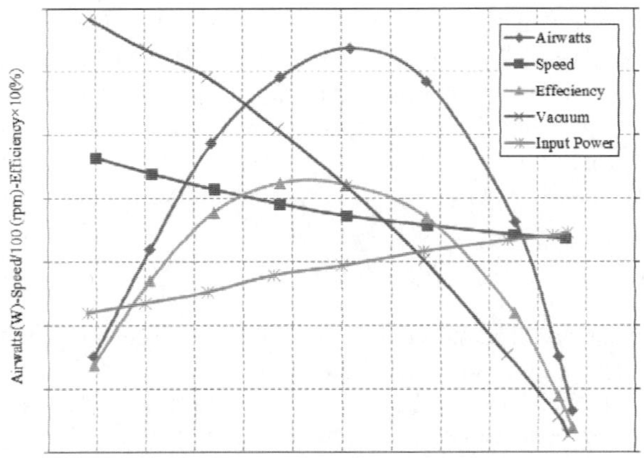

Figure 4. Universal motor supplier datasheet displaying motor performance data.

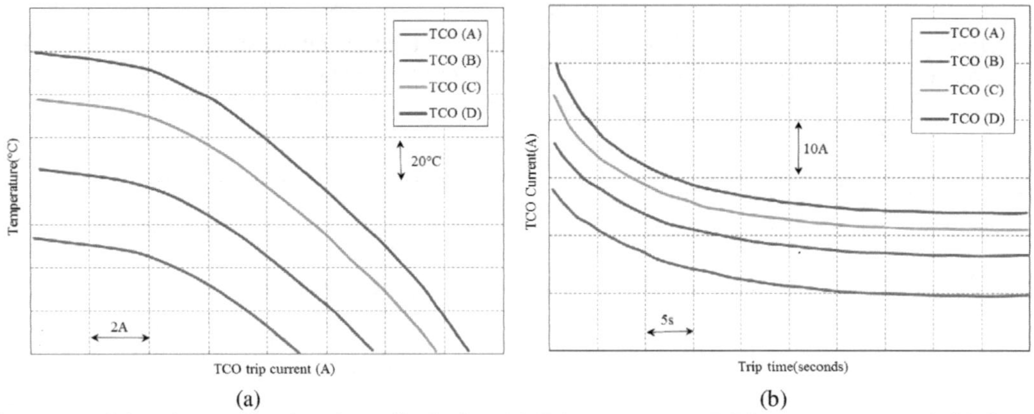

Figure 5. TCO unit supplier datasheet displaying (a) Trip current vs. TCO temperature and (b) Current vs. Trip time.

Based on figure 5 (which contains data taken from a standalone TCO unit), it can be inferred that the TCO unit trips to protect the motor at lower temperature for higher current. This is due to the fact that the TCO unit is using a bimetallic strip circuit breaker that trips when the strip reaches a particular temperature at a certain current. When higher current travels through the strip, the resistive heating due to the current would make the TCO unit to trip at lower operating temperature. Using this knowledge, it is possible to estimate the resistance and temperature coefficient of the TCO component, as shown in equation (7).

$$R_{TCO} = \frac{hA\Delta T}{I^2_{trip\ current}} \tag{7}$$

where $I^2_{trip\ current}$ is the current at which the bimetallic strip snaps and breaks the electric circuit.
Using equation (7), the resistance of TCO can be estimated at all TCO trip temperatures range provided in the supplier datasheet. The temperature coefficient of resistance α is estimated by selecting the highest resistance, R_{high} and lowest resistance, R_{low} values from all the resistance values calculated from equation (7) and using it in equation (8).

$$\alpha = \frac{R_{high} - R_{low}}{R_{high}(T_{high} - T_{low})} \tag{8}$$

The above two information would be stored as reference for each category of TCO variants such that the thermal model would refer the resistance and the temperature coefficient of resistance to estimate the resistance of the corresponding TCO in the available TCO temperature range. Equation (9) will be used to calculate this when the TCO temperature increases from T_1 to T_2.

$$R_2 = R_1[1 + \alpha(T_2 - T_1)] \tag{9}$$

3. Experimental Methodology

The experimental setup for measuring the TCO surface temperature is done based on the schematic shown in figure 6. The TCO connected to the universal motor is placed inside a control volume/appliance. The current passing through the motor is measured using a Yokogowa digital power meter WT210. The resistance of the TCO is checked using a Fluke 8845A 6-1/2 digit precision multimeter before the test starts. One T-type thermocouple is attached to the TCO surface to measure the temperature, which is continuously logged by a Pico Technology USB TC-08 data logger. The flow rate passing through the motor is measured by an orifice plate at the motor upstream. The flow rate is varied by changing the orifice plate diameter.

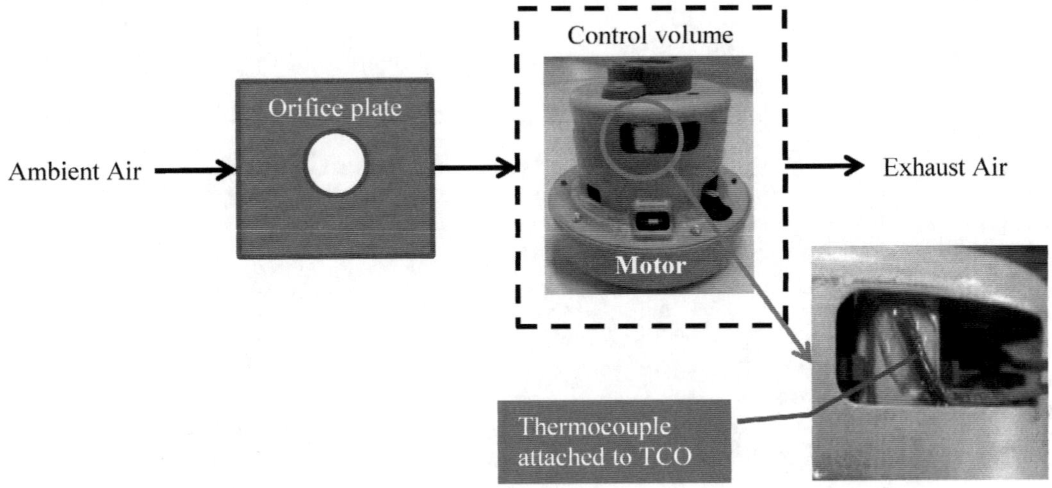

Figure 6. Schematic diagram of experimental test.

An uncertainty budget is made for each of the parameters measured and controlled during the test (as show in table 1). The terms I, R-TCO, Q, and T-Amb are the current passing through the motor, TCO resistance, flow rate and ambient temperature measured during the test.

Based on the uncertainty budget calculated in table 1, the expanded uncertainty of the TCO surface temperature value is ±4.79°C at 95% confidence level. The expanded uncertainty U_{exp} is as expressed in equation (10).

$$U_{exp} = k_c \times U_{comb} \tag{10}$$

where k_c is the coverage factor and U_{comb} is the combined standard uncertainty.

Table 1. Uncertainty budget for the TCO surface temperature measured during the test

Symbol	Source of Uncertainty	Value	±	Probability distribution	Standard Uncertainty
Random	Repeatability		0.2593	Normal	0.26
I	Power meter accuracy	10	0.0100	Rectangular	0.46
I	Resolution error	10	0.0010	Rectangular	0.05
R-TCO	Multimeter accuracy	0.01	0.0002	Rectangular	0.18
R-TCO	Resolution error	0.01	0.0001	Rectangular	0.09
Q	Flow rate Accuracy	10	0.1256	Rectangular	2.82
T-Amb	Calibration uncertainty	40	0.6050	Rectangular	0.35
		Norm/Rect	0.09	Combined Std Uncertainty (±)	2.90
		Coverage factor	1.65	Expanded Uncertainty (±)	4.79

4. Results and Discussions

The proposed thermal model is validated by comparing its prediction with the experimental data. The model is compared with the TCO surface temperature test data from two of the most widely used commercial universal motors.

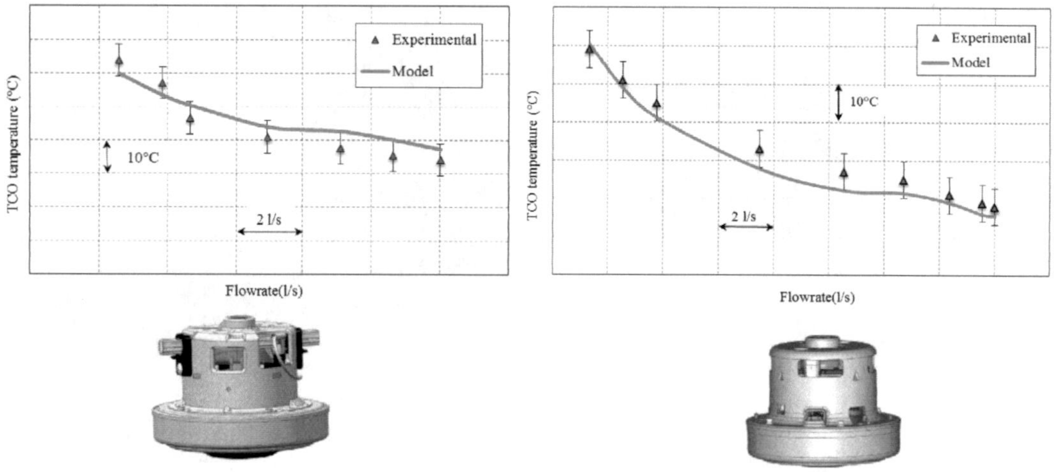

(a) (b)

Figure 7. Comparison of model and actual test data for (a) High power range universal motor (b) Low power range universal motor.

Based on figure 7, we can observe that the model is satisfactorily following the trend of the actual TCO surface temperature. The maximum difference between the test data and model data is 6.4% for the high power range vacuum motors while for the low power range, maximum difference of 7.3% can be observed. The difference in results can be partly explained by the limitations of the model where the prediction is based on a single time constant. In reality, the system would compose of many components, where each component has different thermal time constants and thus achieving different steady state temperature results. This can also be realized from the fact that at short time intervals, the thermal response is dominated by the heat transfer from the stator and rotor conductors to the iron while for longer time intervals, the thermal response is dominated by the heat transfer from the iron to cooling air [9]. Another reason for the discrepancy between the actual test data and the model data can be due to the varying efficiency between different vacuum motor samples.

While checking the logged surface temperature of the TCO component, it was observed that the TCO surface did not immediately cool down after the motor was switched off. In fact it had been heating up due to the phenomenon called thermal or heat soak [10] (as shown in figure 8). In universal motors, heat soak can be theorized to happen due to two reasons. Firstly, it is due to the thermal lag between the heat generating TCO and the cooling flow on the TCO surface. Secondly, it could be due to the relative position of the TCO unit that makes it susceptible to be on the receiving end of the heat dissipated by natural convection from the windings. Both these reasons will lead to an increase in TCO temperature after the motor stops running.

IOP Conf. Series: Journal of Physics: Conf. Series **822** (2017) 012065 doi:10.1088/1742-6596/822/1/012065

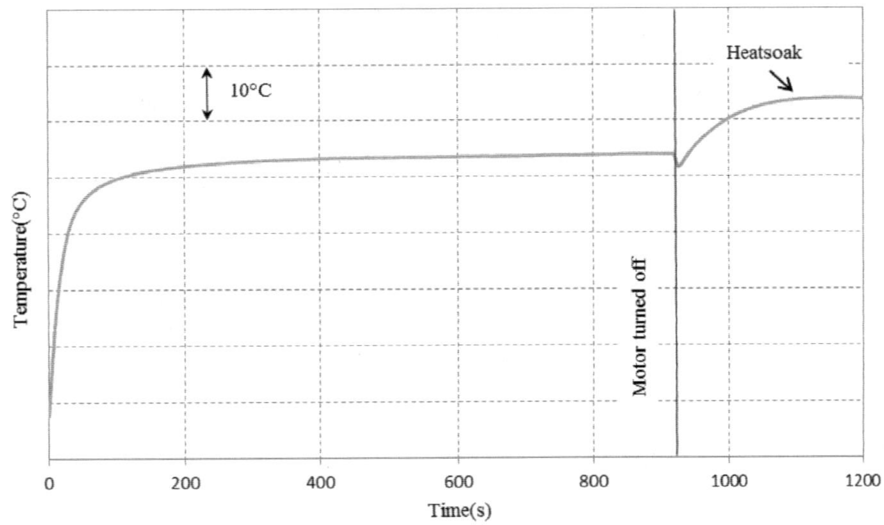

Figure 8. Temperature profile data of TCO surface temperature.

Also, it is was observed that the very high inrush current going to the TCO during start can cause the bimetallic strip to reach the trip temperature at a short moment thereby making steady state comparisons of the model invalid for those scenarios.

The inrush current coupled with heat soak can prove to be a risky combination wherein the TCO unit might trip soon after the motor has reached the heat soak peak during switch off and immediately switched on. This type of scenario can actually cause pre-mature tripping. Thus the model has some real world limitations, which was not addressed in this model. For effective modelling of the TCO response, the response modelling should include the effects of the heat soak and inrush current to provide a better design guideline for using the TCO component in universal motors.

The final result of the model is used to predict the surface temperature of the TCO as shown in figure 9. Figure 9 indicates the temperature profile of TCO surface temperature at 28°C and 40°C environment which are the boundaries for TCO specification validation. Based on this graph, inferences can be made whether TCO surface temperature meets the TCO design criteria. The criteria are based on whether the predicted TCO trip temperature is within the design margin. If the predicted TCO surface temperature is not within the design margin, i.e. the TCO surface temperature is less than the TCO trip specification at the minimum flow rate (defined by the motor supplier as the temperature of the TCO should have reached when the motor is running at minimum flow rate) or the TCO surface temperature is very close to the TCO trip specification at normal flow rate conditions, changes in TCO design criteria has to be made. Changes can be done by lowering the TCO resistance, or by using a TCO with higher trip temperature rating and in some cases both.

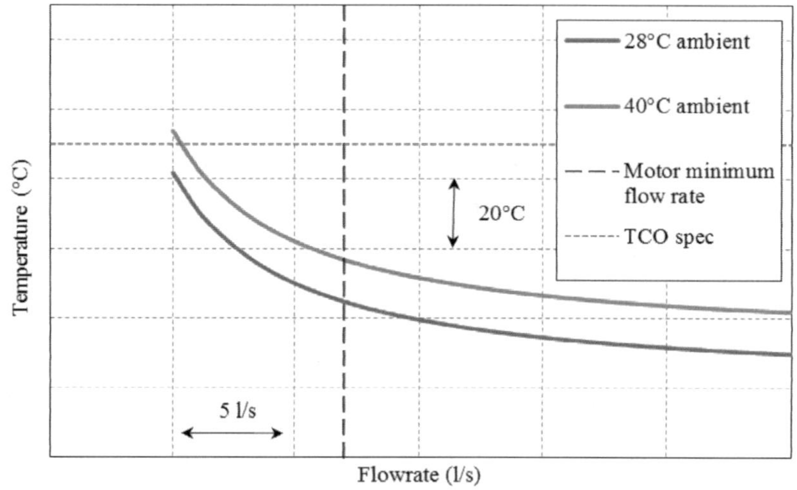

Figure 9. Graphical results of the TCO surface temperature prediction model.

5. Conclusions

This paper presented a mathematical model for predicting the TCO response characteristics for universal motors using a combination of transient and steady state heat transfer model. The thermal interactions within the universal motors which influence the TCO surface temperature is taken into account using a mathematical model that is based on inputs from various universal motors and TCO component supplier datasheets. The model can help to estimate the TCO operating temperature under various conditions, and prescribing a suitable TCO to be used in the motor.

In order to perform a more comprehensive study on the TCO response with the effects of heat soak and inrush current, the entire model needs to be modified to a time dependent one. The more comprehensive model with heat soak and inrush current will be most valuable to accurately predict the working window for TCO at different operating constraints.

References

[1] Venugopal G 2001 Characterization of thermal cut-off mechanisms in prismatic lithium-ion batteries *J. of Power Sources,* **101** 231

[2] Sensata Technologies 17AM *Thermal protector for motor/ballast for fluorescent and temperature sensing controls* [Online] Available: http://www.sensata.com/download/17am.pdf

[3] Thermtrol Corporation *Product Catalogue* [Online]. Available: http://www.thermtrol.com.

[4] Field H L and Solie J B 2007 *Introduction to agricultural engineering technology* in *Electric motors* (New York: Springer) p 353

[5] Ostojic P, Yabiku R, Vico J, Lope P F and Balista A 2012 Improving the usage of temperature sensors for motor thermal protection in *Petrol. & Chem. Ind. Tech. Conf.* **59**

[6] Kanthal AB, Thermostatic bimetal handbook [Online]. Available: http://www.kanthal.com/Global/Downloads/Materials in wire and strip form/Thermostatic bimetal/Bimetal handbook ENG.pdf

[7] Steinmetz J, Patel S C and Zocholl S E 2013 Stator thermal time constant *Ind. & Com. Pow. Sys. Tech. Conf.* **49**

[8] Incropera F P and DeWitt D P 2007 *Fundamentals of Heat and Mass Transfer* 6th ed. (New York: Wiley)

[9] Venkataraman B, Godsey B, Premerlani W, Shulman E, Thankur M and Midence R 2005 Fundamentals of motor thermal model and its application in motor protection *Annual Conf. for Protective Relay Engineers* **58** 127

[10] Khaked M, Elrab M G, Habchi C, Al Shaer A, Elmarakbi A, Harambat F, Peerhossaini H 2015 Analysis and modeling of the thermal soak phase of a vehicle-temperature and heat flux measurements *Intl. J. of Auto. Tech.* **16** 221-229

IOP Conf. Series: Journal of Physics: Conf. Series **822** (2017) 012066 doi:10.1088/1742-6596/822/1/012066

Experimental study to the influences of rotational speed and blade shape on water vortex turbine performance

T C Kueh[1,5] , S L Beh[2], Y S Ooi[1,3] and D G Rilling[4]

[1] Centre for Advanced Materials and Green Technology, Multimedia University, 75450 Melaka, Malaysia
[2] University of Southampton Malaysia Campus, No. 3, Persiaran Canselor 1, Kota Ilmu Educity @ Iskandar, 79200 Nusajaya, Johor.S
[3] Faculty of Engineering and Technology, Multimedia University, 75450 Melaka, Malaysia
[4] Studiendekan School of Engineering & Technology Management, Akad University, 70469 Stuttgart, Germany
[5] Author to whom any correspondence should be addressed.

E-mail: kuehtzecheng@hotmail.com

Abstract. Water vortex turbine utilizes the natural behaviour of water to form free surface vortex for energy extraction. This allows simple construction and ease of management on the whole water vortex power plant system. To our findings, the literature study specifically on water vortex turbine is inadequate and low efficiency was reported. Influences of operating speed and blade shape on turbine performance are the two parameters investigated in this study. Euler Turbomachinery Equation and velocity triangle are used in the improvement analysis. Two turbines with flat blades and curved blades are tested and compared. Both turbines show similar rotational speed at no load condition. This suggested that the circulation force of the water vortex has more dominant effect on the turbine rotational speed, compared to the turbine's geometry. Flat-blades turbine showed maximum efficiency of 21.63% at 3.27 rad/s whereas curved-blades turbine showed 22.24% at 3.56 rad/s. When operating load is applied, the backward-leaning curve helps the turbine blades to reduce the disturbance on the water vortex, and hence provide a better performance.

1. Introduction

1.1. Water vortex power plant

Water vortex power plant (WVPP) is a hydropower system that generates electricity by utilizing the natural feature of water to form free surface vortex. Water vortex creates swirl flow which is beneficial to the rotation of the turbine system for electricity generation purpose. Figure 1 shows an illustration of WVPP.

WVPP can operate under low head condition. It does not need a water dam to operate. Hence, its construction and management can be simple. Not only reduction on financial cost, this can also decrease the damage on the environment. The large cross-section area of flowing water and low runner speed are also doing less harm to under-water creatures that pass through the system. On the other hand, the absence of high head drop causes the efficiency of the system to be low compared to other typical turbine systems. However, WVPP still has its own application especially to act as a decentralized small scale hydropower system in remote area.

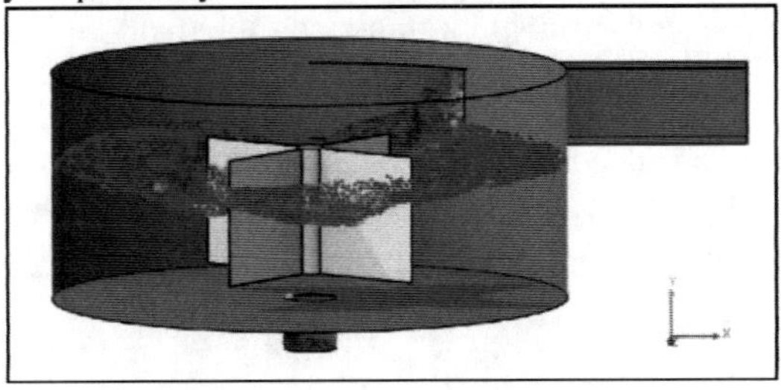

Figure 1. Water vortex power plant.

The number of studies focused on WVPP is quite limited. Yaakob et al. [1] reviewed the studies of free surface vortex and different types of current existing turbine as the early stage of WVPP study. Wanchat et al. [2] studied the WVPP by varying the outlet diameter of the basin and reported an optimum efficiency of 30%. The inadequate of the literatures encourages more studies to be taken into this field to improve the performance of WVPP. With a better efficiency, the application of WVPP can be much widened.

1.2. Turbine optimization

There are a lot of factors that can influence the performance of a turbine. Researchers have been studying turbines' performance from different aspects. Singh and Nestmann [3] [4] compared the influences of the angle, height and number of the runner blades on the axial turbine efficiency. Free vortex theory and velocity triangle were used for the analysis purpose in their studies. Angle at the blades' exit and height of the blades were reported to have a more dominant effect.

Alexander et al. [5] proposed the idea to remove all possible angular momentum in an axial turbine design as the improvement steps. Date et al. [6] studied the efficiency of simple reaction turbine under different operating loads. They suggested that in an ideal case, where assumption of no friction loss is made, the optimum efficiency of the turbine will be obtained when the operating load is half of the maximum shaft torque.

Although water vortex turbine is also classified as a reaction turbine, but its operating concept is different than the typical reaction turbines. Therefore, application of these optimization approaches on water vortex turbine becomes an issue to be investigated in this study.

There are two main objectives to be carried out by this study. First, to investigate the performance of water vortex turbine at different operating speeds, and second, to investigate the performance of water vortex turbine with two different blades' shapes.

2. Methodology

2.1. Theoretical analysis

The output power of the turbine system is defined by equation (1), where T and ω represent the operating load and operating speed of the turbine respectively. The operating load is obtained from the Euler Turbomachinery Equation as shown in equation (2). The term $\dot{m}$ is mass flow rate, r is radius, and V_θ is water velocity in tangential direction. Subscripts 1 and 2 are indicating the location of inlet and outlet.

$$P = T\omega \tag{1}$$

$$T = \dot{m}\left(r_1 V_{\theta 1} - r_2 V_{\theta 2}\right) \tag{2}$$

The tangential velocities at blade inlet and outlet used in equation (2) can be obtained by constructing velocity triangle on the turbine blades. It is shown in figure 2. Since the flowing motion of water vortex is in a swirl direction, velocity triangles at the inlet and outlet have to be constructed in different views.

From the constructions of velocity triangle at the blade inlet and outlet, V_θ in equation (2) can be rewritten in the function of operating angular speed (ω). By substituting this into equation (1), the output power of turbine can be described as a function of operating speed. If a simple shape is used for the turbine blade and assumption of no friction loss is made, the relation between V_θ and ω can be simplified. In this case, the optimum power can be determined by differentiating the equation of power in the function of operating speed.

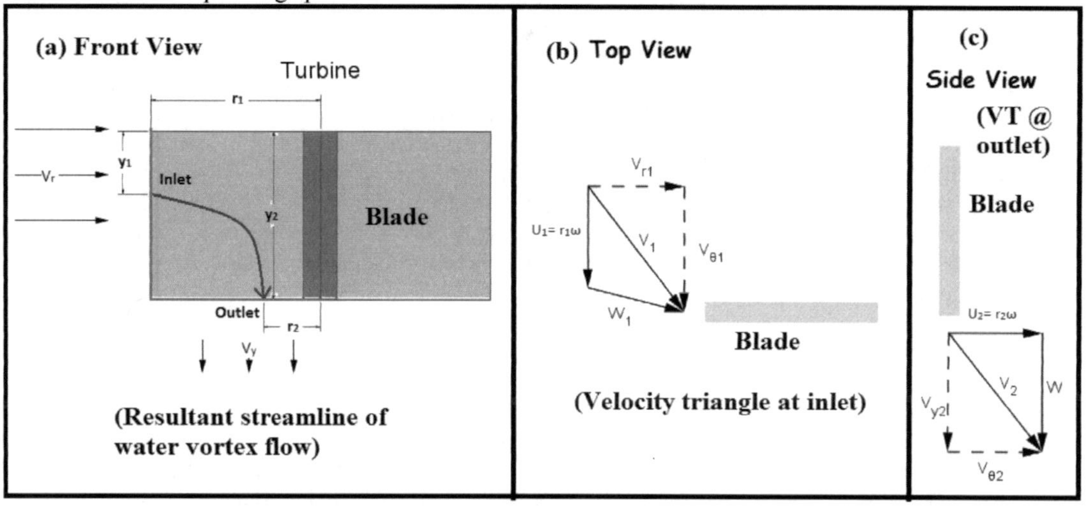

Figure 2. Velocity triangle constructed on water vortex turbine's blade.

However, in real cases, that assumption cannot be true when the turbine blades are often in a more complicated shape. Therefore, a practical way is to observe the output power of the turbine by varying its operating speed, which is one of the objectives in this study.

Besides that, figure 2(c) shows that increment of exit blade angle can reduce the magnitude of $V_{\theta 2}$. Smaller $V_{\theta 2}$ can deliver higher turbine output power as shown in equation (1) and (2). This is another subject of research in the experiment study.

2.2. Experiment setup

An experiment setup system which includes basin and turbine components was built to study the water vortex turbine. Water is circulating within the system by a pump as shown in figure 3. Two turbines are tested in this study. One of them is made of four flat blades while the other one contains four curved blades. Their geometry and dimensions are shown in Figure 4.

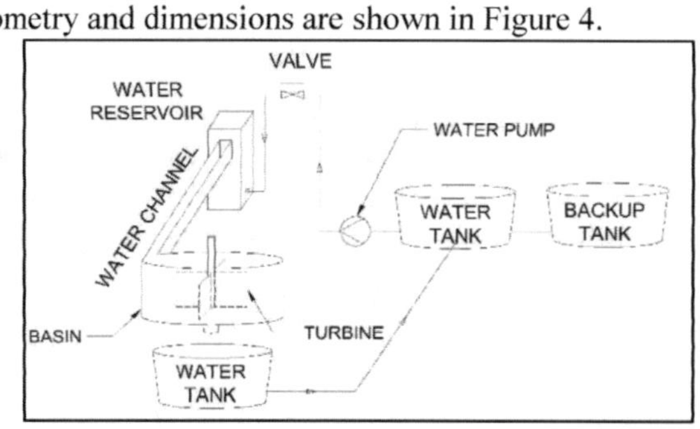

Figure 3. Schematic diagram of experiment setup.

The basin is two meters in diameter with 0.8 m depth. The water inlet is flowing into the basin at 0.5 meter from the bottom of the basin. Hence, 0.5 m is selected as the head drop of the system. Water reservoir is placed to be filled up before the water enters the basin. This can remove the initial water velocity which is imposed by the water pump. By doing this, the input power is only contributed by the water potential energy but not its kinetics energy, therefore equation (3) can be fulfilled.

Input power is shown in equation (3). Gravity acceleration and head drop is known. Mass flow rate of the water is obtained by the portable ultrasonic flow meter. The portable ultrasonic flow meter contains two transducers which are attached at the upstream and downstream of a pipe. The transducers send and receive ultrasound within the pipe. Flowing water as the transmission medium will cause a difference between the times taken for the ultrasound to travel from upstream to downstream and the opposite way. Portable ultrasonic flow meter collects the times difference and obtains the speed of water by a proper calculation.

To measure the output power at different operating speeds, a braking system is designed and located at the turbine shaft. It is shown in figure 5. The braking system is tighten at different strengths to apply different friction torques on the turbine. The friction torque is simulating the operating load of the turbine, and hence relevant operating speed is obtained. Load-cells are connected at the two end of the braking belt to measure the tension force difference. This difference is used to calculate the friction torque.

$$P_{input} = \dot{m}gH \tag{3}$$

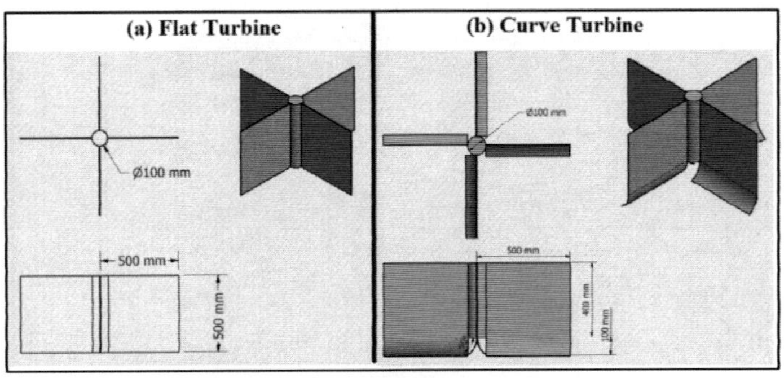

Figure 4. Dimensions of turbines.

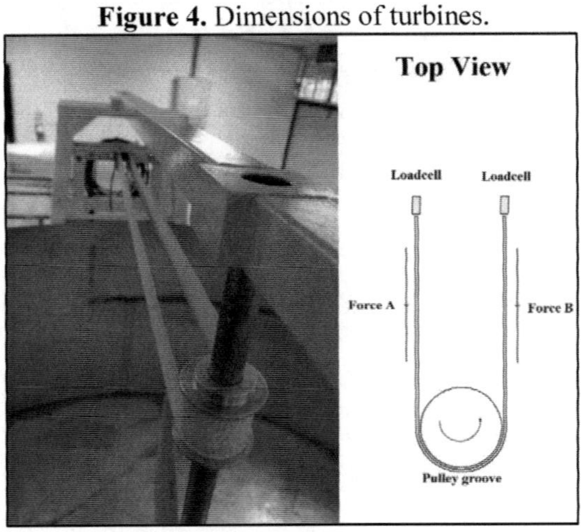

Figure 5. Braking system.

Several friction torques were applied to obtained different operating speeds. Water flow rate, operating torque, and operating speed in these cases were collected and recorded to calculate the input power, output power and efficiency of the turbine. Then the turbine was replaced by another one and the experiment was repeated.

3. Results and discussion

3.1. Results

When the pump was started, water vortex formed inside the basin and the turbine was rotating. Figure 6 shows the illustration of the water vortex turbine system during its operation. Certain strengths of friction torque were applied on the turbine shaft and the turbine was still able to rotate. The values of the friction torques, responding angular speeds and water flow rates were obtained and recorded. From these values, input power, output power and efficiency were calculated and recorded in table 1 and table 2 for flat turbine and curve turbine respectively.

Figure 6. Operation of water vortex turbine setup.

Table 1. Result for flat turbine

Experiment	Flow rate, kg/s	Operating Load, Nm	Operating Speed, rad/s	Input Power, W	Output Power, W	Efficiency, %
1	11.19	0.00	4.08	54.84	0.00	0.00
2	11.98	1.40	3.87	58.76	5.42	9.22
3	12.98	3.21	3.72	63.66	11.94	18.76
4	14.68	4.38	3.43	71.98	15.02	20.87
5	15.47	5.02	3.27	75.90	16.42	21.63

Table 2. Result for curve turbine.

Experiment	Flow rate, kg/s	Operating Load, Nm	Operating Speed, rad/s	Input Power, W	Output Power, W	Efficiency, %
1	10.68	0.00	4.08	52.39	0.00	0.00
2	11.78	1.44	3.93	57.78	5.66	9.79
3	12.08	2.64	3.84	59.25	10.14	17.11
4	12.48	3.23	3.80	61.21	12.27	20.05
5	13.48	4.13	3.56	66.10	14.17	22.24

The maximum efficiency that can be obtained from the flat turbine is 21.63% at 3.27 rad/s of operating speed whereas the curve turbine showed 22.24% of maximum efficiency at 3.56 rad/s. The graph of efficiency versus operating speed for both turbines are plotted in figure 7.

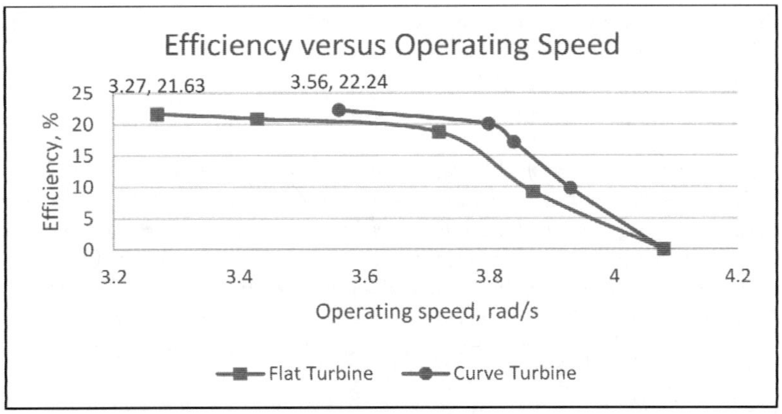

Figure 7. Graph of efficiency versus operating speed.

3.2. Discussion and analysis

For both turbines, generally the efficiency increases when the speed is decreased. The efficiency is expected to drop when the speed is reduced to a certain value. However, those range of experiment cannot be carried out in this study due to equipment limitation. To perform the experiment at such low speed, it requires large friction torque, which is not able to be provided by current braking belt system and load-cells components.

The input flow rate is influenced by the pump capacity and the total water volume in the water tanks. At different turbine speed, the ratio of water volume contained in basin and water tanks is different. This reason causes the input power varies at different operating speeds. Therefore, instead of output power, efficiency as a dimensionless term is taken as the outcome reference of this study.

Figure 7 shows the trend of efficiency changes with respect to the operating speed. The complete graph which is expected in parabolic shape cannot be obtained due to the mentioned reason, and the optimum efficiency is not reached. However the comparison between the performances of two graph is still obviously shown. At any operating speed, the efficiency of curve turbine appears to be higher than the flat turbine's.

From the results, the operating speeds of both turbines are same when no operating load is applied. This shows that initially the speed of the turbine is mostly influenced by the circulation of water vortex only. When operating load is applied, the circulation strength of water vortex is weaken. Therefore, water tends to flow in more axial direction. In these moments, the curves at the turbine's blades can extract the angular momentum better, thus the curve turbine produced a better efficiency.

4. Conclusion

Water vortex forms in the basin to rotate the turbine. With only 5 meters of head drop and flow rate around 15kg/s, the system is able to produce 16.42W of output power. This is an open surface system without a dam involved. Therefore, water vortex power plant is suitable to be used in the remote areas which are off-gridded from the electricity network.

The efficiency was varied at different speeds, but the optimum efficiency of water vortex turbine was failed to be determined in this study. The maximum efficiency obtained is low, at the value of 22.24%. In run-of-the-river type of turbine system, where no water dam is used to control the required flow rate, low efficiency is expected. However, further optimization to increase the efficiency is possible.

Improvement is reported in this study when curves are added to the exit of the turbine blades.

Further study can be carried out to find the optimum speed of the turbines where they produce optimum efficiency. The relation between the optimum speed and the blades' angle can also be studied in more detailed way. The results can be used to adjust the blades' angle to fit different speed requirements of the turbine in different conditions.

References

[1] Yaakob O B, Ahmed Y M, Elbatran A H and Shabara H M 2014 A review on micro hydro gravitational vortex power and turbine systems *Jurnal Teknologi* **69** no. 7 pp 1-7

[2] Wanchat S, Suntivarakorn R, Tonmit S K and Kayanyiem P 2013 A parametric study of a gravitational vortex power plant *Advanced Materials Research* **805-806** pp. 811-17

[3] Singh P and Nestmann F 2009 Experimental optimization of a free vortex propeller runner for micro hydro application *Experimental Thermal and Fluid Science* **33** pp 991-1002

[4] Singh P and Nestmann F 2011 Experimental investigation of the influence of blade height and blade number on the performance of low head axial flow turbine *Renewable Energy* **36** pp 272-81

[5] Alexander K V, Giddens E P and Fuller A M 2009 Axial-flow turbines for low head microhydro systems *Renewable Energy* **34** pp 35-47

[6] Date A, Date A and Akbarzadeh 2013 Investigating the potential for using a simple wter reaction turbine for power production from low head hydro resources *Energy Conversion and Management* **66** pp 257-70

Freak Wave Formation from Phase Coherence

Arnida L. Latifah, E. van Groesen
University of Twente, The Netherlands
Indonesian Institute of Sciences, Indonesia
Labmath, Bandung-Indonesia

E-mail: arni001@lipi.go.id

Abstract. This paper describes freak waves by a (pseudo-)maximal wave proposed in [1]. The freak wave is a consequence of a group event that is present in a time signal at some position and contains successive high amplitudes with different frequencies. The linear theory predicts the position and time of the maximal amplitude wave quite well by minimizing the variance of the total wave phase of the given initial signal. The formation of the freak wave is shown to be mainly triggered by the local interaction of waves evolving from the group event that already contained large local energy. In the evolution, the phases become more coherent and the local energy is focussed to develop a larger amplitude. We investigate two laboratory experimental signals, a dispersive focussing wave with harmonic background and a scaled New Year wave. Both signals generate a freak wave at the predicted position and time and the freak wave can be described by a pseudo-maximal wave with specific parameters.

1. Introduction

Freak waves were first recognized since four and a half decades ago [2]. In general, it is unexpectedly high waves that suddenly appear in a relatively mild background wave field. Many definitions of freak wave have been proposed by researchers [3, 4, 5] and the most common definition is a wave for which the ratio of the maximum crest height and the significant wave height (Hs) is greater than 1.25. Freak waves are also known as extreme wave, rogue wave, giant wave, or monster wave [6]. Understanding the mechanism of freak wave appearance is important as it may cause damages in the ocean or shore. Indonesian region has the largest number of fatalities caused by freak waves [7, 8]. On 1 January 2016, local media reported a freak wave up to 9m high came to the shore, specifically at Bantul Jati and Perawan Desa beach in Malang regency and dragged five people [9, 10].

In the past two decades much research about freak waves have been studied, but the mechanisms of the freak wave appearance are still under discussion. A freak wave can be generated by a simple consequence of linear superposition of waves [11]. Another mechanism was proposed that the interactions between envelope solitons provide an explanation for the formation of extreme waves [12]. Study of [13] gives an overview that the occurrence of extreme waves is highlighted into two scenarios: nonlinear mechanism of second order bound waves and third order four-wave resonance interaction of free waves. In [14] a number of physical mechanisms of extreme waves are described: spatial focusing, dispersive focusing, and nonlinear focusing. Besides these three mechanisms, reference [15] discussed other possible mechanisms, such as nonlinear modulational Benjamin-Feir (BF) instability, wave-current interaction, nonlinear wave group dynamics, and wave-coast interaction. While the

BF instability develops, the population of extreme waves increases, but this only occurs for very long-crested waves [16]. Hence, for waves with broad spectra the linear interaction might be the major cause. Dispersive focusing is the linear mechanism and occurs even in linear Gaussian sea. It was proposed in the eighties [17] that the mechanism of the extreme waves in a Gaussian sea are particular realizations of the evolution of well defined wave groups. The waves are dispersive with phase and group velocity inversely proportional to the frequency at deep water. The dispersive focussing can be explained by creating a long wave group with decreasing frequencies, then the dispersion forces the group to contract to a few wavelengths at a given position. This mechanism is also suggested in [18, 7, 11, 19]. Reference [13] confirmed that extreme waves most likely occur due to the dynamics of a single wave group in agreement with the dynamics imposed by the Zakharov equation [20].

This paper contributes in understanding the appearance of freak waves. We propose one mechanism of freak waves occurrence. We discuss extreme waves that are generated by the linear mechanism without confronting that the nonlinear term will also add some contributions in the wave amplitude. The extreme wave is mainly developed by a group event, part of the initial signal with successive high amplitudes. We give the prediction and description of the extreme waves by a concept of (pseudo-)maximal wave [1], that is based on the phase coherence. We implement it in two different experimental signals, a dispersive focusing wave with harmonic background and an experimental version of the well-known New Year wave. We investigate the dynamics of the group event in both signals in forward time by numerical simulations of the AB-equation [21, 22, 23, 24].

2. Description of a freak wave

One design wave describing a freak wave was proposed by Walker et al. [25]. The design wave has become known as New Wave, which is shown to be an acceptable local linear model for large waves. It is based on the average shape of an extreme wave in a linear random Gaussian process. The full model of New Wave uses Stokes regular wave expansion up to nonlinear fifth order. Another design wave was proposed for a linear description of a freak wave based on the phase coherence by a so-called (pseudo-)maximal wave [1]. In the exceptional case of a fully coherent wave phase, a freak wave can be described well by a so-called maximal wave. The difference with the concept of the New Wave is that the pmw can be designed completely by knowledge of the spectrum, without the necessity as for the New Wave to determine the amplitude based on the probability of appearance.

Suppose $\eta_0(t)$ is a given signal at $x = X_0$ and let $\check{\eta}_0(\omega) = |\check{\eta}_0(\omega)|e^{i\theta_0(\omega)}$ be the Fourier transform of the signal. A *maximal wave* corresponding to $\eta_0(t)$ is the linear wave that is defined as follows:

$$\eta_{max}(x,t) = \int |\check{\eta}_0(\omega)| \cos\left(k(x - X_0) - \omega t\right) d\omega, \tag{1}$$

in which k and ω are related by the exact dispersion relation, $\omega = \Omega(k) = \text{sign}(k)\sqrt{gk\tanh(kD)}$, with water depth D and gravitational acceleration g. In this maximal wave definition, the maximum amplitude is assumed to be at $t = 0$. An irregular wave is obtained in case the phases are uniformly distributed in $(-\pi, \pi]$. Then, a freak wave can be described in between a completely random wave and a fully coherent wave. For a given random phase $\theta(\omega) \in (-\pi, \pi\,]$ as a function of wave frequencies with $\theta(\omega) = -\theta(-\omega)$, we consider a *pseudo-maximal wave* (pmw) corresponding to $\eta_0(t)$, for which the phases are restricted for certain $\alpha \in (0, 1)$ to the phases $\theta_\alpha(\omega) = \alpha\theta(\omega)$, as follows:

$$\left[\eta_{pm}(x,t)\right]_\alpha = \int |\check{\eta}_0(\omega)| \cos\left(k(x - X_0) - \omega t + \theta_\alpha(\omega)\right) d\omega. \tag{2}$$

For given fraction α, the phases of all frequencies are constrained by $|\theta(\omega)| \leq \alpha\pi$. Therefore the maximal amplitude decreases for increasing α. Based on [26], we can also implement a *weak* pmw to describe a freak wave by restricting the phases of only the frequencies of large energy carrying modes. The restriction is typically for frequencies within one (or a half) standard deviation σ_ω, around the mean frequency ω_m, that is $\theta_\alpha(\omega) = \alpha\theta(\omega)$ for $|\omega - \omega_m| < \sigma_\omega$.

3. Coherence of a freak wave

3.1. Coherence and prediction of a freak wave

For a given initial signal $\eta_0(t)$ at a position X_0, the uni-directional linear wave evolution is

$$\eta(x,t) = \int |\check{\eta}_0(\omega)| \cos\left(k(x - X_0) - \omega t + \theta_0(\omega)\right) d\omega.$$

The phase information in the frequency frame is defined by the total wave phase, $\Phi(x,t,\omega) = k(x - X_0) - \omega t + \theta_0(\omega)$ with $\theta_0(\omega)$ is the initial phase. Then we estimate the focussing position and time, (X_{foc}, T_{foc}), at which the variance of the total wave phase is minimal, $PV_{foc} = \mathrm{Var}(\Phi(T_{foc}, X_{foc}, \omega))$. We also measure a degree of coherence, $\Gamma \in (0,1)$ by

$$\Gamma_{foc} = 1 - \frac{3}{\pi^2} PV_{foc}.$$

For random phase $\theta_\alpha = \alpha\theta$, the phase variance and the parameter α is related by $PV(\theta_\alpha) = \frac{(\alpha\pi)^2}{3}$. Therefore, using the spectrum of $\eta_0(t)$, and calculating Γ_{foc}, we obtain the pmw with the phase band α_{foc}. Instead of the pmw that estimates the freak wave in the neighbourhood of (X_{foc}, T_{foc}) only, we can reconstruct a predictive signal which is the signal of the linear wave evolution at position X_{foc}:

$$\eta_{foc}(t) = \int |\check{\eta}_0(\omega)| \cos(\Phi(X_{foc}, t - T_{foc}, \omega)) d\omega$$

To see the performance of the linear-based prediction, we will compare the pseudo-maximal signal (pms) and the signal prediction with the numerical simulation results by the nonlinear AB-equation in Section 4.

3.2. Local coherence

Subsection 3.1 discusses that a freak wave occurs when the total wave phase computed from the given signal has minimal variance. This gives the impression that a freak wave is developed by the whole interval from a given signal. However, most waves will not contribute to develop the freak wave. According to [27, 28, 29, 30, 31], a freak wave can be generated locally from clustered waves. Given a time signal, we investigate parts of the signal that may generate a freak wave based on the amount of the local energy, so-called group events [26]. The local energy is computed using wavelet transform. For a given signal, $\eta_0(t)$, the wavelet transform gives a complex valued function, $\mathcal{W}\eta(u,\omega) = |\mathcal{W}\eta(u,\omega)|e^{-i\Theta(u,\omega)}$. The absolute value represents the distributed local energy and the angle represents the phase in the frame of time and frequency. We introduce a dominating group, that is chosen as the group event containing local energy larger than a certain threshold. Thus we investigate the local behaviour of the dominating group that will generate a freak wave. We compute the local time of the maximal energy in the time interval $u \in [t_1, t_2]$, by

$$\tau(\omega) = \left\{ u \middle| |\mathcal{W}\eta(u,\omega)| = \max_u |\mathcal{W}\eta(u,\omega)| \right\}.$$

Then, we define a local time spreading of waves as a function of frequencies, $\varphi(\omega)$.

$$\varphi(\omega) = \left\{ \varphi_v(\omega) \middle| \overline{\varphi_v(\omega)} = \min_v \overline{\varphi_v(\omega)} \right\} \text{ with } \varphi_v(\omega) = (\tau(\omega) - v) \bmod 2\pi \text{ and } v \in [t_1, t_2]$$

At the focussing time, T_{foc}, the mean of the local time spreading is then minimal, $\varphi(\omega) = \varphi_{T_{foc}}(\omega)$. If the dominating group gives a constant $\tau(\omega)$, all frequencies contribute at the same time which leads to a coherent state. We measure the appearance of a freak wave by local coherence, that is computed from the maximum (M), mean (μ), or standard deviation (σ) of the time spreading. For a dominating group at x position, the local coherence is defined as $\Gamma_{M,\mu,\sigma}(x)$, depending on the choice of the parameters, M, μ, σ:

$$\Gamma_M(x) = 1 - \frac{M}{\pi}, \qquad \Gamma_\mu(x) = 1 - \frac{2\mu}{\pi}, \qquad \Gamma_\sigma(x) = 1 - \frac{\sqrt{3}\sigma}{\pi}.$$

We will show that the evolution of the highest amplitude of the dominating wave group is highly correlated with the local coherence in the restricted frequency interval.

4. Results

This section presents experimental cases, a dispersive focussing wave with harmonic background and the scaled New Year wave. From the measurements, we use the measured elevation data from the first gauge after the wave flap as the influx signal for the numerical simulations. We investigate the coherence of the freak waves and show the evolution of the dominating group.

4.1. Dispersive focussing wave with harmonic background

The experiment was executed at the water depth of 2.12m in the Towing tank 1 at Ship Hydromechanics Laboratory, Delft University of Technology. The performance of the linear and nonlinear AB-simulations is very convincing. Compared to the measurements at seven gauges, the linear and nonlinear AB-simulations give a high correlation, 0.97 and 0.98 respectively. This shows that in this case, the waves are mostly linear and there is only little change by the nonlinear contribution.

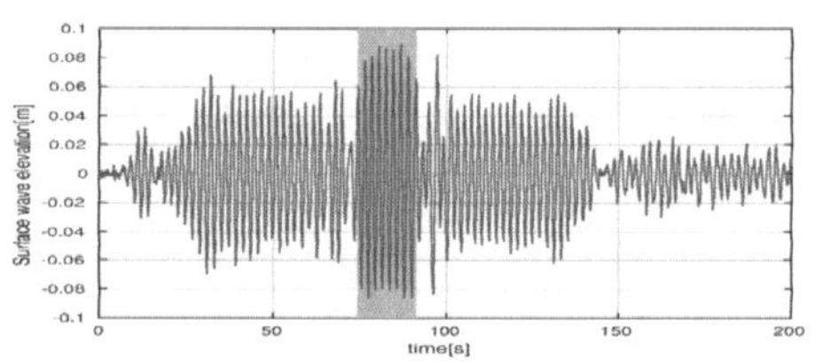

Figure 1. The influx signal of the dispersive focussing wave with harmonic background. The interval of the shaded area shows the dominating group event.

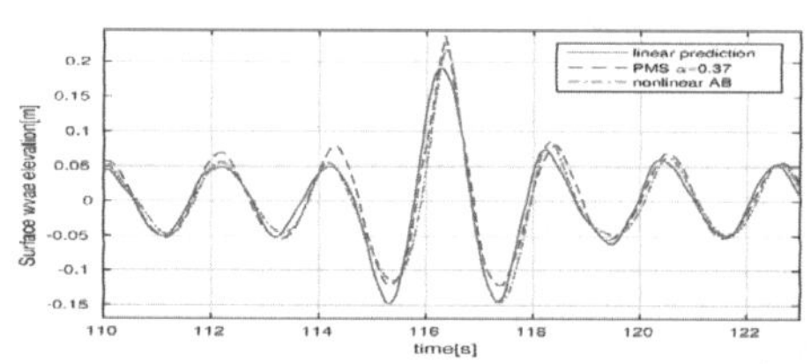

Figure 2. Dispersive focussing wave at the focussing point from the linear prediction (solid), the nonlinear AB-simulation (dash-dotted), and the pms (dashed).

Figure 1 shows that the freak wave is mainly developed by the dominating group event from the influx signal. The linear-based prediction gives a minimum phase variance 0.45, related to a pmw with $\alpha = 0.37$ and a degree of coherence 0.86. It occurs at $t = 116.35$s and $x = 70.2$m from the wave flap, while the nonlinear AB-simulation results a maximal amplitude at $t = 116.72$s and $x = 72$m. The maximum amplitude from the linear prediction reaches $1.27Hs$ and the nonlinear AB-simulation reaches $1.53Hs$. Both can be seen in Figure 2. We also describe the freak wave by the pms with parameter $\alpha = 0.37$. From Figure 2 we can see that the pms can describe the freak wave quite well. The evolution of the local energy of the dominating group is shown in Figure 3; at $x = 40$m the energy is distributed in 10s and at $x = 70$m it is squeezed in approximately 5s. The correlation between the maximal amplitude and the local coherences can be seen in Figure 4, showing that Γ_M and Γ_σ seem to be better than Γ_μ to measure the appearance of the freak wave.

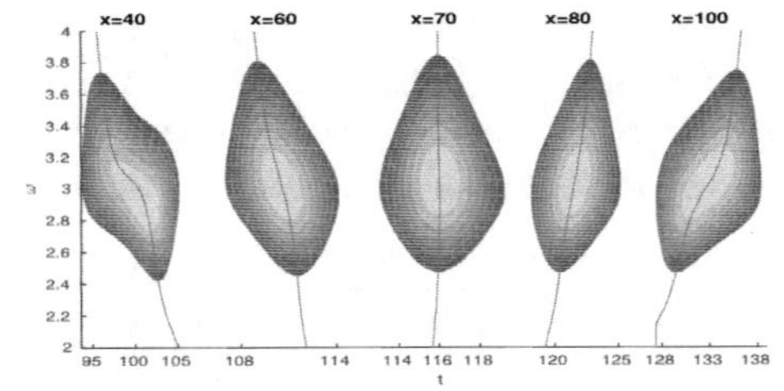

Figure 3. A filled contour plot of the energy distribution of the dominating group at various positions. At each position the solid line shows the $\tau(\omega)$.

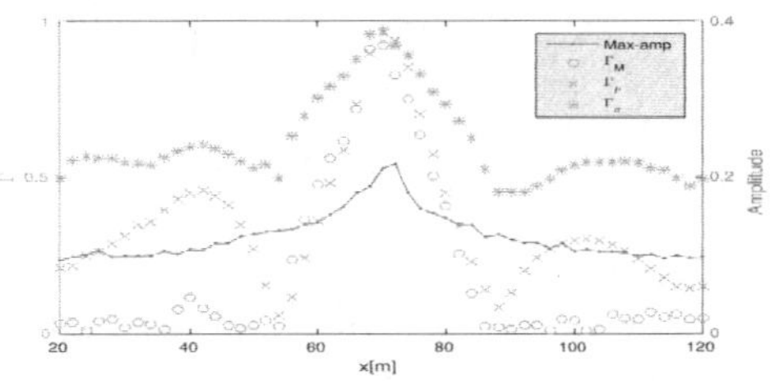

Figure 4. The correlations between the maximal amplitude and the local coherences by the parameters M, μ, σ are 0.93, 0.77, 0.90 respectively.

4.2. Experimental New Year wave

The experiment was conducted at the water depth of 1m, approximately 1:70 of the original New Year wave [3] by MARIN hydrodynamic laboratory with code 204001. The numerical simulations by the linear and nonlinear AB-equation perform well. The mean correlations with the signal measurements at five gauges are 0.78 for the linear and 0.9024 for the nonlinear. This shows that there is significant improvement by adding the nonlinear contribution.

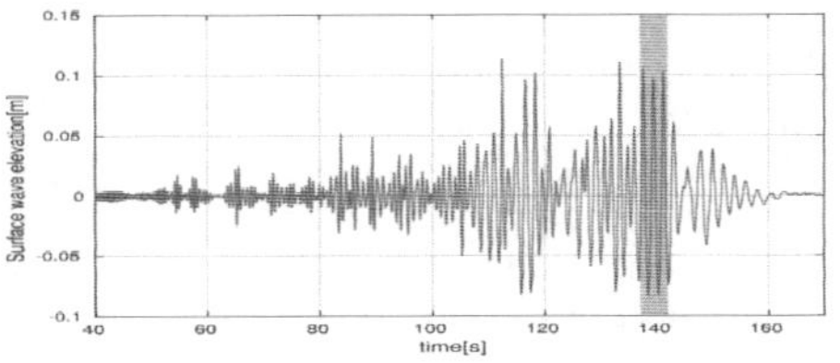

Figure 5. Same as Figure 1, now for New Year wave.

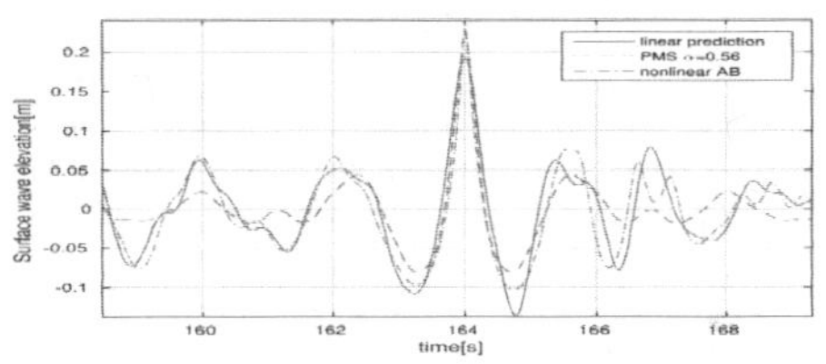

Figure 6. Same as Figure 2, now for New Year wave.

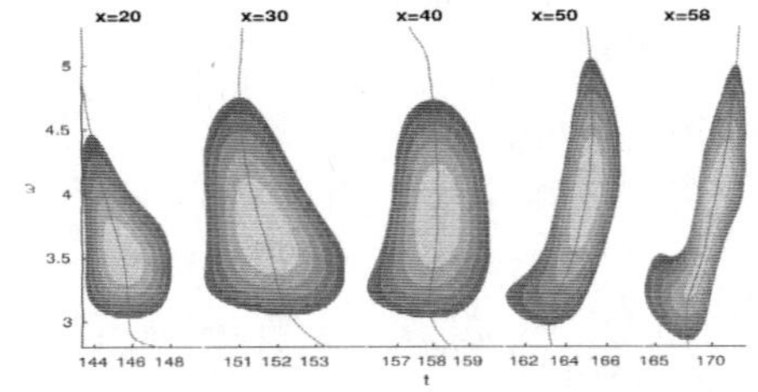

Figure 7. Same as Figure 3, now for New Year wave.

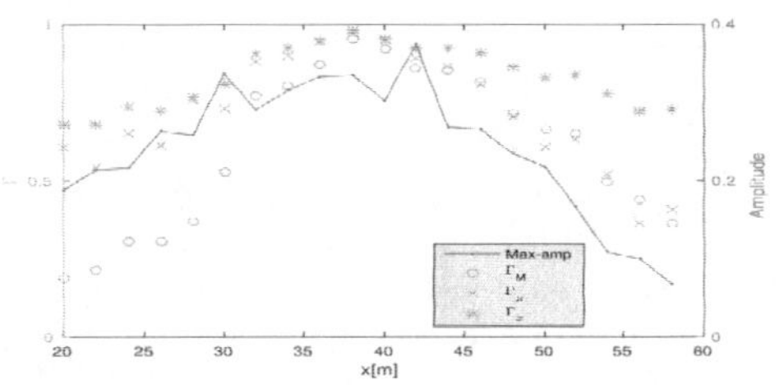

Figure 8. Same as Figure 4, now for New Year wave and the local coherence by parameter M, μ, σ are 0.57, 0.87, 0.67.

Based on the linear prediction, the maximal amplitude reaches $1.58Hs$ at $x = 48.15$m and $t = 162$s, while the nonlinear AB-simulation gives $1.9Hs$ at $x = 49.6$m and $t = 164.5$s. The pmw with parameter $\alpha = 0.56$ describes the freak wave quite well around the focussing time, even though the trough of the freak wave is deeper. Figure 7 shows the local energy at various positions. The main wave frequency becomes higher: at $x = 20$m the main wave frequency is about 3.5 and at $x = 50$m it is about 4. This gives an indication that the waves become steeper. The local phase coherences Γ_M and Γ_σ give quite low correlation and Γ_μ seems to be the best indicator to measure the appearance of the freak wave in this case.

5. Conclusion

Freak wave appears at time and position of high phase coherence. It can be estimated by linear prediction, through minimizing the variance of the total wave phase of a given initial signal. It turns out that the freak wave can be described by a pseudo-maximal wave with a specific parameter that is related to the minimum phase variance. In the two examples presented in this paper, both freak wave positions and times are well estimated by the linear prediction. We remark that the prediction is not a statistical approach, but deterministically based on the phase coherence. The freak wave from the dispersive focussing wave with harmonic background is more coherent than the experimental version of the New Year wave. Both are well described by the pseudo-maximal wave with $\alpha = 0.37$ and $\alpha = 0.56$ respectively. We showed that the freak waves are developed by local waves interactions from a dominating group event that initially contained large energy.

References

[1] Latifah A L and van Groesen E 2012 *Nonlin. Processes Geophysics* **19** 199 – 213
[2] Draper L 1964 *Oceanus* **10** 13–15
[3] Haver S 2004 *Proceeding of Rogue Waves, Brest, France*
[4] Pinho U F D, Liu P C and Ribeiro C E P 2004 *Geofizika* **21** 1931–1943
[5] Olagnon M and van Iseghem S 2000 *Proceeding of the workshop Rogue waves 2000 IFREMER*
[6] Pelinovsky E and Kharif C 2008 *Extreme Ocean Waves* (Springer)
[7] Pelinovsky E and Kharif C 2011 *Natural Hazards and Earth System Sciences* **11** 2043–2046
[8] Nikolkina I and Didenkulova I 2011 *Natural Hazards and Earth System Sciences* **11** 2913–2924
[9] Pitaloka D A 2016 Lima wisatawan terseret ombak di malang, dua meninggal. accessed 20 feb 2016 `http://www.rappler.com/indonesia/117707-lima-wisatawan-terseret-ombak-di-malang`
[10] Arifin Z 2016 Gara-gara selfie, 5 wisatawan terseret ombak di malang. accessed 20 feb 2016 `http://news.liputan6.com/read/2402458/gara-gara-selfie-5-wisatawan-terseret-ombak-di-malang`
[11] Gemmrich J and Garrett C 2008 *Journal of Physical Oceanography* **38** 2330–2336
[12] Kharif C and Pelinovsky E 2006 *Waves in geophysical fluids: Tsunamis, Rogue waves, Internal waves and Internal tides* (Springer) chap Freak waves phenomenon: physical mechanisms and modelling
[13] Fedele F 2007 *Computers and Structures 85 (2007) 291303* **85** 291–303
[14] Dysthe K, Krogstad H E and Muller P 2008 *Annual Review of Fluid Mechanics* **40** 287–310
[15] Slunyaev A, Didenkulova I and Pelinovsky E 2011 *Contemporary Physics* **52** 571–590
[16] Gramstad O and Trulsen K 2007 *J. Fluid Mech.* **582** 463–472
[17] Boccotti P 2000 *Wave mechanics for ocean engineering* (Elsevier Science)
[18] Kharif C and Pelinovsky E 2003 *European Journal of Mechanics - B/Fluids* **22** 603 – 634
[19] Pelinovsky E, Shurgalina E and Chaikovskaya N 2011 *Nat. Hazards Earth Syst. Sci.* **11** 127–134
[20] VE Z 1999 *European Journal of Mechanics B-fluids* **18** 327–34
[21] van Groesen E and Andonowati 2007 *Physics Letters A* **366(3)** 195–201
[22] van Groesen E, Andonowati, Liam L S and Lakhturov I 2010 *J. Comput. Appl. Math.* **234** 1747–1756
[23] Liam L S, Adytia D and van Groesen E 2014 *Ocean Engineering* **80** 73–83
[24] Lakhturov I, Adytia D and van Groesen E 2012 *Wave Motion* **49** 309–322
[25] Walker D A G, Taylor P H and Taylor R E 2004 *Applied Ocean Research* **26** 73–83
[26] Latifah A L and van Groesen E 2016 *Nonlin. Processes Geophys. Discuss*
[27] Sergeeva A and Slunyaev A 2013 *Nat. Hazards Earth Syst. Sci.* **13** 1759–1771
[28] Sergeeva A, Slunyaev A, Pelinovsky E, Talipova T and Doong D J 2014 *Nat. Hazards Earth Syst. Sci.* **14** 861–870
[29] Ruban V P 2013 *JETP letters* **97** 686–689
[30] Cousins W and Sapsis T P 2014 *Physica D* **280-281** 48–58
[31] Cousins W and Sapsis T P 2016 *Journal of Fluid Mechanics* **790** 368–388

IOP Conf. Series: Journal of Physics: Conf. Series **822** (2017) 012068 doi:10.1088/1742-6596/822/1/012068

The effects of the pulsatile period on the size of recirculation bubble in the vicinity of stent struts

B Jiang[1], V Thondapu[1,2], P Barlis[1,2], E K W Poon[1] and A S H Ooi[1]

[1] Department of Mechanical Engineering, Melbourne School of Engineering, The University of Melbourne, Victoria 3010, Australia
[2] Department of Medicine, Faculty of Medicine, Dentistry & Health Sciences, The University of Melbourne, Victoria 3010, Australia

Corresponding author: b.jiang1@student.unimelb.edu.au (B. Jiang)

Abstract. Incomplete stent apposition (ISA) is sometimes found in stent deployment at complex lesions, and it is considered to be one of the causes of post-stenting complications, such as late stent thrombosis and restenosis. The presence of ISA leads to large recirculation bubbles behind the stent struts, which can reduce shear stress at the arterial wall that retards neointimal formation process and thus lead to complications. Computational fluid dynamics (CFD) simulations are performed on simplified two-dimensional axisymmetric arterial models with stents struts of square and circular cross-sectional shapes at a malapposition distance of 120 μm from the arterial wall. To investigate the effects of pulsatile flow period on the dynamics of the recirculation bubbles, high fidelity simulations are carried out with pulsatile flows of period 0.4 s and 0.8 s. Under the condition of the same flow rate, both square and circular strut cases show that shorter period provides greater flow deceleration, leading to the formation of a larger recirculation bubble. With the same thickness, circular strut has a significant improvement over the square strut in terms of the size of the recirculation bubble, and therefore less likely to lead to complications.

1. Introduction

Stenting has been a common intervention treatment for stenosis. Intracoronary stent design has evolved from bare metal stents (BMS) to drug eluting stents (DES) and bioresorbable vascular scaffolds (BVS) over the past 30 years in order to reduce adverse clinical outcomes such as in-stent restenosis (ISR) and stent thrombosis (ST) [1]. However, incomplete stent apposition (ISA) remains common for stent deployment around complex lesions and is considered to be a major cause of post stent complications. In particular, very late ST is often associated with patients who has DES deployed [1].

The presence of malapposed struts influences local haemodynamics, and together with the pulsating effect the human heart recirculation bubbles distal to stent strut are often generated during rapid flow deceleration in a cardiac cycle. These recirculation bubbles slow down local blood flow velocity and yield a lower wall shear stress (WSS) [2]. Low WSS (<0.5 Pa) provides an environment for pro-inflammatory and procoagulant elements to reside longer and leads to less conductivity of endothelialisation [3]. Hence, the long-term outcomes of ISA are possible ISR and/or ST. In this study, we employ computational fluid dynamics (CFD) simulations to show that the size of the recirculation bubble is correlated to the pulsatile nature of the coronary flow (specifically during rapid flow deceleration) and the cross-sectional shapes of the struts.

2. Methodology

Figure 1 (a) presents a commonly available stent model implanted in a 3 mm diameter cylindrical tube. The cylindrical tube represents an idealised segment of the left anterior descending (LAD) artery with an average diameter of 3 mm [4]. We further simplified this axisymmetric stent model into cylindrical ring with both square and circular cross-sections for analyses (see Figure 1(b)) and hence reducing computational expense via performing 2-dimensional axisymmetric CFD calculations (see Figure 1b). We considered stent struts of 120 μm thick, a common thickness of commercial drug-eluting stent strut [1] and a moderate malapposition distance (MD) to be 120 μm [5].

CFD simulations were carried out by directly solving the incompressible Navier–Stokes equations using an open source CFD software, OpenFOAM-2.1.1 (OpenCFD Ltd., ESI group, Bracknell, UK). The temporal term was discretised using a backward Euler scheme whereas standard Gauss linear was used for gradient terms. Gauss linear upwind and Gauss linear corrected discretisation schemes were applied to divergence and Laplacian terms, respectively. Velocity and pressure were coupled through a standard PISO algorithm with a time step of 1×10^{-5} s.

Blood was assumed to be Newtonian with a density of 1000 kg/m^3, and dynamic viscosity of 2.5×10^{-3} Pa·s. In order to capture the boundary layer characteristic near to arterial wall and strut surface, hexahedral elements were cluster to the arterial wall and strut surface (see Figure 1(c)) (first cell element is 7.5 μm normal to the strut surface in square strut model; 6.73 μm in circular strut model). A grid resolution study was performed by quadrupling the number of hexahedral elements, and no significant difference was found in terms of important haemodynamic metrics.

Pulsatile coronary flow velocity from the LAD [6] was applied at the inlet of the axisymmetric domain. A zero-pressure boundary condition was employed at the outlet. The arterial wall and the stent strut surface were assumed to be rigid and non-slip and wedges boundary condition was applied to the remaining patches of the computational domain. The mean blood flow rate was kept constant at 1.3 ml/s (which corresponds to approximately 0.1839 m/s) in this study while increasing the cardiac cycle from 0.4 s and 0.8 s (see Figure 1 (d)). This corresponds to Reynolds number with respect to the arterial diameter, $Re_D = U \cdot D/\nu$, from 107 to 342 throughout the cardiac cycle, and Reynolds number corresponding to the stent strut thickness, $Re_T = U \cdot T/\nu$, from 4.3 to 13.7, where U is the instantaneous inlet velocity, ν is kinematic viscosity, D is arterial diameter and T is the strut thickness.

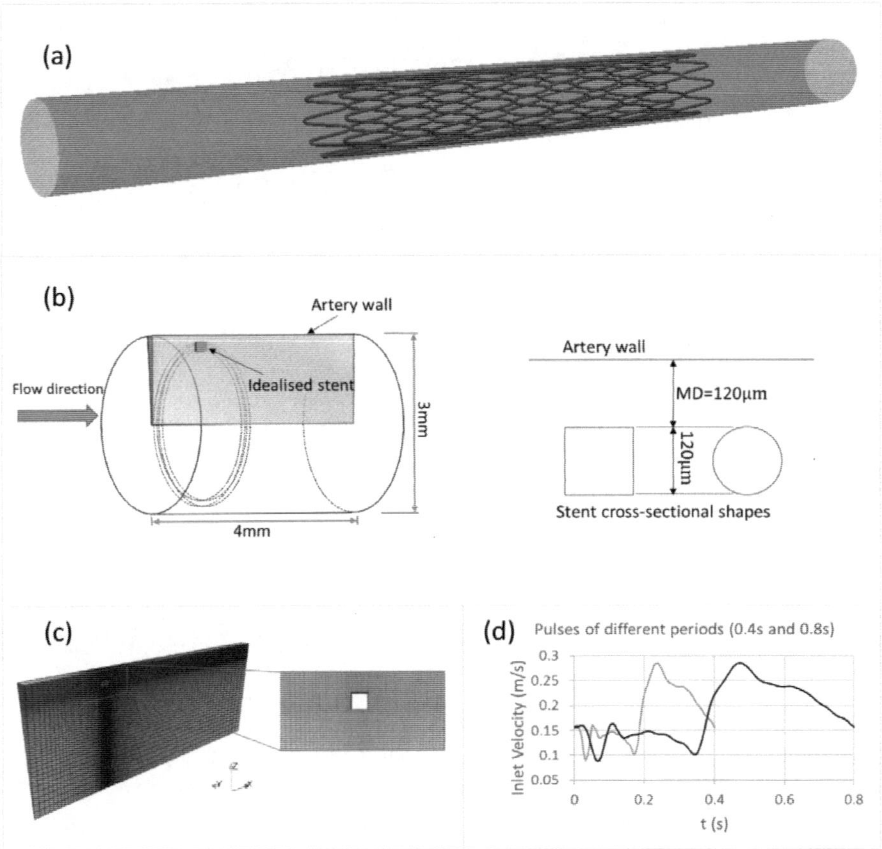

Figure 1. (a) A demonstration of a commonly available stent placed in an idealised segment of a coronary artery. (b) Simplified computation model of the malapposed stent in the idealised arterial geometry for CFD analyses and dimensions of the square and circular cross-sectional stent strut. (c) 2D axisymmetric model (in wedge shape) was generated and hexahedral elements density was increased near the arterial wall and strut surface. (d) A pulsatile coronary flow velocity from the LAD was applied to the inlet of the computational domain as shown in (c). Two different cardiac cycles (0.4 s and 0.8 s) were being considered.

3. Results and discussion

The temporal evolution of recirculation bubble distal to the malapposed strut was studied at various points in time for pulsatile flow with cardiac cycle of 0.4 s and 0.8 s. To ensure that all initial transients have convected out of the computational domain, the simulations were carried out for two cardiac cycles but only the data from the second cycle were used in the analyses below.

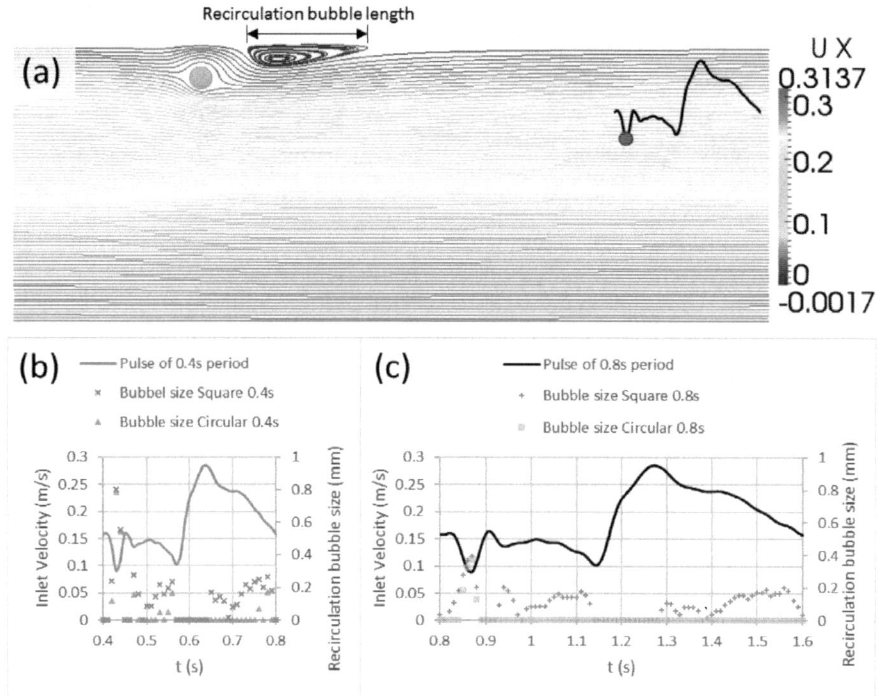

Figure 2. (a) The location and size of the streamline recirculation bubble distal to the malapposed stent strut. (b) & (c) Evolution of the size of the recirculation bubble distal to a malapposed strut (both square and circular cross-section) throughout a cardiac cycle of 0.4 and 0.8 s respectively.

Figure 2 (a) demonstrates the existence of the recirculation bubble distal to a circular stent strut during coronary flow deceleration. The length of the recirculation bubble was measured by identifying the negative WSS region along the arterial lumen. As can be observed in Figure 2 (b), recirculation bubbles exist distal to both square and circular cross-sectional struts. In brief, recirculation bubbles are more prominent during coronary flow deceleration, but disappear as coronary flow increases. Moreover, square strut results in a larger recirculation bubble than the circular strut. In other words, circular strut demonstrates better haemodynamic performance than square strut with identical strut thickness.

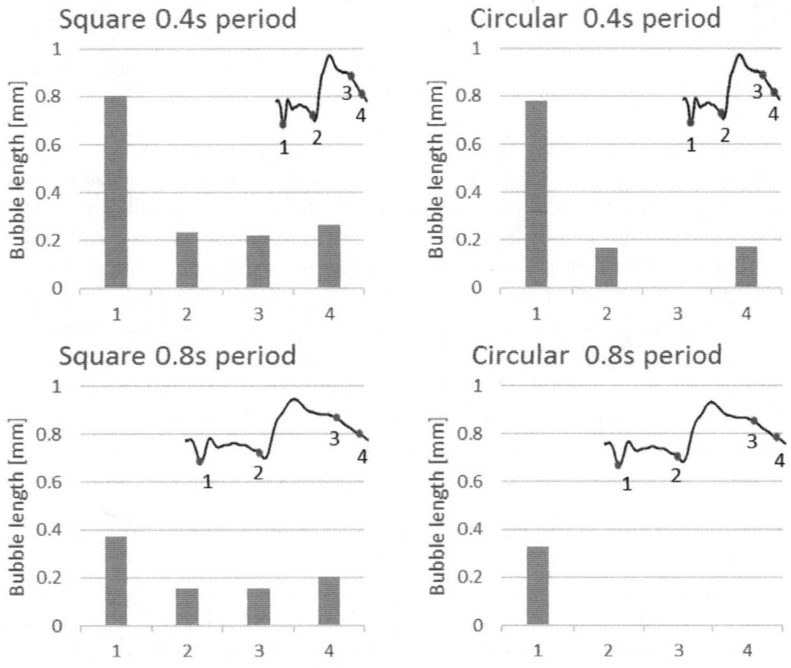

Figure 3. Comparison of recirculation bubble sizes at the same location of the two pulses of different periods.

To further highlight the fact that the size of the recirculation bubble is correlated to rapid flow deceleration, we quantitatively presented the size of the recirculation bubble at different flow deceleration rates and cardiac cycles in figure 3. For a specific strut cross-section profile, there is a reduction of the recirculation bubble size as cardiac cycle increases (from 0.4 s to 0.8 s). In addition, recirculation bubbles are most pronounced at point '1' of both cardiac cycles where coronary flow velocity decelerates at the highest rate. On the other hand, at points '2', '3' and '4', flow deceleration is less steep and the recirculation bubble sizes decrease. Nevertheless, the presented CFD results show that recirculation bubble size is positively correlated to the rate of coronary flow deceleration.

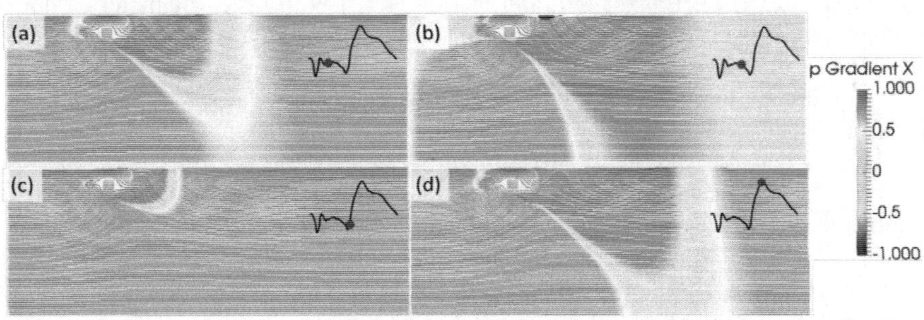

Figure 4. Instantaneous steamwise pressure gradient along with streamlines at different time phase in the 0.4 s period pulse of the square strut case (positive pressure gradient drives the flow from left to right). The phase is indicated in the inset with a dot.

In order to understand the physical mechanism that governs the evolution of recirculation bubbles, we carefully analyse the changes in pressure gradient field at different time for a cardiac cycle of 0.4 s.

It should be noted that the changes in pressure gradient for a cardiac cycle of 0.8 s follow a qualitatively similar trend as those in 0.4 s cycle and thus are not reported here. Pressure gradient range is set from -1 to 1 in order to clearly show the sign of the areas. In Figure 4(a), the change in coronary flow is almost zero (i.e., neither accelerating nor decelerating). The presence of the malapposed strut represents an obstacle in the near wall that affects the pressure recovery behind the stent strut. As a result, there is a small region of negative pressure gradient downstream of the strut. However, as flow starts to decelerate (see Figure 4(b), the negative pressure gradient region behind the malapposed strut gets bigger and begins to push fluid particles backwards. The presence of the malapposed strut prevents the near wall laminar reverse flow (defined by the Womersley profile [7]) and leads to the formation of a recirculation bubble. As inlet flow velocity accelerates again (Figure 4(c)), this creates a substantial positive pressure gradient along the streamwise direction that results in better pressure recovery behind the malapposed strut. This positive pressure gradient in front and behind the malapposed strut helps "wash off" the recirculation bubble. Coronary flow peaks in Figure 4(d) and rate of change of velocity approaches 0 again. As a result, the pressure gradient field becomes qualitatively similar that presented in Figure 4(a) and the pressure gradient field evolves in a similar trend as mentioned above when the flow decelerates again.

4. Conclusion

This study shows that the size of blood flow recirculation bubble expands as cardiac cycle decreases (i.e., a faster deceleration of inlet velocity). When flow decelerates, the malapposed stent strut prevents the laminar reverse flow (Womersley profile) near the arterial wall and forms recirculation bubble. With the same thickness and malapposition distance, circular cross-sectional stent strut always has a smaller size of recirculation bubble than stent strut of a square cross section in both 0.4 s and 0.8 s cardiac cycles, therefore, circular stent strut surpasses the square one in terms of haemodynamic performance.

5. References

[1] Foin N, Lee R D, Torii R, Guitierrez-Chico J L, Mattesini A, Nijjer S, Sen S, Petraco R, Davies J E, Di Mario C, Joner M, Virmani R and Wong P 2014 Impact of stent strut design in metallic stents and biodegradable scaffolds *Int. J. Cardiol.* **177** 800–8

[2] Jimenez J and Davies P F 2009 Hemodynamically Driven Stent Strut Design *Ann. Biomed. Eng.* **37** 1483–94

[3] Sprague E A, Luo J and Palmaz J C 1997 Human aortic endothelial cell migration onto stent surfaces under static and flow conditions *J. Vasc. Interv. Radiol.* **8** 83–92

[4] Dodge J T, Brown B G, Bolson E L and Dodge H T 1992 Lumen diameter of normal human coronary arteries. Influence of age, sex, anatomic variation, and left ventricular hypertrophy or dilation. *Circulation* **86** 232–46

[5] Foin N, Gutiérrez-Chico J L, Nakatani S, Torii R, Bourantas C V., Sen S, Nijjer S, Petraco R, Kousera C, Ghione M, Onuma Y, Garcia-Garcia H M, Francis D P, Wong P, Di Mario C, Davies J E and Serruys P W 2014 Incomplete stent apposition causes high shear flow disturbances and delay in neointimal coverage as a function of strut to wall detachment distance implications for the management of incomplete stent apposition *Circ. Cardiovasc. Interv.* **7** 180–9

[6] Davies J E, Parker K H, Francis D P, Hughes A D and Mayet J 2008 What is the role of the aorta in directing coronary blood flow? *Heart* **94** 1545–7

[7] Womersley J R 1955 Method for the calculation of velocity, rate of flow and viscous drag in arteries when the pressure gradient is known *J. Physiol.* **127** 553–63

IOP Conf. Series: Journal of Physics: Conf. Series **822** (2017) 012069 doi:10.1088/1742-6596/822/1/012069

Analytical solution of subsonic gas flow past a circle

Kamyar Mansour

Department of Aerospace Engineering AmirKabir University of Technology Tehran, Iran, 1587-4413 and Flow Research and Engineering 788 Maplewood Avenue, Palo Alto, CA 94303

E-mail: mansour@aut.ac.ir

Abstract. The Mach-Number series expansion of the potential function for the two dimensional flow of an inviscid, compressible, frozen gas past a circular cylinder are obtained to 29 terms. Analysis of these expansion allows the estimates for the critical Mach-Number M_*, (at which flow first becomes locally Sonic), and the radius of convergence of the series M_c for the maximum velocity. In this case of two-dimensional circular cylinder we obtain consistent results with 29 terms obtained by Van Dyke and Guttmann [19] for the case of diatomic gas with gamma equal to 7/5. For this case of two dimensions, we have been unable to determine the nature of singularities. We hope this work help to shed a light for resolving the 80-years old conterversary which says that whether the airfoil can posses a continuous range of shock free potential flow above the critical Mach-Number ($M_c > M_*$).

1. Introduction

This work is concerned with the two dimensional problem flow of an inviscid, compressible, perfect, frozen gas past a cylinder. The Jansen-Rayleigh approximation to subsonic flow is used. We hope to shed light on the transonic controversy, which may be phrased as: Can an airfoil have a continuous range of smooth shock-free potential flows above the critical Mach number?. Nocilla[11] have numerical results for flow past biconvex airfoils which suggest that shock -free flows can exist for up to 5 percentage. Generally numerical results of any kind like finite volume done by my student for circle case indicates range of shock -free flow above M_*. Inspired by original work of Rayleigh about 60 years ago G.I. Taylor raised the question whether an observed breakdown of flow near the critical Mach number occurs because irrotational flow ceases to be possible or occurs for some other reason connected with viscosity or other properties of the air not considered in the theory.

The goal of this analysis is to provide as complete a description as possible of this controversy. The analysis yields a solution for all values of Mach number from zero to one in a continuous fashion. We hope this work help to shed a light for resolving the 80-years old controversy which says that whether the airfoil can posses a continuous range of shock free potential flow above the critical Mach-Number ($M_c > M_*$). In the case of two-dimensional circular cylinder Van Dyke and Guttmann [19] showed that the radius of convergence for their Jansen-Rayleigh expansion exceeds the critical Mach-Number ($M_c > M_*$) by some 1.1 percentage. Bollman[2] uses series extension for flow of an inviscid, compressible, perfect gas along a sinusoidal wall in the transonic small-disturbance approximation. He generated 36 terms in the series expansion for maximum velocity and finds that the series is a Stieltjes kind. His claim was followed so that the series converge only up to the critical Mach-number.

As it has been illustrated in this example the singularity at Mach number equal to one corresponding to the transonic limit is not an isolated algebraic singularity. This method can be employed very successfully if care be exercised.

In this paper we will show that even in the context of compressible problems computer extension may not easily reveal the flow structure. More work and advanced analysis is required to deal with complications, such as zero radius of convergence, slowly converging type series or series have exponential type singularity. Supercomputers can handle these problems more effectively.

We consider here, that of steady flow of an inviscid, compressible, perfect gas past a circular cylinder without circulation. Following both Rayleigh[13] and Jansen[6], we have used the velocity potential as the dependent variable, rather than stream function, as the simplification of the surface boundary condition obtained by using the stream function as the dependent variable is brought at the price of a more complicated differential equation.

This equation for the velocity potential ϕ is given by Oswatitsch[12] in Cartesian co-ordinates, and re-expressing in the more appropriate cylindrical polar-coordinates ,we have:

$$\nabla^2\phi = M^2\left[\frac{\gamma-1}{2}\left(-1+\frac{\partial\phi}{\partial r}\frac{\partial\phi}{\partial r}+\frac{1}{r^2}\frac{\partial\phi}{\partial\theta}\frac{\partial\phi}{\partial\theta}\right)\nabla^2\phi\right]+$$

$$M^2\left[\frac{\partial\phi}{\partial r}\frac{\partial\phi}{\partial r}\frac{\partial^2\phi}{\partial r\partial r}+\frac{2}{r^2}\frac{\partial\phi}{\partial r}\frac{\partial\phi}{\partial\theta}\frac{\partial^2\phi}{\partial\theta\partial r}+\frac{1}{r^4}\frac{\partial\phi}{\partial\theta}\frac{\partial\phi}{\partial\theta}\frac{\partial^2\phi}{\partial\theta\partial\theta}-\frac{1}{r^3}\frac{\partial\phi}{\partial r}\frac{\partial\phi}{\partial\theta}\frac{\partial\phi}{\partial\theta}\right] \quad (1)$$

$$\nabla^2 = \frac{\partial^2}{\partial r^2}+\frac{1}{r}\frac{\partial}{\partial r}+\frac{1}{r^2}\frac{\partial^2}{\partial\theta^2}$$

Where the angle θ , is measured from the horizontal axis. The corresponding boundary conditions expressing the normal velocity over the surface must be zero:

$$\frac{\partial\phi}{\partial r}=0 \qquad \text{At } r=1$$

$$\phi = r\cos(\theta) \quad \text{At } r=a$$

2. Series derivation and computer extension

The equation for the velocity potential ϕ is just equation (1) in r,θ coordinate, by the requirement that ϕ to be an even function of θ, is enough to ensure the symmetric flow, free of circulation. By neglecting the right-hand side of equation (1), setting $M=0$, gives the basic incompressible solution. Enlightened separation of variables, trying $r^m\cos(m\theta)$ as suggested by the upstream condition, yields $m=1$ and $m=-1$. Hence the first approximation is found as:

$$\phi_1 = (r-\frac{1}{r})\cos(\theta)$$

Frankl and Keldysh[3] have proved that $\phi\left(M^2\right)$ is analytic in the neighbourhood of the origin, therefore one can systematically improve on this approximation by expanding in powers of M according to:

$$\phi = \phi_1 + M\phi_2 + M^2\phi_2 + \ldots = \sum_{n=0}^{\infty} \phi_n M^n \tag{2}$$

Where:

$$\phi_2 = \left(\frac{1}{12r^5} - \frac{1}{2r^3} + \frac{13}{12r}\right)\cos(\theta) + \left(\frac{1}{12r^3} - \frac{1}{4r}\right)\cos(3\theta)$$

$$\phi_3 = \left(\frac{28}{720r^9} - \frac{195}{720r^7} + \frac{414}{720r^5} - \frac{84}{72r^3} + \frac{1563}{720r}\right)\cos(\theta)$$
$$+\left(\frac{-5}{360r^9} - \frac{3}{36r^7} + \frac{171}{360r^5} - \frac{1}{3r^3} - \frac{330}{360r}\right)\cos(3\theta)$$
$$-\left(\frac{1}{10r^5} + \frac{5}{40r^3} + \frac{5}{40r}\right)\cos(5\theta)$$

We have found 12 terms exactly by means of symbolic language, however for saving space only we report the result for the maximum velocity (U is free stream velocity) :

$$q_{\max} = \left(\frac{\partial\phi}{\partial\theta}\right)_{r=1} d\theta$$

$$q_{\max} = 1 + \frac{7}{6}M^2 + \frac{281}{120}M^4 + \frac{9161}{1512}M^6 + \frac{115656949}{6350400}M^8 + \frac{10637691205579}{176033088000}M^{10} + \cdots = \sum_{n=0}^{\infty} q_n (M)^{2n-2} \tag{3}$$

Following the recipes given by Mansour[8] and Mansour[9], we have also obtained 50 terms of the series by writing a Fortran program in quadruple precision. The result for the local variable of maximum velocity is given in Table 1. The critical Mach number M_* is given by the solution of:

$$M_*^2 = \frac{2}{(\gamma+1)q_{\max}^2 - \gamma + 1}$$

The fundamental question we wish to resolve is whether (a) $(M_c > M_*)$ or (b) $(M_c = M_*)$.

3. Analysis of series and Discussion

Pade approximants has been used in original forms to enable us to increase the range of applicability of the series as has been used in the works of Mansour [8] and Mansour[9]. This method does not necessarily require any information about the radius of convergence. The Pade approximants provide an approximation that is invariant under an Euler transformation of the independent variables. The theory of Pade approximants has been used extensively in Mansour [9]. Briefly stated, the Pade approximant is the ratio P(K)/Q(K) of polynomials P and Q of degree m and n, respectively, that, when expanded, agrees with the given series through terms of degree m+n, and normalized by P(Q)=1. Such rational fractions are known to have remarkable properties of analytic continuation. The coefficients of the power series must be known to degree m+n. By equating like power of g(x) and P(K)/Q(K), the linear system of m+n+1 equation must be solved to obtain the coefficients in the functional form P(K)/Q(K).).Pade approximation of orders [1/1], [2/2] and [3/3] for $q_{\max}$ series are respectively:

pade[1/1]:

$$\frac{-2+\frac{299M}{105}}{(1-\frac{281M}{140})}$$

pade[2/2]:

$$\frac{-2+\left(\frac{22264113M}{2876314}\right)-\frac{5601925387M^2}{1208051880}}{1-\left(\frac{38429719M}{8628942}\right)+\frac{13574887007M^2}{3624155640}}$$

pade[3/3]:

$$\frac{\left(-2+\frac{838806830007803891009 11M}{656026242337163518821 0}+\frac{13276598005944335820050 9(M)^2}{626206867685474267965 5}+\frac{2163798482167804440427322 89M^3}{2893075728706891118000610 0}\right)}{\left(1-\frac{2288358062367849087178 9M}{328013121168581759410 5}+\frac{4959394809598266363664843(M)^2}{3673746957088115705397 60}-\frac{11989942158313316555677495 7(M)^3}{18515684663724103155203904}\right)}$$

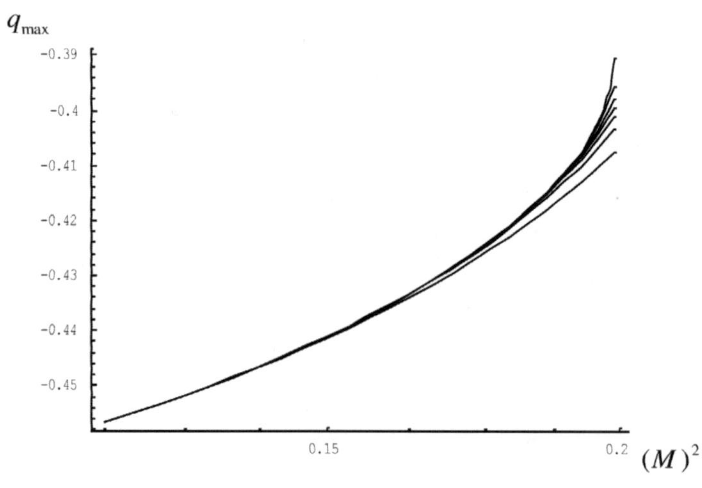

Figure 1. Plots of [8/8], [7/7], [6/6],[5/5], [4/4], [3/3]and [2/2] of the Pade approximants for $q_{\max}$ series versus $(M)^2$

When we form the ratios [8/8], [7/7], [6/6],[5/5], [4/4], [3/3]and [2/2of the Pade approximants, It can be shown, they agreed up to the value $(M)^2$=.17. This conclusion is confirmed as is plotted in Figure 1. This means that we are successful to do analytic continuation for Mach number up to .4

Table 1 . Coefficients of the series for maximum speed ($q_{\max}$)

n	q_n	n	q_n
1	2.00000000E+00	18	1.6242596E+10
2	1.16666667E+00	19	7.4261616E+10
3	2.3416666+00	20	3.4195181E+11
4	6.0588624E+00	21	1.5847778E+12
5	1.8212545E+01	22	7.3879922E+12
6	6.0430066E+01	23	3.4627706E+13
7	2.1495893E+02	24	1.6310689E+14
8	8.0496612E+03	25	7.7180403E+14
9	3.1354777E+03	26	3.6676008E+15
10	1.2598968E+04	27	1.7497171E+16
11	5.1915440E+05	28	8.3781619 E+16
12	2.1841806E+05	29	4.0255125 E+17
13	9.3512994E+06	30	
14	4.0637934E+07	31	
15	1.7889006E+07	32	
16	7.9639262E+08	33	
17	3.5807695E+09	34	

Note: $\boldsymbol{E}$ is a scientific abbreviation of multiplication by, 10 powered by the following number.

Previous works on the problem for the case of diatomic gas, the first attempt was made by O. Janzen [6] and by Lord Rayleigh [13]. Both independently obtained ϕ_1 and ϕ_2 , and Lord Rayleigh expressed the view that $(M_c = M_*)$. Later on Taylor and Sharman [17] and some years later Sauer [14] supported Rayleigh 's view for the cylinder, but pointed out there is some doubt. However

Shapiro stated that ($M_c = M_*$). For both the cylinder and the sphere cases as Lighthill [7] expressed the view which is possible that ($M_c > M_*$). Conclusions of Lighthill[7] presumably relied on the work of Imai[5] and Simasaki[16]. Imai[5] extended the series by obtaining the ϕ_3 term, and in 1941 the ϕ_4 term. Simasaki[16] published the next two terms ϕ_5 and ϕ_6. He estimated $M_* = 0.40$ and $M_c = 0.50$ for $\gamma = 1.405$ approximately. Of course that was a remarkable hand conclusion, but still very inaccurate. Later on by help of computer Hoffman [4] found 7 terms and gave for $\gamma = \frac{7}{5}$ a value of $M_* = 0.3983 + 0.0002$, which compares well with numerical integration technique done by Melnick and Ives[10]. Hoffman [4] finally concluded that ($M_c > M_*$) by 5.5 percentage . Vandyke and Guttmann [19] obtained 29 terms and they gave for $\gamma = \frac{7}{5}$ the estimates $M_* = 0.39823780$ and $M_c = 0.402667605$, so that M_c is larger than M_* by 1.11percentage. We have found 50 terms and agreed well the estimate given by work of Vandyke. Since we have not been unable to found the singularity even 20 more terms can not help. Of course our analysis is based on work by Barber [1] and his transformation is known for losing lots of accuracy, only 20 coefficients by this way can be used. In brief as has been seen in works of Hoffman [4] and Vandyke and Guttmann [19] by having 7 terms and 29 terms, the gap between M_c and M_* has decreased from 5.5 to 1.1 percentages respectively. The author [20] has discussed some aspect of these contradictory conclusions about the appearance of shocks in transonic flow. Question of whether 29 terms obtained in this work with different gamma namely equal one can reduce even further this gap or not?. The answer is, there are some difference between different diatomic gas previously studied and frozen gas we report here, however at this stage we have found no better than 1.1 percentages.

References

[1] Barber M N and Hamer C J 1982 Extrapolation of sequences using a generalized epsilon algorithm *J.Austral.Math.Soc.Ser. B* **23** 229-240
[2] Bollmann G 1982 Potential floew along a wavy wall and the transonic controversy *J.Engrg.Math* **16** 197-207
[3] Frankl F I and Keldysh M V 1934 the exterior Neumann problem for nonlinear elliptic diffrential equations with application to the theory of a wing in a compressible gas (Russin) *Izv.Akad.Nauk SSSR* **12** 561-601
[4] Hoffman H 1974 Extention of perturbation series by computer: symmetric subsonic potential flow past a circle *J.Meqcanique* **13** 433-447
[5] Imai I 1941 On the flow of a compressible fluid past a circular cylinder I and II *Phys. Math. Soc.japan Proc.Ser.3* **20** (1938) 636-645 and **23** 180-193
[6] Janzen O 1913 Beitrag zu einer Theorie der stationaren Stromung kompressibler Flussigkeiten *Phys. Zeits.* **14** 639-43
[7] lighthill M J 1954 Higher approximations in General theory of high speed aerodynamics (ed.W.R.Sears) Princeton Univ.Press,Princeton
[8] Mansour K 1985 Laminar flow through a slowly rotating straight pipe *J.Fluid Mech.* **150** 1-24
[9] Mansour K 1993 Using Stokes Expansion For Natural Convection inside a two-dimensional cavity *Fluid Dynamics Research* 1-33
[10] Melnick R E and Ives D C 1971 Subcritical flows over to dimensional airfoils by a multistrip method of integral relations *Proc.2nd Anternat.Conf.on Numer.Methods in Fluid Dyn* Berkeley CA (Springer-Verlag,Berlin), 243-251
[11] Nocilla S and Geymonat G and Jabutti B 1976 The direct problem of the transonic airfoils on the hodograph *Symposium Transsonicum II(eds.K.Oswatitsch and D.Rues)* Springer-Verlag 134-141

[12] Oswatitsch K 1956 *Gas dynamics* (Academic Press,New York)
[13] Rayleigh L *On the flow of compressible fluid past an obstacle* Phil.Mag.32(1916),1-6
[14] Sauer R 1947 *Introduction to theoretical gas dynamics* (Edwards,Ann Arbor)
[15] Shapiro A H 1954 *The dynamics and thermodynamics of compressible fluid flow* (Ronald,New York)
[16] Simasaki T 1955 *On the flow of a compressible fluid past a circular cylinder* Bull.Naniwa Univ.Ser.A 321-30
[17] Taylor G I and Sharman C F 1928 *A mechanical method for solving problemsmof flow in compressible fluids* Proc.Roy.Soc.London Ser.A 121 ,194-217
[18] Eser F 1943 *Zur Stromung kompressibler Flussigkeiten um feste Korper mit unter-schalgeschwindigkeit* Luftfahrtforschung 20,220
[19] VAN DYKE M D and Guttmann A J 1983 *Subsonic potential flow past a circle and the transonic controversy* J. Austral. Math.Soc. B 24, 243-261
[20] Kwok Y K and Sirovich L 1987 *On Some Aspects of the Transonic Controversy* SIAM Journal on Applied Mathematics, Vol. 47, No. 2, pp. 279-295

Average contra-rotation and co-rotation of line segments for flow field analysis

Jakub Šístek[1,3] and Václav Kolář[2,4]

[1]Inst. of Mathematics, Czech Acad. Sci., CZ-11567 Prague 1, Žitná 25, Czech Rep.
[2]Inst. of Hydrodynamics, Czech Acad. Sci., CZ-16612 Prague 6, Pod Patankou 5, Czech Rep.
[3]sistek@math.cas.cz
[4]kolar@ih.cas.cz, corresponding author

Abstract. The earlier concept of the average co-rotation of infinitesimal radial line segments near a point is extended to the case of contra-rotation. The tensor of the contra-rotation is introduced and averaged over "all planar cross sections" going through the examined point. Both the average contra-rotation and co-rotation, representing shear-free quantities, are applied to describe a complex flow structure.

1. Introduction

The co-rotational approach is based on the analysis of the relative motion near a point which is usually described in terms of the velocity-gradient tensor [1,2]. The relevant quantities under consideration are the vorticity tensor (expressing an average angular velocity of fluid elements) and the strain-rate tensor (dealing with deformation of fluid elements). These quantities determine the local flow kinematics and, in the present context, the local co-rotation and contra-rotation of line segments.

The notion of co-rotation of instantaneous infinitesimal radial line segments near a point has been recently introduced in connection with local vortex-identification schemes [3]. At an arbitrary point, the *vector of local co-rotation* is defined in a planar cross section (of a general 3D flow field) with the orientation of a unit normal vector of the plane. Let us briefly characterize an arbitrary cutting plane at a given point at a given instant of time. The vorticity vector, associated with the cutting plane and aligned with its unit normal vector, represents an *average* angular velocity of all radial line segments near a point. The vector of local co-rotation is the shear-free collinear part of the vorticity vector given by the *least-absolute-value* angular velocity of all radial line segments, according to figure 1.

The least-absolute-value angular velocity in figure 1 is a measure of the local rigid-body rotation of a fluid element. The rigid-body rotation is obtained after eliminating the local shearing motion, characterized by the *shear* vorticity, and quantified by the *residual* vorticity. This part of vorticity is determined through the corresponding decomposition of the velocity-gradient tensor $\nabla \mathbf{u}$ proposed in the frame of the triple decomposition of the relative motion near a point [4]. From the vortex-identification viewpoint, the residual vorticity represents a local vortex intensity.

There exists a number of pointwise vortex-identification methods based on the velocity-gradient tensor. The well-known criteria are Q (Hunt et al. [5]), Δ (Chong et al. [6]), λ_2 (Jeong and Hussain [7]), and λ_{ci} (Zhou et al. [8]). These criteria are briefly reviewed, for example, in Chakraborty et al. [9]. Though the co-rotational approach discussed below is closely related to vortex identification as well, the main difference lies in the ability to characterize also a vortex counterpart, the shear-free deformation, and even shearing effect. A similar but not the same duality property is, from the popular schemes, inherent only to the quantity Q which is just the difference of vorticity and strain-rate tensor magnitudes. The co-rotational approach coincides with the triple decomposition of the relative motion [4] in 2D, distinguishing three elementary motions as indicated in figure 1: rigid-body rotation, shearing motion, and irrotational straining (elongation and/or contraction). However, it differs in 3D.

In 3D, the triple-decomposition method (TDM) becomes computationally expensive. On the other hand, the vortex-identification method based in 3D on the *average co-rotation* vector [3] provides very similar identification results as the TDM at relatively low computational cost [10]. The aim of the present contribution is to show the algorithm for the determination of (i) the (shear-free) *average contra-rotation* and (ii) *average shear*, which is less evident than the algorithm for the determination of the average co-rotation. The new approach motivates an advanced comparison with the TDM.

The application to a wake flow shows the complex flow structure in terms of isosurfaces of magnitudes of the average co-rotation, contra-rotation, and average shear, in direct comparison with the TDM results taken for reference.

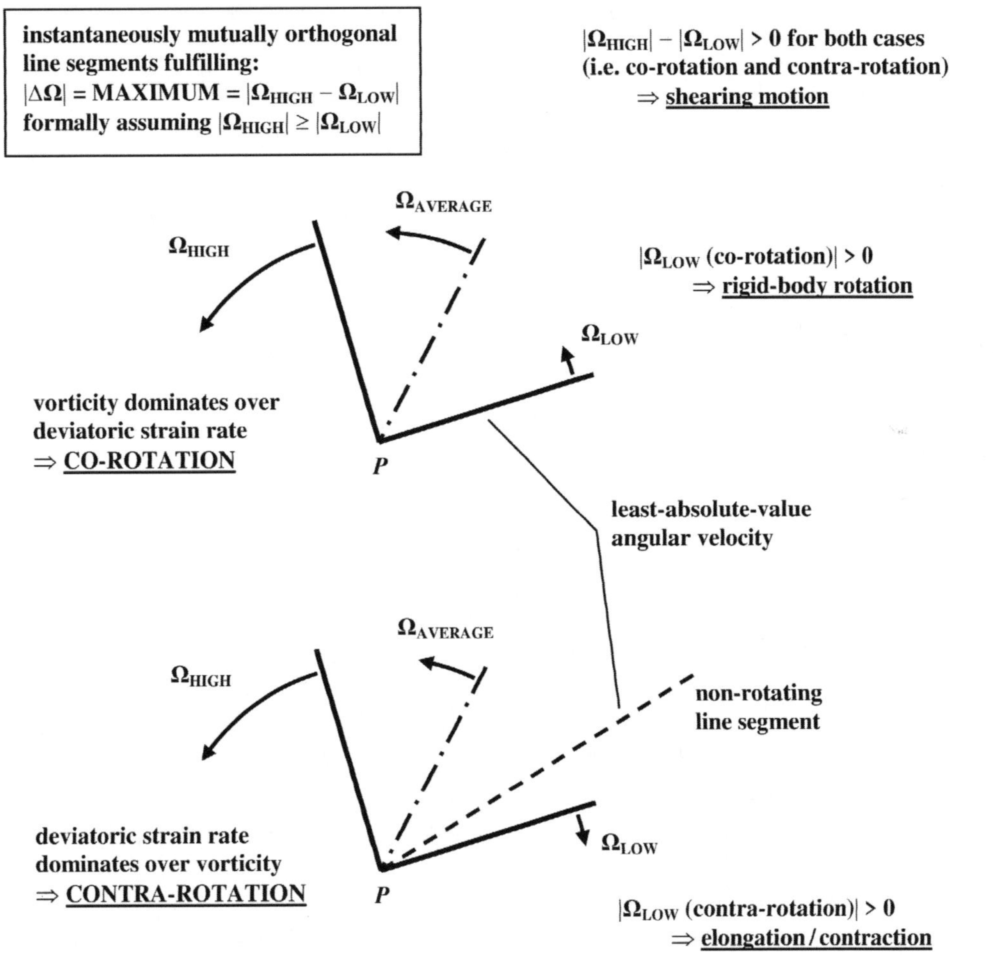

Figure 1. Co-rotation and contra-rotation of radial line segments near a point and elementary motions.

2. Contra-rotation of material line segments and its tensor representation

Recall that, at a given point, the vector of average co-rotation of line segments is defined as the average of the instantaneous local rigid-body rotation over "all planar cross sections" going through the examined point [3]. The procedure consists of two consecutive steps: Firstly, the vector of local co-rotation of material line segments at a point is defined in a plane. This planar concept is directly related to the residual vorticity according to the TDM applied in 2D. Secondly, a proper averaging process is applied to all planar cross sections going through the given point. The averaging procedure is performed as surface integration over a unit sphere, which is approximated using numerical quadrature. Analogous steps — definition in a plane followed by averaging over all planar cross sections — need to be performed to determine the *tensor of local contra-rotation* and *average contra-rotation*.

The necessary condition for the local contra-rotation of instantaneous infinitesimal radial line segments in a plane is that the deviatoric strain rate is greater than vorticity (figure 1). While the co-rotation is, according to the TDM, quantified by the residual vorticity, the contra-rotation is quantified by the residual strain rate s_{RES}. The quantity s_{RES} reads for a 2D velocity gradient [4] (the symbol s_{SH} stands for the shear strain rate associated with the separated shearing motion)

$$s_{\mathrm{RES}} = s_D - s_{\mathrm{SH}} = (\operatorname{sgn} s_D)\left[|s_D| - |\omega|\right] \quad \textit{for} \quad |s_D| \geq |\omega| \tag{1a}$$

$$s_{\mathrm{RES}} = s_D - s_{\mathrm{SH}} = 0 \quad \textit{for} \quad |s_D| \leq |\omega|, \tag{1b}$$

where vorticity ω and deviatoric principal strain rate s_D are given by (see figure 1)

$$\omega = (v_x - u_y)/2 = (\Omega_{\mathrm{HIGH}} + \Omega_{\mathrm{LOW}})/2 = \Omega_{\mathrm{AVERAGE}}, \tag{2}$$

$$|s_D| = \left(\sqrt{(u_x - v_y)^2 + (u_y + v_x)^2}\right)/2. \tag{3}$$

For the co-rotation case (figure 1, $|\omega| > |s_D|$), s_D can be expressed in terms of angular velocities as follows

$$|s_D| = (|\Omega_{\mathrm{HIGH}}| - |\Omega_{\mathrm{LOW}}|)/2. \tag{4}$$

Alternatively, for the contra-rotation case (figure 1, $|\omega| < |s_D|$), s_D can be expressed as

$$|s_D| = (|\Omega_{\mathrm{HIGH}}| + |\Omega_{\mathrm{LOW}}|)/2. \tag{5}$$

Finally, for the shear ($|\omega| = |s_D|$) s_D is given by

$$|s_D| = |\Omega_{\mathrm{HIGH}}|/2. \tag{6}$$

In view of the averaging procedure over a unit sphere and to respect all contributing cross sections we have to construct a tensor representation of the residual strain rate — as a measure of the local contra-rotation — for an arbitrary cross section. This tensor input can then be averaged in the same manner as the vector input in the case of co-rotation [3].

For general angles of spherical coordinates $\varphi \in [0, 2\pi]$ and $\vartheta \in [0, \pi]$, the sequence of two rotational transformations leads to the final 3x3 rotation matrix (orthogonal linear transformation) [4]

$$\mathbf{Q} = \begin{pmatrix} \cos\vartheta\cos\varphi & \cos\vartheta\sin\varphi & -\sin\vartheta \\ -\sin\varphi & \cos\varphi & 0 \\ \sin\vartheta\cos\varphi & \sin\vartheta\sin\varphi & \cos\vartheta \end{pmatrix}. \tag{7}$$

Consider pointwise data input as the general, not necessarily divergence-free, velocity-gradient tensor $\nabla\mathbf{u}$. The transformation (7) of the original coordinate frame results in the rotated tensor components determined by $\mathbf{Q}(\nabla\mathbf{u})\mathbf{Q}^T$. An arbitrarily rotated plane can be identified with the (*x*,*y*)-coordinate plane of the rotated frame. The corresponding leading 2x2 submatrix can be extracted by application of the 2x3 restriction matrix $\mathbf{R} = \begin{pmatrix} 1 & 0 & 0 \\ 0 & 1 & 0 \end{pmatrix}$ on both sides,

$$\begin{pmatrix} u_x & u_y \\ v_x & v_y \end{pmatrix} = \mathbf{RQ}(\nabla\mathbf{u})\mathbf{Q}^T\mathbf{R}^T. \tag{8}$$

Let us now focus on the (x,y)-coordinate plane. Assuming transformation of the 2x2 velocity gradient $\begin{pmatrix} u_x & u_y \\ v_x & v_y \end{pmatrix}$ to the frame of principal axes of the planar strain rate and subtracting half of the trace from the obtained diagonal entries, we find the deviatoric planar strain rate and residual strain rate in the simple 2x2 diagonal matrix form as

$$\begin{pmatrix} s_D & 0 \\ 0 & -s_D \end{pmatrix} \Rightarrow \begin{pmatrix} s_{\mathrm{RES}} & 0 \\ 0 & -s_{\mathrm{RES}} \end{pmatrix} \tag{9}$$

where the value of s_{RES} is given by (1a, b). Denote the 2x2 matrix of eigenvectors of (8), expressed in the original coordinates (x,y), as $\mathbf{W}_2$. This leads to a planar representation of the contra-rotation as

$$\mathbf{S}_{2,\mathrm{RES}} = \mathbf{W}_2 \begin{pmatrix} s_{\mathrm{RES}} & 0 \\ 0 & -s_{\mathrm{RES}} \end{pmatrix} \mathbf{W}_2^T . \tag{10}$$

Here the subscript 2 denotes a 2x2 matrix.

The next step is the search for a proper 3x3 tensor representation of the contra-rotation valid for the original 3D coordinate frame. To this end, the reverse transformation by the matrices **R** and **Q** is used. In particular, the desired 3x3 contra-rotation tensor $\mathbf{S}_{\mathrm{RES}}$ reads

$$\mathbf{S}_{\mathrm{RES}}(\mathbf{x}_0, \mathbf{n}) = \mathbf{Q}^T \mathbf{R}^T (\mathbf{S}_{2,\mathrm{RES}}) \mathbf{RQ} . \tag{11}$$

3. Determination of the average contra-rotation and the scaling factor

We follow the averaging procedure introduced in [3] and apply it to all components of the matrix $\mathbf{S}_{\mathrm{RES}}$. For a given position vector $\mathbf{x}_0$ the averaging procedure is applied over all planar cross sections. The infinite set of all admissible planes can be defined by the infinite set of unit normal vectors which is the same as for a unit sphere $\sigma(0,1)$. Keeping in mind the key role of a normal **n**, the present averaging process translates to an integration over the unit sphere $\sigma(0,1)$. Consequently, the average contra-rotation tensor at $\mathbf{x}_0$ is defined as

$$\mathbf{S}_{\mathrm{RAVG}}(\mathbf{x}_0) = \beta \cdot \frac{\iint_{\sigma(0,1)} \mathbf{S}_{\mathrm{RES}}(\mathbf{x}_0, \mathbf{n})\, d\sigma}{\iint_{\sigma(0,1)} d\sigma} = \frac{\beta}{4\pi} \iint_{\sigma(0,1)} \mathbf{S}_{\mathrm{RES}}(\mathbf{x}_0, \mathbf{n})\, d\sigma \tag{12}$$

where β is a scaling factor. A natural choice $\beta = 5/2$ is derived below. It should be noted that equation (12) represents nine independent surface integrals of scalar functions, one for each component of the average contra-rotation.

A considerable amount of research has been devoted to efficient numerical evaluation of surface integrals over the unit sphere such as equation (12), see the review paper [11]. The approach based on Fibonacci integration from [12] is employed in our computations.

Let us pay attention to the scaling factor β. Suppose a given divergence-free strain-rate tensor $\mathbf{S}_D$ should be reproduced by the given averaging procedure. The coefficient is determined from the requirement that for such input $\mathbf{S}_{\mathrm{RAVG}} = \mathbf{S}_D$. The substitution from equations (10), (11) and from the degenerate relation $s_{\mathrm{RES}} = s_D$ (valid due to the purely irrotational input $\mathbf{S}_D$) into the component-wise integration (12) gives the result $\mathbf{S}_{\mathrm{RAVG}} = 2\beta \mathbf{S}_D / 5$ requiring $\beta = 5/2$ for the reproduction of $\mathbf{S}_D$. A diagonalized input of $\mathbf{S}_D$ makes this evaluation easier.

The average contra-rotation tensor, similarly as the average co-rotation vector, is not affected by a non-zero divergence in the case of compressibility. This fact can be seen from relations (1a, b) dealing with the 2D deviatoric principal strain rate s_D which is free from any impact of a 3D uniform dilatation. Consequently, without loss of generality and applicability to compressible flows, a non-zero divergence can be removed prior to the analysis of $\nabla \mathbf{u}$: only the deviatoric part of $\nabla \mathbf{u}$ is decisive.

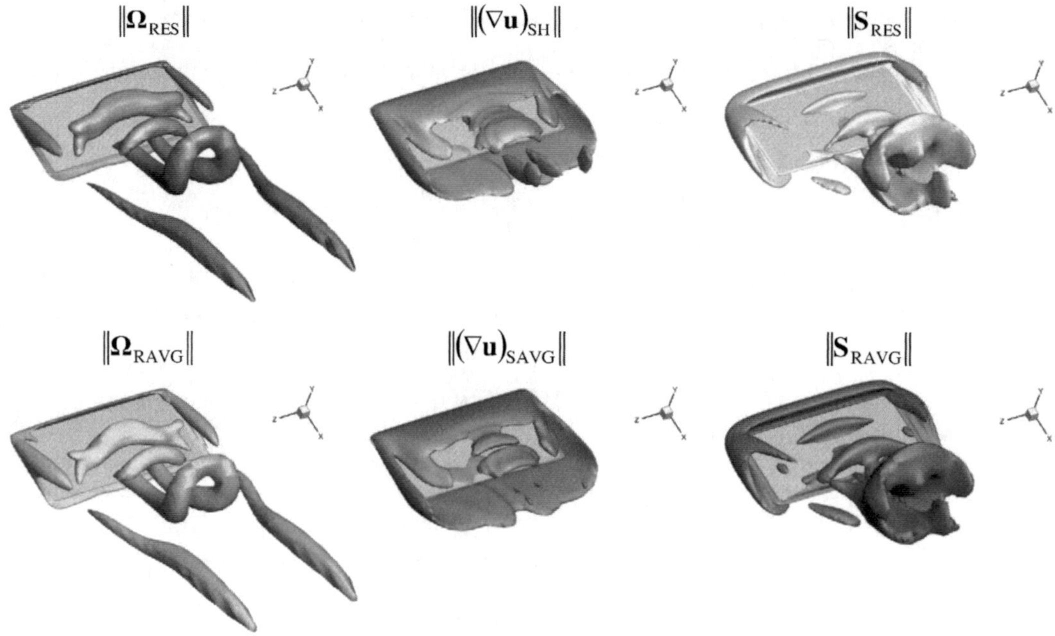

Figure 2. Comparison of isosurfaces for the TDM (top, isovalues from left to right: 1.2, 5.0, 1.2) and for the co-rotation scheme (bottom, isovalues: 1.78, 4.16, 1.26) at *Re*=300.

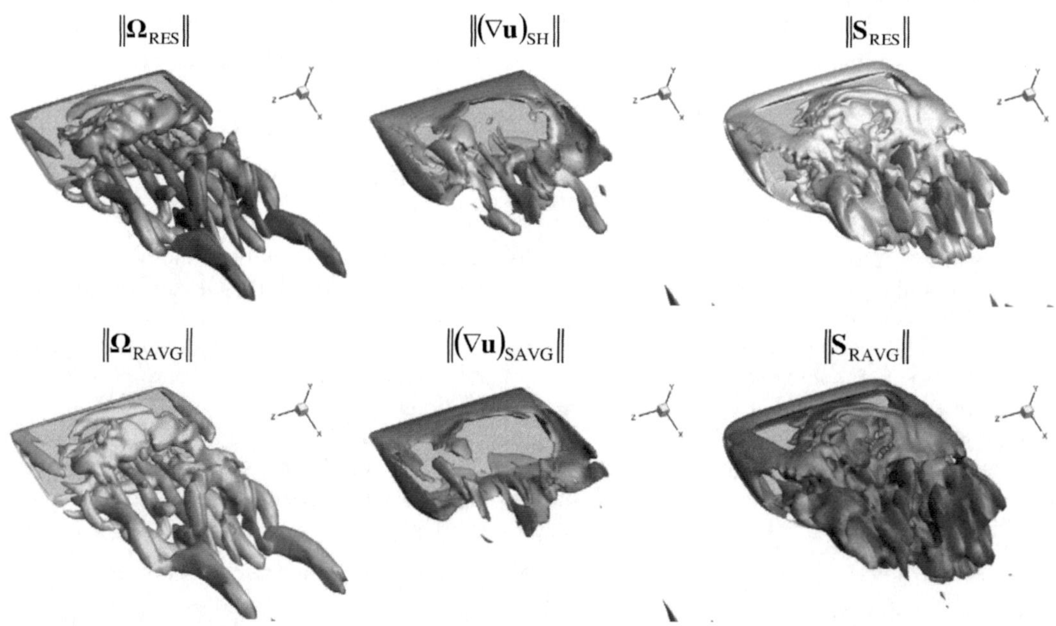

Figure 3. Comparison of isosurfaces for the TDM (top, isovalues from left to right: 1.2, 8.0, 1.2) and for the co-rotation scheme (bottom, isovalues: 1.88, 7.42, 1.28) at *Re*=1200.

4. Average shear

Let us recall that the average co-rotation vector was introduced in [3] as

$$\boldsymbol{\omega}_{\mathrm{RAVG}}(\mathbf{x}_0) = \frac{3}{4\pi} \iint_{\sigma(0,1)} \boldsymbol{\omega}_{\mathrm{RES}}(\mathbf{x}_0, \mathbf{n})\, \mathrm{d}\sigma \qquad (13)$$

where $\boldsymbol{\omega}_{\mathrm{RES}}(\mathbf{x}_0, \mathbf{n})$ denotes the co-rotation vector. The corresponding average co-rotation tensor reads

$$\boldsymbol{\Omega}_{\mathrm{RAVG}}(\mathbf{x}_0) \equiv \Omega_{\mathrm{RAVG},ij} = -\epsilon_{ijk}\, \omega_{\mathrm{RAVG},k} / 2. \qquad (14)$$

A divergence-free analogue of the TDM [4], $\nabla\mathbf{u} = \mathbf{S}_{\mathrm{RES}} + \boldsymbol{\Omega}_{\mathrm{RES}} + (\nabla\mathbf{u})_{\mathrm{SH}}$, which separates the shearing effect and the residual vorticity and strain rate, takes in the present context the form

$$\nabla\mathbf{u} = \mathbf{S}_{\mathrm{RAVG}} + \boldsymbol{\Omega}_{\mathrm{RAVG}} + (\nabla\mathbf{u})_{\mathrm{SAVG}}. \qquad (15)$$

The first two terms on the right-hand side of equation (15) are already known, so the average shear $(\nabla\mathbf{u})_{\mathrm{SAVG}}$ can be determined in a direct manner. However, the quantity $(\nabla\mathbf{u})_{\mathrm{SAVG}}$ is also obtainable by an analogous averaging procedure applied to both its symmetric and antisymmetric parts.

5. Application of the average co-rotation, contra-rotation and shear

The application deals with an impulsively started incompressible flow around a flat plate (aspect ratio 2) at an angle of attack of 30 deg solved numerically for *Re*=300 and *Re*=1200. In figures 2 and 3, isosurfaces of the quantities obtained by the TDM are shown for representative isovalues. Here, the symbol $\|\cdot\|$ denotes the standard Frobenius norm, defined for a tensor $\mathbf{G} = (G_{ij})$ as $\|\mathbf{G}\| = [\mathrm{tr}(\mathbf{G}\mathbf{G}^T)]^{1/2}$. For comparison, isosurfaces of their averaged counterparts are presented, and their isovalues are found by the volumetric matching procedure from [3]. A relatively high level of shear (behind the plate edges) is reflected by a high isovalue adopted for both *Re*.

6. Conclusions

The earlier concept of the average co-rotation of line segments near a point is extended to the case of contra-rotation. The tensor representation of the contra-rotation is introduced and averaged over all planar cross sections going through the examined point. Note that (i) the third elementary motion, shearing, is also "visualized" in terms of the (shear-free) average co-rotation and contra-rotation of line segments, (ii) the volumetric change near a point (compression or expansion) does not affect the co-rotational analysis what results in direct applicability to compressible flows, similarly as the TDM, and (iii) the results obtained for the co-rotational scheme are remarkably similar to those by the TDM.

Acknowledgments
This work was supported by the Czech Science Foundation through Grant 14-02067S, and by the Czech Academy of Sciences through RVO:67985874 and RVO:67985840.

References

[1] Batchelor G K 1967 *An Introduction to Fluid Dynamics* (Cambridge: Cambridge Univ. Press)
[2] Panton R L 1984 *Incompressible Flow* (New York: Wiley)
[3] Kolář V, Šístek J, Cirak F and Moses P 2013 *AIAA J.* **51** 2678–94
[4] Kolář V 2007 *Int. J. Heat Fluid Flow* **28** 638–52
[5] Hunt J C R, Wray A A and Moin P 1988 Eddies, streams, and convergence zones in turbulent flows *Report CTR-S88* (Stanford: Center for Turbulence Research) pp 193–208
[6] Chong M S, Perry A E and Cantwell B J 1990 *Phys. Fluids A* **2** 765–77
[7] Jeong J and Hussain F 1995 *J. Fluid Mech.* **285** 69–94
[8] Zhou J, Adrian R J, Balachandar S and Kendall T M 1999 *J. Fluid Mech.* **387** 353–96
[9] Chakraborty P, Balachandar S and Adrian R J 2005 *J. Fluid Mech.* **535** 189–214
[10] Kolář V and Šístek J 2014 *Proc. 19th Australasian Fluid Mechanics Conf.* (Melbourne: Australasian Fluid Mechanics Soc.) Paper 274
[11] Ehret A E, Itskov M and Schmid H 2010 *Int. J. Numer. Meth. Eng.* **81** 189–206
[12] Hannay J H and Nye J F 2004 *J. Phys. A: Math. Gen.* **37** 11591–601

Development of Flow over Blunt-Nosed Slender Bodies at Transonic Mach Numbers

Gireesh Yanamashetti[1†], G. K. Suryanarayana[2†] and Rinku Mukherjee[3]

[1]PhD Scholar & Senior Scientist, [2]Chief Scientist, [3]Assistant Professor
[†]National Aerospace Laboratories, Bangalore, India 560017
[3]Dept.of Applied Mechanics, Indian Institute of Technology Madras, India 600036

E-mail: [1]gireesh@nal.res.in, [2]surya@nal.res.in, [3]rinku@iitm.ac.in

Abstract. Comparisons of the development of flow over a cylinder with a 20° cone nose and a cylinder with an ogive nose, which represent typical heat-shield configurations are studied using CFD and experiments at transonic Mach numbers. The C_p plots are studied to locate expansion or separation. Experiments are carried out at $M = 0.8$, 0.9, 0.95 and 1.1 and $Re \approx 2.45 \times 10^6$. Computations are carried out using the commercial package, FLUENT 6.3. Inadequate spatial resolution of pressure ports in experiments as well as limitations of the CFD tool result in some differences in experimental and CFD results.

1. Introduction

It is well known that transonic flows are extremely complicated to comprehend because of the mixed nature of the flow involving local pockets of subsonic and supersonic flows. For slender bodies such as launch vehicles, rockets, missiles etc., the payload is placed around the nose, i.e. around the region of the heat shield. Hence, proper understanding of the flow features around the nose is extremely important for successful accomplishment of the mission. Considering geometry of the main rocket engine, volume of the payload and the specific requirements of a mission, use of conical nose shapes along with boat tail [1] are often needed. During the flow, the external structure of the heat shield is subjected to steady as well as unsteady pressure loading [2]. Frequency contents in the external flow can excite the structural modes and result in undesirable levels of vibrations of components mounted inside the heat shield [3].

Since boundary layer separation and reattachments are sources of noise and unsteady forcing of the structure, such features should be avoided in the design of the heat shield. With significant advances made in CFD, it is desirable to exploit the capabilities of CFD tools to obtain deeper understanding of complex transonic flows after validation with experimental data. Hence, in this paper the transonic flow features of two heat-shield configurations, which have nearly identical volume and boat-tail angles but with different nose shapes are compared based on experiments and computations.

2. Experimental Set-Up

The 1.2m Trisonic Wind Tunnel at CSIR-NAL is an intermittent blowdown facility providing maximum run duration of about 40 sec. operating from storage receivers at maximum pressure of 150 psig. The Mach number capability is from 0.2 to 4.0. Subsonic Mach numbers in the test section are achieved by choking the second throat. Supersonic speeds are achieved by variation of nozzle contour using a flexible nozzle Transonic Mach numbers are achieved using a transonic insert with a perforated wall test section. The top and bottom walls of the insert have 0.5 inch diameter inclined holes with 6% open area ratio, the sidewalls have normal holes with an open area ratio of 20%. Acoustic baffles are used in the plenum chamber of the transonic insert to reduce flow unsteadiness in the test section. Test Reynolds number within a certain range can be achieved by variation of the tunnel stagnation pressure. The model incidence may be varied from15^o to 27° continuously or in steps during a run and the model can be rolled and locked from 0° to 360° prior to a run.

2.1. Models used and Test Conditions

The model with a 20° nose consists of a hemispherical nose matched to a 20° cone, cylindrical region and a 20° boat-tail followed by a cylindrical region. The model with an ogive nose is nearly identical, except for the total length, which is greater by 0.16d as shown in Fig. 1. Details of the geometry of the models are shown in Fig. 2.

Experiments are carried out at $M = 0.8$, 0.9, 0.95 and 1.1 and a Reynolds number based on the maximum diameter, $Re \approx 2.45 \times 10^6$. Boundary layer trip was located at 5% of the length of the nose on both the models. The pitch angle was fixed at 0° to ensure axisymmetric flow.

2.2. Model Instrumentation

The models have pressure ports along different generators. Pressure data from the 0° generator and at AOA = 0° alone are discussed here for the purpose of validation of the computations. Surface pressure measurements were carried out using electronically scanned pressure (ESP) scanners (32 ports/16 ports) of range 15 psid housed inside the model. The total tunnel pressure and static pressures were measured in the settling chamber and plenum chamber respectively using 150 psia and 15 psid pressure transducers. All the pressure data from the ESP scanners were acquired at each angle of attack after an initial dwell of 1.2 seconds. At each port, the channels were scanned at the rate of 500 samples per second and a total of 20 samples per channel were acquired at each angle. The averaged pressure data are presented in this paper. Uncertainty in pressure is 0.0045 psi, within 0.1% of range of pressure transducers/scanner employed. Mach number repeatability is within 0.005.

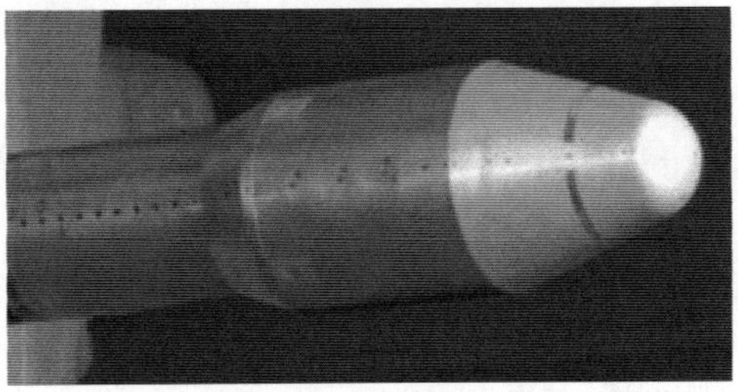

(a) Model with 20° cone nose

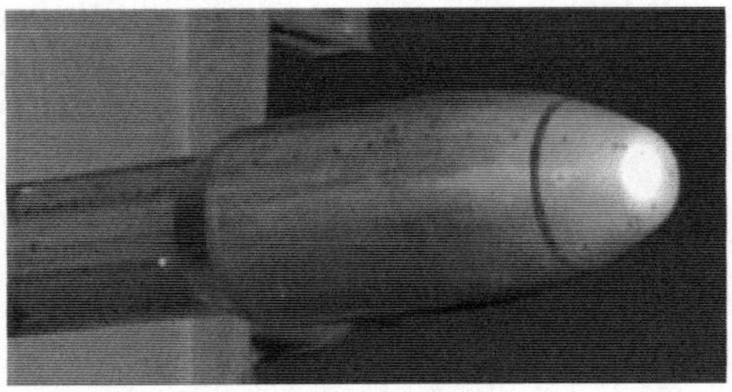

(b) Model with ogive nose

Figure 1: Models

3. Numerical Simulation

Axisymmetric analysis was carried out using Euler equations using the commercial software package FLUENT 6.3. For the study $Re = 2.45 \times 10^6$, based on the maximum diameter.

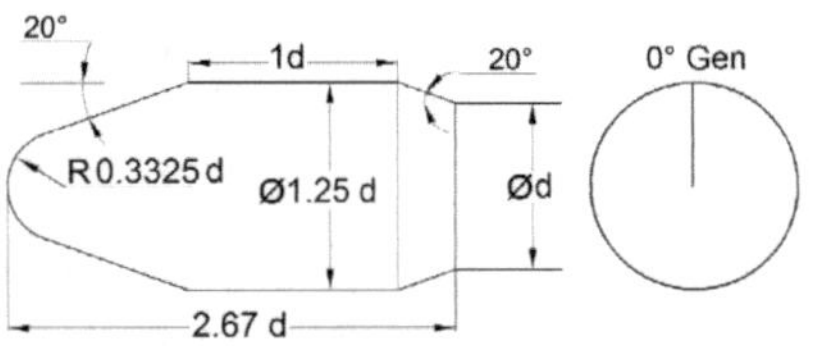

(a) Model with 20° cone nose

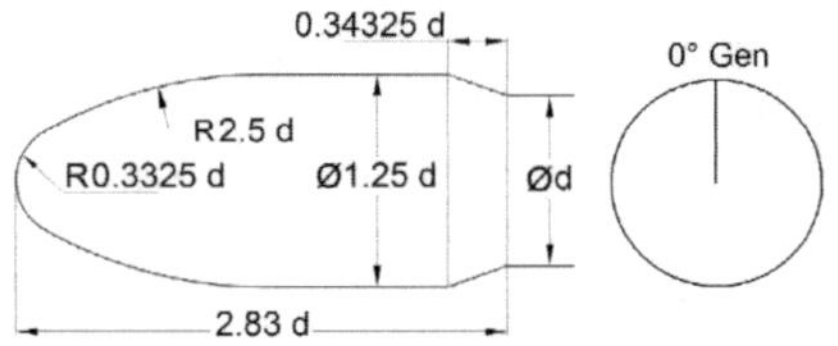

(b) Model with ogive nose

Figure 2: Details of Geometry of Test Models

Ansys Gambit was used for grid generation and Ansys FLUENT for solving the governing Euler equations. Mass conservation equation is the same as for a laminar flow but the momentum and energy conservation equations are reduced due to the absence of molecular diffusion. For 2D and axisymmetric bodies, the continuity equation as given in Eqn. 1. Axial and radial momentum conservation equations are given in Equations 2 and 3 respectively. The equation for conservation of energy is given in Eqn. 4.

$$\frac{\partial \rho}{\partial t} + \frac{\partial}{\partial x}(\rho v_x) + \frac{\partial}{\partial r}(\rho v_r) + \frac{\rho v_r}{r} = S_m \tag{1}$$

$$\frac{\partial}{\partial t}(\rho v_x) + \frac{1}{r}\frac{\partial}{\partial x}(r\rho v_x v_x) + \frac{1}{r}\frac{\partial}{\partial r}(r\rho v_r v_x) = -\frac{\partial p}{\partial x} + F_x \tag{2}$$

$$\frac{\partial}{\partial t}(\rho v_r) + \frac{1}{r}\frac{\partial}{\partial x}(r\rho v_x v_r) + \frac{1}{r}\frac{\partial}{\partial r}(r\rho v_r v_r) = -\frac{\partial p}{\partial r} + F_r \tag{3}$$

$$\frac{\partial}{\partial t}(\rho E) + \nabla \cdot (\vec{v}\,(\rho E + p)) = -\nabla \cdot \left(\sum_j h_j J_j\right) + S_h \tag{4}$$

The computational domain is shown in Fig. 3. The flow is simulated as in the wind tunnel i.e. the top boundary is defined as wall. A structured grid was generated using Ansys Gambit. The present computations were carried out on 470 × 175 grids. Successive ratios of 1.016 to 1.06 are used depending on regions requiring finer mesh. Standard settings of inviscid model of Fluent are applied. First order upwind, implicit time marching schemes were used. The residuals were converged to 10^{-5}. The net mass flow rate balanced till 0.05 Kg/s.

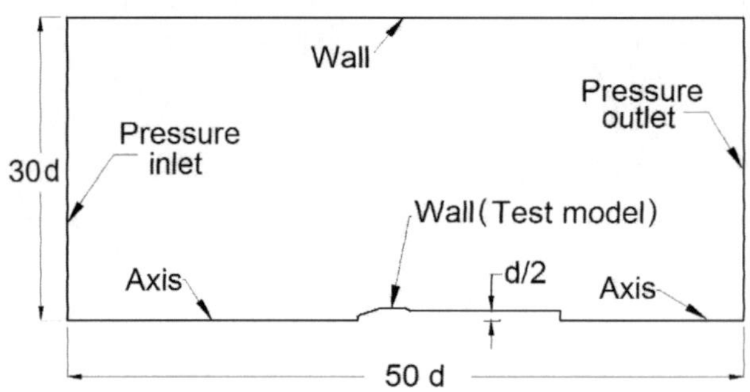

Figure 3: Computational Domain and Boundary Conditions

4. Results

Figure 4 shows a comparison of the experimental and computed C_P distributions on both the models at 4 different Mach numbers, which are in the transonic range. The blue and red dotted lines are model geometry profiles of 20° nose and Ogive nose models respectively. Rapid expansion of the flow as shown by the sharp drop in C_p, starting from the stagnation point ($C_p = 1$) is noted for all Mach numbers. The front of the model is a region of favourable pressure gradient. However, a short separation bubble of length, $x/d \approx 0.1$ exists, which is shown by the static C_p curve at the junction where sphere and cone meet, i.e. at $x/d \approx 0.3$. This may be attributed to the sudden change in curvature although the local slope is continuous. This has also been reported in literature [4]. Stronger expansion of the flow around the shoulder region ($x/d \approx 1.0$) is observed for the conical nose compared to the ogive nose.

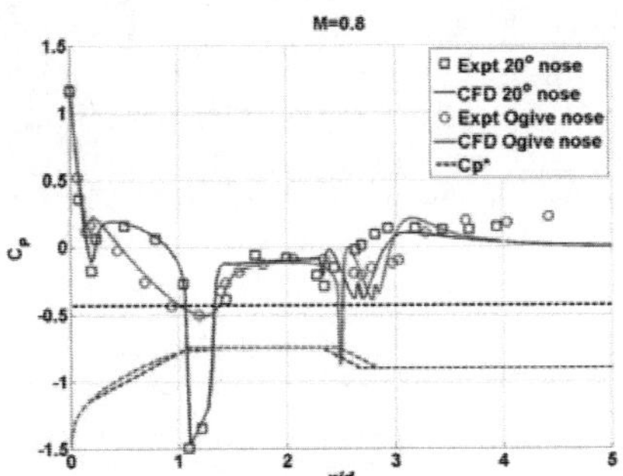

(a) Mach=0.8

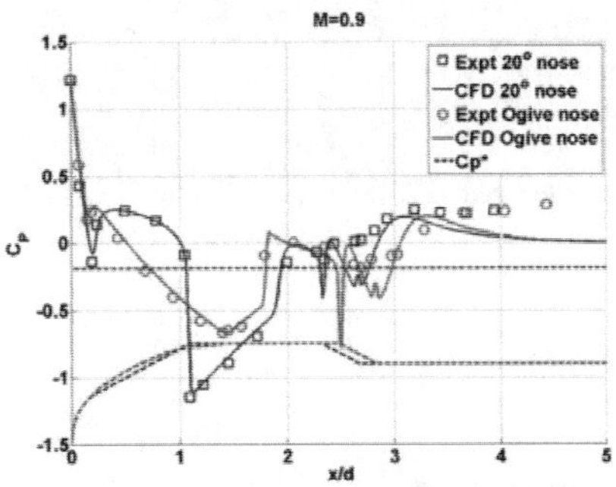

(b) Mach=0.9

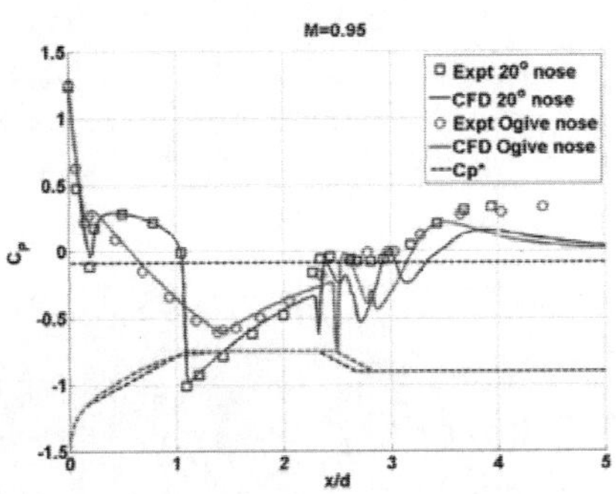

(c) Mach=0.95

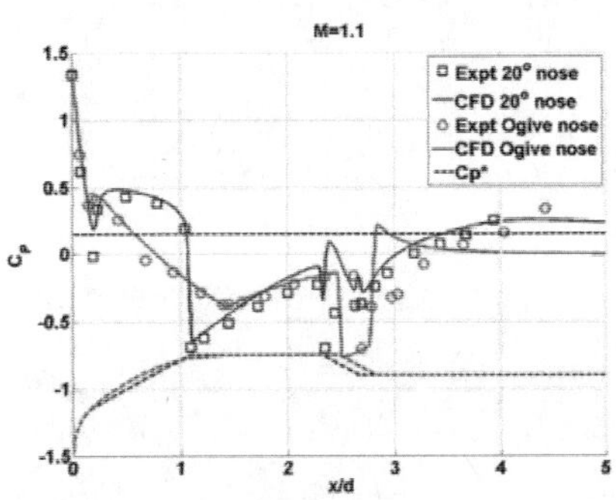

(d) Mach=1.1

Figure 4: Static Pressure Distributions

As shown in Fig. 4a, for the 20° nose, the C_p drops to a minimum of -1.5 while for the ogive cone it is ≈ -0.5. The C_p minima for the ogive cone is also further downstream of $x/d \approx 1.0$. With increase in Mach number as shown in Figures 4b to 4d, the C_p minima for the 20° nose increases, continuing to be located at $x/d \approx 1.0$ while that for the ogive cone decreases and its distance from $x/d \approx 1.0$ also increases. At Mach=0.8 and 0.9, as shown in Figures 4a and 4b, the C_p minima is followed by a shock over the 20° nose whereas there is no shock over the ogive cone. Consequently, shock induced separation occurs on the conical nose as indicated by the pressure plateau in experiments as well as in computations.

Around the boat-tail of the conical nose, computations indicate expansion (drop in C_p) followed by a shock whereas experiments indicate subsonic flow (+ve C_p). This is due to the

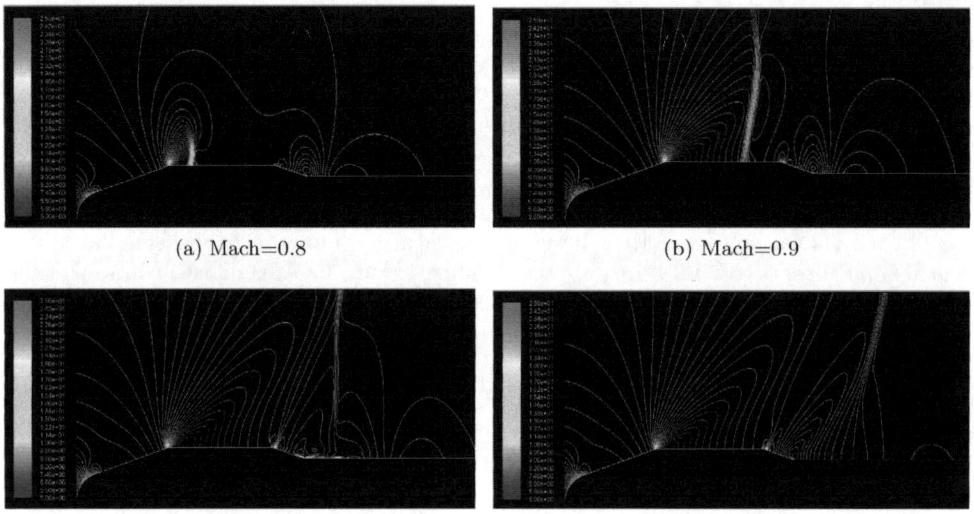

(a) Mach=0.8 (b) Mach=0.9

(c) Mach=0.95 (d) Mach=1.1

Figure 5: Shock Movement on nose with 20° cone

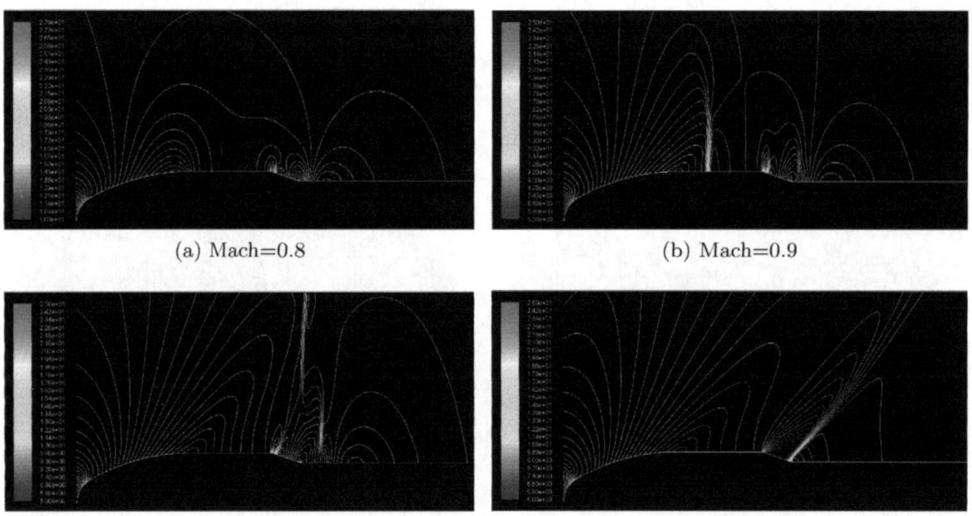

(a) Mach=0.8 (b) Mach=0.9

(c) Mach=0.95 (d) Mach=1.1

Figure 6: Shock Movement on nose with ogive cone

inadequacies in the spatial resolution of pressure ports in experiments as well as in the CFD tools used here, which do not enable a fair comparison. Around the boat-tail of the ogive nose, though both computations and experiments indicate subsonic flow but the agreement in C_p values is not satisfactory. Ambient conditions ($V = V_\infty$, $C_p = 0$) are reached in computation around $x/d \approx 4.0$ whereas in experiments, free-stream conditions are not reached even at $x/d = 4.4$.

The location of the normal shock at these Mach numbers for the 20° nose and the ogive cone is shown in Figures 5 and 6 respectively. The colour key shows bottom most blue for 5 psi and top most red for 25 psi pressure.The comparison of the movement of the shock on both the models is shown in Fig. 7. The normal shock moves downstream as the Mach number is increased, as expected for any configuration. Occurrence of shock is delayed on the conical nose as compared to ogive nose in the transonic range, suggesting that the critical Mach number for the ogive nose cone is higher.

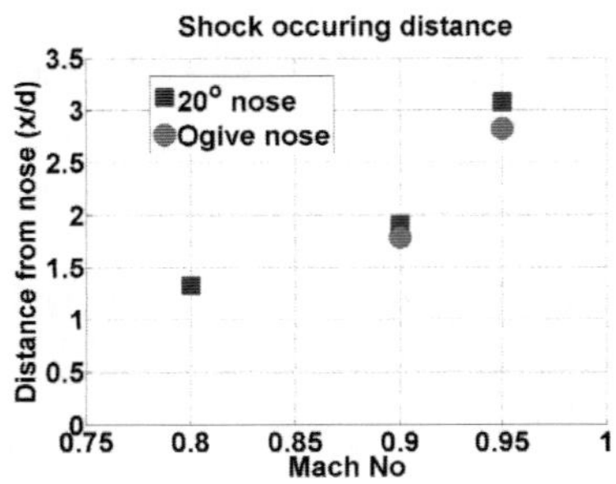

Figure 7: Location of the Normal Shock at different Mach Numbers

5. Conclusion

Comparisons between experiments and computations at transonic Mach numbers on conical and ogive nose-shaped slender bodies with boat-tails is good towards the upstream side. Comparison is not good along the boat-tail, separated flow regions and downstream of the boat-tail for axisymmetric flow conditions. Expansion of the flow around the shoulder region for the conical nose is stronger compared to the ogive nose. Around the boat-tail, both computation and experiment indicate subsonic flow but the agreement in C_p values is not satisfactory. Occurrence of shock is delayed on the ogive nose compared to conical nose. Flow over the ogive nose cone is smoother with much reduced pressure gradients along the body compared to the conical nose. Hence, it seems that the transonic flow over the ogive nose is less prone to flow separation as compared to that over the 20° conical nose.

Acknowledgement

Authors would like to acknowledge the Director, CSIR-National Aerospace Laboratories for grant of the internal research project (N-8-503), which enabled the present study. Support provided by the staff of Design Section, NTAF, Model Shop and 1.2m tunnel is acknowledged. Special thanks to Mr. T. Arun Kumar, Scientist, NTAF for assistance in wind tunnel tests.

References

[1] Charles F. Coe, Steady and fluctuating pressures at transonic speeds on two space vehicle payload shapes, NASA TM X-503, March 1961.

[2] Lawrence, A.G, Gary, A. A., Gallis, M. A. Jr. and George S Deiwert, Comparison of computations and experiments for non-equilibrium flow expansions around blunted cone, AIAA 96-0231, 1996.

[3] R. C. Mehta, Unsteady flow field characteristics over blunt bodies at high speeds, http://dx.doi.org/10.5772/57050, chapter 5, Intech, 2014.

[4] M. A. Ramaswamy and G. Rajendra, Experimental investigation of transonic flow past blunt cone cylinder, Journal of Spacecraft and Rockets, Vol.15, No.2,1978, pp. 120-123. doi: 10.2514/3.28001

Dynamical behaviour of non newtonian spiral blood flow through arterial stenosis

Mohammad Ali[1], **Md. Mahmudul Hasan**[1] **and Mahbub Alam Maruf**[1]

[1]Department of Mechanical Engineering, Bangladesh University of Engineering and Technology, Dhaka-1000, Bangladesh.

E-mail: mali@me.buet.ac.bd

Abstract. The spiral component of blood flow has both beneficial and detrimental effects in human circulatory system. A numerical investigation is carried out to analyze the effects of spiral blood flow through an axisymmetric three dimensional artery having 75% stenosis at the center. Blood is assumed as a non-Newtonian fluid. Standard k-ω model is used for the simulation with the Reynolds number of 1000. A parabolic velocity profile with spiral flow is used as inlet boundary condition. The peak values of all velocity components are found just after stenosis. But total pressure gradually decreases at downstream. Spiral flow of blood has significant effects on tangential component of velocity. However, the effect is mild for radial and axial velocity components. The peak value of wall shear stress is at the stenosis zone and decreases rapidly in downstream. The effect of spiral flow is significant for turbulent kinetic energy. Detailed investigation and relevant pathological issues are delineated throughout the paper.

1. Introduction

At present, congenital and acquired cardiovascular diseases have become one of the most important causes of morbidity and mortality in the world. Among the acquired cardiovascular diseases, atherosclerosis is the most common. These diseases are often characterized by arterial stenosis. Arterial stenosis refers to the localized narrowing of arterial passages. This stenosed artery is responsible to reduce the maximum flow of blood. It happens as a result of coagulation of calcium and fatty materials such as cholesterol and triglyceride. This deposition is named as plaque. In the worst cases plaque may rapture, causing the formation of thrombus cause of death of the tissues. This is the main reason for brain strokes or heart attack and sometimes causes inadequate blood flow in the lower part of the body. In recent years, researchers are interested to simulate the blood flow by computational techniques in a model of three dimensional arteries. The hemodynamical characteristics of flow through arterial stenosis have been continually investigated numerically and experimentally for single as well as multiple restrictions. An experimental investigation of pulsating flow through a smooth constriction was conducted by Ahmed [1]. These experiments were conducted over physiologically relevant mean Reynolds number of 600 based on the tube diameter and the time-averaged value of upstream centerline velocity. A numerical investigation was carried out for laminar sinusoidal pulsating flow through a modeled arterial stenosis with a trapezoidal profile by Hasan and Das [2]. Finite element based numerical technique was used to solve the fluid flow governing equations. The effects of pulsation, stenosis severity, Reynolds number and Womersley number on the flow behavior were studied. Investigation of physiological pulsatile flow in a model arterial stenosis using large-eddy and direct numerical simulations was done by Molla [3]. Physiological pulsatile flow in a 3D model of arterial stenosis was investigated by using large eddy simulation (LES) technique. The effect of spiral flow of blood was studied by Paul [4]. They investigated for 75% reduction in cross sectional area by standard *k*-ω model for Reynolds number 500 and 1000 considering blood as a Newtonian fluid. A numerical simulation of unsteady blood flow through multi-irregular arterial stenosis was conducted by Mustapha [5]. In this paper, effects of spiral blood flow through arterial stenosis are numerically investigated by CFD analysis. Standard *k*-ω model is used to simulate the model. Blood is considered as a Non-Newtonian fluid and Reynolds number of 1000. Artery wall is assumed as rigid.

1.1. Nomenclature

- D: Artery diameter (m)
- ρ: Density (kg/m^3)
- μ: Dynamic viscosity (Ns/m^2)
- V: Stream wise bulk velocity (m/s)
- v: Instantaneous velocity (m/s)
- Ω: Spiral velocity (rad/sec)
- Re: Reynolds number
- R: Radius of blood vessel (m)
- r: Radial location from axis of artery (m)

2. Problem formulations

A 75% axissymetric three dimensional artery is considered here. A parabolic velocity profile with swirling flow is used as inlet boundary condition. The stenosis in the blood vessel is created by using cosine formulae. Both 2D and 3D models of calculation domain are shown in figure 1..The total length of the model is taken as 540 mm (27D) where diameter D=20 mm. Four different spiral velocities are used for investigation. The inlet boundary condition for the stream wise velocity and spiral speed are written in C-language using the interface of User Defined Function (UDF) of Fluent and linked with the solver.

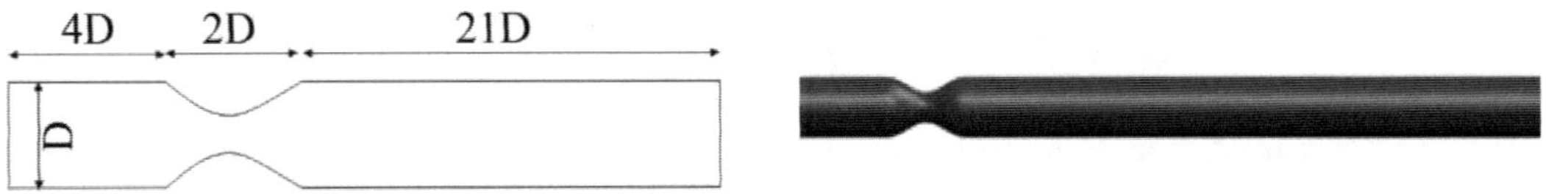

(a) (b)

Figure 1. a) 2D model of stenosed artery (b) 3D model of stenosed artery.

3. Numerical method

The problem is solved numerically with simulating 2D and 3D Navier Stokes equations using proper boundary conditions.

3.1. Governing equations

Navier Stokes equation represents the conservation laws of mass and momentum.

- Conservation of mass

$$\frac{\partial \rho}{\partial t}+\frac{\partial(\rho u)}{\partial x}+\frac{\partial(\rho v)}{\partial y}+\frac{\partial(\rho w)}{\partial z}=0 \tag{1}$$

- Conservation of momentum

$$\frac{\partial(\rho u)}{\partial t}+\frac{\partial(\rho u^2)}{\partial x}+\frac{\partial(\rho uv)}{\partial y}+\frac{\partial(\rho uw)}{\partial z}=\frac{\partial \tau_{xx}}{\partial x}+\frac{\partial \tau_{xy}}{\partial y}+\frac{\partial \tau_{xz}}{\partial z} \tag{2}$$

$$\frac{\partial(\rho v)}{\partial t}+\frac{\partial(\rho uv)}{\partial x}+\frac{\partial(\rho v^2)}{\partial y}+\frac{\partial(\rho vw)}{\partial z}=\frac{\partial \tau_{xy}}{\partial x}+\frac{\partial \tau_{yy}}{\partial y}+\frac{\partial \tau_{yz}}{\partial z} \tag{3}$$

$$\frac{\partial(\rho w)}{\partial t}+\frac{\partial(\rho uw)}{\partial x}+\frac{\partial(\rho vw)}{\partial y}+\frac{\partial(\rho w^2)}{\partial z}=\frac{\partial \tau_{xz}}{\partial x}+\frac{\partial \tau_{yz}}{\partial y}+\frac{\partial \tau_{zz}}{\partial z} \tag{4}$$

3.2. Discretization of the Domain

ANSYS FLUENT is used for the present computation. This CFD tool will solve the governing integral equations for the conservation of mass and momentum. Figure 2 shows the discretization model. The domain consists of 921416 cells, 2793268 faces and 951246 nodes. The geometry is created in solid works and imported in Ansys fluent. Quadrilateral cells were used for this geometry because they can be stretched easily to account for different flow gradients in different directions. The cells near the surface have high aspect ratios. For viscous flow through the stenosis, finely spaced Grid was constructed to calculate the details of the flow near the wall. Three different mesh numbers are used for grid independent test.

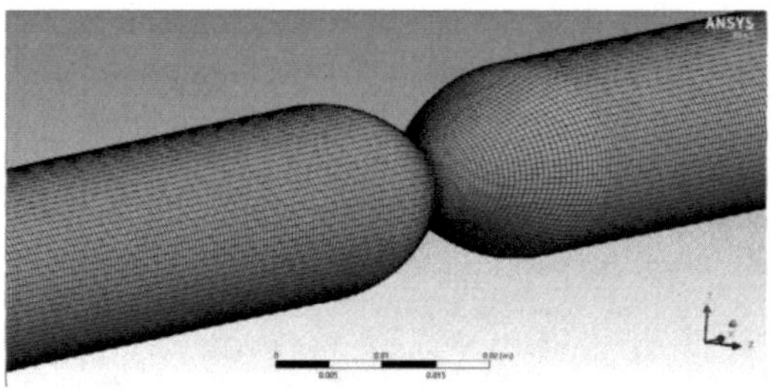

Figure 2. Discretization of the Domain using Fluent meshing tool.

3.3. Boundary Conditions

Inlet-flow: For the geometry, fully developed inlet-flow condition has been assumed as parabolic. The axial velocity distribution is given by,

$$v(r)=2V[1-(\tfrac{r}{R})^2] \tag{5}$$

Where, V= Streamwise mean velocity

Outlet-flow: In case of outlet boundary condition, the gauge pressure is assumed to be zero Pascal.
Wall: At all solid boundaries in the flow geometry, no-slip condition has been used. It can be mathematically expressed as,

$$u_{wall}=0, \text{ and } \quad v_{wall}=0 \tag{6}$$

3.4. Fluid Property

Fluid Material is blood, which is the working fluid in this problem. For this analysis blood is assumed as non Newtonian fluid. Here, typical values of density, ρ=1060 kg/m^3 and viscosity, μ=0.00371 N.s/m^2 are used to calculate velocity from the equation Re = $\rho VD/\mu$. There are several models to solve Non Newtonian fluid problem in CFD. The Carreu fluid model is used here to solve the problem.

3.5. Grid-Independence Test

Since numerical results are greatly dependent on the mesh generation, a grid-independence test is performed to find out the optimum number of elements to discretize the computational domain. Wall static pressure at figure 3 and axial velocity at figure 4 are used for grid independence test. Grid 1 consists of 578835 cells, Grid 2 consists of 921416 cells and Grid 3 consists of 1435082 cells. Grid 2 is giving better results than Grid 1. On the other hand Grid 3 has 55 percent more cells than Grid 2. But, there are no significant changes in results. That's why Grid 2 is good enough for numerical solution.

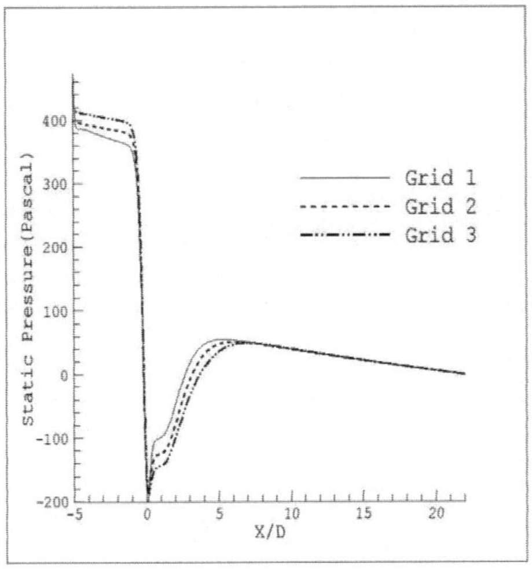

Figure 3. Static pressure for different grids.

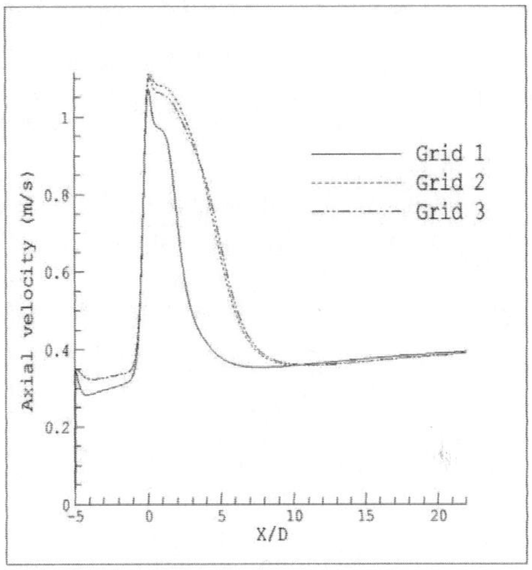

Figure 4. Axial velocity for different grids.

3.6. Validation of results

Before going to the detailed discussion, the computational results have been validated to verify the results. The validation has been done by comparing computational results with the experimental results obtained by Ahmed and Giddens [1]. Single stenosis with 75% severity model is considered for validation. Validation was done for steady inlet velocity condition. Velocity profile, in the post stenotic regions, is compared at Z= 2.5D. Where Z is the normalized distance from the centre of the stenosis. Figure 5 shows that computational results are over predicted at near the artery wall and under predicated near the center of the stenosis than experimental results. But percentage of deviation is too small. However, at middle of radius computational and experimental results are in good agreement. So this model can be used to solve the problem.

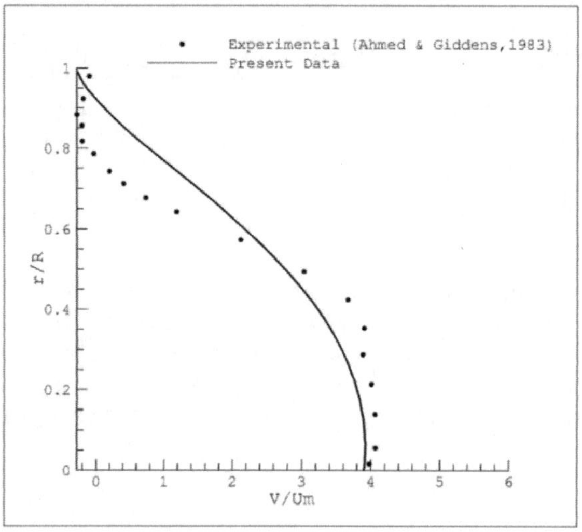

Figure 5. Velocity profile comparison of the computational results and experimental data at Z=2.5D.

4. Results and Discussion

CFD simulation is carried out to observe the effects of spiral flow of blood on different hemodynamic parameter. Figure 6 shows that Spiral flow has a great influence on tangential velocity only, but for axial and radial velocity, its effects are negligible. All velocity components get their maximum value just after stenosis, then gradually decrease to their initial value at downstream.

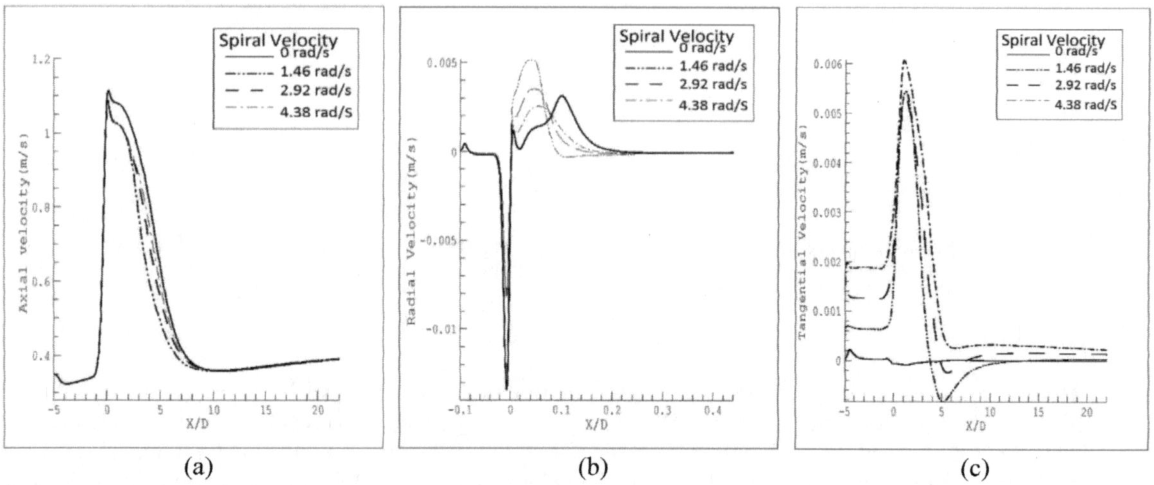

Figure 6. (a) Axial (b) Radial (c) Tangential velocity along the stenosed artery for different spiral speeds.

The centerline total pressure decreases drastically along the length of the stenosed artery which is shown in figure 7. The spiral velocity has little effect on centerline total pressure before stenosis and just after stenosis. But at far downstream spiral flow has no effect on centerline pressure. On the other hand, figure 8 shows that turbulent kinetic energy (TKE) decreases with the increase of spiral speed. TKE is the energy content of eddies of turbulent flow. The spiral flow blood shows the characteristics of turbulent flow. Larger the spiral velocity, higher the energy content of eddies. These eddies extract energy from mean flow. That is why TKE decreases with increase of spiral speed. From figure 9 it is seen that wall shear stress (WSS) tends to rise at the centre of the stenosis. The reason, for this rise of shear stress at the centre of stenosis, is the interaction of vortex with artery wall.

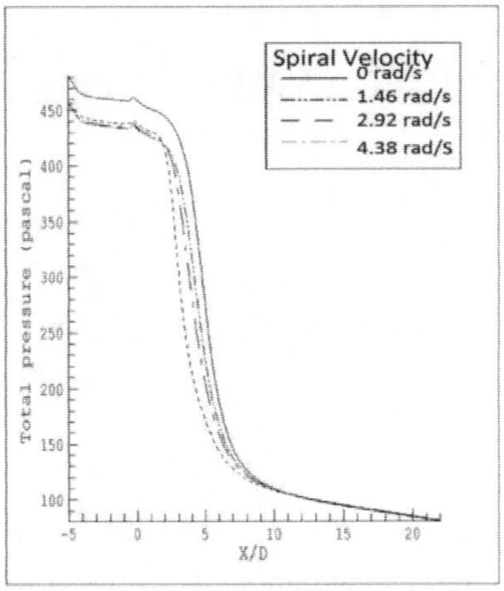

Figure 7. Centreline total pressure for different spiral speeds.

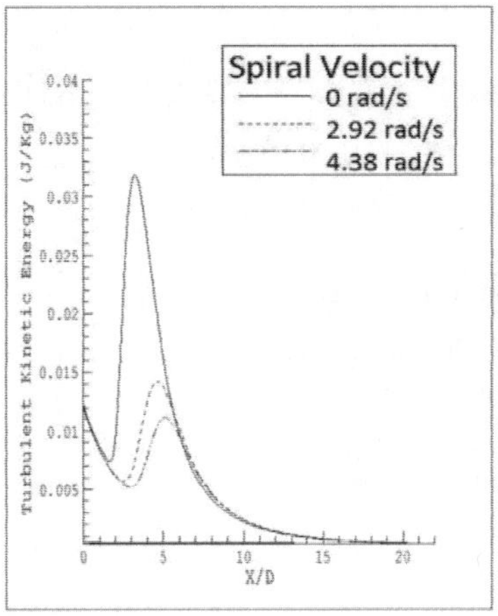

Figure 8. TKE for different spiral speeds.

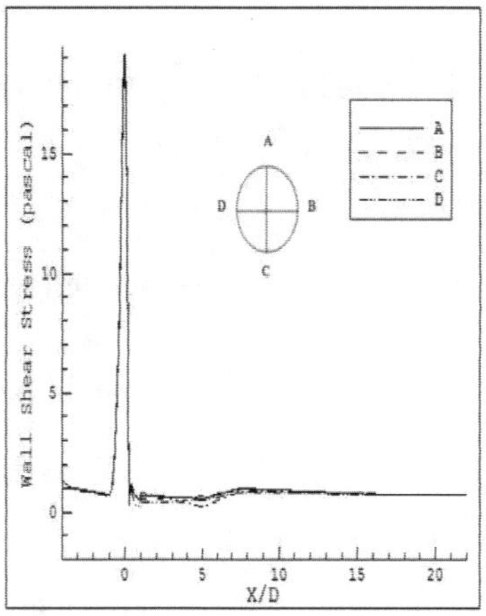

Figure 9. WSS for different spiral speeds.

5. Conclusion

The rise of the turbulent kinetic energy in the post-stenotic region is responsible to cause damage to the blood-cell materials. This activates platelets in the blood, and subsequently they create many pathological diseases. However, the results show that the spiral effect reduces the turbulent kinetic energy, which is a beneficial effect. At the same time spiral flow produces oscillating wall shear stress in the post stenosis, which is a detrimental effect. The oscillating shear stress usually influences to cause potential damage to the inner side of a post-stenotic blood vessel which is kwon as endothelium. In addition, the strong circulation in the post-stenotic region due to the spiral effect is harmful. It may cause the blood to be clotted in the post-stenosis, which is a potential source of stroke. As the spiral flow has potential clinical significance, therefore, the results have some significant impacts on the understanding of blood flow dynamics and its relevance in arterial diseases.

Acknowledgments

This work has been done using different facilities of the Department of Mechanical Engineering, Bangladesh University of Engineering and Technology (BUET).

References

[1] Ahmed, S.A., and Giddens, D.P,1983, *Journal of Biomechanical Engineering*, **16**, 505-516.
[2] Hasan, A.B.M.T., and Das, D.K, 2008, *Journal of Applied Fluid Mechanics*, **1(2)**, 25-35.
[3] Paul, M.C., Molla, M.M., and Roditi, G.2009, *Medical Engineering & Physics*, **31**, 153-159.
[4] Paul, M.C. and Larman, A. 2009, *Medical Engineering and Physics,* **31 (9),** 1195-1203.
[5] Mustapha, N., Mandal, P.K., Johnston, P.R., and Amin, N. 2010, *Applied Mathematical Modelling*, **34(6)**, 1559-1573.

Numerical simulation of parametrically forced gravity waves in a circular cylindrical container

D Krishnaraja and S P Das*

Department of Mechanical Engineering, Indian Institute of Technology Madras, Chennai 600036, India

E-mail: *spdas@iitm.ac.in

Abstract: Parametrically forced gravity waves in axisymmetric mode in a circular cylinder filled with FC-72 with large liquid depth have been studied numerically. The instability threshold and wave breaking thresholds are plotted from the simulated results which show good agreement with the reported experimental and theoretical results. A notable observation is the presence of different time scales of wave amplitude modulations at different regimes. The wave amplitude response exhibits amplitude modulations, period tripling and period quadrupling without breaking of waves. Inertial collapse of the wave trough causes a high velocity jet ejection has also been observed when forcing amplitude crosses the breaking limit.

1. Introduction

Parametrically forced surface waves commonly known as Faraday waves, can be generated by shaking a container vertically inducing oscillation to the interface of two fluids. The Faraday waves have been extensively studied over many decades. Beyond a certain threshold the interface behaviour changes and even leads to various patterns depending on the type of induced vibration. Faraday noticed the waves formed at the interface have half the forcing frequency. This had been confirmed by the Rayleigh through his experiments. Benjamin & Ursell [1] put forward the first theoretical analysis of Faraday waves by carrying out a linear analysis and compared the evolution equation of amplitude to Mathieu Equation. Miles [2] developed a Lagrangian-Hamiltonian formulation to explain the weakly nonlinear surface waves in cylindrical containers. The analytical phase diagram developed based on this theory shows a good agreement with the experimental results.

Large amplitude parametrically forced gravity waves have been investigated by many researchers independently and they observed the period tripling and breaking [3-5]. The different depth ratios, tank base dimensions and its effect in non-linearity have also been investigated [6-7]. Das and Hopfinger [5] experimentally investigated the axisymmetric mode wave motion for parametrically forcing a circular cylinder with different working fluids and determined the stability threshold. They observed period tripling event near the breaking and different time scales of wave amplitude modulations. They reported wave amplitude modulations by period tripling for particular forcing amplitude at a given frequency. Near the resonance they also observed different type of stability behaviour of the wave motion (both subcritical and super-critical). There are also pattern formation studies available in literature at different frequencies [8]. They numerically investigated the dynamics of faraday waves using finite difference method. They reproduced the hexagonal and square patterns using their numerical method for tracking the oscillating interface. The most extensive numerical simulations have been that of Chen and Wu [9] about the nonlinear faraday waves and hysteresis.

In the present paper we focus our study on the axisymmetric mode of wave motion when a circular cylindrical container is forced with parametrically forced gravity waves. We have numerically simulated the flow with single frequency excitation and obtained the stability thresholds. We have also identified different regimes where we observed various time scales. The coexistence of different modes are also observed (neighbouring modes).

2. Theoretical background

The expression for natural frequency of different wave modes for a circular cylinder is developed by Miles (1984) and the dispersion relationship is given by

$$\omega_{mn} = gk_{mn}\left(1+\frac{k_{mn}^{2}\sigma}{g\rho}\right)\tanh\left(k_{mn}h\right)$$

The boundary condition for the container wall, $\left.\frac{\partial\phi}{\partial r}\right|_{r=R} = J_m^{'}(k_{mn}R)=0$ where ϕ is the velocity potential, with axisymmetric modes m=0, n=1 and n=2, $k_{01}R$=3.8317 and $k_{02}R$=7.0156. The asymmetric modes $k_{11}R$=1.841, $k_{21}R$=3.054, $k_{31}R$=4.201, $k_{41}R$=5.318, $k_{12}R$=5.331. The infinite fluid depth limit corresponds to $\tanh(k_{mn}h)=1$, gk_{mn} dominates the sum, the mode is a gravity wave, and when $k_{mn}^{3}\sigma / \rho$ dominates it is a capillary wave. In the present study we used a circular cylinder of radius R=5cm, filled with FC-72 to a depth h=8cm such that $\tanh(k_{01}h)\cong 1$ as used by Das and Hopfinger [5]. Thus the deep water condition is attained, h/R >1. The natural frequency for the axisymmetric mode ω_{01} =27.4719 rad/s. A detailed theory of different sloshing mode and the shape of wave modes for different tank geometries can be found in Ibrahim [10]. The following numerical simulation is done using FC-72 as the liquid of kinematic viscosity ν=0.00406cm^2/s, interfacial surface tension σ =11dyn/cm and density ρ=1690 kg/m^3 at 293K.

3. Computational method

The computational domain is a cylindrical container with FC-72 at bottom and air at top. At the bottom wall we impose the moving boundary and at the cylindrical inner surface no slip condition is applied. The contact angle has been set to 1. The dimensions of the cylinder are radius R=5cm and the height 20cm, filled up to a depth h=8cm such that h/R=1.6. The simulations have been done in ANSYS 15.0. Mesh has been generated in ICEM CFD and for the fluid flow part FLUENT has been used. Both axisymmetric and three dimensional simulations are done for the axisymmetric mode (01). Volume of fluid (VOF) model is used for multiphase modelling. In phase interactions, a formulation of the Continuum surface force (CSF) model is used in conjunction with wall adhesion, where the surface curvature is computed from local gradients in the surface normal at the interface. The interfacial surface tension is set to be σ = 0.011 N/m and the contact angle is taken 1 (FC-72 is fully wetting). Geo-reconstruct scheme is used for tracking the interface. Axisymmetric simulations are done in the stable wave regime and subcritical wave breaking region where pure axisymmetric wave exists. Complete three dimensional simulations are done in the chaotic region and in unstable wave region. Three dimensional simulations are also carried out in the wave breaking and stable wave regions to ensure their axisymmetric nature.

The grid independence test is shown in figure 1. It can be observed from the plot that respectively 21063 and 105336 cells are sufficient for axisymmetric and 3D simulations. We have chosen A/R=0.005 and run the simulation for 3-D case to trace the minimum frequency of instability threshold. Similarly for the two dimensional axisymmetric case we fixed the amplitude ratio as 0.012 and searched the minimum frequency of wave appearance. We have also performed time step independence study and it results in optimum time step size of 0.001s. A different grid is used for the analysis of jet formation, the mesh is chosen based on the last wave amplitude for axisymmetric case (75803 cells) to resolve the jet velocity as accurate as possible. Near the axis mesh is fine and away from axis coarse mesh has been used. The time step used in jetting analysis is 0.0001s.

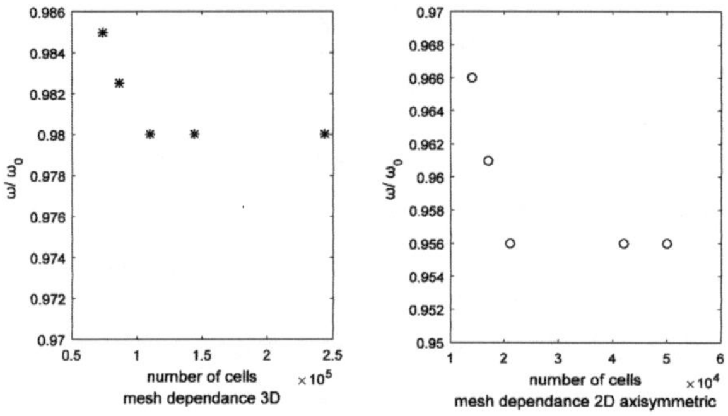

Figure 1. the lowest ω/ω_0 at which the wave appears is plotted against no of cells, for three dimensional case *A/R*=0.005 and for axisymmetric case *A/R*=0.012.

4. Results and Discussions

In this section we discuss the results obtained from numerical simulation of axisymmetric gravity waves in the circular container filled with FC72 up to a depth of *h*=8cm. Throughout the following discussion amplitude corresponds to the non-dimensional amplitude *A/R* and frequency ω/ω_0, where ω is the wave frequency and ω_0 is the natural frequency of 01 mode.

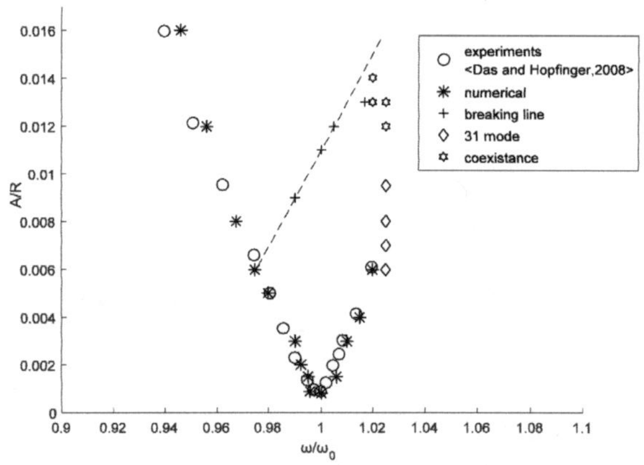

Figure 2. Instability threshold obtained from numerical simulations (*) is compared with experimental results [5]. + shows the region where the wave grows exponentially and breaks. The symbols × and ◊ show the regions for 31 mode and coexistence of 31 and axisymmetric modes respectively. The wave mode for the latter is initially axisymmetric and then coexists with the 31 mode.

4.1. Instability threshold

The dimensionless forcing amplitude as a function of frequency ratio in figure 2 shows the instability threshold for wave appearance and also breaking limit for stable waves. The numerical results show a very good agreement with the experimental results of Das and Hopfinger [5]. The instability threshold indicates required minimum forcing at which waves appear at the interface. Whereas the breaking line shows the region above which the waves will grow exponentially and break as the steepness limited criteria is reached and below the line waves unstable wave motions are observed. In a given range of forcing amplitude (lower that unstable wave and higher than instability threshold) stable waves are observed. The axisymmetric mode 01 lies in the frequency band $\omega/\omega_0 = 0.945$ to $\omega/\omega_0 = 1.025$, and then it follows the 31 mode.

4.2. Amplitude modulations

From the simulation results different types of wave motions are observed in different regimes of the instability diagram (figure 2). Important observations here are the different time scales of wave amplitude modulations. Wave amplitude modulation happens by slow time scale, period tripling and period quadrupling. In some portion of the unstable wave motion (subcritical regime) regime period tripling (here for $A/R = 0.005$) has been observed whereas some part (supercritical regime) shows period quadrupling behavior (here at A/R=0.010).

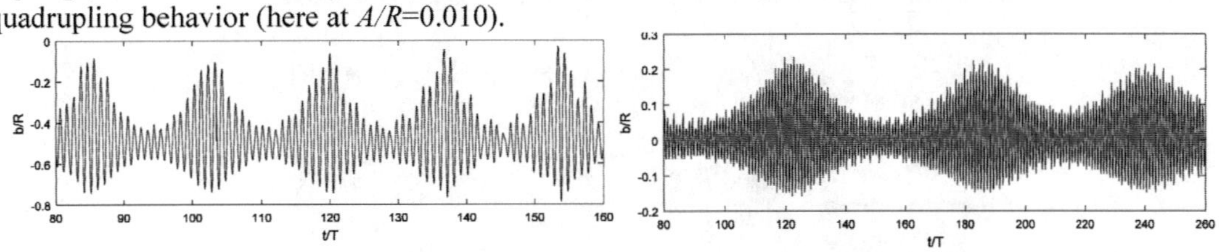

(a) A/R=0.012 (b) A/R=0.0025

Figure 3. Comparison of slow time scale amplitude modulation for different forcing amplitudes (A/R=0.012 and 0.0025). Here b is the wave amplitude.

In figure 3 the wave amplitudes are plotted as a function of dimensionless time t/T, where $T=2\pi/\omega$. Figure shows amplitude modulation occurring in slow time scale for A/R=0.012 and A/R=0.0025 with half period growth in t/T = 63 and 17 respectively. Period of the amplitude modulation decreases with increase in forcing amplitude which is consistent with the slow time similar to Miles [2]. Period tripling is observed at $\omega/\omega_0 = 0.985$, A/R=0.005, where the steep wave crest is seen followed by a flat crest and a slightly higher amplitude wave (figure 4). This instability is also observed with slight

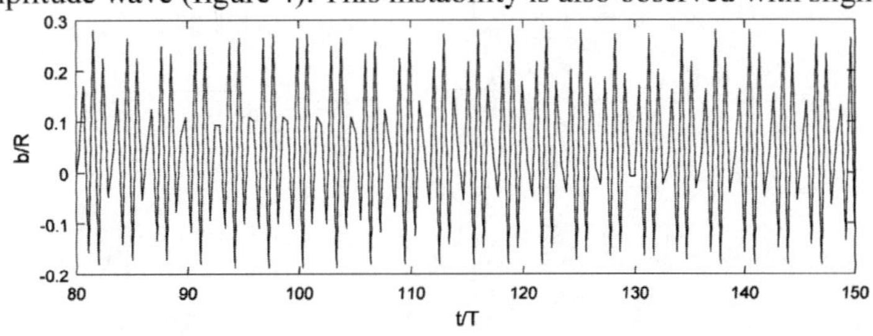

Figure 4. Period tripling cycle, A/R=0.005 and $\omega/\omega_0 = 0.985$, the shorter waves are flat crested and the higher waves appeared with steep crests.

demodulation where the wave initially has an amplitude modulation and eventually shifts to period tripling. In the unstable wave region for A/R=0.005 and $\omega/\omega_0 = 0.9825$, the wave crest is formed slightly off the center and it rotates about the axis of the cylinder. The rotation is observed when the simulation is continued for a longer period of time. Here after t/T=350 rotation starts. Figure 5 shows the images of wave crest for one complete rotation. For A/R=0.010 ω/ω_0 = 1.02, period quadrupling event is observed which is similar to the above mentioned period tripling event (figure 6) at frequencies higher than the natural frequency and in the close proximity of the 31 mode but in the unstable wave motion regime. Below the breaking line, $\omega/\omega_0 = 1.017$, A/R=0.01275 the wave initially shows an amplitude modulation and then slowly demodulate by shrinking the time period of modulation and increasing the amplitude. In some cases at slightly higher amplitude for a given frequency, the wave amplitude demodulates and the wave breaks very near the breaking line for A/R=0.013, $\omega/\omega_0 = 1.017$. This can be explained as demodulation causes increase in the wave height and wave breaks by deep water wave breaking mechanism.

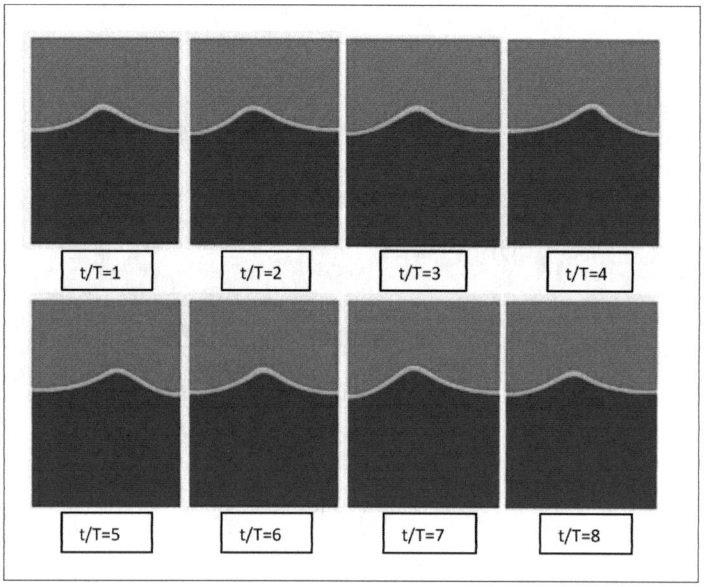

Figure 5. Rotation for A/R=0.005, ω/ω_0= 0.9825 , the time evolution of crest of the wave is shown.

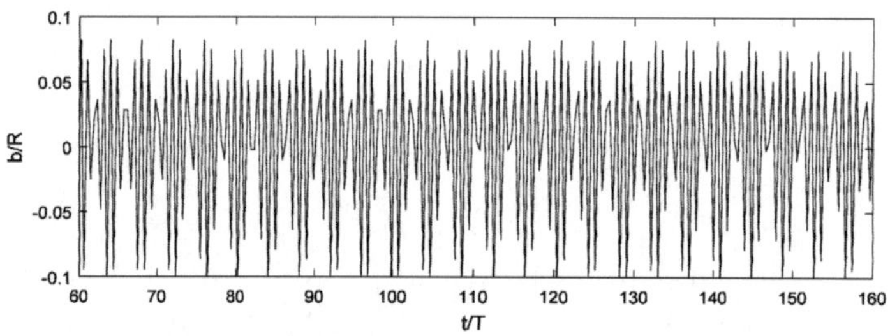

Figure 6. Wave pattern observed near breaking A/R= 0.010 and ω/ω_0= 1.02, it can be seen that each wave packet of envelope carries four waves of the signal

4.3. Coexistence

The axisymmetric mode exists in the range ω/ω_0= 0.94 to ω/ω_0 = 1.025, for amplitude A/R=0.006 and ω/ω_0 = 1.025 shifts to 31 mode. As the amplitude is slowly increased from A/R=0.006 to 0.015 with fixed frequency ω/ω_0 = 1.025 pure 31 mode is seen till 0.012. Above a critical forcing amplitude (A/R = 0.012) coexistence of 01 and 31 modes is observed. Effect of both the modes is substantial to result in a different type of wave pattern. For A/R=0.014 and 0.013 ω/ω_0 = 1.02 the 01 and 31 mode exists together and the wave breaking occurs only on the 31 mode showing dominance of 31 mode. As in figure 7, A/R=0.01225, ω/ω_0 = 1.020, the crest at center is the wave form for axisymmetric mode and the crest in 3 location at the edge is of 31 mode.

4.4. Jet formation

When $A>A_{breaking}$, the wave breaking occurs for axisymmetric mode. The breaking line for this mode has been traced numerically and is shown in instability threshold diagram (figure 2). The deep water wave breaking due to steepness limited criteria H/L=0.142 (for gravity waves in laterally unbounded cases), where H is the peak to peak height and L is the wave length. The maximum steepness numerically obtained is, H/L=0.30 at A/R=0.016, ω/ω_0 = 0.985.

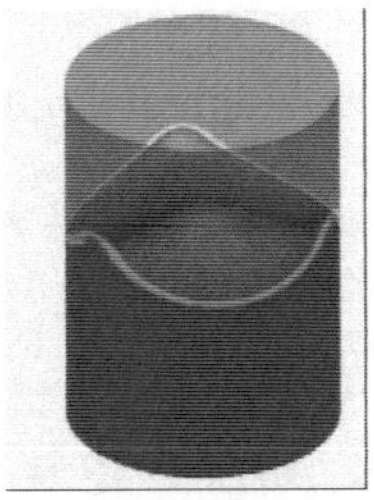

Figure 7. Coexistence of 01 and 31 mode (A/R=0.01225, ω/ω_0 = 1.02).

In the breaking region, exponential growth of amplitude and breaking occurs with the high velocity jet ejection from the fluid interface due to the inertial collapse as observed by previous investigators [5, 11]. The jetting may occur with or without pinch-off. A representative case is shown in figure 8 for A/R=0.016 and ω/ω_0 = 0.995. Here the jet forms with a bubble pinch-off. Jet velocity is measured at a specific point in the path of the jet, the instant at which jet passes the chosen point shows the peak value corresponds to the velocity of jet at that instant. In the breaking region waves are recorded with high steepness H/L>0.3 and the maximum jet velocity is found to be 12m/s in the simulations performed here. The jet velocity depends on the shape of the cavity formed before collapse which also depends on the viscosity and surface tension of the working fluid.

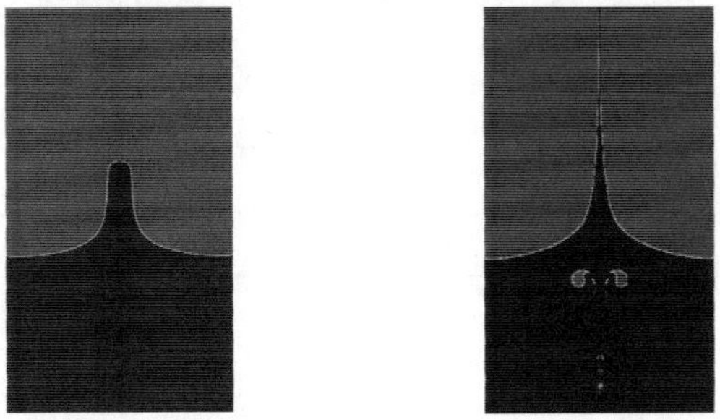

(a) last stable wave amplitude *(b)* jetting

Figure 8. Images showing the last wave amplitude (t=t_0) before collapse and the jetting (t=t_0+0.2 s) (A/R=0.016 with, ω/ω_0 = 0.995).

5. Conclusion

The numerically simulated stability threshold diagram for axisymmetric gravity wave (01 mode) in a circular cylinder for low viscosity and low surface tension fluid (FC72) with infinite liquid depth shows good agreement with the reported experimental investigations. Bifurcation to 31 mode and coexistence with 31 mode of this axisymmetric mode has been found. Amplitude demodulation has been observed near the breaking and the demodulation forms stable wave height.The period tripling event matches with the earlier observations [3, 5]. Wave amplitude modulations occurs at different time scale (slow, period tripling and period quadrupling) without breaking. When the forcing amplitude crosses a critical limit at a given frequency, inertial collapse of the wave trough causes a high velocity jet ejection from the liquid surface with or without pinch-off. The jet velocity obtained from simulation agrees well with the experimental results for same fluid under similar conditions.

References

[1] Benjamin T B & Ursell F1954 The stability of plane free surface of a liquid in a vertical periodic motion. *Proc. R. Soc. Lond. A* **225** 505–515.
[2] Miles J W 1984 Parametrically excited solitary waves *J. Fluid Mech.* **148** 451-460.
[3] Jiang L, Ting C-L, Perlin M & Schultz W W 1996 Moderate and steep Faraday waves:instabilities, modulation and temporal asymmetries *J. Fluid Mech.* **329** 215-301.
[4] Royon-Lebeaud A, Hopfinger EJ & Cartellier A 2007 Liquid sloshing and wave breaking in circular and square-base cylindrical containers *J. Fluid Mech* **577** 467-494.
[5] Das S P & Hopfinger E J 2008 Parametrically forced gravity waves in circular cylinder and finite time singularity *J. Fluid Mech.* **599** 205–228.
[6] Faltinsen O M, Rognebakke O F & Timokha A N 2003 Resonant three-dimensional nonlinear sloshing in a square-base basin. *J. Fluid Mech.* **487** 1–42.
[7] Miles J W 1984 Nonlinear Faraday resonance *J. Fluid Mech.* **146** 285-302.
[8] Nicolas P, Damir J & Tuckerman L S 2009 Numerical simulation of faraday waves *J. Fluid Mech.* **635** 1-26.
[9] Peilong Chen & Kuo-An Wu 2000 Subcritical bifurcations and nonlinear balloons in faraday waves *Phys.Rev.Lett.***85**.3813.
[10] Ibrahim R A 2005 Liquid sloshing dynamics *Cambridge University Press*.
[11] Zeff BW, Kleber B, Fineberg J & Lathrop D P 2000 Singularity dynamics in curvature collapse and jet eruption on a fluid surface *Nature* **403** 401–404.

Performance of nickel-based oxygen carrier produced using renewable fuel aloe vera

NF Afandi[1], D Devaraj[1], A Manap[1*] and N Ibrahim[2]

[1]Department of Mechanical Engineering, College of engineering, Jalan IKRAM-UNITEN 43000, Kajang, Selangor, Malaysia
[2]Faculty of Electrical and Electronic Engineering, Universiti Tun Hussein Onn Malaysia (UTHM), 86400 Parit Raja, Batu Pahat, Johor, MALAYSIA

Email: [1*]Abreeza@uniten.edu.my

Abstract. Consuming and burning of fuel mainly fossil fuel has gradually increased in this upcoming era due to high-energy demand and causes the global warming. One of the most effective ways to reduce the greenhouse gases is by capturing carbon dioxide (CO_2) during the combustion process. Chemical looping combustion (CLC) is one of the most effective methods to capture the CO_2 without the need of an energy intensive air separation unit. This method uses oxygen carrier to provide O_2 that can react with fuel to form CO_2 and H_2O. This research focuses on synthesizing $NiO/NiAl_2O_4$ as an oxygen carrier due to its properties that can withstand high temperature during CLC application. The $NiO/NiAl_2O_4$ powder was synthesized using solution combustion method with plant extract renewable fuel, aloe vera as the fuel. In order to optimize the performance of the particles that can be used in CLC application, various calcination temperatures were varied at 600°C, 800°C, 1050°C and 1300°C. The phase and morphology of obtained powders were characterized using X-ray diffraction (XRD) and Field Emission Microscopy (FESEM) respectively together with the powder elements. In CLC application, high reactivity can be achieved by using smaller particle size of oxygen carrier. This research succeeded in producing nano-structured powder with high crystalline structure at temperature 1050°C which is suitable to be used in CLC application.

1. Introduction

Green house gases (GHGs) are rapidly discharged to the environment throughout these years due to high-energy demands from the industries. The release of GHGs can cause greenhouse effects that bring harm towards the environment. Therefore, many methods were introduced to reduce the emissions of GHGs such as to capture the emissions of CO_2, to increase the efficiency of energy conversion and usage and also to use renewable fuel and nuclear energy [1]. However, capturing and storing CO_2 had gained more interest among researchers due to the inevitable emissions of CO_2 during combustion process [2].

Chemical looping combustion (CLC) was introduced as one of the methods that can capture the CO_2 without requiring high-energy consumptions [3]. In CLC application, oxygen carrier provides oxygen that reacts with fuel to produce CO_2 and H_2O during combustion process. This oxygen carrier will undergo reduction and oxidation process inside the reactors. The regenerability of oxygen carriers is important since it determines the durability of the oxygen carrier [4]. Therefore, efficiency of CLC application highly depends on the performance of oxygen carriers. The oxygen carrier consists of metal oxide and binder material, which provides oxygen and increase the mechanical properties of oxygen carrier respectively [5]. One of the oxygen carriers that can perform good chemical reactivity in CLC application is the $NiO/NiAl_2O_4$ combination [6]. The $NiO/NiAl_2O_4$ had been attracting many researchers due to its properties that can withstand high temperature. The operating temperature of CLC application can be up to 1200°C [7].

Preparation method of oxygen carriers plays an important role in CLC application since it determines the performance of oxygen carriers. Ishida et al. [8] prepared $NiO/NiAl_2O_4$ using a dissolution method and obtained a spherical particle with diameter that is larger than 97 μm. Zhao et al. [9] produce $NiO/NiAl_2O_4$ powder using sol-gel method and obtained particles with sizes ranging from 125-180 μm. However, both methods consume a lot of time (30 – 40 hours). Cho et al. [10] synthesized $NiO/NiAl_2O_4$ using a freeze-granulation method, which reduced the preparation time, but the resulting particles had low mechanism strengths. Rate of redox reaction in CLC application was found to increase when smaller particle size was used due to the decreases in diffusional resistance [8]. Moreover, nanostructured oxygen carrier can improve the CLC performance since it can reduce the formation of sulfur content [11].

Therefore, this research focuses on synthesizing $NiO/NiAl_2O_4$ using solution combustion method with aloe vera as fuel. Solution combustion method is capable to produce nanostructured powder with high purity powder using simple tooling requirements [12]. The usage of aloe vera as fuel is more environmental friendly and capable to produce $NiO/NiAl_2O_4$ powder at low cost and releases non-toxic gases [13]. In this research, various calcination temperatures were varied in order to determine the formation of $NiO/NiAl_2O_4$ since very few extensive studies were found on synthesizing this powder using aloe vera as fuel. Then, the powder was characterized using X-Ray Diffraction, Field Emission Microscopy and Energy Dispersive X-ray to determine the crystalline phase, morphology and elements of the obtained powders.

2. Methodology

In this research, solution combustion method was used to synthesize $NiO/NiAl_2O_4$ powder using aloe vera as fuel. Aloe vera is an environmental friendly source that can act as fuel that produces non-toxic gases during combustion process.

Firstly, few leaves of aloe vera was taken from aloe vera plant and washed gently to remove any impurities and dirt. Then, the top layer of aloe vera leaves were cut in order to obtain the aloe vera gel (5g). After that, this aloe vera gel was placed inside beaker together with 10ml distilled water and stirred for 30 minutes in a room temperature prior to obtain the clear solution of aloe vera. This entire process was known as renewable fuel extraction process.

Then, aluminum nitrate ($Al(NO_3)_3.9H_2O$) and nickel nitrate ($Ni(NO_3)_2.6H_2O$) were mixed together with the aloe vera solution prior the heating process. The mixture was heated on top of the hot plate until the combustion process occurred. This mixture was stirred continuously throughout this heating process. The solution was found to be combust at 300°C about 40 minutes. This heating process was conducted in ambient air. The as-synthesized powders that were obtained after combustion process were collected and grinded by using mortar in order to get fine particle powder prior the calcination process. Then this as-synthesized powders were introduced inside muffle furnace at various calcination temperatures which were 600°C, 800°C, 1050°C and 1300°C.

The crystalline phase of $NiO/NiAl_2O_4$ powder was characterized using X-Ray diffraction (SHIMADZU 6000) at a scan speed 3°/min using CuKα radiation. The crystallite size was calculated by using Scherer equation. The Scherer equation is given below [14],

$$D = z\frac{0.9\lambda}{\beta\cos\theta} \quad (1)$$

where D is crystallite size (in nm), λ is the radiation wavelength (for CuKα radiation, λ 1.5418Å), θ is the diffraction peak angle and β is the broadening of the line ("half width") measured at half its maximum intensity (in radians). The morphology and elements of the $NiO/NiAl_2O_4$ powder were observed using Field Emission Microscopy, FESEM and Energy dispersive X-Ray respectively (Hitachi SU8030).

3. Powder characterizations

3.1. Phase of obtained powder (X-Ray diffraction, XRD)

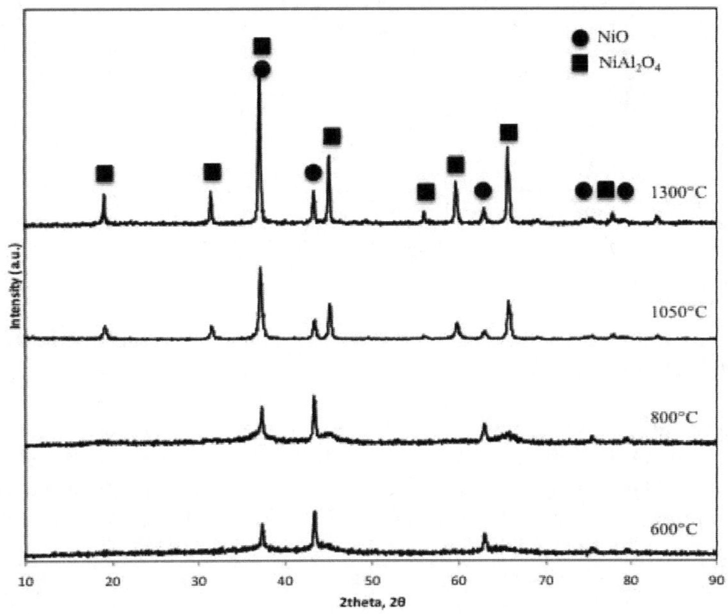

Figure 1. XRD results when $NiO/NiAl_3O_4$ powder calcined at temperature 600°C, 800°C, 1050°C and 1300°C.

Figure 1 shows XRD results of $NiO/NiAl_2O_4$ when calcined at 600°C, 800°C, 1050°C and 1300°C. This figure indicates that only NiO phase was shown at 600°C which corresponds to FCC structure [15]. It is in agreement with Ganachari et al. [15] that the formation of NiO is above than 500°C. Calcination at 800°C produces two more peaks which correspond to $NiAl_2O_4$ peaks (JCPDS 00-010-0339). The formation of $NiAl_2O_4$ was reported to occur at temperature above than 725°C due to the spinel is non-tolerant structure concerning an excess of Ni [16]. Then, more $NiAl_2O_4$ peaks were identified at 1050°C and 1300°C. This result suggested that Al^{3+} ions, Ni^{2+} ions and oxygen had been oxidized to form NiO and $NiAl_2O_4$ phase [17]. However calcination at 1300°C produces sharper peaks than 1050°C which corresponds to a better crystalline phase due to enough heat was introduced at high temperature [18]. In addition, formation of Al_2O_3 was not found at 1300°C even though it is been suggested that Al_2O_3 occurs at temperature above than 1100°C [19]. Formation of Al_2O_3 together with NiO and $NiAl_2O_4$ is not favorable in CLC application since it can degrade the CLC performance [20]. Therefore, NiO and $NiAl_2O_4$ powders were formed at temperature above than 800°C without presence of Al_2O_3.

3.2. *Crystallite size*

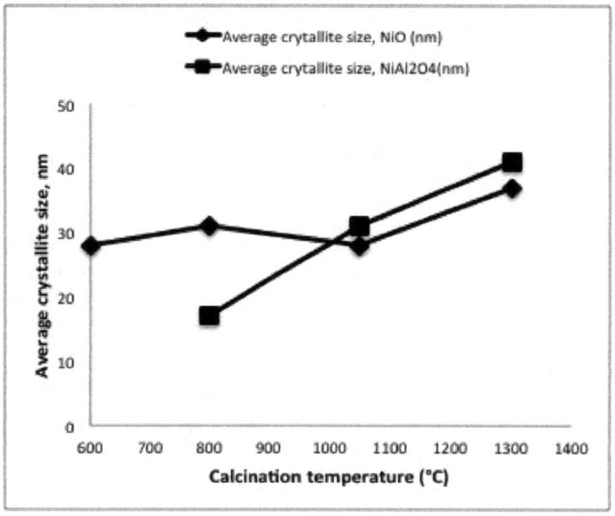

Figure 2. Average crystallite size of the NiO and $NiAl_2O_4$ powder when calcined at temperature 600°C, 800°C, 1050°C and 1300°C.

Figure 2 shows the $NiO/NiAl_2O_4$ average crystallite size as a function of calcination temperature. The crystallite size of $NiAl_2O_4$ increased from 17 nm to 41 nm, respectively when temperature increase from 600°C to 1300°C. The increase in crystallite size resulted from heat presented when increases the temperature that can produce larger crystallite size [18]. However, the crystallite size of NiO starts to decrease from 31nm to 28nm from 800°C to 1050°C. This result suggests that there is a break up in the grain size and causes it to be decreased [21]. However, crystallite size of NiO increases from 28 nm to 31 nm from 600°C to 800°C and 31 nm to 37 nm from 1050°C to 1300°C which in agreement with Leal et al. [18] that increase in calcination temperature causes increases in crystallite size.

3.3. *Morphology of the $NiO/NiAl_2O_4$*

Figure 3 shows FESEM images of the $NiO/NiAl_2O_4$ powders when calcined at 600°C, 800°C, 1050°C and 1300°C. The particles powder calcined at 600°C displays irregular shape with agglomeration and sintered necks with adjacent grain particle due to the large amount of heat occurred during combustion process [22]. At temperature 800°C, the particles started to interconnect with other grain particles to become one large grain size. However, this formation is weak, fragile and could break easily when more heat was introduced since there are still agglomerated particle [22]. However, the particles started to break up to become smaller particle size due to the grain refinement at temperature 1050°C. High crystalline structure started to form at 1300°C. This formation was due to the enough heat to stimulate the formation of crystalline structure [23]. This result also can be seen in figure 1 that sharper peaks were found at 1300°C than other calcination temperature. However, smaller particle size can be obtained at 1050°C, which favorable in CLC application since it could increase CLC performance.

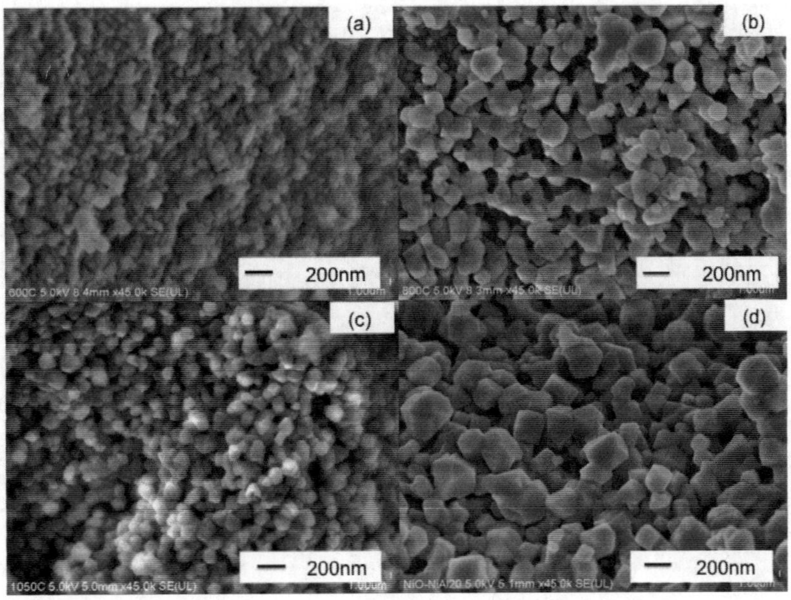

Figure 3. FESEM images of $NiO/NiAl_3O_4$ powder when calcined at temperature (a) 600°C, (b) 800°C, (c) 1050°C (d) 1300°C.

3.4. Elements of the $NiO/NiAl_2O_4$

Figure 4 illustrates elemental analysis of obtained powder using aloe vera as fuel at various calcination temperatures. Only Ni and O elements were shown in figure 4(a). The nuclei of Al_2O_3 does not nucleate yet due to not enough heat presented during synthesization process when calcined at 600°C [24]. Moreover, Al elements start to develop at temperature above than 800°C which can be seen in figure 1. Figure 1 shows that $NiAl_2O_4$ starts to form at 800°C. However, C element also presented in this research, which is not favorable in CLC application due to the carbon inhibit densification during calcination process [25]. However, amount of carbon content can consider as less and this powder would perform comparatively well in CLC application [25].

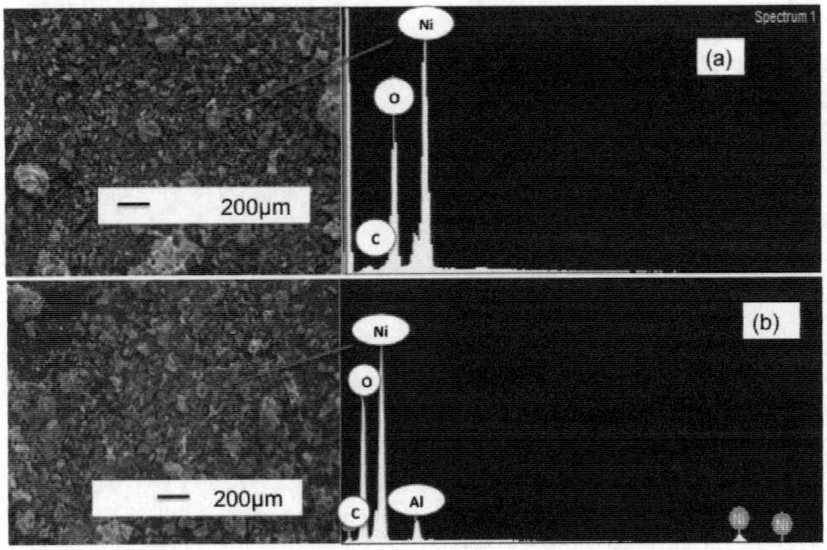

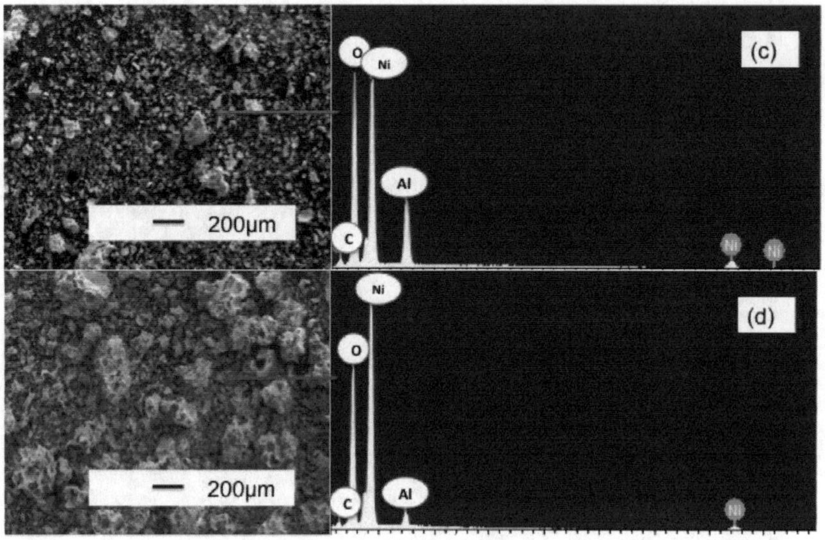

Figure 4. Elemental analysis of $NiO/NiAl_3O_4$ powder when calcined at temperature (a) 600°C, (b) 800°C, (c) 1050°C (d) 1300°C.

4. Conclusions

Effect of the calcination temperature on the $NiO/NiAl_2O_4$ powder synthesized using solution combustion method was studied thoroughly. According to the obtained result, $NiO/NiAl_2O_4$ was successfully synthesized using solution combustion method using aloe vera as fuel at temperature 1050°C, which exhibit smaller size comparing to other calcination temperature and produced oxygen carrier that can be used in CLC application.

Acknowledgements

The authors acknowledge the financial supports by the UNITEN (Grant No. J510050432) and Malaysian Ministry of Higher Education (Grant No. FRGS20130108) and (Grant No. FRGS 20160105).

References

[1] Yamasaki A 2003 An overview of CO2 mitigation options for global warming- emphasizing CO2 sequestration options *Journal of Chemical Engineering of Japan* **36** 361-375.

[2] Sim CY, Brown T, Chen Q, Sharifi V, Swithenbank J, Dennis J and Scott S 2012 Particle characterization in chemical looping combustion *Chemical Engineering Science* **69** 211-224.

[3] Ishida M and Jin H 1994 A New Advanced Power-Generation System Using Chemical-looping Combustion *Energy* **19** 415-422.

[4] Jin H, Okamoto T and Ishida M 1999 Development of a Novel Chemical-Looping Combustion: Synthesis of a Solid Looping Material of NiO/NiAl2O4 *Industrial and Engineering Chemistry Research* **38** 126-132.

[5] Liu LM, Zhao HB and Zheng CG 2006 Advances on oxygen carriers of chemical looping combustion *Coal conversion* **29** 83-93.

[6] Zhao HB, Liu LM, Di Xu, Zheng CG, Liu GJ and Jiang LL 2008 NiO/NiAl2O4 oxygen carriers prepared by sol-gel for chemical-looping combustion fueled by gas *Journal of Fuel Chemistry and Technology* **36** 261-266.

[7] Bhoje R, Kale GR, Labhsetwar N and Borkhade S 2013 Chemical looping combustion of methane: A technology development view *Journal of energy* **Article ID 949408** 1-15.

[8] Ishida M, Yamamoto M and Ohba T 2002 Experimental results of chemical looping combustion with NiO/NiAl2O4 particle circulation at 1200C *Energy Conversion and Management* **43** 1469-78.

[9] Zhao HB, Liu LM, Di Xu, Zheng CG, Liu GJ and Jiang LL 2008 NiO/$NiAl_2O_4$ oxygen carriers prepared by sol-gel for chemical-looping combustion fueled by gas *Journal of Fuel Chemistry and Technology* **36** 261-66.

[10] Cho P, Mattisson T and Lyngfelt A 2004 Comparison of iron-, nickel-, copper- and manganese-based oxygen carriers for chemical-looping combustion *Fuel* **83** 1215–25.

[11] Solunke R and Veser G 2011 Integrating desulfurization with CO2 capture in chemical looping combustion *Fuel* **90** 608-617.

[12] Sutka A and Mezinskis G 2012 Sol–gel auto-combustion synthesis of spinel-type ferrite nanomaterials *Frontiers of Materials Science* **6** 128-141.

[13] Raguphati C, Judith VJ, Surendhar P and John Kennedy L 2014 Comparative investigation of nickel aluminate (NiAl2O4) nano and microstructures for the structural, optical and catalytic properties *Polyhedron* **72** 1-7.

[14] Parra MR and Haque FZ 2014 Aqueous chemical route synthesis and the effect of calcination temperature on the structural and optical properties of ZnO nanoparticles *Journal of Materials Research and Technology* **3** 363-369.

[15] Ganachari SV, Bhat R, Deshpande R and Venkataraman A 2012 Synthesis and characterization of nickel oxide nanoparticles by self-propagating low temperature combustion method *Recent Research in Science and Technology* **4** 50-53.

[16] Salhi N, Boulahouache A, Petit C, Kiennemann A and Rabia C 2010 CO2 reforming of methane to syngas over non reduced spinel catalysts *Journal of the Algerian Chemical Society* **20** 121-131.

[17] Tirsioga A, Visinescu D, Jurca B, Ianculescu A and Carp O 2011 Eco-friendly combustion-based synthesis of metal aluminates MA2O4 (M = Ni, Co) *Journal of nanoparticle Research* **13** 6397-6408.

[18] Chen YF, Lee CY, Yeng MY and Chiu HT 2003 The effect of calcination temperature on the crystallinity of TiO2 nanopowders *Journal of crystal growth* **247** 363-370.

[19] Toniolo JC, Lima MD, Takimi AS and Bergmann CP 2005 Synthesis of alumina powders by the glycine-nitrate combustion process *Materials Research Bulletin* **40** 561-571.

[20] De Diego LF, Garcia-Labiano F, Adanez J, Gayan P, Abad A, Corbella BM and Palacios JM 2004 Development of Cu-Based Oxygen Carriers for Chemical-Looping Combustion *Fuel* **83** 1794-1757.

[21] Afandi NF, Manap A, Misran H, Othman SZ and Pauzi NIM 2015 Characterizations of NiAl-Al_2O_3 produced using gel combustion synthesis method *Advanced mechanics and materials* **761** 457-461

[22] Leal E, de Melo Costa ACF, de Freita NL, Lucena Lira HL, Kiminami RHGA and Gama L 2011 NiAl2O4 catalysts prepared by combustion reaction using glycine as fuel *Materials Research Bulletin* **46** 1409-1413.

[23] Chen YF, Lee CY, Yeng MY, Chiu HT 2003 The effect of calcination temperature on the crystallinity of TiO2 nanopowders *Journal of crystal growth* **247** 363-370.

[24] Taleghani M and Riahi-Noori N 2013 Synthesis of alumina nano powder by a gel combustion method *Journal of Ceramic Processing Research* **14** 17-21.

[25] Lanoş R, Lazău R, Lazău L and Păcurariu C 2012 Chemical oxidation of residual carbon from ZnAl2O4 powders prepared by combustion synthesis *Journal of the European Ceramic Society* **32** 1605-1611.

Rainfall forecast in the Upper Mahaweli basin in Sri Lanka using RegCM model

K M Muhammadh[1], M M M Mafas and S B Weerakoon

Faculty of Engineering, University of Peradeniya, Peradeniya, Sri Lanka.
[1]E-mail: kmmuhammadh@gmail.com

Abstract. The Upper Mahaweli basin is the upper most sub basin of 788 km2 in size above Polgolla barrage in the Mahaweli River, the longest river in Sri Lanka which starts from the central hills of the island and drains to the sea at the North-east coast. Rainfall forecast in the Upper Mahaweli basin is important for issuing flood warning in the river downstream of the reservoirs, landslide warning in the settlements in hilly areas. Anticipatory water management in the basin including reservoir operations, barrage gate operation for releasing water for irrigation and flood control also require reliable rainfall and runoff prediction in the sub basin. In this study, the Regional Climate Model (RegCM V4.4.5.11) is calibrated for the basin to dynamically downscale reanalysis weather data of Global Climate Model (GCM) to forecast the rainfall in the basin. Observed rainfalls at gauging stations within the basin were used for model calibration and validation. The observed rainfall data was analysed using ARC GIS and the output of RegCM was analysed using GrADS tool. The output of the model and the observed precipitation were obtained on grids of size 0.1 degrees and the accuracy of the predictions were analysed using RMSE and Mean Model Absolute Error percentage (MAME %). The predictions by the calibrated RegCM model for the basin is shown to be satisfactory. The model is a useful tool for rainfall forecast in the Upper Mahaweli River basin.

1. Introduction

Extreme weather events, especially heavy rainfall adversely affect the people, services and properties to hinder the societal development. Climate projections show that changing extreme weather patterns are very likely and will have significant consequences for the society and the economy [1]. Prediction of weather phenomenon with higher reliability under changing climate has become a major focus in the world.

The Upper Mahaweli basin is the upper most sub basin of 788 km^2 in size above Polgolla barrage in the Mahaweli River, the longest river in Sri Lanka which starts from the central hills of the island and drains to the sea at the North-east coast. There are two large reservoirs to regulate water and to generate hydropower in this sub basin, and Polgolla barrage in the Central Province diverts part of Mahaweli river water to the North Central Province of Sri Lanka while harnessing hydropower [2]. Rainfall forecast in the sub basin is important for issuing flood warning in the river downstream of reservoirs, landslide warning in the settlements in hilly areas. Anticipatory water management in the basin including reservoir operations, barrage gate operation for releasing water for irrigation and flood control also require reliable rainfall and runoff prediction in the sub basin. Figure 1 shows the location of the basin.

The Upper Mahaweli basin is characterized by predominantly mountainous topography elevations ranging from 400m to 2000m MSL (Figure 2). The basin receives an average annual rainfall of about 2500mm, western slopes in the basin receives higher rainfall up to 6000mm in some years with high intensity causing floods [3 & 4]. Therefore rainfall forecast is important to understand its accompanied impacts and necessary adaptation to minimize its adverse effects and for planning adaptation strategies.

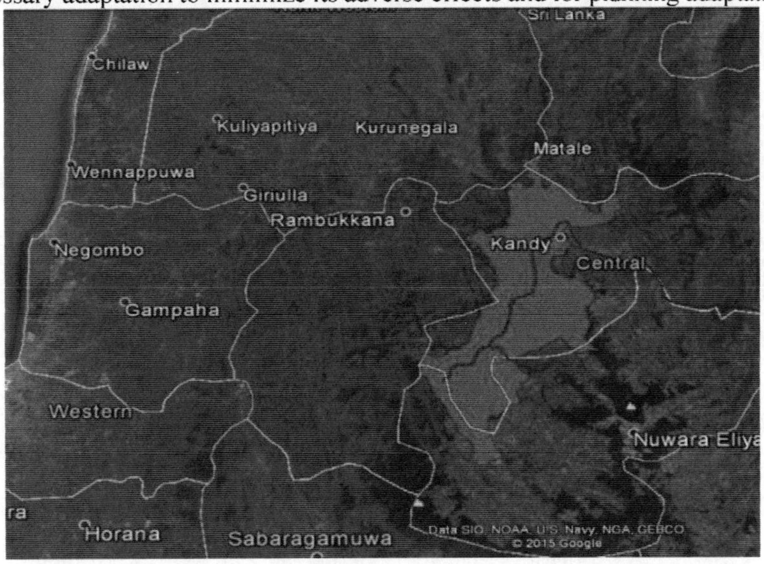

Figure 1. Location of Upper Mahaweli River basin in Sri Lanka

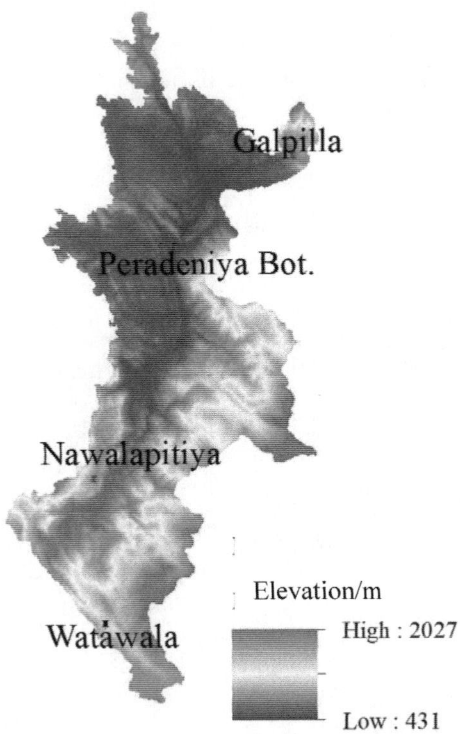

Figure 2. Topography of the basin showing selected rainfall gauging station

Numerical weather prediction (NWP) is based on the solution of atmospheric state variables including temperature, wind, humidity and pressure derived by the numerical solution of mathematical models formed by physical equations describing motion, thermodynamics, continuity, hydrostatic equation together with closure modes [5]. Global Circulation Models (GCM's) and Limited Area Models are two groups of models used for NWP. Global weather simulation, which is computationally expensive, provides predictions at a coarse scale (250 km by 250 km grid) in most cases [6]. GCMs fail to capture sufficient basin properties to provide an accurate description of weather at local or basin scale. Thus, the process of dynamic downscaling of GCM predictions is used to produce the basin-scale weather predictions of the Upper Mahaweli basin. Dynamical downscaling is a computationally intensive technique which makes use of the lateral boundary conditions from GCMs combined with regional-scale forcings such as land-sea contrast, vegetation cover, etc., to produce regional climate models (RCMs) [6]. There are number of dynamic downscaling models and each model has its advantage/disadvantage over the counterparts. WRF [7], RegCM [8], CCSM [9], etc., are among the popularly used RCMs.

RegCM developed and maintained by the Abdus Salam International center for Theoretical Physics (ICTP) is used in this study due to its successful and increase use in Asian countries, and that it has not been used yet in Sri Lanka. The RegCM has been the first limited area model developed for long term regional climate simulation, and it has been applied by a large community for a wide range of regional climate studies, from process studies to paleo-climate and future climate projections [10].

1.1. Model description

RegCM simulation system consists of four modules, including the terrain, the initial / lateral boundary condition (ICBC), the main program (Main) and post-treatment (Post-proc) module (Figure 3). The entire technological process can be divided into pre-processing, simulation and post-processing. The pre-processing stage includes topographic parameters, parameters of the study area as well as the mode setting of the initial and boundary conditions for the setting. Terrain and ICBC are two parts in the pre-processing stage. The main module (Main) is the master control program in the model; post-processing module is to convert the output results for the needs of the average type and data format [11].

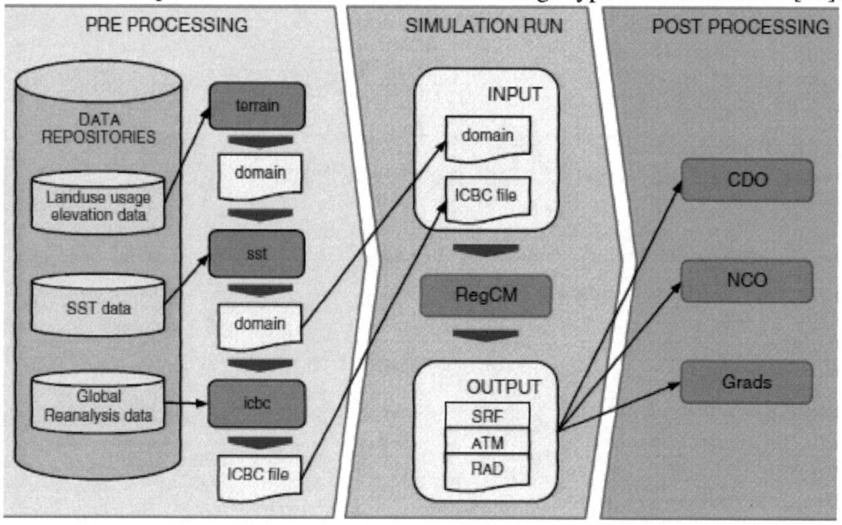

Figure 3. Different components of RegCM modelling system [12]

The RegCM has number of physics options which can be changed to fine-tune the model in order to calibrate the model to the given region. These models include the Biosphere-Atmosphere Transfer scheme(BATS) for surface process representation, The Radiative Transfer scheme of the NCAR Community climate Model, A medium resolution local Planetary boundary layer scheme, the Kuo-type cumulus convection scheme and the explicit Moisture scheme [13]

2. RegCM model application

2.1. Setting up of the model

RegCM model was set up with the centre longitude 80.952, the centre latitude 7.519, resolution of 10 km, which is the lowest grid size possible (RegCM Version 4.4.5.11), and the grids, to cover the Upper Mahaweli basin.

Observed daily rainfall data at four rainfall gauging stations in the basin were obtained from the Department of Meteorology of Sri Lanka to select extreme events for model calibration and validation

GIS based Thiessen polygon method was used to generate average rainfall over the study area and two highest rainfall events were selected as extreme events for model calibration and validation. Table 1 illustrates the details of selected rainfall events. The reanalysed datasets EIN15 at $1.5^0 \times 1.5^0$ resolution ($\approx$ 167 km × 167 km) corresponding to the selected events in 2002 and 2004 were downloaded from the ICTP website: http://clima-dods.ictp.it/data/Data/RegCM_Data/ and used as the initial values and lateral boundary values in the simulation

Table 1. Selected extreme events

Process	Date	Daily Rainfall/(mm)
Calibration	25/03/2004	60
Validation	10/05/2002	74

2.2. Post processing

GrADS tool is used to view the daily rainfall obtained from RegCM model and the observed rainfall. The observed was spatially distributed using GIS Inverse Distance Weighing tool (IDW) in the same grid size of 0.1 degree of the RegCM results using GIS Point to Raster tool. For the purpose of comparison grids in the basin were numbered as shown in Figure 4.

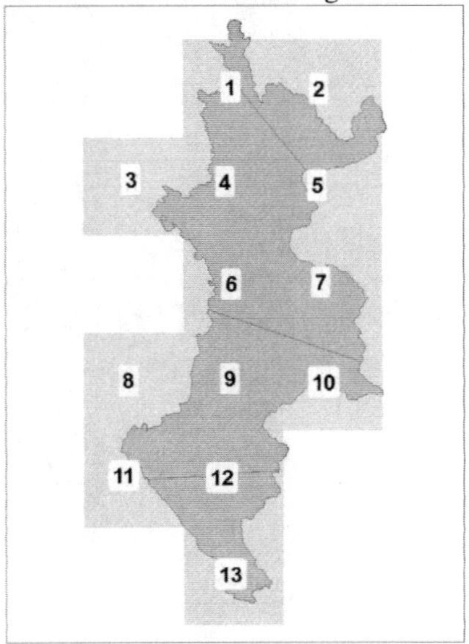

Figure 4. Grid numbering

The Mean Absolute Model Error Percentages (MAME %) and Root Mean Square Error (RMSE) were used to compare the simulated (SR) and observed (OR) rainfall values.

$$\text{MAME \%} = (SR - OR) / OR \; x \; 100 \; \% \qquad (1)$$

IOP Conf. Series: Journal of Physics: Conf. Series **822** (2017) 012075 doi:10.1088/1742-6596/822/1/012075

3. Model calibration and validation

3.1. Model calibration

The RegCM Model was calibrated to forecast the rainfall for the selected rainfall event on 25 March 2004 over the Upper Mahaweli basin by selecting appropriate physics options. Microphysics, Lateral Boundary conditions scheme, Semi-Lagrangian advection scheme for tracers and humidity, Boundary layer scheme, Cumulus convection scheme, Grell Scheme Cumulus closure scheme, Moisture scheme, Ocean Flux scheme were the physics options considered under model calibration.

One physics scheme was changed while keeping other physics schemes at default options. Once the best option was selected, it was kept unchanged for the remaining simulations. This procedure was followed in selecting all physics schemes.

The appropriate physics options that were selected for the Upper Mahaweli basin during calibration are shown in Table 2. Figure 5 shows the RegCM output and the Figure 6 shows the observed rainfall data on same grids for the rainfall event of 25/03/2004.

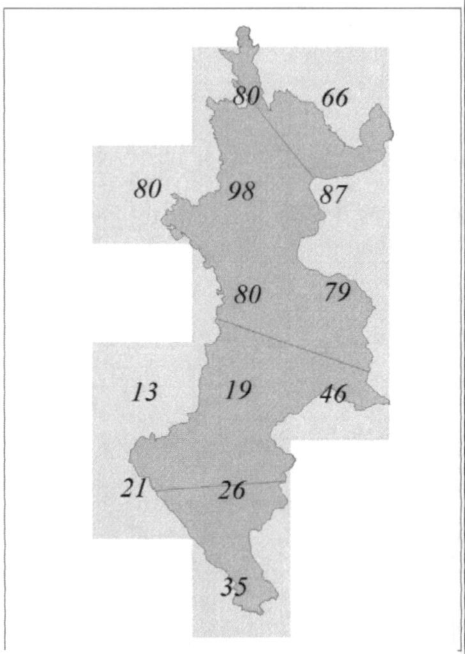

Figure 5. Gridded output of observed rainfall in millimetres on 25/03/2004

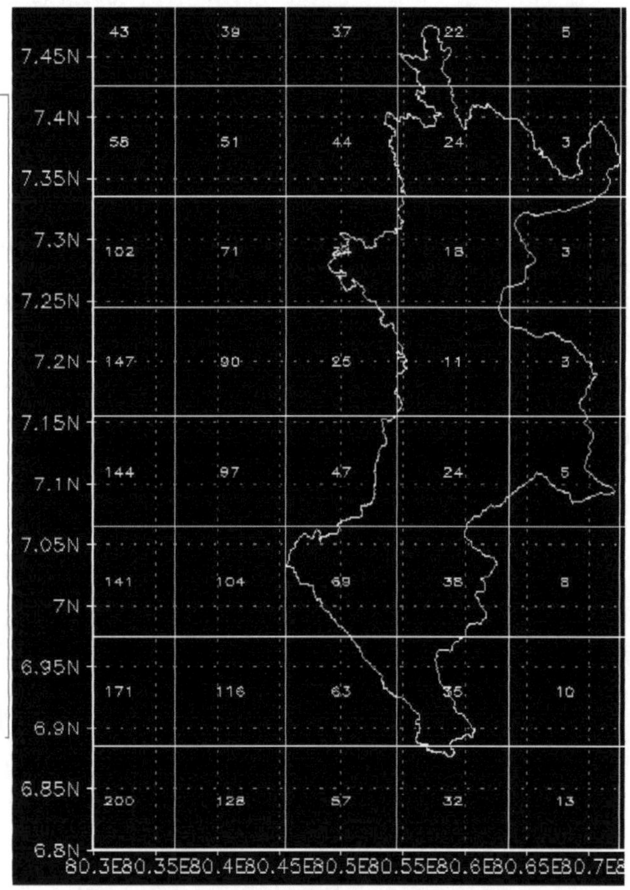

Figure 6. RegCM output of precipitation in millimetres on 25/03/2004

Table 2. Selected physics options based on 25/03/2004 rainfall event

Physics Scheme	Selected options
Boundary layer scheme	Holtslag PBL
Cumulus convection scheme	Emanuel
Moisture scheme	Explicit moisture

Table 3 provides the area percentage corresponding to the MAME % and the cumulative area percentage for the rainfall event of 25/03/2004. It is observed that the MAME % is less than 50%, for about 40% of the basin area for the selected event for calibration.

Table 3. Area percentage corresponding to each MAME % of 25/03/2004

MAME %	Area %	Cumulative area %
0-10	8	8
10-20	8	16
20-30	15	31
30-40	0	31
40-50	8	39
50-60	0	39
60-70	8	47
70-80	0	47
80-90	23	70
90-100	30	100

3.2. Model validation

The calibrated model was applied to simulate the extreme event of 10/05/2002, for model validation. The MAME % and the area percentage respect to each MAME % were calculated to have an assurance about the result for validation. Figure 7 shows the observed rainfall and figure 8 shows the RegCM model predictions for this event

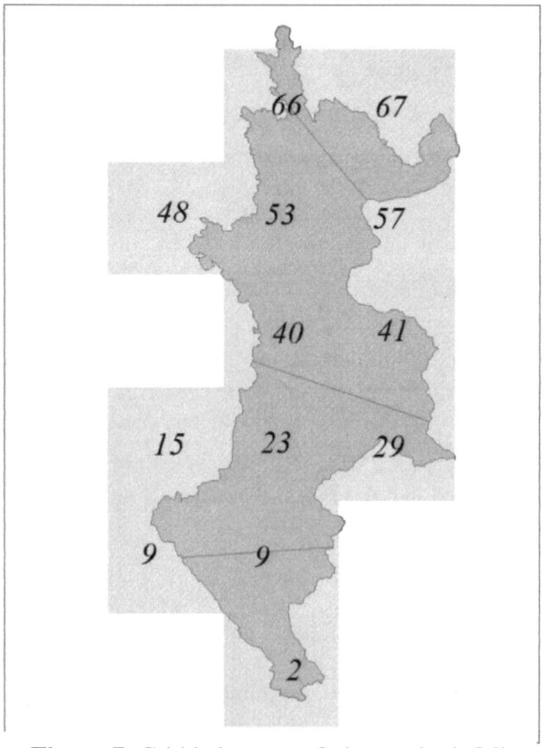

Figure 7. Gridded output of observed rainfall in millimetres on 10/05/2002

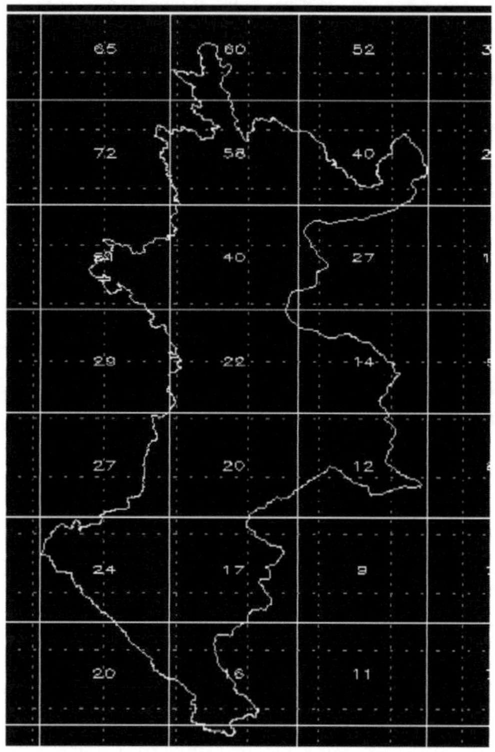

Figure 8. RegCM prediction in millimetres for 10/05/2002 event

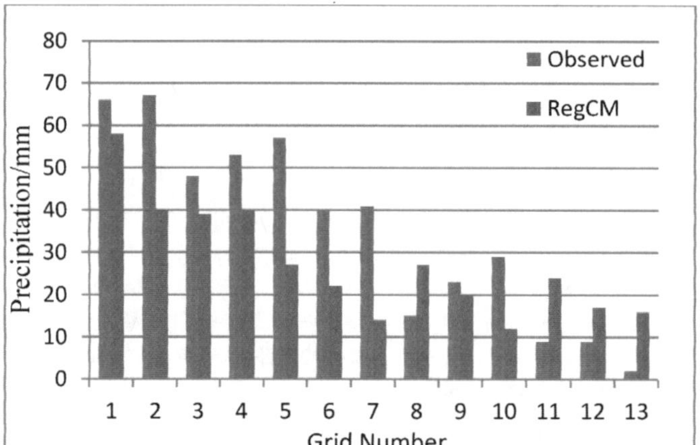

Figure 9. Comparison of RegCM predicted precipitation with the observed precipitation on 10/05/2002

Table 4. Area percentage corresponding to each MAME % of 10/05/2002

MAME %	Area %	Cumulative area %
0-10	0	0
10-20	23	23
20-30	8	31
30-40	8	39
40-50	23	62
50-60	8	70
60-70	0	70
70-80	8	78
80-90	8	86
90-100	14	100

According to Table 4, the MAME is less than 50% over more than 60% of the basin area for the selected extreme event. Further the RMSE value for this date was 17mm, which is an acceptable error. Figure 10 shows the error percentage of the model of the validation results. Therefore the selected physics combination can be applied for the Upper Mahaweli river basin to provide reasonably accurate weather predictions by downscaling the global weather predictions available in coarse grid. However, verifying the model against more events and using more rain gauges would increase the confidence of predictions. Moreover, the resolution of RegCM can be reduced up to 1 km by using the new hydrostatic version of RegCM (version 4.5 of RegCM) which is under development.

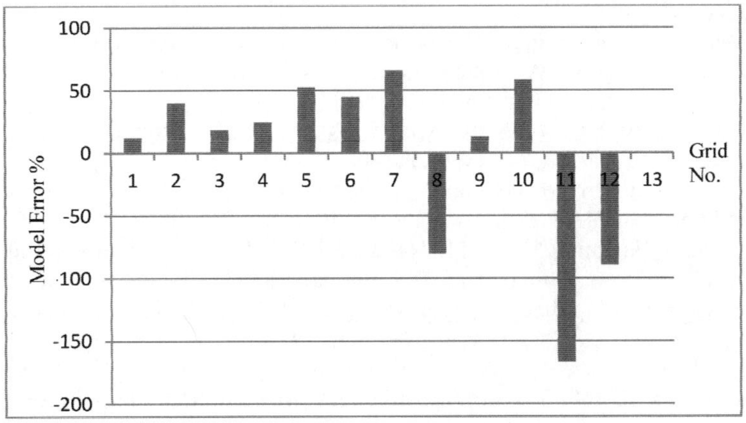

Figure 10. Error percentage of RegCM for 10/05/2002 simulation

4. Conclusion

The RegCM model is proven to be capable of predicting the weather over mountainous Upper Mahaweli River basin. The developed model is a useful tool for extreme rainfall prediction in the basin.

The combination of Relaxation, exponential technique Lateral Boundary conditions scheme, Holtslag PBL (Holtslag, 1990) Boundary layer scheme, Emanuel (1991) Cumulus convection scheme and Explicit moisture (SUBEX; Pal et al 2000) Moisture scheme with other default options were the most suitable physics combinations for the model applied to the Upper Mahaweli basin.

Acknowledgment
Authors are grateful to the RegCM users' forum for sharing the expertise to run RegCM model.

References

[1] Malte Jahn, 2015, Economics of extreme weather events: Terminology and regional impact models. Weather and Climate Extremes (e-journal), Available through: www.sciencedirec.com. [Accessed 20 March 2016]

[2] Mahaweli Hydro Power Complex, Ceylon Electricity Board, 2016, Polgolla Reservoir/Dam. [Online] Available at: http://www.mahawelicomplex.lk/Poldam.html [Accessed 28 September 2016]

[3] Lareef Zubair, 2003. El Nino–southern oscillation influences on the Mahaweli stream flow in Sri Lanka. International Journal of Climatology [e-journal], 23(1). Available through: Wiley Online Library website, http://onlinelibrary.wiley.com/doi/10.1002/joc.865/pdf [Accessed 05 October 2016]

[4] SAARC Disaster Management Centre, 2012, Flood and Risk Management in South Asia. [Pdf] New Delhi: SAARC Disaster Management Centre. Available at: http://www.preventionweb.net/files/31327_31327floodrm.pdf [Accessed 05 October 2016]

[5] Emrullah Sonuc, Baha Sen, Burak Sen, 2012, Verifying regional climate model results with web-based expert-system, Procedia Technology 1(2012) p.24–30. (e-journal), Available through: www.sciencedirec.com. [Accessed 25 March 2016]

[6] Eric M.Laflamme, ErnstLinder, YibinPan, 2015, Statistical downscaling of regional climate model output to achieve projections of precipitation extremes, Weather and Climate Extremes, (e-journal), Available through: www.sciencedirec.com. [Accessed 22 January 2016]

[7] WRF, The Weather Research & Forecasting Model, [online] Available at: http://www.wrf-model.org/index.php

[8] International Centre for Theoretical Physics, Regional Model: REGCM4 - A Regional Climate Model system, [online] Available at: https://www.ictp.it/research/esp/models/regcm4.aspx And at: http://gforge.ictp.it/gf/project/regcm/

[9] UCAR, Community Earth System Model, 2016, CESM Models. [Online] Available at: http://www.cesm.ucar.edu/models/ccsm3.0/

[10] The Abdus Salam International Centre for Theoretical Physics, 2014, Regional Climate Model RegCM User Manual Version 4.4. Italy: ICTP

[11] Hui Liang Duan, Fuxiang Cao, 2011, Numerical Simulation of Regional climate models in a subtropical Region of China, 2011, 3rd International Conference on Environmental Science and Information Application technology (ESIAT 2011), [Online], Available at: www.sciencedirect.com, [Accessed 7 February 2016]

[12] Stefano C., Deepika V., Geol.S, Francesco D.G., Sushil K.D., 2016, Regional Climate Simulations on EU-INDIA Grid Infrastructures: Methodologies and Performance. *Journal of Grid Computing,* [online], Available at: https://www.researchgate.net/ publication/271952252 Regional Climate Simulations on EU-INDIA Grid Infrastructures Methodologies and Performance, [Accessed: 06 October 2016]

[13] Bi Xunqiang, ICTP RegCM and its Regional Climate Simulations, Physics of Weather and Climate Group, Abdus Salam International Centre for Theoretical Physics Trieste, Italy, [Online], Available at: www.ictp.net, [Accessed: 9 February 2016]

IOP Conf. Series: Journal of Physics: Conf. Series **822** (2017) 012076 doi:10.1088/1742-6596/822/1/012076

Streaming flows produced by oscillating interface of magnetic fluid adsorbed on a permanent magnet in alternating magnetic field

S Sudo[1], M Ito[2], Y Ishimoto[3], S Nix[4]

[1] Department of Machine Intelligence and Systems Engineering, Akita Prefectural University, Yurihonjo, Akita, Japan
[2] Graduate School of Systems Science and Technology, Akita Prefectural University, Yurihonjo, Akita, Japan
[3] Department of Machine Intelligence and Systems Engineering, Akita Prefectural University, Yurihonjo, Akita, Japan
[4] Department of Machine Intelligence and Systems Engineering, Akita Prefectural University, Yurihonjo, Akita, Japan

E-mail: sudo@akita-pu.ac.jp

Abstract. This paper describes microstreaming flows generated by oscillating interface of magnetic fluid adsorbed on a circular cylindrical permanent magnet in alternating magnetic field. The interface of magnetic fluid adsorbed on the NdFeB magnet responds to the external alternating magnetic flied as harmonic oscillation. The directions of alternating magnetic field are parallel and antiparallel to the magnetic field of permanent magnet. The oscillation of magnetic fluid interface generates streaming flow around the magnet-magnetic fluid element in water. Microstreaming flows are observed with a high-speed video camera analysis system. The flow pattern generated by magnetic fluid motion depends on the Keulegan–Carpenter number and the Reynolds number.

1. Introduction

The importance of the study on microstreaming flows induced by oscillation of an object is well recognized. Such oscillatory flows have many potential microfluidic applications. Therefore, extensive investigations on the streaming flows have been studied by a number of researchers. For example, Sarpkaya observed the flow around a transversely oscillating cylinder in a fluid otherwise at rest [1]. Dütsch et al performed Time-averaged Laser Doppler Anemometry measurements and time-resolved numerical flow predictions to investigate the laminar flow induced by the harmonic in-line oscillation of a circular cylinder in water at rest [2]. Costalonga et al. reported on an experimental study of streaming flow induced by a vibrating beam in a Hele-Shaw cell of 2 mm span using long exposure flow visualization and particle-image velocimetry measurements [3]. Rigid mechanical direct vibration was used to generate the streaming flows in these studies. On the other hand, Streaming patterns around a oscillating bubble excited at acoustic frequencies were investigated with flow visualization measurements by Tho et al [4]. In this experiment, a thin 12mm diameter lead zirconate titanate (PZT) piezoelectric disk was used to excite the bubbles. The study on microfluidic mixing through electrowetting-induced droplet oscillation was also reported by Mugele et al. They used the wire and the underlying substrate electrode to trigger self-excited oscillations of millimeter-sized sessile droplet [5]. However, the non-contact energy supply method is important in the microfluidic applications.

In this paper, the non-contact energy supply system in order to generate microstreaming flows was proposed. Microstreaming flows were driven by the interface oscillation of magnetic fluid adsorbed on a small circular cylindrical permanent magnet. The magnetic fluid motion was actuated by the external alternating magnetic field. Flow characteristics of water around the magnet-magnetic fluid element were also revealed with a high-speed video camera system.

2. Experimental apparatus and procedures

A block diagram of the experimental apparatus is shown in figure 1. The experimental apparatus is composed of magnet-magnetic fluid element, external magnetic field generation system, and high-speed video camera analysis system. The test magnet-magnetic fluid element is composed of the circular cylindrical permanent magnet and magnetic fluid. The magnetic fluid was adsorbed to the permanent magnet. Letters N and S in figure 1 show the poles of the magnet. This test element was sandwiched between two acrylic plastic parallel plates. Figure 2 shows the test element system. The diameter of the cylindrical magnet is d_p=5mm and the length is ε=5mm. The magnetic flux density of the magnet is B=450mT. The sample magnetic fluid was kerosene-based Ferricolloid HC-50. This magnetic fluid is immiscible with water. The volume of magnetic fluid adsorbed to the magnet is $V_m=2\times10^{-7}\text{m}^3$(200μl). This system was submerged in water as shown in figure 3. The alternating magnetic field was generated by applying alternating current voltage to the Helmholtz coil (Lake Shore MH6) as shown in figure 1. The alternating current signal was supplied from the frequency synthesizer (NF Corporation Multifunction Generator WF1943). The magnet-magnetic fluid element was located on a center part of the Helmholtz coil. The direction of the external magnetic field was parallel and anti-parallel to the direction of permanent magnet. The dynamic behavior of magnetic fluid was analyzed with the high-speed video camera system (Photron FASTCAM-ultima SE). The video camera system is composed a high-speed video camera, a cassette recorder, a video monitor, and a personal computer. The streaming flows generated by the interface oscillation of magnetic fluid were also observed by the high-speed video camera system.

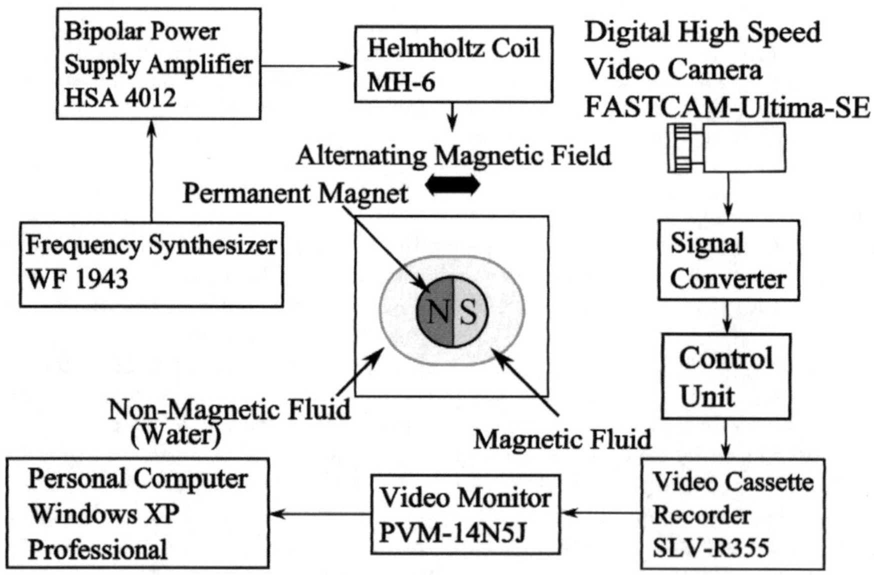

Figure 1. Block diagram of experimental apparatus.

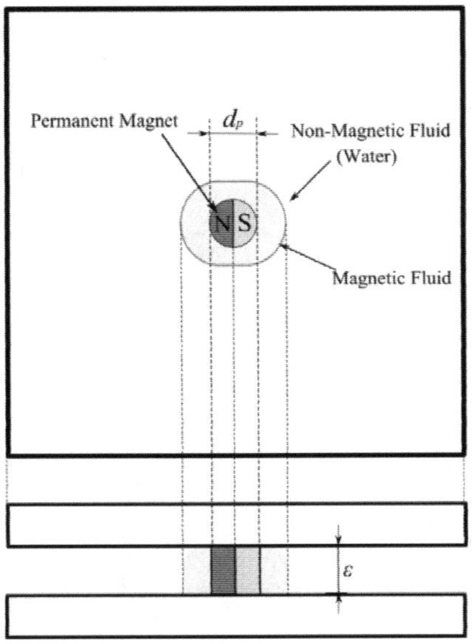

Figure 2. Test element system.

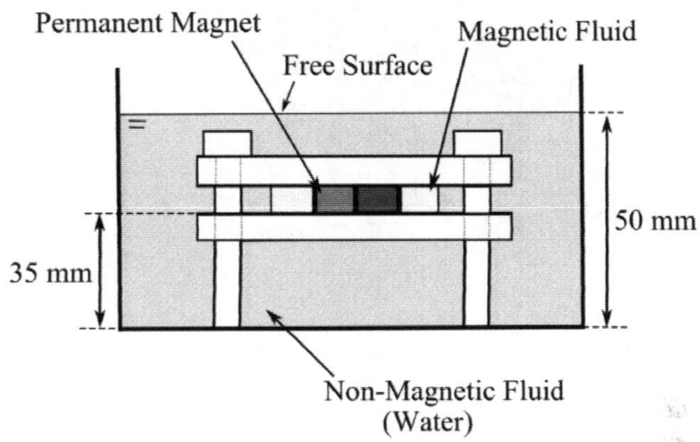

Figure 3. Test system submerged in water.

In this experiment, the applied voltage is given in the sinusoidal form as follows:

$$E_c = \frac{E_0}{2}\sin(2\pi f_0 t) \tag{1}$$

where $E_0/2$ is the amplitude of alternating voltage, f_0 is the frequency, and t is the time. In the experiment, E_0 and f_0 were varied. Most experiments were conducted at the condition of the total voltage amplitude E_0=90V.

In the visualization experiment, milk was spread on the lower flat plate in the test system by a pipette. The experiments were performed under the condition of room temperature.

3. Experimental results and discussion

In this experiment, surface responses of magnetic fluid adsorbed on the permanent magnet were investigated with the high-speed video camera system. Then water flows produced by the magnetic fluid motions were visualized by using the thin milk film spread on the flat plate. Figure 4 shows an example of the visualization photograph of flow produced around the magnet-magnetic fluid element system. This flow was generated by the interface vibration of magnetic fluid subjected to external alternating magnetic field. A coordinate system was set to examine the interface vibration of magnetic fluid and water flows as shown in figure 5.

3.1. Interface oscillation of magnetic fluid

The responses of the magnet-magnetic fluid system in the water subjected to external alternating magnetic field were studied in the first stage. When alternating magnetic field was applied to the magnetic fluid adsorbed on the permanent magnet, the oscillation with small amplitude was generated on the magnetic fluid interface. Figure 6 shows a series of photographs of the interface oscillation of magnetic fluid adsorbed on the magnet. In general, the surface of magnetic fluid adsorbed on the permanent magnet responds to the alternating magnetic field in the elongation and contraction of its length [6]. It can be seen from figure 6 that the horizontal direction length of the magnetic fluid changes with time. Figure 7 shows an example of time variation of the interface displacement at the

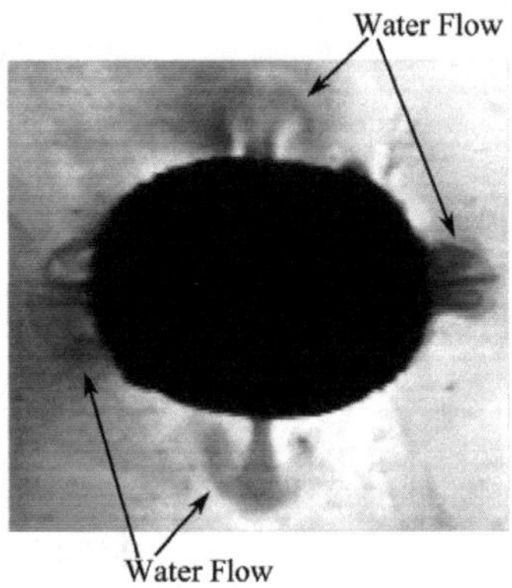

Figure 4. Water flow generated by oscillation.

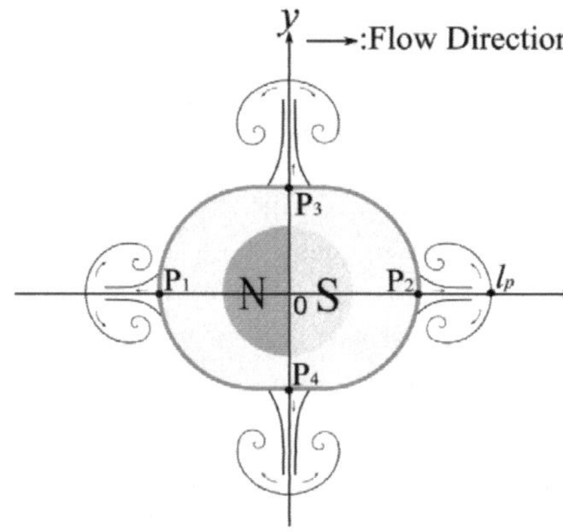

Figure 5. Coordinate system in test system.

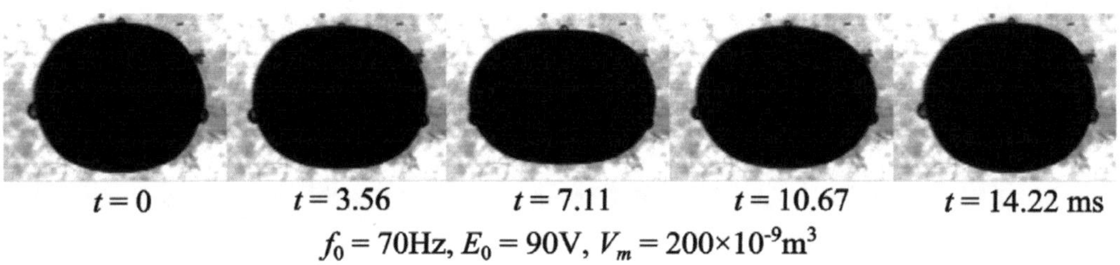

Figure 6. Photographs showing magnetic fluid oscillation produced by alternating magnetic field.

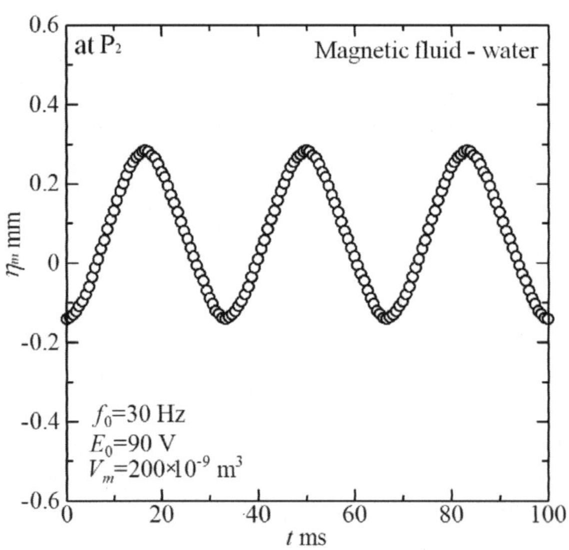

Figure 7. Interface oscillation of magnetic fluid at the horizontal peak position.

IOP Conf. Series: Journal of Physics: Conf. Series **822** (2017) 012076 doi:10.1088/1742-6596/822/1/012076

horizontal peak. In figure 7, η_m is the displacement of magnetic fluid interface in the coordinate system defined as η_m =0 at the interface in the condition E_0=0. It can be seen from figure 7 that the frequency of interface oscillation was precisely same that of the external alternating magnetic field. The amplitude of the interface oscillation depended on the frequency of the external alternating magnetic field at the constant applied voltage. Figure 8 shows the relation between total amplitude $|\eta_m|$ and the excitation frequency f_0. The total amplitude $|\eta_m|$ is defined by the difference between crests and troughs in η_m. Frequency characteristics of total amplitude of magnetic fluid oscillation show peak region. The peak frequency is related to the natural frequency of magnetic fluid oscillation. In figure 8, the peak frequency is f_0=75Hz. In this experimental system, the dominant response of magnetic fluid was elongation and contraction in horizontal length as shown in figure 6. This is the first mode of magnet-magnetic fluid element response.

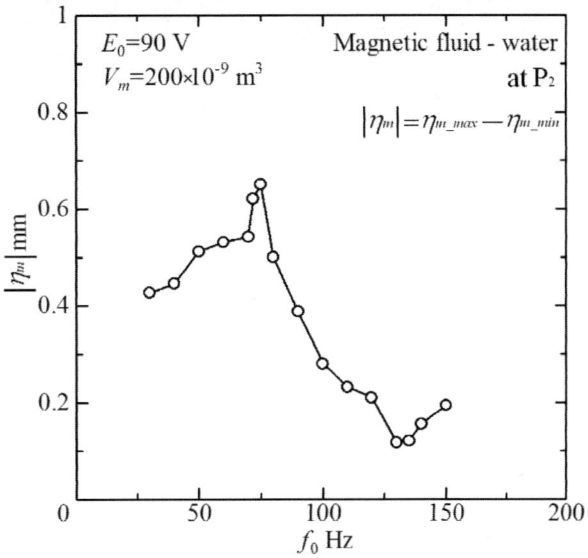

Figure 8. Frequency characteristics of interface oscillation of magnetic fluid adsorbed on permanent magnet subjected to alternating magnetic field.

3.2. Water flows produced around the magnet-magnetic fluid element

When the magnetic fluid adsorbed to the permanent magnet was subjected to alternating magnetic field, water flows were produced around the magnet-magnetic fluid element as shown in figure 4. Water flows around the magnet-magnetic fluid element subjected to external alternating magnetic field were studied in this paragraph. Figure 9 shows a series of photographs of process of water flow generation. Four small flows are visualized in figure 9. Those streaming flow outside as shown in figure 5. Figure 10 shows a position change of the flow tip with time. The flow tip gradually goes away from the magnet-magnetic fluid element with time. The velocity change of the flow tip was calculated by the figure 9. The velocity variation of the flow tip is shown in figure 11. In the beginning of interface oscillation, the speed shows a relatively higher value. After speed reached the peak value, the speed shows gradually decrease. Finally, the speed shows a constant value. The constant speed of the flow is approximately v_p=1m/s in figure 11.

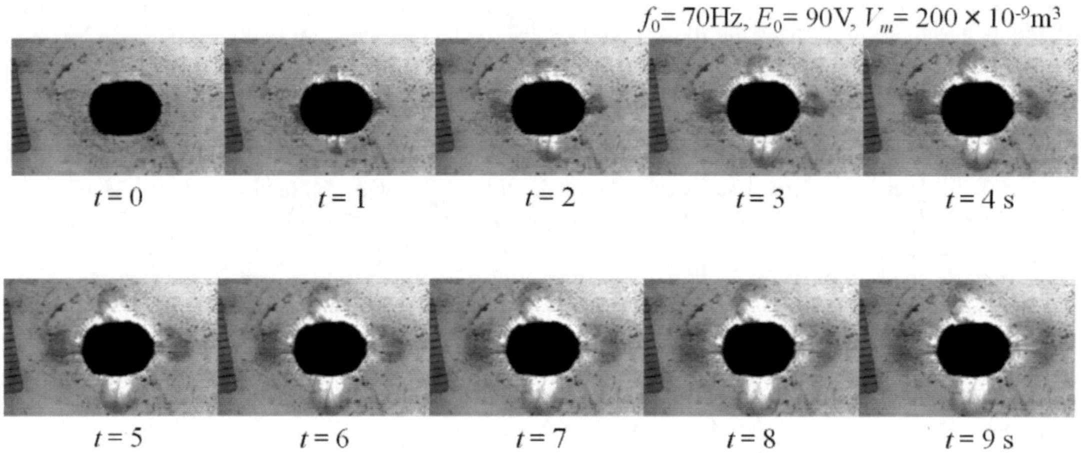

Figure 9. A series of photographs indicating the progress of four flows produced around magnet-magnetic fluid element.

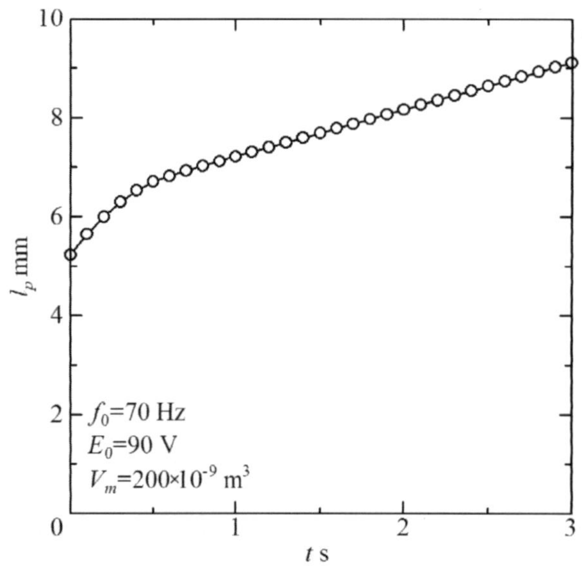

Figure 10. Position change of flow tip after the excitation by alternating magnetic field.

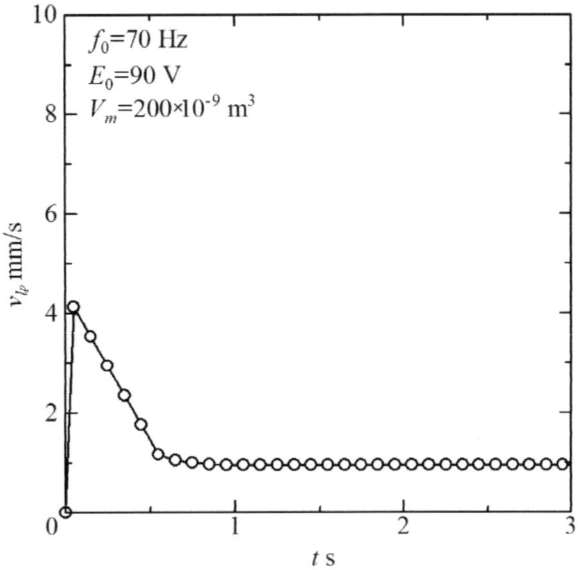

Figure 11. Velocity variation of flow tip after the excitation by alternating magnetic field.

3.3. Effect of frequency in alternating magnetic field

In this paragraph, the influence of frequency in alternating magnetic field on the water flow was examined. Figure 12 shows the visualization photographs of water flow produced by the interface oscillation of magnetic fluid at different frequencies. In figure 12, t is the elapsed time during interface oscillation of magnetic fluid. These photographs show the flow state produced at the interface position P_4 and the elapsed time t=12s. It can be seen from figure 12 that the position of flow tip depends on the frequency f_0. Figure 13 shows the relation between the frequency of alternating magnetic field and the position of flow tip at the constant E_0. In figure 13, the position of flow tip shows a peak at f_0=75Hz.

In figure 12, KC is the Keulegan–Carpenter number, β is the Stokes number, and Re is the Reynolds number. These are key dimensionless parameters. The Keulegan–Carpenter number is defined by

t = 12 s, E_0= 90 V, V_m= 200 × 10^{-9}m^3

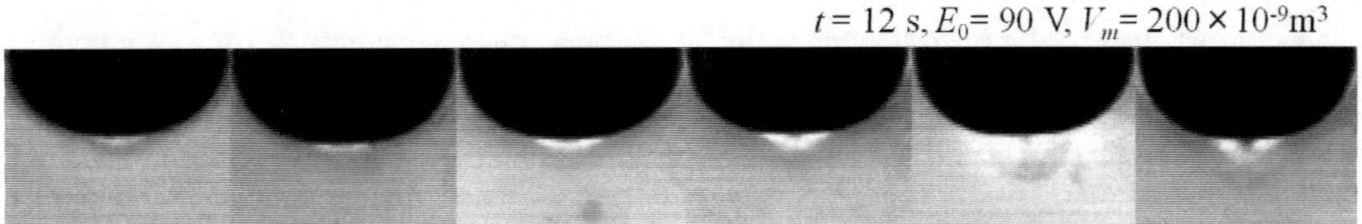

f_0=35	45	55	65	75	85 Hz
KC=0.22	0.24	0.27	0.28	0.34	0.22
β=871.5	1121	1370	1619	1868	2117
Re =189	271	363	450	637	464

Figure 12. Flow patterns produced by interface oscillation at different frequency.

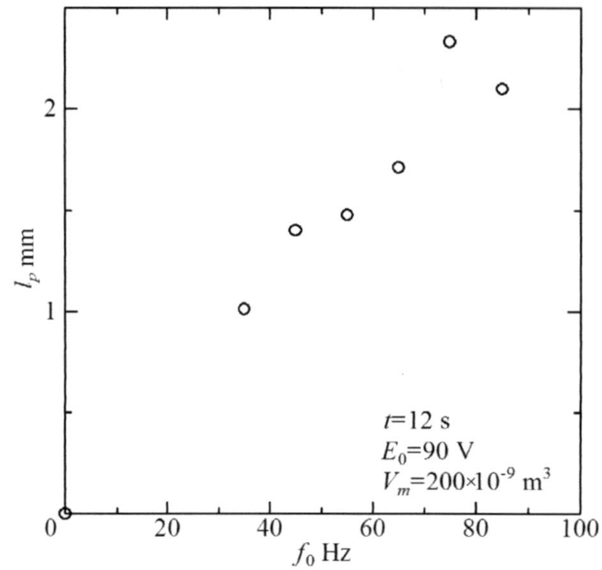

Figure 13. Relation between the tip position of flow produced by the interface oscillation and the frequency of alternating magnetic field.

$$KC = 2\pi A / d_p \tag{2}$$

where A is the amplitude of interface oscillation. The Stokes number β is defined by

$$\beta = f_0 d_p^{\,2} / \nu \tag{3}$$

where ν is the kinematic fluid viscosity. The Reynolds number Re is defined by

$$Re = KC\beta \tag{4}$$

In figure 13, the position of water flow tip shows the peak in the domain that all these parameters have higher value. The flow field produced by the interface oscillation of magnetic fluid depends on three dimensionless parameters.

4. Conclusions

The non-contact energy supply system in order to generate microstreaming flows was proposed. Characteristics of microstreaming flows produced by oscillation of magnetic fluid adsorbed on a permanent magnet subject to the alternating magnetic field were studied with the high-speed video camera system. The results obtained are summarized as follows.

(1) Micro water flows toward the four directions from the element are produced by the interface oscillation of magnetic fluid adsorbed on a permanent magnet subjected to alternating magnetic field.
(2) The amplitude in interface oscillation of magnetic fluid depends on the frequency of alternating magnetic field. Frequency characteristics of the magnet-magnetic fluid element indicates the presence of a natural frequency in the magnetic fluid oscillation.
(3) The flow field produced by the magnetic fluid interface oscillation depends on the Keulegan–Carpenter number, the Stokes number, and the Reynolds number.

Acknowledgment
This work was partly supported by JSPS KAKENHI Grant Number 15K13875.

References

[1] Sarpkaya T 2002 *J. Fluid Mech.* **457** 157
[2] Dütsch H, Durst F, Becker S and Lienhart H 1998 *J. Fluid Mech.* **360** 249
[3] Costalonga M, Brunet P and Peerhossain H 2015 *Phys. Fluid* **27** 013101
[4] Tho P, Manasseh R and Ooi A 2007 *J. Fluid Mech.* **576** 191
[5] Mugele F, Baret J –C, Steinhauser D 2006 *Appl. Phys. Lett.* **88** 204106
[6] Sudo S, Asano D, Takana H and Nishiyama H 2011 *J. Magn. Magn. Mater.* **323** 1314

IOP Conf. Series: Journal of Physics: Conf. Series **822** (2017) 012077 doi:10.1088/1742-6596/822/1/012077

Switching and transport mechanism in the branching tube network due to attached eddies

Yurika Nozaki, Kenta Ohhashi and Osamu Sano

Department of Applied Physics, Tokyo University of Agriculture and Technology, Koganei, Tokyo 184-8588

E-mail: sano@cc.tuat.ac.jp

Abstract. An experimental study was performed on the quasi-two-dimensional symmetric branching tube under oscillatory flow. The velocity field and transport of materials in the Reynolds number $Re = 1500 \sim 3000$ and the Womersley number $\alpha = 17 \sim 23$ were visualized by means of a dye or small spheres as well as measured by using Particle Image Velocimetry. The repeated growth and disappearance of the attached eddies near the junction of the branched channel are recognized, which play a role of the throttle valves to control the transport of newly taken gasses or suspended particles into, and older ones out of, the human lung in spite of zero net volume flux.

1. Introduction

The major function of the respiratory system of human is the gas exchange between the external environment and an organism's circulatory system. This exchange facilitates oxygenation of the blood as well as removal of carbon dioxide and other gaseous metabolic wastes from the circulation [1]-[6]. A no less important application of the respiratory system is to use it as an inhaler, in which nano-particles or liquid droplets that contain medicine are taken into through the lung's air sac, and are transported to the diseased part by means of the blood flow (e.g., [7]). These exchanges occur at the alveoli, which are the tiny sacs to provide very large exchange surface, whereas larger volume of flow is better achieved through larger tubes because it is operated by the pressure difference, which necessitates the branching tube network that becomes narrower, shorter, and more numerous as it penetrates deeper into the lung. According to the anatomical data, the human airway tree consists of the conducting part that starts from the trachea (zeroth generation) to the terminal bronchioles of 14th generation via repeated bifurcation of channels, and the acinar part which consists of the respiratory bronchioles, alveolar ducts and alveolar sacs until the 23rd generation. The length and width L_i and d_i of the i-th generation of the conducting airways follow the geometric ratios: $L_i/d_i \approx 3.5, d_{i+1}/d_i \approx 0.8$, whereas the angle of bifurcation of the branched channel ϕ_i is almost constant ($\approx 70°$) [1]. In normal situation, one breathing is not sufficient to deliver the fresh air to the terminal parts of the lung, which necessitates the transport mechanism that is operated by the repeated application of pressure in spite of the zero net flux of air. Taking into account the fact that oscillatory flow in a straight channel cannot transport fluid or materials after one cycle of oscillation (in the absence of diffusion), and that the tapered or curved tubes do not drastically improve the latter [3], the role of branching conducting airways seems crucial. In general, oscillatory channel flows depend on the Reynolds number $Re \equiv ud/\nu$ and the Womersley number $\alpha \equiv (d/2)\sqrt{2\pi f/\nu}$, where d is the width of the channel (of the zeroth generation), u is the typical velocity there averaged over half period of oscillation, f is the frequency of oscillation, and ν is the kinematic viscosity of the fluid. For the normal respiration $Re \approx 1000$ and $\alpha \approx 3$. In a medical treatment, the "high-frequency respiration" with $Re \approx 2000$ and $\alpha \approx 20$ is conventionally adopted to avoid damaging the diseased part of the lung. However, it is not clear whether or not these conditions are optimum. In order to clarify the latter, we should know the mechanism of the transport of fluid and materials under given conditions. In this paper we shall provide both flow visualization and velocity field measurement in wider range of Re and α, and elucidate the switching and transport mechanism in the branching tubes, which will give deeper insight into more general application of the oscillatory viscous flow to the material transport or ventilation.

2. Experiment

2.1. Experimental apparatus

Figure 1(a) is the block diagram of our experimental apparatus. A branched channel with a branching angle 70° was created up to the second generation between two parallel plates with separation distance 5 mm. The geometric ratios of the lengths and widths between the neighboring ones were adjusted to those in the anatomical data, but the cross section of our channel was chosen rectangular with its aspect ratio about 1/6. This nearly two-dimensional (2-D) setup was used in order to measure the velocity field accurately, with the anticipation that the main contribution comes from the bifurcating plane of the branching tubes of circular cross section (3-D), and that the essential control mechanism will be common to both 2-D and 3-D branching tubes. We used water as a working fluid throughout our experiment. A piston, placed in the zeroth generation channel (first branch), was connected to the motor through a cam, which pushed the fluid sinusoidally, i.e., $s_p = A\sin(2\pi ft)$, where A is the amplitude of oscillation, and s_p is the position of the piston along the channel. The whole system was placed horizontally and was submerged to the water with the outlets of the second generation channels open. Thus the present experimental apparatus is mainly intended to elucidate the structural change of the flow field and the mass transport relation at the *first junction*, which is expected to be applicable to conducting airways up to another few generations. Typical cases reported in this paper are (I) $Re = 1500$ and $\alpha = 17$, (II) $Re = 2100$ and $\alpha = 20$, and (III) $Re = 3000$ and $\alpha = 24$. We have chosen these combinations of Re and α in order to compare the flow fields with the previous works [8][9] as well as to satisfy the fluid mechanical similarity conditions. The material transport was observed by using sodium fluorescein or fine glass beads of diameter 0.1 mm. We also measured the velocity field by means of the particle image velocimeter (PIV). For the latter purpose, nylon beads of a diameter 0.16 mm with a specific density 1.03 were used.

2.2. Material transport from the first branch

We show an example of a sequence of material transport for the case (II) $Re = 2100$ and $\alpha = 20$ in Fig.1(b). These pictures were taken at the same phase of oscillation (i.e., intake phase), $t = 0.5T, 1.5T, 2.5T, ...$ and so on, where $T \equiv 1/f$ is the period of oscillation. Apparently the material (sodium fluorescein) looks transported downward on the average, although the boundary of the dye-containing fluid is of complicated shape. Note that the molecular diffusion of the sodium fluorescein was found to be negligible, which was ascertained by the preliminary experiment of oscillatory flow using a straight channel. We also note that the fluid and materials moved only forward and backward oscillatory within the first branch for flows with Re lower than a certain critical value $Re_c \approx 1000$, so that the average downward transport of the materials was not recognized.

In order to quantify the long-range material transport, we consider a path along the centerlines that connects the centers of the junctions. We divide the channel into rectangular

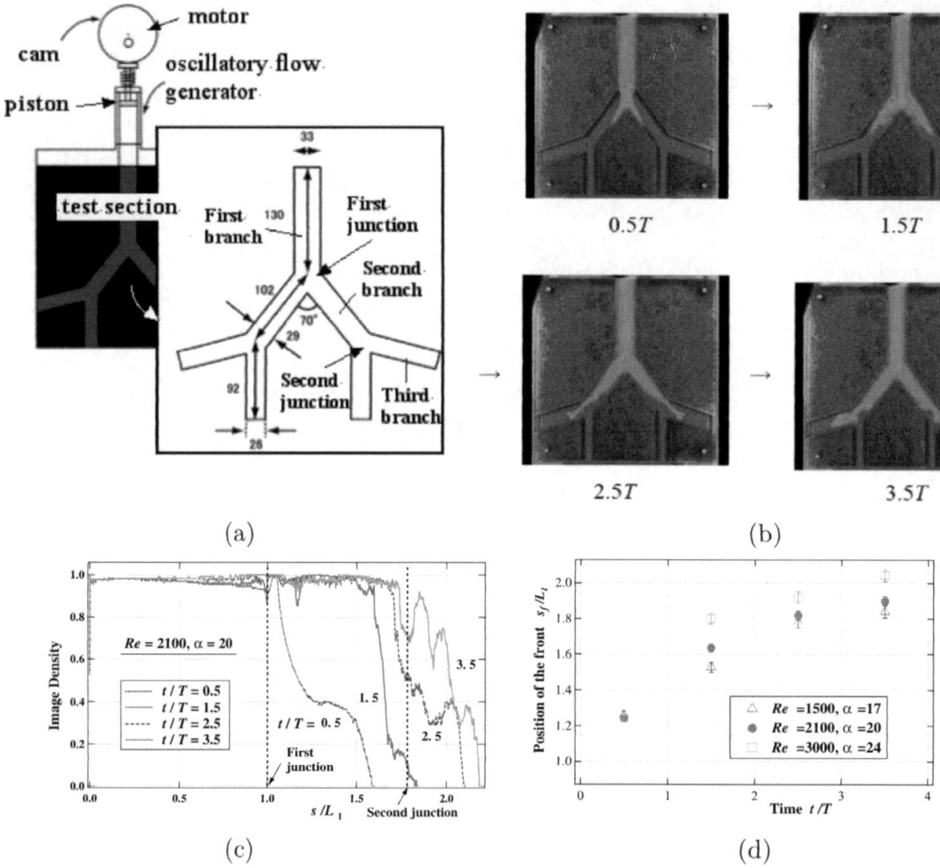

Figure 1. (a) Block diagram of the experimental apparatus. (b) Examples of the visualization of transported material ($Re = 2100, \alpha = 20$). (c) Time sequence of the image density profile of the transported material along the channel, where L_1 is the length of the first branch ($Re = 2100$), and the position of the second branch is $s/L_1 = 1.78$. (d) Positions of the "front" plotted at the maximum intake phase, showing long-term transport of materials: (I) $Re = 1500, \alpha = 17$ (triangles), (II) $Re = 2100, \alpha = 20$ (circles), and (III) $Re = 3000, \alpha = 24$ (squares).

sections, each of which covers each branch, and measure the image density of the sodium fluorescein along these strips. Figure 1(c) is an example of the above analysis for the case (II), where the abscissa is the position s along the centerline of the channel with its origin at the inlet position, i.e., the ranges $0 \leq s \leq 130 (\equiv L_1)$ mm, $130 \leq s \leq 230$ mm, and $s \geq 230$ mm cover the first, second, and third branches. The ordinate is the sum of the image density of the sodium fluorescein over the cross section at s, which is normalized by the maximum value in each branch. The downward transport of the materials is recognized in spite of considerable variation associated with the structural change of the convecting flow near the channel bifurcation. We further consider the "front" s_f of the transported materials defined by the weighted mean of the density near the tip of the fluid that contains sodium fluorescein. Figure 1(d) shows the long-term movement of the front, which is monitored at the maximum intake phase. Material transport is larger for flows with larger Re and α in the tested three cases.

2.3. Velocity field

We show an example of the velocity field ($Re = 2100, \alpha = 20$) measured by PIV in Fig.2(a). Figure 2(b) shows the time variation of the velocity component parallel to the wall near the outlet of the first branch (depicted by the line c-d). During the time $t/T \approx 0.5 \sim 0.6$, the flow in the middle part of the channel is still in the intake phase because of the inertia, whereas the opposite flow grows in the regions near the outer-side of the wall, so that the effective width of the downward flow is reduced. After this phase, the upward flow becomes dominant, but its magnitude decreases toward the end of the period of oscillation of the breathing. Figure 2(c) shows the time variation of the velocity components along the wall measured across the section a-b as is illustrated in the inset. The abscissa of Fig.2(c) is the distance y from the outer-side wall of the second branch (note that the y axis is taken in the negative direction to match the illustration). Here the measured cross section is chosen so as to pass through or nearby the center of the attached eddy, because the latter remains almost at the same position during the intake phase. Further observation reveals the time variation of the attached eddies as follows (see also Fig.3): At first, slow but almost unidirectional flow along the channel is recognized, which develops into faster downstream flow in both sides of the second branch, and the separated flow regions starting from the first corners are recognized. Then the separated flow regions grow and the attached eddies are recognized near the outer-side wall of the second branch (Fig.2(a)), although the net flow is still in the downstream direction. The net flow reverses after that so that the attached eddies become obscure. Finally faster upward stream is recognized in the central region of the first branch as well as in the regions near the outer-side wall of the second branch. The Reynolds number dependence of the position of the center of the eddy y_c from the outer-side wall was also checked, which reveals that the eddy grows faster and larger for larger Reynolds number flows tested in the present experiment.

2.4. Particle path

Figures 2(d)-(f) show the typical particle paths starting from the positions described by the black solid circles. Dotted lines are those in the first cycle $t = 0 \sim T$, whereas solid lines with solid markers are those in the second cycle $t = T \sim 2T$. Generally speaking, particles are pushed downward in the inspiration phase, and are carried back in the expiration phase. Particles that are near the outer-side boundary of the second branch are trapped in the "eddy region", which move further downward in the next breathing.

3. Conclusion

The velocity field and the transport mechanism in the oscillatory flow in the branching tube are summarized as follows (see Fig.3): In the intake phase (Fig.3(a)), the fresh materials flow downward almost uni-directionally in the channel. Separated flow regions start from the outer-side corners of the junction and extend along the outer-side walls of the second branch, which prevents the mixing of the newly coming materials with the older ones that remained in the second branch. The separated flow regions grow in thickness, and the eddies attached to the corner of the first junction are formed (Fig.3(b), Fig.2(a)). Fresh materials are firstly carried by either side of the faster stream between the attached eddy and the inner-side wall of the second branch, and are dispersed laterally throughout the second branch after passing around the eddy region. Immediately before the end of intake phase, the materials are further dispersed by convective diffusion in decelerated flows, whereas the attached eddies are weakened and are wiped upward with their upstream ends pinned at the corners (Fig.3(c)). Then they become thicker and spread into the central part of the second branch, so that they prevent the backward flow of the fresh materials. With the start of expiration phase, the attached eddies are further weakened and the separated flow regions are carried upward (Fig.3(d)). The fluid volumes near the outer-side of the branched channel, which mainly contain older materials, are pushed out

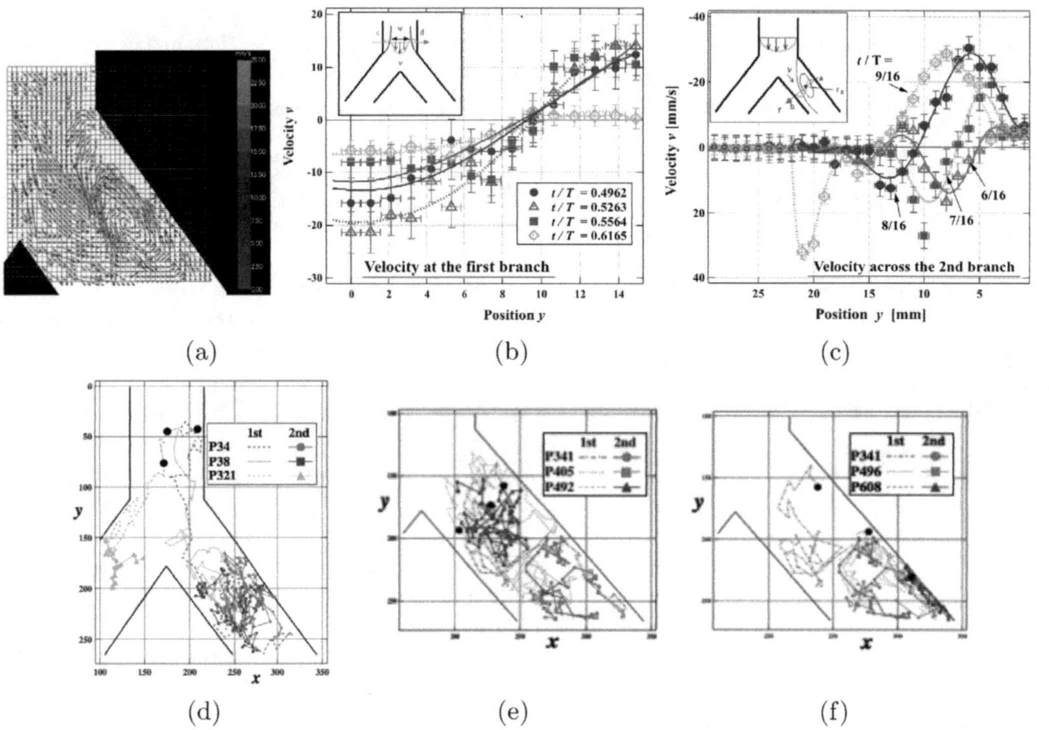

Figure 2. (a) The velocity field at $t/T = 1/2$. (b) Time variation of the velocity profile near the outlet of the first branch at the start of the expiration phase, and (c) the growth of the attached eddy near the second branch in the intake phase. (d)(e)(f) Typical particle trajectories in the first cycle $t = 0 \sim T$ (dotted lines), and those in the second cycle $t = T \sim 2T$ (solid lines with solid markers). Black solid circles describe the initial positions.

along with the fast upward stream in the central region of the first branch (Fig.3(e)). Here, the part of the materials that have been carried far downward remains in the second branch. At the end of one period of oscillatory flow (Fig.3(f)), the older materials are still in the outward motion in the central region of the first branch because of the inertia, but the fresh materials in the next cycle are supplied near the outer-side walls of the first branch because of the conservation of mass. In short, the formation and destruction of the attached eddies play the role of throttle valves, in which newly taken materials are transported far downstream via inner side of the second branch around the eddy regions (Fig.3(b')), whereas the older ones that have been trapped in the separated regions are carried backward to the first junction via outer side of the second branch (Fig.3(e')). These processes eventually transport the newly taken materials downward and the older ones upward, so that the circulation of the materials is achieved in spite of zero net volume of fluid during one period of breathing.

4. Discussion

A lot of studies on the transport of fresh air or suspended particles into the human lung have been made in more realistic geometries (including Pendelluft associated with the asymmetry of the branch) as well as by applying more realistic pressure variation in time, which may better describe the realistic situation. These complexities, however, can mask the essential control

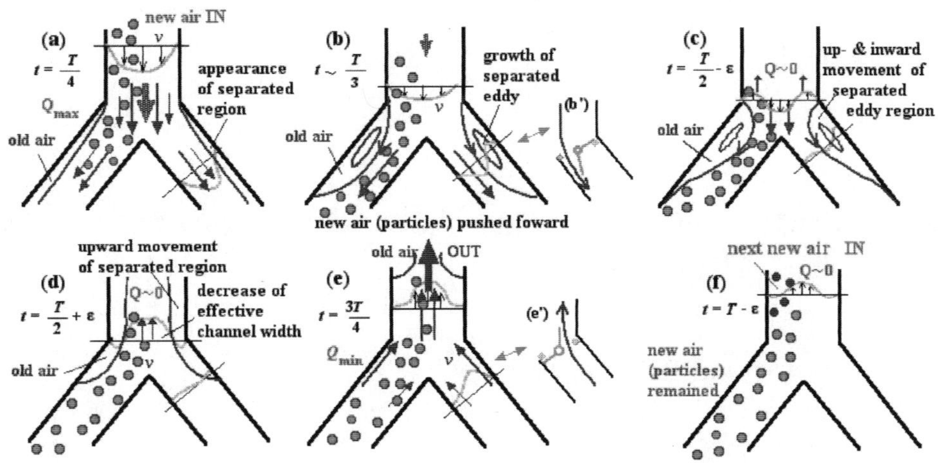

Figure 3. Growth of attached eddies and the control mechanism of material transport.

mechanism. Recently proposed "trap-and-release" mechanism [10], where the formation-and-destruction of the separated regions at the junction leads to bidirectional axial gas-exchange in conductive airways, seems quite promising. Most of their studies, however, are based on flow visualization, and the quantitative understanding of this mechanism is not necessarily clear. On the other hand, we have focused our attention to the simplest possible geometry, and elucidated the role of repeated growth and disappearance of the attached eddies near the junction of the branched channel. The above-mentioned control mechanism is quite simple and fundamental to the transport of materials, and will provide wider application of the oscillatory viscous flow to the material transport or ventilation without using solid valves.

References

[1] Weibel E R 1963 *Morphometry of the Human Lung* (New York: Academic Press)
[2] Pedley T J 1977 Pulmonary fluid dynamics *Annu. Rev. Fluid Mech.* **9** 229
[3] Grotberg J B 1994 Pulmonary flow and transport phenomena *Annu. Rev. Fluid Mech.* **26** 529
[4] Sera T and Tanishita K 2002 Anatomical aspects of airway flow and gas transport in the pulmonary system *Med. Imag. Tech.* **20** (No.6), 654 [in Japanese]
[5] Kleinstreuer C and Zhang Z 2010 Airflow and particle transport in the human respiratory system *Annu. Rev. Fluid Mech.* **42** 301
[6] Choi J, Xia G, Tawhai M H, Hoffman E A and Lin C-L 2010 Numerical study of high-frequency oscillatory air flow and convective mixing in a CT-based human airway model *Ann. Biomed. Eng.* **38** 3550
[7] Daniher D I and Zhu J 2008 Dry powder platform for pulmonary drug delivery *Particuology* **6** 225
[8] Haselton F R and Scherer P W 1982 Flow visualization of steady streaming in oscillatory flow through a bifurcating tube *J. Fluid Mech.* **123** 315
[9] Jan D L, Shapiro A H and Kamm R D 1989 Some features of oscillatory flow in model bifurcations *J. Appl. Physiol.* **67** 147
[10] Mochizuki S, Murata A and Togashi Y 2002 Axial mass transport by reciprocating flow in branching tube network (Gas exchange mechanism in conducting airways of human lung) *Jpn. Soc. Mech. Eng.* **B68** 831 [in Japanese]; Mochizuki S 2003 Convective mass transport during ventilation in a model of branched airways *Proc. PSFVIP-4* (3-5 June 2003, Chamonix, France).

IOP Conf. Series: Journal of Physics: Conf. Series **822** (2017) 012078 doi:10.1088/1742-6596/822/1/012078

Numerical simulation of vortex induced vibration of cylinders with flexible splitter plates

M Furquan, T R Sahu and S Mittal

Department of Aerospace Engineering, Indian Institute of Technology Kanpur, Uttar Pradesh, India

E-mail: mfurquan@iitk.ac.in

Abstract. Vortex induced vibration (VIV) of circular and square cylinders with flexible splitter plates is studied at low Reynolds numbers. Finite element based flow and structure solvers, coupled using a partitioned approach, are used for simulating the fluid-structure interaction. Effect of flexibility of an attached flexible plate on its ability to suppress the VIV of a circular cylinder is considered. Flexibility of the plate is found to adversely affect the reduction in amplitude of the vibration of the cylinder. Next, flow past two square cylinders with deformable splitter plates placed side-by-side is considered. Vibration response of the two plates is studied for different values of flexibility. Initially, the plates vibrate out-of-phase with each other, but eventually settle for an in-phase fully developed response. Large amplitude of vibrations in the fully developed response is observed when its dominant frequency is close to the first natural frequency of the plate vibrations.

1. Introduction

Vortex induced vibrations (VIV) of bluff bodies is a problem of great practical as well as academic interest. There has been a great effort in the direction of understanding and controlling this phenomenon. Splitter plates are a popular device used for modifying/controlling vortex shedding from cylinders and the associated vibrations. The effect of variations in geometry, especially length, of rigid splitter plates on the vortex shedding and cylinder oscillations has been investigated by a number of researchers in past [1, 2]. Kwon and Choi [1] found that the splitter plate length must exceed a critical value, depending on the value of the Reynolds number, to be able suppress the vortex shedding. Wu and Zhao [3] considered the effect of a rigid splitter plate hinged to the rear of a circular cylinder supported on springs. The effect of a flexible splitter on an elastically mounted cylinder is, however, much less explored.

The effect of splitter plates on multiple rigid bodies has also not received great attention. Kim and Durbin [4] studied the effect of a single splitter plate placed along the centerline between two circular cylinders in a side-by-side arrangement. They found the position of the plate required to make the wake steady. A number of researchers have studied the effect of flexible splitter plate on flow past a cylinder, for both circular and square geometries [5-7]. The objective of the present paper is to investigate the effect of flexible splitter plates on aeroelastic response of single and multiple cylinders. While the multiple cylinders are on rigid support, the single cylinder is on elastic support. Both circular as well as square geometries are considered. The effect of flexibility of the plate on the fluid-structure interaction is investigated. In particular, we study the phenomena of lock-in and galloping.

Lock-in occurs when observed frequency of vibration is close to the natural frequency of the structure. This phenomenon is particularly well known in case of vortex induced oscillations of isolated cylinders [8, 9]. Galloping is an instability that occurs with change in angle of attack of flow due to vibration of a radially asymmetric object.

2. Computational Methodology

The fluid-structure system is simulated by solving the coupled equations that govern the flow and structure. The Navier-Stokes equations for incompressible fluid are used to model the flow. The structure is modeled via the force balance along with St. Venant-Kichhoff's constitutive relation. The characteristic non-dimensional parameters for the system are the Reynolds number (*Re*), the ratio of the mass of the structure to that of the fluid it displaces (m^*) and the reduced velocity (U_p^*) calculated using the fundamental natural frequency of the plate vibrations. In the case of an elastically mounted cylinder, reduced velocity (U_s^*) calculated using the natural frequency of the spring-mass oscillator is used for characterizing the compliance of the elastic support for a fixed value of m^*.

The method of Deforming-Spatial-Domain/Stabilized Space-Time (DSD/SST) [10] is used for solving the fluid-flow problem over the deforming spatial domain. DSD/SST utilizes SUPG/PSPG stabilizations in a space-time framework. Mesh is deformed to conform to the evolving spatial domain by modeling it as a linear elastic solid. Smaller elements are made stiff to preserve the quality of mesh in the boundary layer region. The structure dynamics is handled by the open-source program Calculix [11]. It uses standard Galerkin descretization for a total Lagrangian formulation, while the time integration is performed using Newmark's scheme. Forces are transferred from the fluid mesh to the structure mesh using an algorithm that conserves the work done by the interface [12].

3. Results and Discussion

The flow past an isolated square cylinder with flexible splitter plate has been studied by a number of researchers in past [5-7]. We extend these studies to two cylinders placed side-by-side. We also consider the effect of flexibility on the vibration response of the plate, by varying the parameter U_p^*.

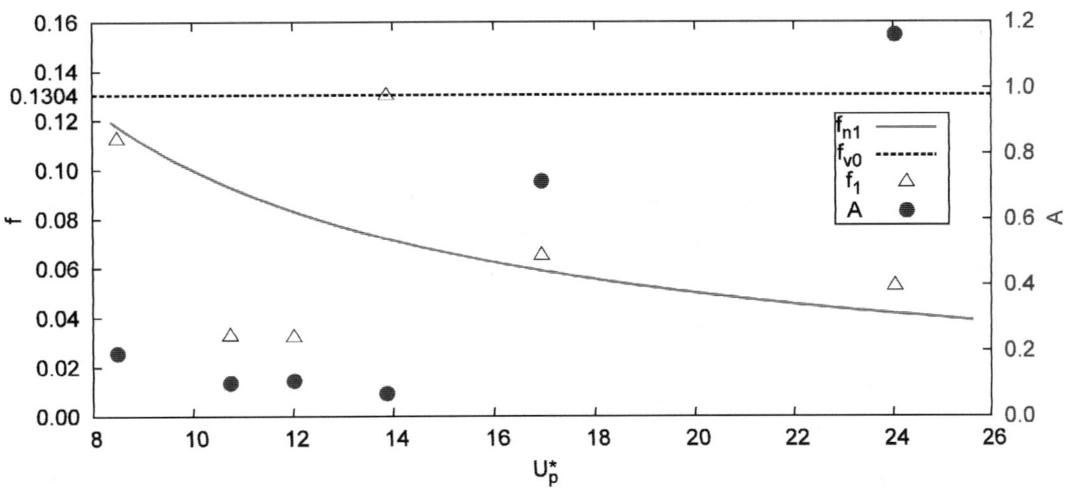

Figure 1. *Re* = 100 flow past two square cylinders with flexible splitter plates: variation of frequency and amplitude of displacement with U_p^*. The frequencies f (plotted on left y-axis) are non-dimensionalized using U_∞/D, while the amplitude A (plotted on right y-axis) is non-dimensionalized by L.

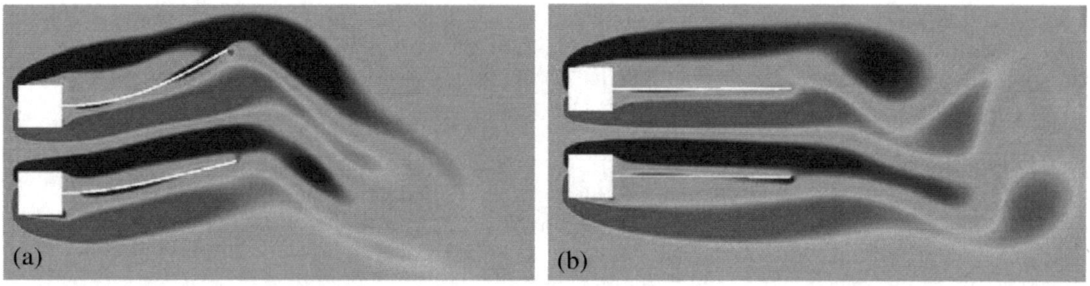

Figure 2. Re = 100 flow past two square cylinders with flexible splitter plates: (a) instantaneous vorticity field for U_p^* = 24 and (b) U_p^* = 12. The plates are in their maximum deflection configuration.

Two square cylinders are placed side-by-side with centre-to-centre separation equal to the twice of their edge length (L). Each plate is $4L$ long and $0.06L$ thick. The Reynolds number of the flow based on the edge length of the cylinder is 100. The flexibility parameter of the plate, U_p^*, is varied from 8.5 to 24, while m^* is fixed at 84.746. For a fixed m^*, smaller values of U_p^* imply greater stiffness of the plates. Larger is the value of m^*, weaker is the coupling between the dynamics of fluid and structure. Initially, the plates vibrate out-of-phase with each other, maintaining the symmetry of the flow field. This happens because, as the fluid is accelerated through the gap between the cylinders, the plates are impulsively pulled towards each other. This results into the vibration response of the plates dominated by their inertia. However, after a period of time, different for different values of U_p^*, the motion of the plates eventually becomes in-phase. The plates now vibrate in unison, more or less behaving like a single cylinder with splitter plate. Figure 1 shows the variation in amplitude and the most dominant frequency of the fully developed in-phase response with U_p^*. Solid circles show the amplitude values, while triangles are used for showing the corresponding frequencies. The variation of the first natural frequency of the splitter plate (beam) vibrations is indicated by the continuous red curve. The broken black line gives the value of vortex shedding frequency for the rigid plate case (f_{v0}). The values of amplitude are normalized by L, and all the frequencies by U_∞/L. Relatively large amplitude of vibration is observed when the dominant frequency in the response is close to the first natural frequency of the plates. This points toward the phenomenon of frequency 'lock-in'. The cases with U_p^* = 24, 17 and 8.5 show lock-in of the dominant frequency in the plate's response with the first natural frequency of the plate bending oscillations. Figure 2 shows the difference between the vorticity fields for a lock-in (U_p^* = 24) and a no lock-in (U_p^* = 12) case, when the plates are at their extreme positions. In contrast with the lock-in case, the shear layers separated from the outer edges of the cylinders, during no lock-in, do not interact with the splitter plates. The vortices are shed much downstream in the lock-in scenario as compared to the case of no lock-in.

Use of rigid plate for suppressing vortex induced vibration of circular cylinder is well known [13]. In this study, we now present the effect of a flexible splitter plate on the vortex induced vibrations of a circular cylinder mounted on a spring. The splitter plate has dimensions $3.5D \times 0.2D$, where D is the diameter of the cylinder. The Reynolds number, based on D, is 150 while m^* has a value of 10. The value of the parameter U_s^* is taken as 6. Computations were carried out for an isolated cylinder, cylinder with a rigid splitter plate and cylinder with a plate having U_p^* = 12.8. The reduction in amplitude due to rigid plate is significantly larger than that due to the flexible one. This can be explained by the ability of the plate to better preserve the symmetry of the flow by restricting the interaction of the flow on the two sides, and thereby suppressing the asymmetric vortex shedding and associated vibrations. The use of rigid plate also reduces the frequency of oscillations appreciably. Figure 3(a-c) shows the differences in the vorticity field observed due to the presence of rigid and flexible splitter plates. The splitter plates, especially the rigid one, delay the shedding of vortices to a much downstream location. The rigid plate, thus, offers a better option for suppressing VIV. To study the effect of stiffness of the spring support, a simulation was carried out at U_s^* = 22. For same value of

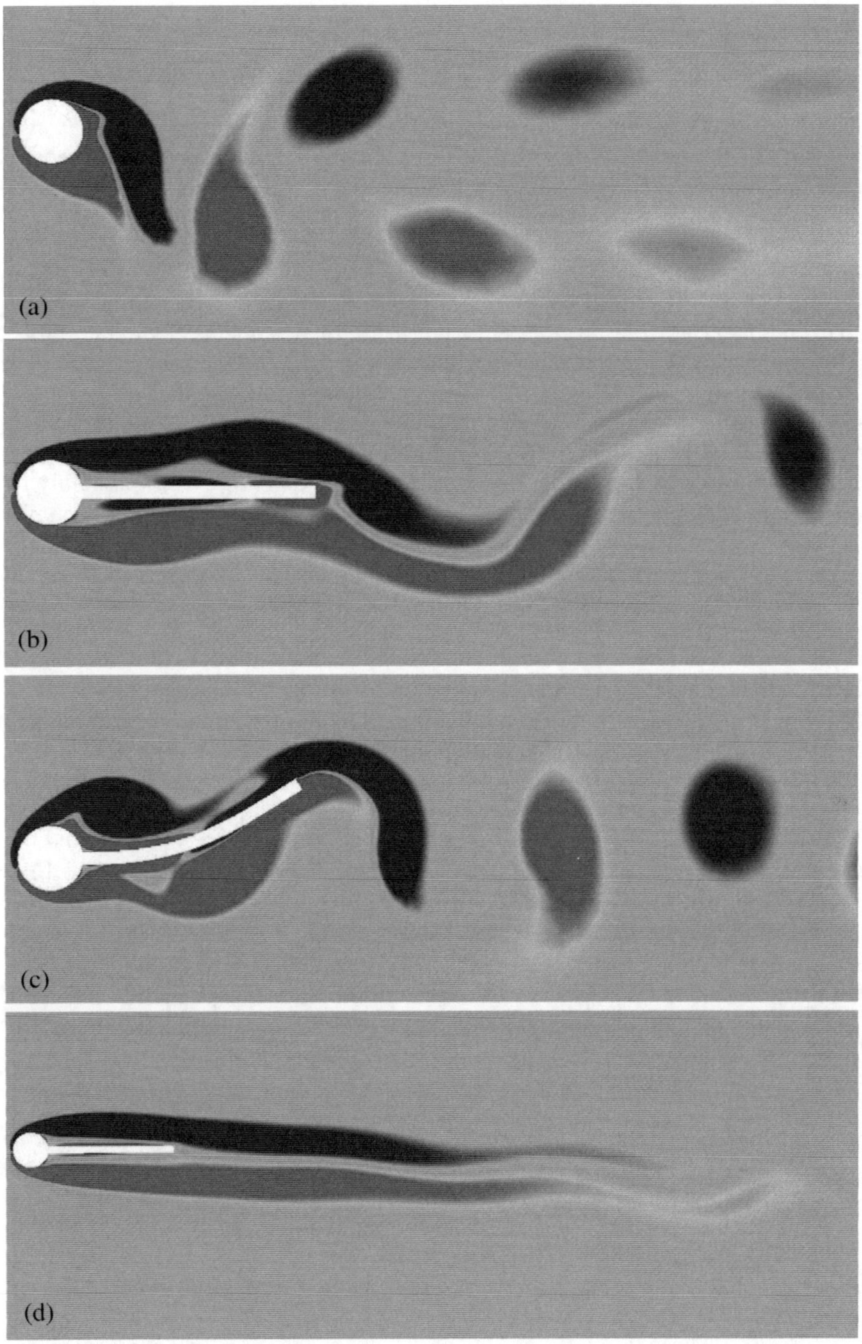

Figure 3. *Re* = 150 flow past an elastically mounted circular cylinder: instantaneous vorticity field for (a) an elastically mounted isolated cylinder with $U_s^* = 6$, (b) a fixed cylinder with a rigid splitter plate with $U_s^* = 6$, (c) a fixed cylinder with a flexible splitter plate having $U_p^* = 12.8$ and (d) an elastically mounted cylinder with a flexible splitter plate for $U_s^* = 22$ and $U_p^* = 4.6$.

m^*, smaller U_p^* corresponds to greater stiffness of the plate. U_p^* was taken to be 4.6, which corresponds to an almost rigid plate. The vorticity field corresponding to this case is shown in figure 3(d). The amplitude of cylinder vibration is quite large, even though it is quite small for the plate-tip vibrations. This shows that for a sufficiently compliant spring galloping may set in. Even though the magnitude of plate oscillations is very small and the near wake appears to be steady, the entire structure undergoes heaving motion. The vortices are shed at a location which is even more downstream as compared to the other cases.

4. Conclusions

A study is presented on the effect of flexible splitter plates on the vortex induced vibration of a single and multiple bluff bodies. The shear layer separated from the bodies interacts with the splitter plates to produce rich dynamical effects. For two square cylinders with splitter plates, initially the plates vibrate out-of-phase with each other. The fully developed response, however, consists of in-phase vibrations of the plates. Lock-in of the response frequency with the first natural bending mode frequency is observed for certain values of flexibility (U_p^*), leading to large amplitude of plate vibrations. For a single elastically mounted circular cylinder, the flexibility of the plate reduces its ability to suppress the vibrations.

References

[1] Kwon K and Choi H 1996 Control of laminar vortex shedding behind a circular cylinder using splitter plates *Phys. Fluids* **8** 479

[2] Assi G R S Bearman P W and Kitney N 2009 Low drag solutions for suppressing vortex-induced vibration of circular cylinders *J. Fluids Struct.* **25** 666

[3] Wu J Shu C and Zhao N 2014 Numerical investigation of vortex-induced vibration of a circular cylinder with a hinged at plate *Phys. Fluids* **26** 063601

[4] Kim H J and Durbin P A 1998 Investigation of the ow between a pair of circular cylinders in the flopping regime *J. Fluid Mech.* **196** 431

[5] Wall W A and Ramm E 1998 *Fluid structure interaction based upon a stabilized (ALE) finite element method* In S.R. Idelsohn and E. Onate, editors, 4th World Congress on Computational Mechanics, CIMNE, Barcelona, Spain, Computational mechanics: New trends and applications, Buenos Aires, Argentina

[6] Hubner B Walhorn E and Dinkler D 2004 A monolithic approach to uidstructure interaction using spacetime finite elements *Comput. Methods Appl. Mech. Engrg.* **193** 2087-2104

[7] Lee J 2013 Study of vortex-shedding-induced vibration of a exible splitter plate behind a cylinder *Phys. Fluids* **25** 110811

[8] Williamson C H K and Govardhan R 2004 Vortex induced vibrations *Annu. Rev. Fluid Mech.* **36** 413-455

[9] Navrose and Mittal S 2016 Lock-in in vortex-induced vibration *J. Fluid Mech.* 794 565-594

[10] Tezduyar T E Behr M Mittal S and Liou J 1992 A new strategy for finite element computations involving moving boundaries and interfaces-the deforming-spatial-domain/space-time procedure: II. Computation of free-surface ows, two-liquid ows, and ows with drifting cylinders *Comput. Meth. Appl. Mech. Eng.* **94** 353

[11] Dhondt G Calculix: A free software three-dimensional structural finite element program http://www.calculix.de/

[12] Samareh J A 2007 Discrete data transfer technique for uid-structure interaction *AIAA Paper* 4309

[13] Roshko 1955 On the wake and drag of blu_ bodies *J. Aeronaut. Sci.* **22** 124-132

Institute of Physics
Dirac House, Temple Back
Bristol BS1 6BE UK

ISSN: 1742-6588
ISBN 978-1-5108-3907-6

9 781510 839076

CURRAN ASSOCIATES INC.
proceedings.com